VEGETABLES AND THEIR ALLIED AS PROTECTIVE FOOD

Other Related Books

- Agroforestry Theory and Practices – *Raj, A.J.*
- Application of Botany in Horticulture – *Genin, A.*
- Bacterial and Viral Disease and their Management in Horticultural Crops – *Reddy, P.P.*
- Biofumigation and Solarization for Management of Soil-Borne Plant Pathogens – *Reddy, P.P.*
- Biointensive Integrated Pest Management in Horticultural Ecosystems – *Reddy, P.P.*
- Chickpea (Gram) Status and Cultivation Technology – *Shekhawat, S.S.*
- Disease Problems in Vegetable Production – *Gupta, S.K.*
- Diseases of Horticultural Crops: Nematode Problems and their Management – *Reddy, P.P.*
- Economics in Horticulture – *Sargent, M.J.*
- Forestry Principles and Applications – *Raj, A.J.*
- Fundamentals of Vegetable Crop Production – *Nair, B.*
- Fungal Diseases and their Management in Horticultural Crops – *Reddy, P.P.*
- Hort-Agro Project Reports – *Sud, R.K.*
- Horticultural Engineering Technology Field Machinery – *Balls, R.C.*
- Horticulture and Livelihood Security – *Nath, P.*
- Industrial Agroforestry Perspectives and Prospectives – *Parthiban, K.T.*
- Insect, Mite and Vertebrate Pests and their Management in Horticultural Crops – *Reddy, P.P.*
- Integrated Nematode Management in Horticultural Crops – *Reddy, P.P.*
- Multipurpose Plants – *Srivastav, P.K.*
- Multipurpose Trees in the Tropics: Management and Improvement Strategies – *Tewari, V.P.*
- Nursery and Plantation Practices in Forestry – *Kumar, V.*
- Ornamental Plants for Gardening – *Chopra, V.L.*
- Plant Nursery Management : How to Start and Operate a Plant Nursery – *Ray, P.K.*
- Plant Pathogens and their Control in Horticulture – *Dixon, G.R.*
- Plant Physiology in Relation to Horticulture, 2nd Ed. – *Bleasdale, J.K.A.*
- Plant Protection in Horticulture Vol. 1-3 (Set) – *Reddy, P.P.*
- Practical Manual on Plantation Forestry – *Panwar, P.*
- Productivity Enhancing Technologies for Horticultural Crops – *Reddy, P.P.*
- Recent Advances in Bamboo Research – *Shanmughavel*
- Salinity Tolerance in Plants: Methods, Mechanisms and Management – *Garg, B.K.*
- Science of Tea Technology – *Ahuja, P.S.*
- The Basics of Human Civilization Food, Agricutlure and Humanity,
Vol. III : Agricutlure – *Nath, P.*
- Tropical Forest Ecosystems Structure and Function – *Singh, V.P.*
- Tropical Planting and Gardening 6th Edition – *Barlow, H.S.*
- Useful Herbs of Planet Earth – *Daniel, M.*

Preface

A diet that harmonizes with natural cycle enhances one's flow of energy and harmonizes the functional activities of all body organs, however, a diet, which is not in harmony with the movements of nature, drains one's energy and brings diseases. Hence, eating healthy food is fundamental to good health and vitality.

For the centuries, fruits and vegetables have been recognized as a good source of minerals and vitamins to which the body needs to perform a variety of functions suitably and keep healthy as they have the ability to prevent vitamins deficiencies in human body. Vitamin A helps to strengthen immune system, B vitamins help to process energy from food, vitamin C helps to keep body cells and tissues healthy, and vitamin D helps to maintain healthy teeth and bones. The steamed carrots and broccoli maintain a higher proportion of vitamins than boiled or fried vegetables. Fruits and vegetables eaten along with skin also contain high amount of fibre, which helps to maintain a healthy gut and digestive system. The vegetables containing starch, also known as carbohydrates, provide most of the energy. The body converts such vegetables into glucose, which is either used as energy immediatcly or stored for later use. Vegetables also contain iron, which is needed to make red blood cells to carry oxygen around the body, hence, eat five kinds of vegetable and two kinds of fruit every day for good health. The National Institutes of Health suggests a minimum of five servings per day of fruits and vegetables combined. To increase consumption of fruits and vegetables, people need motivation. Appropriate communication efforts can often provide motivation and sometimes improve a person's ability to increase consumption of fruits and vegetables.

Fruits and vegetables are part of a well-balanced and healthy eating plan. There are many different ways of losing or maintaining a healthy weight. To lose weight, it is not necessary to eat less food though it is necessary to consume low calorie versions of some of favourite dishes by substituting low calorie fruits and vegetables in place of high calorie food ingredients. It is true that fruits and vegetables are low in calorie. Water and fibre in fruits and vegetables add volume to the dishes, so same amount of food with low calorie can be eaten by consuming fruits and vegetables. Since most of the fruits and vegetables are naturally low in fat and calorie, *per se*, eating more fruits and vegetables helps in controlling weight. Consuming more fruits and vegetables along with whole grains, nuts and beans is safe and healthy since they contain antioxidants or phytochemicals, which help in protecting the body against diabetes, heart disease, cancers and other chronic diseases. Research of the past 20 years has shown that eating fruits and vegetables not only prevents malnutrition but also helps in maintaining optimum health through a host of chemical components that are still being identified, tested and measured. It has been noticed that a high intake of fruits and vegetables could lower the blood pressure in individuals with either high or normal blood pressure.

Scientific research suggests that the role of fruits and vegetables in preventing heart disease is a protective one. Risk reduction has been estimated as high as 20-40% among individuals who consume substantial amount of fruits and vegetables. People who are diagnosed with coronary heart disease are able to reduce blockage modestly through exercise and an extremely low fat, vegan-like diet rich in fruits and vegetables. A diet containing substantial and varied amounts of fruits and vegetables may prevent 20% or more of all cases of cancer. The strongest evidence relates to stomach, lung, mouth, pharynx, esophagus, colon and rectum cancer. Studies show that the patients who take dietary supplements in place of fruits and vegetables die earlier due to a higher mortality rate among the supplement users since dietary supplements do not have the same positive effects as eating real fruits and vegetables.

Delayed development of cataracts is another beneficial effect of fruits and vegetables. A five folds reduction in cataract risk is found in individuals who consume a minimum of one and a half serving of fruits and vegetables each day. Carotenoid-rich fruits and vegetables containing zeaxanthin and lutein have been proved the most beneficial but not all carotenoids offer equal protection. Diets that are high in fibre may be able to help in the management of diabetes and may offer the best protection against diverticulitis. Soluble fibre delays glucose absorption from the small intestine and *per se* may help in preventing spike in blood glucose levels that follow a meal, however, the long-term effect may be insignificant due to many other factors that affect blood glucose. Asthmatics if consume fruits and vegetables more than once a day may have better lung function since higher intake of fruits and vegetables increases ventilation function of the lungs.

The life style diseases, especially cardiovascular diseases, cancer, obesity and type 2 diabetes mellitus, kill more people every year than any other cause of death. The reasons behind these overwhelming diseases are poor diet, physical inactivity, stress and use of tobacco and alcohol. However, fruits and vegetables are an important component of a healthy diet and if consumed daily in sufficient amount, may help in preventing these major diseases. World Health Organization estimated that low fruits and vegetables intake causes about 31% of ischemic heart disease and 11% of stroke worldwide. Overall, it is estimated that up to 2.7 million lives may potentially be saved every year if fruits and vegetables consumption is sufficiently increased. Recommendations in this direction tend to complement and reinforce other valid messages based on the long-known health benefits of consuming vegetables and fruits as dietary sources of fibre, vegetable proteins and protective micronutrients. The FAO/WHO nutrition experts recommended the intake of a minimum of 300 g of vegetables (excluding potatoes and other starchy tubers) and 100 g of fruits per day for the prevention of chronic diseases, such as heart disease, cancer, diabetes and obesity as well as for the prevention and alleviation of several micronutrients deficiencies, especially in less developed countries.

Considering the importance of fruits and vegetables in human diet in view, the author felt obligatory to prepare a book, which may sequentially help the teachers, scientists, graduate and postgraduate students, and of course, the common man to earn adequate knowledge about nutritional and therapeutic significance of fruits and vegetables in maintaining good health. The objective of preparing this book is to make the populace aware about health benefits of fruits and vegetables. This book containing very concise and precise information has been written in a very simple language, which can be explicable even to undergraduate students and common man. The information given in this book are truly based on scientific records of scientists working on particular aspects. In addition, this book may be user-friendly to others who have the concise to expand knowledge concerning human health through fruits and vegetable. Earning scientific knowledge will undoubtedly be rewarding to its users and finally to the nation.

Place: Hisar **–Author**

VEGETABLES AND THEIR ALLIED AS PROTECTIVE FOOD

DR. M.K. RANA

Professor

Department of Vegetable Science

CCS Haryana Agricultural University

Hisar-125 004, Haryana, India

Published by:

Scientific Publishers (India)
5 A, New Pali Road, P.O. Box 91
JODHPUR – 342 001 (India)

Scientific Publishers (India)
4806/24, Ansari Road, Daryaganj
NEW DELHI – 110 002 (India)

E-mail: info@scientificpub.com
Website: www.scientificpub.com

Print : 2021

ISBN: 978-81-7233-899-2

Typeset by Rajesh Parihar
Printed in India

Contents

1. Fundamentals of Nutrition 1
2. Plants as Safeguards to Human Health 13
3. Fruits and Vegetables as Protective Food 29
4. Dietary Fibers 37
5. Wholegrain Food and Health 50
6. Therapeutic Significance of Cereals 60
7. Medicinal Significance of Pulses and Beans 71
8. Coconut: The Miracle Food 87
9. Nuts and their Health Benefit 113
10. Therapeutic Significance of Fruits 139
11. Therapeutic Significance of Minor Fruits 151
12. Therapeutic Significance of Vegetables 160
13. Modern Herbal Uses of Bitter Gourd 177
14. Therapeutic Value of Vegetable Spices 188
15. Spices: A Panacea for Human Health 200
16. Curative Significance of Underutilized Vegetables 215
17. Usefulness of Medicinal and Other Plants 229
18. Herbs That Cure Human Ailments 253
19. Seabuckthorn : A Therapeutic Food 281
20. Tulsi: A Cosmopolitan Herb of Religious and Medicinal Excellence 289
21. Blue Green Algae: Food for the Future 300
22. Mushroom–The Healthy and Healing Food 316
23. The Marine Plants and Human Health 329
24. Aquatic Food and Human Health 341
25. Herbal Significance of Flax Seed 359
26. Therapeutic Significance of Oilseeds 371
27. Organic Farming: Revival of Ancient Practice 385

28.	Plant Pigments and Human Health	406
29.	Polyunsaturated Fatty Acids (PUFA)	421
30.	Green Portals for Vibrant Health	433
31.	Exploring Biocolours for Cosmetic and Textile Industries	445
32.	Free radical Scavenging activity of Some Medicinal Plants	457
33.	Fruits and Vegetables: A Potent Source of Phytochemicals	473
34.	Role of Dietary Antioxidants in Human Health	484
35.	Health Benefits of Plant Enzymes, Juice and Detoxification	498
36.	Fermented Food and Human Health	510
37.	Honey and its Uses	523
38.	Nutritional Aspects of Milk	545
39.	Tea: A Nature's Boon to the Mankind	560
40.	Coffee: Good or Bad for Health	571
41.	Egg in Medical View	583
42.	Red Meat: Facts and Fantasy	593
	Subject Index	604

■■■

1 Fundamentals of Nutrition

M.K. Rana

Department of Vegetable Science, CCS Haryana Agricultural University, Hisar-125 004, Haryana, India, E-mail: <mkrlotus@gmail.com>

Man has always been concerned with food but has never been aware about balanced diet although the intake of nutrients in balanced-proportion is very indispensable for good human health since imbalanced diet, deficiency or excess of dietary components, can cause physiological disorders or chronic ailments. Balanced diet means intake of various kinds of foodstuff in optimum quantity as per requirement of the body. The size of nutrients required by the human body varies with body weight, sex, age, activity and genetic and biochemical make up of an individual. However, in females, pregnancy and lactation increase the requirement of all nutrients. Concurrently, atmospheric temperature also affects the nutritional requirement.

The recommended daily allowance (RDA) should be met by a variety of foodstuffs due to several reasons since most of the food articles contain more than one nutrient but no single food item can supply all the essential nutrients in the amounts that are needed by the body. The greater the variety of foodstuffs, the less likely one can develop a deficiency or an excess of any single nutrient. Consuming variety of foodstuffs also reduces the excessive amounts of toxic substances that occur naturally in food items and additives or contaminants, which may be present in any single foodstuff. A simple approach to adequate nutrition is to consume a variety of foodstuffs.

The total foodstuffs can be divided into four basic food categories, *viz.* (i) food articles related to milk-group supply calcium, protein and riboflavin, while meat group food articles supply protein, fat, iron and other minerals as well as vitamins, (ii) fruits and vegetables are rich in vitamin C and precursor of vitamin A, (iii) while cereals provide carbohydrates, iron and several B group vitamins and (iv) the foodstuffs like tea and certain spices provide vitamins and antioxidants, which are essential for good health. The human body has a powerful ability to take care of itself under certain conditions. It obtains what it needs from widely differing diets and tries to survive by adapting its metabolism when dietary nutrients are short supply. The body has an incredible ability to become accustomed the environment. Biological clocks are set to inform the individual that repletion of nutrients is needed, *e.g.*, hunger, however, if nutrients are not forthcoming, adaptation to endogenous stores of fuel can be accomplished. Hormonal activities are precisely integrated to maintain continuous supply of nutrients.

Nutrients are those chemical substances needed for growth and maintenance of normal cells in human body. In India, generally, diet is entirely based on cereals, which do not supply all the essential nutrients in adequate amount and the availability of calories and protein per capita is probably lowest in the world, which leads to malnutrition. The fruits and vegetables rich in minerals and vitamins are considered the protective food since they develop resistance in the human body against several ailments caused due to the deficiency of certain minerals and vitamins. Under normal conditions, about 10% of the dietary iron is absorbed but when there is deficiency of this nutrient, the absorption is more efficient. On the other hand, when there is an excess of iron in the body, the amount absorbed is curtailed significantly. Almost all the ascorbic acid present in the normal diet is absorbed but consumption of larger doses reduces efficiency of its absorption.

The foodstuffs are made available to an organism for meeting energy requirement, for building and maintaining tissues and for maintaining optimal functional health of an organism. The place of vegetables in human nutrition is well recognized all over the world. Vegetable crops are of comparatively short duration and per unit area give higher yield than the cereal crops. Moreover, vegetables provide all the dietary components, which along with cereals form a balanced diet. Although in comparison to cereals, they provide fewer calories but the calorie requirement can be supplemented or compensated well by vegetables rich in carbohydrates like cassava, taro, potato, sweet potato, elephant foot and yam, hence, by consuming vegetables both the under-nutrition and malnutrition problems can partially be solved.

The consumption of vegetables in sufficient quantity also provides taste and palatability to food, fair amount of dietary fibers and increases appetite. In addition, they provide some specific chemical substances, which are essentially required by the human body for its normal growth, reproduction, maintenance of good health and also found useful in protecting the body against some degenerative diseases. The dieticians advise to consume 125 g leafy vegetables, 100 g root and tuber vegetables and 75 g other vegetables per capita per day to make the diet balanced. In developed countries, the every day intake of vegetables per caput is in the range of 300 to 600 g. The foodstuffs can be categorized into various groups given as under:

Energy food: Vegetables and other foodstuffs rich in carbohydrates and fats are known as energy providing food.

Body building food: Vegetables and other foodstuffs rich in protein are considered as body building food.

Protective food: Vegetables or other foodstuffs rich in minerals and vitamins are known as protective food.

Macronutrients: Macronutrients are those elements, which are required most essentially in huge quantity for the maintenance and normal functioning of human body. These are sodium, calcium, phosphorus, iron, potassium, magnesium, chlorine and sulfur.

Micronutrients: These elements are required in minute quantity for normal functioning of the human body. The important micronutrients are iodine, zinc, fluorine, selenium, copper, molybdenum and chromium.

Undernourished: This term implies to those people whose intake of calories is lower than their actual requirement.

Malnourished: This term is applied to those people who take calorie as per their actual requirement but take diet lacking in minimum daily intake of proteins including essential amino acids, mineral elements and vitamins. It is a condition characterized by inappropriate quality, quantity, digestion, absorption, or utilization of ingested nutrients. It includes under-nutrition, low food intake (calorie deficiency) leading to growth retardation or other deficiency signs and over-nutrition to consume too much food or single nutrient leading to specific toxicity. They meet total calorie from a single kind of food but not from a variety of foodstuffs. The undernourished individuals are always malnourished as well. Dietary essential means that an individual must obtain indispensable chemical substances from the diet. Human body requires at least 45-50 chemical substances, which can be divided into seven main groups, *i.e.*, carbohydrates, fats, proteins, vitamins, inorganic elements, fibers and water.

Malnourishment and undernourishment both increase susceptibility to infection not only in infants but also in adults. Both the conditions also retard physical and mental growth of the children. Brain and nervous system both are affected adversely. Malnutrition during fetus development and during infancy produces an irreversible adverse effect on brain and nervous system and results in lower basic intelligence.

Marasmus: It is a nutritional disorder caused by chronic calorie deficiency due to partial starvation for a long time. A shriveled face just like an old man prior to ones age is a characteristic feature of this disorder. The major symptoms are atrophy of limb muscles and emaciated body.

Based upon food habit, the entire population can be grouped as under:

Vegans/vegetarians: Vegetarians are those individuals who are strictly dependant on food of plant origin.

Loctovegetarians: Loctovegetarians are those individuals who strictly depend on food of plant origin supplemented with milk and milk products.

Ovolactovegetarians: Ovolactovegetarians are those individuals who consume a diet of plant origin supplemented with milk and eggs.

Fruitarians: The individuals are of the type of vegetarians who consume diet consisting of raw, dried fruits, nuts, honey and olive oils, and sometimes, supplemented with grains and legumes.

Non-vegetarian (carnivorous): Carnivorous are those individuals who are completely dependant on meat and never consume food of plant origin.

Principal Nutrient Components in Vegetables

Carbohydrates

Carbohydrates are of different classes such as monosaccharides (glucose and fructose), disaccharides (cane sugar and sucrose) and polysaccharide, polymers or aggregation of simple sugars into complex structures. Carbohydrates are sources of supplying energy to human body, while mono- and disaccharides are soluble in water. Polysaccharides, which are often called starch, are usually insoluble in water. Cooked starches are digested easily, and during digestion, are broken down into sucrose, and then, glucose or any type of monosaccharide, which when absorbed by the body, exerts energy. Cellulose, hemicellulose, gum, pectin and lignin are the complex polysaccharides, which are not always digestible by the human body and designated as *dietary fibre* or *unavoidable carbohydrates*. Their presence in diet prevents constipation. Carbohydrates help in the synthesis of protein. If carbohydrates are adequately ingested then protein is allowed to work for building the tissues. Carbohydrates regulate the fat metabolism, and they supply energy to the microorganisms those synthesize some B-complex vitamins in the intestinal tract.

The primary function of carbohydrates is to provide energy to the body and 50 to 60% energy requirement comes from this source. Carbohydrates such as glucose can be made available in the body from amino acids and from glycerol (fat). Thus, there is no specific requirement for this nutrient in the body through additional carbohydrates intake. If carbohydrates are severely reduced in the diet, fat is metabolized faster, and then, the body can take care of its intermediate products.

The main source of carbohydrates in the diet is vegetable food. In developing countries where vegetable foods are predominant, carbohydrates provide much of the energy, sometimes, as high as 90%, whereas, in developed countries, carbohydrates provide on an average 50% of the total calorie requirement. The available carbohydrates in human diet include starch (polymer of glucose), the disaccharides- sucrose (glucose + fructose), the lactose- milk sugar (glucose + galactose) and maltose (starch converted to malt sugar with the help of yeast).

In human diet, carbohydrates are the major source that supplies energy to the body. According to their structural complexity, carbohydrates can be classified into monosaccharide, *i.e.*, glucose, fructose (laevulose), mannose ($C_6H_{12}O_6$), disaccharide (one molecule of glucose + one molecule of fructose), oligosaccharide, *i.e.*, sucrose, lactose, maltose- $C_{12}H_{22}O_{11}$ and polysaccharide, *i.e.*, starch, cellulose and inulin- $[(C_6H_{10}O_5)\ x\ -]$, which is similar to starch but soluble and available in artichoke. Monosaccharides are the single chain molecules that cannot be further hydrolyzed into simpler form. The union of two or more monosaccharides results in the formation of oligosaccharides and polysaccharides. In vegetables, carbohydrates are present in the form of starch and called homo-polysaccharides. Body tissues store glucose in the form of glycogen $[(C_6H_{10}O_5)\ x\ -]$. Glucose, which is the most favourable fuel, upon oxidation during respiration liberates energy along with carbon dioxide and water as byproduct. Upon oxidation, one-gram glucose provides 4.0 calories energy. A normal weight person daily requires 400-500 g carbohydrates. The chief vegetables those supply carbohydrates are potato, sweet potato, cassava, elephant foot, yam, taro, garlic, onion, pea, bitter gourd, Brussels sprouts, carrot, *etc.* Among them, potato and sweet potato are the important and cheapest source of carbohydrates or calorie.

Protein

Protein, the complex nitrogen containing organic compound, constitutes a major part of the protoplasm. It does not supply energy to the body, however, supplies essential components of the living tissues. Upon hydrolysis, protein forms amino acids. In gastrointestinal tract, the proteins are broken down into individual amino acids, which are absorbed into blood and reached to various body parts. Upon metabolism, 1 g protein provides 4.0 calories energy. An adult requires 60 to 70 g protein per day. In comparison to products of animal origin, the vegetables provide fewer amounts of protein but of quite good quality. The vegetables such as pea, cowpea, broadband, lima bean, fenugreek leaves, mustard leaves, pumpkin, pointed gourd, drumstick, celery, garlic and Brussels sprouts are the good sources of protein.

The importance of protein in the diet is primarily to act as a source of amino acids, and some of which are indispensable dietary constituents. About 20 common amino acids are found in tissue proteins, out of these, eight are identified as essential for adult humans. In addition, infants require arginine and histidine. Prolonged histidine deficiency results in impaired hemoglobin formation. Proteins are essential structural components of all cells and are needed by the body to build up and repair the tissues, which are constantly undergoing breakdown. Body proteins are not static structures but there is a continuous taking up and release of amino acids. In adults, the gains and losses are about equal in the metabolism of proteins. The recommended protein requirement for adults is 0.8 g per kilogram body weight per day, *i.e.*, about 56 g for a person having weight above 70 kg. In general, protein intake should constitute 12-15% of the total calories intake. Although body does not store protein in the sense that it stores fat in the form of glycogen but certain reserves are available from some tissues of the body for use in emergency. A well-fed adult human contains about 11 kilogram of protein so a person can lose about 3 kilogram without serious loss of function or threat to life, however, the use of these reserves eventually requires restoration of these tissues to their normal protein composition.

The nutritional values of proteins depend on constituent amino acids, which are the constituents of proteins, and are of two types, *i.e.*, (i) essential and (ii) non-essential. The former amino acids are not synthesized by the body, while the later ones are synthesized by the body through nutritional conversions from non-protein sources. The essential amino acids are leucine, isoleucine, lysine, methionine, phenylalanine, tryptophan, threonine, valine, arginine and histidine, and the nonessential amino acids are glycine, alanine, serine, tyrosine, proline, asparatic acid, glutamic acid, cysteine, cystine and hydroxyproline. Protein is the main building block of human cells and the vital constituent of muscles, tissues and blood cells. Protein forms antibodies offering resistance to different types of ailment. Dietary protein is broken-down into amino acids, which are absorbed into the body to resynthesize required types of protein. The amino acids, which are not needed for resynthesis of protein, supply energy after breakdown. A protein composed of more amino acids is considered to be of better quality for human nutrition.

Fat and lipids

Fat gives a concentrated form of energy and serves as a carrier of fat-soluble vitamins and certain fatty acids that are essential nutrients. In human body, the energy remains stored in the form of fat. Besides ingested oily food items, carbohydrates and proteins also provide fats during metabolism. Fats upon hydrolysis liberate glycerol [$C_3H_5(OH)_2$] and different saturated palmitic acid [$CH_3(CH_2)_{14}.COOH$] and stearic acid [$CH_2(CH_2)_{16}COOH$] and unsaturated fatty acids [oleic acid ($C_{18}H_{34}O_2$)]. After absorption through blood stream, they are reached the tissues where either burnt or stored for further use. Upon oxidation, 1 g fat provides 9 kilo calories energy. A normal body-weight individual requires 50 g fats per day and most of its requirement is met through cooking of animal fats. Most of the vegetables contain fats in the range of 0.1 to 0.2% vegetables, which contain 0.3% or more than 3% fat, are hot pepper, sweet pepper, sweet corn, Indian bean, cluster bean, globe artichoke, beet leaf, spinach, fenugreek leaves, mustard leaves, coriander leaves, radish leaves and mint. Bengal gram leaves contain 1.4% fat, whereas, bitter gourd contains 1.0% fat.

Lipid is a comprehensive term, which besides lipid includes lipoids natural fat, fatty acids, waxes, lecithin, cephalin, glycolipid (cerebroside), spingomylen, gangliosides, lipoprotein, steroids or sterols and hydrocarbons. All these compounds are categorized into (i) simple lipids, (ii) compound lipids, (iii) steroids or sterols, and (iv) hydrocarbons. Simple lipids are true fats and waxes. True fats are esters of fatty acids with glycerol and are generally called triglycerides. Fatty acids can be obtained on hydrolysis of fat. Waxes are esters of fatty acids with alcohol other than glycerol. Fatty acids are broadly classified into saturated and unsaturated ones. The latter unlike former one contains double bond varying from one to six in number. Fat imparts palatability to the diet and reduces stomach-emptying time. Fat acts as concentrated source of energy. Presence of fat in diet is necessary for the absorption of vitamin A and β-carotene. Fats from vegetable source are called essential fatty acids, which have vitamin like function and form functional and structural components of cells. A diet composed of cereals and pulses is capable of supplying more than 50% requirement of essential fatty acids for an adult individual.

Lipid in diet provides high energy to the body. One gram of lipid on oxidation produces 9.3 kcal energy, which is double the amount that is produced by either carbohydrates or protein of same weight. Lipoid literally means *fat-like*. These fats-like substances are not necessarily related to fat and they do not form soap with alkali like simple lipids. Lecithin, cholesterol, cholesterol esters, and cerebrosides are complex lipoids, which may combine with neutral fat on one side and protein on other side and form a colloidal solution of fat and protein of cytoplasm. Organs like liver, kidney, heart and pancreas contain various amounts of lipoids. Mitochondria are formed of lipoids and cell membranes are composed of cholesterol and cholesterol esters.

Cholesterol and its ester with protein, known as lipoproteins, are of great importance as they are intimately related to good or bad condition of coronary system. However, cholesterol is a natural, normal and essential constituent of many body tissues and fluids. Human liver synthesizes cholesterol from fat taken with food, and such cholesterol is a component of blood and is known as very low-density lipoprotein (VLDL). When this cholesterol passes through the blood serum, the density is increased, and such cholesterol is known as low-density lipoprotein (LDL). The excess of low-density lipoprotein is harmful since it tends to deposit as plaque in inner wall of the arteries, which lead to brain and heart, narrowing down the diameter. Under such conditions, the chances of coronary disease and stroke become a result that is why low-density lipoprotein is called as bad cholesterol. There occurs another type of cholesterol with higher density in human body known as high-density lipoprotein (HDL). Higher the quantity of high-density lipoprotein, higher is the possibility of getting rid of the harmful effect of low-density lipoprotein. High-density lipoprotein gobbles up the low-density lipoprotein, transports back the low-density lipoprotein to the liver, and thereby, clears the inner arterial wall for normal flow of blood to heart and brain. For such beneficial function of high-density lipoprotein, it is called good cholesterol.

Steroids or sterols are complex lipids and occur naturally in edible plants either free or in ester form with fat. Consumption of plant parts having steroids or sterol often offers an ability to spare an administration of synthetic steroids. Cholesterol, coprosterol, ergosterol, male and female sex hormones adrenocortical hormones and vitamin D are the major steroids. Hydrocarbons have no relationship with fatty acids, yet they are grouped as lipids only because of their similar solubility properties. These are squalene, beta-carotene, vitamins A, B and K.

Water

Water is sometimes called a *silent nutrient* but its deficient intake can cause death faster than that of any other nutrient. The total water requirement of human body under normal conditions is about 2450 ml per day. The body has three sources of water, *i.e.*, ingested water and beverages (1100 ml), the water content of solid foodstuffs (1100 ml) and metabolic water (250 ml), which is derived form the oxidation of carbohydrates, fat and protein. According to composite estimates, 1 g of starch yields 0.56 g of water, 1 g of fat yields 1.07 g of water and 1 g of protein gives 0.41 g of water. The body discharges 2450 ml of water of which 1350 ml in the form of waste and 1150 ml in the form of evaporative loss through skin and lungs.

Water is vital to the body as a food solvent and lubricant and as a medium of transporting nutrients and waste. It is essential for maintaining the osmotic balance in the body cells. It also plays an important role in the regulation of body temperature and osmotic balance. A decrease in the content of body water can result either from an inadequate intake or from excessive excretion, *e.g.*, sweetening, vomiting, or diarrhea.

Energy

Energy is well recognized as a prime requirement for a human being. The major dietary sources of energy yielding substrates are carbohydrates, fats and proteins. A part of the nutrients is also used for the growth and maintenance of body tissues. If the food energy intake is inadequate to meet out energy requirements of the body, the loss of body weight occurs, body carbohydrates and fat storages are gradually decreased, and because of the urgent drive for energy, the body protein itself is metabolized to supply energy.

The energy value of food is expressed in terms of a unit of heat, the kilocalorie (kcal) generated from ingested food. This represents the amount of heat required to raise the temperature of 1 litre of water by 1°C. The total calories required by an individual vary considerably with surface area, age, sex and daily activity and may not be same for all periods of life of an individual in question. The surface area is related to the rate of heat loss by the body. The greater the surface area, greater is the heat loss. The energy requirement may also be influenced by genetic differences in the metabolic make up of an individual. The amount of heat required is the sum total of energy needed for basal metabolism, specific dynamic action and physical activity.

Young adult man requires 2000-3000 kcal energy per day, while young adult woman requires energy in the range of 1800-2200 kcal, depending upon work in which an individual is engaged. The higher energy is needed by an individual under more stress or with a higher degree of physical activity. Energy balance occurs when the intake energy is equal to the energy spent for body processes and activities, thus, no weight loss or gain occurs. In order to maintain health and to avoid excessive weight gain, the continued storage of excess food should be discouraged. When the daily diet does not supply adequate calories to meet out the entire energy needs of the body, the body utilizes the stored fat. Rapidly growing children and adolescents require high calories per unit body weight to grow faster. The energy requirements for growth decline with increasing age since the basal metabolic rate and physical activity generally decrease with the increasing age.

During growth, work, or rest, the supply of energy is essential. Even when the body is in rest, the energy supply is essential for blood circulation, reproduction, digestion, assimilation, excretion and maintenance of body temperature, thus, energy is spent, when the body is in complete rest both mentally and physically, is called *Basal Metabolism* or *Resting Metabolism*. In case of any physical labour, heavy or light, done, the body requires additional energy termed as *Maintenance Energy*, which is 1.5 times more than *basal metabolism*. The rate of metabolism depends on age, sex, height and weight of individual as well as nutritional levels of diet. For both types of metabolism, protein, fat and carbohydrates should approximately contribute 10, 30 and 60% of the total energy required, respectively. Generally, the requirement of food is estimated in terms of energy, *i.e.*, calories to be produced in the body.

In human body, food is the important source of supplying energy as it is used as fuel for physiological process of the body. Its requirement depends on the consumption of energy to do work. A person engaged in cumbersome work needs more energy, which is supplied by food. In food, the energy is available in the form of carbohydrates, protein, fats, mineral elements and vitamins. For normal functioning of the body, it is essential to take a balanced diet, which is composed of food containing all the nutrients essentially required by the body to run the physiological processes smoothly.

Source of energy

Fat is a convenient and concentrated source of calorie. Carbohydrates also provide energy. They may be converted into fat and temporarily stored in the liver and body muscles in the form of glycogen. Much of the sugar taken by the body is converted into fat, which the muscles use as fuel for mechanical energy. A considerable amount of energy is used by the nervous system to maintain itself in working order since basal metabolism of about 20% takes place in the brain. The nervous system drives entire energy from carbohydrates. It does not seem to have stored energy, thus, it entirely depends upon carbohydrates supplied by the blood stream.

Factors Influencing Amounts of Nutrients in Vegetables

- **Genetic make up of the plant:** Cultivars may vary in their nutrients content even though these are grown under same soil and environmental conditions. The nutrient content in a particular genetic material can be increased by breeding and selecting material superior in nutrients.
- **Environmental conditions under which the crop is raised:** The following conditions under which the crop is grown are as under:
  - Environmental components such as temperature, light (intensity, duration and quality), wind velocity and relative humidity
  - Atmospheric composition, *i.e.*, gaseous composition
- **Soil factors:** Soil texture, soil structure, soil pH, soil fertility status, soil organic matter content and soil moisture
- **Cultural practices:** Fertilization, attack of insect-pests and diseases and maturity stage

Classification of Vegetables Based on Nutrients Availability

Leafy vegetables alone may never be enough to meet out the entire body requirement for different nutrients. They have very high water content and conversely very low dry matter, thus, for the entire demand of various nutrients, more quantity of leafy vegetables has to be ingested. Carrot is the good source of carotene but not of protein so a small quantity of carrot can supply the minimal daily requirement of vitamin A, but to satisfy the daily requirement of proteins, a large quantity of carrot has to be consumed. The protein demand can be achieved by ingesting a small quantity of food from those plant sources, which are rich in protein, hence, a balance of all the nutrients can be obtained from different plant sources rich in particular-nutrient but the people do not know which plant source is rich in which nutrient and how they can strike this balance. Therefore, vegetables can be categorized into different groups according to their sources and contents of nutrients.

- **High in minerals:** All root crops and most of the leafy vegetables, particularly of cruciferous group
- **High in carotene (vitamin A):** Green leafy vegetables, carrot (yellow or orange colour) sweet potato (yellow or orange fleshed), cucurbits (yellow or orange fleshed), yellow sweet pepper, green beans, green peas and tomato especially yellow types
- **High in ascorbic acid (vitamin C):** Cruciferous group vegetables, peppers, tomato, beans (immature seeds), bean sprouts, melons, most of leafy vegetables and freshly harvested potato
- **High in proteins:** All peas and beans, sweet corn, all leafy vegetables and cruciferous group vegetables
- **High in carbohydrates:** Potato, sweet potato, sweet corn, cassava, dry-beans, taro, yam and arum
- **High in oils:** Soybean, mature vegetable seeds particularly of crucifereae family and cucurbits

Factors Affecting Losses and Stability of Nutrients

Harvesting methods: Losses of nutrients vary with the kind of harvesting method employed. Some vegetables are susceptible to bruising but others are resistant to abrasion. Mechanical harvesting causes more damage to the produce as well as crop than the manual harvesting. Injury to the fruits during either harvesting or transition leads to great loss. Hand harvesting is always better over mechanical harvesting.

Holding of produce prior to utilization: Losses depend on duration between harvesting from the field and the produce finally reaching the consumer end, and temperature at which the produce is held during this period. Vitamins like ascorbic acid are heat labile. If the produce is not stored as per its optimum storage requirement, the loss of weight, appearance and quality will be very high. Humidity is also an important factor to keep the vegetable produce firm. Storage humidity varies with the kind of produce to be stored.

Washing: In some of the vegetables, washing is essential to remove field heat, soil particles adhere to and to improve appearance. This process hardly affects the loss of nutrients from the fruit body. Water used for washing if recycled continuously for long time warms up, and if strong detergents are used, the water soluble substances are leached out.

Peeling: Peeling leads to a great loss of nutrients. The extent of loss depends on peeling intensity and method or instruments used for peeling. Most nutrients are present just below the skin, thus, peeling by any means will certainly remove the nutrients. Ascorbic acid has been reported to be present below the peel of the tomato fruit.

Chopping, dicing, grating, mashing, mincing, or slicing: Any process that breaks vegetable cells is likely to be responsible for significant losses of certain vitamins. The longer the broken process of vegetables are held, the greater is the losses of vitamins, therefore, vitamins may be conserved by (i) cutting/slicing just prior to serving and (ii) using a plastic knife for slicing because metals speed up the destruction of vitamin C. Nutrients are lost more from the cut pieces if washing is done after peeling and chopping, and the loss will be more if shreds are made very small. Vegetables should always be chopped in big size pieces to minimize these losses. Vitamins like vitamin C and minerals like iron are oxidized when the shreds come in direct contact of oxygen, however, addition of acetic acid in shreds cooking vegetables prevents the oxidation and loss of vitamin C.

Blanching: Enzymatic browning occurs when the vegetables are chopped. Enzymes are heat labile and can be made inactive by either steam or hot water. It inactivates the enzymes, which produce undesirable effect on colour of the produce. Some water-soluble vitamins and minerals get lost by leaching when hot water is used for blanching. This process is done to stop the actions of enzymes in vegetables prior to canning, drying and freezing.

Preservation: Vegetables are generally preserved by freezing, canning, fermenting, pickling and dehydration. The degree of nutrients loss depends on the type of processing. The greatest loss takes place when the foodstuffs are processed at higher temperature and also when the duration of processing is longer. High temperature treatment for a short time causes lesser loss of vitamins.

Fermenting/pickling: Vitamins losses in these products may vary widely as a result of factors such as heat developed during fermentation, volume of pickling solution and temperature under which pickles are canned and stored.

Canning: Blanching and strong heat treatments applied to non-acidic vegetables appear to be responsible for hefty losses of vitamins in canning.

Freezing: Most of the losses of vitamins in frozen vegetables occur during blanching since freezing itself is responsible for only small losses. The vitamin losses during frozen storage may be considerably greater when blanching is done prior to freezing.

Dehydration: Moderate amounts of carotene and vitamin C may be destroyed in this process, unless the vegetables are sulfured before drying. However, sulfuring destroys most of the thiamine content. Vitamin losses are greater during slow processes such as sun drying than they are for more rapid processes like freeze-drying.

Packing: Packing affects the nutrients retention. High quality/class packing material can keep the vegetables for longer time and helps in the retention of minerals and vitamins.

Storage prior to processing: Storage temperature, humidity, composition of gases and duration affect the nutrients stability in the produce. Low temperature, high humidity, low oxygen, high carbon dioxide, nitrogen, carbon monoxide, vacuum and darkness are the best conditions for long-term retention of nutrients. Dehydrated food articles can be held safely at low storage moisture. The quality deteriorates with the increase in temperature and storage duration.

Conventional Processing of Vegetables

Vegetables are generally cooked before they are consumed. Cooking methods vary from region to region. The methods include boiling with or without water, pressure cooking, backing, frying, French-frying, steam and microwave cooking. Some vegetables are cooked with or without peel. In some vegetables, peeling is an important traditional process generally employed prior to consumption. The method of peeling significantly influences the percent loss through peeling. The average loss in weight has been accounted from 5 to 24%, varying with the method employed for peeling. Some vegetables such as potato, arum, *etc.* should be peeled off but not before boiling to reduce the losses that taken place due to peeling. The loss of nutrients through peeling removed after boiling will always be lower in all cases than that from which peel is removed before boiling whatever the method of peeling is employed. Some important minerals, trace elements and vitamins present in fruit body just below the rind, hence, peeling results in the loss of some minerals, trace elements, vitamins, amino acids and crude fibers.

Cooking Methods

Baking: Baking is very common in sweet potato. The sweet potato tubers are generally baked with their skin. The fried products retain more ascorbic acid then the baked ones. Cooking method used has a significant influence on the retention of vitamin C. Losses of vitamins are maximum when vegetables are baked since in baking, the temperature of outer surface of the produce goes very high and most of the nutrients remain present just below the rind, thus, they are volatilized due to extreme high temperature near the rind or they are burnt completely. During baking, the minerals like potassium, phosphorus and iron start moving toward the interior.

Frying: Now, a large number of fried products are served in hotels and restaurants. These products are prepared for serving by finish frying in deep fat. French fries are always prepared from good quality produce. The French fries preparation includes washing, peeling, trimming, sorting, cutting, blanching, frying, defatting, cooling, freezing and packaging. Vegetables are first boiled or blanched before converting them into small pieces, and then, fried in shallow oil. Frying decreases the moisture content and makes the nutrients concentrated. Vitamins losses during the frying of vegetables vary greatly as a result of factors such as (i) time and temperature of frying, (ii) type of frying (shallow fat, deep fat, or stir-frying) and (ii) fat or oil that is used (those which break down during frying due to high temperature may increase the rate of vitamin destruction). Fried vegetables retain more vitamin C than that in boiled ones. However, ascorbic acid is oxidized to dehydroascorbic acid more rapidly by boiling than by frying but hydrolysis of dehydroascorbic acid is slowed down by dehydration of the product during frying. Frying decreases mineral content.

Boiling: Boiling is the most common and traditional method of cooking vegetables domestically worldwide. Peeling before boiling results in a higher loss of nutrients. If the loss of nutrients is compared in cooking of vegetables with or without water than it has been shown that the loss will be more if cooking of vegetables is done by boiling with or without water than without water. The cooking of vegetables by boiling increases the digestibility of the starch. Peeled vegetables during boiling lose more nitrogen than unpeeled, and amino acids loss is observed negligible. Significant losses in ascorbic acid have also been noticed due to boiling of both peeled and unpeeled. Vitamin B is water-soluble, thus, it is also lost during boiling. Nutrients losses are found less in boiling than steam cooking. The losses of minerals and water-soluble vitamins increase as the amount of cooking water increases, hence, greater minerals and vitamins retention results in steaming and cooking without water.

Pressure-cooking: This method of cooking vegetables results in a greater retention of minerals and vitamins than boiling in a saucepan, however, best results are obtained when the vegetables are placed on a rack above the water so that they may be cooked only by steam.

Microwave cooking: In conventional cooking, the heat is applied to outside of the fruit body by conduction and then conducted to interior of the food. Heat generated from within the food by a series of molecular vibrations is the basis for microwave cooking. The advantages of microwave cooking are higher energy efficiency, greater timesavings, convenience and easy cleaning. The vitamins losses during this procedure are small and are comparable to those, which occur when vegetables are cooked with little or no water. The disadvantage is less palatability of food.

Steaming: Vegetables cooked with steam but not in contact with water have better nutrients retention than those cooked in boiling water.

Holding on a dinning table: Losses of vitamin C, thiamine and riboflavin appear to increase with the length of time that vegetables are kept warm on a dinning table.

Sulfuring: This process, which consists of either exposing vegetables to fumes from burning sulfur or dipping them in a sulfate solution, is used to prevent discolouration and spoilage. It is generally utilized before the vegetables are dried. Sulfuring helps to prevent the losses of both carotene and vitamin C but destroys thiamine. Fortunately, vegetables are utilized more as a source of the first two vitamins than as a source of thiamine.

■■■

2 Plants as Safeguards to Human Health

Rekha Arya

Head, Department of Botany, Tikaram Kanya Mahavidyalay,
2/365, Vishnu Path, Vishnu Puri, Aligarh-202001, Uttar Pradesh,
E-mail: <aryas3@rediffmail.comient>

Plant kingdom is supporting the very existence of human being since inception of this race. The dependence is increased with the advent of civilizations. Cultivation and improvement of plants were few of the first endeavors of ancient human beings. The history of agriculture dated back to 10,000 years, and now, more than 2000 plants are cultivated to fulfill all the needs and luxuries of modern man. In 21st century, man is utilizing various plants and parts of the plants at every moment of life out of which health and nutrition are of foremost importance.

What is Nutrition and Nutrients?

Nutrition may be defined as the science of food and its relationship with health. Good nutrition means maintaining a nutritional status that enables us to grow well, while nutrients are inorganic and organic complexes contained in food. More than 50 nutrients are supplied through the food. Each nutrient has a specific function. The nutrients are divided in macro- and micronutrients depending upon their requisite amount in diet. Macronutrients include carbohydrates, proteins and fats, while micronutrients consist of minerals, vitamins, antioxidants, *etc.*

Macronutrients: Macronutrients are the chemical compounds, *i.e.*, carbohydrates, protein and fats, which provide bulk energy and are consumed by the humans in large quantity.

Carbohydrates

Carbohydrates are one of three macronutrients that provide the body with energy (protein and fats being the other two). The chemical compounds in carbohydrates are found in both simple and complex forms, and in order for the body to use carbohydrates for energy, food must undergo digestion, absorption and glycolysis. It is recommended that 55 to 60% of the calorie intake come from the carbohydrates. Broadly, the carbohydrates can be divided in two categories, *i.e.*, (i) simple carbohydrates and (ii) complex carbohydrates.

Simple Carbohydrates

Simple carbohydrates or simple sugars are composed of mono- or disaccharide units. Common monosaccharides (carbohydrates composed of single sugar units) include glucose, fructose and galactose. Glucose is the most common type of sugar and the primary form of sugar that is stored in the body for energy. Sometimes, it is referred to as blood sugar or dextrose and is of particular importance to individuals who have diabetes or hypoglycemia.

Oligosaccharides are carbohydrates made of two to ten monosaccharides. Those composed of two sugars are specifically referred to as disaccharides or double sugars. They contain two monosaccharides bound by either an alpha-bond or a beta-bond. Alpha-bonds are digestible by the human body, whereas, beta-bonds are more difficult for the body to break down.

There are three particularly important disaccharides: sucrose, maltose and lactose. Sucrose is formed when glucose and fructose are held together by an alpha-bond. It is found in sugarcane or sugar beets and is refined to make granulated table sugar.

Sugar Comparison

Sugar	Carbohydrate	Monosaccharide or disaccharide
Cane sugar (beet sugar)	Sucrose	Disaccharide (fructose and glucose)
Brown sugar	Sucrose	Disaccharide (fructose and glucose)
Corn syrup	Glucose	Monosaccharide
Fruit sugar	Fructose	Monosaccharide
Honey	Fructose and glucose	Monosaccharides
Malt sugar	Maltose	Disaccharide (glucose and glucose)
Milk sugar	Lactose	Disaccharide (glucose and galactose)
Powdered sugar	Sucrose	Disaccharide (fructose and glucose)
White sugar	Sucrose	Disaccharide (fructose and glucose)

Maltose or malt sugar is composed of two glucose units linked by an alpha-bond. It is produced from the chemical decomposition of starch, which occurs during the germination of seeds and the production of alcohol. Lactose is a combination of glucose and galactose. Since it contains a beta-bond, it is hard for some individuals to digest in large quantities. Effective digestion requires sufficient amounts of lactase enzyme.

Complex carbohydrates: Complex carbohydrates or polysaccharides are composed of simple sugar units in long chains called polymers. Three polysaccharides, *i.e.*, starch, glycogen and dietary fiber, are of particular importance in human nutrition.

Starch and glycogen are digestible forms of complex carbohydrates made of strands of glucose units linked by alpha-bonds. Starch, often contained in seeds, is the form in which plants store energy and there are two types of starch, *i.e.*, amylose and amylopectin. Starch represents the main type of digestible complex carbohydrate. Humans use an enzyme to break down the bonds linking glucose units, thereby releasing the sugar to be absorbed into the bloodstream. At that point, the body can distribute glucose to areas that need energy, or it can store the glucose in the form of glycogen.

Glycogen is the polysaccharide used to store energy in animals, including humans. Like starch, glycogen is made up of chains of glucose linked by alpha-bonds but glycogen chains are more highly branched than starch. It is the highly branched structure, which allows the bonds to be more quickly broken down by enzymes in the body. The primary storage sites for glycogen in the human body are the liver and the muscles.

Another type of complex carbohydrate is dietary fiber. In general, dietary fibers are considered polysaccharides that have not been digested at the point of entry into the large intestine. Fibers contain sugars linked by bonds that cannot be broken down by human enzymes and are therefore labeled as indigestible. Because of this, most fibers do not provide energy for the body. Fibers are derived from plant sources and contain polysaccharides such as cellulose, hemicellulose, pectin, gums, mucilage and lignin.

The indigestible fibers cellulose, hemicellulose and lignin make up the structural part of plants and are classified as insoluble fibers because they usually do not dissolve in water. Cellulose is a non-starchy carbohydrate polymer made of a straight chain of glucose molecules linked by beta-bonds and can be found in whole-wheat flour, bran and vegetables. Hemi cellulose is also a non-starchy carbohydrate polymer made of glucose, galactose, xylose and other monosaccharides. It can be found in bran and whole grains. Lignin, a non-carbohydrate polymer containing alcohols and acids, is a woody fiber found in wheat bran and the seeds of fruits and vegetables.

In contrast, pectin, mucilage and gums are classified as soluble fibers because they dissolve or swell in water. They are not broken down by human enzymes but instead can be metabolized (or fermented) by bacteria present in the large intestine. Pectin is a fiber made of galacturonic acid and other monosaccharides. Since it absorbs water and forms a gel, it is often used in jams and jellies. Sources of pectin include citrus fruits, apples, strawberries and carrots. Mucilage and gums are similar in structure. Mucilage is dietary fiber that contains galactose, manose and other monosaccharides and gums are dietary fibers that contain galactose, glucuronic acid and other monosaccharides. Sources of gums include oats, legumes, guar and barley.

Interesting Facts about Carbohydrates

Fact 1 : Slow digesting carbohydrates are natural.

Slow digesting carbohydrates such as yams, wild rice, beans, red potatoes and fruits all comes from mother nature. They are not produced in food processing plants. Natural carbohydrates are loaded with nutrients and fibers. They produce relatively slow increase in blood glucose and modest insulin release. They should be your first choice as a fuel source for energy and growth.

Fact 2 : Fast digesting carbohydrates are man made.

Refined sugar is quickly digested. Generally, more processing is involved in producing carbohydrate food, which digests faster. White bread, white rice, mashed potatoes, potato chips and fruit juices require one or more processing step in their manufacture. This creates a carbohydrate, which reaches the blood stream quicker than the slow digesting carbohydrates and it produces an insulin spike that is undesirable.

Fats and Oils

Fats and oils are made up of individual molecules called fatty acids. Fatty acids are chains of carbon and hydrogen atoms that have a carboxyl group at one end and a methyl group at other end. Carboxyl groups contain one carbon atom, one hydrogen atoms and two oxygen atoms, and methyl groups include one carbon atom and three hydrogen atoms. The carbon atoms in the fatty acid molecules are linked by single or double bonds.

Fatty acids are either saturated or unsaturated. Saturated fatty acids have no double bonds between any of the carbon atoms in the chain. Unsaturated fatty acids have one or more double bonds in the carbon chain. Monounsaturated fatty acids have one double bond and polyunsaturated fatty acids have two or more double bonds. Unsaturated fatty acids are sometimes named by the position of the double bonds in the carbon chain. The names omega-3, omega-6, or omega-9 refer to the locations of the first double bond in the three different fatty acid molecules.

Most of the fats in the food you eat are called triglycerides. A triglyceride is made up of three fatty acid molecules attached to a glycerol molecule. Human body will use triglycerides as a source of energy or store them as adipose tissue, better known as body fat. The types of fatty acids in the triglycerides have an important impact on physical state of the fats one eats. Fats that are composed of triglycerides with saturated fatty acids are solid at room temperature. Fats that are composed of triglycerides with unsaturated fatty acids are liquid at room temperature.

Good Fats

Monounsaturated Fats: Monounsaturated fats (MUFAs) lower down the total cholesterol and low-density lipoprotein (the bad cholesterol) while increasing high-density lipoprotein (the good cholesterol). Nuts including peanuts, walnuts, almonds and pistachios, avocado, canola and olive oil are high in monounsaturated fatty acids. Monounsaturated fats have also been found to help in weight loss, particularly body fat.

Polyunsaturated Fats (PUFA): Polyunsaturated fats also lower down the total cholesterol and low-density lipoprotein. Seafood like salmon and fish oil as well as corn, soy, safflower and sunflower oils are high in polyunsaturated fats. Omega-3 fatty acids belong to this group.

Bad Fats

Saturated Fats: Saturated fats raise total blood cholesterol as well as low-density lipoprotein (the bad cholesterol). Saturated fats are mainly found in animal products such as meat, dairy, eggs and seafood. Some plant foods are also high in saturated fats such as coconut oil, palm oil and palm kernel oil.

Trans Fats: Trans-fats were invented as the scientists began to hydrogenate liquid oils so that they can withstand better in food production process and provide a better shelf life, as a result of hydrogenation, the trans-fatty acids are formed. Trans-fatty acids are found in many commercially packaged, commercially fried foods such as French fries from some fast food chains, other packaged snacks such as microwave popcorn as well as in vegetable shortening and hard stick margarine.

The foods we eat may also include cholesterol. Cholesterol does not produce any energy like triglycerides but it is important for many biochemical processes and hormone production. Elevated cholesterol levels in human body have been associated with an increased risk of cardiovascular disease. The cholesterol in body is mostly made in liver and there are three different types of cholesterol, *i.e.*, high-density lipoproteins (HDL), low-density lipoproteins (LDL) and very low density lipoproteins. Having higher HDL cholesterol levels can decrease the risk of cardiovascular disease, while elevated LDL cholesterol will increase that risk.

Functions of Fat in the Body

Fats and cholesterol have a number of important functions, which include:

- Lubrication of body surfaces
- Components of cell membrane structures
- Formation of steroid hormones
- Energy storage
- Insulation from cold
- Carrying fat soluble vitamins, *i.e.*, vitamin A, D, E and K

Proteins

Proteins are necessary for building the structural components of the human body, such as muscles and organs. Proteins are needed to keep the immune system healthy, synthesize neurotransmitters, create and signal hormones and much more. A balanced diet supplies all of the proteins, which are needed by the body. Meats, eggs and dairy products are the significant sources of protein but protein can also be obtained from a variety of grains, legumes, nuts and seeds.

Glimpses of Protein Chemistry

Proteins tend to be large molecules made up of several building blocks called amino acids. The general structure of any amino acid molecule includes a carboxyl group of atoms, an amine group and a side chain. The carboxyl group contains one atom of carbon, two of oxygen and one of hydrogen. The amine group contains one nitrogen atom with two hydrogen atoms attached to it. There are 20 different amino acids. Amino acids are linked together to form peptides, which are small chains of amino acids. The peptides are then linked together to form larger proteins.

Protein and Bodybuilding

Protein is probably one of the most important parts of the diet for the bodybuilder to consider. The motor proteins actin and myosin generate the forces exerted by contracting muscles. Current advice says that bodybuilders should consume 25-30% of protein per total calorie intake to further their goal of maintaining and improving their body composition, casein, or whey supplement are most preferred proteins because of their high biological value. Soy is another good source. Nutritionists have divided opinions regarding its use as it has weak estrogenic qualities also.

Protein Functions in the Body

There are many different proteins in your body, and they perform different functions. Proteins functions include:

- Contributing to enzyme activity that promotes chemical reactions in the body
- Signaling cells what to do and when to do it
- Transporting substances around the body
- Keeping fluids and pH balanced in the body
- Serving as building blocks for hormone production
- Helping blood clot
- Promoting antibody activity that controls immune and allergy functions
- Serving as structural components that give our body parts their shapes

Amino Acids

All 20 amino acids have different side chains, which vary in shape including straight chains of atoms, branched chains of atoms and rings of atoms. The side chains may include carbon, hydrogen, sulfur, nitrogen and oxygen atoms. The configuration and molecules found in the side chain is what differentiates one amino acid from another. The branched-chain amino acids are isoleucine, leucine and valine. These amino acids are necessary for muscle structure. Tyrosine, phenylalanine and tryptophan are called aromatic amino acids. Each one contains a side chain with a ring-shaped formation. These three amino acids are needed for neurotransmitter production.

Essential and Non-essential Amino Acids

The amino acids, which are essentially required by the body, are called essential amino acids since they can not be manufactured by the human body, hence, these have to be supplied through diet.

The amino acids, which are essentially required by the body, are called nonessential amino acids. It is not essential to get the 11 amino acids through diet since the human body can build these amino acids from chemicals already present in the body.

The list of essential and non-essential amino acids is given below:

Essential Amino Acids		Non-essential Amino Acids	
Histidine	Phenyl-alanine	Alanine	Glutamine
Isoleucine	Threonine	Arginine	Glycine
Leucine	Tryptophan	Asparagine	Proline
Lysine	Valine	Aspartic acid	Serine
Methionine		Cysteine	Tyrosine
		Glutamic acid	

The amino acids arginine, cysteine, glycine and tyrosine are sometimes also considered to be conditionally essential, meaning that most people manufacture them on their own but some with certain illness or genetic abnormalities do not and need to get them through their diets.

Micronutrients: Micronutrients are nutrients needed throughout life in small quantities. They are dietary minerals needed by the human body in very small quantities (generally less than 100 µg/day) as opposed to macro minerals that are required in larger quantities.

The microelements or trace elements include a least iron, cobalt, chromium, copper, iodine, manganese, selenium, zinc and molybdenum. Plant food sources are main contributors to this part of nutrition especially vegetables, fruits and spices.

Vitamins: A class of organic compounds categorized as essential nutrients. They comes from food and enables the body to use other nutrients. Broadly, they can be categorized in two categories.

Minerals: These are the chemical elements that are required for growth, repair and regulation of body functions.

Key Vitamins and Minerals

Nutrients	Essential For	Found In
Calcium	Building and maintaining bones, muscle and nerve function	Dairy products, leafy green vegetables and calcium-fortified foodstuffs, *e.g.*, cereals and orange juice
Iron	Carries oxygen in red blood cells to body cells	Meats, eggs and dark leafy vegetables
Phosphorus	Supporting tissue growth and repair and a major bone component	Milk, meat, poultry, fish, eggs, legumes and nuts
Magnesium	Energy production and nerve function	Legumes, nuts, whole grains and green vegetables
Potassium	Nerve function and muscle contraction	Fruits, vegetables, meat, poultry, fish and milk
Zinc	Cell reproduction and tissue growth and repair	Meat, eggs, seafood and whole grains
Vitamin A (retinal and retonic acid)	Vision, growth, reproduction, and maintenance of healthy skin	Liver, fish oils, eggs, dairy products. Red, yellow, orange and dark green vegetables and fruits contain alpha- and beta-carotenes, which are converted in the body to vitamin A.
Vitamin B_1 (thiamine)	Carbohydrate metabolism (energy production)	Whole or enriched grain products, fortified cereals, pork and organ meats

Nutrients	Essential For	Found In
Vitamin B_2 (riboflavin)	Metabolism of protein, fat, and carbohydrates into energy	Milk and other dairy foods, organ meats and enriched and fortified grains
Vitamin B_3 (niacin or nicotinic acid)	Needed in many enzymes that convert food to the energy, and helps in maintaining a healthy digestive tract and nervous system	Poultry, fish, beef, peanut butter, legumes and enriched or fortified grain products
Vitamin B_4 (biotin)	Needed for metabolism of glucose and formation of certain fatty acids and essential for proper body chemistry	Meats, soybean, rice bran, peanut, cauliflower, mushroom, carrot, tomato, spinach, beans, peas, orange and apple
Vitamin B_5 (pantothenic acid)	Converts the food to molecular forms and needed to manufacture adrenal hormones and chemicals that regulate nerve function	Yeast, mushroom, peanuts, eggs meat, milk cereals, broccoli and sweet potatoes
Vitamin B_6 (pyridoxine)	Manufacture of amino acids and red blood cells	Fortified cereals, sweet potatoes, chicken and beef liver
Vitamin B_9 (folic acid)	DNA synthesis involved in making protein	Fortified cereals, enriched grains, leafy greens, legumes and asparagus
Vitamin B_{12} (cyanocobalamine)	Responsible for energy and amino acid production, for building up genetic material and helps in formation of red blood cells	Beef, milk, cheese and shellfish
Vitamin C (ascorbic acid)	Acts as an antioxidant, helps in binding cells together, strengthens the blood vessel walls, helps in maintaining healthy gums and aids in the absorption of iron.	Fruits (especially citrus) and vegetables (especially those in the cabbage family)
Vitamin D (D_2- ergocalciferol and D_3- cholecaliciferol)	Proper formation and maintenance of bones and teeth and enhances calcium absorption.	D-fortified milk and cereals, cod liver oil and naturally in the skin when exposed to sunlight
Vitamin E (tocopherol)	Antioxidant action and defends cells against damage by free radicals	Vegetable oils, margarine, wheat germ, nuts, seeds and green leafy vegetables

Nutrients	Essential For	Found In
Vitamin K (K_1- phylloquinone and K_2- menaquinone)	Blood clotting	Eggs, cereals and green leafy vegetables

Nutritional Profile of Plant Food Sources

Cereals: Cereals without any doubt are most important plant food source for man and other lower animals since a very long time. The word cereal comes from the italic word ceres, which means the goddess who gives the grains. Ancient Romans held the festivals at the time of seed harvesting in honour of the goddess, in fact nearly every ancient race has worshipped some deity who presided over the crops. Cereal crops are the members of grass family Poaceae with the characteristic fruit caryopsis. Almost 35 members have been cultivated as cereal crops and only a dozen are important today. True cereal crops are only six in number, *i.e.*, barley, maize, oat, rice, rye and wheat. Wheat, rice and maize are major cereal crops and have played an important role in the development. Millets and sorghums are also important food crops, and sometimes, erroneously mentioned as cereal crops.

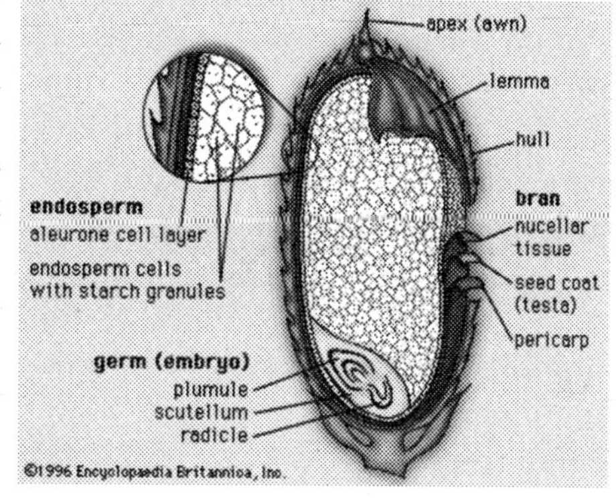

These are chief source of carbohydrates in human diet. Cereals also contain proteins (6-12%), minerals and B-group vitamins and they contribute to the 70-80% of total energy intake and more than 59% of protein intake of Indian diet. The proteins of cereals are lacking in essential amino acid-lysine but if these cereals are eaten with pulses (as popular in Indian diets), they become a complete protein.

Reasons behind the success of cereals as food plants are:

- Easy availability
- Wide range of soil and moisture requirements
- Cultivation involves less labour but more yield
- Easy to handle and store
- High food value

Generalized structure of grain: The seed coat and ovary wall remain fused in a complex structure called bran. It is composed of protein and fat rich layer aleurone. The major part of grain is composed of endosperm rich in starch grain. The region of embryo is particularly rich in proteins, vitamins and fats.

Wheat: It is the second staple food of north Indian families and occupies more than 29 million hectares of land. The wheat grain discovered as a result of Indus valley excavations indicates that northwestern India was one of the ancestral homes of this cereal crop.

Wheat products are probably the most used article of human diet. Worldwide, they furnish almost one third of the total food material of the average family. In Indian cuisine, wheat flour is used chiefly for making *rotis, chapattis, naans, etc*. Western world uses hard wheat (small grain with less starch) for making breads, pasta, noodles macaronis, *etc*. and soft wheat (large grained with more starch) is used for making cakes, pastry and biscuits. Wheat is also extensively used in making alcoholic beverages.

Wheat and Green Revolution

Green revolution was the development and introduction of high yielding wheat varieties over the large areas including multiple cropping, irrigation, crop protection and transfer of technology from laboratories to village farmers. It almost doubles the yield and solves the major food crisis. Professor Norman Borlaug was credited as father of green revolution for his discovery of Norin dwarf varieties and got Noble prize in 1970.

The father of green revolution in India is Prof. M.S. Swaminathan who did pioneering work in mutation breeding of wheat and released the variety *Sharbati Sonora* under his supervision.

Rice: Rice, predominantly the crop of Asia, is the indispensable food of more than half the human race. India is supposed to be the native place for the crop and has a long history of rice cultivation. The area under cultivation (more than 79 million hectares) is largest in world.

Rice is generally eaten with legumes. This starch rich food is taken after boiling. It may also be converted in parched rice, beaten rice, or rice flakes and puffed rice. Mixed with black gram (*urd*) powder, it is used for fermented preparations like *idli* and *dosa*.

Attraction at the cost of Nutrition

White (polished) attractive rice is produced after milling. The process of milling deprives the rice grain of its valuable nutritive elements like vitamin B complex (72%) and proteins (15%). The nutrients are further lost in the process or washing and cooking. Nutrition workers supports parboiled rice cooked in just enough water over white rice.

Nutritive Value of Cereals (values per 100 gram)

Nutrients	Raw rice milled	Wheat whole	Maize dry
Protein (g)	6.8	11.81	11.1
Fat (g)	0.5	1.5	3.6
Carbohydrate (g)	78.2	71.2	66.2
Thiamine (mg)	0.06	0.45	0.42
Niacin (mg)	1.9	5.0	1.8
Riboflavin (mg)	0.06	0.17	0.1
Minerals (g)	0.6	1.5	1.5
Energy (kcal)	345	346	342

Maize: A native of America it ranks next to rice and wheat in world consumption. Maize or Indian corn is utilized in more diversified ways than any other cereal. The grain is very nutritious with a high percentage of easily digestible carbohydrates, relatively high amount of fats and proteins. Mainly a livestock crop it finds a substantial place in human diet also. It is a popular breakfast cereal as corn flakes and the puffed pop corn is so popular that today's generation can be called as pop corn generation.

Millets: The term millet is used for smaller grains, which are ground and eaten without the outer layer removed. They are sorghum, pearl millet, *ragi*, *kodo* and a few other known as minor millets or pseudo cereal crops like *Amaranths* (Chaulai) and *Fagopyrum* (*Kutu*).

Nutritive value of millets (values per 100 g)

Nutrients	Sorghum	Pearl millet	*Ragi*
Protein (g)	10.4	11.6	7.3
Fats (g)	1-9	5.0	1.3
Carbohydrates (g)	72.6	67.5	72.0
Minerals (g)	1.6	2.3	2.7
Calcium (mg)	25.0	42.0	344.0
Iron (mg)	4.1	8.0	3.9
Vitamin B group (mg)	4.7	2.85	2.68
Energy (kcal)	349	361	328

Millets in Indian Culture

Millets are one of the oldest foods known to humans. In India, millets have been used from thousands of years and forms 6th most important grains. There are three main millets in India, *i.e.*, sorghum, pearl millet and *Ragi*. *Ragi* is grown in South India and its flour is eaten as chapattis and boiled cakes. Pearl millet is mainly cultivated in Northern India and eaten as chapattis and *khichdi* and as sweet *tiklis*. Sorghum on the other hand is the crop of Southwest and Central India and mainly eaten as wholegrain preparations. *Hunjas*, inhabitant of remote area of Himalayan foothills, are known excellent health and longevity used millets as their staple diet. They use millets as cereal and soups and also for chapattis. Kutu (pseudocereal) preparations are prevalently used in *Navrartris*- A nine days Hindu ritual ceremony celebrated twice in a year.

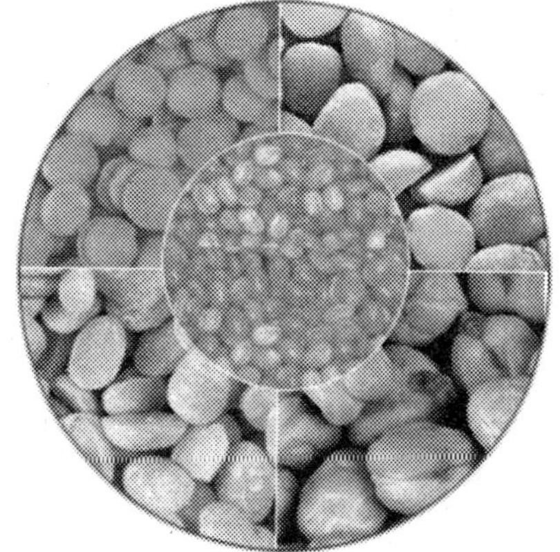

Pulses: An important part of Indian dietary occupies about one seventh of the total cultivated area in India. Seeds are dry with low water content and impervious seed coat. These are easily grown, mature rapidly and are highly nutritious. Together with cereals, they form the essential food adjunct of Indian diet and increase the biological value of food. The protein content is higher than any other vegetable product. Therefore, pulses are sometimes nicknamed as poor man's meat. The high protein content is correlated with the presence of nitrogen fixing bacteria present in roots. They are also rich in minerals, vitamin B, and sometimes, fats. Germinating pulses have also vitamin C and enhanced levels of B-group vitamins, *i.e.*, niacin and riboflavin.

Vegetable protein is not a poor protein. There is simply misconception about the quality of vegetable proteins. They are rather incomplete proteins since cereals are lacking in essential amino acid lysine and pulses in methionine. The common tradition of Indian cuisine of mixing these two in their diet (*dal-chawal/roti*, *khichdi*, *idli-dosa*, *etc.*) makes it complete protein equivalent to animal protein. The common pulses used in India are green gram (*mung*), black gram (*urd*), red gram (*arhar/tuhar*), lentils (*masur*), soybean, cowpea (*ramas*), chickpea (*chana*), kidney bean (*rajma*) and moth bean (*moth*).

Nutritive value of pulses (values per 100 g)

Nutrient	Bengal gram	Black gram	Red gram	Soy bean
Proteins (g)	17.1	24.0	22.3	43.2
Fats (g)	5.3	1.4	1.7	19.5
Calcium (mg)	202	154	73	240
Iron (mg)	4.6	3.8	2.7	10,4
Vitamin B (mg)	3.35	2.62	3.34	4.32
Energy (kcal)	360	347	335	432

Oilseeds and Nuts: This group of plant seeds provides the vegetable fat and oils to the diet. These crops have been the backbone of agricultural economy of India from time immemorial. The area under cultivation constitutes approximately one tenth of total cultivated area. The important crops in this category are coconut, groundnut, rapeseed, mustard, sesame, soybean and sunflower. On world level, cottonseed, corn, olive, *etc.* are the important oilseed crops.

Vegetable fats are rich in unsaturated fatty oils, which are good for health. Some of the important fatty oils are oleic, linoleic, linolenic, palmitic and stearic acid. The measurement index of instauration is iodine number, which refers to the number of grams of iodine absorbed by 100 g of the fats in a medium in which it is soluble. High iodine number indicates high instauration.

Dry Fruits (Nuts): Nuts like almond, cashew, walnut, pistachio and above all peanuts are the source of good quality fats, proteins, minerals like calcium, phosphorus and iron and B-group vitamins. They also contain a good amount of dietary fiber. As a source of omega-3 fatty acids, they are helpful in reducing bad cholesterol, therefore, reducing the risk of cardiovascular heart disease.

Oil Refining: Refining of vegetable oils is usually carried out by adding caustic soda to the oil. Free fatty acids, if any, are removed by this process, and finally, neutral oil is produced. The process includes bleaching, deodorizing, winterizing, and finally, hydrogenation. The unsaturated fatty acid component becomes saturated. The hydrogenated oils are solid at room temperature and have lower iodine number.

Name of Oil	Iodine Number
Coconut oil	7-10
Olive oil	78-88
Peanut oil	85-100
Cottonseed oil	90-117
Rapeseed oil	94-105
Soy bean oil	103-152
Sun flower oil	113-143

Vegetables: The term vegetable is usually applied to edible plants which store reserve food in roots, stems, leaves and fruits and which are eaten cooked or raw as salad. The vegetables rank next to cereals as the source of carbohydrates. The nutritive value is tremendous because of indispensable minerals and vitamins. India grows a large variety of vegetables belonging to the tropical, subtropical and temperate groups. For convenience, the vegetables are divided in three main categories:

***Underground Vegetables*:** These include underground root and stem, *e.g.*, carrot, radish, taro, potato, onion, garlic, beetroot, *etc*. They vary widely in their composition. Some like potato, yam and tapioca are good source of carbohydrates and can serve as subsidiary food at the time of cereal shortage. Carrots are exceptionally good source of beta-carotene, while onion and garlic have organosulphur.

***Green leafy Vegetables*:** The term green designates a number of indigenous leafy vegetables, *i.e.*, leaf mustard (*sarson*), spinach, cabbage, amaranths, chenopods (*bathua*), fenugreek (*methi*), *etc*. consumed by the people. The darker the green colour, the greater is their nutritive value. Green leaves are rich in carotene, calcium, iron and vitamin C. They are fairly good source of riboflavin, folic acid and many other micronutrient. In addition, leaf protein (2-4%) is a good source of lysine. Oxalates present in leaves reduce the bioavailability of calcium and iron. The amount of water and dietary fiber is high. Because of their low calorific value (5-50 kcal/100 g) and large bulk, they have an important role in the dietaries of obese people who wish to reduce their weight.

***Fruit Vegetables*:** They bring variety to the diet. Many of them are good source of minerals and vitamins. The examples of fruit vegetables are tomato, brinjal, cucurbits, beans and peas.

Fruits: Fruits are the protective food. Cultivation of fruit crops is known from ancient times and a large number of cultivated crops had their origin in Asia. The mango is most ancient among the tropical fruits indigenous to India. Later, the civilizations progressed and the nutritive valve and other qualities of fruits were studied. Now, they are invaluable in human nutrition because they are good sources of minerals and vitamins. One special feature that distinguishes fruits from other foodstuffs is that they can be eaten raw and fresh. This makes the minerals and vitamins present in fruits easily available. Fruits are particularly rich in vitamin C.

Phytochemicals : Another good reason to eat fruits and vegetables

Phytochemicals are the natural bioactive compounds found in plant foodstuffs that work with nutrients and dietary fiber to protect against degenerative diseases. Research suggests that phytochemicals, working together with nutrients found in fruits, vegetables and nuts, may help in slowing down the aging process and reducing the risk of many diseases, including cancer, heart disease, stroke, high blood pressure, cataracts, osteoporosis and urinary tract infections.

Pronounced *fight-o-chemicals*, phytochemicals fight to protect the human health. They can have complementary and overlapping mechanisms of action in the body, including antioxidant effects, modulation of detoxification enzymes, stimulation of the immune system, modulation of hormone metabolism and antibacterial and antiviral effect.

Phyto is a Greek word that means plant, and phytochemical are usually related to plant pigments. So, fruits and vegetables that are bright in colours, *i.e.*, yellow, orange, red, green, blue and purple, generally contain the most phytochemicals and the most nutrients.

Colour Wheel of Foodstuffs and Phychemicals

Color group	Phytochemicals	Fruits and Vegetables	Possible Benefits
Red	Lycopene	Tomato	Prevents development of prostrate cancer and down regulate the reactive oxygen species.
	Phytoene	Tomato juice	
	Phytofluene	Tomato soup	
	Vitamin E	Watermelon	
Green	Glucosinolates	Broccoli	Helps in preventing cancer.
	Isothiocyanates	Amaranths	
	Indole-3-carbinol	Cauliflower	
	Folic acid	Cabbage	
Green/yellow	Lutein	Spinach	Prevents age related macular degeneration and cataracts and provides topical pain relief and cancer cell apoptosis.
	Zeaxanthin	Avocado	
		Green peppers	
		Green mustard	
		Kiwi	
Orange	Alpha and Beta	Carrots	Helps in the formation of vitamin A and protects against joint inflammatory diseases (arthritis).
	Carotene	Pumpkins	
	Cryptoxanthin	Mango	
		Apricot	
		Squash	

Color group	Phytochemicals	Fruits and Vegetables	Possible Benefits
Orange/yellow	Vitamin C Flavonoids	Oranges/juices Yellow grapes Peaches Lemons Limes Papaya Pineapples	Treating gall stones, oxidizes LDLs, prevents arteriosclerosis and heart diseases and protects against various diseases.
Red/purple	Anthocyanins Ellagic acid Flavonoides	Grapes/juices Cherries Strawberries Raspberries Plums Raisins Red clover	General antioxidant, easing the symptoms of menopause and prevents arteriosclerosis.
White/green	Allyl sulfides	Garlic Onion Chives	Lowers LDLs and assists in immune system

Spices and Condiments: According to a famous saying, *variety is the spice of life that gives its entire flavour*, and really, spices add the life to food that too in variety of tastes and flavours. The spices and condiments have played a prominent part in all the cultivation of antiquity in ancient India and China. The majority of spices originated in the Asiatic tropics and was among the first objects of commerce between the East and West. The quest for the condiments and spices was once powerful flora in world history. Spices Island of east led to the voyager of Columbus, Vasco-de-gamma, Magellan and many others.

The spices cannot be grouped as food because they contain less nutritive value. However, they give a good flavour and aroma to food, and at the same time, they gently add to taste of the food. They stimulate the appetite and increase the secretion and flow of gastric juices. For this reason, they are commonly known as food adjuncts. The aromatic value of the spices is due the presence of essential oils.

India is known as the home of spices. It earns 3000 million per year by the export of spices. Kerala is the chief state including the peninsular India, which is known as the king of spices and cardamom as queen of spices.

Conclusion

Plants being the producer form the foundation of nutrition. They contain each nutrient with an additional benefit of phytochemicals, which are a good source of antioxidants. As vegetarian diet is an indicative of good health and longevity, and rich and diverse culture of India is the home of vegetarians. The country offers a wide variety of cuisine and can play a vital role in promoting green diet and clean environment. Ancient tradition of *Ayurveda* and *yoga* emphasizes on balanced diet and healthy lifestyle a *mantra*, which is resonating in whole world today. Amongst all the hoopla and chaos in Copenhagen where the world leaders were fighting for their share in global climate, plant kingdom is smiling at the most developed creature of nature- *Homo sapiens.*

■■■

3 Fruits and Vegetables as Protective Food

M.K. Rana

Department of Vegetable Science, CCS Haryana Agricultural University, Hisar- 125 004, Haryana, E-mail: <mkrlotus@gmail.com>

The fruits and vegetables containing high antioxidants appear to stimulate memory and the consumption of fruits and vegetables with high Oxygen Radical Absorbance Capacity (ORAC) helps in slowing down the ageing of both body and brain. The antioxidants present in fruits and vegetables protect the cells and their components from oxidative stress. Consumption of food with high ORAC activity such as spinach, blue berries and strawberries in plenty amount raises the antioxidant power of human blood, prevents some loss of long-term memory and learning ability in middle age, keeps thc ability of brain cells maintained age in middle age and protects the tiny blood vessels capillaries against oxygen damage.

Maladies that come with ageing including cardiovascular disease and cancer are due to the damage caused by oxygen free radicals. Inclusion of fruits and vegetables in diet reduces the risk of cancer and a low intake raises its risk. Diminished brain functions associated with ageing and disorders such as Alzheimer's and Parkinson's may be due to the increased vulnerability to super oxides. In recent years, such evidences have promoted skyrocketing sales of antioxidants to supplement vitamins deficiency in the diet. Individual antioxidant has not been found so effective but combination of nutrients found in various foodstuffs has greater protective effect than each nutrient taken alone.

Different food articles contain more than 4000 flavonoids, which constitute a principal group of dietary antioxidants and these are found accountable for a large part of the protective power of fruits and vegetables. A big number of people are taking antioxidants with their diet to guard themselves against ailments or disorders. Food antioxidants boost up antioxidant power of the blood. People to have protective powers can absorb individual flavonoids.

A daily dose of spinach juice can prevent some loss of long-term memory and learning ability. Spinach has been found the most potent in protecting different types of nerve cells in various parts of the brain against the effects of ageing. The individuals fed with spinach had significantly better long-term memory than the control diets. The vitamin E fed individuals are somewhat less protected against memory loss than the spinach group.

In old age, the cells become sluggish in responding to stimulation of chemicals. The spinach group also scored best among all the fortified foods in a test of nerve cells in the cerebellum (a part of the brain) that maintains balance and coordination. Why spinach is effective more than strawberries? It may be due to some specific nutrients or a specific combination of them in spinach. Fruits and vegetables play an important role in preventing long-term effects of free radicals pressure on the function of brain. Blueberries have been found to be one of the highest in antioxidant capacity.

Table : Fruits & vegetables top in antioxidants value (ORAC units per 100 g fresh weight)

Fruits	ORAC units	Vegetables	ORAC units
Prunes	5770	Kale	1770
Raisins	2830	Spinach	1260
Blueberries	2036	Brussels sprouts	920
Straw berries	1540	Broccoli florets	890
Rasp berries	1220	Beets	840
Plums	949	Red Bell Pepper	710
Oranges	750	Onion	450
Grapes red	739	Corn	400
Cherries	670	Eggplant	390

Carotenoids found in fruits and vegetables are fat-soluble pigments. Beta-carotene is a nutritionally significant carotenoid as it is converted to vitamin A and a principal dietary source of this vitamin. The other provitamin A carotenoids are α-carotene and β-cryptoxanthin. Beta-carotene functions as an antioxidant. A potential anticarcinogenic effect of β-carotene is due to its ability to scavenge reactive oxygen species that cause oxidative DNA damage. It also has a number of functions other than its antioxidant role. It has also been found to influence immune function and enhance gap junction inter-cellular communication. Other cancer preventive mechanisms for this molecule cannot be excluded from its functions.

Fruits and vegetables are rich in other carotenoids and several evidences show that these compounds have the potential anticarcinogenic properties. Carotenoids are the strong quenchers of singlet oxygen. On the other side, lycopene shows the highest singlet oxygen quenching activity and increases communication of growth regulating signals between cells. Carotenoids act as chain breaking antioxidants protecting cells and other body components from free-radical attack. The use of carotenoids in daily diet is associated with a diminished risk of various cancers.

The consumption of lycopene rich diet is also associated with lower risk of prostate cancer. Inclusion of lutein/zeaxanthin in diet has been found associated with a decreased risk of lung cancer. High dietary intake of β-cryptoxanthin has been observed associated with decreased incidence of cervical cancer.

In recent past, the increasing attention has been paid to the role of antioxidants in human health. Such compounds, particularly of plant origin, were previously recognized as factors in food preservation but are now considered important health protecting factors. They play the dual role in human body, *i.e.*, (i) they reduce content of toxic compounds in food and (ii) they supply the human body with exogenous antioxidants.

The naturally present antioxidants in fruits and vegetables could be lost during processing and storage. In any form, thermal treatment causes depletion of naturally occurring antioxidants. On the other hand, heating induces some new compounds with antioxidant properties, which have been evaluated by measuring the chain breaking and the oxygen scavenging activities.

Almost 600 carotenoids from natural sources have been characterized. Green leafy vegetables and many orange yellow colored fruits are considered rich in carotenoids. In humans, these carotenoids are absorbed from the intestine and appear in blood stream. Human serum contains β-carotene, £-carotene, cryptoxanthin, lycopene and lutein. Less than 10% of all the carotenoids serve as precursor of vitamin A.

Flavonoids comprise one of the largest groups of secondary plant metabolites. Flavonoids consist of flavonols, flavones, flavanones, anthocyanidins, catechins and biflavones. Over 4000 flavonoids have been identified. The flavonoid -quercetin- had strong anticarcinogenic effects. Green-yellow vegetables are not important source of quercetin in the human diet.

Impact of Processing on Antioxidants

Dietary and non-dietary nutrients may be an antioxidant. The major antioxidants are vitamin C and vitamin E with fruits and vegetables being major source of vitamin C and vitamin E. These compounds are found in wheat germ, nuts, green leafy vegetables and vegetable oils. Non-nutrients antioxidants include flavonoids, which are found in tea, red wine, onion and apples, polyphenols and terpenes.

Antioxidants help to prevent the occurrence of oxidative damage to biological macromolecules caused by reactive oxygen species. If such oxidative damage occurs due to free radicals, may be a significant causative factor in the development of several diseases.

The protection of foods from oxygen is the basic principle upon which antioxidant protective technologies are based. After microbial spoilage, oxidation leading to overt rancidity is the second most important cause of food spoilage. Maintenance of structural integrity of foods helps to prevent oxidative changes occurring in whole food form, contact of these sensitive compounds with oxygen is reduced, and thus, processes that escape exposure of antioxidants to oxygen have become a requirement. The addition of antioxidants to food helps to preserve constituent of the food by preventing auto-oxidation.

Fats and oils contain natural antioxidants, and provide and carry fat-soluble vitamins in human body. Whatever care is taken in the handling of vegetable oils, rancidity and all the associated deleterious changes always occur sooner, or later. The raw material must be given the requisite protection from elements such as oxygen, heat and variably valent metal ions.

Harvesting and handling of fruits and vegetables bring out many changes and these changes potentially produce impact on their antioxidant status. Intact fruits and vegetables are susceptible to deleterious damages induced by respiratory metabolic and enzymatic activities. Mostly these changes influence the antioxidants status adversely. There is a non-linear relationship between moisture and antioxidant content. Moisture increases the oxidation process rapidly. Temperature control, low oxygen and protection from light constitute other physical procedures from light constitute other physical procedures whose effective control can help to ensure maximum retention of antioxidants. Inactivation of polyphenol oxidase is one example whose inactivation of degerative enzymes can help to maintain antioxidant status.

Controlling food spoilage addition of preservatives such as sodium benzoates, sorbates and metabisulphite and poly phosphate has been found effective. Preservatives also serve as antioxidants to extend shelf life of fruits and vegetables. During processing, several oxidative reactions may occur in which electrons are removed from molecules leading to formation of oxidized form.

Four types of chemical structures, *i.e.*, (i) free radical chain breaking such as tocopherols, (ii) reducing agents and oxygen scavenger such as ascorbic acid and erythorbic acid, (iii) chelating agents such as citric acid and (iv) other secondary antioxidants such as carotenoids are used to stabilize fruits and vegetables. Out of these, the most important compounds used to stabilize fruits and vegetables reducing agents and certain chelating agents that are not actually antioxidants but are important in preventing oxidative damage. The use of sulfites has been restricted, though the combinations of ascorbic acid and derivatives with citric acid and other organic acids are quite effective. Oxygen absorbent sachets containing iron power or to a lesser extent oxygen scavenging matter such as ascorbic acid may be used to prolong the shelf life of various food items.

The link between micronutrients such as minerals and vitamins that are required in minute quantity and various diseases are well recognized. For instance, protein mal-nutrition as well as other vitamin deficiency makes the humans susceptible to several parasitic infections. At least 50% of the pregnant women in India are suffering from anemia due to the deficiency of iron in diet. Deficiency of zinc and selenium affects the sperm production in men. Micronutrient deficiencies and consequent reduced host resistance in HIV infected patients may increase susceptibility to different pathogens.

Vitamin D and calcium supplements in diet reduce the risk of hip fractures considerably, decrease parathyroid hormone secretion and increase mineral density. It is thought that increased parathyroid hormone secretion is due to the deficiency of vitamin D in the body and low intake of calcium in the diet. Six months supplementation with calcium at least one gram a day and vitamin D_2 (ergocalciferol) reduce the incidence of parathyroid. Vitamin D_3 (cholecalciferol) and supplement of calcium decrease frequency of fractures in elderly persons.

Vitamin A reduces mortality from 30 to 34% by supplementing about 2,00,000 IU vitamin A in every six months. An individual who involved himself in physical work for many hours continuously every day requires energy as high as 6000 calories as compared to 250-300 calories for a sedentary person.

Carbohydrates and proteins give about four calories per gram, whereas, fat gives at least nine calories per gram. Ingested fat is rich source of energy when carbohydrates are depleted. Proteins are seldom utilized by the body for energy production. Carbohydrates in the form of glucose and glycogen can be stored, to some extent, for producing instant energy in the body, whereas, dietary proteins can never be stored within the body. Fats can be preserved in the body to an unlimited extent and act as an energy reservoir. Over-weight individuals carry extra fat but not extra carbohydrates and proteins.

Consumption of energy by individuals

Type of work done by the subjects	Energy consumption (kcal/h)
Sleeping individual	65
Lying but awake	75
Sitting in resting position	100
Studying and writing individuals	105
Light exercise	170
Slow walking individuals	200
Fast walking individuals	570
Climbing stairs	1100

Content of β-carotene (in μg/100 g fresh weight basis)

Types of commodity	μg/100 g
Ripe mango	3000
Ripe tomatoes	190
Orange	210
Drumstick leaves	7000-8000
Coriander leaves	7000
Radish leaves	3600
Mint leaves	1800-8000
Carrot	1500-3000
Cabbage	13000

Content of vitamin C in mg/100 g fresh weight basis

Types of fruit	mg/100 g
Aonla fruit	600-700
Guava fruit	200-300
Drumstick/amaranth/chenopod leaves	200-250
Radish leaves	80
Papaya fruit	55-60
Lemon fruit	40-50
Lime fruit	50
Fenugreek leaves	30
Orange juice	30
Beet leaf	30
Tomatoes	25-30

Content of vitamin E mg/100 g fresh weight basis

Types of commodity	mg/100 g
Ground nut oil	261
Wheat germ oil	225
Soybean oil	116
Linseed oil	110
Palm oil	56
Mustard oil	32

In a 70 kg weight adult, the blood glucose pool is around 20 g, liver glycogen 80-120 g and muscle glycogen 300-400 g.

In order to lose a half kilogram body fat, one has to utilize at least 3500 kcal. In other word, one has to walk about 80 km. The better way of reducing weight is to consume calorie as one can and avoid taking diet based on fats and sugary foods, as these are highly calorific food articles.

Production of Free Radicals and Human Health

Free radicals are unpaired electrons, which are considered as fragments of molecules. These free radicals are very reactive in nature. The continuous production of these free radicals is seen in cells as accidental byproducts of metabolism.

Oxygen is required for the conversion of food substrates to release energy, to oxidize endogenous compounds and to detoxify xenobiotics. During this process, oxygen functions as a terminal U-electron acceptor, and eventually, it is transformed to water. In molecule, oxygen and the electrons are distributed in such a way that two of the electrons are unpaired. This nature of oxygen enables it to react readily with many other free radicals. When water is formed from oxygen, super oxide (O_2^-) and hydroxyl (OH^+) are also produced. These are considered the oxygen-derived free radicals. These are highly reactive with DNA, lipids and proteins and lead to cellular damage and tissues or organ failure. In order to survive, aerobic organisms have developed antioxidant systems to protect themselves from the adverse effect of free radicals. This includes enzymes such as super oxide dismutase, catalase and some vitamins like vitamin C and vitamin E. These are said to be free radical scavengers and help in controlling free radical in the body.

When the body suddenly exposed to any kind of stress produces accidentally/deliberately free radicals. Some enzymes utilize free radicals at their active site. The major sources of free radical production are the electron transport chain, auto-oxidation of certain compounds like vitamin C, glutathione, cysteine, adrenaline and flavin.

All biochemicals, like proteins, carbohydrates, lipids and nucleic acids, are attacked by free radicals but lipids are the most susceptible one. Vitamin E in association with vitamin C can be able to terminate this chain reaction by interacting with hydroxyl. Cell membranes are very rich sources of polyunsaturated fatty acids, which are readily attacked by oxidizing radicals.

The preventive defense includes the efficiency of electron transfer. The generation of free radicals may actually release free transition metal ions. Another form of defense is to remove the peroxides that react with transition metal ions to produce reactive free radicals. This includes both hydrogen peroxides and the lipid hydro-peroxides. Catalase and glutathione peroxidase are the enzymes whose role is to decompose produced peroxides safely. Another form of defense is scavenging free radicals. Under this, most are of non-enzymes than enzymes. One of the important enzymes is super oxide dismutase, and non-enzymatic scavengers are vitamin C, vitamin E, uric acid, glutathione, *etc.* All of them are having more or less similar types of action but at different sites in different conditions.

Free radicals are damaging in nature. Their damaging effect can be neutralized by diet containing antioxidants. Fruits and vegetables enriched diets may increase the levels of vitamin C and carotene in plasma. The principal antioxidant vitamin C and E along with carotene, beta-carotene and vitamin A may also reduce the radical related pathogenicity. Supplementary contributions could come from plant components, *e.g.*, carotenoids other than beta-carotene such as lycopene, oxygen radical sensitive B-group vitamins such as folate and vitamin B_6.

Vitamin E

Tocopherol has been considered as the fertility vitamin and the panacea for delaying the onset of a host of age related degenerative diseases. It is even believed that it could play a role in the prevention of cancer. It has the free radicals neutralizing properties. Free radicals are unstable chemicals, toxic in nature, which are released during the normal bodily processes and on exposure to polluted environments and cigarette smoke. When present in excess, these radicals affect the normal cells and predispose them to the development of early heart attack, cataract and some cancers. Free radicals are implicated in inflammation, arthritis, brain damage, circulatory disorders, damages to skin and premature ageing. Fortunately, our natural defense system can minimize production of free radicals or limit the damage. If the system is weakened, supplements like vitamin E, which belong to a class of substances called antioxidants (vitamin C and β-carotene), may be useful.

The important point is that it is very difficult to pick up cases of vitamin E deficiency. Exercise and stress affect the amount of vitamin E a person needs. Those likely to be deficient are smokers, persons suffering from infections or metabolic disorders, patients under going radiation therapy or convalescing after surgery, as well as persons on a restricted diet.

The actual factor chemically called tocopherol, meaning life giving, was isolated from wheat germ oil. Vitamin E is primarily found in foodstuffs of plant origin and vegetable oils.

Wheat germ oil and walnut oil have higher concentration of vitamin E followed by sunflower, cotton seed, safflower, palm, corn, soybean, peanut and olive oil, margarine, butter, nuts, whole grain cereals (fortified breakfast cereals are the good sources) and green leafy vegetables.

How do vitamins work?

The rate of ageing can be slowed down by influencing the factors responsible for the determination of cells. The main factor associated with the ageing process is free radicals. Antioxidants like vitamin E, C and β-carotene react with free radicals and provide protection from their harmful effects. Free radicals are one factor, which may initiate this biological process.

There is strong evidence of vitamin E's role in protection against development of cardiovascular diseases. While the mechanisms are unknown, research suggests vitamin E most likely works by preventing the formation of bad cholesterol known as low-density lipoproteins and its absorption by blood vessels.

Vitamin E supplements are associated with reduced incidence of heart attacks by 75% when taken daily by people with heart disease. Vitamin E supplements for at least two years reduce the risk of heart disease but only up to 40% in healthy men and women. Vitamin E could be used either as a single mode of treatment as a complementary therapy in infertility patients. Women with tubal block and/or problems in egg-production and men with no sperm due to a block need drugs to reproduce the eggs and sperm production, respectively.

It is estimated that 20% of the United States population regularly consumes vitamin E. It is perhaps the fact that vitamin E is a relatively safe drug which has no significant side effects in normal individuals. In practice, vitamin E particularly useful in the treatment of nocturnal cramps in the elderly where the calf muscle goes into severe spasms but its effect on the skin (it is believed that free radicals created by pollution and sun light can affect the skin which is why skin creams containing vitamin E that neutralizes free radicals have hit the market has not been proven. Vitamin E is not advised for patient with vitamin K deficiency as this can complicate the blood-clotting defect associated with vitamin K deficiency. As iron destroys vitamin E, iron supplements should be taken at a different time of the day than vitamin E supplements. It may change the body lipids and alter steroid hormones. The American Heart Association recommends ingesting a variety of food from all of basic food groups to meet ones requirement of vitamin E.

■■■

4 Dietary Fibers

Meenakshi Virmani[1] and Kamla Malik[2]

[1]Assistant Endocrinologist, Department of Animal Production Physiology, College of Veterinary Sciences, Lala Lajpat Rai University of Veterinary and Animal Sciences, Hisar-125004, E-mail: <virmanim2003@yahoo.com>

[2]Assistant Scientist, Department of Microbiology, College of Basic Sciences and Humanities, CCS Haryana Agricultural University, Hisar- 125004, E-mail: <kamlamalik@rediffmail.com>

The term *dietary fiber* refers to the plant cell wall components that are not digestible by human or other mammalian digestive enzymes. They pass undigested through the stomach and small intestine and reach the large intestine virtually unchanged. While passing through the large intestine, some components of fiber are broken down to varying degrees and absorbed by the body, the remaining components are excreted in the faeces. During its breakdown in the large intestine, a small amount of energy is contributed, approximately 3 kcal/g of the ingested dietary fibers.

The development of definition of dietary fiber is fascinating. Earlier in 1953, the term crude fiber was used by Williams and Olmstead for the residue of plant (based) food left after extraction with solvent, dilute acid and dilute alkali. Since crude fiber does not include many of the fiber components normally associated with the physiological activity of fiber, a broader term including all the components was required. The term *dietary fiber* was first introduced by Hipsley in 1953 for the constituents of plant cell wall. It mainly comprised the remnants of plants resistant to hydrolysis by human alimentary enzymes. In 1970s, Dr. Denis Burkitt, a man nicknamed the Fiber Man, and his colleagues made *the fiber hypothesis* that stated that fiber could prevent certain diseases. While the exact mechanism by which fiber might prevent these diseases remained unknown. Burkitt discovered about the beneficial impact of fibers on bowel movements and their relationship with certain diseases. He also proposed that health benefits of fiber intake are due to their ability to increase stool bulk and speed up the movement of stool through the colon. After that, Trowell and Burkitt postulated an inverse relationship between dietary fiber consumption and the incidence of colon cancer and heart disease. In 1976, the term dietary fiber that included all undigestible dietary polysaccharides in addition to other fiber components was broadened by Trowell and his associates since they were found to possess the physiological actions that are characteristic to dietary fibers. This definition gained wide acceptance and became the basis definition for the current AOAC International Official analytical methods.

Dietary fibers constitute an essential part of our daily diet. Daily intake of fibers is a necessity for a healthy lifestyle and possesses unique physiological functions and potential health benefits. The requirement of dietary fiber can be fulfilled by a diet that incorporates food articles of plant origin including fruits, vegetables and grains.

Types of dietary fiber

Soluble fiber: Fiber that is dissolved in water is known as soluble fiber. Soluble fiber is readily fermented in the colon into gases and physiologically active byproducts. After dissolving, it forms a gel, which acts like a plug in the digestive tract, binds the fatty content of the foodstuffs such as bile acids (which contain cholesterol up to 80%) as well as other metabolic products, assists to pass through the digestive system more efficiently and helps in the excretion of these products. This helps in lowering low-density lipoprotein (*bad cholesterol*), while maintaining the high-density lipoprotein (*good cholesterol*). Soluble fiber is found in varying quantities in all plant food items including:

- **Legumes:** Peas, soyabeans, lupins, *etc.*
- **Cereals:** Oats, rye and barley
- **Fruits:** Plums, berries, bananas, apples and pears
- **Vegetables:** broccoli, carrots, sweet potatoes and onions
- **Medicinal herbs:** Psyllium seed husk

Insoluble fibers: This kind of fiber is also known as roughage. The insoluble fiber contributes about two-third of the fiber in most food items. As the name suggests, it does not dissolve in water. It is metabolically inert and more resistant to digestion. It acts as an absorbing agent that absorbs water in the digestive system thereby expanding the volume of matter inside the intestine. This speeds up the bowel movements, decreases the storage period of chyme in the intestine, and thus, it is easily moved through the digestive tract and out of the body. If insoluble fiber is consumed in the right amounts, it can prevent constipation due to its bulking action. The sources of insoluble fiber include:

- Whole grain foods
- Wheat and corn bran
- Nuts and seeds
- Potato skins
- Flax seeds
- Lignans
- Vegetables such as green beans, cauliflower, celery, *etc.*

Composition of Dietary Fiber

Relatively all plants contain both soluble and insoluble fibers, although the proportions vary depending on the type of plant. Chemically, dietary fiber consists of non-starchy polysaccharides such as cellulose, hemicellulose and many other plant components such as resistant starch, lignin, cutins, pectins, β-glucans, gums, waxes, oligosaccharides, *etc.*

Cellulose: Cellulose, the primary component of insoluble dietary fiber, is insoluble in cold and hot water, dilute acid and dilute alkali. It is the most abundant carbohydrate structural material in nature, making up the cell walls of most plant materials, characteristically one-half of the plant's mass. This is a linear polymer of β-D-glucose molecules linked in the 1,4-positions and is indigestible in the human system because humans do not have enzymes that hydrolyze beta linkages.

Hemicellulose: Like cellulose, hemicellulose is also insoluble in cold and hot water, and dilute acid, therefore, hemicellulose is probably considered similar to cellulose. Hemicellulose is distinguished from cellulose due to its solubility in dilute alkali. The name hemicellulose includes a variety of heterosaccharridic polymers, which tend to be small (50-200) saccharide units with branching present usually consisting of more than two sugars. Predominant monomers are xylose, arabinose, mannose, glucose and galactose. Arabinoxylans present in cereal grains are an excellent example of hemicellulose. Part of hemicellulose is quantified as insoluble dietary fiber, part as soluble dietary fiber.

Galactomannan gums: Soluble galactomannan gums are part of the hemicellulose fraction of the food. They are present in leguminous plants such as guar and locust beans. These gums consist of a mannose polymer backbone to which galactose side chains are attached.

Pectins: Pectins (polygalacturonic acids) are the soluble dietary fiber present in food articles. They are found in fruits, vegetables, legumes and roots, *i.e.*, sugar beets and potatoes, as storage polysaccharides. Commercial pectin is isolated from either apple pumice or from citrus peels up to the level of 30% on dry weight basis. The functional groups of the polymer are present as free carboxylic acids, methyl esters, or carboxylate anions (as sodium, potassium, or calcium salts).

Resistant Starch: Resistant starch, when eaten by human passes undigested through the small intestine and into the large bowel where it is fermented or excreted. This fermentation energy source may be significant for maintaining colonic health. The relative amount of indigested starch can vary from food to food and person to person.

Inulin: Inulin is a polymer of fructose monomers and is used to improve taste and mouth feel in certain applications. It has nutritional properties and because of this, inulin products can be used as a replacement for fat or soluble carbohydrates without affecting the taste and texture and still contribute to a foodstuffs nutritional value. Since inulin consists of beta linkages. It cannot be hydrolyzed in the small intestine. Therefore, it enters the large intestine and it is almost completely metabolized by the intestinal microflora. When fermented, they lead to propionate production, which in turn decreases the acetate to propionate ratio resulting in decreased total serum cholesterol and low-density lipoprotein cholesterol.

Beta-glucans: β-glucans are indigestible mixed β-linkage (β-1,3 interspersed with β-1,4) glucose polymers. The alternate positional linkages give water soluble food gums, upon which hydration with water confer high viscosity solutions with relatively little shear and tensile resistance, as compared to cellulose, which is essentially water insoluble and has tremendous shear and tensile resistance, strong rugged and durable enough to use for clothing and shelter. In large intestine, the β-glucans undergo extensive fermentation, whereas, cellulose passes through essentially unchanged.

Cutin: Cutin is a waxy hydrophobic layer made up of highly hydrophobic long chain hydroxyl aliphatic fatty acids polymerized by ester linkages and is resistant to digestion that can be recovered in fecal material.

Lignin: Lignin is formed when polyfunctional phenols are polymerized with ether and ester linkages during plant growth that infiltrate the cellulose of cell walls, resulting in a rigid matrix of tremendous strength. At sufficient lignin concentration, plant tissues become *lignified* or *woody* to the point of being inedible. Lignin is an important component of dietary fiber, making the fiber hydrophobic, resistant to enzymatic breakdown in the small intestine and bacterial breakdown in the large intestine. It is almost completely recovered in the feces. Lignified tissue in the food offers unique textural properties, although they are not always considered desirable.

Various components of dietary fiber along with their source and uses are listed in the table below:

Component	Source	Uses
Cellulosic polysaccharides		
Cellulose	Bran, legumes, peas, root vegetables, cabbage family, outer covering of seeds and apples	▪ Increases bulk of intestinal waste and eases it quickly through the colon. ▪ Prevents constipation and helps in curing diarrhea. ▪ By the above actions, it dilutes and flushes cancer-causing toxins out of the intestinal tract. ▪ Helps in leveling out the glucose in blood. ▪ Helps in curbing weight gain due to its ability of filling without producing fattening.
Non-cellulosic polysaccharides		
Hemicellulose	Bran, cereals and whole grains	▪ Helps in relieving constipation. ▪ Water down carcinogens in the bowel. ▪ Aids in weight reduction.
Gums	Oatmeal, barley, legumes and dried beans	▪ Help in lowering down the cholesterol significantly. ▪ Help the diabetics in handling blood sugar better.
Mucilage	Food products thickenerand stabilizer	▪ Helps in lowering down the cholesterol significantly. ▪ Helps the diabetics in handling blood sugar better.
Pectin	Apples, strawberries, citrus fruits, green beans and carrots	▪ Helps in lowering down the cholesterol significantly. ▪ Aids in the elimination of bile acids ▪ Short-circuiting the development of

Component	Source	Uses
Resistant starch	Unprocessed cereals and grains, ripe bananas, potatoes	▪ Decreases postprandial blood glucose and insulin levels. ▪ Decreases fasting cholesterol and triglyceride levels. ▪ Important to bowel health and may protect against cancer.
Inulin	Onions, garlic, wheat, artichokes and bananas	▪ Functions as prebiotic. ▪ Increases mineral absorption. ▪ Prevents and treats obesity.
Beta-glucans	Endosperm of cereal grains, primarily barley and oats	▪ Decrease serum total and low-density lipoprotein cholesterol ▪ Lower down the post prandial glucose and insulin levels
Non-carbohydrate fibers		
Lignin	Root vegetables, wheat and fruits with edible seeds (such as strawberries, peaches, pears and plums)	▪ Flushes bile acids and cholesterol out of the intestines. ▪ Aids in the prevention of gallstones.

Physiological effects of fiber

Dietary fiber plays an important role in the maintenance of health and prevention of diseases. In most diseases, fat, sugar, or cholesterol has to be restricted but fiber is one constituent in diet, which is allowed in general in all diseases.

Effect on gastro-intestinal tract: In the gastrointestinal tract, some fibers form a matrix with fibrous characteristics, *i.e.*, some fibers because of their ability to swell within the aqueous medium, can trap water and water soluble nutrients such as sugars. The physical characteristics of the gastro-intestinal contents are altered by fiber sources. Since fiber is indigestible, the bulk of material in the gastrointestinal tract is greater during the transition of digesta (digested material) through the small intestine. The increase in volume is due to the water-holding capacity of certain fibers. The viscosity of intestinal contents increases due to the presence of viscous polysaccharides in the fibers. The changes in physical characteristics of the intestinal contents may influence gastric emptying, dilute enzymes and absorbable compounds in the gut, prevent starch from hydrolyzing and slow down the diffusion or mobility of enzymes, substrates and nutrients to the absorptive surface. These effects result in the slower appearance of nutrients such as glucose and some lipid molecules in the plasma following a meal.

- **Bowel regularity:** The fermentation of dietary fiber in large intestine produces short-chain fatty acids, *e.g.*, acetic, propionic and butyric acid. Acetic and propionic acid are used as fuel by the liver and muscles cells. In addition, propionic acid may be partly responsible for the cholesterol-lowering properties of fiber as it inhibits enzyme HMG-CoA reductase, which is involved in the production of cholesterol by the liver.

Butyric acid serves as the primary fuel for the cells of large intestine and helps in maintaining health and integrity of the colon. In addition, fermentable fibers permit healthy populations of friendly bacteria to grow and prevent pathogenic bacteria from surviving in the intestinal tract. This plays an important role in increasing immunity of the body. Fibers that are not fermented in the large intestine help in maintaining the bowel regularity by increasing the bulk of faeces and decreasing the transit time of fecal matter through the intestines. Dietary fibers speed up intestinal transit time when it is slow, slow it down when it is fast and enlarge small stools and moderate high pressure. Bowel regularity is associated with a decreased risk for colon cancer and haemorrhoids.

- **Excretion of faecal energy:** Fiber also helps in the formation of soft stools with bulk, which can be easily evacuated, thus, increasing fiber content of the diet increases the faecal energy excretion, principally in the form of fat and nitrogen.
- **Obesity:** A high fiber diet is recommended for weight reducing regimes. An increased intake of dietary fiber generally leads to an increased satiation and provides fewer calories, thus, helps in the treatment of obesity. This can be attributed to the fact that fibrous food items also known as cellulose and hemicellulose occupy a large space of stomach, thus, making feel full and thereby enable to avoid overeating.
- **Colon disease and hemorrhoids**: An increase in dietary fiber increases stool weight and the number of defecations but decreases the transit time in colon. This results in reduced reabsorption of water from the colon contributing to bulky and moist stools. A soft formed stool can be easily passed without straining and usually produces an improvement in the symptoms resulting from the disorders like hemorrhoids anal bleeding and prolapse. High fiber diet prevents constipation and helps to relieve this condition. Bowel regularity is associated with a decreased risk for colon cancer and hemorrhoids (when the hemorrhoids are related to straining and constipation). Secondly, dietary fibers consist of water-soluble and insoluble plant compounds, which are resistant to digestion by small intestine enzymes but are fermented by colonic bacteria. The products of fermentation like the short chain fatty acids also help to accelerate the movement of the faeces through the colon. Thirdly, dietary fiber increases the bacteria in the large intestines, which require nitrogen for their growth. This in turn reduces the chances of cancerous changes in cells by reducing the amount of ammonia in the large bowel.

Effect on cardiovascular system: Diets rich in fiber have hypocholesterolemic and hypotriglyceridemic effects. This action of dietary fiber is significant in the treatment of atherosclerosis, coronary heart disease, hypercholesterolemia and hyperlipidemia. High fiber content also alters biliary lipid and bile salt metabolism making the bile less saturated with cholesterol. This bile would less likely precipitate its cholesterol and form gallstones.

Effect on diabetes mellitus: High intake of fiber has proved to be useful in the prevention and treatment of diabetes mellitus since dietary fiber has a blood glucose reducing effect as it is manifested by a diminished glycemic index. The rate at which food leaves the stomach is slowed down and this delays the absorption of glucose following a meal. The absorbed glucose is released slowly into the blood circulation, hence, resulting in decreased insulin secretion. Secondly, it promotes a sense of satiety or fullness after a meal, which helps to prevent overeating. This keeps a check on the energy intake, resulting in decreased body weight. As the weight decreases, the number of insulin receptor increases.

Effect on minerals: Purified dietary fibers may reduce acutely the absorption of some vitamins and minerals such as calcium, iron and zinc by binding or entrapping them in the small intestinal lumen. These changes are significantly noticed when the diet rich in fiber is consumed for relatively long periods.

Negative Effects of Dietary Fiber

- While weight management requires feeling of fullness to come at the earliest, it can also be a negative aspect when young growing children need to be fed adequate food to supply energy for their high physical activity.
- Convalescing patients also need enough calories to regain their health. If the diet contains more fiber, they feel full much earlier and may not consume enough of food that may be necessary. Thus, depending on the situation, feeling of fullness can be a benefit or an adverse effect.
- Dietary fiber must also be accompanied by adequate water intake or else they may cause impaction and intestinal obstruction due to lack of fluidity or softness.

  Some other potential adverse effects of dietary fiber include:

  - Bloating caused by the gas formation of the microorganisms while fermentation in the large intestine
  - Additional contamination with xenobiotics
  - Direct epithelial damages (intestinal mucosa)
  - Bonding of calcium, magnesium, iron and zinc, which decrease their absorption rate.
  - Volvulus (twisting of the bowels) as a result of excess filling up of the large intestines.

Requirement of Dietary Fiber and Dietary Guidelines

Modern diet is mostly consisting of products made of refined flours and of animal sources like meat, fish, poultry, eggs and milk products. All of these lack fiber and there are many ailments associated with lack of fiber are quite prevalent in modern society. Various organizations including American Dietetic Association (ADA), National Academy of Science, British Nutrition Foundation, *etc.* have recommended increasing the dietary fiber in the diets recommending up to 35 g fiber per day. Food and Drug Administration of United States has given approval for making health claims for fiber on labels of food products rich in it. The recent trend in dietary consumption pattern in most of the Indian populations, have several dietary imbalances including low intake of monounsaturated fatty acids (MUFA), polyunsaturated fatty acid (*n*-3 PUFA) and fiber, and high intake of saturated fats, carbohydrates and trans fatty acids (mostly related to the widespread use of *vanaspati*, a hydrogenated dietary fiber. Poor intake of fruits and vegetables resulting in low fiber intake was reported in less educated urban in North India. Many studies in rural populations in India including a subset of rural pregnant women have shown a poor intake of fruits and vegetables.

Quantity of Fibers Needed

In general, adult men up to the age of 50 should consume at least 38 grams of dietary fiber each day, while the women up to the age of 50 should consume at least 25 grams of fibers per day. After the age of 50, fiber need is decreased slightly. Unfortunately, most people do not meet these recommendations. In fact, on an average, people consume only 14 grams of fiber per day. Your individual daily fiber intake goal is really based on your calorie consumption. It is recommended that you should get 14 grams of fiber for every 1,000 calories you eat, *e.g.*, if you eat diet containing 2,500 calories a day, you should be getting about 35 grams of fiber a day. Each person's fiber intake needs to be individually assessed for original total fiber content and the contribution of the different fiber sources in your diet needs to be ascertained. This information serves then as a baseline for dietary manipulation. It should be noted that there is a very fine line to be drawn between having the correct amount and overdoing it in terms of fiber intake. Prudent guidelines, *i.e.*, the recommended average for the population, recommend a total fiber intake of 20-35 grams per day of which 6-10.5 grams needs to be soluble.

How to Achieve It

For increasing your fiber intake, the followings ways may be followed:

- **By consuming a healthy natural high fiber diet:** This option is undoubtedly the best, as it will ensure that not only your fiber sources are natural and varied but also your diet will contain the correct balance of other nutrients.
- **By eating fiber-enriched food:** The food items such as specially produced high fiber muffins and biscuits will most certainly increase your insoluble fiber intake but will not rectify any other dietary imbalances. Such items are often too high in fat and sugar to be considered part of a healthy balanced
- **By taking fiber supplements:** Fiber extracts or in the form of bran, pills, tablets, powders and granules. Supervised use of bulk forming laxatives can be beneficial in the early stages of dietary manipulation but should not be viewed as a long-term solution.

Some Tips of Increasing Dietary Fiber Intake

- Eat vegetables and whole fruits, which are good source of carbohydrates but are higher in fiber, low in glycemic load and high in nutrients density.
- Replace white rice, bread and pasta with brown rice and whole-grain products.
- Choose whole-grain and unprocessed cereals for breakfast.
- Snack on raw vegetables instead of chips, crackers, or chocolate bars.
- Substitute legumes for meat two to three times per week in hot pepper and soups.

Precautions with Dietary Fiber Intake

- The persons not using high fiber diets should gradually increase their fiber intake.
- It should always be taken along with adequate amount of water to prevent impaction and intestinal obstruction.
- It should be used with caution in people with gastrointestinal diseases.

Recommended Fiber Intake

A balanced diet is one that provides adequate amounts of energy and nutrients for health and well-being. There are many scientific bodies recommending high intakes of dietary fiber. Indian traditional diets with many vegetables, whole grain cereals products including loafs, chapatti, brown (parboiled) rice, *etc.*, *dals* and other pulses like kidney beans, lentils, Bengal gram, peas, *etc.* along with fruits would give adequate amounts of both soluble and insoluble fibers. However, Indians have been slowly getting away from this traditional diet to more westernized diet with new lifestyle foods that have much less fiber. High fiber foods not only give enough fiber but also contain other nutrients like vitamins and minerals, thus, diet is overall more nutritious and healthful. There is a need for increasing fiber intake not only to prevent certain age-related diseases like diabetes and heart disease but also obesity that affects all ages. Childhood obesity can be curtailed to a great extent, if fiber content of the diet is increased. Children feel full earlier and for a longer period and so tendency of overeating is reduced. Low fiber foodstuffs also have the tendency to increase the blood glucose sharply followed by a sharp decline, compared to high fiber food items of similar carbohydrate contents. This sharp drop also fails to supply energy for a longer time without replenishing again with more food. When enough fiber is there, especially the soluble fiber, absorption of glucose from intestine to blood is more gradual and for a longer time. Hence, there is a continuous supply of energy needed for mental and/or physical activity. Children especially need it, as they are more active as compared to adults. When they feel lacking in energy, they eat or drink beverages supplying enough calories. Thus, total caloric intake is much more and they put on weight.

Indian diets have high dietary fiber content. Therefore, the dietary requirements as recommended by the Indian Council for Medical Research (ICMR) do appear excessive. An average Indian diet contains about 6 to 8.5 g of crude fiber. Today, due to western influence and urbanization, the consumption of highly refined food has increased, thereby, reducing the intake of dietary fiber. Thus, the incidence of diseases coupled with low levels of fiber intake is rapidly increasing in the urbanized class of society. In some parts of India, high fiber diets are common in the poorer class of society. Here, the emphasis must be to look for the healthy diets with addition of minerals and vitamins in adequate amounts.

Dietary Fiber Content of the Indian Diet

Krentler (1980) has listed foods giving their approximate fiber content according to the serving size.

Product	Serving Size	Crude Fiber	Dietary Fiber (g per serving)
Cereal, bread and flour			
Whole grain flour	½ cup	–	13.2
White flour	½ cup	–	4.0
Oatmeal cooked	1 cup	0.5	–
Corn flakes	1 cup	0.2	3.08
Bread whole wheat	1 slice	0.4	2.0

Product	Serving Size	Crude Fiber	Dietary Fiber (g per serving)
Legumes			
Chick peas	½ cup	5.0	–
Beans	½ cup	4.3	–
Lentils	½ cup	3.9	–
Nuts and seeds			
Sunflower seeds	$3^{1/2}$ oz	3.8	–
Sesame seeds	$3^{1/2}$ g	2.4	–
Walnuts	½ cup	2.1	–
Peanuts roasted (with skin)	1 tablespoon	0.5	1.4
Pistachio nuts	30 number	0.3	–
Cashew roasted	6-8 number	0.2	–
Fruits and vegetables			
Peas green	2/3 Cup	2.0	7.0
Pear	1 Medium	–	3.9
Carrot	2/3 Cup	1.5	3.8
Banana	1 Medium	–	3.1
Cabbage (cooked)	2/3 Cup	–	2.8
Apple with skin	1 Medium	2.0	2.6
Strawberries	½ Cup	1.0	1.6
Tomato raw	1 Medium	–	1.4
Plum	1 Medium	–	1.1
Corn (canned)	½ Cup	0.7	–
Raisins (dried/seedless)	½ Cup	0.6	–
Spinach	½ cup	0.4	–
Grapes green (seedless)	18-20 number	0.2	–
Fruit juice	1 Cup	0.2	–

*1 Cup = 150 ml

Dietary Fiber Intake in India

In India, the daily intake of dietary fiber recommendations suggests that adults consume around 40 grams of dietary fiber per day from a variety of sources. In general, this level of dietary fiber intake is easily achieved by Indian diets, being based predominantly on unrefined cereals and plant foods. The dietary fiber contents of several Indian food items have been shown in Table 2. Dietary fiber intake in India by different socioeconomic groups also varies from 60 to 70 g/day, depending upon the type of cereal consumed (Table 1). Dietary fiber intakes from wheat- or millet-based diets are generally higher than from a rice-based diet. Nearly 80% of the fiber content of the diets consumed in India, particularly among the low-income group, is contributed by cereals. Although the fiber intake of Indians through their cereal-based diets is adequate according to the currently recommended desirable intakes but the exact composition of the individual fiber fractions in most Indian foods has not been determined.

Table 1 : Daily requirement of dietary fiber in different age groups

Age	Fiber requirement (g/day)
Children	
1-3 years	19
4-8 years	25
Males	
9-13 years	31
14-18 years	38
19-50 years	38
51+ years	30
Females	
9-13 years	26
14-18 years	26
19-50 years	25
51+ years	21
Pregnancy	
≤18 years	28
18+ years	28
Lactation	
≤18 years	29
18+ years	29

Source: Quarterly medical review report, 2005

Table 2 : Dietary fiber content of some common Indian foods

Foodstuff	Energy (kcal)	Crude fiber (g)	Dietary fiber (g)
Cereals and millets			
Rice	345	0.2	4.1-8.3
Wheat	346	1.2	11.4-17.2
Sorghum	349	1.6	9.7-14.3

Foodstuff	Energy (kcal)	Crude fiber (g)	Dietary fiber (g)
Pearl millet	361	1.6	11.8-20.3
Ragi	328	3.6	11.5-18.6
Pulses and legumes			
Green gram (whole)	334	4.1	15.2
Green gram (dhal)	348	0.8	13.5
Black gram (dhal)	347	0.9	14.3
Red gram (dhal)	335	1.5	14.1
Bengal gram (whole)	360	3.9	26.6
Bengal gram (dhal)	372	1.2	13.6
Nuts and oilseeds			
Groundnut	567	3.1	6.1
Coconut (dry)	662	6.6	8.9
Roots and tubers			
Sweet potato	120	0.8	7.3
Potato	97	0.4	4.0
Yam	79	0.8	5.3
Fruits			
Banana	116	0.4	2.5
Mango	74	0.7	2.3
Vegetables			
Amaranth	45	1.0	3.4
Beet leaf	26	0.6	5.0
Brinjal	24	1.3	2.0
Ridge gourd	17	0.5	5.7
Snake gourd	18	0.8	1.8
Bottle gourd	12	0.6	2.8
Yellow pumpkin	23	0.7	0.5

Table 3 : Dietary fiber intake by urban and rural groups in India†

Group	Energy (kcal)	Dietary fiber‡
Urban		
High income	2604	59.4
Middle income	2230	61.5

Group	Energy (kcal)	Dietary fiber‡
Low income	2008	67.7
Rural		
Average for different income groups	2272	72.7

†Average intake per consumption unit; ‡From cereal-based diet

Source: Anonymous, 1993. Annual Report 1992-93. National Institute of Nutrition, Hyderabad, India.

Conclusion

In a simplified definition, dietary fiber is a carbohydrate that resists digestion and absorption and may or may not undergo microbial fermentation in the large intestine. Dietary fiber consists of many different constituents including arabinoxylan, inulin, beta-glucan, pectin, resistant starches, *etc.* These individual components of dietary fiber play an important role in improving human health. However, excellent sources of fiber include wholegrain cereals, pulses such as peas, beans and lentils, fresh fruits and vegetables, dried fruits and vegetables, dried fruits and nuts that can be added to the diet to increase the fiber intake. For all kinds of disease conditions, there is restriction of food in some form or other but fiber is the only food, which is allowed for almost all diseases.

There exists an inverse relationship between fiber consumption and the risk for coronary heart disease and several types of cancer. This is attributed to several factors including increasing bile acid excretion, decreased caloric intake, increased short chain fatty acid production, carcinogen binding effects and increased antioxidants. Dietary fibers also play a significant role in conditions such as obesity and diabetes. The digestive and viscosity characteristics of dietary fiber are the likely modes of action, which affect diabetes and obesity risk. These result in decrease of nutrient absorption, therefore, decreasing the metabolizable energy.

High-fiber food articles are good for health but adding too much fiber too quickly can promote intestinal gas, abdominal bloating and cramping. The fiber should be increased in diet gradually over a period of a few weeks. This allows the natural bacteria of the digestive system to adapt to the change. In addition, plenty of water should be ingested as the fiber works best when it absorbs water, resulting in soft and bulky stool. Overall, it is better to take a well-balanced high fiber diet with lots of water.

■■■

5 Wholegrain Food and Health

M.K. Rana

Department of Vegetable Science, CCS Haryana Agricultural University, Hisar-125 004, Haryana, India, E-mail: <mkrlotus@gmail.com>

A balanced human diet should have seven basic food constituents, *viz.* carbohydrates, proteins, fats, minerals, vitamins, fibers and water. The nutritionists worldwide generally agree that a quality diet, useful for good health, should consist of eating plenty of fiber-rich food articles, including plenty of whole grains, fruits and vegetables. The carbohydrates of nutritional importance usually present in food items of plant origin including starch (insoluble in water) and sugars are metabolized in human intestinal tract to glucose that serves as a key source of energy. Carbohydrates in foodstuffs contain C, H and O in proportion of $C(H_2O)$. The dietary food of plant origin also contains numerous non-starchy polysaccharides (NSPS), which are not at all metabolized by human digestive system to glucose and not utilized for energy. The animals, particularly the ruminants, can metabolize the non-starchy polysaccharides due to the presence of a definite kind of flora in their digestive tracts and utilize them for energy purposes. Non-starchy polysaccharides, collectively known as food or dietary fibers, are the polymers of glucose. Among the non-starchy polysaccharides, cellulosic material, which in contrast to vegetarian food is completely absent in meat, egg, fish, milk and other dairy products, is water insoluble and rests are soluble in water. These non-starchy polysaccharides jointly act as roughage in the foodstuffs. In spite of having only a minor role as an energy source, non-starchy polysaccharides improve the overall texture of food and play a momentous role in mouth-feel, storage life and in acceptability of foodstuffs by the consumers.

Earlier, it was contemplated that dietary non-starchy polysaccharides provide no energy to the human body and are excreted from the body undigested as such, however, the new concept indicates that non-starchy polysaccharides on their fermentation caused by colon bacteria yield short chain fatty acids (SCFAs), which provide energy to the colon and keep it healthy and prevent it from cancer and other diseases. Having positive health effects, the short chain fatty acids also help in maintaining intestinal electrolytic balance, ensuring smooth bowel movement and in preventing constipation. Low or fiber free diet of modern society is a causative factor for constipation, obesity, cardiovascular disease, diabetes, hemorrhoids and colorectal cancer. A low-fiber diet may also be implicated in the etiology of childhood constipation and appendicitis, thus, inclusion of fiber rich food in diet is now accepted. *Isabgol (Psyllium* seed husk) containing lot of non-starchy polysaccharides with dietary fiber property is usually used to eliminate constipation, and recently, it has been shown to lower blood cholesterol levels, and these properties are thought to be due to its amazing gel-forming characteristic. It is well known all over the world that dietary fiber from grains such as wheat and oats increases stool weight, alters gut transit time, alters activity of colonic microflora, influences appetite, absorbs toxins and modifies the absorption of fats, sugars, minerals and bile acids, however, the extent to which the specific polysaccharides exert their physiological effects is dependent on a complex mixture of structural and physico-chemical properties. The physiological properties of polysaccharides are difficult to predict based on their structure alone. They are partly predictable based on physico-chemical properties such as fermentation, water-holding capacity, viscosity and bile acid binding but water binding has a special important relevance to the physiological actions of dietary fiber.

Definition of Dietary Fiber

Earlier, the term *dietary fiber* was used to explain the remnants of plant components, which were resistant to hydrolysis by enzymes present in human alimentary canal, thus, it is a physiological description related to indigestibility in human small intestine. Plant cell wall components (*i.e.*, cellulose, hemicellulose and lignin) and associated minor substances (*i.e.*, waxes, cutin and suberin) are the major source of digestion-resistant material. Later, the definition of *dietary fiber* was expanded to include all indigestible plant polysaccharides such as cellulose, hemicellulose, lignin, gums, modified celluloses, mucilages, oligosaccharides and pectins, and associated minor substances such as waxes, cutin, lignin and suberin. Finally, the American Association of Cereal Chemists (AACC) concluded that dietary fibers are the remnants of edible part of plants and analogous carbohydrates that are resistant to digestion and absorption in human small intestine with complete or partial fermentation in human large intestine. Dietary fibers include polysaccharides (cellulose and hemicellulose), oligosaccharides (short chain polysaccharides), lignin and associated plant substances.

Generally, the analogous carbohydrates are produced during processing of food by chemical and physical processes affecting the digestibility of starches. For other dietary fibers, the polysaccharides such as β-glucans provide the gummy gelatinous characteristic of soluble dietary fibers. All non-digestible fermentable polysaccharides such as polyglucoses (cellulose or beta-glucans), polyfructoses (arabinogalactans), and analogous carbohydrates are included in the definition of dietary fiber. Though lignin as such is not a polysaccharide, it is intricately tied to the dietary fiber polysaccharides in foods and increases resistance against digestion. Associated plant substances such as waxes, cutin and suberin are indigestible fatty acid derivatives, which like lignin, are intricately tied with the dietary fiber polysaccharides, often serving as chemical cross-links between various other components and increasing resistance to digestion.

This definition reveals several key concepts, *i.e.*, (i) dietary fiber has been defined on the basis of edibility (resistant to digestion and absorption in small intestine, fermentation in large intestine) and behaviour in the body it exhibits; this definition refers those plant components that impart positive health effects, (ii) analogous dietary fiber is defined as those materials, not essentially inherent to a part of a plant as consumed, exhibiting the digestion and fermentation properties of fiber. Analogous fiber, in addition to the requisite digestion and fermentation properties, must also exhibit a positive potential health benefit that has been ascribed to dietary fiber and (iii) the definition obviously describes the meaning of dietary and analogous fiber, and defines the key and relevant functional properties of dietary fiber.

Functions of Dietary Fiber in Human Body

Dietary fibers promote favourable physiological effects and have inverse correlation with the incidence of a range of diseases. Dietary fibers in diet increase faecal bulking and softening and reduce the level of cholesterol and glucose in blood. Increased consumption of fiber-rich food items in place of other foods in a particular meal produces a positive adjustment in levels of serum cholesterol, which is related to the risk of coronary heart disease. An increased consumption of dietary fibers and fiber-rich foods also produces an appreciable reduction in peak level of serum glucose, an effect normally deemed as beneficial to health, however, not all fibers or fiber-rich foods produce beneficial health effects. The highest dietary fiber intakes has been found to be associated with a lower body mass index (BMI), blood systolic pressure, plasma triacylglycerols and plasma glucose in men and lower BMI in women but the most beneficial health effect of a high dietary fibers intake has been found on risk of cardiovascular disease.

Ingestion of fiber-rich diet slows down the absorption of sugars in intestines and reduces hyperglycemia and insulin level in diabetics as well as in healthy individuals. Though it is not precisely implicit how soluble dietary fibers slow down the absorption of sugar but it is alleged that they make the digested food more viscous as a result reduce the absorption of food constituents in the intestines. Dietary fibers also entrap the enzymes, *i.e.*, amylase and trypsin accountable for the hydrolysis of starch and protein, respectively, as a result slow down the digestion of food substrates and further absorption of glucose in intestine. Food also contains soluble dietary fibers that have certain crucial physiological functions and health benefits. The most effective dietary fibers that make the food more viscous in human digestive tract and slows down the absorption of glucose in intestine are acacia gum, deploymerized guar gum, cornhusk and bran of other food cereals, however, guar gum is one of the most widely stipulated food fibers for controlling sugar metabolism in diabetes patients and in those having high serum lipids and high blood pressure. The bran of wheat and corn is also well known to contain oil rich in phytosterols (precursor of vitamin D), antioxidants (vitamin E) and to have cholesterol-lowering property.

Inclusion of food fibers in diet lowers down the blood lipids (triglycerides), particularly low-density lipids and cholesterol, which regulate blood pressure and ultimately reduce the incidence of coronary heart disease. The individuals with hypertension, high blood pressure and arteriosclerosis are advised to include more food fibers in diet. Inclusion of soluble dietary fibers in specifically devised foods can also impersonate the fats, and thus, the fats can be excluded without forfeiting the food texture. Dietary fibers being bulking agent can be customized to impersonate starch, sugar, or lipids, thus, reducing one or more of these food constituents, and replacing with proper fiber can retain the original texture of a food.

In addition to dietary fibers and resistant starch, food grains also contain considerable amounts of oligosaccharides, which are described as carbohydrates with a low degree of polymerization. Common oligosaccharides comprise oligofructose and inulin. The new definition of dietary fibers includes oligosaccharides, which have been shown to have prebiotic effects since fibers increase the growth of lactic acid bacteria (LAB), particularly of bifidobacteria. These bacteria are analogous to the organisms used as probiotics, which have been shown to reduce the incidence of diarrhea. There are potential health benefits of including oligosaccharides, particularly in diet of children. The little intestine of infant may also result in a greater amount of starch entering the colon during weaning. The action of prebiotics in promoting the growth of lactic acid bacteria suggests that in childhood they could be identically useful as the probiotics. Adding prebiotic mixture of oligosaccharides to infant diet can increase the survival of bifidobacteria in gut and can alter faecal flora. Moreover, the ability of wholegrain components to alter the gut microflora has vital role in health promotion and disease suppression. Clinical studies indicate that consumption of fructo-oligosaccharides increases the bifidobacteria population in human gut while decreasing the concentrations of *Escherichia coli*, clostridia and bacteroides.

Dietary fiber generally has inverse correlation with obesity and body fat. Controlling metabolic rates and hormonal responses dietary fiber regulates body weight. Fiber-rich food promotes satiation (lower meal energy content) and satiety (longer duration between meals) due to its bulk and relatively low energy mass, leading to dwindled energy intake. Wholegrain foods provide 20-50% of their fibers in soluble or viscous forms. In contrast to insoluble fiber, soluble fiber results in highly viscous intestinal contents with gel-like properties that may delay intestinal absorption. Grains with high soluble fiber (oats, rye and barley) improve insulin sensitivity by slowing down the absorption of macronutrients. In small intestines, soluble fiber may blunt postprandial glycaemic and insulinaemic responses that are linked to reduction in the rate of return of hunger and subsequent energy intake. It was observed that oat gum added to a glucose solution reduced the glucose and insulin responses of healthy adults. In north India, a nursing mother after childbirth is fed more fiber-rich foods in the form of acacia gum and fenugreek, which prevent obesity and constipation in addition to slowing down the absorption of sugar and lipids.

Dietary fiber also affects the secretion of gut hormones, including cholecystokinin that alters glucose homeostasis. The cells in small intestine on ingestion of fiber-rich food secrete cholecystokinin, which stimulates pancreatic secretion and regulates gastric emptying and central motivation of satiety, while dietary fiber due to its hormonal effects is believed to alter the metabolic fuel partitioning in favor of fat oxidation. Diet with low fiber content and high glycaemic load is positively correlated with risk of non-insulin dependent diabetes mellitus. Hyperinsulinaemia, a sign of insulin resistance, plays a significant role in the progression of diabetes and is further associated with obesity. Chronically high insulin demand induced by high glycaemic load and low fiber may contribute to obesity. The wholegrain bran components due to their insulin sensitizing effects may be involved in the regulation of obesity. It has been reported that consumption of wheat bran for longer period improves glucose tolerance better. Moreover, wholegrain cereals have lower digestion rate than refined ones and have consequently a lower glycaemic index and related increase in plasma insulin concentration. Glycaemic index and glycaemic load have been shown inversely related to the risk of breast, colorectal and gastric cancer. Clinical studies show an inverse correlation between wholegrain cereals and risk of oral, oesophageal and laryngeal, colorectal and gastric cancer, whereas, refined cereals show direct relationship with risk.

Fiber can also act as an antioxidant preventing many infections and slowing down the ageing process. The antioxidants in wholegrain (in bran and germ) also contribute to insulin sensitivity by protecting against oxidative stress, which is associated with reduced insulin-dependent glucose disposal and diabetic complications. Magnesium (Mg) and vitamin E in particular are considered to be involved in insulin metabolism, which may prevent hyperinsulinaemia. In clinical experiments, magnesium supplementation provides beneficial effects on insulin sensitivity among patients with non-insulin-dependent diabetes mellitus.

Dietary Fiber in Naturally Occurring Foodstuffs

Most men and women have low to moderate intakes of food fiber (mean 21.0 and 17.1 g/day respectively), with only 21% of the men and 7% of the women having total food fiber intakes at recommended level (>25 g/day) and soluble fiber accounting for 19% of the total dietary fiber intake for both the genders. The main foods that provide dietary fiber are cereals (30-35%), vegetables (20-24%) and fruits (19-22%). The diet based on cereals, pulses, fruits and vegetables supplies dietary fiber in sufficient amount, and thus, wholegrain cereals along with fruits and vegetables have been the traditional diet in India, however, the recent drastic changes in life styles of developed countries population have led to dietary patterns mostly based on animal foodstuffs and refined cereals with a reduced supply of dietary fiber. Actually a long back, the dietary fibers (bran and germs) were thought to be useless part of food and regarded as waste that could only be fed to the animals, but in late 1970s, this attitude started changing when a lack of food fibers in diet was found concerned with the occurrence of diseases such as diverticulosis, constipation, cardiovascular heart disease (CHD) and colorectal cancer. Epidemiological studies demonstrate a deep correlation between dietary fiber intake and cardiovascular risk and death rates due to cardiovascular heart disease. One of the pioneers showed a 4 folds lower death rate due to cardiovascular heart disease in individuals ingesting high amounts of fiber (≥37 g/day) compared with those ingesting lowest amount (<20 g/day). The surveys confirm the protective effect of dietary fibers against cardiovascular heart disease. Clinical studies have repeatedly shown that only soluble fiber (3 g/day) from various sources has a hypocholesterolaemic effect.

Cereals' bran contains cellulose as food fibers, which are lost to a great extent during manufacturing of white-flour, usually used for the preparation of purified starch and processed packaged food. Normally, dietary fibers are lost during processing and packaging of foods. Some legumes, especially cluster bean (guar) contain polysaccharide mucilage, which is high-quality food fiber. In fruits and vegetables, pectin is the foremost food fiber. In Far Eastern countries, seaweeds containing polysaccharides (food fiber) are customarily employed to thicken the gels. A diet composed of meat and other non-vegetarian is low in food fiber but a conventionally home cooked Indian food by and large contains an essential amount of dietary fiber and does not require any supplementation, but now in India too, the trend of using processed foods, which are usually low in dietary fibers, is bit by bit escalating. Besides guar gum based food fiber, cornhusk and wheat bran are also low viscosity non-starchy polysaccharides, which are recommended additives for fortifying the foods. Now in India, the food fibers based on wheat bran are being marketed as food additives.

Potential Actions of Dietary Fiber in Infants and Children

The great concern about recommending a fiber-rich diet for children is that the dietary fiber would have a major effect on growth by reducing total energy intake. Secondary to this aspect is the concern that the dietary fiber may trap micronutrients within gut, and therefore, reduce bioavailability at a time when demand for these nutrients in the body is high. A study of vegetarian children did find lower growth rates in children less than two years. Vegetarian children are reported to have short stature than normal controls. There are reports that demonstrate growth delay in vegetarian children and children on macrobiotic diets. On the other hand, the other reports show no association between intake of fiber-rich diet and growth rate. However, several studies have shown the role of low-fiber diets in the etiology of childhood constipation. World over, it is well known that a low-fiber diet is the root cause in the development of chronic constipation, and the children with constipation have lower energy and nutrient intakes. Dietary fiber alone is negatively correlated with constipation, with no relationship with age but other studies have not found a strong association between fiber-rich diet and constipation. It has been seen that the children with appendicitis had a lower fiber intake (17.4 g/day *verses* 20.4 g/day) than the controls, and the patients were much more likely to have had a history of chronic constipation than the controls.

Wholegrain Consumption *Verses* Chronic Disease Risk

Wholegrain consumption as essential component of diet is described as the *prudent diet* since whole grains are rich sources of fermentable carbohydrates, insoluble dietary fiber, resistant starch, oligosaccharides, minerals, vitamins and several other phytochemicals such as antioxidants, indoles, phenolic compounds, including flavonoids, isothiocyanates and phyto-oestrogens, which are potential components of diet. The structure of all food grains, which includes endosperm (middle layer), germ (inner layer) and bran (outer layer- seed coat), is similar, and the absolute amount of each of these components varies among grains. Components in wholegrain associated with improved health status include lignans, tocotrienols and phenolic compounds, and anti-nutrients such as phytic acid, tannins and enzyme inhibitors. The bran comprises fatty acids, phytochemicals and fibers, however, during milling (refining) of grains, the bran and germ get separated from the starchy endosperm, which leads to the loss of dietary fiber, vitamins, minerals, lignans, phyto-oestrogens, phenolic compounds and phytic acid, thus, the foods prepared from refined grains are nutritionally inferior to those prepared from wholegrain because they contain higher content of starch and lower amounts of fibers, minerals, vitamins, phenols, phyto-oestrogens and unsaturated fatty acids. Wholegrain foods are said to be protective against a wide range of diseases, and generally, this protectiveness is greater than that seen with any individual ingredient.

Undigested carbohydrates reaching the colon are fermented by intestinal micro-flora to short-chain fatty acids, which include acetate, butyrate and propionate, with butyrate being a preferred fuel for the colonic mucosa cells. These short-chain fatty acids are beneficial to keep the serum cholesterol low and to decrease the risk of cancer. Undigested carbohydrates increase faecal weight and intestinal transit. A comparison of dietary fiber contents of various whole grains reveals that oats, barley, and rye contain about one-third soluble fiber and rest is insoluble fiber. Soluble fiber is associated with lowering of cholesterol and improved glucose response, while insoluble fiber is associated with improved bowel emptying. Wheat is lower in soluble fiber than most grains, while rice contains virtually no soluble fiber. Refining of grains results in low total dietary fiber content and removes proportionally more of the insoluble fiber than soluble fiber. Highly refined food articles contribute to diseases including coronary heart disease (CHD). Conversely, wholegrain foods are associated with a reduced risk for numerous chronic diseases including coronary heart disease, diabetes and certain types of cancer. Foods containing wholegrain protect against chronic diseases by altering serum cholesterol profiles, exerting antioxidant properties and anti-thrombotic action, and through their favorable effects on vascular reactivity and insulin sensitivity.

In contrast to refined-grain foodstuffs, wholegrain foods in general are digested and absorbed more slowly, resulting in smaller postprandial glucose responses and insulin demand on pancreatic cells. Refined-grain foods have more than double the glycaemic and insulinaemic response compared with wholegrain foods. It is well documented that glucose and insulin are associated with chronic diseases, particularly diabetes and colon cancer. Epidemiological studies constantly show that consumption of whole grains decreases the risk for type-II diabetes mellitus. Insulin is an important growth factor of colonic epithelial cells and is a mitogen of tumor cell growth. Epidemiological studies also indicate a similarity of factors that produce elevated insulin levels with those related to colon cancer risk, including obesity and low physical activity. Studies have linked the incident of colon cancer to higher levels of blood glucose and insulin and larger body weight, and the women with diabetes have been found to have higher risk of colorectal cancer.

Consumption of wholegrain food items has been recommended to improve insulin sensitivity and to lower serum insulin as high insulin level may promote obesity by altering adipose tissue physiology, shunting metabolic fuels from oxidation to storage and by increasing appetite. Hyperinsulinaemia has been shown associated with excessive weight-gain, while the intake of wholegrain foods has been hypothesized to protect against obesity and inversely related to body mass index. Studies indicate an association between wholegrain intake and insulin sensitivity, thereby providing indirect support for causal role of wholegrain ingestion on body-weight regulation. Wholegrain intake is negatively related to fasting insulin levels, magnesium and fiber. Similar association has been shown for cereal-bran intake and insulin levels. Intake of wholegrain diet has also been inversely associated with waist : hip ratio, total cholesterol and LDL-cholesterol. Ingestion of wholegrain and legume powder can improve biomarkers relevant to diabetes and coronary heart disease, and reduce insulin demand, lipid peroxidation and plasma homocysteine concentrations in patients with coronary heart disease. Lastly, consumption of wholegrain oats cereal is found to be associated with improved blood pressure control and reduced need for anti-hypertension medicine.

Glycaemic index (GI) is a marker used to compare glycaemic response to foods. The other factors affecting the response include the type of food, cooking, processing, and fat and soluble fiber content of food. Wholegrain foods are known to slow down the digestion and absorption of carbohydrates. Food structure greatly affects the postprandial blood glucose and insulin responses. Any process that disrupts the structure of food components increases the plasma glucose and insulin responses. Food structure has been found to be more important than the presence of dietary fiber in determining glycaemic response. Refining of grains tends to increase glycaemic response, thus, on the contrary, the foodstuffs containing wholegrain slow down the glycaemic response. Intact whole grains of barley, rice, oats, maize, wheat, buckwheat and rye have a glycaemic index value between 36 and 81, with barley and oats having the lowest values. Lower blood glucose levels and decreased insulin secretion have been observed in both normal and diabetic individuals while consuming a low glycaemic index diet containing bread with intact whole grains, parboiled wheat, pasta and legumes compared to a high glycaemic index diet containing white bread and potato. Plasma insulin responses are intensified gradually with whole grains < cracked grains < coarse flour < fine flour. Oat-based diets show lesser glucose and insulin responses than wheat- or maize-based diets. Particle size influences the digestion speed of maize and wheat but not of oats. The increased insulin response to gauzily ground flour could be pertinent to the etiology of diseases associated with hyperinsulinaemia and to the administration of diabetes. Consumption of wholegrain meals improves insulin sensitivity in overweight and obese adults, as insulin sensitivity might be an important mechanism whereby wholegrain foodstuffs reduce the risk of heart disease.

The prospective studies show that a low glycaemic index diet or a diet rich in wholegrain foodstuffs reduces the risk of type-II diabetes. The data available in various journals reveal a negative relationship between serum dietary glycaemic index and HDL-cholesterol. Indeed, the dietary glycaemic index is a potent predictor of serum HDL-cholesterol than dietary fat. Besides, a low glycaemic index diet may reduce the risk of myocardial infarction (heart attack) in women but question arises whether dietary fiber could be a more important deterrent factor than the glycaemic index characteristics. The most commonly consumed wholegrain products such as breakfast cereals and flour-based bread have high rather than low glycaemic index. Cereal fiber may have a preventive effect but the execution of glycaemic index concept in dietary recommendation necessitates a much wider range of low glycaemic index foods. Generally, there is lack of low glycaemic index cereal-based foods. As a result, there is a need for new technologies, which can be used to develop products with low glycaemic index. One obvious option may be to encourage the use of wholegrain bread, as the wholegrain breads are usually a low glycaemic index food, however, the challenge for the food industry will be to find technologies, which can be used to lower the glycaemic index of meal-based bread since the difference in glycaemic index properties between whole meal bread from common cereals and white bread is usually marginal.

Antioxidants in Wholegrain Foodstuffs

The key defensive mechanism of antioxidants in human body is to nullify the actions of free radicals that assault directly on DNA, lipids and protein, the initiating factor for a number of chronic diseases, and the damage to cellular membrane caused by free radical is considered the key causative factor. Cellular damage causes a shift in net charge of the cell, changing osmotic pressure, leading to swelling and eventual death of the cells. Antioxidants in foodstuffs are the phytochemicals that protect the DNA from oxidative damage and mutation, which can lead to cancer. Environmental adversities and normal food metabolism are the major sources for the fabrication of these harming free radicals. Human body has the defense mechanism to prevent itself from the harmful effects of free radicals and to repair oxidative damage caused by reactive oxygen species but numerous ailments may develop when the body defense mechanism is not adequately strong. The disease potential may be diminished to a great extent, if the dietary foods have enough antioxidants to reduce the activity of free radicals in the body.

Meals containing whole grains are high in antioxidant activity. The antioxidants in wholegrain foods are water- and fat-soluble and about half of are insoluble. A lion's share of insoluble antioxidants is bound as cinnamic acid esters to arabinoxylan side chains of hemicellulose. In oatmeal, the soluble antioxidants include flavonoids, tocopherols, phenolic acids, and avenanthramides. Wheat bran insoluble fibers contain around 0.5-1.0% phenolic substances, particularly covalently bound phenolic acids that have good free radical scrubbing capacity. Antioxidant activity is an inherent property of insoluble grain fiber. The diets containing wholegrain provide antioxidant safeguard over extended period through the entire digestive tract. In addition to natural antioxidants, browning reactions during baking and toasting processes increase the total antioxidant activity in grain-based final product as compared to raw ingredients, as a result, the crust of white bread has double the antioxidant activity of the starting flour or crust-free bread. Maillard reactions may increase the antioxidant activity of food.

Phytic acid concentrated in grains is a well-known antioxidant. Colonic bacteria produce free radicals in appreciable amounts and dietary phytic acid could suppress oxidant damage to intestinal epithelium and neighboring cells. In whole grains, vitamin E is another antioxidant that protects polyunsaturated fatty acids from oxidative damage in cell membranes, but unfortunately, that is removed during their refining. Another possible mechanism for vitamin-E is related to its competence to maintain selenium in reduced state. This vitamin inhibits the formation of nitrosamines, especially at low pH. Selenium is another important constituent of wholegrain, which is partly removed during milling. Selenium works as a cofactor for glutathione peroxidase, an enzyme that provides protection against oxidative damage, and at higher concentration, it has a suppressive action on cell proliferation.

Bioactive Compounds in Wholegrain Foodstuffs

Lignans that protect the human against hormones-mediated ailments are hormonally active compounds in grains. Lignans are the compounds that are present in many plants as minor constituents where they form building blocks for the formation of lignin in plant cell wall. The bacteria in digestive tract convert these plant lignans to mammalian lignans. Wholegrain wheat, wholegrain oats and rye meal, flaxseed seeds (the most concentrated source), pumpkin seeds, sunflower seeds and caraway seeds are the concentrated source of lignans. Oilseeds, predominantly the flaxseed flour and meal, turn out the highest concentration of lignans, followed by dried seaweeds, whole legumes, cereal bran, wholegrain cereals, vegetables and fruits. Wholegrain breads and cereals are excellent means of supplying lignans to diet.

Phytosterols and stanols known to reduce serum cholesterol are principally found in oilseeds, dry nuts, cereals' grains and legumes. It is believed that phytosterols are good means to inhibit dietary and biliary cholesterol absorption from small intestine since phytosterols have better solubility than cholesterol in bile-salt micelles. Phytosterols displace cholesterol from micelles, which reduces cholesterol absorption and increases its excretion. To inhibit absorption of dietary cholesterol, the plant sterol must be consumed at the same time as the cholesterol. *Per se* increased wholegrain consumption increases the total phytosterol intake and potentially contributes to cholesterol reduction.

Wholegrain wheat and oat per kilogram contains about 30 and 75 g lipids, respectively. Grain lipids per 100 g comprise about 75 g unsaturated fatty acids, of which there are about equal amounts (1-2 g/100 g) of oleic and linoleic acids, which are regarded as the imperative constituents of a heart healthy diet and known to reduce serum cholesterol. The foodstuffs prepared from wholegrain wheat and wholegrain oat contain about 20 and 55 g unsaturated lipid per kilogram, respectively. There has been considerable emphasis on low-fat diets for reduced cardiovascular heart disease, however, the type of fat consumed is important as well as the amount of fat. If the fat is saturated, LDL-cholesterol and total cholesterol levels increase but these levels decrease when the fat is unsaturated. Individual fatty acids, *i.e.*, oleic acid and linoleic acid have been found associated with lowering total cholesterol and LDL-cholesterol. High-lipid grain/bran products lower down the cholesterol levels in blood stream.

Anti-nutrients found in grains include digestive enzymes, *i.e.*, protease and amylase inhibitors, phytic acid, haemagglutinins, phenolics and tannins. Protease inhibitors, phytic acid, phenolics and saponins have been shown to reduce the risk of colon and breast cancer. Phytic acid, lectins, phenolics, amylase inhibitors and saponins have also been shown to lower the plasma glucose, insulin, plasma cholesterol and triacylgiycerols. In grains, protease inhibitors makeup 5-10% of the water-soluble protein and are concentrated in the endosperm and embryonic parts.

■ ■ ■

6 Therapeutic Significance of Cereals

Anureet Kaur[1], G.P.S. Dhaliwal[2] and Maninder Singh Bons[3]

[1]Assistant Agronomist, P.A.U. Regional Station, Bathinda
[2]Ph.D. Scholar, Department of Plant Breeding, CCS H.A.U., Hisar
[3]Assistant Professor (Plant Protection) KVK, Bahowal (Hoshiarpur),
Punjab Agricultural University, Ludhiana, Punjab, India
E-mail: <anureet_1@yahoo.com>

All cereals are excellent source of carbohydrates, proteins, many essential minerals like iron, magnesium, phosphorus and zinc, vitamins especially B group vitamins and phytochemicals. These are low in fat, high in both soluble and insoluble fibre and free from cholesterol. The wholegrain cereals are also a significant source of antioxidant phytochemicals, including phytoestrogens, which are linked to significant health benefits. These phytochemicals include lignans, phytic acid, saponins, phytosterols, squalene, oryzanol and tocotrienols. Thus, cereals and wholegrain foods can reduce the risk of developing certain diseases, including coronary heart disease, colon cancer, diabetes and diverticular disease. The cereals prepared from rice, wheat, oats, barley and maize are known as wonder cereals. The natural benefits and curative properties of these cereals have been given as under:

Common name: Rice (chawal)

Botanical name: *Oryza sativa*

Family: Poaceae

The nutritional value of rice per 100 grams is given below:

Constituents		Contents
Energy (KJ)	:	1,527
Total dietary fiber (g)	:	1.3
Fat (g)	:	500
Carbohydrates (g)	:	79
Sugars (g)	:	0.12
Protein (g)	:	7.12

Constituents		Contents
Minerals		
Calcium (mg)	:	28 (3%)
Magnesium (mg)	:	25 (7%)
phosphorus (mg)	:	115 (16%)
Potassium (mg)	:	115 (2 %)

Constituents		Contents
Iron (mg)	:	0.80 (6%)
Manganese (mg)	:	25 (7%)
Zinc (mg)	:	1.09 (11%)
Vitamins		
Thiamine (mg)	:	0.0701 (5%)

Constituents		Contents
Riboflavin (mg)	:	0.0149 (1%)
Niacin (mg)	:	1.62 (11%)
Pentothenic acid (mg)	:	1.014 (20%)
Vitamin B_6 (mg)	:	0.164 (13 %)
Total folate (mcg)	:	8 (2%)

Rice is an article of diet for the sick and convalescing, which is often used as medicine. The simplest use of rice is gruel in case of diarrhea. Various other forms as flour, paste, laja (parched rice), boiled, flattened, fried and dried rice and sprouted seedlings are also used for medicinal purposes to treat various diseases. Fried rice is light, suited to invalids and dyspeptics. Flattened rice with curd is often given in dysentery. Rice-water is recommended as an excellent demulcent and refrigerant drink in febrile and inflammatory diseases and dysuria. It is rendered more plalatable, acidulated with limejuice and sweetened with sugar. This is suggested as enema for bowel-related problems.

Uses of rice groups and forms in various ailments

The vaids (traditional doctors) of yore possessed profound knowledge of different effects of rice and were very particular in their prescription. Different groups of rice affect the human body differently, as they possess different inherent qualities to alleviate diseases. *Ayurveda* identified rice groups according to the growing seasons, *i.e.*, transplanted rice is referred as Sali (July to November), broadcasted rice as Vrihi and summer rice maturing in 60 days as Sashtika.

Sali rice: Roots of sali rice are used as a component of Brahma rasayana (*Ayurvedic medicine*). For the cure of piles, Sali and sashtika rice are given along with butter, ghee, goat meat and fresh wine. It checks bleeding from hemorrhoids. For treating burns and fractures, the sali rice is applied to the affected area mixing with ghee. For anemic patients, sali rice is given with barley. The treatment of meno-metrorrhagia is done with powder of red sali mixed in milk and honey and for increasing milk in lactating mothers the powder of sali is given along with milk. The chest, pelvis and head pain can be relieved with liquid gruel of red sali.

Sashtika rice: It is used as aphrodisiac. To treat stomach problems, gruel prepared from sashtika rice soaked in cow urine is given with milk.

Laja rice: For fever, burning sensation, vomiting, debility and thirst, saturated drink of laja rice with sugar and honey is suggested. For diarrhea treatments, pieces of kosakar (a type of sugarcane) fried in ghee and mixed with sugar, honey and powdered laja rice is very effective. For vomiting, powdered laja rice is given along with honey and ghee. Saturated drink of laja powdered with ghee and honey is recommended for intrinsic hemorrhage.

Nivara rice: The oil is used in the treatment of cervical spondilytis, low backache, paralysis, rheumatoid arthritis and neuron-muscular disorders. It is a weaning food for under weight babies. The stomach ulcers can be treated with nivara flour with banana. The Psoriasis and skin lesions can be cured by applying nivara rice paste. This paste also reduces pain in case of snake bite. For general health, body weight and expectant mothers, nivara gruel with meat is given. The paste of coconut, sesame and nivara rice is rubbed over the affected organs, if a person is affected by polio. The hemorrhoids patient is given roasted rice wetted in raw cow's milk for 21 days for early recovery. Urinary complaints in children are overcome by consuming decoction prepared from roots of nivara rice.

Health benefits of rice

Rice meets most of the requirements of good and healthy food and has medicinal properties as follows:

Food/diet for lactating mothers: Rice varieties have been used in increasing milk secretion in lactating mothers. The *Maharaji* and *Bhejri* varieties from Chattisgarh, *Jonga* in Bihar, *Neelam Samba* in Tamil Nadu are used for this purpose.

Post-delivery care: Rice is cooked in excess water and the starch is collected separately. One glass of this with jaggery is taken for five days. As rice has strengthening, hemostatic and expectorant properties, it helps in stopping bleeding.

Skin diseases and care: Rice powder and poultice is used for external application in case of smallpox, measles, prickly heat and other inflammatory infections of the skin, including burns and scalds, as it has cooling and soothing effect. The *Layacha* variety is used to treat boils on the scalp of new-born child. The mother eats cooked rice of this variety and the cure acts through the mother's milk ingested by the infant.

Other diseases: Red colored *Matali* and *Lal Dhan* rices of Himachal Pradesh help in curing blood pressure and fever. *Kafalya* rice from hilly regions of North India cures leucorrhea and helps in treating abortion complications. The *Bora* rice of Assam is given to patients suffering from jaundice. Rice beer is used for stomach and gastric ailments.

Red and black rices are considered more nutritious and have been found to be rich in iron, zinc, minerals and possess antioxidant properties. The parboiled rice of Sri Lanka has lower glycemic index than white rice, therefore, recommended for diabetics. The different researches have concluded that red rice yeast reduces cholesterol and total triglyceride, providing a novel food-based approach to lowering cholesterol. German and Swiss researchers have engineered rice to produce β-carotene, with the intent that it might someday be used to treat vitamin A deficiency. Additional efforts are being made to improve the quantity and quality of other nutrients in golden rice. The addition of carotene turns the rice a *gold* color, hence, the name.

Common name: Wheat (gehun)

Botanical name: *Triticum aestivum*

Family: Poaceae

The nutritional value of wheat per 100 grams is given below :

Constituents		Contents	Constituents		Contents
Energy (KJ)	:	1,506	Iron (mg)	:	6.26 (50%)
Total dietary fiber (g)	:	13.2	Manganese (mg)	:	239 (65%)
Fat (g)	:	9.72	Zinc (mg)	:	12.29 (123%)
Carbohydrates (g)	:	51.8	Manganese	:	13.301
Sugars (g)	:	0.12	**Vitamins**		
Protein (g)	:	23.15	Thiamine (mg)	:	1.882 (145%)
Minerals			Riboflavin (mg)	:	0.0149 (33%)
Calcium (mg)	:	39 (4%)	Niacin (mg)	:	6.813 (45%)
Magnesium (mg)	:	25 (7%)	Pentothenic acid (mg)	:	0.05 (1%)
phosphorus (mg)	:	842(120%)	Vitamin B_6 (mg)	:	1.3 (100 %)
Potassium (mg)	:	892 (19 %)	Total folate (mcg)	:	281 (70%)

The wheat, as produced by nature, contains several medicinal virtues. Every part of the whole wheat grain supplies elements needed by the human body. Starch and gluten in wheat provide heat and energy, the inner bran coats, phosphates and other mineral salts the outer bran, the much-needed roughage the indigestible portion which helps easy movement of bowels, the germ, vitamins B and E and protein of wheat helps in building and repairing muscular tissue. The wheat germ, which is removed in the process of refining, is also rich in essential vitamin E and the lack of which can lead to heart disease. The loss of vitamins and minerals in the refined wheat flour has led to widespread prevalence of constipation and other digestive disturbances and nutritional disorders. The whole wheat, which includes bran and wheat germ, provides protection against diseases such as constipation, ischemic, heart disease, disease of the colon called diverticulum, appendicitis, obesity and diabetes.

Health benefits of wheat

Tooth disorders: Wheat is valuable in the prevention and cure of pyorrhea. It takes time to eat wheat, and as it is generally taken with other foods, it compels the chewing of other foods also. This not only provides the needed exercise for the teeth and gum but also a great aid to digestion. Wheat grass juice acts as an excellent mouth wash for sore throats and pyorrhea. It also prevents tooth decay and tooth aches. Therefore, it is beneficial to chew wheat grass, which draws out toxins from the gums, and thus, checks bacterial growth.

Constipation: The bran of wheat, which is generally discarded in milling of flour, is more wholesome and nourishing than the flour itself. It is an excellent laxative. The laxative effects of bran are much superior to those of fruits or green vegetables as cellulose of the latter is more easily broken down by bacteria while passing through the intestine. The bran is highly beneficial in the prevention and treatment of constipation due to concentration of cellulose, which forms a bulk-mass in the intestines and helps in easy evacuation due to increased peristalsis.

Skin Diseases: It has been scientifically proved that chlorophyll arrests growth and development of harmful bacteria. Wheat grass therapy can effectively be used for skin diseases and ulcerated wounds as by retarding bacterial action, it promotes cell activity and normal re-growth. By drinking wheat grass juice regularly, an unfavourable environment is created for bacterial growth. Poultice of wheat grass juice can be applied on the infected area, as it is an able sterilizer. Externally, wheat flour is useful as dusting powder over inflamed surface as in burns, scalds and various itching and burring eruptions. Whole-wheat flour, mixed with vinegar, boiled and applied outwardly removes freckles.

Digestive System Disorders: Wheat grass juice used as an enema helps in detoxifying the walls of colon. The general procedure is to give an enema with lukewarm or *neem* water. After waiting for 20 minutes, 90 to 120 ml of wheat grass juice enema is given. This should be retained for 15 minutes. This enema is very helpful in disorders of the colon, mucous and ulcerative colitis, chronic constipation and bleeding piles.

Circulatory Disorders: The chlorophyll content present in wheat grass enhances heart and lung functions. Capillary activity also increases while toxemia or blood poisoning is reduced. Due to increased iron content in the blood and hemoglobin, the lungs function better. Oxygenation improves and the effect of carbon dioxide is minimized. It is for this reason that wheat grass juice is prescribed for circulatory disorders.

Wheat grass is grown by soaking a good variety of wheat for eight to 10 hours. The water should then be drained and grains be allowed to sprout for 15 hours. Earthen pots or wooden trays can be used to grow the wheat grass. These trays should be filled with compost manure. The wheat sprouts should be spread and the trays should be covered with a dark cloth or the trays should be kept away from the sunlight in a room. These trays should be sprinkled with water once or twice a day and they should be allowed to stay as such for six to seven days until the grass grows five to seven inches high.

Chewing wheat grass is the easiest method to drink wheat grass juice. As an alternative to chewing, wheat grass should be cut finely and ground in a grinder with addition of water to enable the extract ion of the wheat grass juice. This juice should be drunk within 10 to 15 minutes after extraction. In early stages of wheat grass therapy, about 100 to 150 ml of wheat grass juice should be taken daily. As the body becomes accustomed to its taste, the quantity of wheat grass juice can gradually be increased until the intake is about 250 to 300 ml per day. Wheat grass juice should be mixed thoroughly with saliva before being swallowed slowly. This juice should be drunk an hour before a meal and two to three hours after the meal.

Common name: Oats (Jawi)

Botanical name: *Avena sativa*

Family: Poaceae

The nutritional value of oats per 100 grams is given below:

Constituents		Contents	Constituents		Contents
Total dietary fiber (g)	:	11	Insoluble fiber (g)	:	6
Beta-glucan- soluble (g)	:	5	Fat (g)	:	6

Constituents		Contents
Saturated fat (g)	:	1.217
Monounsaturated fat (g)	:	2.178
Polyunsaturated fat (g)	:	2.535
Cholesterol	:	Nil
Carbohydrates (g)	:	66
Protein (g)	:	17
Minerals		
Calcium (mg)	:	54
Magnesium (mg)	:	177
phosphorus	:	523
Potassium (mg)	:	429
Sodium (mg)	:	2
Iron (mg)	:	4.72
Manganese (mg)	:	4.916
Zinc (mg)	:	3.97
Copper	:	0.626
Vitamins		
Thiamine (mg)	:	0.763
Riboflavin (mg)	:	0.139
Niacin (mg)	:	0.961
Pentothenic acid (mg)	:	1.349
Pyridoxine (mg)	:	0.1119

Constituents		Contents
Total folate (mcg)	:	56
Amino acids		
Tryptophan (g)	:	0.234
Threonine (g)	:	0.575
Isoleucine (g)	:	0.694
Leucine (g)	:	1.284
Lysine (g)	:	0.701
Methionine (g)	:	0.312
Cystine (g)	:	0.408
Phenylalanine (g)	:	0.985
Tyrosine (g)	:	0.573
Valine (g)	:	0.937
Histidine (g)	:	0.405
Arginine (g)	:	1.192
Alanine (g)	:	0.881
Aspartic acid (g)	:	1.448
Glutamic acid (g)	:	3.712
Glycine (g)	:	0.841
Proline (g)	:	0.934
Serine (g)	:	0.750
Energy (kcal)	:	390 or 1630 kJ

In recent years, there has been an increase in the consumption of oat as human food. Oat bran has received considerable attention from the medical community for its role in reducing blood cholesterol. The β-glucans or the water-soluble fibers present in oat bran help in lowering the blood cholesterol, which prevents heart disease. Therefore, daily intake of fiber as in oat bran assists in regulation of gastro-intestinal function.

Along with properties as high fiber and low cholesterol, herbalists have long made use of *avena sativa* grains and straw for their valuable medicinal properties. Oats are an annual grass, which is high in calcium. Calcium rich foodstuffs and herbs are the basis for remedies that relax the muscles and nervous system. Hence, *Avena sativa* is considered as the basis for nerve relaxing.

Oatmeal or gruel is an ideal food for convalescents and can be flavoured with raisins, lemon, butter, or maple syrup. It is easily digested and is a soothing food for those with fever and a good first food for those who have experienced intestinal illnesses or food poisoning.

Oatmeal is an excellent alternative to eggs and sausages or bacon for those desiring a low fat, low cholesterol breakfast routine.

The alcohol tincture of oat straw can be used for nerve complaints and for uterine problems in the standard dose of ten to twenty drops, taken in water, three times a day.

Oat straw tea is recommended by herbalists to soothe chest complaints, especially when mixed with a little lemon and honey. To make the tea, the straw and grains soaked in water are simmered for about an hour. A strong brew can be added to the bath to benefit rheumatism, paralysis, liver ailments, gout and kidney problems. Bladder and bowel conditions, intestinal colic and bedwetting have been helped by soaking in bath of oat straw. These baths are also recommended for tired or chronically cold feet. Oat straw makes a good wash for skin diseases like flaky skin, itching, frostbite, chilblains, wounds and eye problems.

Oats are available in several forms. Oat groats are the whole seed with only the outer hull removed. Rolled oats are the familiar breakfast form. Steel cut oats are oat seeds that have been chopped into pieces. Quick Oats are rolled steel cut oats, and instant oats are pre-cooked, dried and flattened steel cut oats. The finest form is its flour.

The intake of oats can help to prevent many diseases and it works as follows:

Cholesterol and heart: Oatmeal and oat bran are significant sources of dietary fiber. This fiber contains a mixture of about half-soluble and half-insoluble fibers. One component of soluble fibre found in oats is β-glucans- a soluble fiber that has proven effective in lowering blood cholesterol. Soluble fiber breaks down as it passes through the digestive tract, forming a gel that traps some substances related to cholesterol, such as cholesterol-rich bile acids. This entrapment reduces the absorption of cholesterol into the bloodstream. The bad cholesterol (Low-density lipoprotein) is trapped without lowering good cholesterol (High-density lipoprotein). Oat grains are also one of the best sources of compounds called tocotrienols. These are antioxidants, which together with tocopherols form vitamin E. The tocotrienols inhibit cholesterol synthesis and they have been found to lower down the blood cholesterol. The accumulation of cholesterol is implicated in many types of cardiovascular disease. Oats, like all cholesterol-lowering agents, are most effective when consumed as part of a low fat and high-fiber diet taken together with plenty of exercise.

The beneficial health effects of oats are best if half to one cup (one and a half to three ounces) of oats are eaten every day. One study found that the one-tenth ounce (3 grams) of soluble fiber from this amount of oatmeal decreased total cholesterol by approximately 2%, which correlates to a 4% decrease in coronary artery disease. Another study showed that one and a half ounce (43 grams) of oatmeal resulted in a loss of 3% in total cholesterol and a 14% reduction in bad cholesterol after taking two months. Another study found that a 6-8 weeks diet of one and a half to three ounces (43-85 grams) of oat bran daily lowered the total cholesterol by 20% and bad cholesterol (LDL) by as much as 25%. Another study found that consumption of 3.5 ounces (100 grams) of oat bran (one-third of a cup of oat) twice a day lowered down the cholesterol up to 15%. New research has also discovered that the antioxidants found in oats reduce cholesterol by reducing the ability of blood cells to stick to inside of the artery walls.

Blood sugars: Eating oats can spread the rise in blood sugars over a longer time. Control of blood glucose and insulin levels is essential in preventing many of the complications associated with diabetes. Oat β-glucan slows the rise in blood glucose levels following a meal and delays its decline to pre-meal levels. As the β-glucan in soluble fiber of oats is digested, it forms a gel, which causes viscosity of the contents of the stomach and small intestine to be increased. This in turn slows down the digestion and prolongs the absorption of carbohydrates into the bloodstream. Thus, the fluctuations in blood sugar levels are avoided.

Anti-cancer: Oats, like other grains and vegetables, contain hundreds of phytochemicals (plant chemicals). Many phytochemicals are thought to reduce a person's risk of getting cancer. Phytoestrogen compounds, called lignans, in oats have been linked to decreased risk of hormone-related diseases such as breast cancer. Most of the research has been focused on breast cancer but similar effects are expected on other hormone-related cancers such as prostate, endometrium and ovarian cancer. International research has shown that women with a higher intake of dietary fibre have lower circulating oestrogen levels, a factor associated with a lower risk of breast cancer. The insoluble fibers in oats are also thought to reduce carcinogens in the gastrointestinal tract.

Blood pressure: A daily serving of whole oats rich in soluble fibre can reduce hypertension or high blood pressure and so reduce the need for anti-hypertensive medication. It usually has no symptoms but can cause serious problems with the heart and blood vessels, leading to other complications.

Bowel function: Oats have high fiber content. Fiber is necessary in keeping bowel movements regular. Oats are high in both soluble and insoluble fiber. Insoluble fibers are those, which do not dissolve in water. It is spongy and absorbs many times its own weight of liquid. It makes stools heavier and speeds their passage through the gut, relieving constipation.

Weight control: As the soluble fibers of oats are digested, they form a gel, which causes viscosity of the stomach and small intestine contents. As a result, less energy is absorbed by the intestine from the digested food. The gel also delays stomach emptying making you feel full longer which helps with weight loss. New research suggests that children between the age of 2 and 18 years who have a constant intake of oatmeal keep their body weight low. The research found that the children who ate oatmeal were 50% less likely to become overweight, when compared to those children who did not eat oatmeal.

Athletic performance: Oats, like other cereal grains, are valued primarily as a source of carbohydrates, which provide calories for energy needs. Oats have been shown in scientific studies to alter metabolism favorably and enhance performance when ingested 45 minutes to 1 hour before exercise of moderate intensity.

General health and longevity: Oats have a higher concentration of well-balanced protein than other cereals. Oats contain phytochemicals (plant chemicals) which have been associated with protection from chronic disease such as cancer. They contain a good balance of essential fatty acids, which have been linked with longevity and general good health and also have one of the best amino acid profiles of any grain. Amino acids are essential proteins, which help facilitate optimum functioning of the body. Oat β-glucan also appears to help speed up response to infection, which may result in faster healing. According to a new study, it was discovered that beta-glucan could enhance the ability of certain human immune cells to navigate to the site of a bacterial infection, resulting in faster healing.

Common name: Barley (Jon)

Botanical name: *Hordeum vulgare*

Family: Poaceae

Among cereal grains, barley is considered the nutritional powerhouse. It is naturally cholesterol free and low in fat (0.7 g). The nutritional value of per one cooked pearl barley is given below:

Constituents		Contents	Constituents		Contents
Total dietary fiber (g)	:	6	**Minerals**		
Beta-glucan- soluble (g)	:		Zinc (mg)	:	1.2
Insoluble fiber (g)	:		Copper	:	0.16
Carbohydrates (g)	:	44	Selenium (mcg)	:	13.5
Protein (g)	:	3.5	**Vitamins**		
Fat (g)	:	0.7	Vitamin A (IU)	:	11
Minerals			Thiamine (mg)	:	0.13
Calcium (mg)	:	17	Niacin (mg)	:	3.23
Magnesium (mg)	:	35	Pentothenic acid (mg)	:	0.21
Phosphorus	:	85	Pyridoxine (mg)	:	0.18
Potassium (mg)	:	146	Total folate (mg)	:	25
Sodium (mg)	:	5	Vitamin E (mg)		0.01
Iron (mg)	:	2	Vitamin K (mcg)		1.25
Manganese (mg)	:	0.4	Energy (kcal)	:	193

Since barley is a great source of dietary fiber, which is effective in lowering cholesterol, it can reduce the risk of heart disease. It can help in following ways:

Diabetes: Type 2 diabetes and pre-diabetes are common problem these days. However, this disease can be controlled and even prevented by losing weight, increasing physical activity and including plenty of whole grains, which are high fiber foods like barley in daily diet. Barley is an excellent food choice for those concerned about type 2 diabetes or pre-diabetes because the grain contains essential minerals and vitamins and is an excellent source of dietary fiber, particularly beta-glucan soluble fiber.

Research shows that barley β-glucan soluble fiber promotes healthy blood sugar by slowing glucose absorption. In a clinical study reported in the August 2006 edition of the *Journal of the American College of Nutrition,* the data showed that the subjects who ate cookies and crackers made with barley flour enriched with β-glucan soluble fiber also experienced significant reductions in glucose and insulin responses compared to responses after eating the same products made with whole-wheat flour. One more study published in the August 2007 edition of the *Diabetes Research and Clinical Practice Journal* reported a 30 percent decrease in HbA1c (average blood glucose level) in type 2 diabetics who consumed a healthy diet including pearl barley that supplied 18 grams of soluble fiber a day. Regardless of the form of the grain, there is always a ready source of beta-glucan soluble fiber in barley. Unlike many grains, which contain fiber only in the outer bran layer, barley contains fiber throughout the entire kernel. Thus, whether it is whole grain or processed barley products, dietary fiber, including β-glucan soluble fiber, is available in amounts that have a positive impact on improving blood glucose levels.

Lowers cholesterol: The different studies around the world have yielded promising results regarding barley's potential health benefits regarding a significant correlation between barley consumption and cholesterol reduction. The trial was designed to investigate whether barley, as a soluble fiber source, would beneficially change cardiovascular risk factors and included men, pre-menopausal women and post-menopausal women with moderately elevated cholesterol levels. Final data from the trials showed that compared to pre-study concentrations, the total cholesterol levels and low-density lipoprotein (LDL) cholesterol (commonly known as bad cholesterol) levels were significantly reduced after the subjects consumed the modified diets containing low, medium and high levels of soluble fiber from barley.

The researchers reported that the reduction in total cholesterol and LDL cholesterol concentrations was most notable after the subjects consumed the higher levels of soluble fiber from barley. It was concluded that the consumption of barley-containing foods and the associated soluble fiber significantly improved several cardiovascular disease risk factors. They emphasized that the highest barley soluble fiber intake resulted in the greatest reduction in total and LDL cholesterol concentrations. Barley can be included in diet in the form of barley flakes for a hardy cooked breakfast cereal. Pearl or whole grain barley kernels can be added to soups, stews, casseroles and salads. In rural India, it is mostly taken with jaggary or *gur* after roasting. *Sattu* is also commonly used drink in summers prepared after roasting and grinding with *sakkar* (froth of cane juice rich in iron). It is very nutritious and improves digestion.

Common name: Maize or corn (makki or makka)

Botanical name: *Zea mays*

Family: Poaceae

Maize is the most important cereal in the world after wheat and rice. It is unique among cereals. The composition of maize per 100 g edible portion is as follows:

Constituents		Contents	Constituents		Contents
Moisture (%)	:	14.9	Calcium (mg)	:	10
Total dietary fiber (%)	:	0.5	Phosphorus (mg)	:	348
Carbohydrates (%)	:	66.2	Iron (mg)	:	2
Protein (%)	:	3.6	B group vitamins	:	Traces
Fat (%)	:	1.5	Vitamin E (mg)	:	Traces
Minerals (%)	:	2.7	Energy (kcal)	:	342

The bread made from the maize flour is nutritious and palatable. It can be digested without difficulty. Taken at intervals, the corn breads help to keep the colon clean. The dextrose produced from maize is used extensively in medicine. Consumption of maize helps in following ways:

Constipation: Popcorn, made by heating the small grains, is a wholesome cereal food and is easily digested. It has everything in the original grains of corn content. It is not fattening and practically starch free as in the process of popping, its starch is converted into dextrine and intermediate carbohydrates, a digestive product, which is easily assimilable. It is laxative and provides bulk to faeces. It also promotes peristalsis, *i.e.*, wave like movement by its tendency to absorb moisture. It is, therefore, valuable food in constipation.

Other health benefits: The high levels of thiamine and vitamin B in corn can help in preserving memory and keeping the energy levels up. Corn is also high in fiber, which keeps the digestive system functioning well. To check blood sugar levels and weight, use of blue varieties of corns chips and tortillas are the healthiest one, as they contain more antioxidants since blue corn has 8% less starch and 20% more protein than the white and yellow varieties. Corn is also an excellent source of folate, several different B group vitamins and vitamin C. The high amount of folate in corn can help in reducing the risk of a heart attack and colon cancer. It is also high in beta-cryptoxanthin, which is a carotenoid found in orange, red and some yellow vegetables and may help in reducing the chance of developing lung cancer.

Corn oil

Corn oil is extracted from the germ of the corn kernel. It is considered a vegetable oil and has mainly polyunsaturated and monounsaturated fats. It can help in reducing high low-density lipoprotein levels. Oil is composed of 99% triacylglycerols with polyunsaturated fatty acid (PUFA) 59%, monounsaturated fatty acid 24% and saturated fatty acid (SFA) 13%. The polyunsaturated fatty acid is linoleic acid (C18: 2n-6) primarily, with a small amount of linolenic acid (C18: 3n-3) giving n-6/n-3 ratio of 83. The corn oil health benefits also include controlling blood pressure and improving the quality of skin and hair. Linoleic acid is essential for skin and cell membrane integrity and icosanoids production. Icosanoids are necessary for reproductive, cardiovascular, renal and gastrointestinal functions and resistance to disease. Corn oil is also good source of ubiquinone, alpha- and gamma-tocopherols (vitamin E). The consumption of corn oil can replace saturated fatty acids with polyunsaturated fatty acids, as these are present in large amounts in oil. Due to its cholesterol-lowering effects and high smoke point, it is great to use for frying and can be a good substitute for other types of oil.

■■■

7 Medicinal Significance of Pulses and Beans

Anureet Kaur[1], Harsimrat K. Bons[2] and G.P.S. Dhaliwal[3]

[1]Assistant Agronomist, P.A.U. Regional Station, Bathinda
[2]Assistant Professor (Horticulture), KVK, Samrala (Ludhiana)
[3]Ph.D. Scholar, Department of Plant Breeding, CCS H.A.U., Hisar
Punjab Agricultural University, Ludhiana, Punjab, India
E-mail: <anureet_1@yahoo.com>

Beans and other legumes offer complex carbohydrates, which are needed to power bodies and plenty of dietary fiber around fifteen grams per cup. They also contain essential fatty acids (omega-6s) and omega-3s in soybeans. Beans are loaded with protein (about 15 g per cup) and dozens of key nutrients, including calcium, potassium and magnesium. Studies also tie beans to a reduced risk of heart disease, high blood pressure, type 2 diabetes and breast and colon cancers. They are the highest antioxidant food, beating out both blueberries and cranberries. Beans or a lentil with grains like barley, oats and rice, makes a complete food with all vital nutrients for people who should avoid meat or choose not to eat it. The natural benefits and curative properties of these pulses and beans have been given as under:

Common name: Green gram (Moong bean)

Botanical name: ***Vigna radiata***

Family: Fabaceae

The nutritional composition of mature green gram seeds (eaten raw) 100 grams is given below:

Constituents		Contents	Constituents		Contents
Total dietary fiber (g)	:	16.3	Magnesium (mg)	:	189 (51%)
Fat (g)	:	1.15	Phosphorus (mg)	:	367 (52%)
Carbohydrates (g)	:	62.62	Potassium (mg)	:	1246 (27%)
Protein (g)	:	23.86	Sodium (mg)	:	15 (1%)
Sugars (g)	:	6.60	Vitamin C (mg)	:	4.8 (8%)
Calcium (mg)	:	132 (13%)	Energy (kcal)	:	347

The nutritional composition of boiled green gram seeds per 100 grams is given as under:

Constituents		Contents	Constituents		Contents
Total dietary fiber (g)	:	7.6	Magnesium (mg)	:	0.298 (0%)
Fat (g)	:	0.38	Phosphorus	:	99 (14%)
Carbohydrates (g)	:	19.15	Potassium (mg)	:	266 (6%)
Protein (g)	:	7.02	Sodium (mg)	:	2 (0%)
Sugars (g)	:	2.00	Vitamin C (mg)	:	1.0 (2%)
Calcium (mg)	:	27 (3%)	Energy (kcal)	:	105

Moong beans are one of the most cherished foodstuffs in *Ayurveda*. They are very nourishing and easy to digest, as they do not generally create abdominal gas and bloating, which are the drawbacks of larger beans. Persons recuperating are often recommended *khichari*, a combination of rice and mung beans, because of their ability to provide a good level of nourishment without overtaxing the digestion. They offer the astringent taste.

Moong bean sprouts contain a high quantity of vitamin A, B, C and E. They are also known to be an excellent source of many minerals, such as calcium, iron and potassium. This bean is popular as the perfect food for reducing weight. It is recommended as a food replacement in many slimming programs, as it has a very low fat content. It is a rich source of protein and fiber, which help one to lower the high cholesterol level in the blood system. The high fiber content of moong beans yields complex carbohydrates, which aid digestion. Complex carbohydrates are also effective in stabilizing blood sugar and prevent its rapid rise after meal consumption, apart from keeping body's energy at a balanced level. The persons suffering from diabetes or high cholesterol are recommended frequent consumption of moong bean.

Moong beans are also high in fibre, low in saturated fat and sodium and contain no cholesterol. Because of the wide range of nutrients contained in moong beans, they offer a whole host of health benefits for the immune system, the metabolism, the heart and other organs, cell growth, protection against free radicals and diseases such as cancer and diabetes.

The folic acid or folate is contained in moong beans helps in lowering the risk of heart disease, fights birth defects, contributes to normal cell growth, helps in the metabolism of proteins and is essential for the formation of red blood cells and for healing processes in the body. Thiamine is needed to ensure that the nervous system functions properly. It is also important for releasing energy from carbohydrates.

Manganese is a trace mineral and key for energy production and antioxidant defenses. It is also necessary for the metabolism of carbohydrates, fats and proteins and can be helpful for the brain and nerves. Magnesium present in moong beans helps the veins and arteries to relax, lessening resistance and improving the flow of blood, oxygen, and nutrients throughout the body, and lack of magnesium promotes free radical damage to the heart. The body requires copper in order to absorb iron and is also involved in the metabolism of protein. Iron helps to build resistance to stress and disease and it is essential for the formation of hemoglobin, which carries oxygen from the lungs to every cell in the body. It is also part of key enzyme systems for energy production and metabolism. Potassium is necessary for maintaining the acid-alkaline balance in the blood and for muscle contraction and a normal heart beat. Zinc is a well-known immune system booster and can be helpful in fighting male infertility. Zinc aids healing processes in the body, growth and tissue repair.

In Chinese medicine, mung bean sprouts are considered as a cooling food, containing anti-cancer properties. Herbalists use them for all hot, inflammatory conditions, ranging from systematic infections to heat stroke and even hypertension. Moong beans are a nutritional powerhouse and may be able to be defined as a super food

Common name: Black bean (pinto bean or kidney bean)

Botanical name: *Phaselous vulgaris*

Family: Fabaceae

Black beans are a very good source of cholesterol-lowering fiber, as are in most other legumes. In addition to lowering cholesterol, black bean's high fiber content prevents blood sugar levels from rising too rapidly after a meal, making these beans an especially good choice for individuals with diabetes, insulin resistance, or hypoglycemia. When combined with whole grains such as brown rice, black beans provide virtually fat-free high quality protein. Black beans are an excellent source of trace mineral, molybdenum, an integral component of enzyme sulfite oxidasc, which is responsible for detoxifying sulfites. The nutritional composition of black beans per 100 grams is given below:

Constituents		Contents
Total dietary fiber (g)	:	14.96
Fat (g)	:	0.93
Saturated fat (g)	:	0.24
Monounsaturated fat (g)	:	0.08
Polyunsaturated fat (g)	:	0.40
Carbohydrates (g)	:	40.78
Protein (g)	:	15.24
Minerals		
Calcium (mg)	:	46.44
Magnesium (mg)	:	120.4
Phosphorus (mg)	:	240.8
Potassium (mg)	:	610.6
Iron (mg)	:	3.61
Zinc (mg)	:	1.93
copper (mg)	:	0.36
Manganese (mg)	:	0.76

Constituents		Contents
molybdenum (μg)	:	129.0
selenium (μg)	:	2.06
sodium (mg)	:	1.72
Vitamins		
Thiamine (mg)	:	0.42
Riboflavin (mg)	:	0.10
Niacin (mg)	:	0.87
Pentothenic acid (mg)	:	0.42
Vitamin A (IU)	:	10.32
Total folate (μg)	:	255.9
Vitamin E (mg)	:	1.03
Vitamin B_6 (mg)	:	0.12
Amino acids		
Tryptophan (g)	:	0.18
Threonine (g)	:	0.64
Isoleucine (g)	:	0.67

Constituents		Contents	Constituents		Contents
Leucine (g)	:	1.22	Arginine (g)	:	0.94
Lysine (g)	:	1.05	Alanine (g)	:	0.64
Methionine (g)	:	0.23	Aspartic acid (g)	:	1.84
Cystine (g)	:	0.17	Glutamic acid (g)	:	2.32
Phenylalanine (g)	:	0.82	Glycine (g)	:	0.60
Tyrosine (g)	:	0.43	Proline (g)	:	0.65
Valine (g)	:	0.80	Serine (g)	:	0.83
Histidine (g)	:	0.42	Energy (kcal)	:	227.04

Fiber rich: Black beans, like other beans, are rich in dietary fiber and are useful food article for people with irregular glucose metabolism, such as diabetics and those with hypoglycemia, because beans have a low glycemic index rating. The blood glucose (blood sugar) does not raise high after eating beans as it does when compared to white bread. This beneficial effect is probably due to two factors, *i.e.*, (i) the presence of higher amounts of absorption-slowing protein in the beans and (ii) their high soluble fiber content. Soluble fiber absorbs water in the stomach forming a gel that slows down the metabolism of bean's carbohydrates. The presence of fiber is also the primary factor in the cholesterol-lowering power of beans. Fiber binds with the bile acids that are used to make cholesterol. Fiber is not absorbed, so when it exits in the body in feces, it takes the bile acids with it. As a result, the body may end up with less cholesterol. Black beans also contain insoluble fiber, which the research studies have shown not only helps to increase stool bulk and prevent constipation but also helps to prevent digestive disorders like irritable bowel syndrome and diverticulosis.

Loaded with antioxidants: Black beans are as rich in antioxidant compounds called anthocyanins as grapes and cranberries. The researchers analyzed different types of beans and concluded that the beans with darker seed coat had higher level of antioxidant activity. Black beans were found to have the most antioxidant activity, followed in descending order by red, brown, yellow and white beans. The level of antioxidants found in black beans in various studies is approximately 10 times that found in an equivalent amount of oranges and comparable to that found in an equivalent amount of grapes or cranberries.

Promote optimal health: The black beans may help in protecting against cancer but that whole food naturally contains an array of compounds, which work together for our benefit. The researchers fed laboratory animals a 20% black bean diet to check any mutagenic or genotoxic activity. The black beans did not promote cancer though there was a clear reduction in the number of pre-cancerous cells. The reduction was found even in the animals, which were simultaneously given an agent known to promote cancer, the mutagen, cyclophosphamide. In an attempt to identify the bean components responsible for this protective effect, the researchers tested a single commercial anthocyanin, but instead of being protective on its own, the flavonoid, at the highest dose administered (50 mg per kg of bodyweight), actually induced DNA damage.

Lower heart attach risk: Eating high fiber foodstuffs, such as black beans, helps prevent heart disease. Black beans contribution to heart health lies not just in their fiber but in the significant amounts of folate and magnesium also. The folate helps in lowering the levels of homocysteine, an amino acid, which is an intermediate product in an important metabolic process called methylation cycle. Elevated blood levels of homocysteine are an independent risk factor for heart attack, stroke, or peripheral vascular disease and are found in between 20 and 40% of patients with heart disease. Just one cup of cooked black beans provides 64% of the daily value for folate. The supply of magnesium by beans puts yet another addition in the column of its beneficial cardiovascular effects. Magnesium is nature's own calcium channel blocker.

Black beans also contain compounds called polyphenols that are useful for those with elevated cholesterol because they act as antioxidants in the bloodstream, preventing cholesterol from oxidation by free radicals. Only after cholesterol has been damaged by free radicals does it form plaques on blood vessel walls, initiating the development of atherosclerosis. The polyphenols in black beans also have some potentially negative effects. Some of the polyphenols are tannins that can bind some of the protein and iron these beans supply, preventing absorption of these nutrients.

Energy: In addition to its beneficial effects on the digestive system and the heart, black beans soluble fiber helps in stabilizing the blood sugar levels. In case of insulin resistance, hypoglycemia, or diabetes, black beans can help in balancing the blood sugar levels while providing steady, slow-burning energy.

Iron for energy: In addition to providing slow burning complex carbohydrates, black beans can increase energy by helping to replenish iron stores. Though the tannins in black beans may block absorption of some of the iron they contain, a cup of black beans contains 20.1% of the daily value. The black beans are low in calories and virtually fat-free and can be beneficial for menstruating women, as they are at risk of iron deficiency.

Manganese for energy production and antioxidant defense: Black beans are a good source of the trace mineral manganese, which is an essential cofactor in a number of enzymes important in energy production and antioxidant defenses. The key oxidative enzyme superoxide dismutase, which disarms free radicals produced within the mitochondria (the energy production factories within our cells), requires manganese. One cup of black beans supplies 38.0% of the daily value.

Protein power plus: The meat can be replaced with black beans as these are rich in proteins and when combined with a whole grain such as whole-wheat pasta or brown rice, provide protein comparable to that of meat or dairy foodstuffs without high calories or saturated fat found in these food articles. A cup of black beans provides with 15.2 grams of protein, plus 74.8% of the daily value for fiber with virtually no fat.

Common name: French bean (Green bean)

Botanical name: *Phaseolus vulgaris*

Family: Leguminosae

Green beans also known as string beans are one of the only few varieties of beans that are eaten fresh. Green beans are picked while still immature and the inner bean is just beginning to form.

In comparison to the dried bean varieties, the green bean offers less protein but more fiber and other nutrients. The nutritional composition of green bean per 100 grams is given as under:

Constituents		Contents	Constituents		Contents
Total dietary fiber (g)	:	3.6	Iron (mg)	:	1.0(8%)
Fat (g)	:	0.1	Potassium (mg)	:	200 (4%)
Carbohydrates (g)	:	7.1	Vitamin C (mg)	:	16 (27%)
Protein (g)	:	1.8	Energy (kcal)	:	31

Green beans offer cardiovascular protection. The antioxidant in this vegetable helps to prevent cholesterol from becoming oxidized. Oxidized cholesterol is able to stick to and build up in blood vessel walls, where it can cause blocked arteries, heart attack, or stroke. Green beans can help in lowering the high blood pressure. It can help in preventing weakening of the bone. The fiber can help in promoting a healthy colon because of it ability to bind to cancer-causing toxins, removing them from the body before they can harm colon cells. Green beans can help in preventing DNA damage. These have strong anti-inflammatory effects, which make it effective in diseases where inflammation takes place, such as arthritis and asthma. Beans are helpful for menstruating women and for people who are at risk of iron deficiency. Green beans are an energy and diet food, as it contains few calories. Green beans are also important for lactating women.

Common name: Rice bean

Botanical name: *Vigna umbellata*

Family: Fabaceae

Rice bean plays an important role in human health. All varieties seem to be good source of protein, minerals, essential amino acids and essential fatty acids, and its dried seeds make an excellent addition to a cereal-based diet. Rice bean is most often served as a *dal*, either soaked overnight and boiled with a few spices, or cooked in a pressure cooker. Apart from various recipes for *dal* soups and sauces, pulses are also used in a number of other ways, whole, cooked, or roasted, as flour or ground to make various deep fried dishes or snacks. Some recipes are specific to particular pulses but many are open to substitution. The consumption of green pods as a vegetable has been recorded but is not widespread, although the indeterminate growth habit of many varieties is beneficial in providing a steady supply of green pods over a long period of the year.

In South Asia, the idea of a division of foodstuffs into hot, cold and neutral is very common. This has an important bearing on dietary choices, as this perception not only promotes a balance between hot and cold foodstuffs in daily nutrition but also encourages or discourages the consumption of various items according to season and during sickness. An account of the perception of a number of food items in Nepal has been published by Gittelsohn in the year 1997. Their data show that there is hardly any *scientific* basis for the division into hot and cold food articles. For instance, yogurt is cold, while goat milk is hot, buffalo meat is cold, while fish and chicken are hot and black gram is cold, while red gram (cowpea) is hot. In Eastern Nepal, rice bean is considered hot and advised not to consume during hot season by old and sick people, as it is not easily digested and weak people get stomach problems from eating it.

Quantee (*kwati*) is a mixed bean sprout soup served at the *Janai Purnima* or *Raksha Bhandan* festival. Rice bean is one of nine beans prescribed for this recipe. The festival marks the end of the monsoon where people by traditional perception (and probably also in reality) have been weak, undernourished and subject to diseases. In this respect, *quantee* is said *to make one strong* and to purify the stomach as the mixed bean sprouts are hard to digest and so cleans the stomach. In addition, eating *quantee* is said to kill a certain type of mosquito. People often serve it to labourers, while occasionally consuming it themselves in connection with tasks requiring hard work.

In traditional Chinese medicine, the red type rice bean helps to relieve oedema in some illnesses. It is also used in combination with *Angelica sinensis* to remedy red eyes (possibly uveitis) with erosion of oropharyngeal or urogenital mucosae, which is suspected to be Behçet's disease

Common name: Chickpea (chana)

Botanical name: *Cicer arietinum*

Family: Fabaceae

The chickpea also known as garbanzo bean, Indian pea, ceci bean and Bengal gram is an edible legume. It is high in protein and one of the earliest cultivated vegetables since 7,500 years old remains have been found in the Middle East. Mature chickpeas can be cooked and eaten cold in salads, cooked in stews, ground into a flour called gram flour (also known as *besan* and used primarily in Indian cuisine), ground and shaped in balls and fried as falafel, fermented to make an alcoholic drink, stirred into a batter and baked to make farinata, cooked and ground into a paste called hummus or roasted, spiced and eaten as a snack (such as *leblebi*). Chickpeas and Bengal grams are used to make curries and are one of the most popular vegetarian foodstuffs. The nutritional composition of chickpea (cooked with no salt) per 100 grams is given as under:

Constituents		Contents	Constituents		Contents
Total dietary fiber (g)	:	7.6	Potassium (mg)	:	291 (6%)
Fat (g)	:	2.59	Iron (mg)	:	2.89 (23%)
Saturated fat (g)	:	0.269	Zinc (mg)	:	1.53 (15%)
Monounsaturated fat (g)	:	0.583	**Vitamins**		
Polyunsaturated fat (g)	:	1.156	Vitamin A	:	1 (0%)
Carbohydrates (g)	:	27.42	Vitamin B_6	:	0.139 (11%)
Protein (g)	:	8.86	Thiamine (mg)	:	0.116 (9%)
Minerals			Riboflavin (mg)	:	0.063 (4%)
Sodium (mg)	:	7 (0%)	Niacin (mg)	:	0.526 (4%)
Phosphorus	:	168 (24%)	Pentothenic acid (mg)	:	0.286 (6%)

Constituents		Contents	Constituents		Contents
Vitamin C	:	1.3 (2%)	Total folate (μg)	:	172 (43%)
Vitamin E	:	0.35 (2%)	Energy (kcal)	:	164
Vitamin K	:	4 (4%)			

Chickpeas are a helpful source of zinc, folate and protein. They are also very high in dietary fiber, and hence, a healthy source of carbohydrates for persons with insulin sensitivity or diabetes. Chickpeas are highly loaded in folate and protein content, thus, they provide wonderful protection against colorectal cancer in people. People, who are vegetarian, can eat chickpeas with rice or whole grains to include wholesome amount of proteins in their diet. Moreover, the presence of folate also helps in lowering down the homocysteine level in your body. Folic acid is also helpful in preventing neural tube birth defects. Chickpeas are low in fat and most of this is polyunsaturated. The nutrient profile of *desi chana* (the smaller variety) is different, especially in fibre content, which is much higher than the light coloured variety, and hence, a very low glycemic index, which may make them suitable for people with blood sugar problems. Chickpeas also provide dietary calcium (49-53 mg/100 g), as about the same as yogurt and close to milk. The *desi* type is used to make pulse, which is a split chickpea with the skin removed.

Ancient people also associated chickpeas with Venus because they were said to offer medical uses such as increasing sperm and milk, provoking menstruation and urine and helping to treat kidney stones. Wild chickpeas were thought to be especially strong and helpful. Chickpeas do not cause lathyrism, which is caused by consuming other plants of the genus Lathyrism, which contains toxins associated with lathyrism.

Chickpeas contain insoluble fiber content, which helps in improving the stool bulk. This way, it prevents the problem of constipation and other disorders related to digestive tract of body such as irritable bowel syndrome. Being high in both soluble and insoluble fibre and with a low glycaemic index, chickpeas can help you to feel fuller for longer, and thereby, helping to control appetite and manage weight. Eaten regularly they can also help to reduce the risk of chronic diseases such as obesity, diabetes, heart disease and certain cancers. They also contain fructo-oligosaccharides, a type of prebiotic carbohydrate fibre, which supports healthy gut flora. Maintaining a healthy balance of friendly bacteria in the colon is necessary to optimize digestive function and strengthen immunity. Chickpeas are rich in saponins photochemical, which acts as antioxidants. Regular intake of chickpeas in diet helps in lowering down the risk of breast cancer in women along with osteoporosis ailment. It also reduces the hot flushes in women after menopause stage. Chickpeas are also rich source of non-heme iron.

Common name: Soybean

Botanical name: *Glycine max* (L.) Merr

Family: Fabaceae

Soybeans can be broadly classified as vegetable (garden) or field (oil) types. Among legumes, the soybean, also classed as an oilseed, is pre-eminent for its high protein content (38-45%) as well as its high oil content (20%). The beans contain significant amount of phytic acid, alpha-

Linolenic acid and the isoflavones, *i.e.*, genistein and daidzein. They also contain other health promoting compounds including phytosterols, lecithin, isoflavones, phytoestrogens and protease inhibitors. Soy foodstuffs are a source of high-quality protein. In addition, the consumption of soy protein provides health benefits that may help to prevent or treat certain chronic diseases like prostate, colon and breast cancers as well as osteoporosis and other bone health problems and alleviates hot flushes associated with menopause. The nutritional composition of soybeans (mature raw seeds) per 100 grams is given below:

Constituents		Contents	Constituents		Contents
Total dietary fiber (g)	:	9.3	Vitamin B_6 (mg)	:	0.377 (29%)
Fat (g)	:	19.94	**Amino acids**		
Saturated fat (g)	:	2.884	Tryptophan (g)	:	0.591
Monounsaturated fat (g)	:	4.404	Threonine (g)	:	1.766
Polyunsaturated fat (g)	:	11.255	Isoleucine (g)	:	1.971
Carbohydrates (g)	:	30.16	Leucine (g)	:	3.309
Protein (g)	:	36.49	Lysine (g)	:	2.706
Minerals			Methionine (g)	:	0.547
Calcium (mg)	:	277 (28%)	Phenylalanine (g)	:	2.122
Magnesium (mg)	:	280 (76%)	Tyrosine (g)	:	1.539
Phosphorus (mg)	:	704 (101%)	Valine (g)	:	2.029
Potassium (mg)	:	1797 (38%)	Histidine (g)	:	1.097
Iron (mg)	:	15.7(126%)	Arginine (g)	:	3.153
Zinc (mg)	:	4.89 (49%)	Alanine (g)	:	1.915
sodium (mg)	:	2 (0%)	Aspartic acid (g)	:	5.112
Sugar (g)	:	7.33	Glutamic acid (g)	:	7.874
Vitamins			Glycine (g)	:	1.880
Vitamin C (mg)	:	6.0 (10%)	Proline (g)	:	2.379
Vitamin K (μg)	:	47 (45%)	Serine (g)	:	2.357
Vitamin A (μg)	:	1 (0%)	Energy (kcal)	:	446

The beans can be processed in a variety of ways. Common forms of soy or soya include soy meal, soy flour, soy milk, tofu and textured vegetable protein (TVP, which is made into a wide variety of vegetarian food items, some of them intended to imitate meat), tempeh, soy lecithin and soybean oil. Soybeans are also the primary ingredient involved in the production of soy sauce. The use of soybeans and soy products has numerous health benefits as given below:

Childhood obesity: Soy foodstuffs provide critical minerals, vitamins, protein and fiber for growing children. In addition, many soy food items contain fewer calories and fat, making weight loss or maintaining healthy weight much easier. Soy food articles are available in chocolate soymilk, frozen pizza, taco-style meat and chicken-style nuggets. Soy foodstuffs are filled with fiber that provides a feeling of satiety and fullness between meals and reduces hunger cravings. Thus, soy food articles help to promote healthier eating habits with low carbohydrate diets, as they are naturally low in fat and saturated fat, while being high in heart-healthy protein.

Heart health and heart disease: Soy foodstuffs containing soy protein are beneficial against heart disease. Over 40 scientific studies have proven the positive effect of soy protein on lowering cholesterol levels, including the harmful LDL cholesterol, which leads to the decreased risk of heart disease. The Food and Drug Administration recommends consumption of 25 grams of soy protein every day as part of a diet, which is low in saturated fat and cholesterol. Soybeans are one of the best non-fish sources of essential omega-3 fatty acids, which may help reduce the risk of coronary heart disease. Soy protein may also provide positive results for people with high blood pressure. According to recently published scientific study, researchers found that both the systolic and diastolic blood pressure was reduced in middle-aged and elderly women who consumed 25 grams of soy protein daily.

Menopause: Soy protein has proven benefits extending well into post-menopausal years. Research results have shown that consuming soy protein before as well as after menopause may help to protect bones from becoming weak and brittle. As post-menopausal women face an increased risk of osteoporosis, thus, soy protein-rich food items can help.

Pregnancy and omega-3s: A newly released scientific study found that mothers, who eat foodstuffs rich in omega-3 fatty acids during pregnancy and while breastfeeding, might help to reduce significantly their daughter's risk of developing breast cancer later in life. In addition, this study found that including omega-3 rich food articles throughout childhood and teenage years might continue to help in providing protective benefits against breast cancer.

Cancer and soy isoflavones: Foodstuffs rich in fiber, low in fat and high in phytochemicals may help to reduce the risk of developing certain cancers. Including soy food articles during a teenage girl's adolescence years may provide increased protective benefits and reduce the risk of developing breast cancer later in life. Consuming soy foodstuffs may prove beneficial when eating to reduce cancer risk because many soy food articles are not only high in fiber, they are also low in total fat and high in soy protein and phytochemicals called isoflavones. Isoflavones are naturally occurring plant compounds that have been attributed in numerous medical and scientific studies to reduce the risk of colon, breast and prostate cancer. Soy protein may also be protective against prostate cancer by helping to promote healthier prostate tissues. Soy foodstuffs may further more help to protect against colon cancer. The components of soy that may be helping to prevent colon cancer are called isoflavones and saponins. Both are found in soy foodstuffs such as soymilk, soy nuts and green and yellow soybeans.

Diabetes: Soybean is an excellent food for diabetes. Soy foodstuffs have a lower glycemic index, which helps to keep blood sugar levels more stable, making the diabetes much easier to control. Soy food articles like canned yellow soybeans and frozen green sweet soybeans have a lower glylcemic index than other soy foodstuffs. Secondly, many soy food items are high in dietary fiber, which also helps in stabilizing the blood sugar levels.

Other health benefits: It is especially good for growing children to aid growth and development. Soybean has high lecithin content and is therefore excellent for mental fatigue and for protection against cholesterol deposits. Lecithin is also a brain food, a tonic and energizer. It also prevents gallstones. Liquid lecithin application helps to heal bedsores and persistent diaper rashes. It is also reported to prevent pellagra, marked by disturbances of stomach and intestines, skin eruptions and many nervous symptoms such as melancholia. Because of the high content of linoleic and linolenic acids (unsaturated fatty acids), soybeans are conducive to a healthy skin and correct many cases of eczema. The pure pressed oil is used for skin conditions, as it contains the natural lecithin. Lecithin present in soybean has the ability to increase the gamma globulin content of the blood. The gamma globulins are known to be associated with the nature's protective force against the attack of various infections in the body, thus, increasing immunity.

Common name: Horse gram (Kulath)

Botanical name: *Macrotyloma uniflorum*

Family: Fabaceae

Horse gram is legume of the tropics and subtropics, grown mostly under dry-land agriculture. The chemical composition is comparable with more commonly cultivated legumes. Like other legumes, these are deficient in methionine and tryptophan, though horse gram is an excellent source of iron and molybdenum. Comparatively, horse gram seeds have higher trypsin inhibitor and hemagglutinin activities and polyphenols than moth bean seeds. Dehusking, germination, cooking and roasting have been shown to produce beneficial effects on nutritional quality of the legume. Though it requires prolonged cooking, a soak solution has been shown to reduce cooking time and improve protein quality.

Horse gram is used to prepare dishes including *kollu porial*, *kollu avial*, *kollu sambar* and *kollu rasam*. In traditional *siddha* cuisine, horse gram is considered a food with medicinal qualities. In southern parts of India, horse gram is prescribed for persons suffering from jaundice or water retention and as part of a weight loss diet. It is considered helpful for iron deficiencies and is considered helpful for maintaining body temperature in winter season.

The horse gram drink, known as *kollu* is very effective remedy in problems like high cholesterol, obesity, sinus and phlegm in chest. The horse gram is grinded in powdered form and heated in fry pan, and further, the powder in boiling water is added along with cumin and salt. To reduce body fat, the paste of sour buttermilk and horse gram powder is applied on fat deposits on the body and massaged vigorously in upward strokes.

Ayurvedic Physicians of the Gangetic Basin of India use horse gram for various ailments ranging from piles to rheumatism and conjunctivitis of eyes and for eliminating worms. It is widely accepted as a cure for eliminating phlegm in chronic cold and cough, remission of fevers, regulating menstruation in women and controlling post-delivery bleeding. The powdered beans are applied on skin, to minimize heat rash and similar skin eruptions.

It also helps to dissolve kidney stones and expel them through urethra. The stone expeller drink is prepared by soaking one teaspoon of horse gram for overnight in a glass of water. The drink is consumed in the morning after stirring it. The soaked seeds can also be consumed. The kidney stones are dissolved within six to seven days and eliminated through urine. The drink is also a diuretic and regulates ordinary kidney problems, particularly swelling in legs. The old horse gram are boiled and taken to cure rheumatism. It also eliminates germs and worms. It cures impurities of bile and blood, inflammation and checks sweating. Drinking semi-liquid solution of horse gram powder cures flatulence. The horse beans are mixed with powdered dry ginger, asafoetida and *veed salt* and taken to cure the stomach pain. The use of horse gram in diet relieves pain of dry piles and various diseases of stomach. Drinking the semi-liquid mixture of powdered horse gram and powdered black pepper cures sore throat. Its use in diet is good for women as their menstruation is purified.

Common name: Lentil (Masoor)

Botanical name: *Lens culinaris*

Family: Fabaceae

Lentils contain high levels of proteins, including the essential amino acids isoleucine and lysine and are an essential source of inexpensive protein in many parts of the world for those who adhere to a vegetarian diet. Lentils are deficient in two essential amino acids, methionine and cystine. However, sprouted lentils contain sufficient levels of all essential amino acids, including methionine and cystine. Lentils also have anti-nutritional factors such as trypsin inhibitors and relatively high phytate content. Trypsin is an enzyme involved in digestion and phytates reduce the bioavailability of dietary minerals. The phytates content can be reduced by soaking the lentils in warm water overnight. One cup of boiled lentils has 18 grams of muscle building protein, which also helps in keeping the cells functioning normal and the immune system primed to fight off infection. The nutritional composition of lentils per 100 grams is given below:

Constituents		Contents	Constituents		Contents
Total dietary fiber (g)	:	31	Iron (mg)	:	7.5 (60%)
Fat (g)	:	1.0	Thiamine (mg)	:	0.87 (67%)
Carbohydrates (g)	:	60	Folate (μg)	:	479
Protein (g)	:	26	Energy (kcal)	:	353
Sugars (g)	:	2.0			

Cholesterol reduction: Lentils are a good source of soluble fiber to lower cholesterol levels and reduce the risk of heart disease. There is also evidence to prove that **lentils can slow the liver's manufacture of cholesterol**, which similarly helps to reduce levels of cholesterol in the body.

Diabetes: Soluble fiber is particularly important for diabetics since it helps to maintain healthy blood sugars by moderating rapid fluctuations in glucose and insulin levels. The soluble fiber in lentils traps carbohydrates. This in turn slows digestion and absorption, which helps to **prevent wide swings in blood sugar** level throughout the day.

Digestion: Due to their high levels of fiber, lentils increase the size of stool and speed the journey of waste products through the gut, which **help to alleviate constipation.** Fiber can also help both **reduce the risk and the symptoms of diverticulosis**, a condition in which small pouches form in the colon wall (usually from the pressure of straining during bowel movements).

Weight loss: The insoluble fiber is indigestible, and thus, passes through the body and provides few calories. Since the digestive tract can handle only so much bulk at a time, **fiber-rich food articles are more filling than other foodstuffs**, so people tend to eat less.

Cancer prevention: A study carried out by the Department of Nutrition at Harvard School of Public Health, Boston has shown that diets high in lentils and peas (which both contain high levels of flavones) have a reduced risk of breast cancer.

Cardiovascular disease prevention: The intake of dietary fiber, particularly from lentils has been known to reduce the risk of coronary heart disease.

Common name: Lima bean

Botanical name: *Phaseolus lunatus*

Family: Fabaceae

Lima beans, like many other legumes, are a good source of dietary fiber and virtually a fat-free source of high quality protein. These contain both soluble fiber, which helps in regulating the blood sugar levels and lowers cholesterol and insoluble fiber, which prevents constipation, digestive disorders, irritable bowel syndrome and diverticulitis. The nutritional composition of lima beans (cooked 1.0 cup, *i.e.*, 188.00 g) is given below:

Constituents		Contents	Constituents		Contents
Total dietary fiber (g)	:	13.16	Iron (mg)	:	4.49
Protein (g)	:	14.66	Copper (mg)	:	0.44
Manganese (mg)	:	0.97	Tryptophan (g)	:	0.17
Phosphorus (mg)	:	208.68	Vitamin B_1 (mg)	:	0.30
Molybdenum (μg)	:	141.00	Total folate (μg)	:	156.23
Magnesium (mg)	:	80.84	Energy (cal)	:	216.20
Potassium (mg)	:	955.04			

Blood sugar: The high fiber content in lima beans prevents blood sugar levels from rising too rapidly after eating them. This is due to the presence of higher amounts of absorption-slowing protein in the beans, and their high soluble fiber content. Soluble fiber absorbs water in the stomach forming a gel that slows down the metabolism of bean's carbohydrates. Therefore, they can help in balancing the blood sugar levels while providing steady, slow-burning energy, which makes them a good choice for people with diabetes suffering with insulin resistance.

Heart disease: The soluble fiber of beans binds with the bile acids that form cholesterol, and because it is not absorbed by the intestines, it exits in the body taking the bile acids with it. As a result, the cholesterol level is lowered. Therefore, they may help to prevent heart disease, and may reduce the medical dosage required to combat cholesterol in the form of natural food.

Lima beans also provide folate and magnesium. Folate lowers the levels of homocysteine, a bad amino acid, which is an intermediate product in an important metabolic process called the methylation cycle. Elevated blood levels of homocysteine are an independent risk factor for heart attack, stroke and peripheral vascular disease. The magnesium content of lima beans is a calcium channel blocker. When enough magnesium is present, the veins and arteries relax, which reduces resistance and improves the flow of blood, oxygen and nutrients throughout the body.

Rich source of minerals: Apart from providing slow-burning complex carbohydrates, lima beans can increase energy levels by helping to restore more iron. The menstruating women has more risk of iron deficiency, lima beans can provide iron, an integral component of hemoglobin. The hemoglobin transports oxygen from the lungs to all body cells and is also a part of key enzyme systems for energy production and metabolism. Lima beans are a very good source of trace mineral manganese and help enzymes important for energy production and antioxidant defense. Lima beans contain the trace mineral molybdenum, which is an integral component of enzyme sulfite oxidase and which detoxifies sulfites, a preservative often added to prepared foodstuffs.

Raw lima beans and butter beans contain toxic linamarin, a cyanogenic glucoside. One handful of raw beans can make a person violently ill. Therefore, the beans are rendered safe when cooked properly. Low-linamarin varieties are typically used for culinary purposes.

Common name: Velvet bean (Kiwanch, Konch)

Botanical name: *Mucuna pruriens*

Family: Fabaceae

Velvet bean is a reputed *Ayurvedic* remedy for nervous and sexual diseases. Traditionally, it is commonly used as carminative, hypertensive and hypoglycemic agent. Velvet bean has been found to contain L-DOPA, 40 mg/g of the plant. Its plant or seed contains the bioactive alkaloids mucunine, mucunadine, mucuadinine, pruriendine and nicotine, besides B-sitosterol, glutathione, lecithin, oils, venolic and gallic acids.

Velvet bean is sometimes used as a coffee substitute called *Nescafe* (not to be confused with the commercial brand). Cooked fresh shoots or beans can also be eaten. The beans are soaked for at least 30 minutes to 48 hours in advance of cooking, or the water changed up to several times during cooking since the plant can be toxic to humans. The above-described process leaches out the chemical compounds such as levodopa, making the product suitable for consumption.

Use of velvet bean is documented in *Siddha* medicine for a number of uses, including improving sexual function. It has been used as an effective aphrodisiac. It is used to increase libido in both men and women due to its dopamine inducing properties. Dopamine has a profound influence on sexual function. A typical dose for a man is 15 g of ground seeds mixed with cow's milk. The seeds have also been found to have antidepressant properties in cases of depressive neurosis when consumed. Dried leaves of velvet bean are sometimes smoked.

Velvet bean has also recently become popular among lucid dreaming enthusiasts: when combined with other supplements that stimulate the cholinergic system, the dopamine presumably produced from the consumption of velvet bean confers upon the lucid dreamer greater motivation and confidence. The hairs on outside of the pods of velvet beans are a common ingredient in itching powder.

The seeds of in velvet bean contain high concentrations of levodopa, a direct precursor of the neurotransmitter dopamine. L-Dopa is an amino acid that converts into dopamine, which is an essential component required for proper functioning of the brain. The body converts the amino acid tyrosine into L-dopa, which is then converted into dopamine. L-Dopa improves sleep (promotes deep sleep), reduces body fat and decreases cellulite and wrinkles, improves skin texture and appearance, increases bone density and reversal of osteoporosis, increases lean muscle mass, improves mood and sense of well-being, enhances libido and sexual performance, increases energy levels, improves cholesterol profile and regeneration of organs (heart, kidney, liver and lungs) and strengthens immune system. It has long been used in traditional *Ayurvedic* Indian medicine for diseases including Parkinson's disease. The roots are also used against cholera, as a diuretic and for dropsy. The seeds are also used in delirium, impotence, leucorrhoea and spermatorrhoea.

Common name: Tamarind (Imali)

Botanical name: *Tamarindus indica*

Family: Fabaceae

Tamarind is best described as sweet and sour in taste and high in acid, sugar, calcium and vitamin B. The fruit pulp is edible and popular. The hard green pulp of a young fruit is considered too sour and acidic but is often used as a component of savory dishes and as a pickling agent. The ripened fruit is considered the more palatable as it becomes sweeter and less sour (acidic) when it matures. It is used in desserts as a jam, blended into juices or sweetened drinks, sorbets, ice creams and all manners of snack. It is also consumed as a natural laxative.

The tannins, saponins, sesquiterpenes, alkaloids and phlobatamins, which are present in fair amounts in tamarind, are active against both gram-positive and gram-negative bacteria, even at 4-30°C temperature.

Throughout Asia and Africa tamarind is common for health remedies. In Northern Nigeria, fresh stem bark and fresh leaves are used as decoction mixed with potash for the treatment of stomach disorders, general body pain, jaundice, yellow fever and as blood tonic and skin cleanser. In traditional medicines, the Indonesians, Malaysians, Philippines and Japanese use tamarind leaves as an herbal infusion for malarial fever, the fruit juice as an anti-septic and scurvy and used even for the treatment of cough and cold. Fruit of the tamarind is also commonly used throughout South East Asia as a poultice applied to foreheads of fever sufferers.

In Indian *Ayurvedic* medicine, tamarind is used for gastric and/or digestion problems and in cardio protective activity. Tamarind has been found to lower serum cholesterol and blood sugar levels. Due to a lack of available human clinical trials, there is insufficient evidence to recommend tamarind for the treatment of hypercholesterolemia (high cholesterol) or diabetes. It has also been reported that tamarind intake may delay the progression of fluorosis by enhancing excretion of fluoride. Excess consumption has been noted as a traditional laxative.

Other medicinal uses include anthelminthic (expels worms), antimicrobial, antiseptic, antiviral, asthma, astringent, bacterial skin infections (erysipelas), boils, chest pain, cholesterol metabolism disorders, colds, colic, conjunctivitis (pink eye), constipation (chronic or acute), diabetes, diarrhea (chronic), dry eyes, dysentery (severe diarrhea), eye inflammation, fever, food preservative, food uses (coloring), gallbladder disorders, gastrointestinal disorders, gingivitis, hemorrhoids, indigestion, insecticide, jaundice, keratitis (inflammation of the cornea), leprosy, liver disorders, nausea and vomiting (pregnancy-related), paralysis, poisoning (nightshade plant), rash, rheumatism, saliva production, skin disinfectant/sterilization, sore throat, sores, sprains, sunscreen, sunstroke, swelling (joints), urinary stones and wound healing (corneal epithelium).

■■■

8 Coconut: The Miracle Food

Umashashi C. Hegde

Deputy Director, Senior Grade (Retired), National Institute for Research in Reproductive Health (ICMR), Parel, Mumbai-400012, Present Address: Chandrama, Post Kemmannu, Udupi Distt.-576115, Karnataka, Email: <hegdeuc@yahoo.com>

Coconut palm is known in India as *kalpavriksha* or the wish-fulfilling tree. In various cultures in South Asia, the coconut palm provides basic survival needs. In Malaya, coconut is known as *pokok seribu guna*, the tree of a thousand uses. In the Philippines, coconut tree is called the *Tree of Life*.

Coconut is an essential part of Hindu religious ceremonies and is the most auspicious offering to the Gods. It is usually broken in front of deities. *Purnakumbha* is a pitcher or *kalasam* filled with water, topped with fresh mango leaves and a coconut. This is an essential part of Hindu rituals and *pujas*. The fruit believed to represent Lord Shiva, the three eyed God and the three black marks on the coconut shell symbolize his eyes. The fruit is also associated with Goddess *Lakshmi*, and hence, is referred as *shreephal*. Coconuts are used for worship at the time of Kaveri River worship at Tiruchairapalli, India.

Figure 1 : Offering Coconuts to Kaveri River

The coconut palm (cocos nucifera) is a member of the family Arecaceae (the palm family). It is the only species in the genus *Cocos* and is a large palm, growing to 30 metres with pinnate leaves 4–6 m long. Old leaves, fronds, break away cleanly while leaving the trunk smooth. The term coconut refers to seed of the coconut palm. The name of coconut is a bit of a misnomer since it is botanically classified as a drupe, not as a nut. It is the largest seed known.

Figure 2 : Coconut Trees on the Western Coast of India

The flowers of the coconut palm are polygamomonoecious with both male and female flowers in the same inflorescence. Flowering occurs about 13 times annually. Coconut palms are believed to be largely crosspollinated, although some dwarf varieties are self-pollinating. The kernel or the white meat inside shell of the coconut is the edible part or endosperm. Inside endosperm layer, coconut contains coconut water or coconut juice that is slightly sweet and highly nutritious.

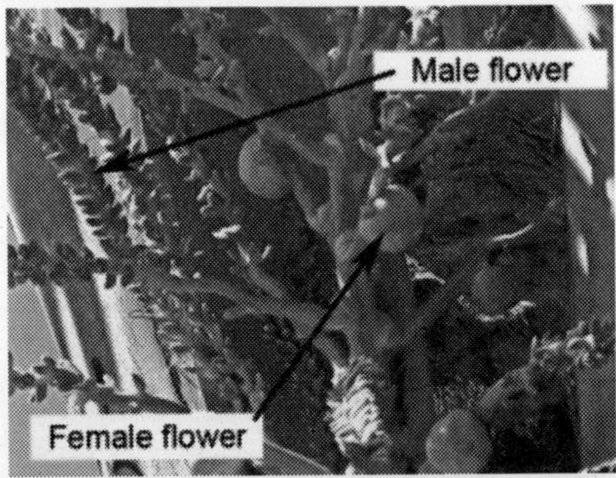

Figure 3 : Coconut Male and Female Flowers

Coconut received their name from Portuguese explorers, the sailors of Vasco da Gama in India, who first brought them to Europe. The brown and hairy surface of coconut reminded them of a ghost or witch called Coco. Before, it was called nux indica, a name given by Marco polo in 1280. When coconut arrived in England, it retained coco portion of the name and nut was added to the end.

Figure 4 : A Bunch of Young Coconuts

Coconuts grow in southern Asia and are prolific in South America, India, the Pacific Islands, Hawaii and Florida. Coconut trees grow on tropical seacoasts and are cultivated in warm and moist climates. In India, coconuts are cultivated in Kerala, Tamil Nadu, Karnataka, Andhra Pradesh, Goa, Maharashtra, Orissa and West Bengal. On Nicobar Islands in the Indian Ocean, whole coconuts were used as currency for the purchase of goods until the early part of the twentieth century.

The trunk of coconut tree is about a foot in diameter. It has no branches but carries a crown of leaves about 70 to 100 feet above the ground. The leaves (fronds) are like feathers and with many segments, which are usually 10 to 20 feet long. Coconuts hang at bases of the leaves in clusters of about 10 to 15. Several clusters ripen during the year. A tree may produce from 20-200 coconuts a year but the average is around 40.

How do coconut trees grow?

Ever wondered, how coconut trees propagated themselves across the world before people started cultivating them for economic reasons? On some islands and coastal regions, the ripe coconuts, which fall from the tree, are frequently washed out to sea. Some may drift for months before they are cast up by waves onto the shore. After 4 or 5 months, the seed sprouts and a new tree begins to grow. In about 6 years, the tree will begin to bear coconuts and it will reach maturity at 20 years. A coconut tree is generally very strong but some older trees are either broken down or uprooted by the strong typhoons. When cultivated as a cash crop, fertilizers are usually applied once a year and trees are watered during summer months to get a good yield of coconuts.

Description of the coconut

Freshly plucked coconuts have a smooth greenish outer covering. Within outer shell is a fibrous husk of one to two inches thickness. The inner shell is brown and hard. When coconut is still young, it contains a large amount of fluid or juice. Coconuts used for coconut water are called tender coconuts. The water of a tender coconut is liquid endosperm. It is sweet, mild and mineral -rich in its stage of development. There is no alternative drink for coconut water. It is a nature's gift, loaded with health-giving nutrients. The meat in a young coconut is much softer and more gelatinous than a mature coconut and is sometimes known as coconut jelly. During this stage, the coconut meat can ideally be eaten as it is, or can be used as part of a salad or blended with milk to make a nutritional drink. When the coconut ripens, the outer husk turns brown, the endosperm thickens and hardens, while the volume of coconut water is considerably reduced. The edible part of coconut is the kernel that can be incorporated into a variety of food preparations. The coconut is considered a functional food as it is rich in numerous essential nutrients with several health benefits. It also has a high content of fiber.

Health benefits of coconut water

During the Second World War, when there was a shortage of blood and plasma, intravenous infusions of coconut water was used to save many lives. It can be injected intravenously in emergency cases as a blood plasma substitute because it is sterile, does not produce heat, does not destroy red blood cells and is readily accepted by the body. It is similar to human blood plasma, and therefore, can function as a universal donor

It is only recently that man has figured out how to package fresh coconut water, preserving its freshness. People in tropical climates have always been able to enjoy fresh coconut water. Now, it is within the reach of everyone. Coconut water has natural preservative qualities so absolutely nothing else has to be added. The shelf life of packed coconut water is an amazing 12 months

Coconut water comes only from young green coconuts. A tender coconut yields about 400-1000 ml of coconut water. As a coconut matures, the natural water inside evaporates into air and is absorbed into the production of endosperm. Coconut water is not to be confused with coconut cream or milk. Some people call the coconut water as *coconut juice*. One should read the labels carefully when buying packaged coconut water.

Figure 5 : Sold as Coconut Water

The coconut kernel has different properties and higher fats. When a fresh coconut is opened, there is the translucent, almost clear coconut water inside that has a slightly nutty flavor. Contrary to popular belief, this is not the coconut milk. The cavity is filled with coconut water. This contains sugar, proteins, antioxidants, minerals and vitamins. Coconut water provides an isotonic electrolyte balance and is a highly nutritious food source. It is used as a refreshing drink throughout the humid tropics and is also used in isotonic sports drinks. Mature fruits have significantly less liquid than young immature coconuts. Barring spoilage, coconut water is sterile until opened. Drinking coconut water has shown numerous health benefits. Coconut water is a natural drink. The water is consumed fresh from the coconut, and it can be used in various recipes. Some of the special properties of coconut water are given as under:

- Coconut water contains four electrolytes- potassium (at about 294 mg, which is more than banana), calcium, magnesium and phosphorus, thus, cures malnourishment. It contains very high chloride (118 mg), less sodium (25 mg) and 5 mg of natural sugars. In addition, coconut water is an excellent tonic and contains organic compounds possessing healthy growth promoting properties that help old, sick, healthy people and children.
- According to Mortin Satin, Chief of the United Nation's Food and Agriculture Organization, it is a biologically pure drink provided by nature, full of natural sugars, salts and vitamins to ward off fatigue and is the best energy drink.
- Coconut water is an isotonic natural rehydration fluid (beverage) and naturally replenishes body fluids after exercise and cools the body to the proper temperature.
- It is very low in fat, in acidity and in sugar and free from cholesterol, thus, it is more nutritious than whole milk.
- It aids in quick absorption of the drugs and makes their peak concentration in the blood easier by its electrolytic effect. It is antiseptic and it detoxifies and eliminates poisons in case of mineral poisoning.
- It is good for feeding infants suffering from intestinal disturbances as it acts as a diuretic and cleans the digestive tract. Coconut water is great at reducing vomiting. People who have typhoid, malaria, fevers, or other ailments, which induce vomiting, drink coconut water to settle the stomach. The presence of saline and proteins makes it a good drink in case of cholera.
- It helps in carrying nutrients and oxygen to the cells, boosts poor circulation and eliminates swelling in hands and feet.
- It helps to relieve hot flushes in menopausal women and helps to get rid of kidney and urethral stones.
- It boosts the immune system in the body, kills intestinal worms and checks bacterial and urinary infections, fights viruses that cause the flu, herpes and AIDs.
- The beneficial effect of feeding coconut water on lipid metabolism is scientifically proven. Administration of coconut water counteracts the increase in total cholesterol, very low-density lipoprotein (VLDL), low-density lipoprotein (LDL) and triglycerides while high-density lipoprotein levels decreased in a high fat diet. Coconut water can also help high blood pressure since it has high concentration of potassium and also helps artery pressure to decrease making it a healthy drink for cardiovascular health.

External uses of coconut water

- It is applied externally to soothe sunburn.
- Application on the body prevents prickly heat and summer boils and helps the rashes caused by small pox, chicken pox and measles to subside.
- It can be directly applied to combat oily skin or to hydrate and moisturize.
- Application to hair, massage and leave in for an hour to reduce breakage helps in preventing hair loss.

With all the reported *health benefits* of coconut and its water, it is no wonder the palm tree is known as the *Tree of Life*.

Most coconut water is still consumed fresh in tropical coastal areas. When it is exposed to air, the liquid rapidly loses most of its organoleptic and nutritional characteristics, and begins to ferment if not stored in sterile conditions. There are reportedly different tastes with different brands of packed coconut water. This phenomenon can be explained by soil conditions and locations. There are many brands of coconut water available. If one is buying packed coconut water, make sure nothing is added to adulterate the natural properties of coconut water.

Table 1 : Nutritional value of 1 cup of fresh coconut water

Constituents	Contents	% daily value	Constituents	Contents	% daily value
Total carbohydrates (g)	8.9	3	Dietary fiber (g)	2.64	11
Protein (g)	1.73	3	Alcohol (mg)	0	0
Total fat (g)	0.48	1	Sodium (mg)	252	11
Saturated fat (g)	0.422	2	Potassium (mg)	600	17
Polyunsaturated fat (g)	0.0048		Calories from water (kcal)	45.6	
Monounsaturated fat (g)	0.0192		Calories from fat (kcal)	4.32	
Cholesterol (mg)	0	0			

Table 2 : Mineral and vitamins contents in coconut water

Constituents	Contents	Constituents	Contents
Calcium	6 %	Riboflavin	7 %
Magnesium	15 %	Folate	2 %
Phosphorus	5 %	Niacin	1 %
Iron	4 %	Vitamin B_6	4 %
Copper	5 %	Vitamin B_{12}	0 %
Zinc	2 %	Vitamin C	10 %
Vitamin A	0 %	Vitamin E	0 %
Thiamine	4 %	Vitamin D	0 %

Table 3 : Analysis of mature and tender coconut water

Constituents	Mature Coconut Water	Tender Coconut Water
Total solids (%)	5.4	6.5
Reducing sugars (%)	0.2	4.4
Minerals (%)	0.5	0.6
Protein (%)	0.1	0.01
Fat (%)	0.1	0.01
Acidity (mg)	60.0	120.0
pH	5.2	4.5
Potassium (mg)	247.0	290.0
Sodium (mg)	48.0	42.0
Calcium (mg)	40.0	44.0
Magnesium (mg)	15.0	10.0
Phosphorous (mg)	6.3	9.2
Iron (mg)	79.0	106.0
Copper (mg)	26.0	26.0

Sugars: Sugars in the forms of glucose and fructose form an important constituent of the tender nut water. The concentration of sugars in the nut water steadily increases from about 1.5 to 5.5% in the early months of maturation, and then, slowly falls reaching about 2% at the stage of full maturity of the nut. In early stages of maturity, sugars are in the form of glucose and fructose (reducing sugars) and sucrose (non-reducing sugar), which increases with the maturity, appears only in later stages, while the reducing sugars fall down. In fully mature nut, approximately 90% of the total sugars are sucrose.

Proteins: Coconut water contains small amounts of protein. The percentage of arginine, alanine, cystine and serene in the protein of tender coconut water are higher than those in cow's milk. Since it does not contain any complex protein, the danger of producing shock in patients is minimized.

Vitamins: Tender coconut water contains both ascorbic acid and vitamins of the B group. The concentration of ascorbic acid ranges from 2.2 to 3.7 mg per ml, which gradually diminishes as the kernel surrounding the water begins to harden.

Minerals: Tender coconut water contains a plethora of minerals such as potassium, sodium, calcium, phosphorous, iron, copper, sulphur and chlorides. Among the minerals, more than half is potassium, the concentration of which is markedly influenced by potash manuring. Tender coconut water, being rich in potassium and other minerals, plays a major role in increasing the urinary output.

Table 4 : Amino acid composition of coconut water (% of total protein)

Constituents	Contents	Constituents	Contents
Alanine	2.41	Leucine	1.95-4.18
Arginine	10.75	Lysine	1.95-4.57
Aspartic acid	3.60	Proline	1.21-4.12
Cystine	0.97-1.17	Phenylalanine	1.23
Glutamic acid	9.76-14.5	Serine	0.59-0.91
Histidine	1.95-2.05	Tyrosine	2.83-3.00

Coconut kernel

The meat in a young coconut is softer and more gelatinous than a mature coconut, so much so, that it is sometimes known as coconut jelly. When coconut has ripened and the outer husk has turned brown (a few months later), it may be plucked or may fall from the palm. At that time, the endosperm has thickened and hardened, while the volume of coconut water is considerably reduced. Mature coconuts have firmer white meat, which tends to cling more stubbornly to the inside of the shell, making it more challenging to remove. The hard shell of the mature coconut is broken with a sickle into two halves. The white kernel is either scraped or scooped out.

People can eat coconut meat plain, or they can grate it and utilize it in a range of dishes. Grated and toasted coconut meat is often used as a garnish, and shavings of coconut meat may be included as decoration on desserts like coconut cream pie, puddings, and cakes. Shredded coconut is also a key ingredient in German chocolate cake. The meat has a distinctive rich, tropical flavor, which some people find very appealing. Although coconut meat contains less fat than other dry nuts such as almonds, it is noted for a high amount of medium-chain saturated fat. About 90% of the fat found in coconut meat is saturated, a proportion exceeding that of foods such as lard, butter and tallow. However, there has been extensive debate as to whether or not the saturated fat in coconuts is healthier than the saturated fat found in other foods. Recent studies have proven the beneficial effects of fats present in coconut and will be explained later. Coconut water also contains less sugar and more protein than popular fruits such as bananas, apples and oranges, and it is relatively high in minerals such as iron, phosphorus and zinc.

This fresh kernel is the base for curries in places growing coconuts. In Western countries, coconuts are used in baked dishes. The coconut kernel, in addition to its nutritional value, is a rich source of fiber. For daily use, people prefer to use coconut kernel because it is a rich source of fiber instead of coconut milk. The coconut is preferable to other nuts because its protein is of a high quality with all the essential amino acids. Its fat content is low and it has lot of roughage, making it an ideal food. The kernel is easily digested. Each globule present in the ground coconut kernel is made up of a spherical structure built of protein, inside which there is the fat particle. This fat does not hinder the digestion of the food in the stomach. When the protein is digested, the oil that is set free goes into the small intestine where it is easily digested. This shows how nature has adapted the food to suit the natural process of digestion. Fried food is just the opposite. The fat is on the outside and the protein or starch on the inside making it unhealthy.

Coconut usage in India

In south India, particularly in Kerala and coastal Karnataka, coconuts are used extensively for curries and various other mouth-watering dishes. A vast majority of dishes, including sweets, have coconut kernel or milk. The most common way is to lightly cook the vegetable and used freshly scraped coconut is used for garnishing. While preparing curries and *sambar*, to enhance the taste of the dish, the coconut is mixed ground and added along with other spices. Coconut milk is used to make fluffy *appams,* rich gravies, puddings and *kheers.* Most South Indian families use at least a coconut a day for their food preparations. They make a wide variety of *chatnis* which involve grinding the coconut kernel with chilly, and various whole spices and salt. These can then be eaten with rice or kanji (rice gruel). Invariably, the *idli, vada* and *dosa* are always accompanied by a side dish of coconut chutney. *Puttu* is a culinary delicacy from Kerala, in which layers of coconut alternate with layers of powdered rice, all of which fit into a metal tube-like container which is then steamed. Different types of breakfast preparations like *dosas* and *appam* include fresh coconut meat or thick coconut milk. The coconut kernel is also used as a snack and is scraped and eaten with jaggery or molasses, especially by children in villages.

South Indian Dish : Idli and Coconut Chutney

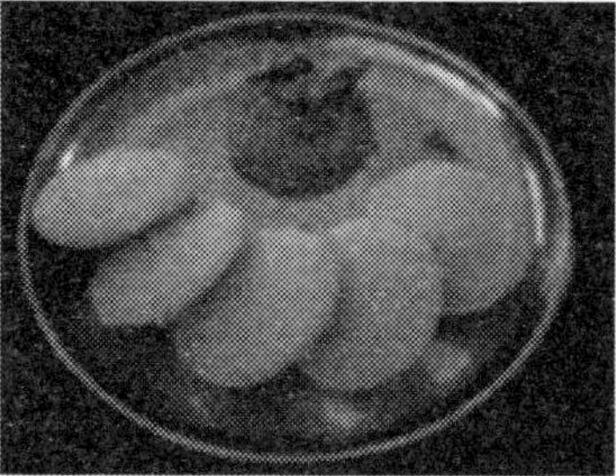

Chicken with Coconut Gravy

Additional information about Coconut

- *Macapuno* nuts are actually aberrant coconuts. Nuts are the fruits whose genetics short-circuited along the way and whose fruit are delightfully different from a regular coconut.

These are also known as sport fruits are harvested, primarily in the Philippines. They are sold in jars as *gelatinous mutant coconut* cut into balls or strands.

- The pleasant smell of coconuts comes from the 6-pentyloxan-2-one molecule, known as delta-decalactone in the food and fragrance industry. .

Table 5 : Nutritional value of raw coconut kernel per 100 g

Constituents	Contents	Constituents	Contents
Energy	350 kcal	Magnesium	32 mg
Carbohydrates	15.23 g	Iron	2.43 mg
Sugars	6.23 g	Zinc	1.1 mg
Dietary fibre	9.0 g	Vitamin B_1 (thiamine)	0.066 mg
Fat	33.49 g	Vitamin B_2 (riboflavin)	0.02 mg
Saturated fat	29.70 g	Vitamin B_3 (niacin)	0.54 mg
Monosaturated fat	1.43 g	Vitamin B_5 (pantothenic acid)	0.300 mg
Polyunsaturated fat	0.37 g	Vitamin B_6 (pyridoxine)	0.054 mg
Calcium	14 mg	Vitamin B_9 (folate)	26 mcg
Potassium	356 mg	Vitamin C	3.3 mg
Phosphorous	113 mg		

The protein content of coconut is of high quality and contains the following amino acids:

Table 6 : Amino acid composition of coconut kernel

Constituents	Contents	Constituents	Contents
Tryptophan (g)	0.039	Methionine (g)	0.062
Threonine (g)	0.121	Cystine (g)	0.066
Isoleucine (g)	0.131	Phenyl alanine (g)	0.169
Leucine (g)	0.247	Tyrosine (g)	0.103
Lysine (g)	0.147	Valine (g)	0.202

Coconut milk

The coconut kernel is separated from the shell and ground with warm water. It is passed through a fine sieve or cheesecloth. This process extracts all the nutrients and aromatic compounds. Coconut milk should not be confused with coconut water and should have a fat content around 17%. This rises to the top and separates from the cloudy watery residue. The first extract or the thick milk is used for making delicious, rich gravies and curries and some special types of pancake or *dosas* as well as sweets including *kheer*. The second extract can be used for soups. The coconut milk is used to produce virgin coconut oil (VCO) which is marketed in western countries. This is produced by controlled heating and removing the oil fraction. Virgin coconut oil is claimed to be superior to the oil extracted from sun-dried coconut. The leftover fiber after extracting coconut oil is used as livestock feed. Coconut milk has all the nutrients of the coconut kernel without the fiber. Packed coconut milk is available in liquid and dehydrated forms and is convenient to use. It is extensively used where coconuts do not grow. Nature cure experts do not consider cow's milk a positive food. It is recommended that coconut milk be used instead of cow's milk wherever possible. On the suggestion of Mahatma Gandhi, a person who had fasted for 28 days, on breaking the fast was given vegetable soup flavored with coconut milk instead of cow's milk and its products. Within a week he regained his vitality, was able to walk to and fro, five miles, both morning and evening. When a working mother cannot breast feed, it is suggested by Dr. Teofilo de la Torre, of Edenia, Central America that she can dilute coconut milk with coconut water and feed the infant. This is definitely better than cow's milk. It is also the best option for infants with lactose intolerance.

How did coconut oil get a bad name?

Tropical oils were very popular in the United States food industry prior to World War II. It is important to realize that around 1960s, the edible oil industry in the United States seized the opportunity to promote its polyunsaturated oils. Supporting evidence came around the same time, when researchers fed laboratory animals with hydrogenated coconut oil that was purposefully altered to make its consistency suitable for mixing with their feed and to make it completely devoid of any essential fatty acids. The animals that were fed the hydrogenated coconut oil as the only source of fat naturally became essential fatty acid deficient and their serum cholesterol levels increased. Diets that cause an essential fatty acid deficiency always produce an increase in serum cholesterol levels as well as an increase in the atherosclerotic indices. Although many studies at the time had also shown research to the contrary, the mud stuck and by the mid 60s, the reputation of all saturated oils in America had been destroyed. This reputation later extended to rest of the western world. Though later studies with unprocessed coconut oil have actually shown beneficial effects in laboratory animals and humans, the commercial interests of the domestic fats and oils industry were successful in driving down usage of coconut oil. The United States of America is the largest exporter of soybeans. The well-oiled marketing machinery funded by the soybean and corn industry and supported by the American Heart Association was committed to change the American Diet, calling among others, for the substitution of saturated fats with polyunsaturated fats. The Prudent Diet, as it was called, left a legacy, which still haunts us today, 40 years later. This conceptual change in the eating habits of Americans has negatively influenced and changed the dietary regimes of societies all around the world that were initially not even affected by America's meat, potato and milk diet. So determined was the pursuit of the American industries in converting their claims into magnificent billboards of health and wealth that even Small Island nations in the South Pacific were converted by this powerful marketing machine to change centuries of dietary traditions of tropical oils to importing polyunsaturated fats. Today, heart disease is still on the increase and obesity, linked to the *new American diet*, is a major social problem worldwide that has governments worried about the health care cost of future generations. The United Kingdom and Australia unfortunately are racing to catch up with their allies with a large percentage of the population being defined as overweight. As a result, in the

United States and in other countries where the influence of the United States is strong, the people have lost the benefit of the lauric acid present in coconut oil.

Types of Fats and oils

- **Mainly saturated:** Butter, hard margarines, cheese, eggs, olives, olive oil, *etc*.
- **Intermediate:** Nuts, coconuts, wheat germ and their oils, *etc*.
- **Mainly unsaturated:** Sunflower seed oil, safflower seed oil, corn oil, linseed oil, soy oil, soft margarines which have high polyunsaturated fats.

Coconut oil

Coconut oil is extracted from the kernel of mature coconut. In India, coconut oil is made using sun-dried coconut kernel (copra). Cold-pressed unrefined coconut oil is an excellent source of good fat. Coconut oil is also made using coconut milk and this is known as virgin coconut oil (VCO). This is believed to be superior to the oil made from copra. The constituents that make the coconut oil different from most other dietary oils is the basic constituents or the fatty acids. Coconut oil has provided the primary source of fat throughout the tropical world for millions of people for generations. It has various applications in food, medicine and industry. While coconut kernel possesses many health benefits due to its fiber, proteins, vitamins and minerals and many other micronutrients, it is the oil that that is truly remarkable. Coconut oil is composed predominantly of a special group of fat molecules. Once mistakenly believed to be unhealthy because of its high saturated fat content, it is now known that the fat in coconut oil is unique and different from most other fats and possesses many health-giving properties. It is now gaining long overdue recognition as a nutritious health food. Coconut oil has been described as the *healthiest oil on earth* by Dr. Bruce Fife, a naturopathic doctor and author of *The Coconut Oil Miracle*, which is quite a remarkable statement. Renee Loux, ecological expert and author of the book *The Essential Elements of Whole Foods and Good Health: The Balanced Plate* says that coconut oil does contain some naturally saturated fats, which is why it turns solid in cold weather. Saturated fats have stronger bonds that make them more stable at higher temperature. Coconut oil is a medium-chain fat, which is very easy to digest and provides adequate energy. The nature of this type of oil does not require the liver to emulsify the fat (break it down to smallest fraction of fat called lipids). Though coconut oil has more saturated fat than unsaturated fat, it is important to understand that it is a medium-chain fatty acid (MCFA), which is directly converted into energy in our bodies and not stored as fat. It is a great source of energy and fuel.

Composition of fats in coconut oil

Of course, some saturated fat is necessary in the diet. As it is absorbed in the intestines, it carries the fat-soluble vitamins A, D, E and K. Without that fat, vitamin deficiency occurs. Saturated fats and cholesterol are also built into cell membranes, strengthening cell walls so they keep their shape. In addition, saturated fat is used by the liver to construct cholesterol, to the tune of 1,000 mg per day. Sunshine converts cholesterol to vitamin D, and cholesterol is the first step in manufacturing testosterone and estrogen. Therefore, saturated fats should not be eliminated. Without saturated fats, the health would be certainly affected. The long-chain saturated fats that make up 20% of coconut oil meet the body's requirements adequately. That is just enough long-chain saturated fat to be healthy but not enough to make you obese.

How is coconut oil different from other oils?

What makes the coconut oil better than other oils? How does it differ from all other oils, especially other saturated fats? The difference is in the fat molecule. All fats and oils are composed of molecules called fatty acids. There are different methods of classifying fatty acids. The first you are probably familiar with is based on saturation. In this category, there are saturated, monounsaturated and polyunsaturated fats. Another system of classification is based on molecular size or length of the carbon chain within each fatty acid. Fatty acids consist of long chains of carbon atoms with hydrogen atoms attached. In this system, we have short-chain fatty acids (SCFA), medium-chain fatty acids (MCFA) and long-chain fatty acids (LCFA). Coconut oil is composed predominantly of medium-chain fatty acids (MCFA), also known as medium-chain triglycerides (MCT). The vast majority of fats and oils in our diets, whether they are saturated or unsaturated or come from animals or plants, are composed of long-chain fatty acids (LCFA). Some 98 to 100% of all the fatty acids consumed are long-chain fatty acids.

Why is the size of the fatty acid extremely important?

That is primarily because our bodies respond to different fatty acids differently. These are metabolized depending on their size. Therefore, the physiological effects of medium-chain fatty acids in coconut oil are distinctly different from those of long-chain fatty acids more commonly found in other vegetable oils. The saturated fatty acids in coconut oil are predominantly medium-chain fatty acids. Both the saturated and unsaturated fat found in meat, milk, eggs, plants, and almost all vegetable oils are composed of long-chain fatty acids.

Medium-chain fatty acids are very different from long-chain fatty acids. They do not have a negative effect on cholesterol and help to protect against heart disease. Medium-chain fatty acids help to lower the risk of both atherosclerosis and heart disease. It is primarily the medium-chain fatty acids in coconut oil, which makes it so special and so beneficial. There are only a very few good dietary sources of medium-chain fatty acids. By far the best sources are from coconut and palm kernel oils.

The primary difference between medium-chain fatty acids and long-chain fatty acids is the size of the molecule, or more precisely, the length of the carbon chain that makes up the backbone of the fatty acid. Medium-chain fatty acids have a chain length of 6 to 12 carbons. Long-chain fatty acids contain 14 or more carbons. The length of the carbon chain influences many of the oil's physical and chemical properties. When consumed, the body processes and metabolizes each fatty acid differently depending on the size of the carbon chain. Therefore, the physiological effects of the medium-chain fatty acids in coconut are significantly different from those of the long-chain fatty acids that are more commonly found in the diet.

Medium-chain fatty acids and long-chain fatty acids can also be classified as saturated, monounsaturated or polyunsaturated fatty acids. Coconut oil contains 92% saturated fatty acids. All of the medium-chain fatty acids in coconut oil are saturated. They, however, are very different chemically from the long chain saturated fatty acids found in animal fat and other vegetable oils. Because coconut oil has a high amount of saturated fatty acids, it also has a relatively high melting point of above 24°C. Coconut oil is a colourless liquid. Below this temperature, it solidifies into a pure, white solid. Coconut oil is very heat stable so it is excellent oil for cooking and frying. It has a smoke point of about 180°C. Because of its stability, it is slow to oxidize and is thus resistant to rancidity, lasting up to two years due to its high saturated fat content.

Coconut oil in traditional medicine

People from many diverse cultures, languages, religions and races scattered around the globe have revered the coconut as a valuable source of both food and medicine. Wherever the coconut palm grows, the people have learnt of its importance as an effective medicine. For thousands of years, coconut products have held a respected and valuable place in folk medicine. In traditional medicine around the world, coconut is used to treat a wide variety of health problems including the following, *i.e.*, abscesses, asthma, baldness, bronchitis, bruises, burns, colds, constipation, cough, dropsy, dysentery, earache, fever, flu, gingivitis, gonorrhea, irregular or painful menstruation, jaundice, kidney stones, lice, malnutrition, nausea, rash, scabies, scurvy, skin infections, sore throat, swelling, syphilis, toothache, tuberculosis, tumors, typhoid, ulcers, upset stomach, weakness and wounds.

Coconut Oil in Modern Medicine

Modern medical science is now confirming the use of coconut in treating many of the above conditions. Published studies in medical journals show that coconut in one or another form may provide a wide range of health benefits. Some of the important health benefits of coconut oil are discussed below in detail:

Table 7 : Fatty acid profile of coconut oil

Fatty Acid	Saturation	Carbons	Percent
Caproic	Saturated	6	0.5
Caprylic	Saturated	8	7.8
Capric	Saturated	10	6.7
Lauric	Saturated	12	47.5
Myristic	Saturated	14	18.1
Palmitic	Saturated	16	8.8
Stearic	Saturated	18	2.6
Arachidic	Saturated	20	0.1
Oleic	Monounsaturated	18	6.2
Linoleic	Polyunsaturated	18	1.6

Coconut oil contains approximately 92.1% saturated fatty acids, 6.2% monounsaturated fatty acids.

Major Constituents with Health Benefits

Historically, coconuts and their extracted oil have served man as important foods for thousands of years. The use of coconut oil for shortening was advertised in the United States in popular cookbooks at the end of 19th century. Both the health-promoting attributes of coconut oil and those functional properties useful to the homemaker were recognized 100 years ago. In addition to some new information, these attributes should be of great interest to both coconut producing and consuming countries. Coconut oil has been a lifesaver for many people. The health and nutritional benefits derived from coconut oil is unique and compelling. The epidemiological studies have shown that medium chain triglyceride present in coconut oil has been identified as an important medically efficacious food. Coconut oil consists of 48% lauric acid (a 12 chain saturated fat), 7% capric acid (a 10 chain saturated fat), 8% caprylic acid (an 8 chain saturated fat) and 5% caproic acid (a 6 chain saturated fat). These make up medium chain triglyceride (MCTs) or medium chain fatty acid (MCFA) molecules, with antimicrobial properties of coconut oil. Human breast milk contains lauric acid. This fatty acid protects the newborn baby from harmful germs. For years, MCFAs have been added to infant formula as protection because they supply easily digestible nutrients. A mother who consumes coconut oil will have more MCFA in her milk to help in protecting and nourishing her baby. Lauric acid, which is the major fat in coconut oil, is used by the body to make disease fighting fatty acid derivative monolaurin that the babies make from their mother's milk. This protects them from getting viral, bacterial, or protozoal infections. Until recently, this important benefit of lauric acid was overlooked by the medical and nutritional fraternity. These are generally absent from all other vegetable and animal fats besides butter. Indeed, diets for critically ill children, premature infants and hospitalized patients include medium chain triglycerides as principle source of fat. Lauric acid is necessary for good health. The references for the research can be found in *The Healing Miracles of Coconut* by Dr. Bruce Fife nutritionist/biochemist and one of the leading authorities on fats. Coconut oil is the healthiest oil on earth according to this naturopath doctor. As a *functional food*, coconut oil is now being recognized by the medical community as a powerful tool against immune diseases. Dr. Fife says *If there was oil you could use for your daily cooking needs that helped protect you from heart disease, cancer and other degenerative conditions, improved your digestion, strengthened your immune system and helped you to lose excess weight, would you be interested? This is what coconut oil can do for you.*

The *Coconut Cures* a valued reference book states that the coconut is one of the best natural defenses against flu. Coconut oil possesses powerful antiviral properties that can kill viruses, including flu virus. Coconut oil is composed of a special group of fats known as medium-chain triglycerides. Our bodies convert these fats into medium-chain fatty acids and monoglycerides both of which have been shown to be effective in killing lipid-coated viruses. Coconut oil also helps boost the immune system enabling the body to fight off infections more efficiently. This is a safe and economical means to protect one from viruses.

Several studies have been done on its effectiveness, and much research is currently being done on the incredible nutritional value of pure coconut oil. Because of its composition and high MFCA content, coconut oil is effective for treating and preventing heart disease as well as AIDs, Chronic Fatigue Syndrome, osteoporosis, gallbladder disease, diabetes, liver disease, Crohn's disease, prostate enlargement and cancer. That is a lot of protection for an inexpensive food that until recently was a standard part of our diets.

Fights Microbes on the Skin and in Digestive Tract

Monolaurin is created when either the saliva or the bacteria on the skin break down the 3 parts of lauric acid triglyceride into single units, called monolaurin. The beauty of monolaurin is that it kills microbial invaders before they even get in the door, so the immune system does not spend all of its time and energy fighting them. Because the immune system only operates *inside* the skin, it can be required to do a lot of work to defend against bacterial and viral invaders that could just as easily have been killed by monolaurin on the skin and in the sinuses. Similarly, lauric acid plays a key role in protecting the digestive tract. Besides coconut oil, the only other good source of lauric acid is mother's milk. Lauric acid is critical for infants because it is the only protection until their immune systems develop. This fact speaks volumes about the power of lauric acid. It keeps babies healthy even in the absence of a well-developed immune system. That is why, coconut oil is a critical ingredient in healthy milk formulas, and that is why, it is effective at treating and preventing AIDS.

AIDS/HIV

Initial trials have confirmed that coconut oil does have an anti-viral effect and can beneficially reduce the viral load of HIV patients. According to Dr. Conrato S. Dayrit, Emeritus professor of pharmacology in the University of the Philippines, a minimum of 50 ml of coconut oil would contain 20 to 25 g of lauric acid. When the oil is metabolized in the body, it releases monolaurin, which functions as an antibiotic and an anti-viral agent. Among the saturated fatty acids, lauric acid has the maximum anti-viral activity. Based on this research, the first clinical trial using monolaurin as monotherapy was conducted on HIV patients. Dr. Dayrit's conclusions after the study is that the initial trial confirmed the anecdotal reports that coconut oil does have an anti-viral effect and can beneficially reduce the viral load of HIV patients. The positive anti-viral action was seen not only with the monoglyceride of lauric acid but also with coconut oil itself. This indicates that coconut oil is metabolized to monoglyceride forms of C-8, C-10 and C-12 to which it must owe its anti-pathogenic activity. On July 19, 1995, Dr. Mary Enig, noted biochemist and nutritionist, was quoted in an article published in The Hindu, India's National Newspaper saying that coconut oil is converted by the body into *Monolaurin* a fatty acid with anti-viral properties that might be useful in the treatment of AIDS. The staff reporter for The Hindu wrote about Enig's presentation at a press conference in Kochi that there was an instance in the United States in which an infant tested HIV positive had become HIV negative and that it was fed with an infant formula with a high coconut oil content gains significance in this context and at present an effort was on to find out how the *viral load* of an HIV infected baby came down when fed a diet that helped in the generation of monolaurin in the body.

Immunity to Bacteria, Virus and Protozoa

Much of the research on the nutritional and medicinal benefits on coconut oil that has surfaced in recent years has been done by Dr. Mary Enig who has classified coconuts as a *functional food*, which provides health benefits over and beyond the basic nutrients. She has specifically identified lauric acid as a key ingredient in coconut products: approximately 50% of the fatty acids in coconut fat are lauric acid. Lauric acid is a medium chain fatty acid (MCF), which has the additional beneficial function of being formed into monolaurin in the human or animal body. Monolaurin is the antiviral, antibacterial and antiprotozoal monoglyceride used by the human or animal to destroy lipid-coated viruses such as HIV, herpes, influenza, cytomegalovirus, various pathogenic bacteria including *Listeria monocytogenes* and *Heliobacter pylori* and protozoa such as *Giardia lamblia*. Some studies have also shown some antimicrobial effects of the free lauric acid. Lauric acid also helps boost the immune system enabling the body to fight off infections more efficiently. By consuming a couple of spoonfuls of coconut oil daily, one can develop an immunity of sorts to the flu. This is a safe and economical means to protection from infectious diseases.

Effect on the Lipid Profile

Regular use of coconut oil in the diet improves cardiovascular health. Coconut oil containing all classes of fat when used in usual diets proves to be anti-cholesterogenic. No topic related to coconut oil has been given more prominence than its relation to cholesterol metabolism. It is claimed that the inhabitants of Thailand have low rate of heart attacks and strokes although coconut oil is their leading dietary fat. In a study on two groups of Polynesians, it was found that the group eating 89% of their fat as coconut oil had lower blood pressure values than those eating 7% and heart attack was not observed in either group. In another study on Polynesians, it was found that consuming large amounts of coconut oil had lower serum cholesterol level and lower incidence of arteriosclerosis than Maoris and Europeans who consumed a European type of diet. However, other studies indicate that a prolonged daily intake of coconut fat, in quantities of 50 g or less produces a sustained rise in blood cholesterol. The International Bank for Reconstruction and Development estimated Sri Lankan consumption to be 1300 nuts per head per year, which is equivalent about 48 g coconut fat per day, whereas, the Central Bank report of 1981 estimates per capita consumption of 990 nuts per year or less than 34 g coconut fat per day. Any increase in Ischemic Heart Disease can not be due to increased coconut consumption. In a study at Philippines, 10 medical students tested diets consisting of different levels of animal fat and coconut oil. When the individuals consumed animal fat and coconut oil in a ratio of 1: 1, 1: 2 and 1: 3 it had no significant change in cholesterol but when animal fat level increased, total calories reached 40% and blood cholesterol increased. This study indicates that not only coconut has no effect on cholesterol levels but also it reduced the cholesterol elevating effects of animal fat.

A review of epidemiological and experimental data regarding coconut eating populations show that the dietary coconut oil does not lead to high blood cholesterol or coronary heart disease. When native people changed their diet and gave up eating coconut oil in favor of polyunsaturated vegetable oils, their risk of heart disease was shown to increase. It also shows that dietary coconut oil does not lead to high serum cholesterol. It has been shown that islanders with high intake of coconut oil showed no evidence of the high saturated fat intake having a harmful effect on this population, however, when these people migrated to New Zealand and lowered their intake of coconut oil, their total cholesterol and low-density lipoprotein increased, and their high-density lipoprotein decreased. In 1989, Mendis and his co-workers reported undesirable lipid changes when young adult Sri Lankan males were changed from their normal diets by the substitution of corn oil for their customary coconut oil. Although the total serum cholesterol decreased from 179.6 to 146.0 mg/ml (18.7%) but the low-density lipoprotein (LDL) decreased from 131.6 to 100.3 mg/ml (18.7%) and the high-density lipoprotein (HDL) decreased from 43.4 to 25.4 mg/ml (41.4%). This brought high-density lipoprotein values below acceptable normal limits and the LDL to HDL ratio increased from 3 to 3.9 (30%). These latter changes are considered undesirable. It was also reported that the addition of coconut oil alone to previously mixed fat diets showed no significant difference.

Numerous studies clearly demonstrated that coconut oil has a neutral effect on cholesterol levels. While reviewing the health and nutritional benefits from coconut oil, some studies reported thirty and more years ago should have cleared coconut oil of any implication in the development of heart disease. A study was carried out in early 90s with 83 adults with normal cholesterol fed 15% of the fat ration as coconut oil. Relative to baseline values, the highest value of total cholesterol increased to 17%, high-density lipoprotein increased to 21.4% and LDL to HDL ratio decreased 3.6%. Those who blindly state that all saturated fats are unhealthy or coconut oil consumption leads to heart disease are ignorant of facts. There is no evidence to support the notion that saturated fat in coconut oil is harmful. Indeed, there is strong evidence now that coconut oil can help to prevent heart disease.

There is another aspect to the coronary heart disease. This is related to the initiation of the atheromas that are reported to be blocking the arteries. It is suggested that there is a causative role for the herpes virus and cytomegalovirus in initial formation of atherosclerotic plaques and reclogging of arteries after angioplasty. What is interesting is that both these viruses are inhibited by the antimicrobial lipid monolaurin a byproduct of lauric acid. Therefore, we could question the recommendations to avoid coconut as a contributing factor for the increased incidence of coronary heart disease.

Effect of coconut oil on hypertension: Dietary fat is one of the many factors that influence blood pressure. Most likely, the cooking oil, shortening and margarine consumed by people allover the world is loaded with omega-6 fatty acids. This is especially true if the individual adheres to the typical Western diet or greatly influenced by the western diet.

Omega-6 polyunsaturated fatty acids present in soybean and corn oils are responsible for the production of prostaglandins, which unfortunately constrict blood vessels and increase inflammatory response and blood platelet stickiness. These changes result in increased blood pressure and encourage atherosclerosis.

Coconut oil is one of the very few fats that are primarily medium chain fatty acids (MCFA). It is actually nature's richest source of MCFAs. When MCFA-rich coconut oil is used in place of other oils in food preparation, the amount of omega-6 fat molecules in the diet is significantly reduced. MCFAs abundantly present in coconut oil are not transformed into prostaglandins. Instead, these fats can weaken the effects of omega-6 fatty acids.

Individuals who had hypertension, (blood pressure around 140/90) before coconut oil was consumed, after just about its use for a month's blood pressure went down to 120/80 and 110/70. With no other dietary modification, the fall in BP is attributed to the incorporation of coconut oil in the diet.

Increased Digestibility in Diabetes

Coconut oil is medicinally used in special food preparations for those who suffer digestive disorders and have trouble in digesting fats. Considerably, more is known about digestibility of coconut oil in men. It is nearly completely absorbed as are most other natural dietary fats. For the same reason, it is also used in infant formulations for the treatment of malnutrition. The absorption of calcium, magnesium and amino acids has been found to increase when infants are fed with diet containing coconut oil, which has been used to enhance absorption of calcium and magnesium when there is deficiency of these minerals. For older people, coconut oil is useful for slowing down the degenerative process by improving mineral absorption. Coconut oil also helps in supplying energy to cells because it is easily absorbed without the need of enzymes and improves insulin secretion and utilization of blood glucose. Those people with diabetes would greatly benefit by adding coconut oil to their daily diet.

Helps in Weight Reduction

Medium chain fatty acid do not circulate in blood stream like other fats but they are sent directly to the liver where they are immediately converted into energy, just like carbohydrates so when coconut oil is consumed, the body immediately converts it to energy rather than converting it to body fat. The weight loss effect of coconut oil has been proven by many researchers. Coconut oil is nature's richest source of medium chain triglycerides (MCTs). Not only MCTs raise the body's metabolism leading to weight loss but they have special health-giving properties as well. The most predominant MCT in coconut is lauric acid. According to Lipid researcher Dr. Jon Kabara, never before in the history of man has it been so important to emphasize the value of lauric acid. The medium-chain fats in coconut oil are similar to fats in mother's milk and have similar nutriceutical effects. Modern research has found a common link between these two natural health products, their fat, or lipid content. The medium chain fatty acids and monoglycerides found primarily in coconut oil and mother's milk have miraculous healing power. Outside of a human mother's breast milk, coconut oil is nature's most abundant source of lauric acid and medium chain fatty acids.

Scientific studies have reported that the fatty acids from medium chain fatty acids (MCFs) or medium chain triglycerides (MCTs) in coconut oil are not easily converted into stored triglycerides and those MCTs cannot be readily used by the body to make larger fat molecules. One animal feeding study evaluated body weight and fat storage for three different diets, low-fat diet, high-fat diet containing long-chain triglycerides (LCTs) and a high-fat diet containing MCTs. All animals were fed the selected diets for a period of 44 days. At the end of that time, the low-fat diet group had stored an average of 0.47 g of fat per day; the LCT group stored 0.48 g per day, while the MCT group deposited only 0.19 g of fat per day, a 60% reduction in the amount of fat stored. The authors conclude that the change from a low-fat diet to a MCT-diet is attended by a decrease in the body weight gain.

This study points out two important facts, *i.e.*, (i) when MCTs are substituted for LCTs in the diet, the body is much less inclined to store fat and (ii) when we eat sensibly, a diet containing MCTs is more effective than a low-fat diet at decreasing stored fat.

Basal Metabolic Rate

In a human study, the researchers compared the metabolic effects of 400 calories meals of MCTs and LCTs by measuring metabolic rates prior to and six hours following the test meals. The results showed that the MCT-containing meals caused an average of 12% increase in basal metabolic rate as compared with a 4% increase with the LCT-containing meal. The authors concluded that replacing dietary fats with MCTs could produce over long period weight loss even in the absence of a reduced intake.

The Miracle Energizer and Fatigue Fighter

Coconut oil energizes for most of the same reasons that it promotes weight loss. It has even been effective in treating Chronic Fatigue Syndrome (CFS).

The increased energy induced by coconut oil occurs for a variety of reasons given below:

- Coconut oil's MCFAs are converted to energy packets in the liver and burned for fuel.
- Since it provides a richer longer-lasting source of energy, it is relatively easy to reduce or eliminate sugar dependence.
- When the sugary snacks and refined flour products are removed from the diet, the sudden drops of energy is eliminated. This makes one hungry and tired enough to go to sleep wherever possible. Sugar and refined flour are considered unhealthy, as they are known to be responsible for sudden increase in blood sugar levels followed by a quick drop. That is why a diet consisting of complex carbohydrates, fruits and vegetables is recommended.
- Fat stores are utilized more readily, which evens out the metabolism.
- Insulin use is stopped and minerals like chromium that are needed to process the sugar and starch, so one does not create a deficiency that incapacitates ones energy-production systems.

Effect on Thyroid

Coconut oil does not interfere with thyroid function unlike soybean oil. The origins of the American obesity epidemic can be traced to the 1970's when a concerted campaign against *harmful tropical oils* by the American Soybean Association led to the elimination of coconut oil in virtually all-commercial cooking. The coconut oil was substituted by partially hydrogenated oil, which interferes with insulin activity and causes other health problems, as explained in *What is Wrong with Partially Hydrogenated Oils*? However, *partially hydrogenated oil* on a food label generally means *soybean* oil, and soybean oil adds an additional problem of its own. It inhibits the thyroid's ability to produce the hormones required to burn fat, generate energy and carry on the processes of life. When the thyroid is not operating properly, the body's metabolism is lowered, so fat is stored instead of burning it. In addition, one does not want to exercise because of low energy. Therefore, the person with hypothyroidism naturally gains weight. Coconut oil is a saturated fat made up primarily of medium chain fatty acids (MCTs). These are known to increase metabolism and promote weight loss. Coconut oil can also raise basal body temperature, while increasing metabolism. This is good news for people who suffer from low thyroid function.

Digestive Disorders

Patients suffering from Crohn's Disease and Irritable Bowel Syndrome reported relief after adding coconut oil into their diet.

Reduces Degenerative Diseases

The biggest advantage of using coconut oil is that it can be substituted for hydrogenated and processed vegetable oils. These are involved in the development of numerous diseases as well as premature aging. While coconut oil may not prevent or cure all these conditions directly, it can limit the amount of potentially toxic hydrogenated and polyunsaturated oil in the body, and thus, reduce chances of suffering prematurely from degenerative disease.

Immune System Support

Polyunsaturated oils put a heavy strain on the immune system. They cause the immune system to shift into feverish activity while at the same time interfering with its ability to form protective compounds. Antioxidants are quickly used up fighting free radicals produced by unsaturated oils. When antioxidants such as vitamin A, C and E are depleted, the immune system slows down and becomes ineffective. Vegetable oils produce an overall depressive effect on immune system. Coconut oil on the other hand supports immune system and it causes no stress. The coconut oil is nontoxic to body so does not burden immune system. Saturated fats are very stable and do not oxidize easily, so antioxidants are not used up. Sixty three percent of coconut oil is composed of antimicrobial MCFA, and therefore, can enhance the power of immune system to fight against microorganisms. Coconut oil is ideal for immune suppressed individuals. For this reason, it is now being studied as a treatment for HIV/AIDS patients whose immune systems are severely compromised. Finally, coconut oil offloads the immune system. Since the immune system needs less fuel, more energy is available for the brain and muscular activity.

Protection against Diseases

Coconut oil may be useful in preventing a wide assortment of diseases. Because of its unique metabolic properties, it can help in shedding unwanted weight, thus, reducing risk of many health problem associated with obesity.

Inhibits Induction of Cancer

Dr. Wickremasinghe of Medical Research Institute, Sri Lanka reported that coconut oil might possess anticarcinogenic properties. Studies have shown that coconut oil inhibits the induction of cancer of colon and mammary by carcinogenic agents in test animal. In chemically induced cancers of colon and breast, coconut oil was by far more protective than unsaturated oil, *e.g.*, 32% of corn oil users got colon cancer, whereas, only 3% of coconut oil eaters got the cancer. A 50 years review of literature shows anti-carcinogenic effects from dietary coconut oil.

Coconut oil is an important component in imitation dairy products like filled cheese, coffee, whiteners, milk shake mix chocolate filled with milk, ice cream, desert topping, spray oil for crackers, cookies and cereals. It is used in these products not only because it resists oxidation but also because it is bland in flavor. It is extremely stable on storage and possesses a unique liquefying property that contributes to feel of the food.

How Much Oil Does an Adult Need?

Based on per capita intake of coconut oil in 1985, the per capita intake of lauric acid can be approximated for those major producing countries such as Philippines (7.3 g), Sri Lanka (4.9 g), Indonesia (4.7 g) and Singapore (2.8 g). In coconut growing areas of India including Kerala and coastal Karnataka, it ranges from 12 to 20 g per day and the average for rest of the country is half a gram. For the Tokelau Islands, the average lauric acid consumption was 68 g. The changes in life style and food habits have changed the situation considerably. In countries like United States of America, there is very little lauric acid in foods. This situation needs positive change. Though researchers have yet to give satisfactory answer, keeping in view of available information, an adult would need about 12 to 24 g of lauric acid a day. This would be available from 2 to 4 tablespoons of coconut oil. Growing children probably need the same amounts as the adults.

Olive oil

In the 1950s, people living in the Mediterranean, especially on the island of Crete, were lean and virtually free of heart disease due to the consumption of olive oil. About 40% of their calorie intake came from fat, primarily olive oil. If we look at the diet, they consumed back then, we note that the Cretans ate mostly fruits, vegetables, beans and some fish. Saturated fat was less than 6% of their total food intake. They certainly consumed lots of olive oil but rest of their food was exceptionally healthy. They also worked hard in the fields walking about nine miles a day, but today, it is changed. With life style change, heart disease has skyrocketed and more than half the population in Crete is overweight. Olive oil is one of the most fattening dense foods on the planet; olive oil packs more calories (4020) per 500 g than butter (3200).

Effects of Polyunsaturated Fats on Health

In view of the increasing incidence of cancer world over, it is of interest to briefly discuss the life style changes in both developed and developing countries during the past 4 decades. One of the major changes is the substitution of coconut oil with polyunsaturated oils even in countries producing coconuts. One study in mice has shown that consuming high amounts of polyunsaturated fat but not of monounsaturated fat may increase the risk of metastasis in cancer patients. The researchers found that linoleic acid in polyunsaturated fats produced increasing membrane phase separation, and thereby, increased the adherence of circulating tumour cells to blood vessel walls and remote organs. According to the report, the new findings support earlier evidence from other research that consuming high amounts of polyunsaturated fat may increase the risk of cancer spreading. The propensity for polyunsaturated fats to oxidize is another possible risk factor. This leads to the generation of free radicles and eventually to rancidity. A high consumption of omega-6 polyunsaturated fatty acids (PUFAs), which are found in most types of vegetable oil, may increase the likelihood of postmenopausal women developing breast cancer. A similar effect was observed on prostate cancer. Other analysis suggested an inverse association between total polyunsaturated fatty acids and breast cancer risk but individual polyunsaturated fatty acids behaved differently from each other. Polyunsaturated fats (PUFs) are greatly immunosuppressive, and anything that suppresses the immune system is likely to cause cancer. The first person to suggest that polyunsaturated fats cause cancer was Dr. R.A. Newsholme of Oxford University, England. What Newsholme wrote was that when the body gets sufficient nutrition, the diet includes immunosuppressive PUFs, which make the body prone to infection by bacteria and viruses. When starved, however, the body's stores of PUFs are depleted. This allows the immune systems to recover which, in turn, allows the body to fight existing infection and prevent other infections. In 1989, there was a report of a ten years trial at a Veterans' Administration Hospital in Los Angeles. In this trial, half the patients were fed a diet, which had twice the amount of PUFs as compared to saturated fats. In half of the patients on high PUF diet, there was a 15% increase in cancer deaths compared to the saturated fat group. The authors reported that the PUFs had been the cause of the increase in cancer deaths. The *British Medical Journal* carried an editorial in its 6 October 1973 issue, which asked if PUFs were carcinogenic. After considering the experimental evidences, it found a direct correlation with PUFs with the increased incidence of cancer. It was concluded that PUFs were indeed carcinogenic. Wayne Martin likes to tell a story, which suggests just how PUFs are cancer causing. In 1930 in the United States of America, eighty percent of men smoked cigarettes and the tar content of cigarettes was much higher than it is today. The death rate at that time from lung cancer was very low. In 1955, doctors decided that PUFs were good in terms of heart disease protection. After this, lung cancer deaths increased dramatically. By 1980, although the number of American men who smoked had dropped to only thirty percent, three times as much PUF was being eaten and there were sixty times as many lung cancer deaths. Since 1974, the increase of polyunsaturated fats has been blamed for the alarming increase in malignant melanoma (skin cancer) in Australia. It is believed that exposure to sun causes it. Are Australians going out in the sun any more now than they were fifty years ago? They are certainly eating more polyunsaturated oils. In Australia in 1995, it was observed that even the cream on milk was removed and replaced with vegetable oil. Victims of the disease have been found to have polyunsaturated oils in their skin cells. Polyunsaturated oils are oxidized readily by ultra-violet radiation from the sun and form harmful *free radicals*. These are known to damage the DNA of cells and this can lead to the deregulation to which we call cancer. Saturated fats are stable. They do not oxidize and form free radicals. Malignant melanoma is also said to be increasing in this country. Does the sun cause this? In Britain, the number of sufferers is so small as to be relatively insignificant. Even so, it is not likely that the sun is to blame since all the significant increase is in the olds over 75 years. People in this age group tend to get very little sun.

It has been known since early 1970s that it is linoleic acid, which is the major culprit. According to Professor Kearney of Sydney University in 1987, many laboratories have shown that a greater proportion of polyunsaturated fats are superior to diets rich in saturated fats in promoting the yield of experimental mammary tumors. In such studies, omega-6 linoleic acid appeared to be the crucial fatty acid and vegetable oils such as corn oil and sunflower oil, which are rich in linoleic acid, are potent promoters of tumor growth.

A study of 61,471 women aged 40 to 60 conducted in Sweden and looked into the relation of different fats and breast cancer. This study found an inverse association with monounsaturated fat and a positive association with polyunsaturated fat. In other words, monounsaturated fats protected against breast cancer and polyunsaturated fats increased the risk.

Hair and Skin Care

The secret of many renowned hairdressers is coconut oil, and it is used in many commercial products. Topically, coconut oil is a beatifying emollient. Because of the small molecular structure of coconut oil, it is absorbed easily by the skin. It protects against damage and aids in skin healing and repair. Its antibacterial properties may be helpful for certain skin conditions including psoriasis, eczema and inflammation. The richness and highly moisturizing properties of coconut oil are said to promote shine, keep hair from turning gray, and believe it or not, are said to help sparse hair grow thicker and fuller, especially if a few drops of Rosemary Essential Oil are added, which increases blood flow in the scalp. The oil also helps repair damaged hair, offering essential proteins to strengthen and nourish it.

A hot coconut oil scalp massage once a week is believed to be the secret of the luxurious hair of Indian women. Follow these steps to do your own treatment at home:

- Heat coconut oil in a pan on the stove, not in microwave.
- Massage the scalp.
- Distribute oil throughout the hair.
- Place a clean towel over your pillow and leave the oil over night.
- Rinse out in the morning.
- Enjoy your lustrous hair.

For a lip balm, use coconut butter on its own or add peppermint or spearmint oil (organic essential oils).

Dottie provided the formulation, *i.e.*, one-third almond oil, one-third beeswax and one-third butter or coconut oil. Stevia can be incorporated into coconut oil for a *sweet taste*. Aqueous extract of Stevia leaves do not mix with oil. It has therefore to be emulsified to make the preparation smooth. Do make sure that the Stevia is dissolved before adding it though or you will get a *grainy feeling*? You could also add a couple of drops of peppermint essential oil for a nice taste.

Body Lotion and Cosmetics

Coconut oil has a natural creamy texture comes from vegetable source and is used in cosmetics. Whether applied topically or internally, coconut oil helps to keep skin young, healthy and free of disease. Polynesian women are famed for their beautiful skin, even though they are exposed everyday to hot blistering sun and the chafing of the ocean breeze. Coconut oil has been used by them to make the skin soft and smooth. It is also used to make natural soaps and other health products, which is said to promote luxurious hair growth and protect the skin from bacterial and viral infection. In *Ayurvedic* medicine, coconut oil is said to nourish the body and increases strength, while application of coconut oil to the skin is said to help fixation of vitamin D in the body. The cosmetic applications of coconut oil include Hair and Skin Oil. Coconut oil mixed with herbal oils and different scents is used as hair oil and is preferred because of its low viscosity. Without chemical modification, it promotes emolliency, gloss lubricity, adhesion and it is believed to prevent dandruff. Different preparations of coconut oil are also used to protect the skin from bacterial, protozoal and other infections in body and baby oils.

Coconut oil is used to prepare natural shampoos, in which the extract of *amla* fruit and soap nut powder are sometimes incorporated to add value.

Herbal/Medicinal Oils

Coconut oil with various herbs/medicinal plants is used for preparing medicated oils such as skin and massage oils. Some people use coconut oil with lime for healing wounds.

Scent-making

Caproic acid, capric acid and caprylic acid obtained from coconut oil are mixed with methyl alcohol, ethyl alcohol, isopropyl alcohol, butyl alcohol and their esters and utilized in scents as well as food products.

Beauty Care Products

Coconut oil has a high level of myristic acid, which in combination with isopropyl myristate is used in many beauty products as an additive. Lauric acid of coconut oil is also used in toothpaste.

Conclusion

Saturated fats and animal fats are usually blamed for all manner of diseases in Western society but look at the facts:

- In the 19th century, mostly animal fats, *i.e.*, lard, tallow and/or butter, were used in western countries for cooking, whereas, coconut oil was one of the major source of fat or oil used for cooking in tropical countries like India. At that time, cancers and heart disease were rare.
- Polyunsaturated fats and oils suppress the immune system, such immuno-suppression is known to cause and promote cancer.
- During the last four decades, there has been a change in favour of polyunsaturated fats and oils, and cancer rates have soared.

Even today, the majority of doctors prohibit the use of coconut and its oil in heart patients. Is it that they are too busy to look up the recent advances on the benefits of using these products? It is like saying mother's milk is bad for her child.

Coconut oil has been called the healthiest dietary oil on the earth. It is being used for thousands of years by Pacific Islanders and in Asia. It is marvelous oil gifted by nature from a perennial palm tree. It is known for its nutrition, health and medicinal value to cure various diseases. It keeps the body shiny without wrinkles and healthy. If someone is not using coconut oil for daily cooking and body care needs, he is missing out one of the nature's most amazing health products. This has proven good for the digestive system in our body, and keeping the intestines regulated every day. It has also been known to kill worms in the intstines over the years.

The immune system is built up by coconut water. Therefore, it can help thc body fight colds, flu and other unpleasant/undesirable illnesses.

Coconut water is high in potassium, which is our central electrolyte and it regulates many of the bodily functions. Drinking coconut water will certainly help in maintaining potassium levels by providing 294 mg per serving.

It is amazing to know that through drinking one cup of coconut water, the individual receives over 2.5 g of fiber, which keeps our intestinal health regulated and adds to colon health.

Kidney stones have been known to help through the drinking of coconut water. There are some unique properties found in this water, which will help in dissolving chronic kidney stones naturally.

The lauric acid found in coconut water helps the body to flush out toxins. This acid has also been known to help benefit those affected by Leaky Gut Syndrome, which is a common stomach disorder. Lauric acid has also been known to help women in the stages of menopause.

If the skin is dry, try tonput some coconut water over it with a washcloth. This water adds moisture to the skin, and since it is so beneficial in hydration to the whole body, it does the skin good too.

Vitamins, which are stored in your body, will be carried to your cells where they need to go. It also helps in carrying oxygen around delivering it into cells.

Coconut oil and water can also help in high blood pressure since it helps artery pressure to decrease. The beneficial effect of coconut water is due to its high concentration of potassium and other health promoting constituents since it helps artery pressure to decrease. This water sometimes helps in transfusions since it can be transfused in place of blood plasma. This is due to the fact that coconut water is not destructive to red blood cells but rather accepted by them and this water, unlike other waters, is sterile and isotonic (clean) without harmful additives.

■■■

9 Nuts and their Health Benefits

Madhusweta Das

Department of Agricultural and Food Engineering, Indian Institute of Technology, Kharagpur 721302, West Midnapur, West Bengal, India,
E-mail: <madhu@agfe.iitkgp.ernet.in>

1. Introduction

Edible nuts, a popular food from prehistoric time, have pleasing flavor and crunch as well as significant health benefits. As a food, nuts can be defined as any of various dry fruits that generally consist of an edible kernel adhered to brownish skin and enclosed in a shell that can range from medium-hard, thin and brittle to woody and tough. Botanically speaking, nuts coming from many different plant families are generally the seeds or extensions of fruits, encased in a hard outer shell and grow on flowering upright trees or shrubs. More commonly, nuts are classified as either tree nuts [seed(s) in a shell inside fruit] or peanuts [seeds that grow inside pods of a member of *Leguminosae* family] growing underground. Globally, the most popular and commercially important edible nuts are peanuts and several tree nuts like almond, cashew, Brazil nut, hazel nut, macadamia, pecan nut, pine nut, pistachio and walnut. Despite their diversity, nut varieties share many common nutritional characteristics.

Long before the development of agriculture, human gathered nuts for food. Recent archeological excavations of a 10,000 years old village in Eastern Turkey have uncovered the existence of a non-migratory society whose economy centered on the harvesting of almond and pistachio. New research suggests that civilizations relied on nuts, even before cereal grains as a staple food. Nuts were able to fill this role probably because they (i) were a more predictable food source than smaller, less-hearty cereal grains that might have been damaged by severe weather, (ii) could be stored through long winters, thus, providing a stable food supply throughout the year and (iii) provided energy and nutrients. Nuts also have been used for centuries as folk medicine, *e.g.*, the pistachio was a cure for liver and other abdominal complaints, tachycardia, nausea, vomiting, toothache, chronic cough, abscesses, sores and vascular insufficiencies. Pistachio leaves are a fertility drug in the folk medicine of Lebanon, and in much of Arabia, the nuts are thought to have aphrodisiacal properties, *e.g.*, in Brazil, a tea made from peanuts is thought to have a calming/relaxing effect. Walnuts are considered an herb in traditional Chinese medicine. They are said to tonify kidneys, strengthen the back and knees, moisten the intestines and move stool. It is also believed to relieve asthma. The ancient Greeks and Romans have used hazel nuts for several medicinal purposes. As cultures evolved so did the role played by nuts in cuisines throughout the world. Cooks from the Mediterranean region, South America and Asia have long used tree nuts and peanuts as ingredients in savory sauces, stuffing, entrees, snacks, appetizers and desserts. Nuts are such an excellent source of protein, so today, nuts are positioned with meat, poultry, fish and eggs as meat alternate group in the United States Food Guide Pyramid (2005), *i.e., MyPyramid* [*MyPyramid*, released by the United States Department of Agriculture (USDA) and produced jointly by the USDA and United States Department of Health and Human Services (HHS), is an update on the American food guidelines]. Nuts also contribute unsaturated fatty acids, important minerals, vitamins and fiber to the diet. In addition to their tastiness, tree nuts and peanuts are both cholesterol-free. With today's busy lifestyles, nuts are a convenient, tasty and easy snack that contributes to a healthy lifestyle. Nuts are also being studied for their potential health benefits. Research suggests that there may be a connection between frequent nut consumption and a reduced incidence of coronary heart disease as well as other diseases. Thus, tradition and promising scientific evidence combine to support the role of nuts in healthful eating.

2. Brief Description of Different Nuts

Different species of nuts are available within a particular genus. The most common types are mentioned in the following discussion.

2.1 Almond

The deciduous almond (*Prunus amygdalus)* tree (about 10 m tall) is commercially grown in large orchards. The hull of the matured fruit opens and seed that contains the nut (the edible part) is exposed. The nut is extracted from the surrounding hard shell of the matured seed by mechanical cracking. Among two varieties (sweet and bitter, with thin/thick skin), sweet variety with thin skin is of commercial importance as food. California shares over 80% of world's almond production. It is followed by Spain, Portugal and Italy, while India produces only 0.22% world's share in restricted hilly terrain of northern India.

Inside fruit

Inside shell

With and without skin

2.2 Brazil nut

Brazil nut (*Bertholletia excelsa*) tree is tall (up to 45 m high) and grows in Amazon basin forest in Brazil, Peru and Bolivia. It produces fruits containing 8-24 triangular shaped (4-5 cm long) seeds (Brazil nuts) packed inside the fruits. The fruit is a large capsule (up to 2 kg and 10-15 cm diameter) having hard and thick, woody cover (8-12 mm thick). The seeds are extricated. Bolivia producing 50% of the world's production, Brazil nut accounts only 1.62% of the total volume on global nut trading.

In fruit **In shell** **Kernel with and without skin**

2.3 Cashew nut

The cashew (*Anacardium occidentale)* tree, an evergreen tropical tree (about 9 m high), produces nut inside a hard shell that grows as an integral bottom part of a fleshy bright yellow to red cashew apple fruit. Because the shell contains a highly irritating caustic resin, cashew nuts are always sold in decorticated form. India is the second largest producer (24% of world production) of cashew nuts.

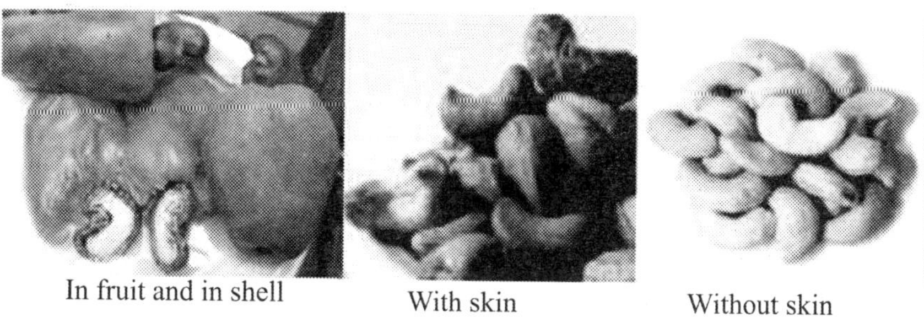

In fruit and in shell With skin Without skin

2.4 Hazelnut

Hazelnut (*Corylus avellana)* tree, typically a deciduous shrub (3-5 m tall), produces fruit (nut) in clusters of one to five together. Each nut is held in involucre of leafy bracts (fuzzy husk). When ripens, the fuzzy outer husk opens up and hard, smooth-shelled nut falls out of it. The nut (also known as cobnut), roughly spherical to oval (15-20 mm long and 12-20 mm broad), is closely related to another more elongated species known as filbert (*Corylus maxima*). Turkey accounts about 70% of the world production of hazelnuts followed by Spain, Italy and Oregon. In India, Pangi valley of Himachal Pradesh produces a small amount of hazelnut.

Shell with husk

Shell and kernel in it

without skin

2.5 Macadamia

Two popular varieties of macadamia nuts (*Macadamia intergrifolia* and *Macadamia tetraphylla*) are grown (evergreen tree up to 30 m height). The spherical fruit comprises of outer fibrous green husk layer (2 to 7 mm thick) that opens up on maturity and releases extremely hard shell from it. Almost round shaped kernel (nut, 1.5-2.5 cm in diameter) is obtained after cracking this shell. Australia is major producer of macadamia nuts (85% of world production), followed by Hawaii, South Africa, Brazil, California, Costa Rica, Israel, Kenya, Bolivia, New Zealand, Colombia and Malawi.

In fruit Kernel with skin Without skin

2.6 Peanut/Groundnut

The peanut/groundnut (*Arachis hypogaea*) is an annual herbaceous plant (30-50 cm tall) and belongs to legume bean family. It is native to Pakistan, India, some African countries, Mexico, South America and Central America. The pods are 3 to 7 cm long, containing 1 to 4 seeds in each pod that gets matured underground. China shares about 32.95% of the world production, while that for India and United States are 18 and 6.8%, respectively.

In shell With and Without skin

2.7 Pecan

Deciduous pecan (*Carya pecan*) tree is tall (20-40 m high) and produces oval to oblong shaped smooth fleshy green fruits (3-6 cm long and 1.5-3 cm broad). It turns into brown and fibrous on maturity, splits naturally into four parts and releases the thin-shelled nut. Edible kernel is obtained after cracking the shell. Seventy five percent of the world production of pecan comes from United States followed by 20% from Mexico and rest from Australia, Israel, Peru and South Africa

In fruit **In shell** **Kernel with skin**

2.8 Pine nut

In Europe, pine nuts come from the most common variety stone pine (*Pinus pinea*). The pine nuts are small (around 12 mm long) and sweet in taste. On drying, the cone scales open up and the nuts fall out automatically. The nuts are roasted lightly to remove *turpentiney* taste followed by cracking by rollers to release edible kernels. Korean pine (*Pinus koraiensis*) in northeast Asia and Chilgoza pine (*Pinus gerardiana*) in the western Himalaya are also used in significant amount. Afganisthan is also an important source of pine nuts.

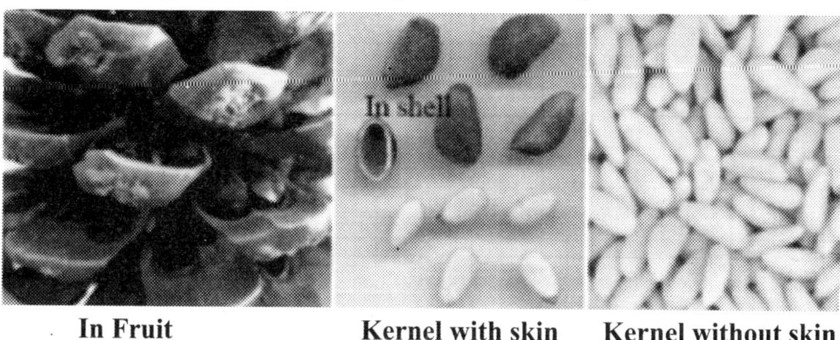

In Fruit **Kernel with skin** **Kernel without skin**

2.9 Pistachio

Pistachio (*Pistacia vera*) tree, a deciduous and salt tolerant tree (average 10 m high) produces hard, whitish exterior shell fruits in clusters like grape. When fully mature, the shell abruptly splits open at one end with audible pop. The seed has a light green flesh (edible part, the kernel) with a distinctive flavour. The mMajor producers of pistachios are Iran (55% of world production), followed by United States, Turkey, Syria and China, while China consumes maximum followed by United States and India.

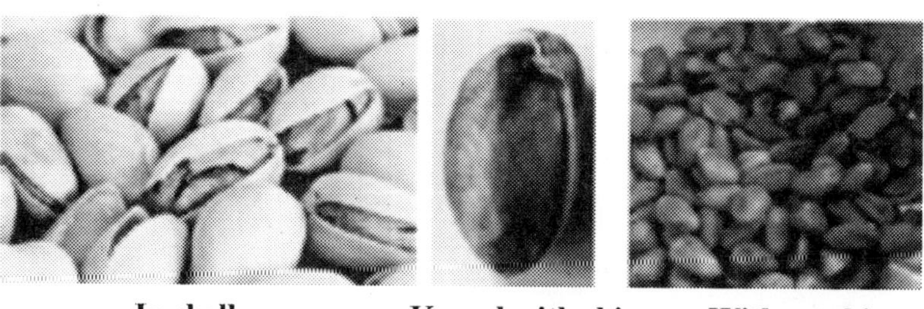

In shell **Kernel with skin** **Without skin**

2.10 Walnut

English walnut (*Juglans regia*), the most common walnut tree is 10-40 m tall deciduous tree. The soft edible kernel is enclosed in brown colour round or oblong shaped thin hard shell. The kernel consists of two bumpy lobes connected to each other. The shell is easily broken with a nutcracker. The leading commercial producers are China (28%), United States (20%), Iran (11%), Turkey (8%), and Ukraine (5%). India, Romania, France, and Egypt, each accounts 2%. In India, it is mainly grown in sub-Himalayan region.

In shell **Kernel with skin** **Kernel colour**

3. Nutrients and Phytochemicals in Nuts

3.1 Nutrients

A nutrient is chemical derived from food during the process of digestion. Absorbed nutrients break down by the process of catabolism and produce energy. The followed by process of anabolism uses the energy to reconstruct carbohydrates, proteins and fats, required in the body. Thus, nutrients give energy and regulate bodybuilding and maintenance. Nutrients needed in large quantities are called macronutrients (carbohydrates, proteins and fat produce the required energy) and those in small quantities are micronutrients (vitamins and minerals), the latter do not directly provide energy but contribute to nutrition by allowing the biochemical reactions to occur. Nut is considered as high energy food and highly nutritious (Table 1).

Table 1 : Macronutrients of raw, unsalted nuts per 100 g

Per 100 g	Energy (kJ)	Digestible Carbo-hydrates (g)	Fiber (g)	Protein (g)	Total fat (g)	Saturated fat (g)	Monoun-saturated fat (g)	Polyun-saturated fat (g)
Almond	2503	4.8	8.8	19.5	54.7	3.7	35.9	12.8
Brazil Nut	2886	2.4	8.5	14.4	68.5	14.8	21.8	29.0
Cashew	2437	16.8	5.9	17.0	49.2	8.4	31.1	7.5
Hazelnut	2689	5.1	10.4	14.8	61.4	2.7	48.8	7.2
Macadamia	3080	7.9	6.4	9.2	74.0	10.0	59.8	3.8
Peanut	2374	16.1	8.5	25.8	49.2	7.1	24.4	15.6
Pecan	2973	4.9	8.4	9.8	71.9	4.5	39.3	25.0

Per 100 g	Energy (kJ)	Digestible Carbo-hydrates (g)	Fiber (g)	Protein (g)	Total fat (g)	Saturated fat (g)	Monoun-saturated fat (g)	Polyun-saturated fat (g)
Pine Nut	2925	4.5	5.1	13.0	70.0	4.2	23.0	39.8
Pistachio	2389	6.8	9.0	19.7	50.6	5.8	26.7	15.8
Walnut	2904	3.0	6.4	14.4	69.2	4.4	12.1	49.6

3.1.1 Carbohydrates

Carbohydrates (sugars joined by glycosidic bond) may be digestible or indigestible. Digestible carbohydrates provide about two-thirds of an individual's energy need. Simple carbohydrates like sugars and sugar enriched foods are quickly digested releasing glucose rapidly into the blood stream (such foods have high glycemic index, GI), whereas, complex carbohydrates like starches take longer time to be broken, hence, provide a slower and more gradual supply of glucose (called low GI food). It is not sensible to eat more starch rich foods since excess sugar rich foods may lead to obesity, high blood sugar, tooth decay and a possible increased risk of developing type II diabetes in later adult life (type I results from the body's failure to produce insulin, and presently, requires the person to inject insulin and type II diabetes also called adult on set diabetes results from insulin resistance, a condition in which cells fail to use insulin properly). Indigestible carbohydrates are the fibers, the cellulosic materials. Fibers in food help in lowering colon cancer, heart disease, type II diabetes, diverticulitis (an inflammation of intestine, one of the most common age-related disorders of colon) and constipation. Nuts are recommended as foods, which are low in digestible carbohydrates but high in fiber content (Table 1).

3.1.2 Proteins

Food proteins on digestion break down into smaller units called amino acids. During the followed by process of biosynthesis of body proteins, the body cannot manufacture eight amino acids, *i.e.*, phenylalanine, valine, threonine, tryptophan, isoleucine, methionine, leucine and lysine (termed essential amino acids), and the diet must supply these. Additionally, cysteine, tyrosine, histidine and arginine are required by infants and growing children. A protein deficiency especially in infants and growing children can cause growth retardation, severe wasting of muscle, *etc.* Animal protein is a good source of essential amino acids and is traditionally regarded as *first class* protein. Nuts are good in protein content (Table 1), and recently, have been placed with animal protein in the same group with respect to its quality. According to a publication of American chemical society, when compared with FAO (Food and Agriculture Organization of the United Nations)/WHO (World Health Organization of the United Nations) recommended essential amino acid requirements for a 2-5 years old child, lysine was the limiting amino acid in Brazil nut, cashew nut, hazel nut, pine nut and walnut, whereas, methionine and cysteine were the limiting amino acids in macadamia, pecan and pistachio, and threonine was the limiting amino acid in Virginia peanut. However, compared to the FAO/WHO recommended, the essential amino acid pattern for an adult, only almond and peanut was deficient in methionine and cysteine, whereas, all other tree nuts seemed to contain adequate amounts of all of the essential amino acids (Table 2).

Table 2 : Essential amino acid composition of nut proteins

Per 100 g	Energy (kJ)	Digestible Carbo-hydrates (g)	Fiber (g)	Protein (g)	Total fat (g)	Saturated fat (g)	Monoun-saturated fat (g)	Polyun-saturated fat (g)
Almond	2503	4.8	8.8	19.5	54.7	3.7	35.9	12.8
Brazil Nut	2886	2.4	8.5	14.4	68.5	14.8	21.8	29.0
Cashew	2437	16.8	5.9	17.0	49.2	8.4	31.1	7.5
Hazelnut	2689	5.1	10.4	14.8	61.4	2.7	48.8	7.2
Macadamia	3080	7.9	6.4	9.2	74.0	10.0	59.8	3.8
Peanut	2374	16.1	8.5	25.8	49.2	7.1	24.4	15.6
Pecan	2973	4.9	8.4	9.8	71.9	4.5	39.3	25.0
Pine Nut	2925	4.5	5.1	13.0	70.0	4.2	23.0	39.8
Pistachio	2389	6.8	9.0	19.7	50.6	5.8	26.7	15.8
Walnut	2904	3.0	6.4	14.4	69.2	4.4	12.1	49.6

3.1.3 Fats

Fats (tryglycerides of fatty acids) should account for 30% or less of the calories consumed daily. Saturated fats (contains only saturated fatty acids) should account for not more than 10% of the total fat intake, as excess of this rise the blood cholesterol and is the main culprit of disease origin. Unsaturated fats (contains fatty acids with one or more double bonds) reduce the blood cholesterol and are good for health. In nuts, saturated fat is much less than unsaturated fat (Table 1). Harmful trans-fat (refers to fat with double bonds in trans-geometric configuration) is absent in all nuts.

Fatty acids increase the thickness and surface area of the digestive-absorptive cells lining the intestinal tract, as well as of the *villi* through which absorption of nutrients takes place. This results in digestion more effectively, better nutrients absorption and better health. The body does not manufacture certain fatty acids (termed essential fatty acids) in the process of anabolism and the diet must supply these. There are two families of essential fatty acids: omega-3 (ω-3 or n-3), *i.e.*, alpha-linolenic acid and omega-6 (ω-6 or n-6), *i.e.*, linoleic acid, both being polyunsaturated fatty acid, PUFA. Omega-3 and omega-6 essential fatty acids are especially important for making prostaglandins (hormone-like substances that regulate many activities including blood pressure, heart and kidney functions, digestive system, body temperature, inflammation, pain and swelling that can relive the symptoms of arthritis and autoimmune diseases) in human body.

Essential fatty acids are also natural blood thinners. They can prevent blood clots, which can lead to heart attack and stroke. *In vitro* studies have documented that omega-3 fatty acids can inhibit the growth of human breast cancer cells. Essential fatty acids assist in the development and function of brain and nervous system, and they help in regulating proper thyroid and adrenal activity. Low levels of omega-3 fatty acid have been linked to an increase in dementia (dementia indicates a group of symptoms caused by disorders that affect the brain), mood changes, memory loss and visual problems. It is also the most abundant fat in breast milk since babies need it to nourish their growing brains and eye development. In a study, the elderly patients with Alzheimer's disease were found to have lower levels of this fatty acid than healthy older people do have. Studies done at Purdue University, Indiana, United States, showed that children diagnosed with attention deficit hyperactivity-disorder had lower levels of these essential fatty acids. Table 3 shows the fatty acid profile for nut. It appears that all the nuts contain considerable amounts of both the essential fatty acids. Moreover, perfect ratio of omega-6 to omega-3 essential fatty acid helps in maintaining good health. Normally, people are consuming higher amount of omega-6 acid. From Table 3, it is observed that macadamia nut, pecan, pine nut, walnut and peanut contain substantial amount of omega-3 essential fatty acid, and among these, walnut is the highest.

Table 3 : Fatty acid composition of nut lipid

g/100 g nut lipid	Almond	Brazil nut	Cashew nut	Hazel nut	Maca-damia	Peanut	Pecan	Pine nut	Pistachio	Walnut
6:0	0.00	0.14	0.02	0.00	0.00	0.00	0.00	0.15	0.03	0.32
8:0	0.00	0.00	0.00	0.00	0.00	0.00	0.00	0.00	0.00	0.00
10:0	0.00	0.05	0.00	0.00	0.00	0.00	0.00	0.07	0.02	0.05
11:0	0.00	0.11	0.05	0.00	0.00	0.00	0.00	0.15	0.08	0.24
12:0	0.00	0.00	0.00	0.00	0.09	0.00	0.00	0.00	0.00	0.00
14:0	0.06	0.05	0.03	0.03	0.90	0.00	0.04	0.00	0.09	0.03
15:0	0.00	0.00	0.00	0.00	0.00	0.00	0.00	0.00	0.00	0.00
16:0	7.36	15.11	10.70	5.78	8.88	6.20	5.90	5.33	12.65	8.14
17:0	0.05	0.08	0.12	0.05	0.00	0.07	0.05	0.05	0.04	0.05
18:0	1.56	9.51	9.33	3.12	4.26	2.06	2.24	2.41	1.09	2.84
20:0	0.06	0.25	0.63	0.14	2.95	1.03	0.12	15.75	0.11	0.10
21:0	0.00	0.00	0.00	0.00	0.00	0.00	0.00	0.09	0.04	0.00
22:0	0.00	0.06	0.12	0.00	0.79	2.22	0.00	0.11	0.08	0.00
24:0	0.00	0.00	0.13	0.00	0.32	1.28	0.00	0.00	0.00	0.00
14:1	0.00	0.00	0.00	0.00	0.00	0.00	0.00	0.00	0.00	0.00
16:1	0.66	0.29	0.54	0.15	18.69	0.07	0.07	0.11	1.18	0.09
18:1	60.93	28.75	61.15	82.95	58.51	81.28	66.66	27.44	50.29	15.19
20:1	0.00	0.00	0.00	0.00	0.00	0.00	0.00	0.00	0.00	0.00
22:1	0.00	0.00	0.00	0.00	0.23	0.14	0.00	0.00	0.00	0.00

g/100 g nut lipid	Almond	Brazil nut	Cashew nut	Hazel nut	Maca-damia	Peanut	Pecan	Pine nut	Pistachio	Walnut
18:2 Linoleic acid (ω-6)	29.21	45.43	16.88	7.55	1.81	3.87	23.68	46.84	33.43	59.79
18:3 α-Linolenic acid (ω-3)	0.10	0.18	0.32	0.24	2.58	1.79	1.24	1.51	0.86	13.17
Total saturated	9.09	25.35	21.12	9.11	18.18	12.86	8.35	24.10	14.24	11.76
Total MUFA	61.60	29.04	61.68	83.10	77.43	81.49	66.73	27.55	51.47	15.28
Total PUFA	29.31	45.61	17.19	7.79	4.39	5.66	24.92	48.35	34.29	72.96

3.1.5 Minerals

The minerals required by human body are supplied from food and drinking water. Minerals form parts of the cellular structure, blood and are major components of the skeletal system. They have functions in enzymes, hormones, muscle actions, *etc*. The minerals calcium, magnesium and phosphorus are needed in higher amounts, while zinc, iron, copper, manganese, chromium, selenium, iodine, potassium, fluorine, molybdenum, sulfur, chlorine, sodium and boron are required in small amounts. The mineral profile of nuts is given in Table 4.

Table 4 : Mineral contents per 100 g of raw unsalted nuts

Nuts	Mg (mg)	Ca (mg)	Fe (mg)	Zn (mg)	Cu (mg)	Mn (mg)	Se (μg)	P (mg)	K (mg)
Almond	260	250	3.9	3.7	1.00	2.29	2.5	484	705
Brazil Nut	350	150	2.2	4.1	1.74	1.22	1917	725	659
Cashew	250	34.0	5.0	5.5	2.20	1.66	20	593	660
Hazelnut	160	86.0	3.2	2.2	1.73	6.18	2.4	290	680
Macadamia	130	85.0	3.7	1.3	0.76	4.13	3.6	188	368
Peanut	168	92.0	4.6	3.2	1.14	1.93	7.2	376	705
Pecan	110	51.0	2.4	3.9	1.20	4.50	3.8	277	410
Pine Nut	230	11.0	4.1	5.3	1.32	8.80	0.7	575	597
Pistachio	100	90.0	3.9	2.3	1.30	1.20	7.0	490	1025
Walnut	150	89.0	2.5	2.5	1.59	3.41	4.9	346	441

3.1.5 Vitamins

With the exception of fat-soluble vitamins D and K, other fat soluble varieties, *i.e.*, vitamin A and D, and all water-soluble vitamins like vitamin C, B_{12}, B_6, B_1 (thiamine), B_3 (niacin), B_2 (riboflavin), B_9 (folate/folic acid), pantothenic acid and biotin, cannot be synthesized by the human body and must be obtained from the diet. Once the supplied vitamins are absorbed into the body, they are assimilated becoming parts of the cells, enzymes, hormones, muscles, blood and bones. Some types of vitamins are retained over a long period of time, while others are utilized as soon as they are assimilated. Nuts can supply the vitamins as shown in Table 5.

Table 5 : Vitamin contents per 100 g of raw unsalted nuts

Nuts	Vit. B_1 (mg)	Vit. B_2 (mg)	Vit. B_3 (mg)	Folate (μg)	Pantothenic acid (mg)	Vit. B_6 (mg)	Vit. A (μg Retinol)	Vit. E (mg)
Almond	0.19	1.4	3.90	29	0.47	0.14	2.0	26.2
Brazil Nut	0.60	0.43	0.60	22	0.18	0.10	2.0	5.7
Cashew	0.64	0.19	1.80	25	0.86	0.42	1.0	0.9
Hazelnut	0.39	0.17	2.20	113	0.92	0.56	3.0	15.0
Macadamia	1.20	0.16	2.50	11	0.76	0.28	0	0.5
Peanut	0.64	0.13	12.06	240	1.76	0.35	0	8.33
Pecan	0.42	0.18	1.30	25	0.86	0.21	4.0	1.4
Pine Nut	0.57	0.19	4.30	34	0.31	0.09	2.0	9.3
Pistachio	0.58	0.29	1.50	81	0.52	1.70	22.0	2.3
Walnut	0.33	0.18	1.40	70	0.57	0.54	4.0	0.7

3.2 Phytochemicals

Phytochemicals are non-nutritive plant chemicals that have protective or disease preventive properties. It is well-known that plant produces these chemicals to protect itself but recent research demonstrates that they can protect human against diseases.

Phytochemicals are broadly classified as alkaloids, carotenoids, organosulfurs, phenols, phytosterols, *etc*., and each works differently. Research suggests that phytochemicals, working together with nutrients found in different plant derived foods promote health and reduce the risk of chronic diseases. These may slow the aging process and reduce the risk of developing cancer, heart disease, stroke, hypertension, cataracts, osteoporosis and urinary tract infections. Phytochemicals may function as antioxidant (protect cells against oxidative damage), antiproliferative (interfere with replication of undesirable cancerous cell) and carcinogen detoxifier, hypocholesterolemic, stimulant of enzymes and hormones, antibacterial and antiviral, anti-inflammatory, ligand to cell wall (some phytochemicals bind physically to human cell walls thereby preventing the adhesion of pathogens) and potential inhibitor of different actions affecting the initiation and progression of several pathogenic processes.

Nuts are a good source of phytochemicals including mainly phytosterols and phenols. Carotenoids have also been reported in some cases.

3.2.1 Phytosterols

Phytosterols – the naturally occurring plant sterols (cholesterol like compounds in animal) – are capable to inhibit the absorption of dietary cholesterol and to lower down the serum cholesterol as well as antagonize selected inflammatory pathways in human. The consumption of 1.5-1.8 g/day of plant sterols reduced cholesterol absorption by 30-40%. At higher doses (2.2 g/day), cholesterol absorption was reduced by 60%. Phytosterol (and other phytochemicals as mentioned below) content of nut is shown in Figure 1. The phytosterol content ranges from 95 mg/100 g in Brazil nuts to 280 mg/100 g in pistachios.

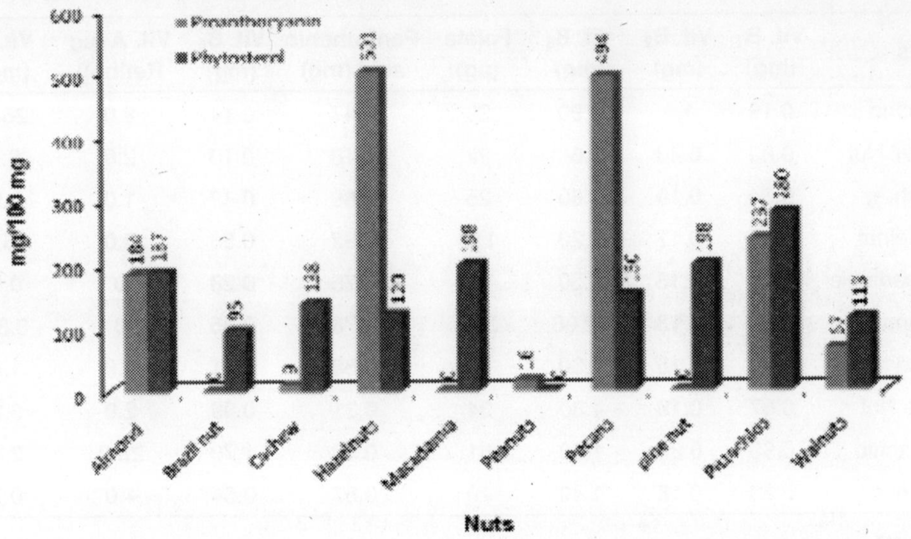

Figure 1 : Phytosterol and proanthocyanin content of various nuts

3.2.2 Carotenoids

Carotenoids intake has been associated with a reduced risk of cardiovascular disease, age-related cataract and macular degeneration (results in a loss of vision in the macula, *i.e.*, the center of the visual field because of damage to the retina) and some forms of cancer. Carotenoids, specifically alpha- and beta-carotene, beta-cryptoxanthin, lutein and zeaxanthin, are found in μg/100 g levels in most nuts but at 0.21–3.0 mg/100 g in pistachios. No carotenoids have been reported in Brazil nuts, macadamias and peanuts.

3.2.3 Phenolics

Phenols/phenolics in nuts consisting flavonoids (comprising anthocyanins, flavanones, flavones, flavanols, flavonols, isoflavones and others), phenolic acids (comprising hydroxy benzoic acids, hydroxy cinnamic acids and others), stilbenes (compounds with 1,2-diphenylethylene back bone), *etc.* may be considered to be the major bioactive compounds for health benefits and have been widely studied.

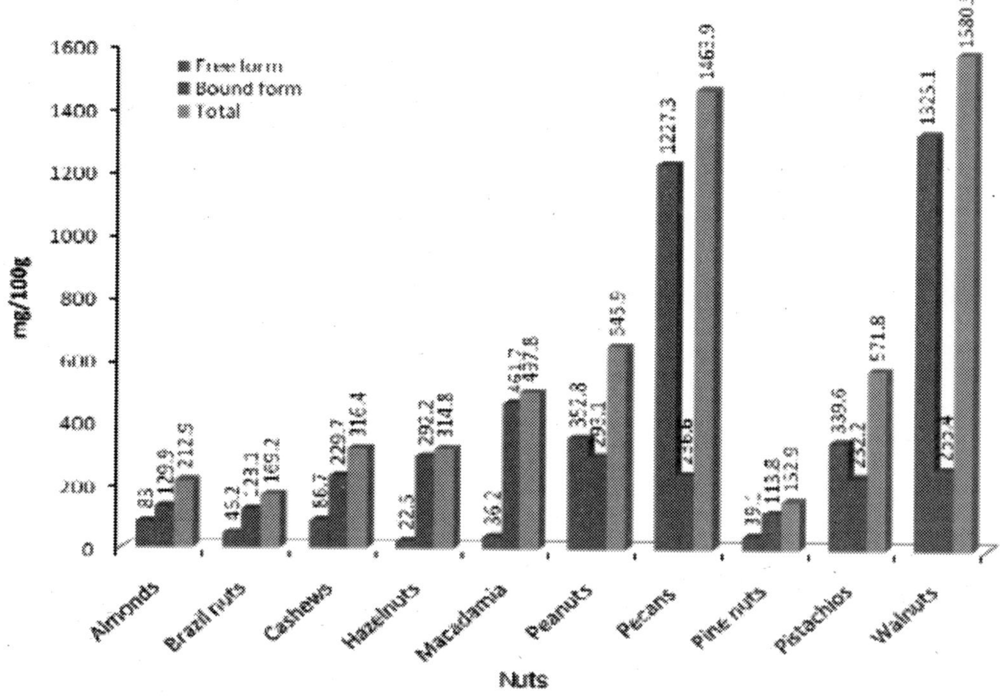

Figure 2 : Phenolic content of different nuts

Phenolics may be present in the cells as both soluble-free and insoluble-bound in the form of glycosides. Nine types of tree nuts and peanuts commonly available in the United States market were evaluated for their soluble-free, bound as well as total phenolic contents, especially the flavonoids (Figure 2 and 3).

Among all the samples analyzed (Figure 2), walnuts contained the highest soluble-free phenolic content followed by pecans, peanuts, pistachios, cashews, almonds, Brazil nuts, pine nuts, macadamia nuts and hazelnuts. The distribution pattern of bound phenolics was different from the free soluble phenolics. Macadamia nuts had the highest bound phenolics followed by peanuts, hazelnuts, walnuts, pecans, pistachios, cashews, almonds, Brazil nuts and pine nuts. The total phenol content among nuts varies widely with walnuts (1580.5 mg/100 g) being the richest source and pine nuts the lowest one.

Among all the samples tested, pecans contained the highest soluble free flavonoid content followed by walnuts, peanuts, pistachios, cashews, almonds, Brazil nuts, hazelnuts, pine nuts, and macadamia nuts (Figure 3). The distribution pattern of bound flavonoids is different from free soluble flavonoids. Walnuts had the highest bound flavonoid content followed by macadamia nuts, hazelnuts, Brazil nuts, pecans, pistachios, almonds, peanuts, pine nuts and cashews. The total flavonoid content was lowest in pine nuts and highest in walnuts (744.8 mg/100 g). The stilbene resveratrol is found in peanuts and pistachios at 84 and 115 mg/100 g. Resveratrol has not been found in other nuts.

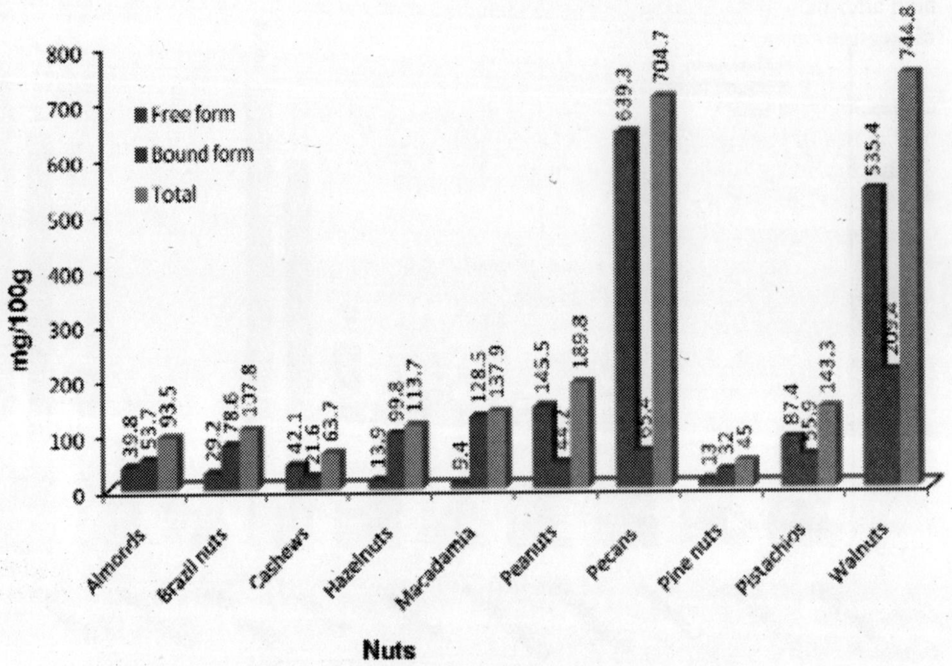

Figure 3 : Flavonoid content of different nuts

Among the flavonoids, pro-anthocyanins (PACs) are short chains of linked flavan-3-ol molecules (catechins). These subsets of polyphenols are recognized for giving nuts an astringent mouth feel. Pro-anthocyanins are found in almonds, cashews, hazelnuts, pecans, pistachios, peanuts, and walnuts, around 500 mg/100 g being present in pecans and hazelnuts (Figure1). Studies show that the antioxidant capabilities of PACs are 20 times more powerful than vitamin C and 50 times more potent than vitamin E. However, all the phytochemicals in nuts appear to contribute substantially to their total antioxidant capacity. Figure 4 compares the lipophilic (L), hydrophilic (H) and total (T) antioxidant capacity of nuts expressed as oxygen radical absorbance capacity (ORAC).

However, the significance of bound phytochemicals in nuts to human health is not clear. Bound phytochemicals, such as glycosides, cannot be digested by human enzymes and could survive in stomach and small intestine digestion to reach the colon. It may be hypothesized that bound phytochemicals can be digested and absorbed at different sites of the gastrointestinal tract and play their unique health benefits. For wheat, it was stated that the colonic microflora might release the phytochemicals from bound state through fermentation, and thus, provide site-specific health benefits in colon or other tissues after absorption.

4. Cardiovascular Health Benefit

Cardiovascular disease (CVD) is any of a number of specific diseases that affect the heart itself and/or the blood vessel system, especially the veins and arteries leading to and from the heart. Over 80% of the cardiovascular disease deaths take place in low-and middle-income countries and occur almost equally in men and women. Cardiovascular risk of women is particularly high after menopause. Proper life style including physical exercise and healthy food can help in preventing cardiovascular disease.

In last two decades, several large prospective cohort Epidemiological studies (an epidemiological study is a statistical study on human populations, which attempts to link human health effects to a specified cause) like Physician's Health Study, the Adventist Health Study, the Iowa Women Health Study, the Nurse's Health Study, and the Cholesterol and Recurrent Events (CARE) Study have examined the relation between nut consumption and the risk of coronary heart disease, coronary heart disease (CHD refers to the failure of coronary circulation to supply adequate blood to cardiac muscle and surrounding tissue), and all have found an inverse association. In addition, several clinical studies (clinical study determines the safety and effectiveness of medications, devices, diagnostic products and treatment regimens intended for human use) have also observed beneficial effects of diets high in nuts including walnuts, peanuts, almonds and other nuts.

Calculations suggest that daily nut eaters gain an extra five to six years of life free of coronary disease. According to a report (2002) released by International Tree Nut Council Nutrition Research and Education Foundation, more than 30 studies have shown that including nuts in the diet can reduce the risk of heart disease regardless of the individual nut studied. In 2003, the United States Food and Drug Administration (FDA) approved the first qualified health claim for a food for use on package labels. A qualified health claim is issued by FDA when it is determined that consumers will benefit from more information on a dietary supplement or conventional food label concerning diet and health even though the claim is based on "somewhat settled science rather than just on the standard of significant scientific agreement, as long as the claims do not mislead the consumers." The claim states that *scientific evidence suggests but does not prove that eating 1.5 ounces (42.5 g) per day of most nuts, as part of a diet low in saturated fat and cholesterol may reduce the risk of heart disease.*

In addition to the benefits in coronary heart disease, the consumption of nuts reduces the risks of other cardiovascular disease like stroke (blocking of blood flow to the brain) and hypertension. Nuts are recommended as part of the Dietary Approaches to Stop Hypertension (DASH) diet, a dietary plan clinically proven to reduce blood pressure significantly. The DASH diet is supported by the United States Department of Health and Human Services and recommends 4 to 5 servings (1 serving is equivalent to 28 g) per week from *nuts, seeds and legumes* grouping. Nuts have been shown to have beneficial effects on inflammation (inflammation is the complex biological response of vascular tissues to harmful stimuli. It is one of the recognized mechanisms involved in the development of atherosclerotic plaque in which an artery wall thickens due to deposition of cholesterol, heart disease and insulin resistance), thrombotic factors (factors that help in formation of thrombosis, *i.e.*, formation of a blood clot/thrombus in heart or a blood vessel, and inflammatory reactions within atherosclerotic plaques can induce clot formation) and endothelial function (the endothelium is the thin layer of cells that line the interior surface of blood vessels. The endothelium, by releasing nitric oxide (NO), promotes vasodilation, *i.e.*, widens blood vessels and inhibits inflammation and thrombosis and prevents hardening of the arteries and vascular smooth muscle cell proliferation).

5. General Health Benefit

Studies show that people who eat more nuts are usually thinner (not obese). Nut consumption lowers the possibility of developing type II diabetes, osteoporosis, neurological disorder, advanced macular degeneration and gallstone. Eating nuts even lowers the risk of cancer. The oil obtained from the macadamia nuts, the only known plant oil, contains a high concentration of palmitoleic acid and is very similar to the sebum of human skin. This substantiates its very good application in cosmetic industry.

Studies on heart disease

Epidemiological study

Physicians' Health Study

The Physician's Health Study carried out by Brigham and Women's Hospital, Boston, USA, followed up for 17 years on 22,000 US male physicians aged 40-84, with no prior history of heart disease found that men who consumed nuts twice or more in a week had a 47% lower risk of sudden cardiac death than men who rarely or never consumed nuts. Nut eaters also had a 30% lower risk of coronary heart disease death.

Adventist Health Study

The Seventh-Day Adventists study, carried out by Loma Linda University, California, USA, over 31,208 ***white*** *persons including men and women,* ***observed*** *frequent nut consumption (comparison: ≥5 times per week nut consumers vs. <1 per week with 6 years of follow-up) was associated with substantial reductions in incidence of both nonfatal myocardial infarction [MI (heart attack, i.e., supply of blood to the heart is blocked, preventing enough oxygen supply to the heart. The heart muscle dies or becomes permanently damaged)] and fatal CHD. In Adventist Study, about 32% of the nuts eaten were peanuts, 29% almonds, 16% walnuts and 23% other types.*

Iowa Women Health Study

The Iowa Women's Health Study conducted by the University of Minnesota, School of Public Health, Minneapolis, USA, over 41,837 Iowa women, white, ages 55 to 69 years, with 5 years of follow-up has recorded that frequent nut consumption (Comparison: 2-4 times per week vs. 0 per week) may offer postmenopausal women a modest protection against the risk of death from coronary heart disease.

Nurse's Health Study

The Nurse's Health Study was conducted together by Harvard Medical School, Harvard School of Public Health, Brigham and Women's Hospital, Dana Farber Cancer Institute, Boston Children's Hospital and Beth Israel Deaconess Medical Center, all from medical community of Massachusetts, USA, over 1,21,700 female registered nurses. While studying the relation between nut consumption and incidence of cardiovascular heart disease among 84,409 women without previously diagnosed cardiovascular disease or cancer, ages 34 to 59 years, with 14 years of follow-up (comparison: ≥5 times per week vs. almost never), it was observed that frequent nut consumption was associated with a reduced risk of both fatal coronary heart disease and non-fatal myocardial infarction.

Cholesterol and Recurrent Events (CARE) Study

The Cholesterol and Recurrent Events Study examined the effects of nut consumption on recurrence of acute myocardial infarction among 3,575 post-myocardial infarction patients. During 4.2 years of follow-up, it was observed that compared with those who rarely ate nuts (never or less than once per month), those eating nuts at least two times per week had significantly lower risk of recurrent coronary heart disease.

Clinical study

The study protocol approved by the Loma Linda University, California, compared the effects of cholesterol-lowering diets either rich in walnuts or without nuts, while maintaining the intake of total dietary fat and calories, on serum lipids. Incorporating walnuts decreased serum levels of total cholesterol and favourably modified the lipoprotein profile in normal men. However, extension of this finding to other population groups deserves further study.

6. How Nuts Bring About Health Benefits

The diseases mentioned above are mostly interrelated and so are the mechanisms involved for the health benefits derived from nut. The actions arise from the contained nutrients and phytochemicals, and work independently and/or synergistically with one other. The contributory role may include improvement in blood lipoprotein profile, antiatherogenic effect (prevents atherogenesis, the process of accumulation of lipid containing plaques on the innermost layers of the arteries that controls inflammation and leads to the disease atheroscelerosis), antiarrhythmic properties (arrhythmic pertains to an absence or irregularity of normal rhythm in the heart's beating), glucose homeostasis (glucose maintenance capacity), insulin sensitivity, antioxidant activity, repair of DNA damage, endothelial function, cholesterol lowering and low-density lipoprotein stability (LDL cholesterol, which is known as *bad* cholesterol, becomes even more dangerous when it is oxidized, and oxidized LDL can produce inflammation in arteries), regulation of cell differentiation and proliferation with a reduction of tumour initiation or promotion, regulation of immunological activity, enzymatic action and regulation of hormones by acting as phytoestrogens (phytoestrogens are non-steroidal phytochemicals that exert estrogen-like beneficial effects on the body). Brief discussion on the mechanisms relating to specific health benefit is as follows.

6.1 Cardiovascular health benefit

6.1.1 Contribution of fatty acid profile

6.1.1.1 Reduction of cardiovascular morbidity and mortality: There are three possible reasons why the fatty acid profile of nuts might contribute to this beneficial effect. These may include:

- Consuming nuts, which are high-energy foods rich in unsaturated fatty acids, suppresses hunger and limits intake of other energy-dense foods due to a satiety effect. This has been explained below (section 6.2.1) for weight management programme. The foods that are displaced by this satiating effect tend to be detrimental for cardiovascular health because they are rich in animal protein, saturated fatty acids, trans-fatty acids and simple sugars.

- Unsaturated fatty acids are cardio protective.
- Omega-3 PUFA in nuts, mainly alpha-linolenic acid in walnuts, due to their anti-arrhythmic properties, protects from fatal coronary heart disease. Epidemiological and clinical studies suggest that omega-3 PUFA also modulates inflammation.

6.1.1.2 Cholesterol-lowering effect: The mono- and polyunsaturated fats in nuts comprising favourable fatty acid composition (Table 1 and 3) can lower low-density lipoprotein level and decreases the ratio of total to high-density lipoprotein probably by regulating gene expression. This has been demonstrated in experimental studies with almonds, macadamia, peanuts, pecans, pistachios and walnuts. It can be stated that daily intake of manageable allowances of a variety of nuts for the period of 4-8 weeks has a clear cholesterol-lowering effect. Based on the data from the Nurse's Health Study, it was estimated that substitution of fat from about 28 g of nuts for equivalent energy from carbohydrates in an average diet was associated with a 30% reduction in coronary heart disease risk and the substitution of nut fat for saturated fat was associated with 45% reduction in the risk. However, the cholesterol-lowering effect observed after nut supplementation has often been higher than that predicted based on fatty acid profiles of the test diets, indicating the synergistic contribution of bioactive components (phytochemicals) present in nuts capable of reducing blood cholesterol.

6.1.1.3 Improvement in vascular activity: A recent feeding trial showed that walnut diet attenuated endothelial dysfunction (endothelial dysfunction is a systemic pathological state of endothelium) associated with hypercholesterolemia and was ascribed in part to their high α-linolenic acid (ALA) content. Recent clinical studies have also shown that diets enriched with α-linolenic acid from walnuts or other sources increase circulating levels of anti-inflammatory cytokines (cytokines are protein materials secreted by specific cells of immune system). This may improve the endothelial dysfunction. Walnut feeding also reduced expression of endothelins (endothelins are proteins that constrict blood vessels and raise blood pressure).

6.1.2 Contribution of amino acid profile

Nuts contain significant levels of plant protein that is rich in arginine (Table 2). Recent research has found that arginine is a dietary precursor of nitric oxide, a potent endogenous endothelium-dependent vasodilator that acts much like nitroglycerin (nitroglycerin is used medically as a vasodilator to treat heart disease). Nitric oxide may have other antiatherogenic properties as well, such as inhibiting platelet aggregation and monocyte adherence (formation of platelet-monocyte aggregates is an established link between inflammation and thrombosis in acute coronary syndromes and related disorder).

6.1.3 Contribution of phytochemical, vitamin and mineral

Oxidative modification of low-density lipoprotein is a key event in early atherogenesis and inflammation leading to cardiovascular disaese. The formation of atherogenic oxidized low-density lipoprotein can be inhibited by dietary antioxidants. Published research suggests that nut's high antioxidant content may be the key to their cardio-protective effects. The antioxidants include both water and lipid soluble components (Figure 4), and presence of phytochemicals imparts such high H-ORAC values. Walnuts and pecans are rich in antioxidant of the tree nuts. However, besides the antioxidant capacity, other actions of phytochemicals towards cardiovascular health benefit include improvement in the functioning of blood vessel walls, suppression of platelet stickiness in arteries and lowering of blood pressure in addition to the cholesterol lowering effect as indicated in section 6.1.1.2.

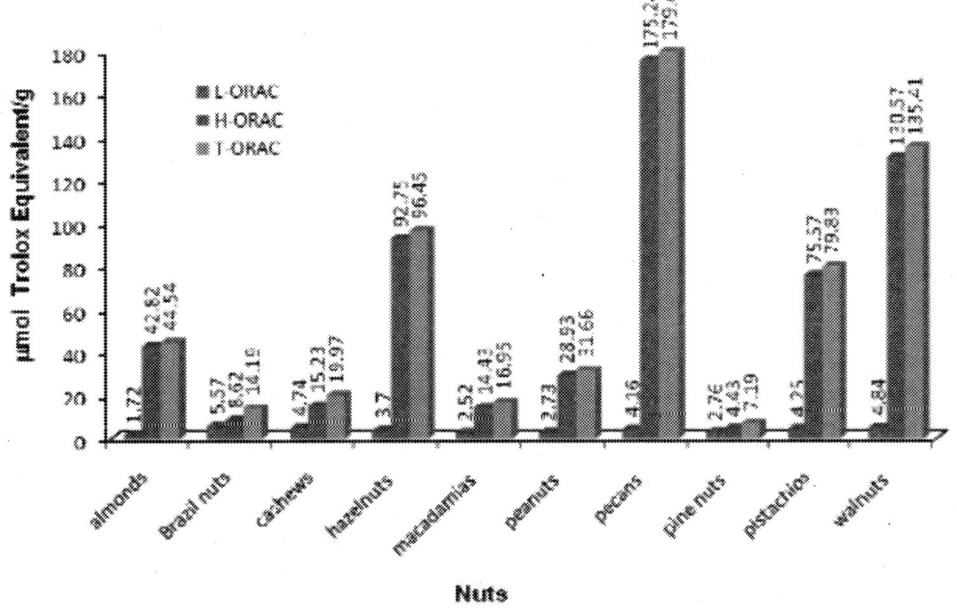

Figure 4 : Lipophilic (L), Hydrophilic (H) and total antioxidant capacity expressed as oxygen radical adsorbents capacity (ORAC) of nuts. All data are expressed as micromoles of Trolox (a water-soluble derivative of vitamin E) equivalence per gram of nuts

Deficiencies of the vitamins folic acid (B_9), pyridoxine (B_6), or cyanocobalamin (B_{12}) can lead to high homocysteine levels (homocysteine is biosynthesized from essential amino acid methionine. High level of homocysteine has been linked to increased coronary heart disease risk). Pistachio is a good source of pyridoxine and peanut a good source of folate (Table 4).

Finally, nuts are an important food source of copper and magnesium. Copper helps to sustain the elasticity of blood vessels maintaining proper blood pressure, and healthy muscle tone which performs vital role in heart. Magnesium intake may have beneficial effects on endothelial function, modulating inflammation and lowering blood pressure. Potassium, an important electrolyte involved in nerve transmission and contraction of all muscles including heart, is another mineral in nut (pistachio) that is essential for maintaining normal blood pressure and heart function.

6.1.4 Contribution of dietary fiber

Foods that reduce glycaemic response to diet can be useful in alleviating inflammation. Recently, nuts have been shown to reduce the glycaemic impact of ingested carbohydrates-rich foods, due to their high fiber and unsaturated fat content (Table 1). A cross-sectional study (which correlates disease and exposure status) performed on diabetic patients from Health Professionals' observed that subjects whose diets had low glycaemic load and high dietary fiber, had in their plasma higher levels of adiponectin, a molecule with anti-inflammatory properties that protects against atherosclerosis. Some intervention studies (meant for improvement in condition of individual) and epidemiological studies also support that dietary fiber can decrease risk of inflammation improving atherosclerosis *vis a vis* cardiovascular health.

6.2 General health benefits

6.2.1 Obesity

It is natural to think that nuts, being high in fat content, would likely cause an increase in weight. However, the epidemiological evidence consistently indicates that nuts consumers have a lower body mass index (BMI) than non-consumers. Actually, an inverse association was observed between the frequency of nut consumption and body mass index. Clinical studies also observed that inclusion of nut in diet leads to little or no weight gain. Three mechanisms have been proposed to account for such benefit.

Firstly, nuts appear to increase fullness imparting satiety. It is posited that inclusion of nuts in diet results in a spontaneous reduction of food intake at other times of the day to offset some or all of energy provided by nuts such that total energy balance is unchanged.

Secondly, nut consumption increases resting energy expenditure (represents the calories required for a 24 hours period by the body during a non-active period) and diet-induced thermogenesis (refers to stimulation of energy expenditure following nutrient ingestion) because of their high-protein content and high polyunsaturated-to-saturated-fatty acid ratio, and thereby, dissipates a portion of energy they provide.

Thirdly, absorption of energy from nuts is incomplete, probably due to the structure of lipid-storing granules that do not completely release the fatty acids during digestion, or to various fiber components, which limit digestion/absorption capacity. These inefficiencies reduce their contribution to energy intake. Actually, some portion of ingested calorie is excreted without absorption. The current best estimates are that 55-75% of the energy contributed is offset by dietary compensation, another 10-15% by fecal loss, and an additional, less well-established estimate of 10% *via* increased energy expenditure. Studies at Brigham and Women's Hospital and Harvard School of Public Health concluded that people on a Mediterranean-style diet, which is low in saturated fat and high in nut, were more successful in keeping weight off for longer than on a traditional low fat diet. Thus, both epidemiological and clinical studies show that moderate consumption of nuts provides health benefits without threatening energy balance and that nuts can be included in weight loss regimen.

6.2.2 Diabetes

A prospective cohort study of 83,818 women participating in the Nurses Health Study followed up for 16 years, reported higher consumption of nuts and peanut butter to be associated with a lower risk of developing type II diabetes. Women who consumed about 28 g of nuts $\geq$ five times per week had 25% lower risk compared with women who never ate nuts. Several other studies also observed such preventive action. Different factors may control the impact of nut consumption on diabetes.

6.2.2.1 Dietary composition: Individuals with diabetes are recommended to follow a low GI diet. A food with a GI of less than 55 is considered a low glycemic food, while 56 to 69 is moderate and over 70 is considered high. Most nuts are low on the glycemic list because of the protein and fat found in them, *e.g.*, cashews are 27 and peanuts are 22.

High-fat diets have been associated with a detrimental effect on glucose homeostasis in epidemiological and intervention studies. Nuts, despite being rich in fat, do not negatively affect glucose homeostasis. Some people with diabetes even benefit from replacing some of the carbohydrates rich foods in their diet with foods rich in monounsaturated fats. Reports are available on the ability of nuts, when eaten with carbohydrates to depress postprandial glycemia. When about 28 g of almond was eaten along with white bread, GI of the meal (105.8) was comparable to eating white bread alone but when 57 g of almonds was consumed with the white bread, GI dropped to 63, and when 86 g almond was eaten, GI was only 45.2, which is less than half the GI of white bread only meal.

A diet high in fiber, particularly soluble fiber, absorbs blood glucose levels. As nuts are a source of fiber, they are especially helpful in controlling blood sugar level.

Scientists opine that high intake of several antioxidants significantly reduced the risk of type II diabetes. People consuming high amounts of vitamin E were 31% less likely to develop diabetes than those consuming low amounts. Nuts particularly almonds, hazelnuts, pine nuts and peanuts are good sources of vitamin E.

Based on current knowledge, clinicians have good reasons to believe that magnesium repletion may play a role in delaying onset of type II diabetes and warding off its devastating complications- cardiovascular disease, retinopathy and nephropathy. This is supported by several prospective cohort studies that found magnesium intake to be inversely associated with incidence of type II diabetes. Hence, magnesium rich foods such as nuts may contribute to reduce the risk of type II diabetes.

6.2.2.2 Insulin sensitivity: Dietary fatty acids influence glucose metabolism through a number of mechanisms including cell membrane function, enzyme activity, gene expression and insulin signaling. Substantial evidence is there that higher intakes of mono- and polyunsaturated fats (found in relatively high amounts in nuts) are associated with improvement in insulin sensitivity, whereas, increasing saturated and trans-fatty acids worsens the situation. A recent study researched the long-term effect of walnut consumption on type II diabetes sufferers. Fifty overweight adults with non-insulin-treated diabetes were monitored to receive low-fat dietary advice and a handful (28 g) of walnuts per day for one year, together with control group (no nut diet). The walnut group showed significantly greater reductions in fasting insulin and glucose levels in blood (glucose and insulin levels in blood are directly proportional).

Besides this, this amino acid arginine and fibre have been shown to help insulin work more effectively in people with diabetes, and nuts are a good source of these ingredients.

6.2.3 Osteoporosis

Bone health requires adequate amount of magnesium, calcium, phosphorous and potassium. About 60% of magnesium in body is located in bones. Magnesium is also necessary for regulating blood-calcium levels. Researchers at the Human Nutrition Research Center at Tufts University in Boston showed that greater intake of foods that are rich in magnesium and potassium led to increased hip and forearm bone density in men and women, aged 69 to 97. All these minerals are present in good quantities in nuts.

Omega-3 fatty acid in nuts also protects bone health and prevents excessive bone turnover by lowering ratio of omega-6 to omega-3 fats in diet. Researchers think, this is because omega-6 fats, which are converted into pro-inflammatory prostaglandins (prostaglandin is a group of lipid compound acting as hormone-like substances that are derived from fatty acids in human body), while omega-3 fats are metabolized into anti-inflammatory prostaglandins. Protective omega-3 levels greatly improve by eating just 4 walnuts a day.

6.2.4 Neurological disorder

Physicians who recommend dietary habits should take heed that the best way for nutritional prevention of Alzheimer's disease is through foods rather than vitamin supplements. Vitamin E and omega-3 fatty acids are the most convincing evidence of neuro-protection to date. Nuts are rich in these components, especially almonds are rich in vitamin E. Almonds are also high in riboflavin which boosts memory. Additionally, the phytochemicals in nuts acting as antioxidant may keep the brain health and delay the onset of dementia.

Nuts are rich in amino acid arginine, which stimulates pituitary gland to release a hormone that declines quickly after the age of 35. This is a real anti-ageing boon to the brain. Nuts also by providing amino acid phenylalanine can do wonders for mental and neurological health. Recently in a medical journal, it has been published that dietary pattern including nut may protect against Parkinson's disease.

Magnesium in nuts keeps the nerves and blood vessels and muscles they enervate relaxed, whereas, calcium provides activation. Magnesium serves as calcium channel blocker preventing calcium from rushing into nerve cells. If diet provides too little magnesium, calcium can gain free entry, and the nerve cell becomes over-activated, sending too many messages to muscles causing excessive contraction and fatigue.

6.2.5 Macular degeneration

Vitamin E and omega-3 fatty acid in nuts have been proven to prevent development of macular degeneration and cataract. Selenium in Brazil nuts renders such health benefit by helping absorption vitamin E and production of body's own antioxidants. Zinc found in nuts helps the body to absorb vitamin A and reduce free radical formation protecting against night blindness and macular degeneration.

6.2.6 Gallstone

Data from Health Professionals Follow-up Study involving 43,823 men (over 40 years) and Nurse's Health Study involving 80,718 women (30-55 years) were analyzed. It was observed that eating approximately 28 g of nuts five or more times per week reduced the risk of developing problems due to gallstones by 27% for men and 25% for women, compared with eating nuts less than once per month.

Most gallstones are formed when excess cholesterol precipitates from bile, which is produced by liver and stored in gallbladder. Previous studies have consistently shown that nut consumption prevents heart disease by reducing total cholesterol and LDL-cholesterol levels. Such effects might also be the reason for their ability to prevent gallstone.

6.2.7 Cancer

An important emerging question is whether nut intake can lower the risk of cancer with the help of possible mechanisms as discussed below:

HER-2/neu oncogene, a particular gene, is known to be over-expressed in many human cancers including cancers of breast, ovary, lung, gastric, oral cavity, *etc*. Repression of *HER-2/neu* over-expression suppresses malignant growth. Oleic acid and ALA, the major fats in nuts, have been observed to induce repression of *Her-2/neu* in cell culture.

Lower than predicted weight gain from nut consumption (section 6.2.1) may provide another mechanism, whereby nuts can reduce cancer risk. Obesity is associated with inflammation and elevated endothelin level. Low endothelin is associated with less tumor growth in ovarian cancer and walnut's ability to reduce endothelin has been mentioned (section 6.1.1.3).

Walnuts also contain a polyphenolic antioxidant called ellagic acid, which blocks the metabolic pathways that can lead to cancer. Ellagic acid not only protects healthy cells from free radical damage but also detoxifies potential cancer-causing substances and prevents cancer cells from replicating. The proliferation of cancerous cells of small intestine, and liver models was significantly inhibited in a dose-dependent pattern after exposure to extracts of nuts, with walnuts and pecans exhibiting high antiproliferative activity.

Though laboratory results suggest a positive indication towards cancer prevention, the epidemiologic data, such as the European Prospective Investigation into Cancer and Nutrition study involving 10 European countries are weak in this regard. The objective of the study was to determine the effects of nut and seed intake on colorectal cancer risk. The data set included 4,78,040 subjects (1,41,988 men and 3,36,052 women) with a total of 855 colon (327 men and 528 women) and 474 rectal (215 men and 259 women) cancer cases. The data showed no association between higher intake of nuts and seeds and risk of colorectal, colon and rectal cancers in men and women combined. However, a significant inverse association was observed for colon cancer in women (comparison: >6.2 g/day versus non-consumers), while no association in men. It is not evident why the association in women or why it may be limited to colon, suggesting that much further research is necessary.

7. Selection of Nut

The definitive answer to this question is currently unknown. In the Adventist Study, the nut consumed was a mixture of different types (vide section 4). These researchers did not ascertain whether the nuts were fresh, oil-roasted, or dry-roasted. The Nurse's Study found that peanuts, which are legumes, appeared to be just as effective in reducing risk of coronary heart disease as tree nuts. Several other experiments where volunteers were fed nuts as part of their diet for several weeks have found that walnuts, almonds, hazelnuts, peanuts, pecans, pistachio and macadamia nuts, all alter the composition of the blood in ways that would be expected to reduce the risk of coronary heart disease. The best advice currently is probably to eat a variety of nuts.

8. Mode of Consumption

After harvesting, the nuts are processed by sun or low temperature drying to reduce the moisture content to avoid post-harvest microbial infection, and these so dried nuts appear in the store as raw nuts. The term *raw* when applied to nuts does not mean just harvested. As snacks, the nuts are usually consumed raw or roasted, with or without salt. As ingredients, they can be mixed to salads, baked goods like breads, cookies, cakes, desserts like candies and ice creams and other food preparation including sweet and savory dishes. However, snacking of nut is the best way to derive maximum health benefit as it can keep the other calorie imparting factors at bay.

Nuts naturally, as plucked from plant, contain some enzyme inhibitors so that they can wait until they come in contact of moisture for germination. These inhibitors act as preservative for the nut, making sure it can stay viable for a long time. If nuts are picked, dried and placed on grocer's shelf, these inhibitors are still intact. When nuts are eaten raw as snacks, these inhibitors actually inhibit digestion. This is why nuts have a reputation as being difficult to digest. Most nutritionists and dieticians suggest eating only limited amounts of nuts, otherwise pancreas need to produce huge amounts of digestive enzymes to counteract the inhibitors. When such a large amount of pancreatic enzymes are needed to digest raw nuts, the enzymes are not available to perform their other regular maintenance of the body. A continuous diet of raw nuts is so taxing to the pancreas that it can actually swell with distress.

Dry roasting, *i.e.*, high temperature processing of nuts will destroy the inhibitor but it also destroys the desirable enzymes in the nut that the body needs to make it easily digestible. Hence, eating dry roasted nuts also puts a strain on the pancreas. Simultaneously, roasting nuts at a temperature higher than 76°C will cause a breakdown of their fats and the production of free radicals. When nuts roasted at high temperature as used commercially are consumed, the developed free radicals can cause peroxidation of blood lipid that can trigger tiny injuries in artery walls- a first step in the build up of plaque and cardiovascular diseases.

In addition to destroying many of the nutrients in dry roasting, oil roasting of nuts usually involves added oil, which is most likely in the form of seed oil. Because of oil absorption, total energy of nuts increases by 30-40 kcal/100 g. Moreover, when heated to a high temperature, frying oils produce an abundance of free radicals. Thus, nutritional value of oil roasted nuts is linked to the quality of frying oil like its fatty acid composition, extent of recycling, contained antioxidant, *etc.* and the applied method of roasting. These factors can modify the fatty acid profile of nuts.

Nuts can be made even more nutritious, delicious and healthy by sprouting or soaking them. When raw nuts are soaked in water, they slowly lose the inhibitor and allow germination to take place. Besides this, minerals, vitamins, essential fatty acids, flavours and enzymes that have been lying dormant in raw nuts are released. Soaked nuts are so easily digested that they can be eaten in large amounts. They have greater nutritional value than raw nuts.

However, roasting brings out the desirable nutty flavour and develops sweetness. Thus, if this flavour is really unavoidable, the cheapest and easiest way to match these preferences is dry roasting. It is safe to roast nuts at 70-76°C in oven for 15-20 minutes in a single layer.

Serving of nuts also makes a difference. Remarkably, for all nuts, more than 50% of the antioxidants are lost when the skin (a thin brownish covering adhered to kernel) is removed.

Hazelnuts with skins intact indicate approximately 10 times the total phenol content and oxygen reactive absorbing capacity than without skin (removed by either roasting or blanching), and approximately three times the proanthocyanin content. Recent studies have shown that almond, peanut and pistachio skins are very high in antioxidants. In walnuts, less than 10% of total antioxidant activity retains after removing the skin. However, they are almost always consumed as the raw product with skins. Salting of nut is also a matter of concern. Though nuts are usually devoid of sodium, a serving of salted nuts are as high in sodium as a small bag of potato chips.

9. Nut Allergy

Some persons including adult and children may be allergic to peanut and tree nuts. An allergic reaction happens when the immune system of any person is hypersensitive to apparently harmless foods that eventually lead to be harmful. The immune system responds by creating specific antibodies to proteins in that food. These antibodies, called immunoglobulin E (IgE), are designed to fight off the *invading* proteins.

Immunoglobulin E antibodies trigger the release of certain chemicals in the body. One of these is histamine. The release of histamine can affect a person's body systems including respiratory system, gastrointestinal tract, skin and cardiovascular system, causing allergy symptoms like wheezing, stomachache, vomiting, itchy hives, and swelling.

Reactions to peanut and tree nuts may be different. It all depends on the person, and sometimes, the same person can react differently at different times. Some reactions can be very mild and involve only one system of the body, like hives on the skin. Other reactions can be more severe and involve more than one part of the body. Most reactions last less than a day. In really bad cases, nut allergies can cause a condition of sudden, potentially life-threatening reaction like trouble breathing, lose of consciousness and blood pressure drop, in addition to the symptoms mentioned above.

However, in some countries, a considerable fraction of population suffers from nut allergy. It is a well-established and conspicuous fact in India that with high peanut intakes, there is virtually no reported nut allergy.

10. Aflatoxins in Nuts

Certain strains of fungi, *Aspergillus flavus* and *Aspergillus parasiticus* rapidly and *Aspergillus nomius* rarely, spread on groundnuts and tree nuts under warm and humid climates in tropical and subtropical regions, even in temperate regions. They produce a group of secondary metabolites called aflatoxins (AF) and designated as B_1, B_2, G_1 and G_2. Among them, aflatoxins B_1 is the most acutely toxic and has been classified as group 1 carcinogen. Recently, the Codex Alimentarius Commission, Joint FAO and WHO Food Standards Program adopted a limit of 10 μg/kg of total aflatoxin for ready to eat nuts. This limit is 15 μg/kg for nuts intended for further processing. However, the committee discusses and evaluates the matter on regular basis.

Aflatoxin contamination in nuts is most severe when insects damage it. To prevent and combat mold contamination and aflatoxin production, the Codex Alimentarius Commission (FAO/WHO) recommended International Code of hygienic practice in the way of good agricultural practice and commercial processing particularly during raw material handling, inspection and sorting, processing (hulling and shelling) and storage. Completely shelled and in-shell, nuts should be dried quickly to a water activity < 0.7 (moisture content 4.5-5.0%) prior to storage to arrest mould growth and toxin formation.

Aflatoxins are highly stable to dry heat. However, roasting of contaminated nuts in presence of salts has been reported to destroy aflatoxins to some extent depending on initial level of contamination, moisture content, temperature and duration of heating.

Rapid and cheap method for aflatoxin detection in nuts is done with high intensity long-wave (365 nm) ultraviolet lamp, called black light test. The sample should be inspected (for screening but not quantitative estimation) for fluorescence in broken and damaged kernels under the specific lamp. Additional confirmatory test is needed for samples that react positively.

11. Storage of Nuts

Rancidity of fat during storage develops off flavour and free radicals that cause health hazard. Because of high content of unsaturated fats, all nuts, whether roasted or raw, are susceptible to rancidity. Walnuts are extremely perishable possibly due to their high delicate polyunsaturated fat content (Table 3). Control of mould growth in nuts is also very critical during storage. These undesirable developments can be reduced by maintaining low moisture content of nut (around 4.5-5.0%), low oxygen tension and temperature (0-10°C) of the environment. Lower these parameters, longer will be the shelf life. The seller maintains low oxygen tension by vacuum or nitrogen packaging. Along with these, light and presence of metal also conspire to spoil nuts. Hence, they are best stored in sealed plastic or glass containers in dark, cool and dry place. Nuts quickly absorb odours from surroundings, which is another argument for keeping them in isolation in cold storage. Generally, if stored in airtight container in refrigerator, nuts will remain fresh for 6 months, whereas, in freezer up to 12 months. Unshelled nuts, however, store better than shelled nut as the latter is directly in contact of air.

It is therefore important to purchase nuts from a well-maintained store with high turnover to ensure freshness. Once purchased, it is advisable for the consumers to keep tightly sealed at low temperature until the stock is finished.

12. Conclusion

Nuts are high-energy food and good sources of excellent nutrients and health promoting bioactive phytochemicals. Eating regularly variety of fresh and unsalted nuts 1.5 oz (42.5 g) per day will provide cardiovascular and other health benefits including maintenance of healthy body weight. As nut skin contains lot of antioxidants, it may be wise to consume nut along with its skin. Therefore, nut consumption can be recommended in the context of a healthy and balanced diet, and snacking on lightly dry roasted or soaked nut is best way to derive the maximum health benefit. Nevertheless, it should not be taken a license to overindulge. When nuts are added to diet, inadvertently calorie is added. Therefore, it becomes imperative to cut down calorie from other foods to avoid weight gain.

■■■

10 Therapeutic Significance of Fruits

M.K. Rana[1] and Harsimrat K. Bons[2]

[1]Department of Vegetable Science, CCS Haryana Agricultural University, Hisar-125 004, Haryana, India, E-mail: <mkrlotus@gmail.com>
[2]Krishi Vigyan Kendra, Samrala, Punjab Agricultural University, Ludhiana, Punjab, E-mail: <harsimratpau@gmail.com>

The importance of fruits in human diet is well recognized allover the world. Fruits being rich in minerals and vitamins are known as protective foods, which help in the maintenance of proper health and resistance to diseases. They also provide minerals like calcium, iron, *etc.* the deficiencies of which may lead to disturbances in the metabolism and can cause several ailments. Fruits are also rich source of organic acids like citric, tartic, malic acid, *etc.*, which stimulate appetite and help in proper digestion. Some fruits also contain protein-digesting enzyme (papain in papaya and bromolein in pineapple). Fruits are also rich in pectin and cellulose, which stimulate intestinal activity and save the human body from various disorders. Most fruits are high in sugars, which are present in easily digestible forms, *i.e.*, dextrose, levulose, fructose and many others. These sugars are very essential for growth and development and supply quick energy to our body. Banana, dates, karonda and raisin are very rich in carbohydrates. Fruits are generally low in fats and oils and are an important food components in diets designed to reduce the risk of coronary heart diseases. Fruits are liked and relished both by rich and poor people because of their delicious taste, refreshing property, bodybuilding, protective nature and calorie yielding value. They have special significance in our country because a sizeable population is vegetarian. Realizing the worth of fruits in human health, Indian council of Medical Research (ICMR) recommends the use of 120 g fruits per capita per day. Consumption of adequate amount of fruits helps in maintaining health and vigour of an individual. Apart from the nutritive value, many fruits are known to possess medicinal properties as well. Many therapeutic drugs in use in modern medicine originated as plant extracts. It is not surprising that certain fruit components exert pharmacological or therapeutic effects. Limonin and nomilin are present in citrus plants such as orange, lemon, lime and grapefruits, are believed to have a role in inhibiting the development of certain forms of cancer. Likewise, many fruits like mango, guava, banana, *bael*, *jamun*, papaya, *etc.* have been valued for certain medicinal properties, which are described below:

Common name: Almond

Botanical name: *Amygdalus communis* var. dulcis and *Amygdalus communis* var. amara

Family: Rosaceae

The almond, known as the king of nuts among dry fruits, is a highly nutritious food since it is rich in almost all the elements needed by the body. Almonds have a special dietary value besides containing 20% of proteins but contain practically no starch. It is an effective health-building food, both for the body and mind and a valuable food remedy for several common ailments. Though pleasant to taste, their nutritive value is diminished unless well masticated, as they are difficult to digest, and are therefore, often made into flour for cakes and biscuits for patients suffering from diabetes. Moreover, they yield nearly half their weight in oil, which is employed medicinally for the treatment of bronchial diseases, nephritic pains, *etc*.

In almonds, the major portion of fat (67%) is mono-unsaturated, which is *good fat* and is valued for its beneficial properties. The fat profile of almonds is even more desirable because it is lower in saturated fat. Monounsaturated fats lower the overall cholesterol, while good cholesterol (HDL) is maintained. The bad cholesterol (LDL) is lowered considerably, and thereby, decreasing the risk of heart diseases. It is also used in medicines, as it is rich in bland oil.

Fresh sweet almonds possess demulcent and nutritive properties, but as the outer brown skin sometimes causes irritation to the alimentary canal, they are blanched by removal of the skin when used as food. However, proper caution must be taken while using the bitter almonds as they have dangerous poisonous properties. The bitter almonds also yield a volatile oil, which is largely employed as a flouring agent. These almonds contain a colourless, crystalline glucoside called amygdalin, which is not present in sweet almonds. This substance is left in the cake obtained after the oil has been expressed and can be extracted from it by digestion with alcohol. Bitter almonds also yield 6-8% of prussic acid. The consumption of the oil extracted from bitter almonds can cause giddiness, headache and dimness of sight. It has been found to be poisonous to animals and birds.

The oil contains chiefly olein with a small proportion of glyceride of linoleic add and other glycerides but contains no stearin, therefore, it is very similar in composition to olive oil but is devoid of chlorophyll. The almond milk prepared by pounding almonds in water is a pleasant drink having a cooling effect. It is prescribed as diluents in acute diseases. Blanched and beaten into an emulsion with barley water, sweet almonds are of great use in the stones and other disorders of the kidneys, bladder and bile duct. Almonds have always been a part of the beauty world, as they not only nourish the skin but also have a gentle bleaching effect, and thus, help in maintaining the colour.

Common name: *Aonla*

Botanical name: *Emblica officinalis*

Family: Euphorbiaceae

Aonla is one of the richest natural sources of ascorbic acid, containing as much as 747 mg/100 g of pulp. It is also packed with other essential elements like calcium, phosphorus, iron and vitamins such as thiamine, riboflavin, niacin, *etc*. The astringency of aonla fruit may be assigned to the presence of polyphenols. Owing to its high vitamin C content, it keeps the gums and teeth healthy and suppresses respiratory disorders. Vitamin C is also a powerful scavenger of free radicals, which cause cell damage leading to several diseases including cancer, cataract, heart disease, diabetes and aging. Dried fruits and fruits powder are also rich in vitamin C. Aonla rebuilds new tissues and increases red blood cells, making the body more vigorous. It helps in anemia by enhancing iron absorption if taken along with iron rich food articles. Amla is also effective in lowering the blood pressure and blood sugar. Regular consumption of *Amla* as part of a healthy diet is known to lower down the blood cholesterol, preventing hardening of arteries and delay ageing. According to research, taking *Amla* is beneficial for gall bladder infections and pregnant and lactating mothers. Its fresh juice mixed with honey is a multipurpose home remedy particularly for improving eyesight and for promoting vitality in men and women both. Its fruits are laxative and diuretic in nature. Thus, remove chronic constipation and other stomach and urinary disorders. A tablespoon of its juice taken with a cup of bitter gourd juice regularly can stimulate insulin secretion in diabetic patients. It is world widely known for stimulating hair growth and preventing premature graying. Its shade-dried pieces are boiled with a little coconut oil unless that becomes paste like and its application to hair makes the hair soft and lustrous and prevents early graying. One tablespoon of paste of *Amla* leaves mixed with honey is an effective treatment for diarrhea.

Common name: Apple

Botanical name: *Malus pumila*

Family: Rosaceae

It is a common saying that the intake of an apple every day keeps the doctor away. Whatever the colour or taste is, an apple makes a very nutritive snack. It is packed with several essential nutrients, like carbohydrates, calcium, phosphorus, iron, *etc*. Most of these nutrients are present just below the peel. Therefore, it is advisable to consume the apple unpeeled. It also provides the sufficient amount of valuable fiber, which helps in eliminating the constipation. It is commonly recommended to the patients in pre- and post-operative phase. It is particularly recommended for post-gastro-intestinal surgery patients because it does not cause any problem to the stitches made during operation. It is very useful in restoring the valuable nutrients lost during the acute forms of diarrhea.

Apples are laxative, increase memory power and are helpful in hypertension induced paralysis of organs. Raw apples are good for treating diarrhea and ripe ones for constipation. Apples are also beneficial for anemia, heart patients and high blood pressure. Apples provide relief from arthritis and gout. They are natural mouth cleansers and teeth protectors. Eating apples before sleeping for 10-15 days continuously are highly beneficial in case of worms.

Common name: Apricot

Botanical name: *Prunus armeniaca*

Family: Rosaceae

The fresh apricot is rich in calcium, phosphorus and vitamin C but dried fruits are rich in carbohydrates. Its kernels are edible and have medicinal, curative and cosmetic properties. Its oil also has the soothing properties and acts as an ointment to heel the wounds. Its fresh juice if applied on the skin cures several skin diseases, like eczema, sunburn, scabies and skin itching.

Being rich in iron, it is recommended to involve in the diet of anemic patients to promote hemoglobin levels in their blood. It is rich in cellulose and pectin, thus, it improves the digestion of other foodstuffs. Its fruits are alkaline in nature so help in neutralizing the acids formed during digestion of proteinous foods, like cheese, meat, egg, mushroom, *etc*. and in maintaining alkaline reserve in the body, as our body tissues are also alkaline in nature.

Common name: *Bael*

Botanical name: *Aegle marmelos*

Family: Rutaceae

Its edible portion is packed with vitamin B, C and E. It is also rich in mineral salts, like calcium, phosphorus and iron. It has a fair amount of tannin, which is responsible for its medicinal value. It may be noted that no other fruit has such a high content of riboflavin (1.19 mg/100 g edible pulp) as *bael*. Marmelosin is therapeutically active principle present in *bael* fruit (0.03 to 0.37%) according to locality of cultivation. The pulp of ripe fruit is sweet and may be consumed either directly or its juice may be extracted and consumed. Mature fruits as astringent, digestive and stomachic are prescribed for diarrhea and dysentery. The ripe fruit is tonic, restorative, laxative and good for heart and brain. It tones up the heart, liver and gastro-intestinal tract. It is helpful in various digestive disorders, like colitis, chronic dysentery and constipation. Its fruits are very nutritious and purify the blood, and its pulp squash is a perfect preventive of heat strokes as it has a cooling effect on the body. Its rich iron content increases hemoglobin content in the blood.

Beverage prepared from *bael* fruit is beneficial in healing ulcers due to the presence of mucilage content Intake of *bael* leaves water in the morning is effective against peptic ulcer. *Bael* leaves are rich in tannin and help in the healing of ulcers. The roots of this tree are used for curing ear problems. Medicated oil prepared from bael leaves gives relief from colds and respiratory disorders. The root, leaves and bark are used in treating snakebite.

Common name: Banana

Botanical name: *Musa paradisiacal* Linn.

Family: Musaceae

Banana makes a complete balanced meal when it is mixed together with skimmed milk. It is rich in calcium, phosphorus, iron and vitamin C and contains traces of vitamin B. It is beneficial in converting harmful *Bacilli* normally present in intestine to the useful ones, thus, it cures several disorders of intestines. It neutralizes the acids secreted inside stomach during the digestion of proteinous food. It forms a thin protective layer against the stomach wall. They should be included in the diet of patients suffering from depression as it contains the neurotransmitter serotonin.

Due to high pectin content, it gives relief to the patients suffering from chronic constipation. It also fights against arthritis and gout and gives immediate relief when its paste spreads over the bums and wounds. A decoction prepared from banana, tamarind, and salt is a useful remedy for the treatment of dysentery. Being a rich source of iron, it increases hemoglobin level in the blood, thus, it is good for anemia patients. The unripe bananas should not be consumed as they contain high starch, which is not easily digestible. The starch present in fruits on ripening is changed to sugars, which are easily digestible.

Common name: Cherry

Botanical name: *Prunus avium*

Family: Rosaceae

Cherry is an excellent source of vitamin C (13 to 56 mg/100 g) and potassium. It is good source of thiamine, riboflavin, niacin, vitamin A, calcium, magnesium and dietary fibre. It has more calorific value than apples. Cherry can be beneficial in reducing the pain of headaches, arthritis and gout cherry contains a red pigment named anthocyanin, which acts as an antioxidant and along with vitamin C helps in strengthening of collagen. Cherry is also beneficial in treating nausea, vomiting bronchitis, asthma and chronic diarrhea. Cherries can help in reducing the risk of heart attack and cancer.

Common name: Coconut

Botanical name: *Cocos nucifera*

Family: Palmae

Coconut is a storehouse of several minerals and vitamins, which are considered most essential for maintaining good health. It is packed with carbohydrates, protein, calcium, phosphorus, iron, copper, vitamin B and folic acid and contains high amounts of fat, which is easily digestible, thus, it is used for cooking foodstuffs. It is believed that coconut is useful in curing heartburn, indigestion and gastric problems. It lowers down the acidity and as a result helps in restoring the body from internal injury. If a spoonful of ground coconut is taken with honey gives relief to the patients suffering from dry cough wracking their chest and soothes the throat.

Its drinking water provides strength to the liver and gives relief to the patients suffering from nausea. It is also used as a skin lotion and helps in removing small pox and scabs. It enriches the health by providing the strength and vigour to the body. It has high therapeutic quality. Its oil is used as an ointment for bums, bruises and cuts and its regular use keeps the skin smooth and soft. It checks the untimely falling and graying of hair if rubbed into the scalp regularly with light fingers. It removes dandruff and provides luster to the hair.

Coconut water is a refreshing drink, which prevents from sunstroke and heatstroke and keeps the body cool. It also purifies the blood. Its water contains electrolytes, like sodium chloride and potassium. It is rich in chlorine too. It reduces both internal and external inflammation and produces soothing effects on digestion system. It maintains the electrolyte balance of the body. It is a good substitute for water and acts as an antidote to counteract the side effects of strong allopathic drugs.

Coconut water taken with some lemon juice is an excellent remedy for dehydration caused by vomiting and diarrhea. It is good for patients suffering from renal, cardiac and liver disorders. In south, coconut water is mixed with milk before feeding the infants who may not digest the pure milk. Coconut water is prescribed extensively to the patients of cholera to recover the dehydration, as it is considered superior to injectable potassium.

Common name: Date palm

Botanical name: *Phoenix dactylifera*

Family: Palmaceae

Being a rich source of common sugars, like glucose and fructose, it provides tremendous energy because our body can assimilate it quickly. It is also rich in calcium, phosphorus, iron, minerals and fibers. The fruits of date palm have a number of natural curative properties. It is very good nourishing food especially for growing children particularly decoction prepared from its fruits and its milk is taken in the breakfast. Being of a good laxative nature and because of high fiber content, it removes chronic constipation often the result of dietary food deficient in fiber. Its deficiency can be cured by a daily intake of dates. For this, 2 to 3 fruits should be taken with lukewarm water before sleeping. This provides roughage to the bowel for its smooth functioning. Its fruit juice is a very good tonic, which provides strength to the heart. A decoction prepared from its fruits soaked overnight in water is also said to be very good for several stomach disorders

Common name: Guava

Botanical name: *Psidium guajava*

Family: Myrataceae

Guava is an excellent source of vitamin C (150-400 mg/100 g) and pectin. The fruit is also rich in minerals like phosphorous, calcium, iron and vitamins like niacin, pantothenic acid, thiamine, riboflavin and vitamin A. Being good source of fibre, it is the best guard against constipation and it paves way for easy bowl with complete cleaning. Guavas consumed continuously for a month help in cleaning impurities of blood and keeps away from skin diseases. Guava helps in reducing high blood pressure and increasing good cholesterol level (HDL). Being low in saturated fat and in cholesterol and rich in dietary fiber, it reduces the risk of coronary heart disease and cancer. It cleanses the mouth and strengthens the teeth and gum. It is also helpful in curing toothache and bleeding of teeth. Green guava peel 500 g with 10 g each of phitkari, black pepper and *sendha namak* powder are mixed firmly and if rubbed the teeth and gums with the mixture twice a day the problem of bleeding of teeth and gums can be cured. Drinking guava water (prepared by cutting the guava into pieces and leaving them into water for about three hours) helps in controlling the blood sugar of diabetic person. Guava baked on gas or coal thoroughly and eaten after meals with salt and black pepper is very effective medicine against cold of any type. The flowers of guava are beneficial for treating bronchitis. Guava leaf extract is effective to treat digestive disorders like diarrhea, vomiting and gastrointestinal due to the presence of quercetin and other flavonoids. The bark extract of guava protects the heart and improves its functionality. It is also used to treat acne and painful menses.

Common name: Grape

Botanical name: *Vitis vinifera*

Family: Vitaceae

Its juice contains moderate amounts of vitamin A, B complex and vitamin C. The main mineral found in grape juice is phosphorus but it also contains several other minerals, like potassium, sodium and calcium. Its juice is also rich in natural sugars, *i.e.*, dextrose and fructose. It suppresses the appetite so it is useful for the women who are figure conscious. Its juice is alkaline in nature, thus, it helps in maintaining the alkaline reserve of the body. Berries are useful in anemia, gallstone, stomatitis and cough and have diuretic properties. The acid present in grapes stimulate liver and bile secretion and prevent liver disorders. Experiments show that grapes have properties that may help in preventing heart disease and certain type of cancer due to the presence of anti-carcinogenic compounds in the seeds.

Because of its high water content, it eliminates the routinely accumulated hardened substances that may have settled in any part of the body. Its juice cures low blood pressure and other blood disorders and improves circulation of blood in the body. Naturopaths recommend it to cure arthritis, cancer, hyper-acidity, jaundice and renal stones. Its regular use in daily diet maintains the youth and slows down the aging process. However, the grape skin adds bulk to the diet but it can cause colic and ulcers. Diabetic patients should not consume its juice.

Common name: Jamun

Botanical name: Syzgium cumini

Family: Myrataceae

The *jamun* fruit is a good source of iron, minerals, protein and sugars. Glucose and fructose are the principal sugars in the ripe fruit and not even a trace of sucrose is detected. The seed is also rich in protein and carbohydrates. *Jamun* fruits, known for their acrid and astringent taste, are useful for curing diarrhea and diabetes. The fruit syrup is very useful for curing diarrhea. It is stomachic, carminative and diuretic, apart from having cooling and digestive properties. *Jamun* seeds are also known for their medicinal properties being high in phenolics. They are prescribed in diabetes, diarrhea and dysentery. Studies have shown that *jamun* markedly lowers blood pressure and it is also used as a lotion for the cure of ringworms. The seeds contain a glucose called jamboline that checks the conversion of starch into sugar in cases of increased glucose production. *Jamun* bark being good astringent is used for gargles and mouth washes. The leaves are antibacterial in nature and used for strengthening the teeth and gums. The ripe *jamun* fruits are widely eaten in India and also used for preparation of different products like ready to serve beverage, squash, syrup and jam and wine. *Jamun* is also dried with salt and preserved as *churan*. Eating a few *jamuns* every morning with either salt or honey is also effective in curing bleeding piles. The ash from burnt leaves of *jamun* is good for gums and teeth. The paste of *jamun* leaves is beneficial to dry the pus-filled wounds. The extract of soft mango and *jamun* leaves mixed in honey two times a day gives relief from vomiting and burning sensation.

Common name: Lemon

Botanical name: *Citrus limon*

Family: Rutaceae

Lemon is a storehouse of several minerals and vitamins particularly vitamin C. Being of a cool nature, lemon protects the body from heat strokes. Besides making the body system immune, it improves the general health. As a mild laxative and a good purifier, it promotes the secretion of bile and is used to prevent the formation of gall stones and acid deposits. Its high potassium content is good for the patients suffering from heart trouble. Lemon juice being full of antioxidants also help to protect our bodies from heart disease, by attacking any free radical cells in our body caused by pollution residue, cigarettes and the suns radiation. In fact, its juice provides detoxifying properties. The toxins, which routinely accumulate in the body, may be eliminated with the intake of lemon juice in a glass of lukewarm water.

The intake of its juice with empty stomach is also recommended to eliminate poison from the body through urine, and honey may be added to make it more beneficial. Its juice is very effective in destroying the bacteria from the intestinal tract. It is also effective in curing coughs and colds, if it is taken daily three or four times with little garlic juice. It provides relief from constipation and body pain if lemon juice is taken with warm water every day. For reducing body weight, lemon juice is mixed with honey in equal proportion and taken in the morning. Gargling for few minutes with lemon juice mixed with water cures mouth ulcer and throat infection.

It controls nausea, vomiting and motion sickness, if it is taken while traveling over high and zigzag mountains. With little ginger juice, it can cure chronic constipation, hyperactive acidity, skin diseases and restore the lost appetite. Its juice is a very good appetizer. It is useful in destroying intestinal worms and expelling excessive gases from intestinal tract. Regular use of lemon juice can also improve vision and face glossiness.

Common name: Litchi

Botanical name: *Litchi chinensis*

Family: Sapindaceae

Lichi fruits are a rich source of vitamin C (40-90 mg/100 g). It is good source of calcium, phosphorous and vitamin B. Litchi helps the body to digest food properly to improve health. According to studies, litchi fruit prevents the growth of cancer cells, particularly breast cancer. Eating litchi can be beneficial for colds, fever, sore throats, digestive complaints and even ulcers. Litchi is a good appetizer, which relieves body ache and improves memory power.

Common name: Mango

Botanical name: *Mangifera indica* Linn.

Family: Anacardiaceae

It is packed with large quantities of calcium, phosphorus, iron and lot of vitamin C. It also contains proteins, fats, minerals, sugars and fibers. Raw mango fruits are rich in vitamin C and can be consumed in different forms such as pickles, sauces and jam. Ripe mango contains vitamin A, thus, it is considered very beneficial in all kinds of eye ailment. If its fruits are consumed raw with a little salt quench thirst and prevent loss of body ions, which are lost during excessive sweating in summers. A decoction prepared from raw mangoes with little salt and honey gives an effective relief in constipation, indigestion and piles. Its use develops resistance in the body against tuberculosis and provides relief in dysentery, cholera and anemia. Its powder generally prepared from immature dried mangoes is a useful remedy for the treatment of scurvy disease. Powdered mango seed along with *juman* seeds are useful in diabetes and diarrhea. Fine mango seed powder is also helpful in curing pellagra and other tooth problems. Similarly, powdered mango leaves are known for the treatment of bleeding piles and stones. Raw mango seed powder if taken twice a day helps in killing worms.

It contains certain body essential nutrients and sugars so if it is taken after blending with milk and honey, it increases the body weight, appetite and memory. However, if it is consumed in excess, it may cause throat irritation, dysentery and colic. Its regular use cures irritability, hearing troubles, arthritis, hepatitis, haemoptosis and nervous disorders, insomnia and loss of smell. It gives strength to the heart, liver and intestinal tract and assists in the multiplication of red blood corpuscles. It also promotes vitality in both men and women. Naturopaths believe that raw mango is a good blood purifier and a perfect preventive against heat strokes, as it produces cooling effect on the body. A good refreshing drink can be prepared from pulp extracted from roasted raw mango, spices and jaggery.

Common name: Orange

Botanical name: *Citrus aurantium*

Family: Rutaceae

Its juice is an excellent source of vitamin C. Besides boosting up resistance to coughs and colds, vitamin C promotes absorption of iron in the body. It contains high amount of beta-carotene, which is a precursor of vitamin A. This valuable vitamin protects against blindness due to nutritional deficiency.

It provides fiber when fruits eaten completely with its thin white peel which adds roughages to diet and eliminates constipation. It also lowers the cholesterol level in the body. Being rich in electrolytes, it helps in maintaining osmotic balance of the body. Its juice is immensely refreshing and appetizing. It stimulates the secretion of gastric juice in the stomach. Its oil is used extensively in pharmaceutical and food processing units.

Common name: Papaya

Botanical name: *Carica papaya*

Family: Caricaceae

Papaya is an invaluable foodstuff as a health tonic. Although it is low in calories but contains ample amounts of vitamin C, proteins, fibers, carbohydrates and minerals, like calcium and phosphorus. It is well suited for obese people because it has very low calorie content (32 k cal per 100 g ripe fruit), and this helps in assuaging hunger while also reducing weight. Papaya is also a very good source of vitamin C (57 mg per 100 g). Beta-carotene is also present in good quantity and it helps in preventing damage caused by free radicals, which might otherwise lead to several diseases including cancer.

The papaya plant has tremendous medicinal values, and it is considered that nearly every inch of the tree possesses medicinal properties. It contains different types of enzyme of which papain is the most important. Papain is similar to the enzyme pepsin found in human stomach. Papain, which improves digestion by assimilating the other foodstuffs, is an excellent digestive enzyme and is useful in digestive disorders. This enzyme increases the release of gastric juice, overcomes the problem of unhealthy mucus in stomach and cures the intestinal disorders. This proteolytic enzyme is present in the latex and it is collected mainly from the green fruit.

Papain has many uses in the beverage, food and pharmaceutical industries. It is used in chill-proofing beer, tenderizing meat and also in the preparation of digestive enzymes and drugs. It is also used in bathing hides, degumming silk and softening wool. Papain also exhibits pain relieving properties, and the Food and Drug Administration (FDA) has approved its medicinal properties to ease the discomfort of slipped discs. Patients with celiac disease, who cannot digest heat proteins, benefit from treatment with crude papain.

Papaya fibre is able to bind to cancer causing toxins in the colon and keeps them away from the healthy colon cells. In addition, papaya's folate, vitamin C, β-carotene and vitamin E have each been associated with a reduced risk of colon cancer. Papaya fibre has also been shown to lower high cholesterol levels. The cooked unripe fruit aids in nursing mothers, while ripe papaya is said to enhance virility. The skin of a ripe papaya immersed in vinegar helps in controlling spleen enlargement. Papaya also aids in anti- aging process.

Carpaine, an alkaloid present in papaya leaves, can be used as a heart-rate depressant, amoebicide and diuretic. In traditional practice, papaya leaves are used in relieving muscle pain and curing festering wounds and ringworm infection. The root has abortificient and diuretic properties. Potassium present in papaya helps in the maintenance of balance between fluids inside and outside the human body. The fruit is both a laxative and a diuretic. The ripe fruit is also beneficial for those suffering from bleeding piles and dyspepsia. The dried and salted fruit apparently has beneficial effects on the enlarged liver and spleen. Papaya seeds have strong antifungal properties. The crude juice extracted from black seeds is administered in case of bleeding piles and enlarged liver. Besides, the use of ripe fruit cures the constipation and chronic diarrhea. Raw papaya juice cures the swellings and wounds infected by pus. It is very effective in healing pimples and freckles if it is applied on the face.

The raw papaya juice is also considered to have anthelminthic properties and kills parasitic worms, like round worms when a tablespoon of juice is mixed with equal amount of honey and is taken with some lukewarm water. In place of honey, caster oil may also be used. Small babies should be given half of the adult dose.

Common name: Pear

Botanical name: *Pyrus communis*

Family: Rosaceae

Pears are fairly good source of fibre. They contain potassium, calcium, phosphorous, copper and iron. Citric acids and malic acids are the principal organic acids present in pears. The pear contains a greater amount of insoluble fibre rich in lignin, which is effective laxative. The pear is also a diuretic, remineralizing, a mild astringent and refreshing. Pear also has capacity to lower down the blood pressure. The higher intake of pear lowers the risk of hypertension due to the presence of potassium in it.

Pear is recommended in cases of renal failure. Pear helps in maintaining a desirable acid-base balance in human body. Pear consumption promotes the elimination of uric acid and other nitrogenated substances through the urine. They have an alkalizing effect on the blood. It can be consumed by diabetic patients as it contains low sugar content. Pear is also beneficial in weight loss and works against the intestinal putrefaction and flatulence in case of colitis and intestinal dyspepsia.

Common name: Pineapple

Botanical name: *Ananas comosus*

Family: Bromeliaceae

Pineapple is fairly rich in calcium, phosphorous, iron and vitamin C (10-25 mg/100 g). It is good source of vitamin A and B. It is also a good source of fibre (0.30-0.61%). Pineapple contains a proteolytic enzyme bromelian, which digests food by breaking down protein. Bromelin helps in the digestion of proteins, stops blood clot formation and has antiviral properties. Eating pineapple after meals reduces gas, blotting, nausea and constipation. Pineapple has been found to be beneficial in treating cough and cold. Pineapple being rich in manganese helps in building bones in young people and strengthening of bones in older people. Pineapple juice also helps in killing intestinal worms. Pineapple is also good for heart patients, as it lowers cholesterol and blood pressure. It is effective in treating goiter. Due to good source of vitamin C, pineapple helps in wound healing and combating memory loss and depression. Pineapple stem is beneficial in case of certain types of cancer such as ovarian, breast, lung and colon cancer.

Common name: Pomegranate

Botanical name: *Punica granatum*

Family: Punicaceae

Its seeds along with the fleshy portion are rich in minerals, like calcium, phosphorus, iron, *etc.* and vitamin C and vitamin B complex. In spite of sweet taste, it is low calorie food. Its 100 g edible seeds provide about 65 calories so figure conscious persons may safely include this fruit in their daily diet. Pomegranate is digestible, carminative, enhances semen formation, activates memory, improves formation of hemoglobin and is a very good blood purifier and unripe fruits and flowers are beneficial in inducing vomiting and the pomegranate fruit is used to cure diarrheoa and dysentery. Pomegranate juice is mainly used as a health drink. It increases the body resistance against infection particularly tuber culosis. Pomegranate is used to cure liver ailments, anemia, sore throat, sore eyes, stomatitis (inflammation of mouth), fever, spleen complaints, chest troubles, earache, brain and heart diseases.

The bark of the stem and root is used for controlling dysentery, diarrhoea and killing of tapeworms. The dried bark prevents gum disease, if applied on teeth. The powder of seeds helps to cure the gall bladder and kidney stones. A paste of the leaves when massaged into the scalp is reported to reverse baldness. The arils from sour varieties of pomegranate are dried as *anardana*, a condiment commonly used as acidulant in Indian cooking. The fruit rind and seeds are also used for preparing a cosmetic cream for oily skin.

The seed oil has skin moisturizing properties, which makes the skin bright, properly hydrated, elastic and silky smooth. The seed oil being strong antioxidant in nature protects the skin from free radicals and reduces the visible signs of aging. Pomegranate extract has skin whitening property caused by inhibition of proliferation of melanocytes and melanium synthesis. The intake of powdered flower buds with water 2 to 3 times daily improves memory. The flowers are useful in nose bleeding, ulcers, sore eyes and sore throat. Pomegranate also protects the heart and reduces the risk of arteriosclerosis and cancer. Pomegranate may be beneficial in preventing prostate cancer or slow its growth. Drinking pomegranate juice during pregnancy may help in reducing the risk of brain injuries in babies. A squash prepared from its juice flavoured with saffron can cure the ailments, like gastric disorders and asthmatic problems.

Common name: Strawberry

Botanical name: *Fragaria chiloensis*

Family: Rosaceae

Strawberries are excellent source of vitamin C (89 mg/100 g) and potassium (161 mg/100 g). In fact, they have more vitamin C than citrus fruit. It contains good amount of phosphorous, calcium and magnesium. Strawberry is also a good source of vitamin A, thiamine, riboflavin and nicotinic acid and dietary fibre, vitamin E, vitamin A and vitamin K. Strawberries have an ability to lower the risk of gastrointestinal track cancers. Strawberry is an excellent antioxidant and good source of flavonoids, which protect artery walls from damage by cholesterol. Strawberry reduces the risk of heart disease by decreasing blood pressure. Strawberries are found to be beneficial in enhancing memory.

■■■

11 Therapeutic Significance of Minor Fruits

G.S. Unnikrishnan Nair

Assistant Director of Agriculture, Farm Information Bureau, Department of Agriculture, Government of Kerala, Kowdiar, P.O. Thiruvananthapuram, E-mail: <vanchiyurunni@gmail.com>

Minor fruits are the natures' blessing to the mankind but the man hardly realizes the significance of these underexploited fruit species. Due to this, these fruit trees and shrubs have been neglected by the man and many of these plants are facing extinction. However, the ancient system of medicine like *Ayurveda* has recognized the importance of these fruits in human life and included them in many medicinal formulations. Modern studies are also revealing the properties of these fruits of miracle nature. Another important aspect is that these fruit plants mostly grow naturally in the nature, and hence, their fruits are fully organic. The therapeutic significance of some important minor fruits is discussed below:

Common name: West Indian Cherry

Botanical name: *Malpighia punicifolia*

Family: Malpighiaceae

One of the richest known sources of vitamin C is the West Indian cherry fruit. It is grown mostly as an ornamental shrub and people are rarely aware of the nutritional qualities. West Indian cherry is a medium-sized tropical evergreen fruit-bearing shrub belonging to the family Malpighiaceae. It is native to the West Indies and North and South America. West Indian cherry grows up to 3 meters, with dense crowns of leaves that are ovate.

The flowers are formed in umbels (a cluster of flowers with stalks of nearly equal length, which spring from the same point, like the ribs of an umbrella). Each flower is small having diameter 1-1.5 cm with five pink petals. The ripe fruit is bright, containing seeds. It has a sweet and sour taste. West Indian cherry can be propagated using seedlings and by air layering. Pruning once a year prior to monsoon and application of organic manure is recommended. Layers bear fruit in 6 months, while seedlings bear fruit after 2 years of planting. Fruiting is noticed from August to November and April to May.

The juice of the fruit is used as a nutritional supplement in baby foods and in canning and processing industries. With added pectin, excellent jelly, jam and other preserves can also be prepared.

The vitamin C in the fruit acts as an antioxidant. According to the study conducted at Michigan State University, West Indian cherry contains powerful anthocyanins and flavanoids, which can inhibit colon cancer and heart diseases.

Common name: Star Gooseberry

Botanical name: *Phyllanthus acidus*

Family: Euphorbiaceae

Star gooseberry (Otaheite gooseberry or Tahitian gooseberry) is a curious deciduous fruit tree that was once frequently found in households of South India. It belongs to the family *Euphorbiaceae* and is locally known by the names arinelli, seemanelli and pulinelli. Native to Madagascar, it was spread long ago to India, Southeast Asia and Pacific islands.

The Star gooseberry prefers tropical climate and is grown from seeds and by air layering. It grows 6-9 m in height with spreading branches. The branchlets bear alternate leaves that are ovate or lanceolate in form with short petioles and pointed ends. Leaves are 2-5 cm long, green and smooth on the upper surface, and blue-green on the underside. There are two tiny pointed stipules at base of the each leaf.

Small rosy flowers are male, female and some hermaphrodite. They are borne collectively at leafless parts of the main branches in little clusters. The tree often bears twice a year in South India, first during April-May and second during August- September. The fruits are numerous and oblate with 6 to 8 ribs. They develop very densely to form spectacular masses giving the tree an ornamental appearance. Fruits are pale yellow, waxy, crisp, juicy and sour. There is only one seed in each fruit.

The fruit is eaten raw and pickled in Kerala. In Indonesia, Philippines, Surinam and Bahamas, the fruit flesh is used for flavoring dishes and for making cold drinks, syrup and vinegar. The fruit pulp cooked in sugar yields ruby-red coloured jelly. The young leaves are cooked and eaten in Indonesia. The wood is lightly brown, fine-grained, attractive, moderately hard, strong, tough and long lasting if seasoned.

According to the report published in *Molecular Pharmacology*, a combined study conducted by University of Lisboa, University of Regensburg and Prince of Songkla University proved that an extract from the *Phyllanthus acidus* and its isolated compounds induce airway chloride secretion. A potential treatment for cystic fibrosis (hereditary disorder causes the body to secrete abnormally thick and sticky mucus that clogs the pancreas and the lungs, leading to the problems with breathing and digestion.

In India, Star gooseberry fruits are taken to enrich the blood and as liver tonic. Fruit syrup is prescribed as a stomachic and the seeds promote bowel movement. Because of the mucilaginous nature of leaves, they are taken as a demulcent (having soothing effect on the skin) in cases of gonorrhea. Root poultice is applied to cure psoriasis on the soles of feet.

Common name: Durian

Botanical name: *Durio zibethinus*

Family: Malvaceae

Durian is considered as the King of Tropical Fruits in countries like Malaysia and Singapore. According to them, durian fruit warms up the body. The Japanese believe durian to have aphrodisiac qualities.

Durian means thorny in Indonesian dialect. Thailand and South Vietnam are chief producers. Durian fruits are ovoid-oblong, brownish green, covered with sharp spines and weighs 2.5 to 4 kg. Handling the fruits without gloves can be painful. Inside the fruit, there are five compartments containing the creamy-white flesh and 1 to 7 chestnut-like seeds. The flesh is mostly eaten fresh but is best chilled. The taste is comparable to French custard but flesh of durian has a very nasty odour as that of dirty socks. In Singapore, it is forbidden to carry durian in public transport. Durian fruit is used to flavour a wide variety of dishes like traditional Malay candy- dodol, milk shakes, ice cream, custard, moon cakes and cappuccino. In Java, the sliced seeds are cooked with sugar as a confectionery. In Indonesia, fermented durian is covered in palm leaves and served. The fruit is mixed with rice and sugar to make a dessert called *lempog*. Durian flesh is canned in syrup for export.

Durian fruit contains a high amount of sugar, vitamin C, potassium and the amino acid tryptophan. It is a good source of carbohydrates, proteins and fats and has no cholesterol. The odour of the flesh is believed to be due to indole compounds, which are antibacterial. Eating durian is believed to clear the lungs and breathe pathways. In Malaya, decoctions of leaves and fruits are applied to swellings and skin diseases. Eating durian is assumed to restore the health of sick persons.

Durian is an evergreen tree and grows 80-100 feet in height. There are over 300 named varieties of durian in Thailand, most valued one being *Monthong Durian*. Seedling trees generally come into bearing 7 to 12 years after planting, while grafted trees bear in 4 years. Durian generally blooms in March and April. The fruits mature in July and August. A tree yields about 100 fruits in a year. Malaysia is the largest exporter of fresh durian. Fresh durians are usually shipped to nearby countries such as Singapore, Hong Kong and Taiwan where it is in high demand. Frozen durian from Thailand is shipped to the United States, Australia and Canada.

The latest news is that DKT Indonesia, an anti-AIDS organization sold 150,000 durian-flavoured condoms as part of a campaign to fight AIDS. The group is targeting sales of 600,000 durian-flavoured condoms in a year. Though it has a pungent smell, it remains the king of fruits for people in some parts of the world.

Common name: Sour sop

Botanical name: *Annona muricata*

Family: Annonaceae

The Sour sop is usually processed into ice creams, sherbets and drinks but fibers-free varieties are often eaten raw. According to latest research conducted at Catholic University of South Korea, Sour sop also called Graviola fruit is a miraculous natural cancer cell killer. Sour sop was shown to target the cancer cells selectively, leaving healthy cells untouched. Sour sop is a broad-spectrum antimicrobial agent for both bacterial and fungal infections, is effective against internal parasites and worms, lowers high blood pressure and is used for depression, stress and nervous disorders.

A study at Purdue University recently found that leaves of Sour sop tree killed cancer cells among six human cell lines and were especially effective against prostate, pancreatic and lung cancers. Much of the cancer research on graviola focuses on a novel set of phytochemicals called *Annonaceous acetogenins*. Graviola produces these natural compounds in its leaf and stem, bark and fruit seeds. Three separate research groups have isolated these acetogenin compounds in graviola, which have demonstrated significant antitumorous and anticancerous properties and selective toxicity against various types of cancer cells, without harming healthy cells. Purdue University has filed nine United States and international patents on their work around the antitumorous and insecticidal properties and uses of these acetogenins.

All parts of the graviola tree are used in natural medicine in the tropics. Fruit and fruit juice are taken for worms and parasites to cool fevers and as a lactagogue. The crushed seeds are used as a vermifuge and anthelmintic against internal and external parasites, head lice and worms. Bark, leaves and roots are considered sedative, antispasmodic, hypotensive and nervine. Graviola has a long, rich history of indigenous medicinal use. Leaf tea is used for catarrh (inflammation of mucous membranes) and the crushed seed is used to kill parasites. Bark, roots and leaves are also used for diabetes and as a sedative and antispasmodic. Leaf and bark tea is used as a sedative and heart tonic. In the Brazilian Amazon, a leaf tea is used for liver problems. In Jamaica, the fruit and fruit juice is used for fevers, parasites and diarrhea and as a lactagogue.

Common name: Passion Fruit

Botanical name: *Passiflora edulis*

Family: Passifloraceae

Passion fruit is a vigorous vine, found especially in the tropics. The Catholic Bishops who believed that the peculiar shape of its flower is synonymous to the crucifixion of Christ named it.

The corona threads of the passion flower is seen as a symbol of the crown of thorns, the five stamens for wounds, the five petals and five sepals as the 10 apostles (excluding Judas and Peter) and the three stigmas for the nails on the cross. The Hindus of South India consider the stamens of the Passionflower as a symbol of the *pancha pandavas*.

Propagated using vine cuttings, it can grow over 15 feet in a year. The tendrils need support for climbing. There are two types of passion vines, *i.e.*, yellow-fruited and purple-fruited. While both passion fruits possess an oval shape and contain a yellowish flesh, the yellow passion fruit is larger and contains more fruit juice. However, the purple passion fruit is sweeter in taste than the yellow fruit and has a sweet odour. Profuse fruiting is seen during the months of September to November and May to June.

The fruit can be eaten and is often placed as a topping for ice cream, custards and fruit salads. The undiluted juice is an excellent additive to other fruit juices, or it may be drunk as such with water and sugar added to it. Passion fruit can also be used to make jellies and jams. Passion fruit is also used as an ingredient in ice creams, candies, fruit juice concentrates and alcoholic beverages, such as cordials.

Passion fruits provide a good source of fibers, are low in sodium and contain no cholesterol. They also provide a high amount of the vitamins A and C and offer an excellent source of potassium, calcium and iron.

The fruit juice and the leaves of passion fruit contain alkaloids, including Harmine, Harmalol and Harmol, which has blood pressure lowering, sedative and antispasmodic action. It is used in the treatment of insomnia and depression. Passion fruit juice is used for urinary infections and as a mild diuretic.

Common name: Bilimbi

Botanical name: *Averrhoa bilimbi*

Family: Oxalidaceae

Native to Indonesia, Bilimbi is a tree, which grows about 5-10 m height and is propagated using seeds. The bilimbi prefers a tropical environment and grows well in backyards of South India. The most common use for the fruits is as flavouring for fish and meat dishes. Fruits are also used for beverages and preserves. Half-ripe fruits are salted, set out in the sun and kept for longer use. Squash, wine and pickle are made using bilimbi fruit.

The fruit is a rich source of vitamin C. It fights cholesterol and is used as a tonic and laxative. Syrup made from the fruit is used to cure illness arising from jaundice. The fruit is also known to stop internal bleeding in the stomach. In the Philippines, the leaf paste is applied on itches, swellings, rheumatism and on skin eruptions. Leaf infusion is a remedy for coughs and is taken after childbirth as a tonic. The fruit is given to children as a protection against coughs. Flower infusion is said to be effective against coughs. In Java, the fruits combined with pepper are eaten to cause sweating during fever. The fruit conserve is administered as a treatment for coughs, beriberi and biliousness. Syrup prepared from the fruit is taken as a cure for fever and inflammation, to stop rectal bleeding and alleviate internal hemorrhoids. The fruits and leaves are applied on bites of poisonous insect. The juice is useful for bleaching stains from white clothe and also tarnish from brass.

Common name: Cashew Apple

Botanical name: *Anacardium occidentale*

Family: Anacardiaceae

Cashew apple (Cashew Tree-*entalis*) after separating cashew nut is usually wasted but the fact is that it contains 5 times vitamin C as compared to a lemon fruit. Cashew juice, syrup, candy, jam, vinegar, *etc.* are made using this. Cashew apple is a digestive and aids in control of intestinal worms, diarrhea, body pain and scanty urination.

Cashew apple juice without removal of tannin is prescribed as a remedy for sore throat and chronic dysentery in Cuba and Brazil. In 16th century Brazil, cashew fruits and their juice were taken by Europeans to treat fever, sweeten breath and for stomach. The Tikuna tribe in northwest Amazonia considers the fruit juice medicinal against influenza. A wine made from the fruit is used for dysentery in other parts of the Amazon rainforest. In 1999, researchers reported that cashew fruit exhibited antibacterial activity against the Gram-negative bacterium *Helicobacter pylori*, which is now considered to cause acute gastritis and stomach ulcers. Its effectiveness against ulcers also was documented in two clinical studies. Because of its high amount of vitamin C and mineral salts, cashew fruit is used to remineralize the skin and as a catalyst in the treatment of premature ageing of skin. It is also an effective scalp conditioner and tonic and is often used in shampoos, lotions and scalp creams for conditioning activity of its proteins and mucilage.

Common name: Mangosteen

Botanical name: *Garcinia mangostana*

Family: Clusiaceae

The Mangosteen believed to have originated in the Sunda Islands and the Moluccas is a tropical evergreen tree. The tree grows from 7 to 25 meters tall. This exotic edible Mangosteen fruit is deep reddish purple when ripe. Mangosteen is known as the *Queen of Fruits*. The Mangosteen tree requires a warm and very humid equatorial climate all the year round. The pericarp of Mangosteen is a source of a class of naturally occurring *polyphenolic* compounds known as xanthones and other bioactive substances.

People in Indonesia, Malaysia, Sri Lanka, Philippines and Thailand use pericarp (peel, rind, hull or ripe) of Mangosteen as a traditional medicine for the treatment of abdominal pain, diarrhea, dysentery, infected wound, suppuration and chronic ulcer. These unique antioxidants, known as Xanthones, have properties, which help to heal cells damaged by free radicals, slow aging and ward off degenerative diseases and physical and mental deterioration. The latest in scientific research shows that Xanthones provide beneficial effects on cardiovascular diseases, including ischemic heart disease, atherosclerosis, hypertension and thrombosis. Xanthones demonstrate particularly powerful antioxidant properties. Mangosteen fruit shows vasorelaxant properties, causing dilation of blood vessels and increased blood flow. Xanthones are among the most powerful antioxidants to be found in nature.

According to research reported in professional journals such as *Free Radical Research* and the *Journal of Pharmacology*, these amazing Xanthones have a remarkably beneficial effect on cardiovascular health. They are also naturally antibiotic, antiviral and anti-inflammatory, anti-fatigue, anti-obesity, anti-depression, anti-anxiety, anti-vertigo (dizziness), anti-Alzheimer's, anti-Parkinsonism (Parkinson's Disease), anti-allergenic, anti-seborrheic (skin disease), anti-glaucoma (eye disease), anti-pyretic (anti-fever), anti-diarrhea, anti-periodontic (gum disease), anti-pain (such as dental pain), anti-neuralgia (nerve pain), anti-arthritis, anti-inflammatory and anti-ulcer. Mangosteen has shown inhibitory action against these harmful bacteria organisms. It is highly effective in boosting weak immune systems.

Common name: Rambutan

Botanical name: *Nephelium lappaceum*

Family: Sapindaceae

Commonly known as rambutan, it is botanically known as *Nephelium lappaceum*. The rambutan is a tropical tree, which grows to about 10 to 20 m in height. Rambutan is widely distributed throughout Indonesia, Malaysia and Southeast Asia.

The fruits of rambutan are rich in fat, calcium, iron, protein, nitrogen, ash, sodium, zinc, magnesium, manganese, potassium, phosphorus, vitamin C, vitamin A, thiamine, riboflavin and fibers. It has been determined in studies that vitamin C is a key antioxidant, which protects cells from free radicals and helps with the absorption of iron. It is also a source of copper, which is needed for the production of healthy red and white blood cells, and along with manganese, it manufactures and activates enzymes in the body.

A study released by the University of Malaysia named Rambutan as a fruit that had both antiseptic qualities and an antimicrobial food. It means that the fruit is useful in fighting infections both as a topical ointment produced from the juice but also good for fighting bacteria inside the body as well. The fruit is astringent, stomachic, acts as a vermifuge, febrifuge and is taken to relieve diarrhea and dysentery. The leaves are poultice on the temples to alleviate headache. In Malaya, the dried fruit rind is sold in drugstores and employed in local medicine.

Common name: Custard Apple

Botanical name: *Annona squamosa*

Family: Annonaceae

The custard apple plant is an evergreen shrub and reaches a maximum of 6 m height. Since it tastes like custard, the fruit is known as custard apple, while fruit pulp contains many black shiny seeds in it. It is commonly found in deciduous forests and also cultivated in many parts of India. Pulp of the fruit is eaten fresh or converted into juice or shake. Fruits are normally eaten fresh. It is known for various medicinal properties too.

Custard apples are a well-balanced food having protein, fibre, mincrals, vitamins, energy and little fat. They are an excellent source of vitamin C, a good source of dietary fibre, a useful source of vitamin B_6 with some vitamin B_2. The presence of riboflavin and vitamin C content in custard apple is suggested to help in fighting free-radical damage caused to the cells. These can help in maintaining good eyesight.

Custard apple contains reasonably high amount of magnesium and potassium. Magnesium can relax the smooth muscles of the heart and soothe spasms or cramps, thereby preventing heart attack and strokes. On the other hand, the rich content of potassium may regulate blood pressure, thereby maintaining cardiovascular function. In addition, the presence of magnesium can help in smooth muscle relaxation of the alimentary canal. This might help in treating the problems of dysentery and diarrhea. Experts suggest that high in magnesium content, custard apple may regulate water balance. This might help in removing acids from the joints, thereby alleviating symptoms of arthritis and rheumatism. The presence of potassium can also help in fighting muscle weakness.

Being rich in complex carbohydrate, custard apple is a rich source of energy. This can help in fighting fatigue and weakness. Custard apple is considered a natural source of dietary fiber. This can help in improving digestion by making stools softer and aiding in peristalsis.

The fruit also serves as an expectorant, stimulant, coolant, haematinic (haematic an agent that stimulates the production of red blood cells or increases the amount of haemoglobin in the blood) and is useful in treating anemia. The decoction made out the leaves of this plant serves as vermifuge. Paste of flesh of this plant or crushed leaves of the plant can be used for local application on ulcers, abscesses and boils. The fruit of the plant, in its unripe form, can be dried, crushed and used for treating diarrhea and dysentery. Decoction of the plant bark could be used in the form of tonic for dysentery and diarrhea remedy. The root bark of the tree is used for relief from toothache. The decoction extracted from the root serve as febrifuge. The seeds of the plant have insecticidal and abortifacient properties.

Common name: Atemoya

Botanical name: *Annona atemoya*

Family: Annonaceae

The atemoya is a medium sized tropical fruit, which is produced from the atemoya tree. It is a cross between the tropical fruits cherimoya and custard apple, the latter of which the atemoya resembles in appearance. The compound fruit, 8-16 cm in diameter, may be symmetrically heart-shaped, lopsided, irregular, nearly round, or oblate with a deep or shallow depression at the base. The skin, thin but tough, may be yellow or brownish when ripe with a pink, reddish, or brownish-red blush, and faintly, moderately or distinctly reticulated. There is a thick, creamy white layer of custard like, somewhat granular, flesh beneath the skin surrounding the moderately juicy segments, in many of which, there is a single hard dark-brown or black glossy seed. Atemoya fruit has a sweet taste, which is similar to a *fruity vanilla*. Because of this, the flesh of the atemoya fruit is typically used as an ingredient in desserts, including ice creams. Atemoya fruit is popular when eaten raw with its flesh scooped out, then chilled and served.

Though the atemoya fruit has an appealing, sweet taste, it can also provide a healthy addition to the daily diet. This is because atemoya fruits are low in fat and contain no cholesterol. Additionally, they provide an excellent source of potassium and vitamin C. The fruit is an outstanding source of many health-benefiting groups of vitamins such as niacin, riboflavin, pantothenic acid, pyridoxine and thiamine.

Common name: Jamun

Botanical name: *Syzygium cumini*

Family: Myrtaceae

The Jamun or jambolan (*Syzygium cumini*) is indigenous to India, Burma, Ceylon and the Andaman Islands and was once common throughout Kerala. It is an evergreen tropical tree 50 to 100 feet tall with oblong opposite turpentine-scented shiny leaves. Propagated from seeds, Jambolan starts bearing fruit in 4-7 years. It has scented white flowers in branched clusters at stem tips and purplish-black oval edible berries. The berry has only one seed. The fruiting season is available from March to mid-June. The fruits appear in clusters of 10 to 40 numbers and are oblong and ½ to 2 inches long. Usually, the fruits turn from green to light magenta, and then, dark-purple or nearly black when they ripen. Having a sweet, sub-acid flavour and least astringency, the fruits can be eaten raw. Hundred grams of edible fruit portion contain 48 IU of carotene, 18 mg vitamin C, 55 mg potassium, 15 mg each of phosphorus and calcium and 26 mg of sodium.

Sauces, jam, syrup and squash are made from the fruit. In Goa and the Philippines, jambolans are a chief source of wine. In the Philippines and Suriname, the wine made from the fermented fruit bark, leaves and flowers is used against diabetes. Jambolan vinegar is clear purple, with a pleasing aroma and gentle flavour.

In *Ayurvedic* texts, the fruit is stated to be astringent, stomachic, carminative, antiscorbutic and diuretic. The fruit is used against diarrhea and abdominal pains. Cooked to a thick jam, it is eaten to relieve acute diarrhea and dysentery. The juice of the ripe fruits is administered in India in cases of enlargement of the spleen, chronic diarrhea and urine retention. Eating jamun fruit in empty stomach is believed to be a cure for piles.

Extracts of fruits and seeds, especially the seeds, in liquid or powdered form, are given orally to patients with diabetes mellitus or glycosuiria. Three grams of seed powder 4 to 5 times a day is recommended. In most cases, the blood sugar level is quickly reduced without any side effects. In Suriname's traditional medicine, the leaves are used against diarrhea and fever. Tender leaf juice of jambolan is used as a cure for bleeding gums and bleeding piles. Leaves are also helpful as poultices on skin diseases.

Jambolan tree is of importance in Apiculture also. The flowers have abundant nectar, thus, they are visited by Rock bees. The essential oil distilled from the leaves is used to perfume soap and is blended with other materials in making perfume. The wood is commonly used for beams and rafters, posts, bridges, boats, oars, agricultural implements, carts, solid cartwheels, *etc.*

Common name: Canthium

Botanical name: *Canthium parviflorum*

Family: Rubiaceae

A small thorny shrub grows up to 3 meters in height. Leaves are simple, opposite, small, acute, with axillary spines, and flowers are white, small in axillary cymes. Fruits oblong to chambered drupe become yellow when ripe. *Canthium parviflorum* plant pacifies vitiated cough, diarrhea, fever, leucorrhoea, worm infestation and general debility

Common name: Java Apple

Botanical name: *Syzygium javanica*

Family: Myrtaceae

The tree, 5-15 m tall, has a short trunk 25-30 cm thick and open wide spreading crown, and pinkish-gray flaking bark. The wood is red, coarse, hard and used for constructing huts in the Andaman and Nicobar Islands.

The flowers are astringent and used in Taiwan to treat fever and halt diarrhea. Investigators have found their principal constituent to be tannin. They also contain desmethoxymatteucinol, 5-*O*-methyl-4'-desmethoxymatteucinol, oleanic acid and β-sitosterol. They show weak antibiotic action against *Staphylococcus aureus*, *Mycobacterium smegmati,* and *Candida albicans.*

Common name: Pomelo

Botanical name: *Citrus grandis*

Family: Rutaceae

A small sized evergreen tree grows up to a height of 15 meters. Leaves are simple with winged petiole, oblong-ovate and acute or obtuse. Flowers are whitish, fragrant and fruits are large, oblong, inner part fleshy with several compartments, containing embedded white seeds.

Plant pacifies vitiated pitta, burning sensation, febrile fits in children, epilepsy, cough, asthma and general weakness.

■■■

12 Therapeutic Significance of Vegetables

M.K. Rana

Department of Vegetable Science, CCS Haryana Agricultural University, Hisar- 125 004, Haryana, India, E-mail: <mkrlotus@gmail.com>

The medicinal importance of vegetables in human nutrition is largely recognized. They are known as the cheapest source of natural protective foods and have been considered the essential articles for a balanced diet, for reproduction and for the maintenance of some of the important physiological activities in human body since they are the principal source of providing carbohydrates, fats, proteins, minerals and vitamins without which one cannot maintain proper health and resistance to diseases. They are packed with specific mineral salts of which other food articles are deficient and whose deficiencies lead to the disturbances of metabolism resulting in ill health. Besides this, pectin and cellulose, which stimulate the intestinal activity and protect the body from various disorders, are also present in several foods. Vegetables in particular are also essential for maintaining the body alkaline reserve by neutralizing the acids produced during the digestion of eggs, meat, cheese and other fatty foods and provide valuable roughage, which prevents constipation.

The tremendous potential of vegetables can be exploited on large scale in order to expel these dietary deficiencies, which are generally prevalent among our poor masses. Because of these specific medicinal properties, vegetables have been considered as protective food. However, still the nutritional and medicinal importance of vegetable crops has not yet received due attention. Many therapeutic drugs being used in modern medicines have been extracted from the vegetables. In countries like India where most of the people are vegetarian, use of vegetables in daily diet is very essential to meet the body requirements. In many parts of the world, vegetables form a major part of the daily diet, but in India, per capita consumption of vegetables is very less, thus, most of the people in India suffer from diseases due to deficiency of vitamins and minerals. Besides, vegetables have been valued for certain medicinal properties, which are given below:

Common name: *Bathu* (Lamb's quarters)

Botanical name: *Chenopodium album*

Family: Chenopodiaceae

This vegetable contains high amount of β-carotene and vitamin C. Quercitrin present in *Bathu* is an important cardiotonic compound to control diabetic cataracts by inhibiting aldose reductase (AR). Aldose reductase catalyses the metabolism of sugars to their respective polios such as glucose and galactose reduced to sorbitol and galactiols, respectively. These polios in diabetic patients and galactosaemic patients accumulate in very high level in the lens leading to osmotic over-hydration of tissues that leads to the events, which follow some physiological steps that ultimately end in the formation of cataract, thus, aldose reductase has direct impact in the estimation of cataract formation.

Quercitrin inhibits completely 3',5' adenosine monophosphate-phosphodiesterase, thereby, elevating cellular cyclic adenosine monophosphate concentration by inhibiting phosphodiesterase and affecting the permeability to all organelles to Ca^{2+} ions too. This compound also has strong insecticidal property. This species contains (+) catechin, which is a strong liver protective. It prevents infection or renders recovery from hepatitis B virus. Consumption of (+) catechin imparts cell mediated resistance to hepatitis B infection.

Ferulic acid, (+) catechin, quercitrin and caffeic acid present in *Bathu* are anti-mutagens, which operate within the cell to counter the mutagens. Caffeic acid, quercitrin and (+) catechin are antiviral against several viruses. Caffeic acid has antiviral property for influenza, herpes simplex, vesicular stomatitis, polio, adeno and vaccinia viruses. (+) Catechin is antiviral for herpes simplex, respiratory syncytial and hepatitis B viruses. Quercitrin is anti-pseudorabies, anti-aujeszky and anti-herpes simplex viruses.

Common name: Beet leaf

Botanical name: *Beta vulgaris* var. orientalis

Family: Chenopodiaceae

Beet leaves and root contain caffeic acid and two of its derivatives, chlorogenic acid as well as oxidized product (caffeic acid oxidation product), are virucidal. These compounds are, in general, inhibitors of cell growth and cellular DNA metabolism, thus, offering the property of anti-cancer or anti-tumor. Caffeic acid alone has been shown to possess antiviral properties in many test viruses, including influenza, herpes simplex, vesicular stomatitis, polio, adeno and vaccinia viruses. Caffeic and chlorogenic acids act as anti-hypercholesterolemia and anti-ulcer compounds.

Further, caffeic acid is found to have anti-tumor promotion, anti-mutagenic, antioxidative, antiviral (HIV) and chromosome stabilization activities. Beet contains caffeic, chlorogenic, ferulic and citric acids, which are all anti-mutagenic. The red pigment of beet is due to betaine and betalains. They consist of β-cyanine and partly β-xanthenes. Cyanin is an anti-polio virus.

Common name: Bitter gourd

Botanical name: *Momordica charantia*

Family: Cucurbitaceae

It is said to be very beneficial for patients suffering from diabetes, arthritis, rheumatism and asthma complaints. Bitter gourd contains linoleic acid, which modulates the immune system of the human body. With oral administration of bitter gourd fruit juice 73% of the diabetic patients showed improved glucose tolerance. A polypeptide–P, a 17-amino acid, did not cross react in an immuno-assay for bovine insulin.

Charantin in both *Momordica charantia* and *M. foetida* (bigger fruited) is a mixture of two steroid glycosides, *i.e.*, β-sitosterol- D-glucoside and 5, 25-stigmastadien-3- β-ol- D-glucoside, hypoglycemic in action. Inhibition of glucose uptake by intestinal fragments was also observed and attributed to glucoside constituents of the fruit juice. Thus, constituents in bitter gourd fruit have pancreatic and extra pancreatic effects with therapeutic potential for diabetic patients.

Bitter gourd contains some anti-mutagenic compounds such as cryptoxanthin. Purified lectin II of bitter gourd was found to inhibit vescicular-stomatitis virus plaque formation as well as found to be cytostatic for leukemia-cell culture lines. An antiviral factor has been found to inhibit the plaque formation of vesicular stomatitis virus. The compound is also inhibitory to cellular RNA and protein synthesis, hence, the antiviral activity appears as a result of non specific inhibition of cellular macromolecule synthesis.

The fertility regulating substances, α- and β-momorcharin, glycoprotien being proteinaceous in nature, their administration causes problem of immunogenecity. It seed and outer rind contain lectin, which interferes with protein synthesis.

Common name: Bottle gourd

Botanical name: *Lagenaria siceraria*

Family: Cucurbitaceae

It contains protein, fat, minerals, like phosphorus, iron and potassium and many of the vital amino acids required in our diet. Its grated and boiled pulp added to curds makes very nutritive, refreshing and cooling drinks in summer months. It is often included in the diet of old and convalescing patients. It has certain curative properties, like cooling, diuretic and sedative. It is known to calm the nerves and removes chronic constipation. It is especially recommended in the diet of patients suffering from high blood pressure. It is a useful remedy to cure the urinary disorders. A glass of its juice mixed with 5-10 ml juice of lemon if taken daily can relieve the burning sensation in the urinary passage often caused by hyperacidity because it serves as an alkalinizer. If it is taken with a pinch of salt, it can control severe diarrhea.

Common name: Brinjal

Botanical name: *Solanum melongena* var. esculentum (round shaped)
Solanum melongena var. serpentinum (long type)
Solanum melongena var. depressum (dwarf type)

Family: Solanaceae

Contrary to the common belief, brinjal is quite rich in nutrients. On an average, oblong brinjals are rich in total water-soluble sugars, whereas, long brinjals contain a large amount of free-reducing sugars, amide proteins and solasodine, an alkaloid, which is very useful in family planning. It has certain *Ayurvedic* medicinal properties. It is believed that white brinjals are good for diabetics. Brinjals if consumed after frying in until oil can cure toothache. It is also an excellent remedy for those suffering from liver troubles.

The plant parts are antiviral and protect the poultry birds from *Ranikhet* viral disease. It contains caffeic and chlorogenic acids, which are anti-tumor promotion and anti-mutagenic. Caffeic acid commands chromosomes to behave normally and acts as strong antioxidant. It is antiviral against influenza virus, herpes simplex virus, vesicular stomatitis virus, poliovirus, adeno virus and vaccinia virus. Brinjal contains 4 types of phenylamines that are well-known cardiotonic compounds. Brinjal contains oxidative product, which is antiviral as this compound inhibits cell growth and cellular DNA metabolism. Chlorogenic acid in brinjal is antiviral to poliovirus, herpes simplex virus 1 and herpes simplex virus. Both caffeic acid and chlorogenic acid inhibit the infectivity of H1 virus, which being retrovirus produces DNA templete during active infection with the help of an enzyme-reverse transcriptase, chlorogenic and caffeic acids inhibit the said enzyme, and thereby, inactivate infectivity of HIV. Vanillin in brinjal is antiviral to poliovirus 1. Catechol present in brinjal is anti-vesicular stomatities virus.

Common name: Cabbage and other species

Botanical name: *Brassica oleracea* var. capitata

Family: Brassicaceae (Cruciferae)

Cabbage is an acknowledged source of vitamin A, B, C and U and a storehouse of invaluable minerals like phosphorus, potassium, calcium, sodium and iron. All of which are very beneficial for human health. Vitamin C or ascorbic acid in the body offers protection to the liver and kidney and it acts as antioxidant and guides the chromosomes to act normally. Its use fights ailments such as gout, diarrhea, stomach trouble, colic, deafness and headache and removes constipation, improves digestion and appetite. Cabbage juice is said to be a remedy against poisonous mushrooms and is used as a gargle against throat hoarseness. Its leaves are used to cover wounds and ulcers. Cabbage, in particular, has the protective properties especially against human bowel cancer and peptic ulcer. It is also good for persons suffering from diabetes.

This specie contains caffeic acid, a lignan, which is anti-tumor promoting, anti-mutagenic, antioxidant and antiviral, particularly HIV. Fair amount of β-carotene is present in this species, which in the body acts as anti-carcinogenesis, anti-cellular transformation, anti-mutagenic, gap junction enhancement and chromosome behaviour stabilizer. This specie contains chlorogenic acid, caffeic acid related compound, which imparts in the body a power to resist tumor formation, gene mutation and HIV (virus) infection. The species of this group contain a flavoanol called rutin, which exerts following actions in the body such as aldose reductase inhibition, anti-inflammation, anti-mutagen, anti-tumor production, carcinogen inhibitor, smooth muscle relaxant and oestrogen receptor binding. It contains citric acid, which acts as antibacterial agent in urinary tract.

Crocetin present in these vegetables has been found to increase oxygen diffusion in blood serum by 80%. It also binds serum albumin and lowers serum cholesterol. This species along with its varieties contain desmutagen factor. It is anti-mutagenic substance, which activates enzyme known as super oxide desmutagenase, which in turn detoxifies pro-mutagenic compound and drives out of the body. Another factor known is as desmutagenic enzyme known as cytochrome. The latter enzyme functions like super oxide desmutagenase. The leaf extracts have been found to be anti-mutagenic. Rutin, kaempferol and quercetin present in these vegetables are partial cardiotonic in their properties. Chlorogenic acid is anti-polio, anti-herpes simplex 1 and anti-herpes simplex viruses.

Caffeic acid is anti-vaccinia, anti-herpes simplex, anti-influenza, anti-vesicular stomatitis, anti-polio and anti-adeno viruses. Quercetin present in these subspecies of *B. oleracea* is anti-parainfluenza-3, anti-aujeszky, anti-pseudorabies, anti-herpes simplex, anti-herpes hominis and suis, anti-polio and anti-respiratory syncytial viruses. These subspecies contain allyl-isothiocyanate, benzyl isothiocyanate, 3-benzyl isothiocyanate and phenyl isothiocyanate, although they are goitrogenic compounds in human beings are anti-mutagens. Besides containing ascorbic acid, β-carotene, caffeic acid and chlorogenic acid, this species also contains some other biologically active compounds such as citric acid, crocetin, desmutagenic factor, ferulic acid, kaempferol, nonacosan-15-one, η-nonacosan and pheophytin, all of which are anti-mutagen and anticancer compounds. The varieties of this species contain goitrogen in aerial parts.

Common name: Carrot

Botanical name: *Daucus carota* var. carota (Wild)
Daucus carota var. sativus (European type)
Daucus carota var. atrorubens (Asiatic type)

Family: Umbelliferae

Carrot contains high amount of β-carotene, which exerts a number of beneficial effects such as anti-carcinogen, anti-mutagen, anti-abnormal cellular transformation, stable chromosome behaviour and enhancement of gap junction facilitating movement of growth factors through tissues. It contains caffeic acid, which is functionally anti-ulcer, anti-hypercholesterolaemia, anti-tumor promotion, anti-mutagen, antioxidant and antiviral (HIV). It contains chlorogenic acid, which offers the effect such as anti-tumor promotion, anti-mutagenic and antiviral (HIV). A monoterpene, limonene, in carrot is anticarcinogenic and antibacterial. It contains linoleic acid, a fatty acid, which offers immunomodulation in human body. Psoralens -5 methoxypsoralen and 8-methoxypsoralene are present in carrot. These are photoactive compounds.

The common pharmocological use of psoralen is in rectifying vitiligo, an idopathic de-pigmentation of skin due to tyrosinase enzyme inactivation. Not only that it will be effective on an exposure to ultraviolet ray also. Besides this, psoralene is a preventive of tumor formation. Flavonols such as kaempferol and quercetin and flavones such as apigenin, luteolin and luteolin 7 glucoside in carrot are partial cardiotonic. Methyl eugenol present in carrot acts as sedative. Eugenol has anti-platelet aggregation action and inhibits arachidonic induced platelet aggregation but does not inhibit adenosine diphosphate induced platelet aggregation. It is further observed that eugenol inhibits thromboxane B_2 for activating anti-platelet formation.

Kanji prepared from the juice of carrot roots is an appetizing drink. Its roots are the rich source of vitamin A, which cures the night blindness. An infusion of its seeds has long been used as a folk medicine for threadworms and oil extracted from its roots has an antibacterial property. Its use increases the flow of urine and helps in elimination of uric acid. The essential oil extracted from carrot seeds is being used in pharmaceuticals. Carrot seeds are aromatic, stimulant and carminative and are useful in kidney diseases.

Carrot contains coumarin, which also has anti-platelet forming property and inhibits platelet aggregation factors such as ADP, adrenaline and collagen. It is an effective compound in controlling blood sugar level towards normal one in human beings. Though this compound is hepatotoxic in lower mammals, it is not so in human beings. Carrot contains crocetin, which increases oxygen diffusion in blood plasma by 80%. It binds with serum albumin and reduces serum cholesterol. Bornyl acetate present in carrot is strong antibacterial compound. It contains β-asarone, which has reputed insecticidal property. So far, there are three types of anti-mutagens present in edible plants. They are called desmutagen, anti mutagen and bio-anti-mutagen. Desmutagen (D) interacts directly with mutagens outside the cell and deactivates them through chemical or enzymatic reaction. Anti-mutagens (A) are agents that act within the cell through a wide variety of mechanisms such as inhibition of pro-mutagens, metabolic activation and detoxification by enzymes, chemical reaction with mutagen, scavenging of the reactive free radicals and protection of nucleophilic sites in DNA. Bio anti-mutagens act by increasing fidelity of DNA replication (favoring the repair of DNA damage or inhibiting error prone repair system). Anti-mutagen and bio-anti-mutagen are expressed anti-mutagen, however, for simplification anti-mutagenic compounds are presented here as desmutagen and anti-mutagen.

It contains citric acid, aesculetin, chrysin, crocetin, ferulic acid, kaempferol, luteolin, umbelliferone, vanillin and coumarin, which are anti-mutagens that act within the cell. Apigenin, quercetin, chrysin and ferulic acid also inhibit platelet aggregation factors. Some of the above-mentioned compounds are antiviral. Apigenin is anti-herpes simplex and anti-aujeszki viruses. Caffeic acid is anti herpes simplex, anti-vesicular stomatits, anti-polio, anti-adeno and anti-vaccinia viruses.

Common name: Celery

Botanical name: *Apium graveolens*

Family: Umbelliferae

Due to the presence of β-carotene, ingestion of celery has the properties of anti-carcinogenesis, anti-cellular transformation, anti-mutagenesis, intracellular gap enhancement facilitating movement of growth factors and chromosome stabilization and due to the presence of ascorbic acid in raw leaves exerts anti-hepatotoxicity, anti-nephrotoxicity, antioxidant effect and induces stable chromosomal function. It contains limonene, which is anticarcinogenic and anti-influenza A and B viruses. It contains psoralen, a coumarin substance, which helps in skin pigment formation. This compound is cytostatic, anti-mutagenic, immunomodulator and bronchodialator. Rutin, a flavonol, effects aldose reductase inhibition. It is anti-inflammatory, anti-mutagenic, anti-tumor, carcinogen inhibitor, smooth muscle relaxant and oestrogen receptor binding. Celery seed contains 3-η-butylphthalide, which is found to be anti-epilepsy induced by conaria lactone in rats. This compound has the ability to counteract the learning and memory impairment induced by coriaria lactone in rat.

Clinically celery juice has been found effective in lowering the blood pressure in patients with high blood pressure. Aqueous extract of celery stem acts as anti-inflammatory and anti-rheumatoid arthritis. Celery contains carvone, which induces enzyme glutathione-S-transferase, which detoxifies the harmful chemical compounds in the body, thus, carvone acts as potential inhibitor of carcinogenesis and anti-mutagen. Alpha-pinene is anti-influenza A and B viruses. Celery contains coumarins of acyldihydropyrano and furano types, which are typical to the species of the umbelliferae. These coumarins have been observed to possess anti-platelet activities. They inhibit adenosine diphosphate, adrenaline and collagen induced platelet aggregation, but they are weak in imparting anti-platelet aggregation. Vitamin E present in celery has been shown to reduce the levels of glycosylated haemoglobin in diabetic complications. It is also anti-mutagen. 3-carene present in celery has antimicrobial property. Isoquercitrin in celery inhibits angiotensin-converting enzyme, which if not inhibited, induces angina pectoris. Isoquercitrin, an analogous compound of zoapatle with similar activities of zoapatle, has properties of inducing labor pain, increasing the tone and the frequencies of uterus constriction during labor and of decreasing post partum bleeding. This compound is also used as fertility regulating agent

Common name: Chilli

Botanical name: Hot pepper: *Capsicum annuum* var. longum

Sweet pepper: *Capsicum annuum* var. grossum

Cayenne pepper: *Capsicum annuum* var. acuminatum

Bird pepper: *Capsicum annuum* var. minimum

Family: Solanaceae

Chillies when taken with food articles increase the flow of saliva by stimulating the taste buds. Saliva contains amylase enzyme, which helps in the digestion of starchy foods. The green chillies are packed with vitamin C and A and act as an appetizer. Green chillies also contain vitamin P, which is very useful in providing protection against damage due to secondary irradiation caused by atomic explosion. Addition of chillies in food increases the flow of saliva that contains enzyme amylase helps in the digestion of starchy foods. Its paste is used externally as rubefacient and as a local stimulant for the tonsils in tonsillities. Its application in diphtheria is said to hasten the separation of false membranes, but its large doses produce gastric enteritis and irritation in intestine.

Medicinally chillies have been used internally as a cure for several maladies including colic, diarrhea, dropsy, black vomit, gout and toothache. It is used as a tincture to stimulate digestion, appetite and mitigate flatulence. Chillies are used as counter irritants in neuralgia, lumbago and rheumatic disorder and are especially useful in dyspepsia. It has a tonic and carminative action when taken with food and may be added to rose gargles for pharyngitis and sore throat. Its paste is externally applied as a rubefacient and as local stimulant for tonsils in tonsillitis. In diphtheria, its application hastens the separation of false membranes.

Capsaicin, an alkaloid responsible for pungency in chilli, which has significant physiological action, is used in many pharmaceutical preparations like tincture, ointment for cold, chest congestion, sore throat, pain balms, vaporous, linaments, plaster, medicated wool *etc*. It is also used in cosmetics, like prickly heat powders and skin ointments.

Common name: Cluster bean

Botanical name: *Cymopsis tetragonoloba*

Family: Papilionaceae or Fabaceae

Seeds of cluster bean are used as medicine for reducing human blood sugar. In addition to direct hypoglycaemic effects of their constituents, dietary effect is also important. High legume diets have improved many of the facets of blood sugar (glucose) and have controlled, specially, postprandial levels of sugar. A non-cellulosic guar gum (galactomannan) is present in this vegetable. This compound is absorbed slowly in the intestine in the form of glucose by slowing gastric emptying and by thickening the unstirred water layer adjacent to the intestinal villi. Leguminous gum modifies the physiological and chemical characteristics of intestinal content and may modify the release of gastro-intestinal hormones, which in turn influences insulin secretion and gastro intestinal motility.

Tablets of guar gum when taken, the blood glucose is not reduced, though significant reduction takes place as regards Low Density Lipoprotein (LDL) cholesterol. Pure guar gum being unpalatable tends to cause abdominal distension and diarrhoea but intake of natural gum with high carbohydrate food has been shown to reduce postprandial glycaemic profile without gastric distress. Guar gum lowers the serum and liver cholesterol levels and postprandial glucose level in human beings. Guar gum, as long-term dietary supplement, controls hyperchlo- lesterolaemia in diabetic patient. Long-term intake of guar gum (21 g/day) produced an improvement in the control of non-insulin dependent diabetic patients with significantly lowering total and low density (LDL) cholesterol concentration in blood. This fruit contains gallic acid, which is anti-influenza A and B, anti-adeno and anti-polio 1 viruses. Gallic acid and β-carotene are the direct anti-mutagen, while ascorbic acid is anti-mutagen acting indirectly.

Common name: Cucumber

Botanical name: *Cucumis sativus*

Family: Cucurbitaceae

It produces cooling effects during hot summer and it prevents constipation. It is very useful food for jaundice patients. Cucumber contains caffeic acid and its related compound, chlorogenic acid that are very useful in the human body. The above-mentioned two compounds along with ferulic acid, present in cucumber, prevent gene mutation. Cucumber contains linoleic acid, an unsaturated fatty acid, which offers immunomodulation in the body. Patients suffering from rheumatoid arthritis, after being treated with linoleic acid, showed reduction of inflammation due to this disease, however, such result was observed after three months of treatment. Caffeic and chlorogenic acids are a class of lignan, which acts as scavenger of free radicals. They bind with protein and other large molecules and base compounds. This binding participates in the anti-diarrhoeal event. One of the most useful effects of the plant-lignan is to offer protection of the digestive system from injurious attack. Cucumber contains proteolytic enzyme that helps in digestion.

Common name: Drumstick

Botanical name: *Moringa oleifera*

Family: Moringaceae

Vitamin C and β-carotene present in leaves and pod are biological active. Benzyl isothiocyanates, though goitrogenic, are bioantimutagens. The ripen seeds contain alum like substance. The powder of dry seed clears by precipitation of dusty or muddy substances nearly cent percent and little chlorination renders the water drinkable. It contains indole acetic acid and indole acetonitrite, which are plant growth promoting compounds and hypoglycaemic in human being. Gum and polysaccharide are present in pod, and hence, they have a role in inducing hypoglycaemia in human beings. Flower contains kaempferol and quercetin, action of which has already been described.

Common name: French bean

Botanical name: *Phaseolus vulgaris*

Family: Leguminosae

French bean contains a number of anti-mutagenic compounds such as aesculetin, caffeic acid, (+) catechin, chlorogenic acid, coumarin, diadzein, β-carotene, ferulic acid, genistein, kaempferol, limnonene, quercetin, robinin, rutin and umbelliferone. Quercetin and coumarin inhibit platelet aggregation. Umbelliferone is the coumarin, which inhibits ADP, adrenaline and collagen induced platelet aggregation but the inhibition is relatively weak. French bean contains isoflavonoids like diadzein and genistein. They have proliferative actions on human body. They produce significant oestrogenic potential effects, however, are less potent than capacity to influence oestrogenic action through variety of pathways and may exert multisite action in cell system. They perform various functions such as binding of the oestrogen site, transcription of protein synthesis, tissue proliferation and uterine growth. Ferulic acid, besides having anti-platelet activity, has anti-allergic, broad cardiovascular simulative and inhibitory effect on lipid peroxidase. Many of the compounds in French bean are strong antiviral. Antiviral properties of these compounds have already described. Quercetin has therapeutic potential as anti-cancer drug either alone or in combination with an anti-tumor antibiotic such as Adriamycin in multi-drug resistant breast tumor cells. Flavonoids are pro oxidant in nature and they act as antioxidant and anticancer compounds. (+) Catechin is a liver protective compound and anti-hepatitis B virus.

Common name: Lettuce

Botanical name: *Lactuca sativas*

Family: Compositae

Lactuca indicates its relation with lactation or flow of breast milk in women. Psychological and emotional factors such as stress, tension, fear, anger and anxiety can have an inhibitory effect on lactation, particularly on milk ejection reflex. Lettuce has anxiety relieving property. Ingestion of lettuce reduces anxiety and calms the individual female leading to lactation without inhibition. This is an indirectly active galactogenic and acts as a tranquilizer for persons who suffer from tension, anxiety, anger and fear. Since lettuce is taken raw as salad, vitamin C, nearly the whole amount of it will be consumed, thus, vitamin C will exert beneficial effects such as anti-inflammation, antipyretic, analgesic, anti-hapatotoxicity, anti-nephrotoxicity, antioxidant and chromosome stabilization. Lettuce is very rich in β-carotene, thus, intake of raw lettuce will exert anti-carcinogenesis, anti-mutagenic, anti-cellular transformation and enhancement of gap junction in the tissue system. It contains caffeic acid, which exerts in the body beneficial effects such as anti-hypercholesterolomaeia, anti-ulcer, anti-tumor promotion, anti-mutagenic, antioxidant and antiviral HIV. This compound also controls normal behaviour of chromosomes. It contains chlorogenic acid. The biological activities are anti-tumor promotion, anti-mutagenic and anti-HIV. It contains rutin, the biological effects of which in the human body are aldose reductase inhibition, anti-inflammatory, anti-mutagenic, anti-tumor, carcinogenesis inhibition, smooth muscle relaxation and binding of oestrogen receptor. Rutin and kaempferol are also moderate cardiotonic in action. Lettuce contains flavonoids of quercetin. They suppress the cytotoxicity of active oxygen species on Chinese hamster V79 cell. The potential therapeutic application of quercetin as anticancer drug either alone or in combination with anticancer antibiotic such as Adriamycin is found effective in the case of multi-drug resistant breast cancer cells. The pro-oxidant properties of flavanone such as quercetin are antioxidant in nature. This vegetable contains luteolin-7 -glucoside which is partial cardiotonic. This compound is anti-herpes simplex virus.

Common name: Musk melon

Botanical name: *Cucumis melo*

Family: Cucurbitaceae

In muskmelon, β-carotene, which essential for eyesight, is 95% of the total. Beta-carotene, cryptoxanthin, ascorbic acid and β-sitosterol are anti-mutagenic, which act as desmutagens. Vitamin C (ascorbic acid) present in muskmelon has also numerous pharmacological actions. Caffeic acid and chlorogenic acid present in the fruit are antioxidant, bioantimutagens, desmutagens, anti-tumor promotion, anti-ulcer and anti-hypercholesterolemia. Caffeic acid is anti-influenza, anti-herpes simplex, anti-vesicular stomatitis, anti-adeno and anti-vaccinia viruses. The fruit contains some compounds, which are anti-mutagens. In addition, the compounds like apigenin and luteolin are of similar type anti-mutagens. Ferulic acid present in the fruit is anti-mutagen, which acts as intracellular-bio-antimutagen. Apigenin is anti-aujeszki and anti-herpes simplex viruses. Chlorogenic acid is anti-polio, anti-herpes simplex and anti-herpes simplex 1 viruses. Luteolin is anti-aujeszki, anti-polio, anti-parainfluenza-3 and some bird infecting viruses. Apigenine and luteolin are partially cardiotonic compounds.

Common name: Okra

Botanical name: *Abelmoschus esculentus*

Family: Malvaceae

It is rich in vitamins, calcium, potassium and other mineral matters. In *Ayurveda*, it is considered a virilific and a very good remedy for the people suffering from kidney stone. Its fruits give vitality to the body and increase semen content if consumed fresh raw on an empty stomach every morning. The powder of its roots is given with sugar for leucorrhoea backache in women. Okra acts as a tonic for both men and women and enables them to increase vigor. It is an excellent source of iodine, which is considered very beneficial for the control of goiter. Its roasted seeds are useful against genito-urinary disorders and chronic dysentery.

This fruit contains very high amount of mucilage, which is hypoglycaemic and hypolipemic. Folic acid present in this fruit prevents anaemia. Carotene exerts protective actions against carcinogens, tumor formation, abnormal chromosome behaviour, abnormal cell transformation and narrowing gap junction, which allows movement of growth factors. In seeds, gossypol that has both good and bad effects is present. This compound acts as anti-fertility, promoter of calcium metabolism, modulator of gap junction through which vital fluid of the body travels, genotoxicity and cell cycle disruption depending on dose, anti-oestrogenic, menstrual inhibition, prostaglandin inducer, spermatocidal and toxic. Derivatives of oleanolic acid present in seeds have the properties like body-immunomodulation, hypolipemic, anti-inflammation and anti-atherosclerosis. Gossypol is anti-influenza A, anti-herpes, anti-parainfluenza 3, anti-Japanese encephalitis, anti-vesicular stomatitis, anti-rabies and human anti-acute encephalitis viruses in human beings. A remarkable action of gossypol is that it has virucidal effect on HIV at a dose of 100 mg concentration.

Common name: Parsley

Botanical name: *Petroselinum crispum*

Family: Umbelliferae

Our kidneys are filtering the blood by removing salt, poison and any unwanted entering our body. With time, the salt accumulates and this needs to undergo cleaning treatments and how are we going to overcome this? It is very easy, first take a bunch of parsley and wash it clean then cut it in small pieces and put it in a pot and pour clean water. Boil it for ten minutes and let it cool down and then filter it. Pour it in a clean bottle and keep it inside refrigerator to cool. Drink one glass daily and you will notice all salt and other accumulated poison coming out of your kidney by urination also you will be able to notice the difference, which you never felt before. Parsley is known as best cleaning treatment for kidneys and it is natural

Common name: Pea

Botanical name: *Pisum sativum*

Family: Leguminosae

Pea contains β-carotene, which is more in green than in dry one and anti-carcinogenic and anti-mutagenic and it resists cellular abnormal transformation and enhances tissue gap junction facilitating flow of growth fluid. Linoleic acid in pea acts as immunomodulator in the body. Pea contains trigonelline, an alkoloid, which is present in high quantity in fenugreek, is an antidiabetic in effect. Beta-sitosterol present in pea prevents genesis of breast cancer, detoxifies 7,12-dimethylbenz (α) anthracene induced carcinogenesis and also prevents the mutation of genes. This compound activates enzymes, which prevent mutation.

Quercetin found in pea has strong virucidal and anti-mutagenic property and acts against viruses such as mengo, encephalomyocarditis virus, aujeszky virus and para influenza-3-virus, pseudorabies, herpes simplex, herpes hominis and suis, polio and respiratory syncytial viruses. Ferulic acid and kaempferol in pea are anti-mutagenic, Pea oil has been found to reduce the rate of pregnancy in females and reduce spermatozoa count in males by 50-60% without any toxic effect.

Common name: Potato

Botanical name: *Solanum tuberosum*

Family: Solanaceae

Potato is a nature's great gift to the humankind as it is a powerhouse of dietary constituents like protein, carbohydrates, vitamins (vitamin A, B and C) along with traces of thiamine, iron, potassium, calcium, magnesium and various minerals, which are lacking in milk. However, it is low in protein content but contains all the eight essential amino acids, which are not only important for general health but also for nervous system, normal digestion and health of skin. Potatoes are rich in potassium but poor in sodium, thus, potato is a great boon to those who are suffering from blood pressure and want to reduce blood pressure by limiting intake of salt. Being a rich source of iron, it prevents anemia. It has high content of starch, so diabetics are advocated not to use potato in large quantity, but potatoes are generally recommended for the people who require a carbohydrates rich diet and for some convalescing nutrients. Potato has been considered the protective food. Its consumption in sufficient quantity plays an important role in preventing teeth decay because it produces cleaning effects. Potato also helps in maintaining the alkaline reserve of our body tissues. Thus, it plays an important role in human diet in India. However, some believe that it is harmful and causes obesity. Actually, it is not the potato itself but it is the frying medium, which causes obesity and defames the potato as one of the most fattening foods of all times. The calorific content shoots up only when they are fried deep in the oil. Potatoes should be consumed without peeling off because many of the essential nutrients are present in the potato just below the peel.

Potato contains bioactivities such as ascorbic acid, caffeic acid, chlorogenic acid, flavonoids kaempferol, quercetin, rutin and luteolin that exert partial cardiotonic action by inhibition of cellular phosphodiesterase and by elevation of cyclic adenosine monophosphate concentration. They affect permeability of cell organelles to Ca^{2+} ions. Potato contains linoleic acid, which exerts immunomodulation in human body, reduces the frequency and intensity of heart attack in persons having multiple sclerosis and inhibits histamine release in the body.

Myricetin and quercetin present in potato increase prostaglandin I_2 stimulated rise in cyclic adenosine monophophate and both these compounds inhibit primarily lipooxygenase activity. Such action of these compounds is through inhibition of phosphodiesterase resulting reduction in platelet formation. In action, these compounds are hypoglycaemic. Paradoxically both are known to be potent inhibitors of protein tyrosinekinase, activity of which is essential in the post receptor binding activity of insulin.

The coumarins, umbelliferone and scopoletin, present in potato are possessing anti-platelet aggregation activity as they inhibit platelet aggregation factors (PAF) such as adenosine diphosphate, adrenaline and collagen. Riboflavin present in potato safeguards the health of liver and increases consumption of oxygen in human body efficiently and is an anti-mutagen. Potato contains rutin, which has several biological actions such as aldose reductase inhibitor, anti-inflammatory, anti-tumor promotion, anti-carcinogenesis and smooth muscle relaxant and oestrogen binder.

Citric acid present in potato keeps the urinary tract disinfected from bacterial infection. Compounds such as L-ascorbic acid, citric acid, riboflavin and rutin in potato are anti-mutagens of desmutagen type. Caffeic acid, chlorogenic acid, kaempferol, luteolin, myricetin, quercetin and umbelliferone in potato are anti-mutagen. Caffeic acid, chlorogenic acid, luteolin, quercetin and rutin are antiviral to various types of viruses. Potato contains high amount of choline, which is involved in the production of acetylcholine that acts in transmitting nerve impulse.

Common name: Pumpkin

Botanical name: *Cucurbita maxima*

Family: Cucurbitaceae

Pumpkin contains β-carotene, which prevents sister chromatid exchange, carcinogenesis, mutagenesis, cellular transformation and narrowing tissue gap junction. Ferulic acid present in pumpkin has anti-allergy, broad cardiovascular and anti-platelet aggregation and anti-thrombin properties. Quercetin present in this vegetable has partial cardiotonic, insecticidal and antiviral properties. This compound is anti-aujeszki, anti-parainfluenza-3, anti-pseudorabies, anti-herpes simplex, anti-herpes hominis and suis, anti-polio and anti-respiratory cyncytial viruses. This compound is anti-platelet aggregation in property.

Gourd fruit contains several important compounds, which prevent induction of gene mutation. They are caffeic acid, cryptoxanthin, ferulic acid, quercetin, p-cresol, transcinnamaldehyde and β-carotene. Of these compounds, cresol, ferulic acid and trans-cinnamaldehyde prevent gene mutation within the cell, whereas, other compounds do so from outside the cell. Cinnamaldehyde has insecticidal property for stored grain pests.

Common name: Radish

Botanical name: Asiatic type: *Raphanus sativus* var. sativus
European type: *Raphanus sativus* var. radicula
Rat-tail radish: *Raphanus sativus* var. caudatus
Chinese radish: *Raphanus sativus* var. longipinnatus

Family: Cruciferae

Its leaves are a good source for the extraction of proteins on commercial scale. Its use in diet increases appetite and prevents constipation. It is a very good natural alkalinizer. Its preparations are a useful remedy for liver and gall bladder troubles and for the patients suffering from jaundice. In homeopathy, it is used for neurological headaches, sleeplessness and chronic diarrhea. Its roots are said to be useful in urinary complaints, piles and gastric troubles. The juice of its fresh leaves is consumed as diuretic and laxative. The seeds are said to be digestive, expectorant and carminative.

Radish root juice is virucidal. Radish root contains ascorbic acid, which acts as chromosomal stabilizer among others stated under different items discussed. It contains caffeic acid, which functions in chromosome stabilization, anti-tumor promotion, antioxidant and antiviral to HIV and to other viruses mentioned earlier. Caffeic acid also exerts actions such as anti-ulcer and anti-hyperchole-sterolaemia. Caffeic acid serves as chemical trap for highly electrophilic agents such as DMBA and other similar compounds. Radish contains malic acid, which is antibacterial and keeps the urinary tract disinfected from the bacteria usually attack the same and p-coumaric acid, which exerts in human body an antipyretic effect. It contains crocetin, which increases oxygen diffusion in blood plasma by 80%, binds serum albumin and lowers the serum cholesterol. Ferulic acid present in radish has various bioactions such as anti-platelet forming action, stimulation on broad cardiovascular system, anti-allergic and restriction of damaging action of lipid peroxidase. Cyanin present radish is antiviral against poliovirus.

Common name: Sprouting broccoli

Botanical name: *Brassica oleracea* var. italica

Family: Brassicaceae (Cruciferae)

A compound, which is powerful anti-cancer, has been detected from sprouting broccoli. Scientists of John Hopkins University School of Medicine have isolated an ingredient from sprouting broccoli that makes certain critical enzymes active in the cells, which guard against tumors.

Common name: Sponge gourd

Botanical name: *Luffa aegiptiaca*

Family: Cucurbitaceae

From the buds and vines of the sponge gourd, an antiviral fraction is obtained. This fraction was found to be active *in vitro* and *in vivo* against vesicular stomatitis virus. This fraction acts as an interferon inducer against the virus. *Luffa* charcoal is effective in the treatment of herpes zoster in the face and eye regions. Decoction of *Luffa* sponge exhibited marked anti-inflammatory, analgesic and tranquilizing effects in mice.

Common name: Spinach

Botanical name: *Spinacia oleracia*

Family: henopodiaceae

Spinach leaves contain very high amount of β-carotene and vitamin C, which are very useful to human body. Spinach leaves contain α-tocopherol, a provitamin E, which reduces the levels of glycosylated haemoglobin in diabetic patients independently of changes in plasma glucose, which may help to reduce the incidences of diabetic complications. Caffeic acid, β-carotene, citric acid, α-tocopherol and quercitrin present in spinach leaves are anti-mutagen of desmutagen type. It contains linoleic acid, an unsaturated fatty acid that exerts immunomodulation in human body.

Quercitrin present in spinach leaves is found effective in lowering the blood sugar levels to about half the levels of control subjects, not given quercitrin, thus, prevents cataract formation. Spinach contains histamine, which when exceeds tolerable limit may cause allergy. Quercitrin is anti-pseudorabies, anti-aujeszki and anti-herpes simplex viruses. Spinach contains high amount of dopamine, which in human body converts to L-dopa. This compound is a strong cardiotonic. However, person having high blood pressure on consumption of spinach is likely to suffer from hypertonic crisis. Quercitrin among the flavonoids is the best cardiotonic in property, however, Rutin, comparatively less cardiotonic.

Citric acid in spinach keeps the urinary tract infection less from infective bacteria. Rutin present in spinach leaves has number of biological activities such as aldose reductase inhibition, anti-inflammation, anti-tumor, carcinogenesis inhibition, smooth muscle relaxation and oestrogen receptor binding. Rutin is antiviral to a number of viruses infecting some birds. Chlorogenic acid and rutin present in spinach leaves are anti-mutagens of both types of functions.

Chlorogenic acid is anti-hypercholesterolemia, anticarcinogenic, anti-ulcer, anti-polio, anti-herpes simplex-1 and anti-herpes simplex viruses and anti-HIV but slightly less active than caffeic acid. Caffeic acid has been shown to have antiviral properties against influenza, herpes simplex, polio, vesicular stomatitis, adeno and vaccinia viruses. Ferulic acid is bioanti-mutagen.

Common name: Sweet potato

Botanical name: *Ipomoea batatas*

Family: Convolvulaceae

Sweet potato contains β-carotene, caffeic acid and chlorogenic acid, which are very useful for human body. Besides, it contains aesculetin and ellagic acid and both are anti-mutagenic and caffeic acid (D) and chlorogenic acid (D) are lignan type, which are anti-mutagens. Beta-carotene (D) is a carotenoid and provitamin A. 'A' means anti-mutagenic and 'D' means desmutatigen. Sweet potato is also a galactogogue, which may promote the secretion and flow of breast milk of a mother directly or indirectly. Many of the above-mentioned compounds present in sweet potato have biological activities such as anti-carcinogenesis antiviral, *etc*.

Common name: Sweet corn

Botanical name: *Zea mays* var. rugosa

Family: Gramineae

Beta-carotene present in corn regulates normal chromosome behaviour, opposes carcinogenesis, mutagenesis and cellular undesirable transformation and enhances cellular gap junction. Alpha-tocopherol is an antioxidant (free radical scavenger), anti-lipid peroxidative and it protects vitamin A, β-carotene and ascorbic acid from destruction in the body. Though corn silk is not a food item, its medicinal values are quite appreciable. Com silk contains, cryptoxanthin, vitamin C and K, sitosterol and stigmasterol. An aqueous extract of com silk has diuretic, hypoglycaemic and hypotensive activities. Alpha-tocopherol offers protection to diabetic patients from having other complication due to diabetes.

Linoleic acid in corn keeps the immune system of the human being in alert condition. Irrespective of stage, corn contains caffeic acid, coumarin, crocetin, eugenol, ferulic acid, gallic acid, lignin, limonene, α-tocopherol and vanillin. All these compounds are anti-mutagenic. Eugenal, gallic acid and lignin are desmutagenic and the above remaining compounds act directly as antimutagen. Some of these compounds are virucidal. Caffeic acid has been shown to possess antiviral properties in many test viruses including influenza, herpes simplex, vesicular stomatitis, polio, adeno and vaccania viruses. Gallic acid is antiviral to the viruses such as influenza A, B, adeno and polio 1. Limonene is anti-influenza A and B viruses. Lignin is anti-herpes simplex virus. Vanillin is anti-polio virus. Ferulic acid present in corn has anti-platelet activity as it inhibits platelet aggregation factor, thrombin. This acid has inhibitory effect on lypooxygenase dependent peroxidation. This acid stimulates broadly cardiovascular system and acts as anti-allergen. Caffeic acid has more stronger inhibitory effect on lipooxygenase dependent peroxidation.

Caffeic acid inhibits the histamine release. Gallic acid inhibit lipid peroxidation caused by adenosine diphosphate and ascorbic acid in liver cell mitochondria and the same caused by adenosene diphosphate and nicotinamide adenosine dinucleotide phosphate (in reduced form) in microsome of liver cell. Such inhibition is mainly due to free radical scavenging action of gallic acid. Crocetin present in corn increases the diffusion of oxygen flow in blood by 80% and lowers the serum cholesterol.

Coumarin in corn, a good hypoglycaemic agent, has anti-platelet activity, inhibits the platelet aggregation factors such as adenosine diphosphate, adrenaline and collagen. Eugenol present in corn has anti-platelet activity. This compound inhibits platelet aggregation factor, arachidonic acid.

Common name: Tomato

Botanical name: *Lycopersicon esculentum* Mill.

Family: Solanaceae

Tomato is well known for its nutritive value. It is a rich source of vitamins, especially vitamin A and C and minerals, like calcium, magnesium, phosphorus, iron and potassium. The important amino acids present in tomato are glutamic acid, aspartic acid, triptophane and tyrosine. It is the most important protective food. Tomato has certain medicinal properties. It increases appetite and removes constipation. It is one of the best vegetables, which keeps our stomach and intestine in good condition.

The pulp and juice are digestible, mild aperient, a promoter of gastric secretion and blood purifier. It is also considered a good antiseptic for the intestines. Tomato is said to be useful in curing cankers and sores in mouth. It stimulates torpid (sluggish, dull, inactivated) liver and is good in chronic dyspepsia. The people suffering from hyperacidity and from stones in gall bladder should avoid the excessive use of tomato in daily diet.

Biological activities of β-carotene are anti sister chromated exchanges to encourage more and more recombination for genetic variation, anti carcinogenesis, anti-cellular transformation and gap junction enhancement. Caffeic acid in tomato exerts biological activities such as anti-tumor promotion, chromosome stabilization from breakage by mutagenic compounds, antioxidant and antiviral to HIV. Vitamin C is analgesic, anti-inflammatory, anti-pyretic, anti-hepatotoxicity, anti-nephrotoxicity, antioxidant and chromosomal stabilization. Lycopene in tomato inhibits malignant carcinoma. Tomatin compound of tomato inhibits further growth of malignant transplanted tissue. This compound increases corticosteroid and neutrophils levels in blood. Tomatin is antihistaminic as well as acts similar way of digitonin. Tomato in kaempferol, quercetin, rutin and naringenin are cardiotonic. Quercetin inhibits platelet aggregation by inhibiting three of platelet aggregation factors (PAF) such as phosphodiesterase, adenosine diphosphate and prostaglandin. Quercetin increases the prostaglandin-stimulated rise in platelet cyclic adenosine monophosphate and inhibits primarily lipooxygenase activity.

Eugenol present in tomato inhibits mainly platelet forming agent-arachidonic acid but does not inhibit adenosine diphosphate induced platelet aggregation. Bomyol present in tomato has strong antibacterial activity against *Staphylococcus aureus, Escherichia coli* and *Candida albicans.* Citric acid is a mild laxative and diuretic. It is reported that citric acid acts as disincentive compound in urinary tract. Caffeic acid besides being anti-HIV is anti-influenza, anti-herpes simplex, anti-vesicular stomatitis, anti-polio, anti-adeno and anti-vaccinia viruses.

Common name: Water melon

Botanical name: *Citrullus lanatus* Manst

Family: Cucurbitaceae

The juice of watermelon contains plenty of fructose, which refreshes instantly. It is alkaline in nature; thus, it helps in reducing acidity of body fluids, however, its juice is rather better in reducing acidity than its pulp. The redness of fruit denotes a high level of vitamin A.

As per the recommendation of Naturopaths, the unstrained juice is more beneficial than strained ones. It also has a good amount of potassium, calcium and iron. Its juice keeps the body cool and calm in hot summer and prevents the body from heatstroke. Being an alkaline drink, it acts as a digestive and rejuvenating tonic. It increases the flow of urine and eliminates the body wastes. However, the Naturopaths never recommend this drink to asthmatics. Its juice has some soluble chemicals, which reach the various body organs quicker, thus, it deposits the chemicals needed our body to carry away the waste matter. The water distilled from the juice of watermelon contains a wealth of minerals.

■■■

13 Modern Herbal Uses of Bitter Gourd

S.K. Gupta

Assistant General Manager, National Seeds Corporation Ltd., Beej Bhawan, Pusa, New Delhi-110 012, India
E-mail: <skguptanscplan@yahoo.co.in>, <skguptaji@gmail.com>

The plant kingdom has treasure of medicinal plants, and since antiquity, diseases cum deficiencies have been treated with plant medicines. Recent scientific investigations have confirmed the efficacy of many, some of which including bitter gourd are remarkably effective, relatively non-toxic and have substantial documentation. Considered one of the most nutritious gourds, it has been used extensively in folk medicines.

Bitter gourd

The original home of the bitter gourd (*Momordica charantia* of the family Cucurbitaceae) is not known, however, it is considered a native of the tropics. It is widely cultivated as vegetable in India and other parts of the Indian subcontinent, Southeast Asia, China, Africa and the Caribbean. It is the most bitter of all the vegetables that is why the name in any language derived from its taste. The English names for the plant and its fruit include bitter melon or bitter gourd (translated from Chinese: *Pinyin* or *kugua*) and *goya* from Japanese). It is also known as bitter cucumber, balsam pear or balsam apple, boston apple and in Hindi karela.

Bitter gourd vine with tendril, leaves, flower and fruit (Courtesy: webindia123.com)

Description

Bitter gourd is a fast growing warm season climbing annual plant. The plant vine grows up to 3-5 meters long. The vine has a slender hairy stem with numerous branches and dense foliage. It bears simple, alternate leaves 4-12 cm across, with 3-7 deeply separated lobes. Each plant bears separate yellow male and female flowers.

A plant on pole **Male Flower** **Female Flower**

Immature Fruit **Fruits** **Fruit (Oblong)**

Fruits (Pointed) **Ripe Fruit** **Dried Seeds**

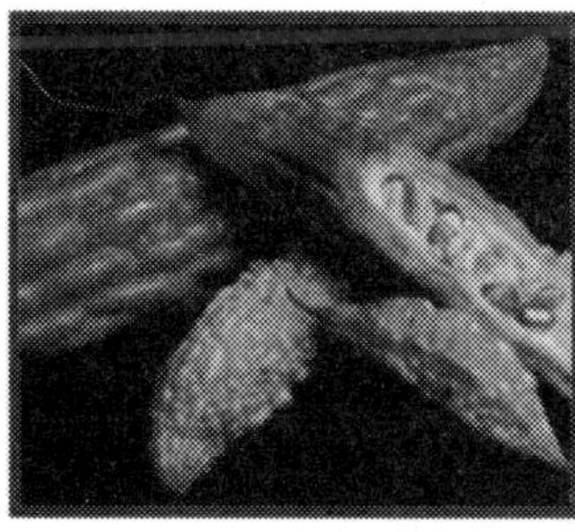

Cut Bitter gourd Fruits

Cooked Bitt er Gourd

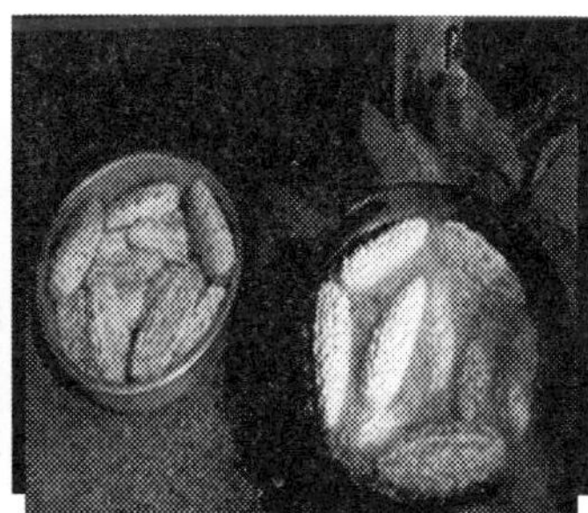

Soaking in Brine Solution

Bitter melon comes in a variety of shapes and sizes. The fruits are green usually oblong or narrower shape with pointed ends. They have an irregular surface with warts and 8-10 vertical ridges. There are several varieties, having fruits 3-4 to even 12 inches long. The typical Chinese phenotype is 20-30 cm long, oblong with bluntly tapering ends and pale green in colour with a gently undulating, warty surface, whereas, the more typical Indian bitter melon has a narrower shape and pointed ends and a surface covered with jagged, triangular *teeth* and ridges. Colouration is green or white. Between these two extremes is any number of intermediate forms. Some bear miniature fruits of only 6-10 cm in length, which may be served individually as stuffed vegetables. These miniature fruits are popular in South Asia as well as in India. When ripe, the fruits turn yellowish orange in colour. It is hollow in cross-section, with a relatively thin layer of flesh surrounding a central seed cavity filled with large flat seeds and pith. Seeds and pith appear white in unripe fruits and that turn red when they ripe. They are not bitter intensely and can be removed before cooking. However, the pith will become sweet when the fruit is fully ripe, and the pith's colour will turn red. Ripe bitter melons can be excessively bitter.

Propagation and Planting

Bitter gourd thrives in hot and humid climate. Propagation is through direct seeding and transplanting both. The best medium for the seeds to germinate is a fertile well-drained soil with a pH ranging from 5.5 to 6.7, enriched with organic matter, such as compost or dried manure. But it will tolerate any soil that provides a good drainage system. The soil must be prepared well by adding organic matter before planting. Two or three seeds can be sown together in a pit 1.5 cm deep. For better results, the seeds are soaked in water for 24 hours before sowing. The seeds will germinate in 2-3 days. The germinated seeds can be replanted on raised beds 45-50 cm apart. Transplanting should be done in such a way as to avoid disturbance to the root system. As the plants grow, the poles of 2 m height are placed and wire or twine supports are given in rows across the poles. Regular watering is essential for its better growth. Flowers will start appearing in 5-6 weeks and fruition will occur between two to four months. Vines should be pruned at the tips when female flowers start developing to encourage branching and fast bearing. Regular fertilizing is essential for its growth and water immediately after applying fertilizers.

Mature fruits are ready to be picked within 3 months from planting and they will be light green and juicy with white flesh but bitter. The fruits are picked at 2-3 days intervals when they are still at the tender stage. Regular picking is important as fruits will become more bitter as they mature and it can also hamper the growth of new fruits. Leave some fruits to reach full maturity if their seeds have to be reserved for subsequent crops. When fully mature, the fruits will break open them selves and release brown or white seeds, which can be collected.

Insect-pests, Diseases and Care

Bitter gourd is susceptible to many insect-pests and diseases. It is susceptible to watermelon mosaic virus, other cucurbit viruses and powdery mildew, which can be controlled by sulfur dust. Rust disease is controlled by spraying foliage with oxycarboxin. The fruits are subject to attack by various fruit flies and fruit rots. Pests attack can be prevented by wrapping fruits with newspapers, when they are about a few centimeters long.

Nutritional Value

Bitter gourds are very low in calories but dense with precious nutrients. It is an excellent source of magnesium, folic acid, zinc, iron, phosphorus, manganese and vitamins A, B_1, B_2 B_3 and C and has high dietary fiber. Bitter gourd has an affluent amount of carbohydrates, 5.6 g protein, 290 mg calcium, 5 mg iron, 5.1 mg vitamin A and 170 mg vitamin C, 44 kcal, per 100 g serving (webindia123.com). It contains twice the beta-carotene of broccoli, twice the calcium of spinach and twice the potassium of a banana.

Table 1: Nutrition Facts: Serving 1 Cup (93 grams) 16 Calories

Daily Values are based upon a 2000 calorie diet					
Constitutents	**% Daily Value↓**	**Constitutents**	**% Daily Value↓**	**Constitutents**	**% Daily Value↓**
Total Fat 0 g		Zinc	5	Folate	17
Saturated Fat 0 g		Magnesium	4	Vitamin A	9
Trans Fat 0 g		Phosphorus	3	Thiamine	2
Cholesterol 0 mg	0	Iron	2	Riboflavin	2
Total carbohydrates 3 g	1	Copper	2	Niacin	2
Dietary fiber 3 g	13	Calcium	2	Vitamin B_5	2
Sugars 0 g	0	Sodium 5 mg	0	Vitamin B_6	2
Protein 1 g	1	Vitamin C	130		

Source: http://www.foodsdatabase.com

Chemical Constituents

Balsam pear contains an array of biologically active phytochemicals including triterpenes, proteins and steroids. Alkaloids, such as charantin, charine, cryptoxanthin, cucurbitins, cucurbitacins, cucurbitanes, cycloartenols, diosgenin, elaeostearic acids, erythrodiol, galacturonic and gentisic acids, goyaglycosides, goyasaponins, guanylate cyclase inhibitors, gypsogenin, hydroxytryptamines, karounidiols, lanosterol, lauric-, linoleic- and linolenic-acids, momorcharasides, momorcharins, momordenol, momordicilin, momordicins, momordicinin, momordicosides, momordin, multiflorenol, nerolidol, myristic-, oleanolic-, oleic- and oxalic- acid, pentadecans, peptides, petroselinic acid, polypeptides, proteins, ribosome-inactivating proteins, rosmarinic acid, rubixanthin, spinasterol, steroidal glycosides, stigmasta-diols, stigmasterol, taraxerol, trehalose, trypsin inhibitors, uracil, vacine, v-insulin, verbascoside, vicine, zeatin, zeatin riboside, zeaxanthin and zeinoxanthin are found in it.

Traditional Uses

It has a long history of use by the indigenous tribes and local peoples of the Amazon and Asia, who grow it in their gardens for food and medicine. Bitter melons are seldom mixed with other vegetables due to strong bitter taste, although this can be moderated to some extent by salting the pieces, and then, washing before use. However, bottle gourd has extensive use. Indian cooking often prepares curries with potatoes. The fruts are stuffed with spicy ingredients and then fried and often used in China in stir-fries, soups and also as tea, likewise, different types of dishes in different countries.

Culinary Uses

Young immature bitter gourds, having skin of bright green colour, white flesh inside and small tender seeds, are the best for cooking. Bitter melons are seldom mixed with other vegetables due to strong bitter taste, although this can be moderated to some extent by salting the pieces, and then, washing before use. However, bitter gourd has extensive use. If it is difficult to eat bitter gourd by itself, one can add some to other dishes such as onion, potato, lentil, raw mango, spices, *etc*. The fruits are some times stuffed with spicy ingredients and then fried. People add the fruits and leaves to beans and soup for a bitter or sour flavour. Likewise, different types of dishes are prepared in different countries. To reduce bitterness, sometimes, it is salted and exposed to direct sunlight for few hours. After a few hours, its bitter salty water is reduced by squeezing out the excess by hand, and thereafter, it is rinsed with water a few times. Pickles are also prepared from tender fruits of bitter gourd. It is often used in China, Japan, Vietnam, *etc* typically in stir-fries

Natural Benefits and Curative Properties

From the *Ayurvedic* perspective, bitter gourd is excellent for balancing Kapha. It helps in purifying the blood tissue, enhances digestion and stimulates the liver. It is antidotal, antipyretic, tonic, appetizing, stomachic, antibilious and laxative. Based on clinical and traditional reports, the fact is that bitter melon is also a demulcent and at least mild inflammation modulator, meaning that it rarely does have these negative effects. Over the years, scientists have verified that many of the tribes of Asia, Africa and America use this bitter plant as an important natural remedy in herbal medicine systems. In Myanmar, China and India, it is highly prized for its anti-diabetic activity, where traditional physicians prescribe to diabetics. It is used topically for sores, wounds and infections and internally and externally for worms and parasites. In Brazil, it is used for tumors, wounds, rheumatism, malaria, vaginal discharge, inflammation, menstrual problems, diabetes, colic fevers, worms, abortions and as an aphrodisiac. It is prepared into a topical remedy for the skin to treat vaginitis, haemorrhoids, scabies, itchy rashes, eczema, leprosy and other skin problems. In Mexico, the entire plant is used for diabetes and dysentery, and the roots are a reputed aphrodisiac. In Peru, the leaves or aerial parts of the plant are used to treat measles, malaria and all types of inflammation. In Nicaragua, the leaves are commonly used for stomach pain, diabetes, fevers, colds, coughs, headaches, malaria, skin complaints, menstrual disorders, aches and pains, hypertension, infections and as an aid in childbirth.

Main Actions	Other Actions
Kills bacteria	Reduces inflammation
Kills viruses	Fights free radicals
Expels worms	Enhances libido
Treats diabetes	Cleanses blood
Reduces blood sugar	Detoxifies
Reduces blood pressure	Balances hormones
Lowers body temperature	Enhances immunity
Lowers cholesterol	Kills cancer cells
Mildly laxative	Kills leukemia cells
Controls piles	Prevents tumors
Controls diarrhoea	Promotes milk flow
Prevents cholera	Prevents respiratory disorders

Health Benefits

A majority of diabetics usually suffer from malnutrition as they are usually under-nourished. Being rich in all the essential vitamins, especially vitamin A, B_1, B_2, C and minerals, especially iron, regular use of bitter gourd prevents many complications such as hypertension, eye complications, neuritis and defective metabolism of carbohydrates. It increases body's resistance against infection. The applications of bitter gourd in some common ailments are explained below:

Diabetes mellitus: Folk wisdom states that bitter melon helps to prevent or counteract Type-2 diabetes. In a scientific study at Jawaharlal Institute of Post Graduate Medical Education and Research, Puducherry, India has proved that bitter melon increases insulin sensitivity. The Philippine Department of Health issued a circular in 2007 that bitter melon, as a scientifically validated herbal medicinal plant, could lower down the elevated blood sugar levels. The study revealed that a 100 milligram per kilo dose per day is comparable to 2.5 milligrams of the anti-diabetes drug *Glibenclamide* taken twice per day. Bitter melon is sold in the Philippines as a food supplement and marketed under the trade name Charantia. Charantia capsules and tea are being exported to the United States, Canada, Europe, Mexico, Japan, Korea and parts of the Middle East.

Lowering down the blood glucose levels: There are a myriad of phytochemicals present in bitter gourd and at least three different groups of extracted components have been shown to increase the number of beta-cells, those which produce insulin in the pancreas and to regulate and lower down blood glucose levels. These involve glucoside, a steroidal saponin-like substance called charantin, alkaloids called momordicine that suppress neural response to sweet taste stimuli and peptides mimicking the action of animal insulin, may be able to help with insulin secretion, glucose oxidation and other processes. It has been confirmed that charantin have a hypoglycemic effect. There is also another insulin-like compound known as polypeptide P that has been suggested as insulin replacement in some diabetic patients. This is still a matter of research on which of these is most effective, if not all working synergistically. Therefore, it should be included liberally in the diet of the diabetic people. The oral administration of 50-60 ml of juice has shown good results in clinical trials.

Bitter melon contains very promising four bioactive compounds, which activate a protein called AMPK, which is well known for regulating fuel metabolism and enabling glucose uptake, processes that are impaired in diabetics. *We can now understand at a molecular level why bitter melon works as a treatment for diabetes*, said David James, director of the diabetes and obesity program at the Garvan Institute of Medical Research in Sydney. *By isolating the compounds we believe to be therapeutic, we can investigate how they work together in our cells.* Bitter melon contains a lectin that has insulin-like activity. The insulin-like bioactivity of this lectin is due to its linking together two insulin receptors. This lectin lowers the blood glucose concentrations by acting on peripheral tissues and similar to insulin's effects in the brain suppressing appetite. This lectin is likely a major contributor to the hypoglycemic effect that develops after eating bitter melon and why it may be a way of managing adult-onset diabetes. Lectin binding is non-protein specific and this is likely why bitter melon has been credited with immunostimulatory activity by linking receptors that modulate the immune system, thereby stimulating said receptors.

Constipation and piles: Like most bitter-tasting foods, bitter melon stimulates digestion. While this can be helpful in people with sluggish digestion, dyspepsia and constipation, it can sometimes make heartburn and ulcers worse. The cellulose content of the fruit acts as a bulk residue and increases peristalsis. Its regular use prevents habitual constipation. A decoction of root is highly valuable in chronic constipation and haemorrhoids, *i.e.*, piles. It may be given one cup at bedtime, the juice of fresh leaves is also valuable and three teaspoonfuls of leaf juice mixed with a glassful of buttermilk should be taken every morning for about a month in this condition. Juice of bitter melon leaves (3 teaspoonfuls) is mixed with a glassful of buttermilk and taken every morning on empty stomach for about a month gives relief in constipation and piles. To hasten the healing, the paste of bitter melon roots may be applied over the piles.

Cholera: Fresh juice of leaves is effective in early stages of cholera and other types of diarrhoea during summer. Two teaspoonfuls of this juice mixed with equal quantity of white onion juice and a teaspoonful of limejuice should be given in these conditions.

Blood disorders cum skin diseases: Bitter gourd is highly beneficial in the treatment of blood disorders like blood boils, scabies, itching due to toxemia, psoriasis, ringworm and other fungal diseases. A cupful of fresh juice mixed with a spoon of lime juice should be taken sip by sip on empty stomach daily for four and six months and see improvement in the condition. Its regular use in endemic regions of leprosy acts as a preventive medicine.

- **Toxemia:** Bitter gourd contains beneficial properties that cleanse the blood from toxins. Sipping two teaspoonfuls of the juice daily is helpful in cleansing the liver and also is helpful in ridding jaundice for the same reasons.
- **Psoriasis:** Regular consumption of this bitter juice has also been known to improve psoriasis condition and other fungal infections like ringworm and athletes feet.

Respiratory disorders: Bitter gourd roots are used for respiratory disorders from ancient times. A teaspoonful of root paste with equal amount of honey or basil (*Tulsi*) leaf juice or two ounces of fresh bitter melon juice with a cup of honey diluted in water, given once every night for a month, acts as an excellent medicine for asthma, bronchitis, pharyngitis, colds and rhinitis.

Malaria: Though it has been claimed that bitter melon's bitterness comes from quinine, no evidence could be located supporting this claim. Bitter melon is traditionally regarded by Asian as well as Panamanians and Colombians as useful for preventing and treating malaria. Laboratory studies have confirmed that various species of bitter melon have anti-malarial activity, though human studies have not yet been published.

In Panama, bitter melon is known as Balsamino. The pods are smaller and bright orange when ripe with very sweet red seeds but only the leaves of the plant are brewed in hot water to create a tea to treat malaria and diabetes. The leaves are allowed to steep in hot water before being strained thoroughly so that only the remaining liquid is used for the tea.

Eye problems: The high β-carotene and other properties in bitter gourd make it one of the finest vegetable fruit that helps to alleviate eye problems and improve eyesight.

Alcoholism (hangover): Leaf juice is also beneficial in the treatment of alcoholism. It is used as an antidote for alcohol intoxication and also in liver damage due to alcoholism. Bitter melon juice may be beneficial in the treatment of a hangover for its alcohol intoxication properties. It also helps to cleanse, repair and nourish liver problems due to alcohol consumption.

Energy: Regular consumption of bitter gourd juice has been proven to improve energy and stamina level. Even the sleeping patterns have been shown to be improved or stabilized.

Immune booster: This bitter juice can also help to build immune system and to increase body's resistance against infection.

Human Immunodeficiency Virus (HIV): It is the human immunodeficiency virus (HIV), which causes acquired immune deficiency syndrome (AIDS). Laboratory tests suggest that compounds in bitter melon might be effective for treating HIV infection. As most compounds isolated from bitter melon that impact HIV have either been proteins or glycoproteins (lectins), neither of which are well-absorbed, it is unlikely that oral intake of bitter melon will slow HIV in infected people. It is possible oral ingestion of bitter melon could offset negative effects of anti-HIV drugs, if a test tube study can be shown to be applicable to people. In one preliminary clinical trial, an enema form of a bitter melon extract showed some benefits in people infected with HIV. Advertently, more researches are necessary before this could be recommended.

The other realm showing the most promise related to bitter melon is as an immunomodulator. One clinical trial found very limited evidence that bitter melon might improve immune cell function in people with cancer but this needs to be verified and amplified in other research. If proven correct this is another way, bitter melon could help people infected with HIV.

Parts of Plant Used: Mostly fruits are used but leaves, vines and even roots are also used.

Standard Dosage: Decoction 1 cup 1-2 times daily, tincture 1-3 ml twice daily and capsules 1 g twice daily.

Methods of Use

The herb may be prepared in a variety of different ways, however, the appropriate one is to make the fresh juice or extract from the fruit, vine and leaves, and either drink it, or administer it as a retention enema. Bitter gourd capsules and tablets with concentrated fruit and seed extracts, powder and tinctures with whole herb/vine extracts are becoming more widely available in United States of America and are employed by natural health practitioners for diabetes, viruses, colds and flu, cancer and tumors, high cholesterol and psoriasis. A leaf tea is used in diabetes to expel intestinal gas, to promote menstruation and as an antiviral for measles, hepatitis and feverish conditions.

Figure: Bitter melon transformed into capsule form and sold as a food supplement

It would be better to take the fresh fruit juice/extract, half to full cup and once twice or thrice a day depending on the sugar level, preferably in morning on an empty stomach. Diabetics can also use bitter gourd in the form of decoction by boiling the pieces in water or in the form of dry powder. The appropriate method of administering is to make the extract from the fruits, vines and leaves. To make juice more effective add tender leaves of *Jambolana*, in Hindi *Jamun*, (*Syzygium cumini* syn. *Myrtus cumini* and *Eugenia jambolana* of family Myrtaceae) and Emblic, in Hindi *Amla*, (*Phyllanthus emblica* synonym *officinalis* of family Euphorbiaceae) at the ratio of 6-8 leaves of *Jamun* and one or half fruit of *Amla* to per fruit of *karela*. The seeds of bitter gourd can be added to food in the powdered form. Drink it as a juice, tea, or administer it as a retention enema. Some say that drinking the juice or tea will result in the breakdown of the active components by stomach acids. Therefore, a retention enema would be a more efficient route.

Biological Activities and Clinical Research

The ability of fresh juice of the unripe fruit to enhance cells' uptake of glucose, to promote insulin release and to potentiate the effect of insulin in lowering down the blood sugar has been clearly established in both experimental and clinical studies. In 1999, a Bangladeshi clinical trial examined the effect on 100 patients with Non-insulin Dependent or Type 2 Diabetes mellitus and recorded the patients' sugar levels both without food intake for 12-24 hours and after taking 75 g of glucose. They administered a bitter melon extract and 86 out of 100 responded to it, showing a significant reduction (14%) in fasting and post-meal serum glucose levels. Similar study in India at Devi Ahilya University in 2004 proved to have the same positive effects, where 15 men and women between the age of 52 and 65 with Type- 2 Diabetes took 200 mg extract together with half doses of either Metformin or Glibenclamide or a combination of both. The result was hypoglycemia than what may acquire from taking full doses of medicine.

Charantin, extracted by alcohol composed of mixed steroids is more potent than the drug tolbutamide. In 2006, Japanese researchers confirmed that the major pure cucurbutanoid compounds possess similar effects. Recent researches have also established that it contains an insulin-like principle or substance, polypeptide-P, which lowers down the blood and urine sugar levels even into type1 diabetics when injected subcutaneously. It was likewise concluded that the vegetable might enhance the hypoglycemic effect of the drugs. Should they continue to use these prescription drugs? Several rat and hamster trials taking extract alone also yielded good results although animal studies may not always hold true in humans.

Around 100 *in vivo* studies have demonstrated the hypoglycemic effect to reduce total cholesterol, the antitumorous activity and also to increase resistance to viral infections and an immunostimulant effect in humans and animals, increasing interferon production and natural cell killer activity. It has clinically demonstrated the ability to inhibit the enzyme guanylate cyclase that is thought to be linked to the cause of psoriasis and also necessary for the growth of leukemia and cancer cells. In addition, a protein, momordin, has clinically demonstrated anticancerous activity against Hodgkin's lymphoma in animals. Other proteins in the plant, alpha- and beta-momorcharin and cucurbitacin B have been tested for possible anticancerous effects. The extract blocked the growth of rat prostate carcinoma, while hot water extract of entire plant inhibited the development of mammary tumors in mice. Numerous *in vitro* studies have also demonstrated the anticancerous and antileukemic activity including liver cancer, human leukemia, melanoma and solid sarcomas and also antiviral activity against numerous viruses, including Epstein-Barr, herpes and HIV. In addition to these properties, leaf extracts have demonstrated broad-spectrum antimicrobial activities like antibacterial against *E. coli*, *Staphylococcus*, *Pseudomonas*, *Salmonella*, *Streptobacillus* and *Streptococcus*, the entire plant extract have antiprotozoal activity against *Entamoeba histolytica* and the fruit juice have shown similar properties like against the stomach ulcer-causing bacteria *Helicobacter pylori*.

Many *in vivo* studies have demonstrated relatively low toxicity (even safe during pregnancy) of all plant parts when ingested orally. However, toxicity and even death in animals have been reported when extracts are injected intravenously. The seeds, however, have shown the ability to induce abortions, while the root as a uterine stimulant and the fruit and leaf as antifertilitant in rats and mice.

A chemical analog of proteins has been developed, patent and named *MAP-30*. Its developers reported that it was able to inhibit prostate tumour growth. Two of them, *i.e.*, alpha- and beta-momorcharin, have been reported to inhibit HIV in test tube studies and showed a nearly complete loss of viral antigen, while healthy cells were largely unaffected. They also filed another patent stating it was useful for treating tumours and HIV infections.

Caution

Various cautions are indicated. Small children or anyone with hypoglycemia should not take bitter gourd since this herb could theoretically trigger or worsen low blood sugar. Furthermore, diabetics taking hypoglycemic drugs (such as chlorpropamide, glyburide, or phenformin) or insulin should use bitter gourd with caution, as it may potentiate the effectiveness of the drugs, leading to severe hypoglycemia so one should be careful in a case of active diabetes, as it masks the sugar output in the urine. It should not be used except under close medical supervision. Pregnant women should avoid taking too much bitter gourd or its juice as it may stimulate the uterus that may lead to preterm labour. It has been used as an ingredient in some abortifacent mixtures. Excessively high doses of juice can cause abdominal pain and diarrhoea. The seeds contain vicine, and therefore, can trigger symptoms of favism in susceptible individuals. In addition, the red arils of the seeds are reported to be toxic to children, and the fruit is contraindicated during pregnancy. Though numerous 'miracle herbal cure' companies exist and champion the ability of herbal compounds to supplement insulin as a treatment, these should not be taken at face value without thorough research and consultation with experts.

Conclusion

The Indians, among other Asians and Latin Americans, have been planting and nourishing themselves of natural insulin vegetables and glucose regulating food for a long time, and bitter gourd is one of them. Bitter gourd is believed to work in a holistic manner. Being a part of naturopathy, naturally, it will work best in natural atmosphere, which one should maintain in life, like in- food, water, thinking, behaviour, or other daily routine. It would not work so well in case of high intake of fried foods or foods with a lot of chemicals, fats, or spices. In case of high toxins in the body, there is need to reduce it, use sparingly at first, as purification reactions may be strong. This is why no chemical extracts or synthetics should be made from bitter gourd. Incidentally, it is so easily available with an affordable cost that people of all the strata can easily use it; especially the rural folk have edge over it to get fresher, cheaper and without any time factor. Secondly, due to its herbal medicinal value, its popularity and consumption have been increased several folds that has created a round the year demand of bitter gourd especially in urban population for greens as well as in *Ayurvedic* preparations. The efficacy of its bitterness has thus become a boon for this botanical herb that is now giving a good economic return to the farming community.

■■■

14 Therapeutic Value of Vegetable Spices

M.K. Rana[1], Harsimrat K. Bons[2] and S.K. Sanwal[3]

[1]Department of Vegetable Science, CCS Haryana Agricultural University, Hisar-125 004, Haryana, India E-mail: <mkrlotus@gmail.com>

[2]Krishi Vigyan Kendra, Samrala, Punjab Agricultural University, Ludhiana, Punjab, India, E-mail: <harsimratpau@gmail.com>

[3]Division of Crop Improvement, Indian Institute of Vegetable Research, Varanasi-221305, Uttar Pradesh, India,E-mail: <satishsanwal@rediffmail.com>

Spices play a pivotal role in human life. A large quantity of spices is being consumed in our country for flavoring and seasoning foods and is used in medicine, pharmaceutical, perfumery, cosmetics and several other industries. Spices are well known appetizers; some of them have antioxidant properties, while others are added as preservatives in some foods and are added to insipid foods to improve their flavor and taste and to suppress unpleasant odor. Some spices possess strong anti-microbial and antibiotic effects; many of them possess medicinal properties and have grand significance in human life since they affect many functional processes. Spices promote flow and secretion of saliva, amylase, neuraminic acid and hexosamines. They keep the bacteria under control by cleansing oral cavity from food adhesion and help in checking infection and in protecting the mucous membrane against thermal, mechanical and chemical irritations. Spices inhibit formation of thrombus and accelerate thrombosis. Hence, the physiological and medicinal importance of spices deserves serious consideration. The therapeutic significance of spices in human life is given as under:

Common name: Bishop's weed (Ajwain)

Botanical name: *Trachyspermum ammi, Synonym Carum copticum*

Family: Umbelliferae

The composition of the *ajwain* is as follows: moisture 7.4%, protein 17.1%, fat 21.8%, fibre 21.25%, carbohydrate 24.6%, calcium 0.015%, iron 0.0275%, sodium 0.056%, potassium 0.0014%, vitamins (per 100 g) include thiamine 0.21 mg, riboflavin 0.28 mg, nicotinic acid 2.1 mg and carotene 71 mg. Its calorific value is 363. About 27 compounds have been identified in *ajwain* of which thymol 61% is present in the highest concentration. Among the other major components are *p*-cymene 15.6%, gamma terpinene (10%), *a*-pinene and limonene.

Ajwain owes its characteristic odor and taste to the presence of an essential oil called *ajwain* oil. The essential oil known as oil of *ajwain* is obtained by steam distillation of the fruits. It is used in medicines. The oil of *ajwain* is a principal source of thymol, which is valued considerably in medicine. It is said to be very rich in calcium and iron. *Ajwain* oil is colorless or brownish-yellow liquid possessing a characteristic odor of thymol and a sharp burning taste. The principle constituents of the oil are the phenols and some carvacrol.

Thymol is easily crystallized out from the oil and is known as *sat-ajwain.* The rest of the oil is called thymene, which constitutes about 45% of the oil has the following composition: para-cymene 50-55% used as germicide, gamma terpinene 30-35% used in cheap-perfumes for soaps, gamma-beta-pinenes 4-5% and dipentene 4-6%.

Oil of *ajwain* is used in surgery as an antiseptic and has been found very effective in the treatment of hookworm. Leaves of *ajwain* are used as a vermicide. Aqueous solution of thymol is an excellent mouthwash and it is a constituent of many toothpastes, *Ajwain* seeds are reported to be useful in flatulence, colic, atonic dyspepsia, diarrhoea and spasmodic affection of the bowels. *Ajwain* oil is added in ointments usually applied externally on muscles to relieve rheumatic and neuralgic pains. A teaspoonful of the seeds with a little salt is a common home-remedy for indigestion. For stomachache, cough and indigestion, the seeds are thoroughly masticated prior to being swallowed and this is followed by a glass of hot water.

Ajwain can check profuse expectoration from bronchitis. A tablespoonful of seeds crushed and tied up in a hanky can be used for inhalation particularly when nasal passages are congested. This is an effective remedy since ajowain seeds have the remarkable power to open up clogged and congested nasal passages. In case of adults, a teaspoonful of the seeds can be put in boiling water and the vapors inhaled. It is also beneficial in the treatment of migraine pains. Even the roots of *ajwain* plant are reported to be diuretic and carminative. *Ajwain* seeds are used along with *Kalmegh* (*Andrographis paniculata*) to treat malaria.

According to *Ayurveda*, *ajwain* seeds are hot, bitter, pungent, appetizer, aphrodisiac, antihelminthic, carminative, laxative, digestive and useful in the treatment of ascites, abdominal tumors, enlargement of spleen, piles, vomiting, abdominal pains and diseases of the heart and mouth. Both *ajwain* seeds and oil have been reported to be useful in atonic dyspepsia, cholera, diarrhea, spasmodic, hysteria and flatulence-colic. Its seed powder with a pinch of black salt is a common folk medicine for stomachache, cough and indigestion due to irregular diet.

It is also effective in skin diseases. The seeds are used in conjunction with asafoetida and rock salt. A hot poultice of ajowain seeds is used as a dry fomentation to chest in asthma and to hands and feet in fainting and cholera. They also check chronic discharges such as expectoration from bronchitis. Even the roots of *ajwain* plant are diuretic and carminative.

Ajwain is one of the most valuable spices used as a chief ingredient in veterinary medicines. According to the *Unani* system of medicine, it is useful in the treatment of weakness of the limbs, chest pains and diseases of liver, spleen, hiccup, dyspepsia, kidney troubles and inflammations. Indian traditional healers use all the parts of the plant as medicine.

Of late emphasis has been laid on the identification of pesticides of herbal origin, which are not easily degradable and which can be used with the minimum of side effects. These bio-pesticides can be obtained from available sources and are generally inexpensive compared to chemical pesticides. *Ajowain* has been found to be greatly effective in inhibiting microbial growth. It is useful in a variety of fungal infections of the skin.

The different components present in *ajwain* oil inhibit the growth of many deadly bacteria and fungus. This includes some of the virulent strains of (i) *Escherichia coli* that cause infections, septic abscesses and gastroenteritis, (ii) *Klebsiella pneumoniae* that causes respiratory tract infections, (iii) *Staphylococcus aureus* that cause pneumonia, endocarditis and meningitis, (iv) *Agrobacterium tumefaciens,* pathogen of fruit trees and crop plants, (v) *Alternaria brassicicola* plant pathogen, (vi) *Aspergillus niger* that causes allergy and lung infection, (vii) *Helminthosporum oryzae* that is a plant pathogen causing great damage to paddy and (viii) *Macrophomina phaseolina,* which is a plant pathogen causing root and stem rot of jute besides being harmful to cotton, tobacco, potato, *etc*.

Many more deadly plant and animal pathogens are killed by extremely low doses of *aiwain* oil. The pungent *ajowain* therefore surely has great prospects as an herbal antibiotic.

Common name: Cumin

Botanical name: *Cuminum cyminum*

Family: Umbelliferae

Its seeds have an aromatic odor; in *Ayurveda* and Unani system of medicine, they are considered astringent, carminative, stimulant and stomachic. It may be a very useful remedy in diarrhea and dyspepsia. A medicine prepared from mixture of a teaspoon of fresh coriander leaf juice and a teaspoon of cumin seeds boiled in a glass of water is beneficial for diarrhea if taken daily twice after meals. Intake of 3 g of powder prepared from 30 g roasted black cumin seed and 30 g without roasting is beneficial for treating piles. Powder prepared from cumin seeds (20 g), caraway seeds (12 g) and black salt (6 g) mixed with a little vinegar is useful in renal colic if taken about 3 g every hour until relief is obtained. Cumin water is an antiseptic beverage highly beneficial in treating common colds and fevers. It is prepared by boiling a teaspoon of cumin seed in water for few seconds. To this, ginger can be added for soothing throat irritation. Black cumin seeds mixed with honey are beneficial to improve dullness of memory. Paste of black cumin is applied over the boils. Moreover, paste of cumin seed prepared with onion juice retards the frequency of upbeats of scorpion sting. During conception and after delivery, cumin decoction is very popular. It is good to improve health and beauty of women and for proper growth of their breasts. It cures menstrual disorders and is said to correct hormonal imbalance. Half a teaspoon of cumin powder mixed with a teaspoon of tamarind pulp and honey taken before breakfast is a good remedy for morning sickness, nausea, vomiting and jaundice. Cumin seeds are chiefly used in veterinary medicines.

Common name: Coriander

Botanical name: *Coriandrum sativum*

Family: Umbelliferae

All parts of coriander plant have a pleasant aromatic odor. Thus, the principal use of coriander is in flavoring foods and a limited use in pharmaceuticals and perfumery. Its leaves are the richest source of vitamin C and A and used fresh in the form of chutney as an appetizer. Its seeds are considered aphrodisiac, anti-bilious, carminative, diuretic, laxative, refrigerant and stomachic and a very good tonic and are chiefly chewed to conceal the odor and griping qualities of other foods and medicines and to lesson the intoxicating effects of spirituous liquors.

An infusion of coriander seeds is taken thrice a day to lower down the blood sugar and its infusion in combination with cardamom and caraway seeds is a useful medicine in indigestion, flatulence, vomiting and other intestinal disorders. Fresh buttermilk added to coriander juice is highly beneficial in treating digestive disorders, such as indigestion, nausea, dysentery, hepatitis, ulcerative colitis and typhoid fever. Coriander is helpful in treating acidity, diarrhea and chronic dysentery. Chutney made from dry coriander, green chilies, grated coconut, ginger and black grapes without seeds, is beneficial for abdominal pain. A teaspoon fresh coriander juice mixed with 1 or 2 seeds of banana and given once daily regularly for a week is a very effective preventive measure against small pox. Putting coriander leaf juice in the eyes during an attack of small pox prevents eye damage. It is also used in bleeding piles, mucous diarrhea, constipation, rheumatism, neuralgia, cephalalgia and locally in eye infection and as a carminative in human and veterinary medicines. Crushed seeds tied up in a cloth, kept in milk or rose water for few days and dropped in eyes are excellent eyewash in conjunctivitis to reduce burning, pain and swelling.

Regular intake of coriander water prepared by boiling dry seeds of coriander helps to lower blood cholesterol since it is a diuretic in nature. Coriander seeds check excessive menstrual flow. Six grams of the seeds boiled in half a litre of water until half the water remains after adding sugar gives relief to the patient after taking the medicine for 3 or 4 days. A teaspoon of coriander juice, mixed with a pinch of turmeric powder, is an effective remedy for pimples, blackheads and dry skin. Coriander leaves can help in accelerating the excretion of mercury, lead and aluminium from the body.

Common name: Fennel

Botanical name: *Foeniculum officinale*

Family: Umbelliferae

The leaves have been reported to have diuretic and purgative properties. Its dried immature seeds are masticated alone to overcome the foul breath. Fruits are aromatic, carminative and stimulant and are considered useful in ailments of chest, spleen and kidney. A hot infusion of its fruits is used to increase lacteal secretion and cold infusion is employed as an enema for the expulsion of flatus in infants. It is the main ingredient of *Fennel Water* given to the infants suffering from constipation; also useful in infantile colic and flatulence and checks griping in purgatives. Seeds are considered a good vermicide especially against hookworm. Fruits and oil are added as corrective in drugs to conceal the odour and to cover the taste of less pleasant medicines. Intake of fennel relieves menstrual and menopausal problems in women. It promotes lactation and helps in smoothing milk flow in nursing mothers. Fennel is also believed to help in reducing obesity. Almond oil after adding few drops of fennel oil is used as massage for the abdomen. This relives nausea and vomiting. Fennel oil is inhaled from the bottle or in steam to relieve hiccoughs.

Common name: Fenugreek

Botanical name: *Trigonella foenum–graecum*

Family: Leguminosae

The seeds of fenugreek are a very good source of protein, starch, sugars, mineral matters, vitamins, enzymes and amino acids. The leaves and stems are also rich in carotene (vitamin A), ascorbic acid (vitamin C), calcium and iron. Its addition makes the food more nutritive so this is of considerable importance for the people who are purely vegetarian due to religious reasons. Its leaves are appetizing and diuretic and stimulate digestive and metabolic processes. The seeds are used as aphrodisiac, carminative and tonic. Its infusion is given to small pox and dysentery patients as a cooling drink and served to the women during postnatal period. The seeds are also used in chronic cough, colic flatulence, diabetes, diarrhea, dropsy, dysentery, dyspepsia with the loss of appetite, enlargement of liver and spleen, gout and rickets and good for the elimination of bad breadth and body odor. It reduces the blood glucose, urinary sugar excretion and cholesterol levels significantly. Its seeds contain steroid- diosgenin, which is used as an instituting material in the synthesis of sex hormones and manufacturing of oral contraceptives. It is particularly useful to cure disturbances after delivery and difficulty in breathing. The seeds are taken by the Indian women to promote lactation and fed to milch cows and buffaloes to increase the flow of milk and lactation period. It is externally applied on burns, wounds and used in hair tonic preparation to cure baldness in men swellings. It helps to combat dandruff.

Fenugreek seeds are not only very nourishing but they have been found to lower blood cholesterol levels and even inhibit cancer of the liver. The seeds are harvested when fully ripe and dried for later use. The seeds are popular as a spice in curry and it is a source of yellow dye too. Its seeds yield a strong mucilage and are, therefore, useful in the treatment of inflammation and ulcers of the stomach and intestines. These are also used in the treatment of adult-onset diabetes, poor digestion, insufficient lactation and painful menstruation. Externally, the seeds can be ground into a powder and used as a poultice for abscesses, boils, ulcers and burns. The hypoglycemic substance, isolated from water extract of seeds of fenugreek, has a noticeable effect on lowering of the fasting blood glucose in diabetic rabbits who were given the fenugreek seed extract for about 30 days.

Similar effect of fenugreek on the glycogen content of tissues and on the key enzymes of carbohydrate metabolism was noticed. The aqueous extract and a gel fraction isolated from fenugreek seeds have shown significant ulcer protective effects. Histological studies reveal that the soluble gel fraction derived from the fenugreek seeds is more effective than the anti-ulcer drug, Omeprazole, in preventing lesion formation. A compound, protodioscin, purified from fenugreek has a strong growth inhibitory effect against certain cancer cells. Fenugreek extracts have effective nematicidal properties, as the aqueous, methanol and chloroform extracts of fenugreek could kill the larvae of *Meloidogyne javanica*.

Fenugreek seeds contain the saponin namely, diosgenin, an important substance in the synthesis of oral contraceptives. An essential oil is obtained from the seed that has both culinary and medicinal purposes. Other sapogenins found in fenugreek seed include yamogenin, gitogenin, tigogenin and neotigogens. Besides, fenugreek seeds also contain mucilage, bitter fixed oil, volatile oil and the alkaloids choline and trigonelline. The seeds also have cosmetic values. The crushed seeds, mixed with oil and massaged into the scalp, are recommended for glossy hair. An infusion of the seeds is also used as a skin lotion since it is good for the complexion.

Common name: Garlic

Botanical name: *Allium sativum* L.

Family: Alliaceae

The good reputation of the strong-smelling bulb is not only based just on its alleged ability to ward off the bloodthirsty Count Dracula, but also on its many health-inducing qualities. Present-day researchers have accepted these qualities and given scientific basis to long hold folk-remedies. Ancient Egyptians appreciated the qualities of garlic more than any others did. An Egyptian medical text from about 1550 B.C. lists some 22 prescriptions, which call for garlic application against heart complaints, insect bites, headaches, worms and bowel problems. The Greek physician Hippocrates, known as the Father of modem medicine used garlic for a variety of infections and intestinal disorders. The ancient Romans nicknamed garlic as *theriaca rusticorias*, meaning a cure for all poor. Ancient Hindus knew the medicinal uses and pharmacological action of garlic. Known as a natural antibiotic, garlic was used during both World Wars to disinfect wounds.

Garlic contains volatile garlic oil, proteins, phosphorus, iron and calcium, vitamins B_1, B_2, C and nicotinic acid. This wondrous white bulb is said to cure various diseases. It helps in digestion, improves appetite, voice, complexion, inflammation of eyes, bronchitis *etc.* Laboratory investigations have shown that garlic is indeed a mild antibiotic. Garlic may work against some intestinal parasites. It appears to have roughly 1% the strength of penicillin against certain types of bacteria. This means it is not a substitute for antibiotics, but it can be considered as a support against some bacterial infections. *Candida albicans* growth is inhibited by garlic and garlic has shown long-term benefit for recurrent yeast infections.

Experiments indicate it to be anti-coagulant, preventing blood clotting inside veins. It also prevents deposition of cholesterol. A German doctor claimed that garlic consumption prevents hardening of arteries thus preventing heart attacks. Garlic has a preventive effect on the development of hardening of the arteries. However, garlic only keeps clotting in check- it cannot effectively replace stronger anti-clotting drugs. Its primary value is as a preventive.

Eating 2-3 cloves of fresh garlic daily really seems to work for hypertensive. Eating a clove every day for several months apparently has a significant blood-thinning effect. Some lesser ailments for which garlic is said to be beneficial include beriberi and constipation. Garlic may have a generalized mechanism of action to fight an array of cancers. These chemicals suppress runaway growth of tumors by forcing the cancerous cells to commit suicide (apoptosis). The compounds extracted from garlic could also put the DNA repair pathway, which goes awry in cancer- back on the right track. Allyl sulfur tucked away in bulbs of garlic could be that sought- after drug.

The ability of garlic to decrease the concentration of serum lipids and serum fibrinogen has been attributed to ajoene. Clinical studies have shown that extracts of garlic can decrease serum levels of cholesterol and triglycerides. Fresh garlic extracts have been found to inhibit many fungal species, bacteria and parasites. Garlic also possesses anticancer activity. One underlying mechanism of action for these activities is stimulation of the immune system, both systemically and locally. Lymphocytes and macrophages are attracted to the site where garlic is injected; therefore, injections near the sites of the localized disease can mobilize the body's immune system. In case of cancer, besides stimulating the immune system, it may directly inhibit guanylate cyclose activity, which would interfere with tumor metabolism. Garlic reduces platelet aggregation and blood glucose level and increases blood insulin levels.

It has since long been recognized all over the world as a popular remedy for various ailments and human physiological disorders. Garlic has possessed a place of king among the natural folk medicines. According to Ayurveda, garlic is carminative, gastric stimulant, anthelmintic, antiseptic and strong natural antibiotic. It aids in digestion and absorption of food and is given in flatulence, breathing problems, worms-infection and leucoderma and skin diseases. It is good to cure epilepsy and convulsive *vata* diseases and is especially good to contract the uterus after delivery; thus, the Indian women take preparations with garlic after delivery. Naturopaths believe that use of garlic in diet cures mucous in sinus cavities, in bronchial tubes and lungs, deafness and even gall bladder disorders. Chest diseases may also be cured effectively by the use of garlic. One gram of garlic clove, boiled in a quarter liter of milk until a quarter of decoction is left, is regarded as a wonderful medicine for tuberculosis if taken thrice a day. This is also useful remedy to cure asthma.

The anti-bacterial action of garlic, its healing capacity and effectiveness against cholera and several other bacteria have also been noticed from early days. The active principle in garlic is an antibiotic allicin and one milligram of allicin has been reported equivalent to 15 oxford units of penicillin. The inhalation of garlic oil or juice has been found very effective in controlling pulmonary tuberculosis, rheumatism, sterility, impotency, cough and sore eyes. Garlic juice is used in various ailments of stomach, as a rubefacient in skin diseases and as eardrop in earache. The juice diluted with water can be used against peptic ulcers. Allicin present in aqueous extract of garlic reduces cholesterol level in human blood.

Use of garlic in diet eliminates obnoxious waste matter from the body. The digestive disorders can be cured if crushed garlic cloves are taken with milk. Use of garlic this way kills the harmful bacteria in the intestines by leaving the beneficial ones untouched. A regular use of garlic capsules can eliminate the internal infection and purify the blood. Because of its antiseptic properties, garlic juice mixed with boiled water in the ratio of 1: 3 is an effective lotion for cleansing infected wounds. A formulation containing 1% garlic extract gives protection to persons against mosquitoes and black fly for about 8 h. The aqueous garlic bulb extract inhibits the growth of many zoo pathogenic fungi and has inhibitory and fungicidal effects on a number of medically important fungi colonies of food poisoning bacteria are unable to grow if concentration of garlic extract in the medium is more than 12%.

Common name: Onion

Botanical name: *Allium cepa* L.

Family: Alliaceae

Onion is packed with minerals like phosphorus, calcium, carbohydrates, protein and vitamin C and keeps the body calm and cool, besides preventing sunstroke and heatstroke. Its bulbs contain volatile oil, *i.e.*, allyl-propyl disulfide, which rejuvenates the body, removes toxins, revitalizes the blood and stimulates blood circulation in the body. Freshly extracted juice has antiseptic and moderate bactericidal properties due to presence of several sulfur compounds. It also possesses several medicinal properties like diuretic, expectorant, stimulant *etc.* It is considered useful in flatulence and dysentery. Crushed onion can be applied on eruptions, bruises, boils and wounds. Its juice is used as inhaler on hysterical fits in faintness and used to relieve sore throat. Equal amounts of onion juice and honey should be mixed and taken daily in case of cold, cough and bronchitis. In these conditions, onions are also beneficial in the treatment of anemia. Cotton wool dipped in onion juice and put into the ear is a popular Russian remedy for ringing sound in the ears. Dropping hot juice of onion in the ear relives earache. Onion is an effective remedy for cholera. About 30 g of onion and powder of black peppers should be given to the patient of cholera. It allays thirst and restlessness. Moreover, it lessens vomiting and diarrhea immediately.

Onion is used to treat various kinds of ailments like abdominal pain, hypertension, heart disease, rheumatism, stomach disorders, *etc.* The medicinal properties of this plant are usually attributed to sulfur containing compounds. The flavonoids, which occur in low concentrations, are thought to add to the antioxidant properties of *Allium* spp., increasing quantity of onion in diet could decrease the chances of heart diseases. Onion is said to be very effective in the prevention of atherosclerosis and coronary heart disease. Its extract can inhibit the aggregation of human blood platelets for arterial blocking. Inclusion of onion in diet results in fewer tendencies to form blood clots and lower levels of cholesterol and lipoproteins in blood serum, which are appeared to be associated with heart disease. Its extract is also beneficial in the treatment of diabetes, cancer and asthma. Doctors also advise the use of onion for people suffering from high cholesterol, lethargy, feebleness, weariness and lack of vitality.

Common name: Ginger

Botanical name: *Zingiber officinale*

Family: Zingiberaceae

Ginger is considered carminative, stimulant and extremely useful in dyspepsia, flatulence, colic, vomiting, spasms and other pains of the stomach and bowels. It is also an excellent remedy to cure cough and cold. Ginger juice mixed with equal honey is taken at least thrice a day to cure cough and cold. It is added in appetizing tonics and stimulating remedies and used in veterinary medicines as stimulant and carminative, which are given in indigestion of horses and cattle, in spasmodic colic of horses and in preventing griping by purgatives. It can cure all types of pain happen in the body. In headache, an ointment prepared by rubbing dry ginger with little water is applied to forehead that gives relief. In case of earache due to infection, a few drops of its juice give relief. In toothache, it gives relief when applied externally to the face. It is also used to promote digestion and as an antiflatulence or carminative to reduce gas and bloating, for temporary relief and protection against gastrointestinal ulcers to improve blood circulation, to lower blood glucose in the treatment of diabetes, to act as an anti-inflammatory against rheumatic pain and arthritis, to treat cancer, due to its possible antitumorigenic effects. *In vitro* trials support the use of ginger preparations in the prevention of vomiting in pregnancy. Several studies suggest that ginger may be more effective than placebo in reducing symptoms associated with motion sickness.

Common name: Turmeric

Botanical name: *Curcuma longa*

Family: Curcumaceae

Turmeric, which is a mild aromatic stimulant, is derived from the thick, rounded underground stems or rhizomes of a large-leaved herb. Turmeric has held a special place of honor in both Indian and Chinese systems of medicines for the thousands of years. Considering it as a cleansing herb for the whole body turmeric was used in ancient times as a digestive aid and treatment for fever, infections, dysentery, arthritis, jaundice and liver problems. It was also used to make yellow-orange dyes.

Curcumin, which gives curry its yellow colour, is the main active ingredient of turmeric. Its oleoresin is a mixture of compounds, namely curcumin, volatile oils and resinous material, obtained by solvent extraction of the ground spice. It is normally mixed with non-volatile edible solvents like vegetable oils, for commercial use. The volatile oil present about 3-5% in turmeric, which is obtained by steam distillation of turmeric powder, is pale yellow in colour and has peppery aroma. It contains turmeron and zingiberene as its major constituents.

Turmeric is endowed with enormous medicinal properties. It has anti-bacterial and anti-fungal effects, besides being well known for its anti-inflammatory and cholesterol-lowering properties. In a clinical study, oral curcumin administration was shown to reduce the serum levels of cholesterol and lipid peroxides dramatically. Curcumin acts as a chemo-preventive substance against arterial diseases and reduces platelets from clumping together, which in turn, improves circulation and helps protect against atherosclerosis. Phenlybutazone and curcumin produced a better anti-inflammatory response than placebo.

The yellow color of turmeric is due to a chemical substance- curcumin, which is chiefly responsible for most of its therapeutic properties. Its anti-inflammatory action is as good as aspirin. Being cytotoxic, curcumin checks growth of cells in culture and tumor in animals. Turmeric can reverse the cancer in early stages. Application of either curcumin or turmeric has been found effective for reducing itching and inflammation in patients having cancer in most advance stage. Turmeric due to its non-toxic effects can be taken up to 50 g/day without any adverse effects to prevent cancer. Turmeric and curcumin can also inhibit liver fibrosis caused by alcohol and other chemicals. These substances also reduce the level of cholesterol in blood by reducing its uptake from the gut and are especially useful in increasing the flow of bile juice in infected bile ducts.

A pinch of turmeric powder acts as an appetizer, aids in digestion and is a very good tonic and blood purifier. It is a useful remedy for diabetes and leprosy and prescribed as an anti-periodic alternative. It relieves sore throat, common cold and cough when taken internally with warm milk or used as an inhalation from boiling water. Burnt turmeric if used as tooth powder relieves teeth diseases. Turmeric is very useful in gallstones and other gall complaints. Addition of high amounts of red pepper in food causes blindness but use of turmeric in diet checks the effect of red chillies. In rural, women are usually exposed to wood smoke and suffer with eye opaqueness, the use of turmeric in diet can reduce the effect of smoke on eyes.

Turmeric acts not only as an excellent natural antibiotic, but also strengthens digestion and helps improve intestinal flora. Curcumin also protects the liver by detoxification and stimulates gall bladder. Numerous studies have also shown the cancer-preventing effects of turmeric. This may be due to curcuminoids, the other pigments present in turmeric, which have a powerful antioxidant activity in the body. In fact, the antioxidant property of turmeric has been experimentally proved, as it has been found to be effective in preventing the development of peroxides or the free radicals.

Turmeric and its preparations are traditional beauty aids for Indian women. Its antiseptic and healing properties are said to be a preventive cure for adolescence pimples. It is used on large scale in cosmetics and several skin ointments. Smearing turmeric paste on the face and limbs during bathing is found very effective to clear the skin, beatify the face and to discourage unwanted hair on the famine skin. In small pox, its paste prepared with gingelly oil and *neem* leaves is applied as face pack.

Turmeric has an important place as an ingredient in the preparation of medicinal oils, ointments and poultice and has vermicidal properties. It is anthelmintic and inhibits the growth of parasites not only inside but also outside the body. The juice extracted from raw rhizomes is applied as an anti-parasitic ointment for skin infections. Its application in foods inhibits the growth of a fungus *Aspergillus parasiticus*, which produces mycotoxin aflatoxin- a major food contaminant and serious liver diseases including liver cancer and death.

Several *in vitro* and *in vivo* studies have established the effectiveness of curcumin and total extracts of turmeric against many microorganisms such as *Micrococcus pyogenous, Staphylococcus* sp., *Paramecium caudatum*, *Mycobacterium tuberculosis*, *Salmonella typhi*, *Vibrio cholerae* among others. Studies have also shown that the aqueous extract of turmeric and its essential oils are credited with biopesticidal properties. The efficacy of turmeric constituents against various plant pathogens has also been experimentally established. The turmeric oil has been seen to be effective as a mosquito repellant and for controlling housefly and aphid populations.

Common name: Mango ginger

Vernacular name: Sanskrit: Ama haldi, Hindi: Amada, Bengali: Mangai inchi Tamil: Manga inch and Malyalam: Amiya haldi

Botanical name: *Curcuma amada* Roxb.

Family: Zingiberaceae

Mango-ginger is botanically related to neither mango nor ginger but it is related to turmeric (*Curcuma longa*). Morphologically mango-ginger plant is similar to turmeric but has shorter crop duration of six months. The rhizomes are pale yellow from inside with lighter colour outside and have sweet smell of unripe mango when crushed. The rhizomes of mango ginger is used as a food spice due to its mango flavour especially for the preparation of pickles, chutney, candy, preservatives, sauce, salad and in other culinary preparations. The rhizome has acquired excellent medicinal properties. Part of the rhizome is applied for the skin diseases and itching. Tuber pastes made with the spirit and egg white is used as applicant for chronic rheumatism and bruises. Oil is applied on the sprains, reducing muscular swellings, *etc*. It has received a reputed position as a traditional medicinal herb. *Ayurveda* considers it as a sweet, bitter, cooling, appetizer, antipyretic, aphrodisiac, carminative, stomachic, *etc*. It is also used as one of the ingredients of *rasayanams* and number of carminative and digestive *churnams*. The plant is getting importance due to its antifungal properties found in volatile oil. It is reported that the extract of rhizomes show anti-inflammatory activity when it is tasted in albino rats. On stem distillation of the dried rhizomes, it yields 1.1% volatile oil, which constitutes d-U-pinene, ocimene, linalool, linalyl acetate and safrole.

Common name: Black ginger

Vernacular name: Hindi, Bengali, Gujarati: Karkashur, bitter ginger, or shampoo ginger

Botanical name: *Zingiber zerumbet* Smith (Syn. *Amomm zerumbet* L.)

Family: Zingiberaceae

Black ginger is known throughout most of the world as the *shampoo ginger* due to the milky substance in the flower cones being used as a shampoo by some cultures. It is a wild herb available in warm and moist areas of the interior hilly regions. It is reported that the herb possesses medicinal values. Hence, it is cultivated as a medicinal plant since old time. The rhizomes are much similar to ginger rhizome except its shape and size with deep blue in colour. The rhizome is also very bitter with disagreeable taste. The structure of leaves is just like that of turmeric. However, its leaves are small with light purple colour in midrib portion.

The rhizome is the principle product like its cousin ginger. However, it is not popular as a spice due to its bitter taste. However, after certain traditional treatment, it is used as spice by certain tribes in North Eastern India but it is grown basically for its medicinal properties and being popularly used as medicine by various tribal groups of Northeastern India like the Naga group, the Meiteis, the Kuki-Chin-Mizo group, the Bodo group and so on.

The extract of rhizome is conventionally used in cough, asthma and stomach complaints. Some of the tribal people in Churachandpur district of Manipur used black ginger rhizomes in the treatment of colic pain. It is also used as an anti-pyretic and analgesic and in the treatment of leprosy and other skin disease. The rhizomes produce an essential oil, which is used in the preparation of soap and other toilet items. The rhizome yields xeruminoe, a monocyclic sequiteopena ketone, which acts against a number of bacteria and viruses. Rhizome also yields an essential oil. Zerumbone can be isolated from rhizome and tetrahydrozerumbol and tetrahydric zerumbone can be prepared.

In Hawaii, the spicy-smelling fresh rhizome were pounded and used as medicine for indigestion and other ailments. The rhizomes can be stored in a cool, dark place to keep for use when needed. In traditional use, the rhizome was ground in a stone mortar with a stone pestle, was mixed with a ripe Noni fruit, and then, used to treat severe sprains. The pulp was placed in a cloth and loosely bound around the injured area. For a toothache or a cavity, the cooked and softened rhizome was pressed into the hollow and left for as long as was needed. To ease stomachache, the ground and strained rhizome material is mixed with water and drunk. Similarly, Awapuhi Pake or ginger root is widely cultivated and eaten, or made into a tea for indigestion as well as increased circulation of the blood and an increased sense of well-being.

Common name: Long coriander

Vernacular name: English: Puerto Rican coriander, black Benny, saw leaf herb, Mexican coriander, spiny coriander, fireweed, Hindi: bhandhania or bhandhanya

Botanical name: *Eryngium foetidum* L.

Family: Apiaceae

Long coriander is a tropical perennial as well as annual herb. It grows best under moist, shaded conditions near cultivated areas. The herb is rapidly becoming an important import item in the United States mainly due to the increasing ethnic immigrant populations who utilize it in their many varied dishes from around the world. It is closely related botanically to coriander but has a distinctly different appearance and much more potent volatile leaf oil. It is rich in iron, carotene, riboflavin and calcium. This plant has a strong aroma similar to fresh coriander leaves. The essential oil from the leaves is rich in aliphatic aldehydes. Another essential oil can be obtained from its roots of which unsaturated alicyclic or aromatic aldehydes dominate here. In essential oil from the seeds, sesquiterpenoids and monoterpenes are found.

Long coriander is widely used in seasoning and marinating in the Caribbean. It is also used extensively in Thailand, India Vietnam and other parts of the Asia as a culinary herb. This variety of coriander dries well, retaining good colour and flavour and making it valuable in the dried herb industry. It is sometimes used as a substitute for coriander but has a much stronger taste. Like coriander, many people find the leaves distasteful, due to a genetic trait.

Medicinally, the leaves and roots are used in tea to stimulate appetite, improve digestion, combat colic, soothe stomach pains, eliminate gases and used as an aphrodisiac. The plant is used in traditional medicines for fevers and chills, vomiting, diarrhoea and in Jamaica for colds and convulsions in children. The leaves and roots are boiled and the infusion is drunk for pneumonia, flu, diabetes, constipation and malaria fever. The roots can be eaten raw for scorpion stings, and in India, the root is reportedly used to alleviate stomach pains. The leaves themselves can be eaten in the form of chutney as an appetite stimulant.

■■■

15 Spices: A Panacea for Human Health

M.K. Rana

Department of Vegetable Science, CCS Haryana Agricultural University,
Hisar-125 004, Haryana, India, E-mail: <mkrlotus@gmail.com>

Spices and condiments are very popularly world widely known for their extra-ordinary flavoring, medicinal and coloring properties and are extensively used in medicine, pharmaceutical, perfumery, cosmetics and several other industries. Spices and condiments exhibit strong anti-microbial and anti-biotic activities, thus, have a profound importance in human life. They are added in savory dishes and in other products for stimulating action on the digestive organs to produce an increased flow of saliva and gastric juices. They are well known appetizers and are considered indispensable in the culinary art all over the world and widely used by meat, sauce, canning, frozen food industries and food manufacturing industries. They are used as spices and condiments on large scale in pickles, preservatives and in vegetable dishes. They promote the taste of the baked and cooked foodstuffs and produce salivation and numbness of mouth, thus, gaining importance in food industry, pharmaceuticals, preservatives and body-care products.

Spices keep the bacteria under control by cleansing oral cavity from food adhesion and help in checking infection and in protecting the mucous membrane against thermic, mechanical and chemical irritations. They affect many functional processes and improve human health in a number of ways. The medicinal importance of an individual spice in human life is given individually as under:

Common name: Allspice

Botanical name: *Pimenta officinalis*

Family: Myrtaceae

Its dried berries contain fat, carbohydrates, fiber, calcium, phosphorus, sodium, potassium, vitamins and are thus, used as a condiment and flavoring agent in several food preparations and preservatives. Allspice is used as a stimulant in digestive disorders. The powdered form of its berries is used in diarrhea, dyspepsia, flatulence, neuralgia and rheumatism. It is also added in tonics and purgatives as a carminative. Formerly, it was considered carminative and used in medicines as an adjuvant to tonics and purgatives. It is an anodyne against rheumatism and neuralgia. The berry oil is used as a flavoring condiment in food preparations, which avoids microbial contamination. The oil is also used as a carminative and stimulant in pharmaceutical preparations. It also adds bactericidal, fungicidal and anti-oxidant properties.

Common name: Anardana

Botanical name: *Punica granatum* Linn.

Family: Labiatae

It is generally used as an agent for acidification of foodstuff. The seeds are said to be stomachic, while the pulp is both cardiac and stomachic. The powder of its rind and cloves is very useful remedy for diarrhea and dysentery. The seeds contain steroidal estrogens, thus, they produce estrogenic activity in the body. Its stem and root barks contain alkaloids and have astringent and anthelmintic properties and are specifically useful in tapeworm's infection.

Common name: Angelica

Botanical name: *Angelica archengelica* Linnaeus

Family: Umbelliferae

The dry root, rootstalk and fruits possess stimulant, expectorant and diaphoretic properties. Use of angelica root and seed in pharmacy is probably a survival of those ancient beliefs, which attributed medicinal properties to many aromatic herbs.

Common name: Aniseed (Vilaiti sounf)

Botanical name: *Pimpinella anisum* Linn.

Family: Umbelliferae

It is used for flavoring various food preparations. The seeds are widely known for their medicinal properties. Seeds and oil both are added into the composition of several indigenous preparations. Its oil or emulsion in water is said to be an aromatic carminative especially beneficial in colic and flatulence. The fruits are reported carminative, diaphoretic, diuretic, mild-expectorant and stimulant and are used in the preparations of powders for asthma, flatulent colic and in veterinary medicines. The alcoholic extract of the fruits possesses fungicidal properties. It is also used in the preparations of toothpastes, mouthwashes and gripe water. The oil of the fruit is also used as carminative in flatulence as an ingredient of cough syrups as an antiseptic and as a remedy of cholera. Its essential oil possesses insecticidal and fungicidal properties, thus, it is used externally as an insecticide and fungicide against small insects and fungal infections, respectively.

Common name: Asafoetida (Hing)

Botanical name: *Ferula asafetida* Linn.

Family: Umbellifereae

Asafetida's use in food stimulates the intestinal and respiratory tracts and the nervous system. It is very beneficial in asthma, chronic bronchitis and whooping cough. It is also used as an enema to relieve intestinal flatulence and a useful medicine for relieving spasms, indigestion and colic and in cholera. It is also often used in veterinary medicine. Its oil possesses antibiotic and sedative properties. It inhibits the growth of microbes and its use has been suggested in heart diseases. An *Ayurvedic* medicine *Hingastik churan* contains *asafoetida* as one of the major ingredients. It is applied externally on the stomach to stimulate the functioning of the intestine.

Common name: Basil (Tulsi)

Botanical name: *Ocimum basilicum* Linn.

Family: Labiatae

Tulsi, a household name, is revered as a sacred plant in India since time immemorial. This truly gifted herb holds a special place in ancient *Ayurvedic* system of medicine due to its unique medicinal properties. All parts of this plant, which are used as powder or boiled as an infusion, are medicinally useful. *Tulsi* has anti-microbial as well as anti-inflammatory properties. The anti-microbial activity of the essential oil of *Tulsi* has been shown against *M. tuberculosis* and *Staphylococcus aureus*. Ursolic acid present in *Tulsi* has anti-allergic properties. *Tulsi* also has anti-stress properties as its essential oil shows to increase physical endurance and prevent stress-induced ulcers. It is protective against histamine-induced bronchospasm and commonly used with black pepper in bronchial asthma. Eugenol, a major component of volatile oil and some of other compounds show good antioxidant activity at 10 μM concentrations. Anti-inflammatory activity or cyclooxygenase inhibitory activity of these compounds is also observed.

The tea prepared from its leaves is a popular home remedy for treating ailments affecting the stomach and considered highly beneficial in relieving cough and cold. The basil leaves have been reported alexipharmic, anthelmintic, anti-pyretic, carminative, diaphoretic, expectorant, pectoral, stimulant and stomachic. Its leaves, roots and bark are cyanogenetic. This wonderful plant is used in homeopathic medicine. An infusion of leaves is given for cephalalgia and gouty joints and is used as a gargle for foul breath.

The juice of its leaves is a common remedy for coughs and used as a nasal douche and as a nostrum for earache and for ringworm. Alcoholic leaves or flower extract has antibacterial properties and the seed possesses diuretic, demulcent, diaphoretic, stimulant and cooling properties. Its seeds are given to the patients suffering from chronic constipation and piles. An infusion prepared from seeds is useful in chronic diarrhea, dysentery and gonorrhea. The aqueous extract of seeds is effective against gram +ve bacteria and mycobacterium. Poultice is prepared from seeds is used for sores and sinuses. Its roots are given to the children in bowel complaints and its oil possesses bactericidal, insect repellent and insecticidal properties, thus, it is effective against houseflies and mosquitoes. Flowers are carminative, diuretic, demulcent and stimulants.

Amazing results of *Tulsi* leaves have been in patients with non insulin-dependent diabetes mellitus. Leaf extract of *Tulsi* has hypoglycemic effect. Significant decrease in fasting and postprandial blood glucose levels is noticed during treatment with basil leaves compared to treatment with placebo leaves. Fasting blood glucose falls by 21.0 mg/dl and postprandial blood glucose by 15.8 mg/dl and similar trend is observed in urine glucose levels. The mild reduction in total cholesterol levels is noted in these patients. Basil leaves also have the immuno-stimulatory effect.

Common name: Bay leaf

Botanical name: *Laurus nobilis* Linn.

Family: Lauraceae

Leaves and fruits both possess stimulant and narcotic properties and were formerly used for amenorrhoea, flatulent colic and hysteria. They are chiefly used in veterinary medicine. The oil of its berries is externally applied as a stimulant in sprains. Fruits are used in buer, diarrhea and dropsy. Its aromatic fat is used to some extent in pharmacy.

Common name: Black cumin (Kalaunji)

Botanical name: *Nigella sativa* Linn.

Family: Umbelliferae

The seeds of black cumin have been reported to be carminative, diuretic, emenagogue, galactogogue and stimulant and are used as a remedy of mild puerperal fever. It is applied externally for treatment the skin eruptions. They are used as an antidote to scorpion venom. Alcoholic extracts of its seeds possess anti-bacterial properties against *Escherchia coli* and *Micrococcus Pyogenes* var. aureus. Its seeds act as a stabilizing agent for edible fat. The seeds if scattered between thc folds of linen or woolen clothes can protect them from insect attack.

Common name: Black pepper (Kalimirch)

Botanical name: *Piper nigrum* Linn.

Family: Piperaceae

Black pepper is referred to as the King of spices. Pepper is unique in the spice world as the pepper fruits are marketed in four different versions: black, white, green and red peppercorns. By choosing time of harvest and processing done thereafter, all four types could be produced from the same pepper plant. However, it is the black pepper, which is the most popular. This valuable spice is produced by a woody, evergreen vine, widely cultivated in the Malabar region of South India. Cochin is the pepper trade centre in India. The pepper grades are identified by their origin, as the most important Indian grades are Malabar and Tellicherry.

The Malabar grade is regular black pepper with a slightly greenish hue, while Tellicherry is very aromatic and pungent. The fruits that develop upon these spikes are generally harvested while still unripe and green in colour. These green fruits are then dried at moderate temperature until the flesh around the single hard seed is wrinkled and turned almost black. It is now that these fully ripened seeds are ground into black pepper or are even sold as whole seeds. The milder white pepper is made from the same plant; but the fruits are allowed to ripen and the flesh is removed before the seed are ground. White pepper retains the full pungency of black pepper, but it has an altered flavor due to partial loss of aroma compounds. Green pepper is early-harvested pepper, which is processed by pickling the freshly harvested peppercorns in salt or vinegar, or by quick drying at elevated temperature or in a vacuum (lyophilization). It is mildly pungent and has a fresh-herbal flavor. The same kind of pickling procedure when applied to ripe peppercorns yields red pepper, which is more pungent and aromatic than green pepper and has the spicy flavor of black pepper.

The edible part of black pepper is its dried fruit, which is highly aromatic due to the presence of volatile oils that contain monoterpenes hydrocarbons. Sesquiterpenes constitute about 20% of the essential oils. Besides flavoring, the volatile oils are also used in perfumes. The seeds contains the alkaloids namely, piperine and piperidine along with some resins, which are the main active constituents. The bioavailability-enhancing activity of piperine results in increased intestinal absorption of several therapeutically useful drugs. Piperine increases intestinal brush border membrane-fluidity and stimulates the activity of certain membrane enzymes. This clearly suggests the role of piperine in altering, membrane dynamics and its permeation characteristics, thus, causing an increase in the absorptive surface of small intestine. Piperine influences the bioavailability of curcumin, ingredient of *Curcuma longa* or *Haldi*, in healthy human volunteers.

In India, black pepper has been used as a medicine for a variety of ailments. Besides its use as a cooking spice, black pepper is known for its curative effects in conditions such as chronic indigestion, sinus congestion, fever, colic, hemorrhoids, worms' infection, toothache and sore throat. It is even considered effective in treating boils and skin diseases. It is a powerful remedy for various disorders of the anatomical system and recommended as an effective cure for aphrodisiac delirium tremens, dyspepsia, hemorrhoids and malaria. It may be used for embalming. The homemakers use it as an insect repellent and moth killer. It possesses some feeble anti periodic properties.

Common name: Caper (Kabra)

Botanical name: *Capparis spinosa* Linn.

Family: Caparidaceae

Caper buds as well as the fruits are considered useful in scurvy disease. Its macerated leaves are used as a poultice in gout. The bark is bitter, aperient, diuretic, expectorant, emmenagogue and tonic. It is used in rheumatism, paralysis, toothache, affections of liver and spleen and tubercular glands.

Common name: Caraway (Siya zira)

Botanical name: *Carum carvi* Linn.

Family: Umbellifereae

Caraway is a mild carminative and stomachic, thus, occasionally used in flatulent colic but its oil is used more often than the fruit itself. Carvene isolated from its oil is used as anthelmintic and in hookworm. Its oil is chiefly used in medicines as a carminative and also useful to correct nauseating and griping effects of medicines.

Common name: Celery seeds (Ajmud)

Botanical name: *Apium graveolens* var. chanu celery

Family: Umbelliferae

The dried ripe fruits are used as a pre-dinner appetizer, stimulant and tonic in asthma and in liver complaints. Its essential oil helps in contraction of gravel and virginal uterus. In celery tonics, its seeds are the main ingredients. The decoction prepared from its seeds is a domestic remedy for rheumatic pains. Its seeds are also used as nerving sedative medicine and its oil is used as an anti-spasmodic and nerve stimulant. It is a useful remedy for rheumatoid arthritis and acts as an intestinal antiseptic. The powder of its root's is considered diuretic and is given in colic.

Common name: Cinnamon (Dal chini)

Botanical name: Cinnamomum zeylanicum Blume

Family: Lauraceae

Every part of this tree, leaves, bark, buds, flowers, fruits and roots is useful in one or other way in human life. Its bark's powder is aromatic, astringent, carminative and stimulant and has the potential of controlling nausea and vomiting. The bark oil is extensively used in pharmaceutical and dental preparations due to its high germicidal activity. It also acts as a fungicide and possesses the carminative and cordial properties, thus, added as adjuvant in carminatives and stomachic medicines. It is some times recommended in flatulent colic, gastric debility and gastrodynia. Cinnamon leaf oil is used as an embrocation in rheumatism.

Common name: Clove (Laung)

Botanical name: *Eugenia caryoehyllus* Sprengel

Family: Myrtaceae

It is extensively used in pharmaceuticals as aromatic, carminative and stimulant and used in dyspepsia and flatulence. It acts as a digestive agent and it is added in several medicines especially for toothache because of its antiseptic and antibiotic. It is an ingredient of many tooth powders, toothpastes and mouthwashes. Its oil is extensively employed as a cleansing agent in histological work. It imparts warming qualities. It is also used as flavoring agent in oral preparations, *i.e.*, dentifrices and gargles and in chewing gums.

Common name: Curry leaf

Botanical name: *Murraya Koenigii* (Linn) Sprengel

Family: Rutaceae

The leaves, roots and bark of this plant are used in medicine as a carminative, stimulant, stomachic and tonic. A decoction of roasted leaves is used in vomiting. They are also externally used to cure eruptions and the bites of poisonous animals. The green tender leaves are eaten raw for controlling dysentery. Sometimes, a decoction of its leaves is given with bitters as a febrifuge and the leaves chutney with peppermint is used to stop vomiting. It has also been used as an anti-periodic, and many a time a dried leaf powder mixed with honey and betel nut juice is recommended in *Ayurvedic* system of medicine.

The leaves, fruits, bark and roots have been reported to contain a glucoside, koenigin. Aqueous leaves extracts, when administered parenterally female guinea pigs, not only raise the phagocytic index but also mobilize a greater number of leucocytes to take part in phagocytosis but the effect is not long lasting. The roots juice is taken in relieving pain associated with kidney.

Common name: Dill (Sowa)

Botanical name: *Anethum sowa* Roxb. Ex flem

Family: Umbelliferae

The seeds due to their essential oil are well known for their medicinal properties. Both seeds and oil are used as an ingredient in a variety of indigenous medicinal preparations. The essential oil, dill oil, or its emulsion in water, dill water, is considered an aromatic carminative specially useful in flatulence, colic and hiccups of infants and children.

Common name: Greater cardamom (Bari elaichi)

Botanical name: *Amomum aromaticum* Roxb.

Family: Zingiberaceae

Greater cardamom seeds are a good tonic to the heart and liver, an astringent to the bowels, an appetizer and a hypnotic and cause belching. The outside coverings of seeds heal stomatitis and are good for headache and teeth. The decoction of seeds is good for gums and teeth if used as a gargle. The decoction of its seeds prepared with melon seeds is used as a diuretic in case of kidney stone. It may cure certain digestive disorders by promoting secretion from intestine and is useful in congestion of liver. Large doses of grains (30 grains) in conjunction with quinine have been found very useful in neuralgia. The seeds are used in gonorrhoea and as an aphrodisiac. They are an antidote to either snake venom or scorpion venom. Oil extracted from seeds is applied to allay inflammation. It is aromatic, stomachic and stimulant

Common name: Large cardamom (Elaichi)

Botanical name: *Aframomum* species

Family: Zingiberaceae

Large cardamom is used in several human medicines, which are taken as a sedative against toothache, to guard against rheumatism and migraine, to cure fever, dysentery *etc*. Its seeds are chewed to correct the foul breath and the pulp of the plant is often applied externally to relieve pain. The stem extract is used as a purgative to cure the tapeworm infection. An extract from young leaves is used as a styptic. Before the seeds mature, the pulp is chewed as a stimulant. The seeds are taken to make aphrodisiac and abortifacients more palatable. Both its leaves and seeds are used in veterinary medicines too.

Common name: Lesser cardamom (Chhoti elaichi)

Botanical name: *Elletaria cardamom* Maton

Family: Umbelliferae

True cardamom is popularly known as the *Queen of Spices* and used to prepare tinctures, which are used in medicines for stomachic or windiness. Powdered seeds in combination with ground ginger, cloves and caraway are very useful in combating digestive disorders. In medicines, it is added as a powerful aromatic stimulant, carminative, diuretic and stomachic. It is reported to be a cardiac stimulant; its seeds are masticated to prevent foul breath, indigestion, nausea and vomiting, morning sickness and excessive watering in mouth. Gargling with cardamom and cinnamon decoction cures the sore throat and pharyngitis. Cardamom seeds powder is used as a medicine for scanty urination, diarrhea, dysentery and palpitation of heart, exhaustion due to over work and depression. A single cardamom if taken daily with a tablespoon of honey can improve eyesight, strengthen the nervous system and keep the subjects healthy. However, its excessive use causes impotency in human beings. A good nasal drop is prepared from cardamom extracts, *neem* and myrobalans in combination with animal fat and camphor.

Common name: Galangal

Botanical name: *Alpinia galangal* Linn.

Family: Zingiberaceae

The galangal rhizomes are used in indigenous medicine for the treatment of rheumatism and catarrhal affections, bronchial catarrh. The rhizomes are also carminative and stomachic. The drug is a depressant of the cardio-vascular system. Small doses stimulate the respiration but larger ones depress it. It has important action on the bronchioles. The rhizome and its essential oil are useful in respiratory troubles, especially of children.

Common name: Horseradish

Botanical name: *Armoracia rusticana*

Family: Brassicaceae (Cruciferae)

It is considered diaphoretic, digestive, diuretic and stimulant. It is used as a counter irritant in lumbago and similar painful affections. Crushed-form of horseradish produces an inhibitory effect on the growth of microorganisms. This effect is due to the presence of an alkaloid allyl isothiocynate.

Common name: Hyssop

Botanical name: *Hyssopus officinalis* Linn.

Family: Labiatae

Hyssop is carminative, pectoral and stimulant and used in colds, coughs and lung complaints. An infusion of hyssop plant is effective in nervous disorders, toothache and in pulmonary, digestive, uterine and urinary troubles. A plant decoction is useful in asthma and coughs. Leaves are stimulant, stomachic, carminative and useful in hysteria and colic. The leaf juice is applied for the expulsion of round worms. The macerated leaves are applied as a resolvent and vulnerary. The steeped leaves in hot water is used as fomentation for wounds, sprains and strains, muscular rheumatism and for clearing skin discoloration due to blows. It is also used as a salve in catarrhal ophthalmic. Hyssop oil is used as a tonic. Small dose of its oil (1 or 2 drops) promotes expectoration in bronchial catarrh and asthma and induces clonic convulsions when injected intravenously.

Common name: Juniper (Aarhaar)

Botanical name: *Juniperus communis*

Family: Pinaceae

The fruit and its oil both have carminative, diuretic and stimulant properties and are useful in different forms of dropsy, in disorders of urino-genital tract such as gleets, gonorrhea and leucorrhoea and in certain cutaneous diseases. The wood and its berries are blood purifier, diuretic and sudorific. It is employed in gout, rheumatism and cutaneous diseases.

Common name: Kokam

Botanical name: *Garcinia indica* Choisy

Family: Guttiferae

The fruits of kokam are anthelmintic and cardiotonic and considered useful in piles, dysentery, pains, heart complaints and any kinds of tumor and its roots powder is considered strong astringent. Kokam butter has been reported to be very nutritive, astringent, demulcent and emollient and used for several ailments, suppositories and other pharmaceutical purposes. It is applied as local application to fissures of lips, hands and ulcerations.

Common name: Lemon balm

Botanical name: *Mellissa officinalis* Linn.

Family: Labiatae

Its leaves and flowering tops are used in medicine. It is used to strengthen the gums and to remove bad taste of mouth. Lemon balm possesses stomachic, anti-tubercular and anti-pyretic properties. The fruits are considered a brain tonic, which is useful in hypochondriac conditions. Leaves and stems are useful in brain, liver and heart ailments and in bites of poisonous insects. Oil of lemon balm is said to be sedative and its properties are similar to those of peppermint oil. Balm distillates act as a mild spasmolytic agent.

Common name: Mace (Jaivitri) and Nutmeg (Jaiphal)

Botanical name: *Myristica fragrans*

Family: Myristicaceae

Both are used chiefly as drugs than a condiment only because of their valuable medicinal properties. Both the spices are astringent, aphrodisiac, carminative and stimulant, thus, they are used on large scale in pharmaceutical preparations for dysentery, flatulence, leprosy, malaria, nausea, rheumatism, stomachache, sciatica, vomiting and early stage of leprosy, however, high doses cause narcotic effects. Symptoms of delirium and epileptic convulsions appear after 1-6 h. Nutmeg is used in tonics and electuaries. Mace is masticated to correct the foul breath.

The use of nutmeg is comparatively greater in the medicine than the mace. Their volatile oil is employed as a flavourant in dental creams and perfumery. It is used in liniments and hair lotions. It has been prescribed for the treatment of inflammations of urinary bladder and tract. The oil is irritant and toxic so it should be used with proper care. Nutmeg butter is used as a stimulant in external ointments, hair lotions and plasters. It also forms a useful application in paralysis, rheumatism and sprains. Its alcoholic extract possesses antibacterial activities against *Micrococcus* var. Aureus.

Common name: Marjoram (Marwa)

Botanical name: *Marjoram hortensis* Moench

Family: Labiatae

Its leaves and seeds are considered astringent, carminative, expectorant and used as a tonic. The decoction of plant is used as stimulant, emmenagogue, glalactagogue and sudorific. It is said to be useful in asthma, hysteria, colic and paralysis. Its oil is applied externally for bruises, hot fomentation in acute diarrhoea, paralytic and stiff limbs, sprains and toothache.

Common name: Mint or Japanese mint

Botanical name: *Mentha arvensis* Linn.

Family: Labiateae

Azulene, a compound, has anti-inflammatory and anti-ulcer properties. Mint possesses α-tocopherol, choline and niacin. Choline acts as good neurotransmitter. Niacin when administered orally through ingesting mint leaves exerts weak and transient hypoglycaemic effect in diabetic patients. Nicotinic acid is hypoglycaemic if administered parenterally, outside the intestine. Alfa-tocopherol is shown to reduce the levels of giycosylated haemoglobin, which may help to reduce the incidences of diabetic complications. Mint contains biologically active compounds such as apigenin, anisaldehyde, caffeic acid, chlorogenic acid, coumarin limonene, eugenol, luteolin, *o*-cresol, *p*-cresol and vanillin, which are all anti-mutagens and luteolin and apigenin, are cardiotonic. Eugenol, apigenin and rosmarinic acid have anti-platelet aggregation function. Apigenin inhibits platelet aggregation and depresses the cyclooxygenase pathway. It also reduces platelet c-adenosine monophate response to protaglandin I_2 probably mediated by inhibition of adenylate cyclase. Peppermints contain carvone, which has chemo-protective action. Carvone excites an enzyme, glutathione-S-transferase that detoxifies the toxic substances having specific organ targets to induce cancer. Some anti-mutagenic compounds mentioned above present in mint are virucidal. Caffeic acid is anti-influenza, anti-herpes simplex, anti-HIV, anti-vesicular stomatitis, anti-polio, anti-adeno and anti-vaccinia viruses.

Dementholized oil is used in mouthwashes, toothpastes and pharmaceutical preparations for flavoring. The major use of mint is the extraction of volatile oil, which contains menthol used in medicine for stomach disorders, in ointments for headaches, rheumatism and other pains, in cough drops and in inhalers. The oil and the dried plant are antiseptic, carminative, refrigerant, stimulant and diuretic. The dried plant is expectorant, emmenagogue, tonic to the kidney, useful in liver and spleen diseases and in asthma. It also possesses antispasmodic properties and it is used in jaundice and frequently given to stop vomiting. A leaves and stems infusion is used as antispasmodic carminative and sudorific and given in fever and indigestion. The plant is also considered as an excellent diaphoretic.

Common name: Origanum (Sathra)

Botanical name: *Origanum vulgare* Linn.

Family: Labiatae

Origanum oil has carminative, diaphoretic, diuretic, emmenagogue and stomachic properties. It is taken as a stimulant and tonic in colic and diarrhea. It is mixed in wound healing ointments and applied externally in chronic rheumatism, earache and toothache. Because of its spasmolytic action, it is used in bronchitis and whooping cough and as homeopathic medicine for hysteria. Its oil is employed in veterinary liniments. It stimulates the growth of hair. It is used in gargle and bath water. The oil stimulates the growth of hair.

Common name: Parsley (Ajmood)

Botanical name: *Petroselinum crispum*

Family: Umbelliferae

Parsley plant possesses anti-pyretic, carminative, diuretic, ecbolic and emmenagogue properties. It has long been used in uterine troubles. The crushed leaves are applied to remove the biting effect of poisonous insects. The juice of fresh leaves has properties of insecticide, thus, its mericarps are used to kill the lice and skin parasites. Extracts and infusions of leaves and roots on subcutaneous administration in mice produced a depressing effect on the nervous system.

Common name: Peppermint

Botanical name: *Mentha piperita*

Family: Labiateae

The herb is said to be stimulant, stomachic and carminative and used for allaying nausea, flatulence and vomiting. Macerated leaves are employed as an external application for relieving local pains and headache. A hot infusion is taken to allay stomachache and colicky diarrhoea. Peppermint oil employed for flavoring pharmaceuticals, cough drops, chewing gums, mouthwashes and dental preparations. It is valued in medicine both for internal and external uses. It is widely used in flatulence, nausea and gastralgia. The oil has mild antiseptic and local anesthetic properties. It is used as an external application in rheumatism, neuralgia, congestive headache and toothache.

Common name: Rosemary (Rusmary)

Botanical name: *Rosmarinus officinalis* Linn.

Family: Labiatae

The oil of rosemary is used as in medicinal preparations and as an ingredient in rubefacient liniments. The oil is useful in atonic dyspepsia, headache and in tardy menstruation. The oil possesses carminative, diaphoretic, stimulant, anti-bacterial and protistocidal properties. It is employed with hot water in chills and colds. The oil emulsion is employed for blending in Eau-de-colonge. All plant parts are astringent and used as a nerving tonic and an excellent remedy for stomach troubles. An infusion of plant with borax is said to be very useful for preventing pre maturing baldness if employed as a hair wash. Its tea is used to cause temporary sterility. Its flowering tops and leaves are reported to be aperient, carminative, diaphoretic, diuretic, emmenagogue, stimulant and stomachic. They are employed for vapor bath for the relief of incipient catarrhs, paralysis and rheumatism. Dried leaves are smoked for relief of asthma. Juice of the leaves has strong anti-bacterial action against *E. coli*, *Bacillus subtilis* and *staphylococcus aureus.*

Common name: Saffron (Kesar, Jaffran)

Botanical name: *Crocus sativus* Linn.

Family: Iridaceae

Saffron has been a popular spice since ancient times known for its unique medicinal properties besides its use in culinary proposes. It not only imparts yellow colour to food but it has a unique aroma that has no substitute. It possesses stimulant and stomachic properties. It is used in catarrhal affections of children, fever, enlargement of liver and spleen, melancholia and in the treatment of coughs, gastrointestinal colic and insomnia is well known. It is an important ingredient of several *Ayurvedic* and *Unani* medicines.

It is employed as an abortifacient and useful in curing urinary, digestive and uterine troubles. It enables the patient suffering from urine trouble to pass the urine freely when used with honey after soaking overnight in water. It is added in other medicines to help in normal menstruation. It is popularly known as dry, warm and stimulant in action, helping diabetic, digestive, urinary and uterine problems. It has been reported to provide strength to the heart and brain when used in large quantity. Besides, saffron has been used in making perfumes and dyes. Recent pharmacological studies have shown that saffron extract has anti-tumor as well as hypolipidemic effects. It also has memory-improving properties.

Saffron contains cancer-inhibiting constituents, namely crocin, safranal and picrocrocin. These compounds of saffron could inhibit cell growth of human tumor cells. This has opened the door for developing new saffron-based drugs for the treatment of cancer since saffron extract (dimethyl-crocetin) has the anti-cancerous activity against a wide spectrum of murine tumors and human leukemia cell lines. Scientists believe this anti-cancer property of saffron is due to its rich carotenoid content.

Traditionally, saffron is known to have sedative properties. Saffron has been reported to have some behavioral effects on the central nervous system. The aqueous and ethanolic extracts of saffron have anti-convulsing effects in mice. In mice, saffron was shown to delay the onset of tonic convulsions but it failed to produce complete protection against mortality. The extracts of saffron may be beneficial in seizures.

Saffron extract affects learning and memory in experimental animals. Ethanol-induced impairment of learning behaviour in mice showed improvement with saffron extract. This effect of saffron extract is attributed *to* its constituent namely, crocin. Saffron has the potential of being employed for treating memory impairment associated with certain neurodegenerative disorders. Saffron extracts have a strong anti-inflammatory as well and this property of saffron has been assessed in experimental animals, as formalin-induced edema in the rat paw has been shown to improve with saffron extracts.

Common name: Sage (Salvia)

Botanical name: *Salvia officinalis* Linn.

Family: Labiatae

It is used as an astringent, carminative and mild tonic. An infusion of leaves is recommended as a gargle for curing sore throat and hot infusion is said to be diaphoretic. A strong infusion is used for drying up of the breast milk for weaning children. Extract of leaves is believed to be antipyretic. It is prescribed to cure female disorders. Its plant is used in gargles, hair tonics, hairdressings, poultices, mouthwashes, tooth powders and toothpastes. Leaves are rubbed on teeth as a good dentifrice. Dried leaves are used as a fumitory. Sage and sage oil exhibit antioxidant properties

Common name: Savory

Botanical name: *Saturcia hortensis* Linn.

Family: Labiatae

Savory is said to be carminative and stimulating. The antioxidant property of the herb is reported to be due partly to labiatic acid.

Common name: Shallot

Botanical name: *Allium ascalonicum* Linn.

Family: Liliaceae

The shallots are used to cure earache, a small piece being placed in the meats. It is also fried in butter and preserved in honey as an aphrodisiac. The bulbs are ground and rubbed on the skins of feverish children. This is said to cure them. Sometimes, they are mixed with palm vine and large pepper and heated in the sun, the mixture being used to cure fever. They are also used as antidote for snakebite and poisoning.

Common name: Spearmint

Botanical name: *Mentha spicata*

Family: Labiateae

Spearmint is considered stimulant, carminative ant antispasmodic. An alcoholic beverage is prepared from leaves and used as an antidote for poison. A sweetened infusion of the herb is given as a remedy for infantile troubles, vomiting in pregnancy and hysteria. The leaves are used in fevers and bronchitis. Spearmint oil is used for flavoring chewing gums, toothpastes and pharmaceutical preparations.

Common name: Star-anise

Botanical name: *Illicium verum* Hook.

Family: Umbelliferae

Star-anise fruit is chewed to sweeten the breath and to help digestion. It is stomachic and carminative and considered useful in flatulence, spasmodic affection of the intestinal canal and dysentery. It is used as an adjunct to cough mixtures and as a corrective of taste. Star-anise oil is used as flavouring agent in chewing gum and pharmaceutical preparations. The oil is stimulant, stomachic, carminative, mildly expectorant and diuretic. It relieves colic and is an ingredient of cough lozenges. The oil is employed as massage in rheumatism and otalgia and as an antiseptic. It is useful against body lice and bed bugs and forms an ingredient of cattle sprays. It is used in favus (honeycomb ringworm) and scabies. Star-anise plant is considered useful for colic, constipation and insomnia.

Common name: Sweet flag or calamus

Botanical name: *Acorus calamus* Linn.

Family: Araceae

The medicinal properties of the calamus plant have been known since the earliest dawn of human history. Its medicinal action is attributed to its essential oil content. The roots also contain a glucoside called *acorin*. The oil is reported to cure gastritis and is supposed to promote digestion. It is also administered to animals as a stomachic. The essential oil is also used in stomach and skin diseases as a vermifuge and antiseptic as well.

Common name: Tamarind (Imli)

Botanical name: *Tamarindus indica* Linn.

Family: Leguminosae

In *Ayurvedic* system, its ripe fruits are said to be anthelmintic, appetizing, digestive, heating, laxative and refrigerant. It cures cough and colds and heals wounds and fractures. It is useful in diseases caused by deranged bile such as intoxication from liquors and burning of the body. A gargle of tamarind water is useful in the treatment of sore throat. Its leaves are used in a number of medicines. They are employed externally to reduce inflammatory swellings, ringworm and a number of other diseases. The bark possesses astringent and tonic properties. It is applied topically for loss of sensation in paralysis. Its ash is given for normal urinary discharges and gonorrhea. It is very useful in ulcers. The seeds are considered astringent, aphrodisiac and useful in giddiness and vertigo and applied externally in inflammations and liver complaints.

Common name: Thyme (Banajwain)

Botanical name: *Thymus vulgare* L. or *T. serpyllum* L.

Family: Labiatae

The oil of thyme is very popular for its medicinal properties. It is employed in the preparations used for the treatment of bronchitis and whooping cough. It is useful in toothache. This herb is given in weak vision, stomach and liver troubles, suppression of urine and menstruation. It is considered anti-spasmodic, carminative and tonic. An infusion is given in flatulence, convulsive coughs, whooping coughs, catarrh and sore throat. It is a good remedy for nervous or hysterical headaches. The leaves are laxative, stomachic, tonic and blood purifier and are good for kidney and eyes. The herb is alexiteric, emmenagogue, anthelmintic, good in liver complaints, chest pains, liver and spleen problems, useful in asthma and bronchitis. Seeds infusion is used in skin eruptions diseases.

Common name: Vanilla

Botanical name: *Vanilla fragrans* (Salisbury) Ames.

Family: Orchidaceae

The medicinal uses of vanilla have gradually reduced. It is no longer regarded as a drug. However, it is still the most lovable flavourant. The remaining quantity was used in deodorants, perfumes, odor fixatives and as a masking agent in pharmaceutical and vitamin preparations. The greatest use for the technical grade vanillin (98% pure) is a chemical intermediate in the production of a number of pharmaceutical products. Foaming in lubricating oils can be prevented by the use of vanillin. It can also be used as solubilizing agents for riboflavin (vitamin B_2) as vanillin is an anti-oxidant.

■■■

16 Curative Significance of Underutilized Vegetables

S.K. Sanwal[1], Harsimrat K. Bons[2] and M.K. Rana[3]

[1]Division of Crop Improvement, Indian Institute of Vegetable Research, Varanasi-221305, Uttar Pradesh, E-mail: <satishsanwal@rediffmail.com>
[2]Krishi Vigyan Kendra, Samrala, Punjab Agricultural University, Ludhiana, Punjab, India, E-mail: <harsimratpau@gmail.com>
[3]Department of Vegetable Science, CCS Haryana Agricultural University, Hisar-125 004, Haryana, India, E-mail: <mkrlotus@gmail.com>

Many neglected and underutilized species are nutritionally rich and are adapted to low input agriculture. The erosion of these species can have immediate consequences on nutritional status and food security of the poor. Wild, semi-domesticated, or fully cultivated, these species are important for the well-being of local communities. Their enhanced use can bring about better nutrition and fight hidden hunger, *e.g.*, the vitamin C in the underutilized fruit Barbados cherry (*Malpighia glabra*) is ten times higher than the kiwi fruit, which is notably very rich in this micronutrient. Tree tomato has more acidity and vitamin C and beta-carotene content than normal tomato. The nutritional value of the Himalayan chenopod grains, *Chenopodium* spp., sometimes called pseudocereals is superior to that of most major cereals.

The vitamin A content of many underutilized leafy vegetable species used in India and Sub-Saharan Africa is comparatively higher than other well-established vegetable crops such as spinach. Another example is that of Bambara groundnut (*Vigna subterranea*), a neglected pulse from Africa, whose seeds are rich in protein (24%) with levels of the essential amino acid methionine higher than that found in most other grain legumes. Broad bean is also a very rich source of protein (24-28%) and replaces meat from human diet. Many underutilized leafy vegetables are rich source of folic acid, commonly referred to as folate, which is helpful in reducing the risk of cardiovascular disease. It is found in abundance in poi/Indian spinach (123 µg/100 g), chenopods and leaves of drumsticks. Folic acid is crucial in foetal development in the early stage of pregnancy. This chapter highlights the medicinal and therapeutic significance of underutilized vegetable crops.

Common name: Chow-chow or chayote

Botanical name: *Sechium edule*

Family: Cucurbitaceae

The fruit of chow-chow are rich source of carbohydrates, proteins, fat, minerals and vitamins, particularly vitamin A and vitamin C. Owing to carbohydrates and minerals, chow-chow fruits can also be used for the preparation of pickles, candy and flour products. The fruit and particularly the seeds are rich in amino acids such as aspartic acid, glutamic acid, alanine, arginine, cysteine, phenylalanine, glycine, histidine, isoleucine, leucine, methionine (only in the fruit), proline, serine, tyrosine, threonine and valine. The extracts of leaves and shoots show transient depresses or effects when injected in laboratory animals.

The chayote has many medicinal uses. The infusions of leaves are used to dissolve kidney stones and to assist in the treatment of arteriosclerosis and hypertension and the infusion of fruits is used to alleviate urine retention. The cardiovascular properties of leaves infusions have been tested in modern studies, while their great effectiveness in curing kidney diseases has been known since colonial times on the Yucatan peninsula, where these ailments are very common. Leaf ethanolic extracts and leaf and seed water extracts showed high antioxidant activity. Leaf and seed extracts may be exploited as biopreservatives in food applications as well as for health supplements or functional food to alleviate oxidative stress.

Common name: Sweet gourd

Vernacular name: Kakrol, baby jackfruit, spiny bitter gourd

Botanical name: *Momordica cochinchinensis* Spreng

Family: Cucurbitaceae

Sweet gourd is a highly nutritious and delicious vegetable that has many industrial and medicinal uses. It is a good source of carbohydrates, ascorbic acid and beta-carotene. The fruits are used as vegetable when they are green in colour and just before attaining full maturity. The unripe fruits act as appetizer and astringent. It is a good medicine for those who suffer from cough and other digestible problems. Young leaves, flowers and seeds of sweet gourd are also edible. Its seeds contain oleostearic acid, a kind of fatty acids. Roots of sweet gourds are used in bleeding piles, bowel infection and urinary complaints. When powder or infusion of dry fruits is introduced into nostrils, it produces erring effect and provokes a copious discharge from schneiderian membrane. The powder is indicated to promote the subsidence of nodulation and swelling, combat poisons and cure sores and is orally consumed for the treatment of sores and inflammatory swellings, mastitis, scrofula, hemorrhoids, anal fistula, chronic eczema, neurodermatitis and scalds.

Sweet gourd is valued for many important health benefits and is prized for its ability to promote longevity, vitality and health. The seeds, known in traditional Chinese medicine (TCM) as *Mubiezi*, are thought to have resolvent and cooling properties and are used for liver and spleen disorders, wounds, haemorrhoids, bruises, swelling and pus. The seed membranes are used to make a tonic for children and lactating or pregnant women and to treat *dry eyes* (xerophthalmia) and night blindness.

Common name: Snake gourd

Vernacular name: Chichinda, serpent gourd and padwal

Botanical name: *Trichosanthes cucumerina* L.

Family: Cucurbitaceae

Snake gourd is the most valuable home remedy for common illness. The fruits are rich in carbohydrates, protein, minerals and vitamin A, B and C. The predominant mineral elements were potassium (121.60 mg/100 g) and phosphorus (135.0 mg/100 g). It improves appetite, acts as a tonic and stomachic and cures biliousness. The roots and seeds are anthelmintic and are also used in the treatment of diarrhea, bronchitis and fever. Snake gourd is useful in restoring the disordered processes of nutrition. It creates a cooling effect in the body. Being a low-calorie food, diabetics can safely include it to reduce weight while getting enough nutrition. Its leaves are used in indigenous medicine in India. Combination of leaves is beneficial in the treatment of jaundice, if given in 30 to 60 gram doses with a decoction of coriander seeds thrice daily. Snake gourd is the best household remedy for heart problems. The juice of the fresh leaves is useful in heart disorders like palpitation and pain in the heart on physical exertion. It also plays a vital role in curing fever also. A decoction of snake gourd is useful in bilious fevers, as a febrifuge or thirst reliever and laxative. Its efficacy increases provided it is given with *chiratta* and honey. In obstinate cases of fevers, a combined infusion of this plant and coriander is more beneficial. About 30 grams of each should be infused in water overnight. The strained liquid should be given in two doses the next day. The leaf juice is used to induce vomiting. The latter is also applied locally as a liniment in cases of liver congestion. In remittent fevers, it is applied over the whole body. The leaf juice is beneficial in the treatment of alopecia, a disease of the scalp resulting in complete or partial baldness.

In *Ayurveda*, the snake gourd is a folk cure for intestinal infestations. It is also used to induce vomiting after poisoning. Snake gourd's root serves as purgative and tonic, whereas, its juice is a strong purgative, and an infusion of dried fruit is a mild purgative. It also aids digestion. Its leaves are useful as an emetic and purgative in children suffering from constipation. A teaspoon of fresh juice can be given early in the morning for the ailing children.

Common name: Kachari

Vernacular name: Kachariya, pathya, or sengha

Botanical name: *Cucumis melo* var. calosus

Family: Cucurbitaceae

Kachari is consumed in various forms as cooked vegetable, chutney and pickles. Besides, the dehydrated *kachari* is also used for making chutney and vegetable. About 20-25% of *kachari* is used for making powder like *amchur* and approximately 75% of the total produce is used as part in cooked vegetable. Powdered kachari is used in various types of spices on commercial scale. Dried *kachari* is also a component of the *panchkuta*, a blend of five dried fruits/seeds, *viz.* dehydrated pods of *Khejri* (*Prosophis cineraia*), ker (*Capparis decidua*), fruits of lasora (*Cordia dichotoma*) and seeds of *kumat* (*Acacia Senegal*). *Kachari* supplies fairly good amount of carbohydrates, proteins, minerals and moisture. The dehydrated kachari is coughcide, vermicide, cooling refrigerant, diuretic and gastric stimulant. It is useful in digestive disorders, facial paralysis, cough, coryza, bile, obesity, piles and constipation.

Common name: Stuffing cucumber (meetha karela)

Vernacular name: Slipper gourd, sparrow gourd, or wild cucumber

Botanical name: *Cyclanthera pedata* Schard

Family: Cucurbitaceae

The *Cyclanthra pedata* is a vine, grown for its small fruits, which are used as a vegetable. It is a rich source of minerals such as calcium, magnesium and iron. Ripe fruits contain peptin, galacturonic acid, dihydroxitriptamine, pierine, resins, minerals (phosphorus), vitamins (thiamine and vitamin C), lipoproteins and steroidal compound (systosterol and 3 β-D-glucoside) with hypoglicaemic and antilipemic (against cholesterol LDL) action and low-density lipoproteins. Investigations into the chemical constituents of *Cyclanthera pedata* seeds and fruits have revealed the presence of 6 flavon glycosides, 9 triterpenoid saponins and 6 cucurbitacin glycosides.

The immature fruits may be eaten raw or pickled. Unlike many cucurbitaceous fruits, the stuffing cucumber becomes quite hollow as it matures, and the mature fruits are often eaten stuffed. The young shoots and leaves may also be eaten as greens. In case of disease and/or obesity, it can be eaten as juice extract during a fast. It is also effective as dehydrated powder. It has various traditional medicinal usages, mainly to control cholesterol, reduce obesity and control high blood pressure.

The stuffing cucumber has long been utilized for weight loss and problems associated with arteriosclerosis. It is thought to be one of the most potent natural fat absorbers and is traditionally taken together with Hercampuri to rejuvenate and reduce cellulite. Clinical studies have demonstrated that the regular consumption of this fruit regulates the metabolism of lipids and sugar in the blood stream, controlling cholesterol levels by reducing low-density lipoprotein (the bad cholesterol) while increasing high-density lipoprotein cholesterol (the good type). Flavonoids and triterpenoid saponins have been isolated from the fruits. A preclinical study has also shown stuffing cucumber to be a more potent antiinflammatory than ibuprofin. Stuffing cucumber weight reducing and cholesterol-balancing effects are most noted when consumed 2-4 teaspoons daily for several weeks or months. Due to its pleasant taste, it is mixed well with foods or can be dissolved in liquids or taken in capsules.

Common name: Ash gourd

Vernacular name: Petha, white gourd or wax gourd

Botanical name: *Benincasa hispida*

Family: Cucurbitaceae

Ash gourd is wholesome and nutritive. Being low in calories, it is particularly useful for diabetic and obese people. It is cooling and laxative. This herb increases the secretion and discharge of urine. It promotes libido and is useful in periodic attacks of hysteria and convulsions. The dilute juice of ash gourd is beneficial in the treatment of peptic ulcer. The juice, squeezed out of grated ash gourd with an equal amount of water added to it, should be taken daily in the morning on an empty stomach. This also relieves inflammation in the alimentary canal. It also acts as a blood coagulant. From ancient times, its fresh juice mixed with a teaspoon of Alma or limejuice is used as a specific medicine to stop profuse bleeding from lungs and nose, in piles and haematuria, a disease characterised by the presence of blood cells in the urine. A delicious sweet prepared from pulp of the fruit by boiling its pieces in water and adding sugar syrup to it is used as a medicine to increase weight, in tuberculosis, weakness of the heart, heat in the body, thinness of semen and anaemia. Shelled seeds of ash gourd are anabolic that promote tissue growth-especially when taken with coconut milk. They expel tapeworm and other worms from the intestine.

Common name: Ivy gourd

Vernacular Name: Bimba- Sanskrit; Konduru - Hindi; Telakuchu- Bengali; Tondale- Marathi; Ghole - Gujarati; Tondekayi-Kannada; Kovakai- Tamil; Dondakayi-Telugu

Botanical name: *Brassica oleracea* var. rubra

Family: Cucurbitaceae

The fruits contain protein 1.2%, vitamin A 260 IU per 100 g and vitamin C 28 mg per 100 g. The roots, stems and leaves are used to cure skin diseases, bronchitis and diabetes. The fresh juice extracted from the leaves and roots either by itself or in combination with certain metallic preparations is administered for diabetes. The various parts of the plants are used in the indigenous system of medicine in both *Ayurvedic* and *Unani* practices. The plant is reported to reduce the amount of sugar in the urine and improve the general metabolism of patients suffering from diabetes. Some parts of the plant particularly the leaves and roots are used by the *Ayurvedic* and *Unani* practitioners in the treatment of conditions similar to diabetes and claims are often made of cures affected by such treatment when insulin treatment has failed. The cooked leaves are used externally in eruptions of the skin and mixed with ghee are applied as ointment to sores, boiled in ginger oil, they are applied externally in the treatment of ring worm, psoriasis, itch, *etc*. The root pounded with juice of leaves and smeared over the whole body induces respiration in fever. The young bitter fruits have also been reported to contain cucurbitacin 'B' in the form of a glycoside. The bitterness rapidly reduces during ripening and ultimately becomes sweet to test and scarlet in colour. The water soluble alkaloid fraction from the leaves possessed hypoglycalonic B-sitosterol activity for a brief duration, B-sitosterol was the main constituent of the petroleum ether extract.

Common name: Winged bean

Vernacular name: Goa bean, Manila bean, princess pea, asparagus pea and four angled bean

Botanical name: *Psophocarpus tetragonolobus* (L.) DC.

Family: Fabaceae/Leguminosae

The nutty tasting root contains about 20% protein. The seeds possess a high nutritional value and contain about 40% protein. A winged bean milk and flour are used as dietary treatments for protein-deprived children.

The immature pods are used as vegetable in almost all parts of the world. The plant produces pea-like beans with four winged edges. Almost every part of this unique plant is tasty and edible. The fresh young pods are similar to green beans with a chewy texture and a slightly sweet taste. The leaves are edible and rich in vitamin A, which prevents night blindness, the tendrils taste like asparagus and the flowers like steamed mushrooms. This remarkable bean could become one of the most important crops for underdeveloped countries because it offers a high source of protein. The dried seeds can be useful as flour and also to make a coffee-like drink. Each of these parts of winged bean provides a source of vitamin A and other vitamins.

Common name: Velvet bean

Vernacular name: Cowitch, Cowhage

Botanical name: *Mucuna pruriens* Linn.

Family: Fabaceae

Velvet bean is one of the popular and important medicinal plants of India. It is a constituent of more than 200 indigenous drug formulations. All parts of velvet bean possess valuable medicinal properties, and there is a heavy demand of *Mucuna* in Indian and International drug markets. This plant is widely used in *Ayurveda*, which is an ancient traditional medicinal system in India. The seeds of velvet bean are high in protein, carbohydrates, lipids, fibers and minerals. They are also rich in novel alkaloids, saponins and sterols. The seeds of all mucuna species contain a high concentration (7-10%) of L-dopa. Concentrations of serotonin also have been found in the pod, leaf and fruit. The stinging hairs of the seedpods contain phytochemical mucunain, which is responsible for causing skin irritation and itch. It is considered a diuretic, nerve tonic and aphrodisiac. Externally, it is applied to ulcers. Velvet bean has a long history of use in Indian *Ayurvedic* medicine, where it is used for worms, dysentery, diarrhoea, snakebite, sexual debility, cough, tuberculosis, impotence, rheumatic disorders, muscular pain, sterility, gout, menstrual disorders, diabetes and cancer. In India, it is considered an aphrodisiac, menstrual promoter, uterine stimulant, nerve tonic, diuretic and blood purifier.

Traditionally, velvet bean has been used as a nerve tonic for nervous system disorders. Due to the high concentration of L-dopa in seeds, it has been studied for its possible use in Parkinson's disease. It is associated with progressive degeneration of dopaminergic neurons in specific areas in the brain. Dopamine does not cross the blood-brain barrier, and therefore, cannot be used directly as a treatment. However, L-dopa (levodopa) does gain access to the brain where it is converted to dopamine. Several *in vivo* studies have been conducted on the blood-sugar-lowering effect of velvet bean. These studies all validate the traditional use of the plant for diabetes. An ethanol-water extract of the root, fruit and seed dropped blood sugar levels in rats by more than 30%. At 200 mg an ethanol extract produced a 40% fall in blood glucose within one month and a 51% reduction at four months. The root, fruit, leaf and seed have shown significant *in vivo* antispasmodic, anti-inflammatory, pain-relieving and fever reducing activities in various clinical research with animals. Traditionally, the seed has been used by indigenous people throughout the world for snakebite and several *in vivo* studies validate this traditional use.

Common name: Broad bean

Vernacular name: Faba bean, Windsor bean, or horse bean

Botanical name: *Vicia faba*

Family: Fabaceae

It is used as green-shell, the seeds removed from the pod before maturity, or as dry beans. It is the principal protein source for poor people in North Eastern Hilly region. The protein content of bean seeds is high, amounting to about 20-25%. Being a high protein crop, it is used to replace meat in human diet. However, presence of anti-nutritional factors [tannins, lectin, protease inhibitor and glucocide vicine that induce favism (a haemolytic anemia) in susceptible individuals] in seed has limited its widespread use. Broad beans are rich in L-dopa, a substance used medically in the treatment of Parkinson's disease. L-dopa is also a natriuretic agent, which might help in controlling hypertension. Some also use faba beans as a natural alternative to drugs like Viagra, citing a link between L-dopa production and the human libido. The green plants are used as cattle feed and in green manuring.

Common name: Lima bean

Vernacular name: Butter bean, Burma bean, Madagascar bean, prolific bean, or sugar bean

Botanical name: *Phaseolus lunatus*

Family: Leguminosae

Lima bean is a good source of minor elements. Immature seed is cooked and used like peas in soups, stews, *etc*. The mature seed is dried and stored for future use. It must be thoroughly cooked before being eaten. The dried seed can be ground into a powder then used as a thickener in soups or can be mixed with cereal flours when making bread. Young seed/pods are steamed and used as a side dish with rice, or added to soups, stews, *etc*. The cooked leaves often have a bitter taste. Lima beans are a very good source of cholesterol-lowering fiber as are most other legumes. In addition to lowering cholesterol, lima bean's high fiber content prevents blood sugar levels from rising too rapidly after a meal, making these beans especially a good choice for individuals with diabetes, insulin resistance, or hypoglycemia. When combined with whole grains such as rice, lima beans provide virtually fat-free high quality protein. Lima beans and butter beans contain linamarin and acyanogenic glucoside, although the beans are rendered safe when cooked, and low-linamarin varieties are typically used for culinary purposes. The seeds are astringent and are used in the diet of people with fevers.

Common name: Tree tomato

Vernacular name: Tamarillo

Botanical name: *Cyphomandra betacea* Cav., Syn. *Solanum betaceum* Cav.

Family: Solanaceae

Tree tomato fruit tastes sour due to high citric acid content. The fully ripe fruits are slightly sweet in taste due to accumulation of sugar during fruit maturity. It also contains phosphorus, calcium, potassium, nitrogen, iron, beta-carotene (pro-vitamin A), vitamin B_6, vitamin C (ascorbic acid), vitamin E and carbohydrates.

It has a unique flavour and consumed as delicious chutney when raw or after roasting and peeling off the skin. The peeled fruits may be pickled completely or may be substituted for tomatoes in a hot chilli sauce. Being high in pectin, the fruit is easily made into jelly but the fruit oxidizes and discolours without special treatment during processing. Peeled, sliced and seeded tree tomatoes with lemon rind, lemon juice and sugar can also be cooked to a jam. The peeled fruit can then be sliced and the slices added to stew or soups, or served with a sprinkling of sugar and perhaps with a scoop of vanilla ice cream. Chopped slices are blended with cream cheese and used as sandwich spread. The peeled, stewed fruits are combined with gelatine, milk, sugar and lemon juice to make a dessert, which is then garnished with fresh tree tomato slices. The peeled fruits can be pureed in a blender or by cooking, strained to remove the seeds, and then, packed in plastic containers and frozen. Lemon juice may be added to the puree to enhance flavour.

The fruit is eaten by scooping the flesh from a halved fruit, but in New Zealand children, palpate the ripe fruit until it is soft, then bite off the stem end and squeeze the flesh directly into their mouths. When lightly sugared and cooled, the flesh makes a refreshing breakfast dish. In addition, they give a unique flavour when added to stews (such as Boeuf Bourguignon), hollandaise, chutneys and curries. They are also tasty and decorative in, *e.g.*, radicchio salads. In Colombia and Ecuador, fresh tamarillos are frequently blended together with water and sugar to make a juice. It is also available as a commercially pasteurized puree. As a natural remedy, the tree tomato is considered a fruit that relieves stress and helps to cure migraines and severe headaches. It is used to treat obesity- because it is low in calories- as well as high blood pressure. It is used for colds, sore throats, liver diseases and diabetes, and it is said to improve the immune and circulatory systems.

Common name: Bird eye pepper

Vernacular name: Malayalam: Kanthari mulagu, Khasi: Sohmynken khnai, Assamese: Khut jolokia

Botanical name: *Capsicum frutescens* (Syn. *C. fastigiatum*)

Family: Solanaceae

The name bird's eye pepper, whose small fruits separate easily from the calyx and are dispersed by birds, is often used to denote small sized, pointed chilli of high pungency. The common features of all *frutescens* cultivars are their greenish flowers and their upward directed pedicels.

Like all other peppers, bird's eye pepper is also rich in beta-carotene, ascorbic acid and contains tocopherol, which is a well-known vital antioxidant. Most bird's eye peppers are processed to extract the oleoresins (colouring matter and capsaicin) for sale to the food and pharmaceutical industries due to its high pungency. It eliminates microbial contaminations, provides uniformity of colour, flavour and economical transport and storage over bulky whole or powdered spices. Capsaicin is used as a chemical in many pain balms, carminative tonics, *etc.* and as colour additives in food industry. A powerful local stimulant with no narcotic effect is largely used in hot climates as a condiment and most useful in atony of the intestines and stomach. For persons addicted to drink, it seems to be useful possibly by reducing the dilated blood vessels and thus, relieving chronic congestion. It is also used to cure Ranikhet disease of poultry at initial stage. It is used extremely for the treatment of tonsillitis, diphtheria and snake bites. It is also used in lozenge form, which helps in curing sore throat and hoarseness. In Assam, it is traditionally used to treat mouth blisters by rubbing it on the blisters. Indonesians use it to treat acute stomachache, loss of appetite and low blood pressure. It is also used in the treatment of cancer patients, under going chemotherapy to reduce nausea. It relieves (but temporarily) of blocked sinuses, a congested throat or lungs. Studies have also shown that capsaicin works with protein molecules to decrease cholesterol absorption in the body and to speed up the metabolism but probably the best thing about eating bird's eye pepper is that it enhances the release of endoephins, a substance in the brain that causes one to feel pleasure. A non-conventional use of it is in the self-defence sprays gaining popularity among working women who have to travel along and work late and senior citizens of United States of America. The spray consists of oleoresins of hot pepper at ultra high emission rate, which temporarily immobilizes the attacker. It is also becoming an indispensable war-weapon that is why the Defence Research Laboratories are assigned to find out its improvement in all aspects.

Common name: King pepper (World hottest pepper)

Vernacular name: Bhut jolokia, bih jolokia (Assamese), naga jolokia, naga mircha (Nagamese), raja mircha

Botanical name: *Capsicum chinense*

Family: Solanaceae

The king pepper is a pepper that grows in northeastern India (Assam, Nagaland and Manipur), Sri Lanka and Bangladesh. King pepper or bhut Jolokia has created history when it captured attention of the world as the world's hottest pepper variety. The certificate of *Guinness World Records* for the hottest of all spices brought more laurels for India as a predominant pepper growing and exporting country.

King pepper is loaded with vitamin A, a potent antioxidant, which boosts up the immune system. As the pods mature and darken high quantities of vitamin C are gradually replaced with beta-carotene and the capsaicin levels are at their highest. Due to these capsaicin levels, some believe that eating peppers may have an extra thermic effect, temporarily speeding up the metabolic rate hence, burning off calories at a faster rate. The alkaloids from the capsaicin stimulate the action of stomach and intestine improving the whole digestion process.

King chilli is used as a spice in food or eaten alone. One seed from a Bhut Jolokia can sustain intense pain sensations in the mouth for up to 30 minutes before subsiding. Passionate care should be taken when ingesting the pepper and its seeds so as not to get it in the eyes. It is used as a cure for stomach ailments. It is also used as a remedy to summer heat, presumably by inducing perspiration. In northeastern India, the peppers are smeared on fences or used in smoke bombs as a safety precaution to keep wild elephants at a distance.

The hot flavour of chillies is due to the presence of a group of seven closely related compounds called capsainoids but capsaicin (8-methyl-N-vanillyl-6-nonenamide) and dihydrocapsaicin are responsible for approximately 90% of the pungency. Due to its extra-ordinary pungency level and irritating properties, it has also been used as lachrymatory agent (a chemical compound that irritates the eyes to cause tears, pain and even temporary blindness). Nagas are known to have used this pepper as biological weapon in ancient warfare to get rid of enemies and also used to smoke out fox and rodents in their fields. The oleoresin capsicum (OC) extract from pepper is popularly in use for riots control, crowd control and personal self-defense. The general concept is that irritating compounds from the natural product will be most acceptable from the human point of view and environment-friendly than synthetically produced compounds CN (1-chloroacetaphenon), CS (O-chlorobazyladene malononitrile and CR (Dibenz (b, f)-1,4-oxozepine.

Common name: Bitter brinjal

Vernacular name: Jilo

Botanical name: *Solanum gilo* Raddi

Family: Solanaceae

Bitter brinjal, producing red tomato like fruit, is a kind of brinjal, which has enlarged calyx almost covering the long fruit and has the appearance between a brinjal and a tomato. It is important among the tribes of northeastern region. They grow it in the kitchen garden. The bitter brinjal is a prized vegetable particularly of Khasi and Mizo tribes of Meghalaya. It is highly resistant to shoot and fruit borer, the most destructive pest of eggplant in India and other tropical and sub-tropical parts of the world. The green fruits of bitter brinjal are a good source of protein. The high phenolics content is beneficial in the diet, as they are known to minimize the toxicity and also have antiseptic and bacteria static action and regulate the pH in the intestine. In African countries, these bitter fruits are liked for flavour and are believed to confer various medicinal properties such as prevention of malaria or increasing lactation.

Common name: Curry leaf

Vernacular name: Assamese: Narsinghe, Bengali: Barsanga,
Punjabi: Currypatia

Botanical name: *Murraya koenigi* L. Spreng

Family: Rutaceae

The tree is valued for its leaves that are used to flavour and season the dishes. The curry leaves tree adorns every household of south India. It is rich source of vitamin A. Leaves contain moisture (63.8%), protein (6.1%), fat (1%), carbohydrates (16%), fibre (6.4%), minerals (4.2%), vitamin A (12,600 IU/100 g), nicotinic acid (2.3 mg/100 g) and vitamin C (4 mg/100 g). The leaves are used mainly for seasoning and flavouring of dishes. Ground curry leaf with matured coconut kernal and adequate spices forms an excellent preserve. Leaves, bark and root of *M. koeingii* are used in ayurvedic and yunani medicines. The plant is tonic, stomachic, stimulant and carminative. Leaves and roots are bitter, arid, alexetric, anthelmintic, allay heat of body and cure itching and inflammation. They are also useful in leucoderma and blood disorders. The leaves are used as a febrifuge in fevers and are prescribed to eat raw for the cure of dysentery. The roots are purgative and the juices are used for relieving pain associated with kidneys. External application of the root and pulp of bark are used to cure eruptions and bites of poisonous insects and animals. Decoction of bark is prescribed for checking vomiting. The aromatic taste, flavour and its value as a piece is due to the presence of an essential oil in the leaves. Fresh leaves on stem distillation under pressure (90 lb/sq in) yield 2.6 per cent of volatile oil. The volatile oil of curry leaf is used as a fixative for heavy soaps and perfumes. The stem distillates and extracts of different part of the plant have own anti microbial and pesticidal activities. It has been observed that shrubs impart relative immunity to be spike disease of sandalwood. Grinimbine, murrayanine and mahanimbine from stem bark show antifungal properties against some dermatophytes like *Trichophyton rubrum, Microsporum gypseum etc.* Root extract displayed significant in vitro cytotoxic activity against cultured KB cells. Plant extract displayed juvenile hormone activity of crumpling of wings against *Dysdercus arigulatus.* Powdered leaves and their solvent extracts showed growth inhibitory effect against *Spilosoma oblique.* Efficacy of powdered leaves is for controlling *Phthorimaea operculella* on stored potatoes. The germplasm of *M. koenigii* showed resistances to citrus leaf minor *Phythenistic citrella.* Volatile oil of leaves of *M. koenigii* is extracted by hydrodistillation and solvents and evaluated for its anti microbial insecticidal and pesticidal value. It was found effective against *Pseudomonas* sp., *Staphyllococcus aureus, Bacillus subtilis, Pleurotus sajor-caju, Odontotermes* sp. (termite) and *Callosobruchus chinensis,* a stored grain pest.

Common name: Drumstick

Vernacular name: Horseradish or ben oil tree

Botanical name: *Moringa oleifera* Lam.

Family: Moringaceae

Drumstick leaves are good source of protein. The leaves, flowers and pods are used as significant sources of vitamins A, B and C, riboflavin, nicotinic acid, folic acid, pyridoxine, alpha-tocopherol, calcium and iron. The drumstick leaves and pods are used as vegetables, aphrodisiac, food and medicine in Asia. Fresh tender leaflets are stripped and used as a leafy vegetable and a potherb that is cooked either alone or with lentils. Tender leaves are also used to flavour *ghee* and to enhance the shelf life of *ghee.*

Almost all parts of the tree have been utilized in Indian *Ayurvedic* traditional medicine and other folk remedies. Aqueous leaf extracts are reported to regulate thyroid hormone and can be used to treat hyperthyroidism. Leaf extracts are also used to treat ulcer. Reportedly, drumstick leaves and pods also have a positive effect in reducing blood cholesterol and anti-tumor promoting activity. Furthermore, it is an important source of glucosinolate, which is a precursor of isothiocyanate group of chemo-preventives that can inhibit carcinogenesis.

Drumstick is also being studied for its anti-inflammatory, antimicrobial, diuretic, antibiotic and antimicrobial properties. A compound found in moringa roots and flowers (pterygospermin) has powerful antibiotic and fungicidal effects. The root bark contains two alkaloids, *i.e.*, moringine and moringinine. Moringinine acts as cardiac stimulant, produces rise of blood pressure, acts on sympathetic nerve-endings as well as smooth muscles all over the body and depresses the sympathetic motor fibers of vessels in large doses only.

Common name: Chenopods

Vernacular name: Goosefoot, pigweed, fat-hen, or bathua

Botanical name: *Chenopodium album*

Family: Chenopodiaceae

The leaves contain about 3.9% protein, 0.76% fat, 8.93% carbohydrates and 3% ash, and the seed contains about 49% carbohydrates, 16% protein and 7% ash. Each plant produces tens of thousands of black seeds that are high in protein, vitamin A, calcium, phosphorus and potassium.

The leaves and young shoots may be eaten as a leaf vegetable, either steamed, or cooked like spinach. One report says that when eaten with beans, the leaves act as a carminative to prevent gas synthesis and bloating. It is better not eat the leaves raw. The leaves are generally very nutritious but very large quantities can disturb the nervous system and cause gastric pain. Seeds are dried and ground into a meal and eaten raw or baked into bread. The seed can also be sprouted and added to salad. The seed is very fiddly to harvest and use due to its small size. The seed should be soaked in water overnight and thoroughly rinsed before being used in order to remove any saponins. The young inflorescences are also cooked as a tasty broccoli substitute.

The leaves are anthelmintic, antiphlogistic, antirheumatic, mildly laxative and odontalgic. An infusion is taken in the treatment of rheumatism. The leaves are applied as a wash or poultice to bug bites, sunstroke, rheumatic joints and swollen feet, while a decoction is used for carious teeth. The juice of stems is applied to freckles and sunburn. Incorporation of dehydrated chenopods leaves in various conventional food items can improve the nutritional quality of the products as well as add variety in the diet.

As the common names suggest, it is also used as food (both the leaves and the seeds) for chickens, hens and other poultry. However, the nitrates in the plant can be converted very efficiently to nitrites in the rumen of cattle, leading to changes in haemoglobin and reducing the ruminant's oxygen binding capacity.

Common name: Multiplier onion

Vernacular name: Potato onion

Botanical name: *Allium cepa Group aggregatum*

Family: Liliceae

Bulb has nice onion flavour, when chopped up finely they makes a nice addition to salads and can also be used as a cooked vegetable or as flavouring in cooked foods. Leaves are raw or cooked have a strong onion flavour. Flowers are used as a garnish on salads. The flowers are somewhat dry and are less pleasant than many other species. Besides, it has medicinal uses. Although rarely used specifically as a medicinal herb, the onion has a wide range of beneficial actions on the body and when eaten (especially raw) on a regular basis will promote the general health of the body. The bulb is anthelmintic, anti-inflammatory, antiseptic, antispasmodic, carminative, diuretic, expectorant, febrifuge, hypoglycaemic, hypotensive, lithontripic, stomachic and tonic. When used regularly in the diet it offsets tendencies towards angina, arteriosclerosis and heart attack. It is also useful in preventing oral infection and tooth decay. Baked onions can be used as a poultice to remove pus from sores. Fresh onion juice is a very useful first aid treatment for bee and wasp stings, bites, grazes or fungal skin complaints. When warmed, the juice can be dropped into the ear to treat earache. It also aids the formation of scar tissue on wounds, thus, speeding up the healing process, and has been used as a cosmetic to remove freckles. The juice of the plant is used as a moth repellent and can also be rubbed on the skin to repel insects and promote the growth of hair and to be a remedy for baldness. The plant juice can be used as a rust preventative on metals and as a polish for copper and glass. A yellow-brown dye is obtained from the skins of the bulbs. A spray made by pouring enough boiling water to cover 1 kg of chopped unpeeled onions is said to increase the resistance of other plants to diseases and parasites.

Common name: Broccoli

Botanical name: *Brassica oleracea* var. *italica*

Family: Cruceferae

Broccoli is most nutritious of the cole crops, especially in vitamin content, iron and calcium. It contains twice as much vitamin C as spinach and as much as lemons, making it a great immune booster. Besides, it is also rich source of sulphoraphane, which is a compound associated with reducing the risk of cancer. It also purifies the liver and is rich in cellulose, which promotes intestinal transit. Furthermore, its high fiber content (3%) lessens the risk of cancer, particularly colon cancer. This effect is a result of the particular substances found in cruciferous vegetables (indoles and sulfurous compounds) that broccoli contains naturally. The effectiveness of its fiber is also augmented by its specific glucidic compounds, which have a slight laxative effect. Broccoli is highly recommended for all its healthful effects. It helps fight heart disease, promotes good eyesight by combating presbyopia, cataracts and conjunctivitis, and defends against flu and bronchitis, Parkinson's disease and aging. Rich in vitamin B_9, broccoli is also advised in cases of anemia. Broccoli is the only food to contain almost all the major antioxidant agents to be found in food, sulforaphane, indoles, glutathione, quercetin and beta-carotene. Its antioxidant properties are most effective when sprouted broccoli is eaten raw every day. It is easily digestible. Only kidney stones are a contraindicating factor.

Common name: Horseradish

Botanical name: *Armoracia rusticana*

Family: Cruceferae

Horseradish is grown for its pungent root. Roots are ground for widespread culinary use as a condiment and spice. It is a powerful stimulant and used against facial neuralgia, rheumatic limbs, indigestion, sciatica, joint-ache, hard swelling of the liver, whooping cough and curing acne. It is employed to induce vomiting, sweating and to stimulate the entire nervous system.

Additionally, horseradish has been shown in laboratory tests to be antibiotic, active against a variety of bacteria, so this can benefit a sinus infection. It has a high sulphur content, which may contribute to its antibiotic properties. Pungent oil in the root contains these properties. Horseradish has long been considered a powerfully effective diuretic, used by herbalists for centuries to treat kidney stones and like conditions. For bladder infections, mix 3-4 tablespoons of the fresh grated root with apple cider vinegar and honey to taste.

The root has also been used for indigestion and putrefaction in the digestive tract. Horseradish has long been taken with fish or meats to cut the fat and counteract spoilage. Horseradish preparations can also clear lung problems, coughs and asthma. Try it for such conditions and you will see that it is an immediate and very effective expectorant, cutting mucus and allowing you to eliminate it. Similarly, it works well in respiratory ailments related to allergies, such as hay fever. Horseradish can also be used for rheumatic and arthritic conditions. It stimulates the mucus surfaces throughout the entire body. Horseradish is noted as a skin treatment, to remove spots and blemishes from the skin. Horseradish is often used as a tonic for a tired or deranged system. It is known as a successful vermifuge, helping to kill and expel worms from the system. It is an important immune stimulate which is used in herbal formulas to increase the number of white blood cells in the bloodstream. It is also a known antioxidant, helping to counteract the bad effects of stress and pollution in the environment. Horseradish is known to be a specific tonic for the spleen and most especially the liver, which it helps to detoxify and stimulate.

Common name: Red cabbage

Botanical name: *Brassica oleracea* var. rubra

Family: Cruceferae

Recently, there has been an increased interest in red cabbage pigments as a possible source of natural colourants. Presently, a red cabbage colourant is marketed under the name San Red RC, recommended for use in beverages, chewing gums, candies, sherbets, dressings, yogurts and other fermented products. Flavour in red cabbage is mainly due to the presence of 2-propenyl isothiocyanate and 3-butenyl isothiocynate. Nutritional composition of the red cabbage is given in per 100 g of edible portion as: water 90%, protein 2.0 g, fat 0.2 g, carbohydrates 6.9 g, energy 31 kcal, calcium 42 mg, phosphorus 35 g, iron 0.8 mg, sodium 26 mg, potassium 268 mg, vitamin 'A' 40 I.U., thiamine 0.09 mg, riboflavin 0.06 mg, niacin 0.4 mg, ascorbic acid 61 mg. The presence of phytocides, which have antibacterial properties, in red head cabbage, is a valuable means of tuberculosis prophylaxis. They also contain a large amount of cyanide a special substance influences the regulation of permeability of blood vessel walls, which contributes to the prevention of vascular diseases. The anthocyanin pigment, which causes the red colouring of the head, has antioxidant properties.

■■■

17 Usefulness of Medicinal and Other Plants

M.K. Rana[1], Nidhi Dubey[2] and C.S. Raghav[3]

[1]Department of Vegetable Science, CCS Haryana Agricultural University, Hisar- 125 004, Haryana, India, E-mail: <mkrlotus@gmail.com>
[2]Indian Pharmacopoeia Commission, Government of India, Rajnagar, Ghaziabad, Uttar Pradesh-201002, E-mail: <nidhiknj@gmail.com>
[3]National Bureau of Plant Genetic Resources, Pusa Campus, New Delhi-110012, E-mail: <raghav63@yahoo.com>

The use of medicinal herbs on the earth is as old as the origin of mankind and has been documented in history of all civilizations. For thousands of years, the man has been using herbs for treating various diseases and aliments. The mythological references in several religions of the world indicate that the use of herbs is even older than the origin of mankind. The use of herbs by the God and Goddesses are such examples. As per Indian mythology, the holy basil existed on the earth before the origin of man and *parijat* tree existed in the era of Lord Krishna. In modern era, also the medicinal plants find application in pharmaceutical, cosmetic, agricultural and food industry. Man in the pre-historic era was probably not aware about the health hazards associated with irrational therapy. With the onset of research in medicine, it was concluded that plants contain active principles, which are responsible for curative action of the herbs. Before the onset of synthetic era, man was completely dependent on medicinal herbs for the prevention and treatment of diseases. With introduction of scientific procedures, the researchers were able to understand about toxic principles present in green flora. The scientists isolated active constituents of the medicinal herbs, and after testing, some were found to be therapeutically active. Aconitine, atisine, lobeline, nicotine, strychnine, digoxin, atropine and morphine are the some common examples. The efficacy of some herbal products is beyond doubt, the most recent examples being *Silybum marianum* (silymarin), *Artemisia annua* (artemisinin) and *Taxus baccata* (taxol). On the other hand, randomized controlled trials have proved the efficacy of some established remedies, for instance, *Ginkgo biloba* is a reputed remedy for tinnitus and *Hypericum perforatum* for depression. In *Hypericum*, some researchers are of the view that hypericin is the active principle of the herb, and some believe that hyperforin is responsible for antidepressant action of the herb. Recently, the research has supported biological activities of some medicinal herbs. Cancer is such a segment where researchers are expecting new molecules from herbs that can provide us with tools for fighting this dreaded disease. *Allamanda cathratica* (allamandin), *Elephatopus elatus* (elephantpoin), *Helenium autmnale* (helenalin) *Vernonia hymenolepis*, *Heliotropium indicum* (Indicine-N-oxide), *Daphne mezereum* (mezerien) and *Stereospermum suaveolans* (laphacol) are medicinal plants that have shown significant tumor inhibiting effect. Diabetes mellitus is another area where a lot of research is going on. *Ajuga reptans* (the active principle is said to be potentiate effects of insulin), *Galega officinalis* (galagine), *Bougainvillea spectabilis* (pinitol), *Momordica charantia* (chirantin) and *Gymnema sylvestre* (gymnemic acid) are some medicinal herbs that have shown effectiveness in non-insulin dependent diabetes. Recently, the extract of *Tecoma stans* has shown potent anti-diabetic activity. Alkaloid tecomonine is considered active principle of the herb. Arthritis is another potential disease where no satisfactory answer is present in modern medicine. *Commiphora mukul* (guggulsterones), *Boswellia serrata* (boswellic acid), *Withania somnifera* (withanolides), *Ruscus aculeatus* (ruscogenin) and *Harpagophytum procumbens* (harpagoside) are prominent plants with anti-arthritic activity. Harpagoside is a precious constituent as it has anti-rheumatoid activity. Rest of all natural products has anti-inflammatory activity. The concept of antioxidants is fast catching up and latest research has shown that a number of herbal derivatives have excellent antioxidant action. *Bacopa monnieri* contains bacosides A and B, and bacoside A is a strong antioxidant, which reduces several steps of free radical damage. *Coleus forskohlii* (forskolin), grape seed (proanthocyanidins), *Camellia sinensis* (polyphenols), *Huperzia serrata* (huperzine), *Pinus maritima* (pycnogenol), *Borago officinalis* (gamma linoleic acid) and *Vinca minor* (vinpocetine) are potential antioxidants. The plant is a biosynthetic laboratory, not only for chemical compounds but also for a multitude of compounds like glycosides, alkaloids, *etc.* These exert physiological and therapeutic effects. The compounds that are responsible for medicinal property of the drug are usually secondary metabolites. A systematic study of a crude drug embraces through consideration of primary and secondary metabolites derived as a result of plant metabolism. The plant material is subjected to phytochemical screening for the detection of various plant constituents. With the onset of scientific research in herbals, it is becoming clearer that the medicinal herbs have a potential in today's synthetic era, as numbers of medicines are becoming resistant. Under such situation, ancient knowledge coupled with scientific principles can come to the forefront and provide us with powerful remedies to eradicate the diseases. Here is an account of some potential herbs having proven therapeutic properties.

Common name: Arjun tree

Botanical name: *Terminalia arjuna* (Roxb. ex DC.) Wight and Arn.

Family: Combretaceae

Arjuna is a dense and tall deciduous tree and has long cone-shaped round leaves and thick white to pinkish-grey bark used in medicines. It grows up to a height of 15 to 30 meters. The tree is indigenous to India though it is found also in Myanmar and Sri Lanka. The bark of this tree strengthens the heart muscles, increases the circulation of blood and supplies important nutrients to the heart. It has stimulant effect on the heart. However, it strengthens the heart muscle unlike *Digitalis*, or *nitroglycerine* that only restores the heartbeat with no nutritive effect on heart muscles itself. Several clinical studies reveal its effects on various cardiac disorders including congestive heart failure, coronary artery disease and hypertension. Improvement of cardiac muscle function and subsequent improved pumping activity of the heart seems to be the primary benefit of its bark. The bark is highly useful as an anti-ischemic and cardio protective agent in hypertension and in ischemic heart disease. It induces a drug-dependent decrease in blood pressure and heart beat rate. It also relieves symptomatic complaints related with essential hypertension such as giddiness, insomnia, exhaustion and headache. It also lowers the level of cholesterol and triglycerides, while enhancing the good cholesterol (HDL). The suppression of hepatic cholesterol biosynthesis by Arjun is the mechanism responsible for a significant lowering of β-lipoprotein lipids and the recovery of HDL components in hyperlipidemia. Besides having the astounding properties of reducing blood cholesterol levels, the bark-powder is a good antioxidant. In addition to its benefits for the heart, this tree has also been shown to have anti-mutagenic and anti-carcinogenic effects. An ethanolic extract of its leaves has been shown to have some cytotoxic activity against human breast carcinoma cells. The active constituents of Arjun tree include tannins, triterpenoid saponins (arjunic acid, arjunolic acid, arjungenin and arjunglycosides), flavonoids (arjunone, arjunolone and luteolin), gallic acid, ellagic acid, oligomeric proanthocyanidins (OPCs), phytosterols, calcium, magnesium, zinc and copper. A new ellagitannin named arjunin-l, four known tannins and two phenolic acids have been isolated from Arjun tree.

Common name: Arnica

Botanical name: *Arnica montana* L.

Family: Compositae

Arnica montana is endemic to Europe, from southern Iberia to southern Scandinavia and the Carpathians. It grows in nutrient-poor silicaceous meadows up to nearly 3,000 meters. It is rare overall but may be locally abundant. It is naturalized in North America. It is becoming rarer, particularly in the north of its distribution, largely due to increasingly intensive agriculture. In more upland regions, it may also be found on nutrients-poor moors and heaths. It has tall stems, 20-60 centimeters high, supporting usually a single flower head. The flowers heads are yellow, approximately 5 cm in diameter and appear from May to August. Arnica, a popular herb used in several homeopathic preparations, is well known to promote the healing of wounds and bruises and relieve abdominal pains when used as a poultice. It is usually used for external applications. It is most commonly prepared as a tincture, which can also be used as the base for creams, ointments, compresses and poultices. As a dilute tincture, it works as an excellent anti-inflammatory agent. This tincture could also be helpful in treating the inflammations of mouth and throat. Besides healing of wounds and bruises, several herbal preparations containing oil are used for a wide range of conditions including sprains, muscle aches, acne, rheumatic pain, inflammation from insect bites and even swelling due to fractures. Though the topical use of Arnica is generally safe, its prolonged use may irritate the skin, causing eczema, peeling, or blisters. It is, however, important that arnica should not be used on broken skin, such as to treat an ulcer. It is not taken internally except in the form of herbal tea. Internal use of arnica as an herb is not recommended for children. It is rarely used as an internal herbal remedy because it can cause dizziness, tremors and heart irregularities. It may also irritate mucous membranes and cause vomiting. Arnica has been shown to be an immuno-stimulant, as both the sesquiterpene lactones and the polysaccharide fraction present in arnica stimulate phagocytosis, which involves the ingestion of invading foreign particles.

Sesquiterpene lactones are known to have anti-inflammatory activity and their biological effects appear to be mediated through processes related to the body's defense machinery. Helenalin is recognized to be one of the most active sesquiterpene lactones, which bestows this herb with unique properties to relieve pain and inflammation. The chemicals constituents of plant include sesquiterpene lactones, flavonoids, volatile oil, mucilage and polysaccharides, tannins and miscellaneous substances. The flowers are said to contain more arnicin than the rhizome but no tannin. Helenalin and dihydrohelenalin produce anti-inflammatory and analgesic effects.

Common name: *Ashwagandha*

Botanical name: *Withania somnifera*

Family: Solanaceae

The rejuvenating effects of *Ashwagandha, p*opularly called the Indian ginseng, have been known to ancient Indians for centuries. The name *Ashwagandha* comes from the peculiar odour of this herb, which is quite akin to that of a sweaty horse. The active compounds of *Ashwagandha* comprise mainly the alkaloids and withanolides. The latter are steroidal, and interestingly, these compounds bear a resemblance to the active constituents of Asian ginseng *(Panax ginseng)* known as ginsenosides.

Ashwagandha is one of the most famous *Ayurvedic* herbs due to its strong tonic action on the entire body. The whole plant, especially the leaves and the root bark, are enriched with medicinal properties. Mainly acting on the reproductive and nervous systems, it has strong tranquilizing effects as well. This uniquely gifted plant also stimulates the immune system. It has been shown to inhibit inflammation and improve memory while counteracting the effects of stress. Its external application as a poultice helps to provide relief to boils, swellings and other painful ports, and it is said to be an excellent tonic for reproductive system.

Ashwagandha is famous as a sexual tonic known to increase libido and the power of retention of sperm, thereby reducing premature ejaculation. It increases sperm count up to 10 times. In women, it is used to tone the uterus after a miscarriage and in treating post-partum difficulties. This wonderful herb has been found useful for curing manic depression, alcoholic paranoia and even schizophrenia. It is also used to treat nervous exhaustion and insomnia. Memory enhancing effects of plant extracts of *Ashwagandha* have been assessed in context with the neuro-chemical alterations of specific transmitter systems. This investigation throws light on the effect of *Ashwagandha* on increasing the cerebral receptor capacity to the neurotransmitter called acetylcholine, which explains, to some extent, the plant's cognition-enhancing and memory-improving properties.

The alcoholic root extract of *Ashwagandha* is even helpful in regression of tumors and in reducing inflammation in rheumatoid arthritis. The clinical treatment with *Ayurvedic* formulation that mainly contained *Ashwagandha* along with *Boswellia serrata* (guggul), *Curcuma longa* (turmeric) and zinc for three months showed a significant relief of rheumatoid arthritis pain and decreased morning stiffness. The fruit and seed are diuretic. Among other uses of this plant, its seeds are used to curdle plant milk in order to make vegetarian cheese. Since the fruits are rich in saponins, they can be used as a soap substitute.

Common name: Asiatic Ginseng/Ginseng

Botanical name: *Panax ginseng* C.A. Meyer

Family: Araliaceae

Ginseng is a short, perennial woodland plant of the Araliaceae botanical family. It is indigenous to the mountain forests of Asia from Nepal to Manchuria and is cultivated primarily in Korea. It grows naturally on the slopes of ravines and shady, well-drained mountainous hardwood forests. The plant grows from 7 to 21 inches in height when wild. The roots are creamy yellow or white resembling a parsnip with rootlets that branch off, sometimes, taking the shape of a human body. Ginseng's value is mainly as a preventative. However, the root is consumed for its medicinal and nutritive properties as well. It is used to increase physical and mental endurance, boost energy, normalize body functions, reduce cholesterol and prevent cancer. Traditionally, ginseng has been used to enhance sexual desire by promoting sex hormone production. Ginseng supports the natural balance of your body to combat fatigue and it strengthens and protects the nervous system. Recent research shows that regular use of ginseng is helpful in the treatment of memory loss (Alzheimer's), balance of blood sugar levels (diabetes), slowing down the aging process, helping the immune system, *etc*. Ginseng may be effective for treating colds, coughs, rheumatism, neuralgia, gout, anemia, insomnia, stress, headache, backache and double vision. Studies in China showed that ginsenosides also increase the protein synthesis and the activity of neurotransmitters in brain. The plant is used to restore memory and enhance concentration and cognitive abilities, which may be impaired by improper blood supply to the brain. Women find it helpful in normalizing menstruation and easing childbirth. Ginseng also counteracts the effects of physical and emotional stress, counteracts fatigue without caffeine and improves stamina. Medically, ginseng stimulates the immune system by spurring the production of body's own virus fighting chemicals, helps in reducing the cholesterol levels in blood, has anti-clotting effects, reducing risk of arterial blood clots, is known as an antioxidant, preventing the cumulative cell damage, researchers believe that it cumulates in cancer, protects liver from the effects of drug, alcohol and toxins, minimizes cell damage from radiation and increases intestinal absorption of nutrients. It is used as a decoction, liquid extract, or as a powder. Ginseng has a deep influence on metabolism and prevents the development of atherosclerosis. It may be taken over a long period to stimulate rejuvenation and virility. The best long-term benefits are achieved with a dosage of between 500 and 1500 mg per day. For therapeutic use, a dosage of 2000 mg per day is generally recommended. The dried roots of ginseng species contain complex mixture of over 30 therapeutic elements saponin triterpenoid glycosides termed as ginsenosides. It contains vitamins A, B_6 and the mineral zinc, which aids in the production of thymic hormones, necessary for the functioning of the defense system. These steroid-like ingredients provide the adaptogenic properties that enable it to balance and counter the effects of stress. The glycosides appear to act on the adrenal glands, helping to prevent adrenal hypertrophy and excess corticosteroid production in response to physical, chemical, or biological stress. The other important compounds include panaquilon, a bitter substance, resin, tannin, choline, *etc*.

Common name: Ephedra (Somlata)

Botanical name: *Ephedra gerardiana* Wall. Ex. Stapf.

Family: Ephedraceae

Ephedra is an evergreen herb found in high mountain regions of the Himalayas, specially cold and dry areas. In India, it is found in Laddakh cold desert and Sikkim. The herb is 1-2 feet high. Its stem is slender, erect, grooved, green and branching. Branchlets are green, ascending and smooth. Leaves are small, scale like, articulate and sheathing at each joint, and fruits are ovoid and red. Ephedra is a dioecious plant, in which male and female reproductive organs are born on separate plants. The tincture of plant is cardiac and circulatory system stimulant. Liquid extract of plant is used for controlling asthmatic attack. Decoction of stem and root is considered a remedy for rheumatism and syphilis in Russia. In China, it is used to treat bronchial asthma, cold, flu, edema, typhoid, cough and various pains. In Pakistan and India, it is used to relieve bronchial asthma. It is said to improve the lung condition and control wheezing. It also promotes sweating and urination. It is used in cold and flue preparations, energy products and weight management formulas. The multifarious effects of this plant are related to the stimulation of sympathetic nervous system through α- and β-receptors, however, persons with heart problems and high blood pressure should be especially careful when using these preparations. The whole plant possesses oxytoxic, diuretic and anti-fertility properties. The major chemical constituent of this plant is ephedrine.

Common name: Foxglove/Digitalis

Botanical name: *Digitalis lanata* Ehrh

Family: Scrophulariaceae

Foxglove is an annual herb characterized by a thick, cylindrical and downy stem that reaches a height of up to 2 m. Some species of foxglove are biannual and perennial also. Leaves form a thick rosette during the first year of growth. The leaves, which are woolly, veined and covered with white hairs on the underside, have a very bitter taste. Flowers grow in the first or second year depending on the species and are tubular and bell-shaped growing to 8 cm in length. Many colours of flowers have been bred from digitalis and they are rarely white. European herbalists used the leaves of foxglove purely for the healing of wounds and skin diseases. An English doctor William Withering discovered that foxglove was the active herb in a folk recipe for dropsy and also recognized its value in the treatment of heart disease. Its plant contains a number of therapeutically active groups of compounds such as flavonoids and chalcones. The seeds also contain digitalis glycosides, while steroidal saponins, flavones, the flavonoid, chrysoeriol, anthraquinones and organic acids have been identified in the leaves. High performance liquid chromatography and mass spectroscopy have been used to identify and quantify glycoside composition. Today, glycosides are extracted from foxglove by the pharmaceutical industry to produce the cardioactive glycosides such as digitoxin, digitalin and digoxin. Digoxin from *D. lanata* is widely prescribed in orthodox medicine for heart failure and irregular heart beat. Digitalis leaves have a narrow therapeutic index, requiring close medical supervision for safe use. Traditional dosage starts at 1.5 g of leaves divided into 2 daily doses. Purified digoxin is typically used at daily doses of 0.125 to 0.25 mg. All parts of the plant are toxic. The incidence of digitalis toxicity in therapeutic use has been estimated to a range from 5 to 25%. Ingestion of extremely small amount of the plant may be fatal to humans, especially children and to animals, and its toxicity is cumulative. The ornamental strains of *D. purpurea* typically have low concentrations of active compounds. Leaves of wild varieties that have been used for medicinal purposes contain at least 30 different glycosides in total quantities ranging from 0.1 to 0.6%. These consist primarily of purpurea glycoside A (yielding digitoxin) and glycoside B, which are the precursor of gitoxin. The main glycosides of *D. lanata* are the lanatosides, designated A through E. Removal of acetate groups and sugars results in the formation of digitoxin, gitoxin, digoxin, digitalin and gitaloxin. It also serves as a major source of lanatoside C and digoxin (300 times more potent than the powder prepared from *D. purpurea*). The isolated digitoxin is 1,000 times more potent than the whole powdered leaves and is completely and rapidly absorbed from the gastrointestinal tract.

Common name: German chamomile/Babuna

Botanical name: *Matricaria chamomilla* L. Sensu

Family: Compositae

German chamomile is a wonderful aromatic herb with a very beautiful daisy-like flower. Many people know it as a relaxing tea but do not realize how valuable it is. It can be distinguished from lawn chamomile or Roman chamomile by its hollow conical flower heads. It has a slightly bitter taste, which is reminiscent of the apple taste. The plant contains several pharmaceutically active compounds such as sesquiterpenes, flavonoids, coumarins, *etc*. These drugs are used for treating the travel sickness, inflammations of the skin and bacterial diseases of the skin, gums and respiratory tract. The oil of chamomile is used as a flavouring agent in liquors particularly of the French type. When eaten by cattle, the flowers have a deleterious effect on milk, and in some cases, the butter provided from it. The herb is used to treat various nervous tensions and conditions of irritability or stress. Topical use of the German chamomile involves its use as a cream in the treatment of sore skin and disorders such as eczema. German chamomile is similar to the Roman chamomile, a close herbal relative, which is also used to treat many of the same disorders and conditions, which can be treated using the German chamomile. The herbal remedies made from the chamomile are very gentle acting and efficient, this is the reason that the chamomile is considered ideal for the treatment of conditions in children. Disorders such as abdominal pain, persistent indigestion, problems related to acidity, excess gas and the conditions such as gastritis, abdominal bloating and colic are normally treated using German chamomile. In addition, the remedies made from the chamomile are used in the treatment of conditions such as hiatus hernia and disorders like peptic ulcer. It is used in the treatment of Crohn's disease and the symptoms related to the irritable bowel syndrome. The active chemical compound in the German chamomile is the compound known as spiroether, which has a very strong antispasmodic action. It soothes aching muscles and helps in relaxing muscular tension and in easing the menstrual pain in women affected by such disorders during menstruation. Irritability is also relieved by the chamomile herb, which also promotes sleep, particularly in affected children. Disorders such as the hay fever and conditions like asthma are also treated using the remedies prepared from the German chamomile. When put under steam distillation, it produces deep blue coloured essential oil and compounds known as proazulenes in the chamomile herb-produce chamazulene. This secondary compound is markedly anti-allergenic in its action and helps in alleviating all the symptoms associated with allergies in the body. Topical treatments can also be prepared from the chamomile, and remedies prepared from the herb can be applied on to sore and itchy skin. It can also be used in the treatment of disorders such as eczema. Eyestrain and other optic disorders can also be relieved using herbal remedies made from the chamomile herb. This essential oil has a calming effect on the mind and body and is excellent in treating any type of inflammation, *i.e.*, internal or external, and is very effective on urinary stones (bladder gravel) as well. It stimulates the liver and gall bladder, thereby improving digestion and is valuable in treating menstrual and menopausal problems. On the skin, it is a miracle worker and calms red, dry and irritated skin, as well as calming allergies, boils, colic, cuts, cystitis, dermatitis, eczema, psoriasis and all other flaky skin problems. It is high in (-*a*)-bisabolol, which promotes granulation (healing) and is a great tissue regenerator. The main chemical components of essential oil of Roman chamomile *a*-pinene, camphene, *b*-pinene, sabinene, myrcene, 1,8-cineole, *y*-terpinene, caryophyllene and propyl angelate and butyl angelate, while the main constituents of German chamomile oil are chamazulene, *a*-bisabolol, bisabolol oxide A, bisabolol oxide B and bisabolone oxide A.

Common name: Glory lily/*Kalmari*

Botanical name: *Gloriosa superba* L.

Family: Liliaceae

The name gloriosa comes from the word *gloriosus*, which means handsome and *superba* form the word superb clearly alluding to its beautiful flowers. Glory lily is a climbing herb with wavy-edged 3-5 inch long yellow and red splendid flowers, which is a remarkable sight. Glory lily is a twining vine, which is able to climb up with tendrils formed at tips of the leaves. Leaves are bright green and lance shape, 2-3 inches long. The leaf tip is modified into tendrils that coil around nearby supports to get a grip. Due to its thin shape, it is difficult to locate the plant in the forest. However, it is visible in the monsoons, especially when the plant is in flower. In bud, the pale green petals face downward. As the blossom matures, the petals elongate, wrinkle and gradually arch backward while sequencing through a spectrum of colour from green to yellow and yellow to scarlet. The stamens are extremely prominent and spread outward in graceful curves that follow the petals in their backward progression. This is an important medicinal plant, used as an antidote for snake poison, which is in demand commercially. The plant grows wild throughout tropical parts of India. A postal stamp was issued by the Indian Postal Department to commemorate this flower. It is also the state flower of Tamil Nadu. Root of this plant mixed with other herbs is used as good anti-venom. The tuber is poisonous, when consumed in high quantities. Colchicine is the important alkaloid extracted from the seed and used in modern medicine. The colchicine is effective in the treatment of gouty arthritis. The whole rhizome is oxytoxic and antibiotic but poisonous in large doses. Fresh juice of plant is uterine stimulant. The other properties of the plant are tonic, antiperiodic, cholagogue, anthelmintic, alternative and purgative. It is grown commercially for its chemical compound. They are also used by certain people to treat intestinal worms, bruises, infertility, skin problem and impotence. It is also said that sap from the leaf tip is used for pimples and skin eruptions. It is used as an *ayurvedic* medicinal herb to cure diseases in various parts of Africa and Southeast Asia. The plant was under threatened category due to its imprudent harvesting from wild as it is extensively used by medicinal industries for its colchicine content. The plant is used to cure arthritis, gout, rheumatism, inflammation, ulcers, bleeding piles, skin diseases, leprosy, impotency, snakebites, *etc*. Various compounds, *viz.* colchicine, colchicoside (its semi-synthetic derivative- thiocolchicoside), superbine, gloriosine, lumicolchicine, 3-demethyl-N-deformyl-N-deacetylcolchicine, 3-demethylcolchicine and N-formyl deacetylcolchicine, have been isolated from the plant parts mainly tubers and seeds.

Common name: Periwinkle

Botanical name: *Catharanthus roseous* (L.) G. Don

Family: Apocynaceae

A widely growing evergreen shrub periwinkle has had the reputation of being a magic plant for its beautiful flowers and leaves endowed with unique chemical compounds that have high medicinal value. The plant has been found to be a useful remedy for not only treating headache, diarrhea and bleeding gums but also for improving memory, lowering blood sugar, lowering blood pressure and specially for combating cancer. Periwinkle contains about 70 alkaloids and tannins, some of which are useful in the treatment of abnormal uterine bleeding and vaginal discharge. The main active compounds of periwinkle are alkaloids. One of its major alkaloids is known as vincamine. A closely related semi-synthetic derivative of vincamine, widely used in medicine, is known as ethyl-apovincaminate or vinpocetine, which has vasodilating, blood thinning and memory-enhancing properties. In clinical studies, this compound has been shown to help alleviate a type of dementia known as vascular dementia where the arteries supplying blood to the brain develop atherosclerotic plaques. Vinpocetine actually improves circulation by keeping acetylcholine, a primary neurotransmitter, active for a longer time. In one study, vinpocetine was shown to prevent the decline in short-term memory induced by anti-anxiety benzodiazepine drug, flunitrazepam. Similarly, vinpocetine has been shown to have a beneficial effect on hearing loss due to ageing, besides reducing the symptoms of tinnitus (ringing in the ears) due to impaired blood flow to the inner ear. The tannins present in periwinkle have a mild astringent effect, which explains the traditional use of this plant in relieving pain from canker sores or sore throat. Other alkaloids that have been discovered in aerial parts of periwinkle include catharanthine, leurosine sulfate, lochnerine, tetrahydroalstonine, vindolinine and vindoline, some of which have the property to lower the blood sugar level. Others lower blood pressure and some have hemostatic property that aids in arresting bleeding. Besides, vincristine and vinblastine, isolated from periwinkle are today well known to have anti-cancer properties. Periwinkle also contains the alkaloids reserpine and serpentine, which are powerful tranquilizers. The two alkaloids vinblastine and vincristine of this plant are popular in cancer chemotherapy. Oncovin is a cancer chemotherapy drug, containing vincristine, which is used for treating different types of cancer. This drug interferes with the growth of rapidly dividing cells. The anti-tumor activity of another its alkaloid vinorelbine is thought to be due to inhibition of mitotic-cell division at metaphase-stage through its interaction with microtubulin. The Food and Drug Administration approved vinorelbine for the treatment of advanced non-small cell lung cancer. This compound has been already launched in more than 70 countries of the world as a therapeutic agent of non-small cell lung and breast cancer.

Common name: Himalayan fir/Birmi

Botanical name: *Taxus baccata* L.

Family: Coniferaceae

Himalayan fir is a tree with 40 to 50 feet height and very stout trunk covered with red-brown, peeling bark and topped with a rounded or wide-spreading head of branches. The leaves are dark, glossy, almost black-green above, grey, pale-green or yellowish beneath, 1/2 to 1 and 1/2 inch long and 1/16 to 1/12 inch wide. Flowers are unisexual with the sexes invariably on different trees. The fruits are bright red, sometimes yellow, juicy and enclose the seed. The tree is deeply associated with the history and legends of Great Britain. Before Christianity was introduced, it was a sacred tree favoured by the Druids, who built their temples near these trees- a custom followed by the early Christians. The association of the tree with places of worship still prevails. Many cases of poisoning amongst cattle have resulted from eating parts of the Yew. The fruit and seeds seem to be the most poisonous parts of the tree. In the ancient Celtic world, the yew tree had extraordinary importance. A passage by Caesar narrates that Catuvolcus, chief of the Eburones, poisoned himself with yew rather than submit to Rome. Similarly, Florus notes that when the Cantabrians were under siege by the legate Gaius Furnius in 22 BC, most of them took their lives either by the sword or fire or by a poison extracted from yew tree (*Arboribus taxies*), which is from the yew tree. In homoeopathy, a tincture of young shoots and also of berries is used in a variety of diseases, such as cystitis, eruptions, headache and neuralgia, affections of the heart and kidneys, dimness of vision and gout and rheurmatism. In 1021, Avicenna introduced the medicinal use of *T. baccata* for phytotherapy in *The Cannon of Medicine*. He named this herbal drug as *Zarnab* and used it as a cardiac remedy. This was the first known use of a calcium-channel blocker drug, which was not in wide use in the Western world until the 1960s. In Central Himalayas, the plant is used as a treatment for breast and ovary cancer. The chief active principle of various *Taxus* sp. (*T. baccata* and *T. brevifolia*) is paclitaxel, which is an anti-cancer compound. The precursors of chemotherapy drug Paclitaxel can be derived from the leaves of European Yew, which is a more-renewable source than the bark of Pacific Yew (*Taxus brevifolia*). A commercial drug Taxol is isolated from the bark of this plant. The other pharmacological actions of this plant are emmenagogue, antispasmodic, carminative, expectorant, stomachic, sedative and tonic. Paclitaxel belongs to a small class of taxanes that possess four member's ether (also called an oxirane) and a complex nitrogen-containing ester side-chain. Both of these functional groups are essential for anti-tumour activity. Yew leaves contain diterpene esters, taxine A, taxine B, taxol and flavonoids, including sciadopytisin, ginkgetin and sequoia-flavone. The pigments rhodoxanthin, lycopene, beta-carotone and zeaxanthin are present in the fruit. The presence of eschescholtzxanthone has been reported in the freshly ripe red fruit. The alkaloid taxine has also been obtained from the seeds. This is a poisonous white crystalline powder, only slightly soluble in water, and another principle Milossin has also been found.

Common name: Indian Belladonna/Belladonna

Botanical name: *Atropa belladonna* L.

Family: Solanaceae

Belladonna, also known as deadly nightshade, is a perennial herbaceous plant, 1 m to 2 m high with a many-headed cylindrical rhizome. The woody stem is erect, branched, bluntly angular and hairy. The leaves are ovately pointed with entire-margined up to 15 cm long. The lower ones are alternate. Near the inflorescence, they are in pairs of 1 large and 1 small. The plant has a strong narcotic smell, a sharp and bitter taste and is poisonous. The other species *Atropa acuminata* is endemic to India, while *Atropa belladonna* Linn. is the European species. Indian *Belladonna* is found in the Western Himalayan Ranges, extending from Kashmir at the altitude of 1800-3600 m to adjoining hills of the Himachal Pradesh up to 2500 m. *Atropa belladonna* has been introduced in Kashmir and the Himachal Pradesh. Belladonna use may result in muscular tremor or rigidity due to the effects on central nervous system. The preparations from this plant have a positive dromotropic as well as a positive chronotropic effect on the heart. The drug has always been important in folk medicine for its hallucinogenic effect. Belladonna accumulates the tropane alkaloid hyoscyamine, which together with its racemic form (atropine) are used as a remedy against spasms in skeletal muscles and as an antidote against cholinesterase inhibitors (organophosphates) poisoning. The epoxidated derivative scopolamine is also an anti-cholinergic agent but unlike hyo-scyamine it is primarily a CNS depressant. Scopolamine is used prophylactically against motion sickness. The products from the leaves of this plant are well known to produce hallucinogenic effects and to cause dilations of the pupils. The alkaloids responsible for these activities, especially atropine and scopolamine are thought to be responsible for the use of plant in the treatment of toothache and for inducing sleep. Atropine has been used for the effective treatment of people exposed to cholinesterase inhibiting nerve gases. The roots of belladonna are narcotic, sedative, diuretic, antispasmodic, mydriatic and anodyne. *Atropa belladonna* is one of the most poisonous plants known with all parts of the plant containing toxic tropane alkaloids. *Atropa belladonna* extracts have been used since ancient times as both a poison and as a cosmetic agent for dilation of the pupils. A number of the *A. belladonna* alkaloids are used in modern medicine as anti-cholinergic agents. Atropine is used for a wide variety of conditions, including bradychardia and Wenckebach block. Scopolamine (levoduboisine and hyoscine) is used to treat motion sickness in the form of a patch. Hyoscyamine alleviates the symptoms of various gastrointestinal disorders. Donnatal combines *A. belladonna* alkaloids hyoscyamine, atropine and scopolamine with phenobarbital to treat intestinal cramping.

Common name: Indian Cotton/Kapas

Botanical name: *Gossypium herbaceum* L.

Family: Malvaceae

Cotton is a seasonal crop producing natural fiber of vegetable origin, like linen, jute, or hemp. It thrives in dry tropical and subtropical areas. The cotton flower has five large petals (showy, white, white-creamy, or even rose in colour), which soon fall off, leaving capsules or *cotton bolls*, having a thick and rigid external layer. The capsule bursts open upon maturity, revealing the seeds and masses of white/creamy and downy fibers. Cotton fibers of the *Gossypium hirsutum* species range from about 2 to 3 centimeters in length, whereas, *Gossypium barbadense* cotton produces long-staple fibers up to 5 centimeters length. Their surface is finely indented and they become kinked together and interlocked. The cotton plant is almost exclusively cultivated for its oleaginous seeds and for the seminal fibers growing from them. Besides being a major natural fiber crop, cotton also provides edible oil and seed by-products for livestock food. Cottonseed oil is a vegetable oil ranking fifth in world use among edible oils (accounting for about 4% of the world consumption of vegetable oil). The cottonseed meal is usually used as roughage in the diet of cattle for its high protein and energetic value. The cottonseed oil contains a yellowish phenolic compound known as gossypol. Gossypol has been shown to be an effective anti-fertility agent in humans with an efficacy of 99%. The aerial part of plant shows diuretic and oxytoxic activities, while the leaf is antibacterial. The other reported properties of the plant are demulcent, laxative, expectorant and aphrodisiac. This has also been indicated as an emmenagogue where there is backache, with a sense of dragging in the pelvis. A feeling of fullness and weight in the bladder, accompanied with difficult micturation, meaning the act of urination. It is useful in suppression of menses from any cause and in hysteria of females. As a parturient, it is one of our best remedies producing firm, regular and strong uterine contraction. In uterine inertia, it will increase expulsive power and prevent danger of postpartum hemorrhage, may be successfully combined with cimicifuga and caulophyllum, useful to control hemorrhage of cancer and uterine fibroids and no danger of gangrene from it as from ergot. As a parturient, it is much safer than ergot, as it brings on regular intermittent contraction, while ergot produces tonic contraction.

Common name: Indian hemp/Bhang

Botanical name: *Cannabis indica* L.

Family: Calmabinaceae

Cannabis is an annual crop plant propagated from seed. It can reach up to 5 m in height in 4 to 8 months from spring to autumn growing season. It is frequently found in association with human habitation, disturbed lands, farm fields, roadsides, railways, trash piles and exposed riverbanks cannabis. Cannabis is normally dioecious (male and female flowers developing on separate plants) and the gender of each plant is anatomically indistinguishable before flowering. Cannabis, also known as hemp or marijuana, has been used by humans for fiber, medicine and as a psychoactive for at least 4,500 years. Its biological name is *Cannabis sativa* with a subspecies, *Cannabis sativa* var. indica, used most often for recreational or medicinal purposes. The earliest milestone in the history of cannabis is charred cannabis seeds dated to 2500 BCE found on a ritual brazier in Romania. In 2008, nearly two pounds of cannabis was found buried in the tomb of a Gushi shaman who lived about 700 BCE in the Gobi Desert, northern China. Genetic analysis of the plants found that they were cultivated rather than from the wild. There are three types of cannabis used in India. The first, *Bhang*, consists of leaves of the marijuana plant. It is usually consumed as an infusion in beverage form and is the weakest form of the drug. The second, *Ganja*, consisting of the leaves and the plant tops, is smoked. The third, called *Charas* or *Hashish*, consists of the resinous buds and/or extracted resin from the leaves and stems of the marijuana plant. Typically, Bhang is the most commonly used form of cannabis in religious festivals. Cannabis was used in Hindu culture as early as 1500 BCE and its ancient use is confirmed within the Vedas (*Sama Veda*, *Rig Veda* and *Atharva Veda*). Cannabis or *ganja* is associated with worship of the Hindu deity Shiva, who is popularly believed to like the hemp plant. *Bhang* is offered to Shiva images, especially on *Shivratri* festival. This practice is particularly witnessed at temples of *Benares*, *Baidynath* and *Tarakeswar*. Bhang is not only offered to the deity but also consumed by *Shaivite* (sect of Shiva) yogis. *Charas* is smoked by some *Shaivite* devotees and cannabis itself is seen as a gift (*prasad* or offering) to Shiva to aid in *Sadhana*. Some of the wandering ascetics in India had known as *sadhus* smoke *Charas* out of a clay chillum. During the Hindu festival of Holy, people consume a drink called *bhang*, which contains cannabis flowers. According to one description, when the elixir of life was produced from the churning of the ocean by the *Devas* and the *Asuras*, Shiva created cannabis from his own body to purify the elixir. Another account suggests that the cannabis plant sprang when a drop of the elixir dropped on the ground. Thus, cannabis is used by sages due to association with elixir and Shiva. According to religious rites, a wise drinking of *bhang* is believed to cleanse sins, unite one with Shiva and avoid the miseries of hell in the after life. In contrast, foolish drinking of *bhang* without rites is considered a sin. Researchers claim that in the 5th century BCE Siddhartha ate only hemp seeds for six years, prior to becoming the Buddha. Cannabis continues to play a significant role in the meditation ritual of Tibetan, Tantric Buddhism and it has been a practice since 500 BCE when cannabis was regarded as a holy plant. Cannabis is known to provoke several behavioural and physiological effects, some considered both positive and negative. Behavioural effects include euphoria, relaxation, changes in perceptual and cognitive function, impaired memory and time perception, improved mood and a sense of disinhibition characterized by giggling. Physiological effects of cannabis use include increased heart rate, decreased blood pressure and increased appetite. Recreational cannabis smokers receiving chemotherapy have been known to report to their doctor that smoking cannabis relieved their nausea. The wealth of research provides support for the use of both synthetic and natural cannabis as a treatment for nausea and vomiting in cancer chemotherapy patients. There are many uses of medical marijuana. Medical marijuana or cannabis is used to treat cancer. One of the most common uses of medical marijuana is antiemetic, which minimizes the sick feeling and let patients to eat and live normally. It is safer, more cost-effective and often more effective than standard synthetic antiemetics. Smoking cannabis is more effective than taking ingesting it (or its synthetic derivatives such as Marinol), as patients find it tough to keep anything down long enough for it to have an effect. Smoking cannabis produces an immediate effect and patients find it easier to manage the dosage. In addition, the euphoric components act as an anti-depressant and the hunger and enjoyment of food properties (the munchies) make weight gain easy and increasing the chances of recovery. Cannabis based drugs are used for the treatment of pain, asthma, epilepsy and anorexia. The chief active principles of the plant are cannabinoids, which have hallucinogenic properties. The plant possesses other therapeutic properties such as stomachic, antispasmodic, analgesic, stimulant, aphrodisiac and sedative.

Common name: Snakeroot/Echinacea

Botanical name: *Echinacea purpurea* L.

Family: Compositae

Echinacea or snakeroot is an ornamental wild plant with 3-5 feet height. The dark pink petals that surround an orange center make this a superb cut flower. It attracts butterflies and small birds and is famous for its medicinal uses. It was and still is a widely used medicinal plant of the Indians. The roots and upper parts of the plant are used in medicinal preparations. Since olden days, Echinacea has been used to treat various diseases ranging from scarlet fever to venereal diseases such as syphilis, malarial infections, blood poisoning and diphtheria. In recent times, it is used to reduce the effects of common cold and flu and it soothes the symptoms of painful diseases such as sore throat or pharyngitis, severe cough and cold and reduces fever. Most herbalists are of the opinion that *Echinacea* be recommended to help in building the immune system and boost the body's fight against infections. Recent laboratory and animal research studies show that *Echinacea* possesses active chemical compounds that help to boost the activity of immune system, alleviate pain, cut down inflammation and bring about positive hormonal, antiviral and antioxidant changes to treat disorders. Some professional herbalists advocate the use of *Echinacea* in treating urinary tract disorders, yeast infections, ear infections or otitis media, tinea pedis, inflammation of the sinus and allergic rhinitis. It is also used to treat certain wounds that are slow in healing. Although *Echinacea* is known to be a safe herb, patients experiencing tuberculosis, blood disorders, diabetes, multiple sclerosis, autoimmune disorders and liver disorders should refrain from consuming Echinacea preparations. It is possible that Echinacea could bring down the potency of medications that inhibit the immune system. Moreover, since *Echinacea* has the ability to boost immune function, the people who are under treatment for organ transplant with immunosuppressive medications must strictly avoid taking the herb, as it contradicts the traditional medication. Herbalists usually recommend the use of *Echinacea* in boosting general immunity in the event of colds, flu, respiratory tract infections and mild bladder infections. *Echinacea purpurea* or purple coneflower is usually administered in the form of dried root or herb as tea, standardized tincture extract, powdered extract, tincture and as stabilized fresh extract. It was used as a painkiller and for a variety of other ailments, including toothache, coughs, colds, fever, headache, sore throats, infection, rheumatism, hemorrhoids, *etc*. It alleviates skin rashes and boosting the immune system. *Echinacea* has been used since 1831 in Europe and America for its immune stimulating effects and skin repairing properties. *Echinacea purpurea* has a long history of medicinal use. Native peoples used it to treat sore throats, toothaches, inflammation, coughs and to dress wounds and treat infections. Modern science recognizes the benefits of *Echinacea* as an immune booster. The main indication for *Echinacea* is helping the body to recover in the early stages of infection. It is especially useful for sore throats, colds and other respiratory infections. The herb works best when taken at the first signs of illness. *Echinacea* can be taken in tea, tincture (alcohol extract of the plant), or capsule form. The root is the part most commonly used but some herbalists will use the stems, leaves and flowers as well. *Echinacea* is generally used for short periods, from a few days to a few weeks, when combating infection or for the prevention of recurrent infection. There is little benefit to taking this herb for *immune support* in the absence of current or threatening illness.

Common name: Indian rhubarb/Ravand Chini

Botanical name: *Rheum emodi* Wall. Ex Meissn

Family: Polygonaceae

Rhubarb is an herbaceous perennial plant that grows from thick short rhizomes. Leaves are roundish with a heart-shaped base. Basal leaves can be quite large up to 60 cm across. Though the leaves of rhubarb are toxic, the plant, especially its stalk has medicinal uses. The plant is 1.5-2 m high and has stout streaked green and brown stem. The flowers are small, greenish-white, or dark reddish-purple and born in large compound leafy inflorescences. It is grown throughout the valleys in Himalayan Region at altitude of 3200-5200 m. It is native to India, Myanmar, Nepal and Pakistan. Rhubarb has a long and proven history of herbal usage, its main effect being a positive and balancing effect upon the whole digestive system. It is one of the most widely used herbs in Chinese medicine. Small doses act as an astringent and tonic to the digestive system, whilst larger doses act as a mild laxative. The root is taken internally in the treatment of chronic constipation, diarrhoea, liver and gall bladder complaints, haemorrhoids, menstrual problems and skin eruptions due to an accumulation of toxins. They have an astringent-like effect on mucous membranes of the mouth and nasal cavity. This remedy is prescribed neither for pregnant and lactating women nor for the patients with intestinal obstruction. Externally, the root is used in the treatment of burns. The root of the plant is also used in traditional Chinese medicine. It was also used in the medieval Arabic and European prescriptions. Rhubarb root is alterative, anthelmintic, antibiotic, anti-inflammatory, antiseptic, antispasmodic, aperient, bitter, cholagogue, diuretic, emmenagogue, haemostatic, sedative, sialagogue, anticholesterolemic and antitumor. Rhubarb also strengthens the liver and is useful in phlegmatic cough, cirrhosis, jaundice and oedema. Rhubarb also promotes blood circulation in the pelvic cavity and is, therefore, useful as a treatment for dysmenorrhoea, especially when there is heavy clotting. Rhubarb roots have a purgative action for use in the treatment of constipation but also have an astringent effect too. Therefore, it has a truly cleansing action upon the gut, removing debris and then astringing with antiseptic properties as well. The primary chemical constituents of rhubarb include rhein anthraquinones, chrysophanol, emodin, physcion, sennidine, rheidine, palmmidine, tannins, catechin, gallic acid, oxalic acid, rutin, phytosterol and calcium oxalate. It is the anthraquinones, which contribute to its laxative and purgative properties, yet the tannin content helps in balancing those properties and even stops diarrhoea. Though no specific mention has been seen for this species, the leaves of some if not all members of this genus contain significant quantities of oxalic acid and should not be eaten in any quantity. Oxalic acid can lock up certain minerals in the body, especially calcium, leading to nutritional deficiency. The content of oxalic acid will be reduced if the plant is cooked. People with a tendency to rheumatism, arthritis, gout, kidney stones, or hyperacidity should take especial caution if including this plant in their diet since it can aggravate their condition. The roots are also used to make a strong dye, which creates a golden hair colour and is suitable for those who have light brown or blondish hair. Simmer three tablespoons of rhubarb root in 2 cups of water for about 15 minutes, leave overnight and strain in the morning. First, test on a few strands of hair to see the effect and then pour through the hair for a rinse.

Common name: Milk thistle/Silybam

Botanical name: *Silybam marianum* L

Family: Asteraceae

Milk thistle is native to the Mediterranean Region, but now, it is found throughout the world. This stout thistle usually grows in dry and sunny areas. The spiny stems branch at the top and reach a height of 4 to 10 feet. The leaves are wide with white blotches or veins. Milk thistle gets its name from the milky white fluid, which comes from the leaves when they are crushed. The flowers are reddish purple. The small, hard-skinned fruit is brown, spotted and shiny. Milk thistle has been used for 2,000 years as an herbal remedy for a variety of ailments, particularly liver and gall bladder problems. Several scientific studies suggest that the substances in milk thistle, especially a flavonoid called silymarin, protect the liver from toxins, including certain drugs such as acetaminophen (Tylenol), which can cause liver damage in high doses. Silymarin has antioxidant and anti-inflammatory properties and it may help the liver repair itself by growing new cells. Milk thistle is often suggested as a treatment for hepatitis and alcoholic cirrhosis. The active ingredient that protects the liver in milk thistle is known as silymarin. It reduces the inflammation (which is why it is often suggested for people with liver inflammation or hepatitis) and is a strong antioxidant. Therefore, milk thistle protects and regenerates the liver in most liver diseases such as cirrhosis (hardening of the liver), jaundice and hepatitis, (inflammation of the liver) and Cholangitis (inflammation of bile ducts, resulting in decreased bile flow). Milk thistle is a gentle and mild laxative due to its ability to increase bile secretion and flow in the intestinal tract. This herb can have actions ranging from lubrication and softening of the stools to a mild laxative effect to actually balancing individuals that alternate between diarrhea and constipation. Since milk thistle is a demulcent and it stimulates bile flow, it is a natural remedy for treating a wide array of gall bladder diseases and symptoms. When using milk thistle to combat gall bladder symptoms, it is important to verify that there are no gallstones present, which would be too large for the gall bladder to pass.

Common name: Harsingar/Night-flowering Jasmine

Botanical name: *Nyctanthes arbortristis* L.

Family: Verbenaceae

Harsingar, also known as Parijat in India, is a species of Nyctanthes, native to southern Asia, from northern Pakistan and Nepal south through northern India and southeast to Thailand. It is a shrub or tree growing to 10 m tall with flaky grey bark. The leaves are opposite, simple, 6-12 cm long and 2-6.5 cm broad, rough and hairy with an entire margin. The small white star shaped sweet scented flowers have five to eight lobed white corollas with an orange-red centre. They are produced in clusters of two to seven together with individual flowers opening at dusk and finishing at dawn. The fruit is a flat brown heart-shaped to round capsule having diameter 2 cm with two sections, each containing a single seed. In India, they grow in the outer Himalayas at the height of about 1400 m and are found in tracts of Jammu and Kashmir, Nepal to East of Assam, Bengal, Tripura and extended through the Central region up to Godavari in the South. Besides India, they are found in Thailand, Indonesia, Nepal and Pakistan.

Parijat appears in several Hindu myths. In one myth, *Parijat* appeared as the result of churning of the milky ocean. In another myth, *Parijat* was brought to earth by Krishna from the garden of Indra Dev. In Hindu mythology, there is a story involving Lord Krishna about a parijat and Krishna's two wives, Satyabhama and Rukmini. Satyabhama wanted this *Parijat* tree from the heaven to be planted in her garden. Rukmini too took a fancy to the flower. Krishna, wanting to keep both his wives happy, planted this tree so that the flowers fell in Rukmini's garden, while the tree remained in Satyabhama's garden. The tree was planted in the garden of Indra, the Lord of Heavens. Even as Krishna stole a branch of the tree, he was spotted by Indra. However, Indra desisted from placing a curse on Krishna since he was an incarnation of Vishnu. Still, Indra put forth a curse on the stolen branch that it will never bear fruit even though the flowers may bloom on the tree. Apart from contributing to the beauty of gardens, *Parijat* also has medicinal properties. The juice of its leaves is bitter and saline in taste. Taking its juice is considered useful in the treatment of fever. Its bark, if taken as medicine, is known to give relief in and swelling of lungs. The oil produced from its bark brings relief to pains the eyes and used for hair tonics. Fresh juice of its leaves mixed with honey provides effective relief in the treatment of chronic fever. The decoction of its leaves prepared over low fire gives relief in the treatment of chronic *Citica pain.* The juice of its leaves mixed with little sugar is used for the treatment of stomach diseases of children. Application of six to seven of its tender leaves with little flesh ginger juice, churned in water brings relief in case of high fever. Its flowers, though bitter in taste, act as appetizer and soothe the stomach. The hot infusion of flowers has the sedative effects, which appears to result mainly by antioxidant and membrane stabilizing. Its seeds are used in the treatment of skin diseases and piles. Pharmacological studies suggest anti-inflammatory, analgesic, antipyretic and ulcerogenic (tending to cause ulcer) activities of water-soluble portion of an ethanol extract of its leaves. The extract of stem is used to relive headache. The extract exhibited aspirin like antinociceptive activity but failed to produce morphine like analgesia (a deadening or absence of sense of pain without loss of consciousness). Widely known since ancient times, the *Harsingar* or *parijat* tree has been used by *Ayurvedic* physicians for the treatment of arthritis, intestinal worms, liver disease, piles sciatica, various skin diseases and as laxative.

Common name: Passionflower

Botanical name: *Passiflora incarnate* L.

Family: Passifloraceae

The passionflower, which is native to the southeastern areas of the Americas, is a perennial climbing vine. Folklore states that passionflower gets its name from the corona feature that closely resembles the crown of thorns worn by Jesus at his crucifixion. Some varieties of passion fruit are edible and the leaves, stems and flowers are used for medicinal purposes. The passionflower is sometimes classified as a weed due to its aggressiveness in growing once established and difficulty in exterminating. The passionflower possesses a sturdy woody stem with herbaceous shoots. Passionflowers can grow to a length of about 32 feet. Leaves are three-lobed and measure 2 to 6 inches in length. Individual flowers have five white petals and five sepals, a modified leaf in the outermost section of the flower, which vary in colour from magenta to blue. The flowers are 3 inches in diameter and 2 to 4 inches long. The passionflower has two rings of thin filaments that circle the reproductive organs. There are three stigmas, which are part of the pistil and five pollens, bearing stamens. The fruit of the passionflower is called maypops and is a fleshy, green, egg-shaped berry that can be eaten fresh. Passionflower is famous for its beauty and fragrance. This calming herb has been a valuable sedative since ancient times. Due to its tranquilizing-properties, it is frequently used in the treatment of anxiety, insomnia, epilepsy and hysteria. The active substances in flowers, leaves and stems have sedative and vasodilating effects, thus, this herb finds use in the treatment of neuralgia, irritable bowel syndrome and premenstrual tension. An extract of the herb depresses the motor nerves of the spinal cord, slightly reduces blood pressure and increases the respiratory rate. A poultice of roots is applied to boils, cuts, earaches and inflammation. In the United States, passionflower is available as an over the counter sedative in combination with other calming herbs such as valerian and lemon balm. It is also used in German homeopathic medicine to treat pain, insomnia and nervous restlessness. The extracts of passionflower have also been used in making herbal formulations for the treatment of attention-deficit hyperactivity disorders. Passionflower is effective as a leading anti-anxiety medication when taken for one month. It may also relieve anxiety in people who are recovering from heroin addiction. It enhances the effects of pentobarbital, a medication used to promote sleep and for controlling seizures. The alkaloids of passionflower, *viz.* passiflorine, harmine, harmaline and flavone glycosides act as effective non-addictive sedatives, which do not cause drowsiness. Normally, the passionflower preparations are made from its fresh or dried flowers and other aerial plant parts. The available forms include infusions, teas, liquid extracts and tinctures. In general, passionflower is considered safe and non-toxic, however, it must not be taken by pregnant or breastfeeding women.

Common name: Poppy seed

Botanical name: *Papaver somniferum* L.

Family: Papaveraceae

The opium poppy is an annual that grows 30 to 150 cm high. It is a single-stemmed blue-gray frosted plant. The stem is erect, straight, or branched and produces white milky latex, as does the whole plant. The leaves are entire, glabrous, serrated, or crenate at the margin and clasping. The flowers are of white, purple, red, or pink colour. The cultivation of plant, extraction and sale of opium is banned in many countries. However, it is cultivated in many countries, including India, Turkey, Australia, Canada, Poland, Romania and in Central and South America The plant originated in western Asia and southeast Europe. The poppy seeds are demulcent and are used in the form of emulsion as an emollient and as specific against obstinate constipation and in catarrh of the bladder. The white seeds are sometimes used in pharmaceuticals. The oil is useful in the preparation of ointments, emulsions and compositions for skin care. It is used against diarrhoea, dysentery and scalds. An infusion of poppy capsules is used as a soothing application for bruises, inflammatory swellings and sometimes for painful conjunctivitis and inflammation of the ear. A hot decoction of capsules is applied as an anodyne. Leaves containing morphine are smeared as anodyne. Capsules are also used in the form of syrup or extract as a sedative against irritant coughing and sleeplessness. Red poppy flowers are used in medicine for making syrup.

Common name: *Sarpgandha*

Botanical name: *Rauwolfia serpentina* (L.) Benth

Family: Apocynaceae

Sarpagandha or snakeroot is gifted with unique serpentine roots that have extraordinary medicinal properties and a special place in *Ayurvedic* system of medicine for centuries. The roots of this plant contain several alkaloids. The ajmaline group (ajmaline, ajmalicine, isoajmaline and neoajmaline) acts as a general depressant to the heart, respiration and nerves. The serpentine group (serpentine, serpentinine and sarpagine) paralyzes respiration and depresses the nervous system but stimulates the heart. The whole root of *Sarpgandha* is hypnotic, sedative, hypertensive and a remedy in painful affections of bowels. It is useful in insomnia, hypochondria and highly effective in providing relief to various central nervous system disorders like nerve impulses along with certain pathways that affect heat and blood vessels, thus, lowering blood pressure. They also deplete catecholamine and serotonin from nerves in the central nervous system.

This small wood evergreen plant is useful in treating anxiety, insomnia, epilepsy and even abnormal behaviour associated with psychosis and schizophrenia. Besides, due to its anti-hypertensive and sedative properties, this wonderful plant has won much acclaim worldwide. The root extract has been a useful remedy for poisonous snakebites and stings of insects. The juice of its leaves also has therapeutic properties, as it is useful in treating opacity of cornea.

This herbal drug rapidly replaced the electric shock treatment for a certain types of mental illness. It is the first therapeutically useful tranquilizer of natural plant origin. Reserpine in snakeroot depletes the neuro-transmitter, *viz.* dopamine. The snakeroot constituent *ajmaline* not only lowers the blood pressure but also has a potent anti-arrhythmic effect. Ajmaline specifically depresses intra-ventricular conduction, suggesting its role in the treatment of ventricular arrhythmias. Besides, it also decreases the elevated serum cadmium levels. Though snakeroot has numerous medicinal properties, it may produce many dangerous side effects if taken indiscriminately. These adverse side effects include depression, decreased heart rate, low blood pressure, decreased sex drive, weight gain, gastrointestinal ulcers and decreased motor coordination.

Common name: Seabuckthorn

Botanical name: *Hippophae rhamnoides* L., *Hippophae salicifolia* and *Hippophae tibetana*

Family: Eleagnaceae

The berries of this species are known to be a storehouse of more than 100 different nutrients and bioactive substances, more than any other cultivated vegetable and fruit on earth. Sea-buckthorn contains 8 vitamins, 24 mineral compounds, 18 amino acids, organic acids, antioxidants, unsaturated fatty acids and other compounds that are essential to our health and well-being. Sea-buckthorn seeds, fruit pulp and juice have been reported to contain more than 190 compounds such as vitamins (A, B_1, B_2, C, E, K and folic-acid), 22 fatty acids (90% unsaturated fats), 42 lipids, organic acids, amino acids, carbohydrates and flavonoids, phenols, terpernes and tannins. It is high in both linoleic acid and linolenic acid, both of which are essential to human health.

The ratio of these fatty acids in seabuckthorn is almost perfect one at roughly 1:1 ratio and can most certainly play a beneficial role in decreasing heart disease. In addition to vitamins, high contents of flavonoids, oils, oil soluble bioactive compounds like carotenoids, plant sterols, lipid classes and their fatty acids, sugars, aromatic compounds and elements such as iron, calcium, magnesium, potassium, zinc, copper and manganese, have been reported in the berries of sea-buckthorn. Sea-buckthorn fruit oil is rich in palmitic and palmitoleic acids. The latter is a rare fatty acid that is a component of skin fat and can support cell tissue and wound healing.

The leaves of male plants have 18-22% crude protein and 4-5% fat and other micronutrients like phosphorous (1.52-2.23%), calcium (10.02 mg/kg), copper (23.52 mg/kg) and zinc (0.18-0.94%). Fruits of seabuckthorn are rich source of vitamin C (87-1400 mg/l00 g), vitamin E (162-255 mg/l00 g), vitamin A (11 mg/l00 g), vitamin B_1 (0.04 mg/l00 g), B_2 (0.56 mg/l00 g) and organic acids (2-4%). It also contains essential amino acids such as lysine, *etc.* It possesses various properties like anti-stress, antioxidants and contains flavonoides. Seabuckthorn fruit pulp, seeds and leaves have valuable medicinal oil, which is rich in β-carotenoides, vitamin E and anti-radiation and anti-aging properties.

The mature seeds contain 8-20% oil, whereas, dried fruit pulp yields about 18-26% oil. Pomace, the residue after pulping of fruit, contains 15-22% oil. Out of total fatty acids of the oil, more than 80% is unsaturated fatty acids. Oil has medicinal properties because of having vitamin E, antioxidants, essential fatty acids and have anti-radiation and anti-ageing effect, thus, it is widely used in making ultraviolet protective creams and beauty products. In Ladakh, traditional doctors (Amchies) use it for the traditional treatment of sunburns and regeneration of damaged skin. Plant leaves, fruits and bark is used as medicine for treatment of various ailments like cold, cough, malnutrition and skin diseases by local traditional doctors (Amchies). Bark of seabuckthorn is an effective blood purifier. Leaves having antioxidants, flavonoides, and other medicinal components are used for production of herbal tea.

Every part of the seabuckthorn plant has some pharmacological effect on the body and these have been thoroughly documented since antiquity. Historically, the Chinese were the first to use seabuckthorn as a drug. This wonder plant is capable of curing a number of human diseases and that it is an excellent medicinal food containing many kinds of vitamins, trace elements and amino acids. This combination of beneficial fatty acids, vitamins and antioxidant phytochemicals can be effective in maintaining a healthy immune system, important for all ages but particularly for the aged. Its products have been reported to have beneficial effects on blood pressure, and in respiratory infections, heart problems, gastric ulcers, cancer, bums, scalds, blood deficiency, digestive disorders, cardiovascular, memory loss and some brain disorders. It has also been proved to have good anti-inflammatory, anti-senility and anti-radiation properties, probably because the bark and fruits contain five hydroxy tryptamine (a rare occurrence in plant kingdom), which can act as a neurotransmitter. It is reported to regulate human emotions, hormone levels and neuro-endocrine disorders, age related problems, loss of memory and inhibit cancer cell growth.

Cosmetic companies claim that not only the seabuckthorn oils do have anti-aging properties but also they are quite good as sun-blocking agents. Cosmonauts have used seabuckthorn products for protection against radiation in outer space. Seabuckthorn containing beauty creams have been reported to have anti-wrinkle and anti-skin sclerosis properties. These also apparently improve microcirculation, nourish skin and hair, and cure scabies.

Common name: Sweet wormwood/Artemisia

Botanical name: *Artemisia annua* L.

Family: Asteraceae

Sweet wormwood is a common type of wormwood that grows throughout the world. It has fern-like leaves, bright yellow flowers and a camphor-like scent. It averages about 2 m tall and has a single stem, alternating branches and alternating leaves, which range 2.5 to 5 cm in length. Sweet wormwood was used by Chinese herbalists in ancient times to treat fever but had fallen out of common use, to be rediscovered in 1970 when the *Chinese Handbook of Prescriptions for Emergency Treatments* (340 AD) was found. This pharmacopeia contained recipes for a tea from dried leaves, prescribed for fevers (not specifically malaria). In 1971, scientists demonstrated that the plant extracts had antimalarial activity in primate models, and in 1972, the active ingredient, artemisinin was isolated and its chemical structure described. The plant has also been shown to have anti-cancer properties. It is said to have the ability to be selectively toxic to breast cancer cells and some form of prostate cancer. There have been exciting preclinical results against leukemia and other cancer cells. Popularly called sweet wormwood has been a traditional Chinese herbal medicine for centuries. Though this aromatic herb was already well recognized for having anti-bacterial properties, it has now earned the reputation of being effective against malaria and even cancer. Several flavonoids in crude extracts of *Artemisia annua* exhibit antiplasmodial activity. Other properties include antiperiodic, aperient, stomachic, tonic, anthelmintic, cardiac stimulant and expectorant. An infusion of the leaves of sweet wormwood has been a popular remedy for treating fevers, cold and diarrhoea. Externally, the poultice of leaves has been found to be effective in controlling nosebleeds, boils and abscesses. Artemisinin, an active ingredient of wormwood, is found to be effective in treating multi-drug resistant malaria, particularly caused by the protozoan parasite, *Plasmodium falciparum.* This substance selectively kills all the human breast cancer cells exposed to it within 16 hours. More importantly, this substance was found to be highly toxic to just cancer cells, as it had a marginal impact on the normal breast cells. Artemisinin is a sesquiterpene lactone, which occurs in several aerial parts of the plant. *Artemisia annua* also has the potential against tumor cells.

Common name: Witch hazel

Botanical name: *Hamamelis virginiana* L.

Family: Hamamelidaceae

Witch hazel, also known as winter bloom happens is a deciduous shrub or a small tree, which is grown prominently in Atlantic North America. Today, the plant is grown in Europe and Asia as well. On an average, witch hazel plants touch a height of 3 to 8 meters, rarely going beyond 12 meters. It bears rich yellow and orange-red fragrant flowers in the autumn and continues throughout winter. A versatile plant, witch hazel has many uses, ranging from facial care to relaxing aching feet. Earlier, the leaves and bark of the plant were used by people as poultice to treat swelling and inflammation. People have now discovered various more ways of making use of it as well. Read on further to know about the uses and benefits of witch hazel plant. The bark and leaves of witch hazel are endowed with remarkable medicinal properties. This plant has been used for treating bruises, varicose veins, hemorrhoids, sore nipples and eye problems since ancient times. The bark is about 30 times richer in hamamelitannin than the leaf extract. Witch hazel contains large quantities of tannins, which have a drying, astringent effect, causing the tightening up of proteins in skin and across the surface of abrasions. This creates a protective covering that increases resistance against inflammation and promotes healing of broken skin. It also appears to help damaged blood vessels beneath the skin. A homeopathic medicine is prepared from fresh bark, which is used internally in the treatment of diarrhea, colitis, dysentery, hemorrhoids, piles, varicose veins, vaginal discharge, excessive menstruation, internal bleeding and prolapsed organs. An infusion of leaves is used to reduce inflammation, internal hemorrhages, treat pile and eye inflammations. A refreshing tea, which helps to contain bleeding and excessive mucous discharge from the alimentary canal, is prepared from its leaves and twigs. A cotton plug moistened in its extract is quite useful in relieving pain and swelling of varicose veins. It is also used topically for treating external hemorrhoids, varicose veins, bruises, sprains and localized inflamed swellings, spots and blemishes. It can also be applied to insect bites and minor burns such as sunburn. Witch hazel is used as an ingredient in eye drops, after-shave lotions and a wide range of cosmetic preparations. A lotion, decoction, or tincture can also be used as a douche for vaginal discharge and irritation, or as a gargle for sore throat and infections. The bark and leaves are used for the treatment of mild diarrhea, inflammation of gums and mucous membranes of mouth, hemorrhoids and varicose veins. Witch hazel is believed to have a slight action on the muscular fibre of veins and strengthens the veins due to its anti-inflammatory property. A cream prepared with phosphatidylcholine and witch hazel has been found to be as effective as 1% hydrocortisone in topical management of eczema. Some proanthocyanidin fractions inhibit inflammatory mediators derived from arachidonic acid and inhibit the formation of platelet-activation factor, a chemical mediator of inflammatory processes. Besides, a strong antioxidant activity against the free radical and super oxide, released by several enzymes during inflammatory process may also play a role in witch hazel's anti-inflammatory effects. A specially filtered fraction of witch hazel extract containing mostly proanthocyandins has significant anti-viral activity against Herpes *simplex* virus type 1.

Common name: *Indian Wild Rice/Nivara*

Botanical name: *Oryza nivara,* S.D. Sharma and Shastry

Family: Gramineae

Oryza nivara is a wild progenitor of the cultivated rice *Oryza sativa.* It is found growing in swampy areas, at edge of pond and tanks, beside streams, in ditches and in or around rice fields. It grows in shallow water up to 0.3 m, in seasonally dry and open habitats. It is an annual, short to intermediate height (usually <2 m) grass. Its panicles are usually compact, rarely open and spikelets are large, 6-10.4 mm long and 1.9-3.4 mm wide, with strong awn (4-10 cm long) and anthers (1.5-3 mm long). Its distribution includes Bangladesh, Cambodia, China, India, Laos, Malaysia, Myanmar, Nepal, Sri Lanka, Thailand and Vietnam. Nivara is the same specie as the cultivated rice. However, its occurrence is limited to Kerala perhaps due to the climatic conditions, the heavy monsoon rains and the acidic soil. In ancient *Ayurvedic* texts such as the *Ashtanga Hrudayam* and the *Sushrutha Samhita, Njavara* is referred to as *shashtika* rice, implying that it matures over a period of sixty days. In *Ayurveda,* the *Njavara* grain is referred to as *tridosha hara,* which is capable of eliminating the three ailments from the human body. Being easily digestible, it now forms an ingredient of several baby foods that are available in the market but the plant is famous mainly due to *Njavara Kizhi,* a special massage treatment program of Kerala. *Njavara Kizhi,* a *rasayana* treatment, is considered very effective in the treatment of wasting muscles, nervous weaknesses and rheumatic diseases. The *Njavara* grain is well cooked in milk and herbal decoctions and is filled inside linen poultice, which is applied gently on the body on which oil has been previously applied. The deficiency of vitamin B_1 or thiamine causes beriberi disease and the chief symptoms of which are muscular weakness, neuritis, *etc.* Indigenous medicine prescribes *Njavara* for the relief of these very symptoms. Though the actual medicinal ingredient has not been identified, evidence abounds that vitamins and proteins are responsible for healing properties of *Njavara.* Very high total free amino acid content in the grain justifies the classification of this plant as a protein-cereal. However, there are significant variations in the free amino acid content of the various *Njavara* types grown in different localities, and the black-glumed *Njavara* records the highest content.

Common name: Oat (spring oat or white oat)

Botanical name: *Avena sativa* L.

Family: Gramineae

Oat is the most commonly cultivated cereal crop. Though oat is a minor cereal compared to wheat or rice, today it is believed as a major protective food against a number of health afflictions like hypercholesterolemia, heart disease, diabetes, hypertension and even cancer. Oat has many vital micronutrients and phytochemicals, however, the energy is slightly lower (3 kcal/g), which is a favourable trait for its effective use as a food with relatively low glycaemic index for most of the population, except in cases where quick energy yield is called for. Worldwide recognition about its health-protecting ability reveals that the dietary fiber component and a plethora of phytochemicals present in it have multidimensional functions in human body. The dietary fiber, which is present in oat, especially beta-glucan, has been proved to reduce blood serum cholesterol levels significantly. This is brought about by the ability of oat fiber to bind with cholesterol precursors in intestine to form gel like structures and excrete it from the body. This necessitates drawing out the cholesterol from the body for the digestion of fatty foods, thus, reducing its effective level in body system. Such an action reduces the potential for cholesterol to form plaques on the arterial walls and lowers the risk of myocardial infarction, stroke and clot formation elsewhere in the body by 8-23%. Reducing the cholesterol in the load by 1% reduces the risk of a heart attack by 2%. A combination of oat and tocopherol prevents restriction of blood vessels after a fatty meal, whereas, the diameter of the arterial walls is normally restricted by about 13% after 3 h of ingesting such a meal, while tocopherol reduces the number of free radicals. The soluble fiber in oat prevents fewer of these radicals from forming in the first place by blocking the gastrointestinal absorption of fats and carbohydrates into the blood stream. Also found in oat are the unique phenolic substances avenanthramides, which are known to reduce the number of molecules that stick to cells in human aorta, and consequently, lessen the chances of formation of plaques that stick to the artery walls constricting the blood flow from the heart. These flavonoids also appear to have anti-inflammatory effect by reducing the production of cytokines, which are proteins that play a prominent role in the immune system of human body. Oat extract inhibits the oxidation of low-density lipoprotein, which contributes to the formation of plaques in the heart arteries. Oat has also been shown to prevent in building up of blood sugar level, and consequently, lowering the demand for insulin production. Oat has relatively low glycemic index again due to the presence of high levels of fibers both soluble and insoluble types that help in slowing down the glucose absorption from the intestines. Oat's suitability in food formulations for overweight people is due to its role in creating satiety even when small quantity is consumed, and thus, the resultant feeling of fullness reduces the urge to eat frequently, one of the important reasons for obesity. Similarly, the fiber component in oat can imitate the role of fat in many foodstuffs, and oat-rich food products help in reducing the calorie intake significantly, arresting the tendency to put on unwanted weight. Dietary fibers in oat are recognized to have the ability to prevent constipation, varicose veins, piles, colon cancer and some gastrointestinal disorders. Lysine, the first limiting amino acid in oat, averages about 4.2% of the protein, which is higher than that in other cereals. Oat is also a significant source of other amino acids essential for balanced growth. Oat lipids are nutritionally important owing to high concentration of polyunsaturated fatty acids, especially linoleic acid, the essential fatty acid, required in the synthesis of prostaglandin, the versatile metabolic regulator vital for many body functions. Oat is also a significant source of antioxidants like tocotrienol and lignan precursors for the formation of mammalian lignans by friendly intestinal bacteria. The mammalian lignan enterolactone is one of the key substances conferring protection against breast cancer and other hormone-dependent cancers and the heart. Selenium is universally acknowledged as a trace mineral essential for all antioxidant activity in human body system and oat is a good source of this trace-nutrient. It also acts as a co-factor in all important antioxidant enzyme system glutathione peroxidase and in synergy with tocopherol. Selenium performs more efficiently in almost all antioxidant systems through out the body systems. Selenium is also vital to repair the DNA and to reduce the risk of several types of cancer, especially colon cancer.

■■■

18 Herbs That Cure Human Ailments

C.S. Raghav[1] and Nidhi Dubey[2]

[1]National Bureau of Plant Genetic Resources, Pusa Campus, New Delhi-110012, E-mail: <raghav63@yahoo.com>

[2]Indian Pharmacopoeia Commission, Government of India, Sector-23, Rajnagar, Ghaziabad, Uttar Pradesh-201002, E-mail: <nidhiknj@gmail.com>

Our limited knowledge about the medicinal properties of herbs compels us to choose a title as that of this chapter or else there is hardly any herb on the earth, which has no medicinal value. When *Jeevak*, a great exponent of the *Ayurveda*, had nearly completed his long training in the *Gurukul*, the Guru called all his disciples and asked them, *Go unto the forest and fetch me a plant that has no use*. The disciples went away to the forest. By the evening, every one returned with one or the other plants but *Jeevak* did not return that night or even the next. On the third day, *Jeevak* walked back with a tired gait, unbathed, unshaved and unfed. He approached the great Guru empty-handed and fell at his feet. He lamented, O Guru, I am good for nothing. I am not worthy of being your student. You had given a simple command but I failed you. I could not fulfill it even after three days' untiring toil. *Gurudev*, I could not, in these three days, find a single plant that is useless in the vast forest. The great Guru raised *Jeevak* from his feet and embraced him. He said, oh *Jeevak*, you are the fulfillment of my life's purpose. You alone have understood that nothing in God's creation and Nature is useless. You alone, of all my disciples, have realized this. Go, my son, I have nothing more to teach you. Go and serve the mankind.

Herbs have been used by man since time immemorial for treating the illness. Various mythological testaments throughout the world mention the use of herbs for wellness. Though lacking material evidences, the Hindu mythology refers to the oldest references of use of herbs for the welfare of mankind. The birth of God Brahma, the creator of the universe, is said to occur through the lotus sprigged from the navel of God Vishnu. Chandan (*Santalum album*), ketki (*Agave americana*), datura (*Datura metel*), ashok (*Saraca indica*), peepal (*Ficus religiosa*) bargad (*Ficus bengalensis*), somlata (*Ephedra gerardiana*), kapoor (*Cinnamomom camphora*), tulsi (*Ocimum sanctum*), neem (*Azadirachta indica*), bhang (*Cannabis sativa*), bael (*Aegle marmelos*), *etc*, are the plants referred in various testaments related to Hindu mythology and spread through the entire time span since the origin of universe till date. Similar reference could be seen in mythological testaments of various religions of the world. Ancient Greece, as with Ancient Rome and Egypt played an important role in medical history. The most famous of all Ancient Greek doctors was Hippocrates. The Greeks were practicing medicine 1000 years before the birth of Christ. The use of herbs in health care existed even about 60000 years ago when Neanderthal man used herbal preparations to cure animal and insect bites and injuries. As civilizations came about, each society or civilization started practicing separate forms of Herbalism. This depended on the locally available materials and individual cultures. Thus, different forms of medicine, like Chinese, Indian and Greek originated. During 14th century in the European countries, a spirit of curiosity prevailed as to 'How' and 'Why' things worked. This saw the birth of Allopathy or modern medicine as we know it today.

Both, Natural remedies and Allopathy have their own basis and principles. Natural remedies use whole herbs or herbal extracts in different forms along with some ingredients of animal or mineral origin. Allopathy on the other hand uses naturally or synthetically engineered molecules. Allopathy or modern system of medicine has many examples of plant products used successfully in their natural or synthetic form. An allopathic drug that provides symptomatic relief can be given along with a natural remedy intended for root cause of the disease, *e.g.*,, in the treatment of a bronchial problem, allopathic drugs can be given to reduce the symptoms of wheezing and a natural remedy can be given to treat the bronchitis itself. Another example could be about using a natural remedy to reduce the side effects of an allopathic drug leading to comparatively less stress on the body. It is definitely possible to achieve a synergy with the world becoming one global entity and the intermingling of cultures and knowledge from different parts of the world. Ancient knowledge coupled with scientific principles can come to the forefront and provide us with powerful remedies to eradicate the diseases. Hopefully, this will percolate down to scientific medical research as well.

Research in medicine has been established that plants contain active principles, which are responsible for curative action of the herbs. The scientists isolated active constituents of the medicinal herbs and after testing, some were found to be therapeutically active. Aconitine, atisine, lobeline, nicotine, strychnine, digoxin, atropine and morphine are some common examples. Today, there are at least 150 distinct chemical substances derived from plants that are considered important drugs and are currently in use in one or more countries in the world. Recently, research has supported biological activities of some medicinal herbs. Cancer is such a segment where researchers are expecting new molecules from herbs that can provide us with tools for fighting this dreaded disease. *Allamanda cathratica* (allamandin), *Elephatopus elatus* (elephantpoin), *Helenium autmnale* (helenalin), *Heliotropium indicum* (Indicine-N-oxide), *Daphne mezereum* (mezerien) and *Stereospermum suaveolans* (laphacol) are medicinal plants that have shown significant tumour inhibiting effect. Diabetes mellitus is another area where a lot of research is going on. *Ajuga reptens*, *Galagea officinalis* (galagine), *Bougainvillea spectabilis* (pinitol), *Momordica charantia* (chirantin), *Gymnema sylvestre* (gymnemic acid) and *Tecoma stans* (tecomonine) are some medicinal herbs that have shown effectiveness in non-insulin dependent diabetes. Arthritis is another potential disease where no satisfactory answer is present in modern medicine. *Commiphora mukul* (guggulsterones), *Boswellia serrata* (boswellic acid), *Withania somnifera* (withanolides), *Ruscus acueleatus* (ruscogenin) and *Harpagophytum procumbens* (harpagoside) are prominent plants with anti-arthritic activity. *Chrysanthemum parthenium* (parthenolides) traditionally known as few has shown promising results in migraine, a disease that has eluded the researchers from centuries.

Hepatoprotective action of certain botanicals deserves attention. *Sedum sarmentosum* (sarmentosin) and *Schisandra chinensis* (waweizichun and schisantherin) have shown their ability to lower down the raised liver enzymes in viral hepatitis. *Croton sublyratus* (plaunotol) has potent and wide spectrum anti-peptic ulcer action. A number of plant derivatives have shown anti-AIDS activity. *Ancistrocladus korupensis* (michellamine-b), *Caulophyllum langigerum* (calanolide-a), *Caulophyllum teymani* (costatolide-a), *Homalanthus nutans* (prostratin) and *Conospermum* sp. (concurvone) are the medicinal herbs from African countries that are being employed in research for finding a suitable cure for Aids. The concept of antioxidants is catching up fast and latest research has shown that a number of herbal derivatives have excellent antioxidant action. *Bacopa monnieri* contains bacosides A and B, and bacoside A is a strong antioxidant, which reduces several steps of free radical damage. *Coleus forskohlii* (forskolin), grape seed (proanthocyanidins), *Camellia sinensis* (polyphenols), *Huperzia serrata* (huperzine), *Pinus maritima* (pycnogenol), *Borago officinalis* (gamma linoleic acid) and *Vinca minor* (vinpocetine) are potential antioxidants.

To conclude, the medicinal value of herbs is unlimited, and of course, there are unlimited herbs having medicinal value. The need is to explore them. Increasing scientific research has strengthened our belief in use of herbs validating several traditional claims. This chapter describes medicinal properties and other useful information about few medicinal plants.

Common name: Indian Atees

Botanical name: *Aconitum heterophyllum* Wall. Ex Royle

Family: Ranunculaceae

Vernacular Names: English- Indian Atees; Sanskrit- Ativisha; Hindi- Atis; Bengali- Ataicha; Tamil- Athividayam, Telagu- Ativasu; Gujarati- Atavasa, Kannada- Ativisa, and Kashmiri-Atis

Distribution: This herb is usually found on humus-rich soils in the Alpine and sub-alpine zone of the Himalayas from Indus to Kumaon and from 1800 to 4500 m above mean sea level.

Plant details: It is a perennial herb with simple or branched, long, erect stem, leaves broad, ovate or orbicular, 5-lobed and toothed and flowers helmet shaped, 2-2.5 cm long, blue or greenish with purple veins.

Chemical Composition: The roots contain various alkaloids such as heterophyllisine, hetrophylline and heterophyllidine, heteratisine, hetidine, hetisinone, histisine, benzoylheteratisin, atidine, F-dihydroatisine, isoatisine, dihydroatisine, hetisine, hetidines and hetisinone. However, the main constituent of the plants is an amorphous nontoxic diterpenoid alkaloid atisine.

Medicinal Properties and Uses: Its roots contain nontoxic amorphous alkaloids and are one of the best bitter tonics for children. The dried roots are analgesic, anti-inflammatory, antipyretic, antiperiodic, aphrodisiac, astringent, cholagogue, febrifuge and tonic. It is used in the treatment of liver disorders, dyspepsia, diarrhoea, indigestion, nausea, vomiting, throat pain, anorexia, piles and coughs. Seeds are stimulant, aromatic, emmenaguogue, stomachic, carminative, anti-pyretic, hyperacidity, useful in hiccup and bad breath. It is regarded as a valuable tonic and digestive. It is also used in the treatment of rheumatism, intermittent fevers and acute inflammatory infections.

The extracts from its plants, as well as purified compounds, are used in stomachache and as an aphrodisiac. Its roots contain alkaloids and they are used in the treatment of hysteria and throat diseases. They are considered astringent and are prescribed in diarrhoea, especially in infants and for dyspepsia, vomiting, and cough. The root is also considered antidiabetic and very efficacious for irritability of stomach and in abdominal pains.

Indigenous Knowledge: The plain powder of the tuberous root mixed with honey is given to children for cough, cold, fever and vomiting. In Uttaranchal, the roota are used to treat fever, cold and cough in children.

Common name: Sweet flag

Botanical name: *Acorus calamus* Linn.

Family: Araceae

Vernacular Names: English- Sweet flag; Sanskrit- Vacha; Hindi- Bach; Tamil-vashambu, Telugu- vasa vadaja and Kannada- bajai

Distribution: It is cultivated in damp, marshy places in India and Burma. It is quite common in Manipur and the Naga Hills on the edges of lakes and streams, also growing wild, ascending to 3000 to 6000 ft in the Himalayan ranges.

Plant details: A rhizomatous, perennial semi-aquatic plant grows up to 40 cm in height. Leaves are simple, bright green, distichous, thickened in the middle and ensiform flowers seen in densely packed spandex inflorescence and fruits oblong berries with seeds few suspended from the apex of cells. Underground rhizomes are creeping and branched about 1 cm in diameter

Chemical Composition: The dry rhizomes contain 1.5 to 3.5% of a yellow aromatic volatile oil. It has a mellow odour resembling to that of patchouli. The chief constituent of the oil is asaryl-aldehyde. There are also present a bitter glycoside named acorin and some other substances such as eugenol, methyl eugenol, calamol, calamine, azulene, sesqui terpene ketone like acorone, calarene, calacorene, acolamone, isoacolamone, epyshyobunone, shyobunone, isoshyobunone, acoragermacrone and alcohol like isocalmendiol, preisocalmendiol, pinene and camphene. The Indian oil has much higher asarone content than the European commercial oil.

Medicinal Properties and Uses: The rhizome is an aromatic, stimulant, bitter, tonic, carminative, antispasmodic, emetic, expectorant, emmenagogue, aphrodisiac, laxative and diuretic. Alcoholic extract of the plant has been shown to possess sedative and analgesic properties. It causes a moderate depression in the blood pressure and respiration. The water-soluble fraction of the dealcoholized extract produced relaxation of isolated intestine and negative isotropic action on frog's heart. The insecticidal activity of the solvent extracts and steam-distilled volatile principle of rhizome against common houseflies is quite marked. The insecticidal activity of the oil and the extract appears to be due to the presence of the trans-isomer of asarone.

Indigenous Knowledge: Its infusion is given in diarrhea, dysentery, bronchial and chest affections and epilepsy. The burned root mixed with some bland oil is applied over the abdomen in flatulent colic. The poultice is also usefully applied to paralyzed limbs and rheumatic swelling.

Common name: Indian aloes

Botanical name: *Aloe vera* Linn.

Family: Liliaceae

Vernacular Names: English- Indian aloes; Sanskrit- Kumari; Hindi- Ghikanuar; Bengali- Ghrit Kumari, Tamil- Rattabolam, Telugu- kalabanda and Kannada- Lolesara.

Distribution: Many varieties are cultivated throughout India of which some grow wild on the coasts of Bombay, Gujarat and South India.

Plant details: It is a perennial shrub with fleshy leaves and short stem. Its leaves are fleshy and sessil, with prickles on margins, flowers in long spikes and fruits capsules, often sterile. Reproduction is by budding. Leave juice dried in air is called aloin.

Chemical Composition: Aloe contains C-glycosides and resins. They are anthraquinone-producing plants and the content of anthraquinones is subject to seasonal variation. The leaves contain barbaloin, chrysophanol and aloe emodin. The inside mucilaginous part contains glucose, galactose, mannose and galacturonic acid. The barbaloin content of Indian aloe was found to be as low as 4.24%. Purgative properties of aloes are due to presence of three pentosides, *i.e.*, barbaloin, isobarbaloin and B-barbaloin.

Medicinal Properties and Uses: The use of Aloe dates back to ancient times in Egypt and Greece where it was used for skin infections and wound healing. The leaf contains three medicinally important and distinct parts, *i.e.*, the leaf exudates, the leaf epidermis and the leaf pulp. The leaf exudates, a bitter yellow liquid, can be heated, concentrated and dried to a black powder. This is the source of the drug, aloe, which is used as purgative. The whole leaf of *Aloe vera* or its product's has been used directly on radiation burns, thermal bums, wounds, stasis ulcers and diabetic ulcers. The whole drug is stomachic, tonic in small doses and purgative in large doses, emmenagogue and anthelmintic. Dried juice of *Aloe vera* is cathartic and given in constipation. Fresh juice is cooling. Juice of *A. vera* roots in the form of an ointment in *vaseline* has been found to hasten healing of wound of thermal burns and radiation injury in albino rats. It has been found that aloe compound is very useful in cases of functional sterility and disturbed menstrual function. The *Ayurvedic* drug known as *kumari asava* is useful in general debility, cough, asthma, consumption, piles, epilepsy and colic.

Indigenous Knowledge: Pulp is useful in menstrual suppression. Root is given in colic. It is a favorite remedy for intestinal worms in children. After dissolving in alcohol, it is used as hair dye to stimulate hair growth.

Common name: Java glangal

Botanical name: *Alpinia galanga* Willd.

Family: Scitaminiaceae

Vernacular Names: English- Java glangal; Sanskrit- Malayavach; Hindi- Kulanjan; Bengali- Kulinjan; Tamil- Arathai; Telugu- Pedda-dumparash-tram; Kannada-Dumba-rasmi and Malyalam- perraratta.

Distribution: It is distributed in Eastern Himalayas, South India Bengal and Malabar region.

Plant details: The herb grows to a height of about 12 cm, the leaves being long rather narrow blades and the flowers, of curious formation, growing in a simple, terminal spike and the petals white with deep-red veining distinguishing the lip petal. The branched pieces of dark reddish-brown rhizome are from 4 to 7 cm in length and seldom more than 2 cm thick. Their odour is aromatic and their taste pungent and spicy.

Chemical Composition: Rhizomes contain a pale yellow volatile essential oil with a pleasant odour obtained on distillation. This oil contains 48% of methyl cinnamate, 20 to 30% of cineol and other includes camphor, galangin, alpinin, kaempfedine, 3-dioxy 4-methoxyflavone and pinene. Alcoholic extract of the rhizome showed the presence of tannin, phlobaphenes and starch. The ash contained chloride, phosphate, sulfate, potassium and manganese.

Medicinal Properties and Uses: It is aromatic, stimulant and bitter stomachic and carminative. Steam-volatile portion of extract of rhizome stimulates the bronchial glands directly but the nonvolatile portion acts reflexly through the gastric mucosa. The alcoholic extract of rhizome produces hypotemia in mice. It also potentiates amphetamine toxicity in grouped mice. The rhizome is very useful in rheumatism and catarrhal affections, especially bronchial catarrh. The seeds have the same uses as the rhizome. It is also used as flavoring agent.

Indigenous Knowledge: Its paste made with bland oil is applied over acne and other skin diseases. It is given in doses of 5 to 10 g in controlling urine and in diabetics. The rhizome is used as deodorizer of foul smells from the mouth and other parts of the body.

Common name: King of bitters

Botanical name: *Andrpgraphis paniculata* (Burm.f.) Nees

Family: Acanthaceae

Vernacular Names: Sanskrit- Kirta; Hindi- Kalmegh; English- King of bitters; Bengali-Kalmegh; Tamil- Nilavembu; Malyalam- Nilaveppu; Telugu- Nelavembu and Kannada- Kreata

Distribution: Found throughout the plains of India.

Plant details: An annual herb grows up to 50 cm in height, having slender quadrangular branches, leaves opposite, lanceolate, sub-sessile, flowers small bluish white, numerous, whitish, fruits minute pointed capsules and seeds are many

Chemical Composition: The main constituent is andrographolide and other constituents include neoandrographolide, andrographin, panicalin, panaculoside, apigenin, 7-4-dimethyl ether, kalmeghin, traces of an essential oil and 5 hydroxy 1-7, 8, 2' 3'-tetramethoxy flavone.

Medicinal Properties and Uses: Plant pacifies tridoshas, skin diseases, burning sensation, cough and bronchitis. Plant stimulates liver and purifies vitiated blood. Roots and leaves are stomachic, tonic, antipyretic, alterative, anthelmintic, febrifuge and cholagogue.

Indigenous Knowledge: The leaf juice is a household remedy for flatulence, loss of appetite bowel complaints of children, diarrhea, dysentery dyspepsia and general debility. It is preferably given with the addition of aromatics. Decoction or infusion of the leaves gives good results in sluggish liver, neuralgia, in general debility, in convalescence after fevers and in advanced stages of dysentery.

Common name: Wormseed

Botanical name: *Artimisia annua* Linn.

Family: Compositae

Vernacular Names: Sanskrit- Chauhar; Hindi- Kirmala; English- Wormseed; Tamil- Machipattari; Malayalam- Tiruni-tipacha; Telugu- Davanamu and Kannada- Gopata puttiopalu

Distribution: Many species grow abundantly in the western Himalayas from Kumaon to Kashmir, in Kurran Valley, Baluchistan and in Afghanistan.

Plant details: It is an annual or perennial sub-shrub. Its roots are usually taproots, sometimes fibrous. Stems are usually erect, sometimes prostrate to ascending (underground stems are sometimes woody caudices or rhizome, sometimes fleshy). Leaves are usually alternate or opposite, sometimes in basal rosettes, rarely in whorls, rarely stipulate, usually etiolate, sometimes sessile and fruit are exalbuminous and embryos straight.

Chemical Composition: Important constituents of *A. annua* are an essential oil having an odour like cajuput oil and camphor, containing cineole, thujone, thujyl alcohol, cadinene, phellandrene and pinene, camphene, santonin, *etc.* The santonin is responsible for anthelmintic properties. The herb also contains the bitter glucoside absinthin, absinthic acid, together with tannin, resin, starch, nitrate of potash and other salts.

Medicinal Properties and Uses: The plant is chiefly used for its anti-malarial activity. It is antiperodic, aperient, stomachic, tonic, anthelmintic and cardiac stimulant.

Indigenous Knowledge: Its infusion is given as an anthelmintic and cardiac stimulant. As a disinfectant and antiseptic, a fomentation of the herb is applied over inflammations, tumors and foul ulcers.

Common name: Asparagus

Botanical name: *Asparagus racemosus* Willd.

Family: Liliaceae

Vernacular Names: Sanskrit- Shatavari; Hindi- Satavari; English- Asparagus; Bengali- Satmuli and Malyalam- Satavari

Distribution: Found all over India, especially in North India.

Plant details: It is a thorny, climbing under-shrub with woody stems. Young stems are very delicate, brittle and smooth and leaves are reduced to minute scales and spines. Curved cladodes replace the leaves. Flowers are white fragrant in simple or branched recemes. Fruits are globular, or vaguely 3 lobbed, pulpy berries, purplish black when ripe and seeds with hard and brittle testa. Roots are succulent and tuberous from 30 cm to 1 m in length, smooth and tapering at both ends.

Chemical Composition: The alcohol, ethyl acetate and acetone extract of powdered dry roots of *A. racemosus* yielded pharmacologically active (antioxytocic) saponins. The main constituents include inulin, asparagusic acid and eight fructo-oligosaccharides. The two glycosidic bitter principles, *i.e.*, officinalisnin-I and -II are isolated from the dried root and yield *b*-sitosterol, sarsasapogenin and nine steroidal glycosides (named asparagosides A to I, in order to increase polarity). Other constituents include asparagine, tyrosine, succinic acid, arginine, *a*-aminodimethyl-*g*-butyrothetin (a methylsulfonium derivative of methionine), fat and sugar. The roots are thought to possess diuretic and hypotensive properties and to enhance the renal elimination of water.

Medicinal Properties and Uses: Roots are antidiarrhetic, refrigerant, diuretic, nutritive, tonic, galactagogue, aphrodisiac and antispasmodic. Alcoholic extract of the aerial parts of *A. racemosus* showed anticancer activity in human epidermal carcinoma of the nasopharynx in tissue culture. The aqueous solution of the crude alcoholic extract of the roots also had an inhibitory effect on the growth of *E. histolytica in vitro*. It is useful in dysentery, diarrhea, tumors, inflammations, biliousness, diseases of the blood, kidney and liver complaints. It is also useful in rheumatism, dyspepsia, gleets and gonorrhea. The roots have been found to have galactogogue action in buffaloes in which the milk yield was found increased significantly after the uses of drug. The aqueous extracts of both fresh and dried roots were found to have amylase and lipase activities.

Indigenous Knowledge: The juice of this drug taken with milk is useful in gonorrhea. It is also used as tonic.

Common name: Periwinkle

Botanical name: *Catharanthus roseus* (L) G. Don.

Family: Apocynaceae

Vernacular Names: English- Periwinkle; Hindi- Sadabahar; Sanskrit- Sadampuspa; Tamil- Nithyakalyani and Telugu- Billaganneru

Distribution: Throughout India

Plant details: An erect annual herb grows up to a height of one meter. Leaves are deep green oblong or obovate flowers in cymose, axillary clusters, deep rose, or white in colour and fruits cylindrical follicle.

Chemical Composition: The stem of Madagascar periwinkle produces a milky sap, which is the source of over 70 different indole alkaloids. Some of the constituents found in various parts of plant include Ajmalicine, catharanthine, catharicine, Catharine, catharocine, quercetine, reserpine, serpentine, vinblastine, vincamine, vincarodine, vincristine, vindoline, vindolinine, vindilicine, *etc*.

Medicinal Properties and Uses: This plant, commonly known as periwinkle, contains more than 100 alkaloids, two of which, vinblastine and vincristine, have potent activity in the treatment of different forms of cancer. Vinblastine is used for the treatment of Hodgkin's disease, whereas, vincristine is primarily used for the treatment of paediatric leukaemia. They are also used in the treatment of breast, ovarian and renal cell cancer, squamous cell's cancer of the head, *etc*. The herb is found growing throughout India. These vinca alkaloids bind to tubulin dimers and inhibiting microtubule structures of the cells, thus, inhibiting the metaphase of cellular mitosis. Their main side effects of these drugs are peripheral neuropathy, hair loss, hyponatremia and constipation.

Indigenous Knowledge: The whole plant of *Sadabahar* is used to treat diabetes, menorrhagia, leukemia and stomachache.

Common name: Black Oil Tree

Botanical name: *Celastrus paniculata* Willd.

Family: Celastraceae

Vernacular Names: Sanskrit- Jyotishmati; Hindi- Malakanguni; English- black oil tree; Bengali- Latafarki; Tamil- Valuzhuvai and Malayalam- Palulavam

Distribution: It is found all over the hilly district of the Himalayas up to an altitude of 6000 ft. It grows in holly parts of Bombay, Gujurat, Central India and Madras

Plant details: It is a large climbing shrub with woody stem and elongate branches. Leaves are simple, alternate, ovate-obovate, crenulated and coriaceous and flowers greenish white, in terminal panicles. Fruits are capsules yellow coloured, become bright red when ripe and contain 4 brown seeds.

Chemical Composition: The seeds yield brownish-yellow oil (52.2%) with an unpleasant taste. An appreciable proportion of acetic and benzoic acids in the oil of seeds and fruit coat has been reported, in addition to the usual higher fatty acids. Scientists have isolated a crystalline material believed to be a tetracasanol and sterol. They reported two alkaloids from oil cake, *viz.* celastrine $C_{19}H_{25}NO_3$ melting point (mp) 260°C and paniculatine. An orange-red semisolid fat (30%) is obtained on extraction of the fleshy arils with petroleum. This contains 6.46% unsaponifiable matter from which a phytosterol, celastrol $C_{27}H_{46}O_3$ mp 142°C has been isolated. In addition, a highly coloured resinous substance is present in the unsaponifiable matter, which has not been identified.

Medicinal Properties and Uses: Oil is rubefacient. Seeds are alterative, stimulant, and nervine tonic. Seeds and oil stimulate intellect and sharpen memory. The seed oil produces sedation and anticonvulsant activity in rats and produces a gradual fall in blood pressure in cats. Aerial parts of the plant indicated antiviral activity against *Ranikhet* disease virus. An active fraction isolated from the oil by the countercurrent distribution method had a tranquilizing effect on rats, mice, monkeys and cats. Oil with benzoin, cloves, nutmeg and mace is a sovereign remedy in beriberi and a powerful stimulant in doses of 10 to 15 minims. The leaves are emmenagogue, and the leaf sap is used as an antidote for opium poisoning. The bark is abortifacient and the seeds are bitter, laxative, emetic and tonic.

Indigenous Knowledge: Decoction of the seeds with or without the addition of some aromatics is recommended in rheumatism, gout, paralysis and leprosy.

Common name: Safed Musli

Botanical name: *Chlorophytum borivilianum* Santapau.

Family: Liliaceae (Anthericaceae)

Vernacular Names: Hindi- Safed Musli; Tamil- Tiruvanticam and Kannada- Dravanti

Distribution: As safed musli grows naturally in most parts of the Central Region of India, the normal climate of the Central region suits the crop most and it can grow successfully in the wide range of temperature and rainfall. Sandy loam soil with proper drainage system facilitates its growth.

Plant details: It is an annual herb. Roots are usually thick or slightly fleshy, leaves basal, sub-distichous or fasciculate, sessile or petiolate, usually linear to elliptic-lanceolate, centuplicate and base sheathing. Scape is axillary, proximally with bract like cauline leaves. Inflorescence is a terminal raceme or panicle and bracts small. Flowers are bisexual and pedicel articulate. Perianth is usually white, tepals 6, free, 3-7-veined, persistent, or marcescent. Stamens are 6, inserted at base of the tepals, filaments filiform, usually slightly widened near middle, anthers nearly basifixed and introrse. Ovary is 3-loculed and ovules 1 to several per locule. Style is slender, and stigma small. Fruit is a capsule, acutely 3-angled, loculicidal. Seeds are black coated and flattened.

Chemical Composition: The roots contain carbohydrates (35-45%), proteins (5-10%), fibre (25-35%), saponins (2-20%), 25 different alkaloids (15-25%), minerals and vitamins. The saponins and alkaloids present in the plant are the primary source of their significant medicinal properties.

Medicinal Properties and Uses: In *Ayurvedic* literature, safed musli is celebrated as a *Divya Aushadi* with unparalleled medicinal properties. It is a chief ingredient in the preparation of over a hundred *Ayurvedic* formulations. Safed musli is also gaining increasing acceptance as a vitalizer and health-giving tonic, a curative for pre- and post-natal problems, a restorative for immunity-improvement and as a remedy for diabetes and arthritis. The recent discovery of Safed musli as a natural and safe aphrodisiac agent has also opened up new channels of application and new demand for it. By virtue of being an herb, safed musli has been found to be an ideal aphrodisiac with no negative side effects associated with chemical-based aphrodisiac. It is also found to be effective as an energetic one in cases of cough and asthma and chronic leucorrhoea. Post-partum it replenishes lost fluids, *prana*, *ojas* and improves the quantity and flow of breast milk.

Indigenous Knowledge: It is used as aphrodisiac and tonic and during pregnancy as a nutritive tonic for mother and foetus.

Common name: Coleus

Botanical name: *Coleus forskohlii*

Family: Lamiaceae

Vernacular Names: English- Coleus; Sanskrit- Pashan Bhedi; Hindi- Patharchur; Gujarati- Garmai; Kannada- Makandiberu and Tamil- Koorkan Kilangu

Distribution: It grows on dry slopes of the Indian plains and in foothills of the Himalayas. It is under cultivation in parts of Rajasthan, Maharashtra, Karnataka and Tamil Nadu in large scale.

Plant details: Coleus is a crop of tropics. It prefers humid climate with a relative humidity ranging from 60 to 85% and a temperature range between 10 and 25°C for its successful growth. It is also found to perform well in less humid and warmer regions when grown as an irrigated crop. It thrives better in porous and well-drained soils with a pH ranging from 5.5 to 7. It does not require very fertile soils though it can be economically grown even on the soils with marginal fertility. The red sandy soils are ideal for the cultivation of this crop.

Chemical Composition: Barbatusine, barbatusol, coleol, coleonols, coleons, coleosol, cyclobutatusin, deacetyl-forskohlin, forskolin, *etc.*

Medicinal Properties and Uses: Scientists have studied coleus and forskolin, a root extract of coleus, for the various health problems. A small number of studies suggest that forskolin may improve cardiovascular function in patients with cardiomyopathy. It may improve breathing in patients with bronchial asthma due to relaxation of airway smooth muscle.

In last 20 years, forskolin has been successfully evaluated in numerous clinical studies, including bronchial asthma, cardiovascular health, glaucoma, sexual dysfunction in men and dermatological diseases. In cited clinical studies, no adverse effects of forskolin administration were reported in spite of the fact that different doses, forms of administration and routes of administration were employed in these studies.

Coleus and forskolin have been suggested for many other uses, based on tradition or on scientific theories. However, these uses have not been thoroughly studied in humans, and there is limited scientific evidence about safety or effectiveness. Some of these suggested uses are for potentially serious and even life threatening conditions.

Indigenous Knowledge: As part of a long-standing tradition in India, *C. forskohlii* roots have been used as a marinated food or pickle, which is commonly eaten as part of a vegetarian meal. Coleus has been used as a breast milk stimulant for hundreds of years, however, the traditional use has not been well documented, and scientific evidence is limited. The other traditional uses include abdominal pain, allergies, antimicrobial, antiviral, cancer, cataract, diabetes, liver diseases and malabsorption.

Common name: Babchi Seeds

Botanical name: *Cullen corylifolia* Linn.

Family: Papilionaceae

Vernacular Names: Sanskrit- Vakuchi; Hindi- Babchi; English- Babchi Seeds; Bengali- Bavachi; Tamil- Karpokarisi; Telugu- Pilligaddalu; Kannada- Aheruballi and Malayalam- Chatawali

Distribution: This plant is found as a weed in waste places allover the plains of India. It is also cultivated in some other places.

Plant details: It is an erect herbaceous annual with 60-120 cm height. Its leaves are simple, broadly elliptic, rounded, clothed with white hairs on both surfaces, covered with numerous black dots. Flowers are blue in dense, axillary, 10-30 flowered racemes, fruit pods sub-globose, slightly compressed and black and seeds oblong, flattened, dark brown with an agreeable aromatic odour and taste.

Chemical Composition: The seeds contained an essential oil, nonvolatile terpenoid oil, a pigment, a resin, fixed oil, raffinose, two coumarin compounds, *viz.* psoralen and isopsoralen, psoralidin, corylifolean and isopsoralidin, chalcones, flavones, coumesterol. The seed oil yields limonene, β-caryophyllene, 4-terpineol, linalool, geranylacetate, angelicin, psoralen and bakuchiol. From the petroleum extract of roots, daidzein, trilaurin and coumasterol were isolated for the first time. Angelicin, psoralen and sitosterol were also isolated from the extract.

Medicinal Properties and Uses: Seeds are laxative, anthelmintic, diuretic and diaphoretic. Oil has a powerful effect against skin streptococci. Studies have showed that psoralen significantly accelerates the photooxidation of DOPA (diydroxyphenylalanine) under sunlight as well as photoflood lamplight. The alcoholic and aqueous extracts of seeds exerted antibacterial activity. Volatile oil of seeds inhibited the growth of staphylococci. Ether and chloroform extracts also showed similar activity. Essential oil of *P. corylifolia* showed a moderate selective antifungal activity. Petroleum ether extract of seeds exhibited anthelmintic activity against earthworms. The extract also produced a rise in blood pressure on anesthetized dogs and caused a stimulation of the intestinal smooth muscle. Literature shows that isopsoralene exhibited tranquillo-sedative, anticonvulsant and central muscle-relaxant properties in rats, mice and rabbits. Bavachinine, a flavonoid isolated from *P. coryliflora*, showed anti-inflammatory activity against carrageenin-induced edema in rats. It has been reported that oleoresin extract of the seeds (containing most of the essential oil) was found to be a most effective preparation when applied locally on the patches of leucoderma. Seed powder is useful in leprosy and leucoderma internally and is also applied in the form of paste or ointment externally. As a laxative, it is particularly used in bilious disorders.

Indigenous Knowledge: Traditionally, the paste of seeds is used in the treatment of leucoderma.

Common name: Black musalie

Botanical name: *Curculigo orchioides* Gaertn.

Family: Amaryllidaceae

Vernacular Names: Sanskrit- Talamulika; Hindi- Kali musli; English- Black musalie; Bengali- Talamusli; Tamil- Nila panaikelangu; Malayalam- Vella Panakelanga; Telugu- Nela tadi kelangu and Kannada- Neladaali.

Distribution: It is found occurring wild in sandy regions of hotter parts of India and Sri Lanka.

Plant details: It is a perennial herb with tuberous roots. Its leaves are simple, crowded in the base, short petioled, long, acute, lanceolate and often produce buds from the tip when it contact with the wet soil. Flowers are bright yellow, fruits capsule, 1-4 seeded and seeds black and oblong.

Chemical Composition: Resin tannin (1.45%), mucilage, fat starch (48.48%) and ash (8.60%) containing oxalate of calcium

Medicinal Properties and Uses: Bitter aromatic tonic, demulcent, diuretic and aphrodisiac Roots are alterative. As an aphrodisiac, the powder is given with sugar and ghee also to remove stones from the urinary system. Rootstock is alternative and carminative. It is useful in biliousness, diseases of the blood, bronchitis, ophthalmia, lumbago and in gonorrhea.

Indigenous Knowledge: It is given with some aromatics in piles, diarrhea, jaundice, and asthma and used as a poultice for itch and in some skin diseases. One to two ounces of powdered rhizome in warm milk and sugar is prescribed during convalescence after acute illness.

Common name: Lemon grass

Botanical name: *Cymbopogon citrates* (DC.) Stapf.

Family: Poaceae

Vernacular Names: Sanskrit- Bhustrina; English- Lemon grass; Hindi- Gandhatrina; Bengla- Gandhabena; Telugu- Nimmagaddi; Tamil- Vasanappillu; Kannnada- Majjjgehullu and Malayalam- Vasanappulla

Distribution: Its species cultivated as a culinary herb and for medicinal use.

Plant details: Lemongrass is a clump forming grass native to India and Sri Lanka. It forms dense clumps up to 1.8 m tall and 1.2 m wide. The leaves are tapered at both ends and are up to 1 m long by 1.2 cm wide with scabrous margins.

Chemical Composition: Citral, geranyl acetate, linalyl acetate, geranial, neral, limonene, myrcene, beta-caryophyllene, isopulegol, citronellic acid, geranium acid and *a*-camphorene.

Medicinal Properties and Uses: The entire plant possesses antiseptic, antifebrile and stomachic properties. It is used in treating coryza, influnza and pyrexia. It is also prescribed against dyspepsia and vomiting and as a carminative, using three to four drops of the essential oil, diluted in water. It is used externally to treat eczema. The essential oil is used as an insecticide against mosquitoes and as a deodorant. Lemongrass oil revitalizes the body and relieves the symptoms of jetlag, clears headaches and helps to combat nervous exhaustion and stress-related conditions. It is a great overall tonic for the body and it boosts the parasympathetic nervous system, which is a boon when recovering from illness, as it also stimulates glandular secretions. It is useful with respiratory infections such as sore throats, laryngitis and fever and helps prevent spreading of infectious diseases. It is helpful with colitis, indigestion and gastro-enteritis. Lemongrass oil helps in toning the muscles and tissue and relieves muscle pains by making the muscle suppler. It helps with correcting poor circulation and as an insect repellent. It helps to keep the pets clean of fleas, ticks and lice. It is also used for clearing up oily skin and acne as well as athlete's foot. It alleviates excessive perspiration. It blends well with basil, cedar wood, Clary sage, lavender, geranium, fennel, orange, chamomile, eucalyptus, tea tree, patchouli, marjoram, elemi, sandalwood, ylang ylang and other citrus oils.

Indigenous Knowledge: The leaves of lemongrass are used as tealeaves in headache and fever.

Common name: Rosha grass

Botanical name: *Cymbopogon martini* (Roxb.) Wats.

Family: Poaceae

Vernacular Names: Sanskrit- Dhyamakah; English- Rosha grass or Palmarosa; Hindi- Gandhabel; Kannad- Harehullu, Malyala- Sambharappullu; Tamil- Kavattampillu and Telugu- Kamakchi

Distribution: It is found throughout India

Plant details: A perennial sweet scented grass grows up to 2 meters in height. Its leaves are linear-lanceolate, acute and flower in spikelets, in long panicles, turning reddish brown when mature. A well-drained sandy loam soil with soil pH 7.5-8.5 is ideal for cultivation of Palmarosa. It can also be cultivated in well-drained clay loam soil, which is free from water logging. Warm tropical climate up to 300 m elevation in the foothills is suitable for cultivation of Palmarosa. Temperature ranging from 10 to 36°C with annual rainfall around 1000 mm and ample sunshine are congenial for its growth. Moist and warm climate throughout the year accelerates its growth. Areas, which are affected by severe frost, are not suitable as the frost kills the grass and reduces the oil content.

Chemical Composition: Geraniol, geranyl acetate, limonene, methyl heptenone, pipritone and farnesol

Medicinal Properties and Uses: The plant is acrid, bitter, thermogenic, appetiser, carminative, digestive, cardiotonic, depurative, galactagogue, diuretic, sudorific and febrifuge and is useful in vitiated conditions of *kapha* and *vata*, neuralgia, bronchitis, cough, catarrh, helminthiasis, anorexia, dyspepsia, colic, cardiac debility, leprosy, skin diseases, agalactia, strangury, epileptic fits in children, pharyngopathy and fever.

It is most useful during convalescence and cools the body of fever, while aiding the digestive system, helping to clear intestinal infection, digestive atonia and anorexia nervosa. It is effective in relieving sore, stiff muscles. Palmarosa oil moisturizes the skin, while balancing the hydration levels and stimulating cell regeneration. It balances production of sebum to keep the skin supple and elastic and is valuable for use with acne, dermatitis, preventing scarring, rejuvenating and regenerating the skin, as well as fighting minor skin infections, sore tired feet and athlete's foot.

Palmarosa is used for many kinds of skin issues including dry skin, swellings, dermatitis, eczema, psoriasis, skin infections and fungus. It helps to balance sebum production and stimulate cell regeneration. The effect of Palmarosa on the nervous system is calming, yet uplifting. Palmarosa is an excellent immune stimulant and it helps the body heal from all sorts of infections. It is also used for heart palpitations, spasms, pain, insomnia and to reduce anxiety.

Indigenous Knowledge: Traditional healers use Palmarosa oil for its calming and mood uplifting effect, while clearing muddled thinking. It is used to counter physical and nervous exhaustion, stress-related problems and nervousness.

Common name: Citronella grass

Botanical name: *Cymbopogon nardus* (L.) Rendle.

Family: Poaceae

Vernacular Names: Sanskrit- Bhustrina; English- citronella grass; Hind- Gandhatrina; Bangla- Gandhabena; Telugu- Nimmagaddi; Tamil- Vasanappillu; Kannnad- Majjjgehullu and Malyalam- Vasanappulla.

Distribution: It is found growing throughout India

Plant details: It flourishes on a wide variety of soils ranging from loam to poor laterite with best growth on well-drained sandy-loam soil. The ideal elevation for commercial cultivation ranges from sea level to 300 m. However, in Kerala, it grows well at altitudes between 900 and 1200 m, generally on poor soils along with hill slopes. Lemongrass requires warm and humid climate with sufficient sunshine and rainfall, ranging from 250 to 300 cm, uniformly distributed throughout the year. In the hilly areas of Kerala receiving heavy rainfall, the plant grows luxuriantly and it is harvested more frequently but the oil and citral content are less as compared to the plants growing in the regions of less rainfall.

Chemical Composition: Alpha- and beta-pinene, alpha-terpeneol, camphene, citronellol, citronellal, elemol, caryophyllene, limonine, farnesol, methyl heptinone, methyl iso eugenol, alpha-phellandrene, *etc.*

Medicinal Properties and Uses: The entire plant possesses antiseptic, antifebrile and stomachic properties. It is used in treating cold, influnza and pyrexia. It is also prescribed against dyspepsia and vomiting and as a carminative, using three to four drops of the essential oil, diluted in water. It is used externally to treat eczema. The essential oil is used as an insecticide against mosquitoes and as a deodorant. Citronella oil's most useful quality is that of it being an insect repellent. Citronella oil helps to clear the mind and has a general toning and tonic effect on the body. It is helpful with colds, flu and minor infections and also has deodorizing qualities. In studies using laboratory animals, the oil of citronella showed little or no toxicity. The only concern is skin irritation because some products are applied to human skin.

Indigenous Knowledge: Citronella oil is a popular mosquito repellant in many rural areas of the country. People in villages plant the grass near the houses to ward off the mosquitoes.

Common name: Kokam butter tree

Botanical name: *Garcinia indica*

Family: Guttiferae

Vernacular Names: Sanskrit- Brikshamla; Hindi- Kokam; English- Kokam butter tree; Tamil-Murugal; Malayalam- Punampuli; Kannada- Murgala; Bangla- Kokam; Gujarati- Kokan and Marathi- Amsol, Katambi.

Distribution: The tree is oriental in origin, found in Southern India, particularly in the tropical rain forest of Western Ghats of Ratnagiri, Konkan, Coorg and Wynadu region. It is also found in the evergreen forests of Assam, Khasi, Jantia hills, West Bengal and Gujarat. The crop prefers warm and moderately humid tropical climate with a total rainfall range of 2500-5000 mm grows under a mean annual temperature of 20-30°C, relative humidity 60-80% and up to an altitude of 800 m above mean sea level.

Plant details: A moderate to large sized evergreen tree grows up to 40 meters in height. Its leaves are simple, opposite, ovate, acute and shining, flowers unisexual, found in separate tree and fruits red, ovoid, smooth, containing 6-8 seeds, covered with whitish sweet pulp. Kokam fruits are harvested when ripe.

Chemical Composition: The seeds contain 30% fat, yielding pale yellow concrete oil known as kokam oil. Fruits contain cellulose, an extractive and an insoluble reside. Fruit rind contains commercially important hydroxyl-citric acid.

Medicinal Properties and Uses: Fruit is cooling, cholagogue, demulcent, emollient, and antiscorbutic and bark is astringent. Oil is soothing, used in skin diseases. Kokam oil is a specific remedy in dysentery, mucus and diarrhea. It is also useful in phthisis and some skin diseases. Externally, oil has a healing property and usefully employed as an application to ulcerations and fissures of the lips. It is considered an excellent substitute for animal fat as a basis for ointment. The fruit is refrigerant, demulcent, anticorbutic and astringent. A drink of infusion and its local application allover the body are prescribed in urticaria. The rind has antioxidant property

Indigenous Knowledge: The ripened rind and juice of kokam fruit are commonly used in cooking. The dried and salted rind is used as a condiment in curries. It is also used as a garnish to give an acid flavour to curries and for preparing attractive, red and pleasant flavoured cooling syrup. Kokam butter used as an edible fat is nutritive also.

Common name: Miracle fruit

Botanical name: *Gymnema sylvestre* R. Br.

Family: Asclepiadaceae

Vernacular Names: Sanskrit- Meshasringi; Hindi- Gurmara; English- Miracle fruit; Bengali- Meshasringi; Tamil- Sirukurinjan; Telugu- Podapatri and Kannada-Sannagerasehambu

Distribution: A climbing plant is very common in central and southern India, Western Ghats and in the Goa territory. Specific states include Uttar Pradesh, Madhya Pradesh, Maharashtra, Punjab, Haryana and Tamil Nadu, Andhra Pradesh, Kerala, Karnataka, Bihar and Bengal.

Plant details: It is a large, woody climber. Its leaves are simple, opposite, elliptic, pointed at the end, flowers small, yellow in umbellate cymes and axillary and fruits slender, brownish follicles. The plant grows in a variety of soil and agro-climatic conditions in tropical and sub-tropical regions up to 600 m. Its mature seeds are collected between October and December, and seedlings raised in the nursery are transplanted in field during February-March. The plant grows well with the on-set of rainy season.

Chemical Composition: Its dried leaves contain two resins, *i.e.*, one is insoluble in alcohol forming the large portion and another is soluble in alcohol. There are also a bitter neutral principle, albuminous and colouring matter, calcium oxalate, pararobin, glucose, carbohydrates, some tartaric acid- an organic acid said to be a glucoside and possessing antisaccharine property; gymnemic acid 6%, cellulose, ash and quercitol. Air-dried leaves yielded after ignition 11.45% of inorganic matter consisting of alkali, phosphoric acid, ferric oxide, manganese and two hydrocarbons. Literature shows the presence of lupeol B-amyrin, stigmasterol and also the gymnemic acid (+) quercitol in *G. sylvestre*. Saponins are present in the alcoholic extract of the plant. Research findings indicate the presence of alkaloids betain, choline and trimethylamine in the leaves. Bark contains a large amount of calcium salts and other crystalline concretions. Gymnemic acid resembles chrysophanic acid and forms insoluble salts with alkaloids.

Medicinal Properties and Uses: It has a stringent, stomachic, tonic refrigerant and diuretic properties. *Gymnema sylvestre* caused insignificant reduction in blood sugar in normal rats but produced marked and significant reduction in anterior pituitary treated hyperglycemic animals. The body weight of rats treated with *G. sylvestre* increased as compared with controls. The urine output was also greater in rats treated with the drug. The leaf material stimulates insulin secretion and later studies revealed that it has active principles, which have blood sugar-reducing properties. Oral administration caused inhibition of hyperglycemic response of anterior pituitary extract in rats. It is used as a destroyer of glycosuria and in other urinary disorders. Because of its property of abolishing, the taste of sugar, it has been given the name of gurmar (sugar destroying), and it is believed to neutralize the excess of sugar present in the body in diabetes mellitus.

Indigenous Knowledge: Roots are very useful remedy for snakebite. Their powder is dusted upon the wound or made into a paste with water and applied. Their decoction is given internally. A decoction of the leaves is given in fever and cough. Leaves triturated and mixed with castor oil are applied to swollen glands.

Common name: Seabuckthorn

Botanical name: *Hippophae rhamnoides* L.

Family: Elaegnaceae

Vernacular Names: English- Seabuckthorn and Hindi- Dhurchuk

Distribution: Seabuckthorn is one of the major plants of the Ladakh region, found in four of the five valleys. Seabuckthorn is locally called as *Tsermang* and the fruits are called as *Tsestalullu*. It is generally found in the altitude range of 8000-13500 ft mol. Ladakh is an old desert in the Trans Himalayan range of 32°N to 76°E. The region is characterized by extreme climatic conditions like high wind velocity, high rate of soil erosion and extreme temperatures, *i.e.*, –30°C in winter to +35°C in summer. Seabuckthorn is generally found in the cold and hilly areas of Ladakh in Jammu and Kashmir, Lahaul-Spiti in Himachal Pradesh and parts of Arunachal Pradesh and Sikkim.

Plant details: Flowering takes place in April and the seeds ripen from September to October. The flowers are dioecious (individual flowers are either male or female but only one sex is found on any one plant so both male and female plants must be grown if seed is required) and are pollinated by wind. The fruit is very freely borne along the stems and is about 6-8 mm in diameter. The plant is not self-fertile. It can fix nitrogen. The plant prefers light (sandy), medium (loamy) and heavy (clay) soils and can grow in nutritionally poor soil. The plant prefers acid, neutral and basic (alkaline) soils. It cannot grow in the shade. It requires dry, moist, or wet soil and can tolerate drought. The plant can tolerate maritime exposure.

Chemical Composition: The main constituents are 5-hydroxytryptamine, ascorbic acid, harmalol, harmol, hemin, hippophain, hippophainum, kaempferol, paeonin, physaleiene, querbrachitol, serotonin, *etc*. The fruits of some species and cultivars contain up to 9.2% oil. The twigs and leaves contain 4-5% tannin. The tender branches and leaves contain bioactive substances, which are used to produce oil that is quite distinct from oil produced from the fruit. Yield of around 3% of oil is obtained. It is also a fairly good source of essential fatty acids, which are fairly unusual for a fruit.

Medicinal Properties and Uses: Twigs are astringent and vermifuge. Berries are very high in vitamin C (120 mg per 100 g), vitamin A and E. They have principally been used to help improve resistance to infection. It is being increasingly used in making fruit juices, especially when mixed with other fruits because of its reputed health benefits. Oil is used as an ointment for treating burns. High-quality medicinal oil is made from the fruit and used in the treatment of cardiac disorders. It is also said to be particularly effective when applied to the skin to heal burns, eczema and radiation injury and is taken internally in the treatment of stomach and intestinal diseases.

Indigenous Knowledge: The fruit is astringent and used as a tonic. The freshly pressed juice is used in the treatment of colds, febrile conditions, exhaustion, *etc.*

Common name: Cow-itch plant

Botanical name: *Mucuna pruriens* (Stickm.) DC.

Family: Papilionaceae

Vernacular Names: Sanskrit- Kapikachchha; Hindi- Kavach; English- Cow-itch plant; Bangla- Alkusi; Malyalam- Naicoma; Tamil- poonaykali; Telugu- Dulagondi; Kannada- Nayisoonnu-balle; Marathi- Kanchkuri and Gujarati- Kiwatch, kandchan.

Distribution: It occurs commonly in the tropical parts of India. It is also found in the forests of Andeman and Nicobar islands.

Plant details: It is an annual or perennial climber with hairy branches. Leaves trifoliate, leaflets ovate or rhomboid ovate, membranous, pubescent above, flowers purple in axillary drooping pendulous racemes, fruits curved pods, densely clothed with pale brown irritant hairs and seeds black, ovoid and 4-6 per pod.

Chemical Composition: *M. pruriens* has been reported to be a good source of 3,4-dihydroxyphenylalanine (l-DOPA). It also contains a few amino acids. The seeds gave four alkaloids, namely, mucunine, mucunadine, prurienine and prurieninine. Four indole-3-alkylamines and choline have been isolated from different parts of the plant. The seed oil contains steric, palmitic, myristic, arachidic, oleic and linoleic acids and a sterol.

Medicinal Properties and Uses: Its seeds are astringent, anthelmintic, nervine tonic and aphrodisiac and roots are tonic, stimulant, diuretic, purgative and emmenagogue. When tested on frogs, purieninine slows down the heart, dilates the blood vessels, depresses blood pressure and increases peristaltic action of the intestine. *Ayurvedic* physicians claim that *M. pruriens* is very effective in the treatment of Parkinsonism but the latest studies on powdered seed extract showed that it was devoid of anticholinergic activity. The extract showed hypotensive action in dogs and spasmodic action in guinea pig preparation. The extract had no effect on frog's rectus abdominis muscle but revealed a histaminergic activity. Seed powder may act by some mechanism other than through anticholinergic property in Parkinsonism. Studies indicate that the extracts of seed had no anticholinergic activity but had a potent anti-Parkinsonian effect. *Mucuna pruriens* seed diet showed a hypocholesterolemic effect in rats. In another study, the protein isolate of the seed in the diet led to a reduction in cholesterol content of the liver and blood in rats. Studies indicate that the total alkaloids of seeds showed a weak neuromuscular blocking effect on frog's rectus abdominus. Bufotenine showed anticholinesterase activity *in vivo* and *in vitro* similar to but 20-30 times weaker than that of Physostigmine.

Indigenous Knowledge: The plant is traditionally used for diseases of the nervous system, kidney and dropsy. An ointment prepared from the roots is applied for elephantiasis. The leaves of the plant are applied to ulcers. An infusion of hair is used in diseases of the liver and gall bladder and applied externally as a local stimulant and mild vesicant. Seeds are also used in the form of powder in leucorrhea, profuse menstruation, paralysis, spermatorrhea and in cases requiring aphrodisiac action.

Common name: Holy basil

Botanical name: *Ocimum Sanctum* Linn.

Family: Labiatae

Vernacular Names: Sanskrit- Tulsi; Hindi- Kali tulsi; English- Holy basil; Bengali-Krishna tulsai and Tamil- Thulasi

Distribution: This is a small herb found throughout India and cultivated near temples and gardens.

Plant details: An erect much branched under shrub grows up to 1 m in height. Its leaves are pale greenish in colour, simple, opposite, elliptic, oblong, obtuse, or acute, serrate, entire and pubescent on both sides, petiole slender and hairy, flowers purplish in elongate racemes and fruits nutlets, smooth and not mucilaginous when wet

Chemical Composition: The leaf contains the highest percentage of essential oil, followed by inflorescence and stem but roots are devoid of any oil. Seeds contain a large amount of mucilage. Studies indicate the presence of 70% eugenol as a major constituent in the essential oil. The other components identified were nerol, eugenol methyl ether, caryophyllene, tepinene-4-ol, decylaldehyde, y-selinene, ox-pinene, B-pinene, camphor and carvacrol. The leaves also yield ursolic acid, apigenin, luteolin, apigenin 7-*o*-glucuronide, luteolin, orientin and molludistrin. Old leaves of *Ocimum sanctum* contained 3.15% calcium, 0.34% phosphorus and 4.97% insoluble oxalate.

Medicinal Properties and Uses: It is demulcent, expectorant and antiperiodic. Root is febrifuge and seeds are mucilaginous and demulcent. Dried plant is stomachic and expectorant. Leaves are anticatarrhal, expectorant, fragrant and aromatic. *Ocimum sanctum* leaves are reported to show abortifacient and antifertility activity. Ethanolic extract (50%) of leaves showed a hypoglycemic effect in rats and antispasmodic activity against spasmogen-induced spasms in isolated guinea pig ileum. Crude watery extract of leaves showed a transient hypotensive effect in anesthetized dogs and cats and a negative inotropic and chronotropic effect on rabbit's heart. The extract inhibited the spasm of smooth muscles induced by acetylcholine, carbachol, and histamine and potentiated hexobarbitone-induced hyposis in mice. The essential oil showed larvicidal activity against *Culex pipiens-fatigans*. The seed mucilage of *O. sanctum* also showed marked larvicidal activity against *C. pipiens-fatigans* and *Aedes aegypti*. Antifungal activity of this plant has also been reported. The seeds exhibited anticoagulose activity as evidenced by the suppression of coagulose activity and mannitol fermentability of pathogenic staphylococci. The essential oil also showed antimicrobial and antimycotic effects *in vivo*. Ethanolic extract of leaves showed fungitoxic activity against *Rhizoctonia solani*.

Indigenous Knowledge: Infusion of the leaves is given in malaria and as a stomachic in gastric diseases of children and hepatic affections. Juice of the leaves should be taken internally and is very effective in skin diseases such as itches, ringworm and leprosy and in impurities of the blood. Dried plant in decoction is a domestic remedy for catarrh, bronchitis and diarrhea. Fresh leaves also cure chronic fever, hemorrhage, dysentery and dyspepsia. With honey and ginger juice, it is a good expectorant, useful in cough, bronchitis and children's fever.

Common name: Kutki

Botanical name: *Picrorhiza Kurroa* Royle ex Benth.

Family: Scrophulariaceae

Vernacular Names: Sanskrit- Katula; Hindi- Katki, Kuru; Bengali- Kuru and Tamil- Kadukurokani

Distribution: A perennial herb is found only in the higher mountains at 9,000 to 12,000 feet in the northwestern Himalayas from Kashmir to Sikkim

Plant details: The herb is more or less hairy, with perennial woody bitter rhizome 15 to 25 cm long clothed with dry leaf bases. Leaves spoathulate, serrate, 5 to 10 cm long, rather coriaceous with rounded tips and the base is narrowed into winged sheathing petioles. Flowers are small, in spikes, dimorphic, bracteates and bracts oblong or lanceolate as long as the calyx, sepals 5, lanceolate, 6 mm long and ciliate. They are of two kinds, *i.e.*, some have 8 mm long filament and others have 2 cm long filaments. Fruits are 1 to 3 cm long. The rhizomes of the plant are cylindrical with an average diameter of 2 cm.

Chemical Composition: Roots yielded kutkin, a glucosidal bitter principle, which was reported to be a stable mixture of two glycosides, *viz.* picroside-1 and a new glycoside, kutkoside. The petroleum ether extract also yielded d-mannitol, kutkiol, and kutkisterol, and a ketone, which was found to be identical with apocynin. Picorhizin also reported to be its bitter principle, as glucosidovanilloyl glucose was isolated from the plant.

Medicinal Properties and Uses: In small doses, it is a bitter stomachic and laxative and in large doses, a cathartic. It is reputed as an antiperiodic, cholagogue. It is a bitter tonic and is reported to have beneficial action in case of dropsy and is also used in snakebite and scorpion sting. In bilious fever, a compound decoction of its root, liquorice, raisins and *Neem* bark is very curative. In dyspepsia with severe pains, the powder of kutuki, *Acorus calamus*, *Chebulic myrobalan* and plumbago root in equal parts is given in doses of 28 ml with cow's urine. The drug is also used in scorpion sting. Recently, it has been tried and found beneficial in many cases of ill-defined fever, such as low fever with constipation, symptomatic fever of elephantiasis and fever of material origin, which resisted other home remedies.

Indigenous Knowledge: The root is also used in diseases of liver and spleen including jaundice and anemia. Root powder when given with sugar and hot water acts as a mild purgative; 1 to 1.5 g of the powder with aromatics or pepper, asafetida, triphala and salts is very useful in constipation due to scanty intestinal secretions.

Commercial value: Kutki is in high demand through out the world. Since it is in the endangered category in India, it is imported from Nepal and other adjoining countries. Due to its high commercial value, its cultivation has been started in India. The kutki is an important ingredient in several Ayurvedic drugs.

Common name: Indian Mayapple

Botanical name: *Podophyllum hexandrum* Royle.

Family: Berberidaceae

Vernacular Names: English- Indian Mayapple; Hindi- Papra, Bankakdi and Sanskrit-Vantrrapusli

Distribution: It found in Himalaya region at an altitude from 2000 to 3500 metre height.

Plant details: A perennial herb commonly known as Himalayan or Indian mayapple is a native to the lower elevations in and surrounding the Himalaya. It is an erect, succulent herb with toothed, glossy green and purple spotted leaves, deeply divided in 3 to 5 lobes. It has flower bearing erect branches leafy at top. The flowers are white or pinkish, cup shaped, and fruits are bright red-orange bulbous. The attractive berries come during the winter. It has a creeping rootstalk. The ornamental appearance of the plant makes it a desirable addition to woodland-type gardens. It can be propagated by seed or by dividing the rhizome. Himalayan Mayapple can survive very cold winters with annual averages as low as -40°F. The optimal soil texture for this species is one that is fertile. Partial to full shade is ideal for this plant. It usually does best in moist well-drained soils.

Chemical Composition: The active principle of podophyllum is contained in the resinous mixture known as podophyllin. Other constituent of the root is podophyllotoxin. The rhizomes yield podophyllol, a sticky resin and quercetin.

Medicinal Properties and Uses: The herb podophyllum is useful as a drug to correct disordered processes of nutrition and to restore normal function of the system. It is highly beneficial for treating chronic constipation and is used as a purgative. The safe single dose is 0.01 g. Its action is slow but strong. In large doses, it can cause acute irritation and gripping. It should therefore be administered either in combination with belladonna or Indian aloe. It is used as a hepatic stimulant and as an agent to promote the flow of bile. Podophyllum is useful in treating many skin diseases and tumorous growths. It has acquired importance in recent years for its possible use in controlling skin cancer.

Indigenous Knowledge: It has been historically used as an intestinal purgative and emetic, salve for infected and necrotic wounds and inhibitor of tumor growth.

Common name: Indian Serpent wood

Botanical name: *Rauwolfia serpentine* (L) Benth ex Kurz.

Family: Apocynaceae

Vernacular Names: English- Indian Serpentwood; Hindi- Chandrabhaga; Sanskrit-Chundrika; Tamil- Chivanamelpodi; Malyalam- Chuvannavilpuri; Telugu- Patala-agandhi; Kannada-Patalagarud and Bangla- Chandra

Distribution: It is found in the sub-Himalayan tracts from Sirhind eastward to Assam, especially in Dehradun, Siwalk range, Rohilkhand Gorkhpur, ascending to 4000 ft. Also in Konkon, Kanara and Maharashtra, Eastern and Western Ghat of South India, Bihar and Bengal.

Plant details: A small erect shrub grows up to 60 cm in height. Leaves are simple, in a whorl of three, thin, acuminate and oblong-lanceolate, flowers white, streaked with violet, found in cymes and fruits green drupes and become dark purple when ripe. Roots are cylindrical, yellowish brown and tortuous.

Chemical Composition: The alkaloids, *viz.* reserpine, rescinnamine, reserpinine, serpentinine, sarpagine, raubasine, yohimbine, ajmaline, ajmalinine, ajmalicine, isoajmaline, neoajmaline, rauwolfinine, rauhimbine, the baine and papaverine occur in the roots of *R. serpentina*. The drug also contains in oleoresin, a sterol starch and calcium oxalate. The yield of total alkaloids in *R. serpentine* ranges from about 0.8 to 1.3%.

Medicinal Properties and Uses: Root is bitter tonic and possesses well-marked sedative properties. It acts also as febrifuge. The ajamaline group acts as a general depressant to the heart, respiration and nerves and the serpentine group paralyzes respiration and depresses the nervous system but stimulates the heart. Root is hypnotic, sedative, specific, or insanity, reduces blood pressure, remedy in painful affections of bowels, *etc.* Reserpine is a Central Nervous System (CNS) depressant, sedative hypotensive with bradycardia and used as tranquilizer. Deserpidine modifies the hypophysiogental system of the rat. In females, it suppresses vaginal keralinization, induces permanent estrus and stimulates the mammary gland, and in the male, it dissociates the germinal and endocrine functions of testicles. Reserpinine has hypotensive, sedative and bradycardiac activity similar to that of reserpine. Sarpajmaline an alkaloid complex is a much more potent hypotensive than reserpine without the sedative and CNS depressant action of the latter. Ajmalicine is a CNS depressant and adrenergic blocking agent. Serpentine is the major yellow anhydronium base of *R. serpentine*. It is twice as hypotensive as ajmaline and shows synergistic activity when administered with resperine.

Indigenous Knowledge: Decoction of roots is employed to increase uterine contraction and promotes expulsion of the fetus. Root is a valuable remedy in dysentery, painful affection of the bowels, and recently, it has attained prominence as a remedy for insomnia, hypochondria and irritating condition of the central nervous system. Powdered root in doses of 20 to 30 g given daily produces not only sedative effects but also a reduction of the blood pressure. The hypnotic action the drug appears to have been known to the poorer classes in Bihar, and the practice of putting children to sleep by this drug is stated to be still present in certain parts of the province. In Uttar Pradesh and Bihar, the drug is sold as pagal-ka-dawa (insanity specific), and its use is common among practioners of indigenous medicine.

Common name: Stevia

Botanical name: *Stevia rebudiana*, Bertoni.

Family: Asteraceae

Vernacular Names: Since it is an exotic plant, it is commonly known as stevia through out India.

Distribution: It is a native to Paraguay where it grows wild in sandy soils and it has been introduced in India a decade back.

Plant details: A small herbaceous plant grows up to 50 cm in height. Leaves are simple, opposite, ovate-lanceolate, membranous, short petioled and very sweet to taste, flowers white, small and seen in terminal inflorescence.

Chemical Composition: The leaves contain diterpene glucosides such as stevioside (6-8%) and rebaudioside

Medicinal Properties and Uses: Stevia is 30 times sweeter than sugar but it does not have the neurological or renal side effects of some of the artificial sweeteners. It is especially popular as a sweetener for coffees and teas. Stevia can also be used in cooking and baking, helping you reduce your calorie intake and stay healthy. It also helps to keep the body's blood sugar in balance. Hence, it is beneficial for diabetics. It is also low caloric and aids weight management. Stevia possesses anti-fungal and anti-bacterial property also, and it can be placed directly to cuts and wounds for more rapid healing, without scarring. It has shown effective results when applied to acne, seborrhoea, dermatitis, eczema, *etc*. Stevia also helps to improve digestion, increase energy levels and mental activity. It also helps to reduce desire for tobacco and alcoholic beverages. Mild Stevia leaf tea offers excellent relief for an upset stomach. It can be safely used in herbal medicines, tonics and also in the daily usage products like mouthwashes and toothpastes.

Indigenous Knowledge: Being a recently introduced plant in India, the indigenous knowledge is based on the literature only.

Common name: Charayatah

Botanical name: *Swertia chirayita* (Roxb. Ex Fleming) Karsten

Family: Menispermaceae

Vernacular Names: Sanskirit- Kirata tikta, Bhunimba; Hindi- Charayatah; Marathi-Cherayata, Chiraita and Kiraita; Gujarati- Chirayatin; Bengla- Chireta; Kerala- Nilaveppa; Urdu-Chiaravata; Deccan- Charayatah; Kannada- Nelabevu and Tamil- Nenilawandi, Nilavembu, Shirattakuchi

Distribution: The plant is a native of temperate Himalayas and found at an altitude of 1200-3000 m from Kashmir to Assam and in the Khasi hills at 1200-1500 m.

Plant details: It can be grown in sub-temperate regions between 1500 and 2100 m altitudes. The genus *Swertia* consists of annual and perennial herbs. There is no consistency in the literature citing the habit of *Swertia chirayita*. Some authors have described chirata as an annual and others as biennial or pluri-annual. It is not clear whether the plant behaves differently due to climatic conditions or varying genotypes. The plant can be grown in a variety of soils with sandy loam rich in carbon and humus. It is also found in open ground and recently slash and burnt forests. *Swertia chirayita* has an erect, about 2-3 ft long stem and the middle portion round, while the upper is four-angled with a prominent decurrent line at each angle. The stems are orange brown or purplish in colour and contain large continuous yellowish pith. The root is simple, tapering and stout, short, almost 7 cm long and usually half an inch thick.

Chemical Composition: 1,3,5,8-tetrahydroxyxanthone, 1,3,7,8-tetrahydroxyxanthone, 1,3,8-trihydroxy-5-methoxyxanthone, 1,5,8-trihydroxy-3-methoxyxanthone, 1,8-dihydroxy-3,5-dimethoxyxanthone/swerchirin, 1,8-dihydroxy-3,7-dimethoxyxanthone/7-*O*-methylswertanin, 1-hydroxy-3,5,8-trimethoxyxanthone, 1-hydroxy-3,7,8-trimethoxyxanthone, 2,5-dihydroxyterephthalic acid, 21-α-H-hop-22(29)-en-3-β-ol, (Triterpenoid), Amarogentin, Amaroswerin, Chiratanin, Chiratenol, Chiratol/1,5 dihydroxy 3,8-dimethoxyxanthone, Decussatin, Enicoflavine, Episwertenol, Erythrodiol, Gammacer-16-en-β-ol (Triterpenoid), Gentianine, Gentiocrucine, Kairatenol, Lupeol, Mangiferin, Mangostin, Oleanolic acid, Pichierenol, Sweroside, Sweroside-2′-*O*-3″,5″-trihydroxy biphenyl-2″ carboxylic acid ester, Swerta-7,9(11)-dien-3-β-ol, Swertanone, Swertenol, Swertianin/1,7,8-trihydroxy-3-methoxyxanthone, Syingaresinol, Taraxerol, Ursoilic acid, β-Amyrin, β-Sitosterol-3-β-D-glucoside, Ø-Taraxasterol or heterolupeol, tannin and 4 to 8% of ash.

Medicinal Properties and Uses: It is bitter tonic, stomachic, anthelmintic, laxative, alterative, antidiarrheic and antiperiodic. It is a well-reputed drug for intermittent fevers, skin diseases and intestinal worms. It is given in the form of infusion or tincture in bronchial asthma. The infusion made in hot water with cloves, cinnamon, *etc.* is given in doses of 14 to 28 ml as a tonic. To check hiccup and vomiting, the decoction of root is taken in doses of 0.5 to 2 g with honey. Chireta is used in scorpion sting also. It is given as tonic to a gouty person.

Indigenous Knowledge: The plant powder is used to cure fever and stomach related problems.

Common name: Giloe

Botanical name: *Tinospora cordifolia* (Willd) Miers.

Family: Menispermaceae

Vernacular Names: English- heart leaf moonseed; Hindi- Giloe; Sanskrit- Amrita; Tamil-Amridavalli; Malyalam- Amrytu; Telugu- Jivantika and Kannada- Amrytaballi

Distribution: The giloe is found throughout tropical regions of India.

Plant details: It is a spreading perennial deciduous twinner with succulent stems and easily pealing off papery bark. Leaves are simple, alternate, cordate, entire and glabrous with 7-9 nerves. Flowers are yellow in racemes, arising from nodes on the old wood. Male flowers usually are in clusters and female solitary. Fruits are drupes, red when ripe. Mucilaginous greenish crust is covered with grey coloured coat studded with tubercles.

Chemical Composition: The chemical constituents reported from this shrub belong to different classes such as alkaloids (Berberine, Palmatine, Tembetarine (0.012%), Magnoflorine (0.075%), Choline, Tinosporin, Isocolumbin, Palmatine, Tetrahydropalmatine, Magnoflorine); diterpenoid lactones (Furanolactone, Clerodane derivatives and [(5R,10R)-4R-8R-dihydroxy-2S-3R:15,16-diepoxy-cleroda-13 (16), 14-dieno-17,12S: 18,1S-dilactone] and Tinosporon, Tinosporides and Jateorine, Columbin; glycosides (18-norclerodane glucoside, Furanoid diterpene glucoside and Tinocordiside,Tinocordifolioside. Cordioside, Cordifolioside A, Cordifolioside B, Syringin, Syringin-apiosylglycoside, Palmatosides C, Palmatosides F, Cordifoliside A, Cordiofoliside B, Cordifoliside C, Cordifoliside D, Cordifoliside E); Steroids (b-sitosterol, d-sitosterol, hydroxy ecdysone, ecdysterone, Makisterone A, Giloinsterol); sesquiterpenoid (Tinocordifolin); phenolics; aliphatic compounds (Octacosanol, Heptacosanol,

Nonacosan-15-one); polysaccharides and Miscellaneous compounds 3, (a,4-dihydroxy-3-methoxy-benzyl)-4-(4-hydroxy-3-methoxy-benzyl)-tetrahydrofuran, Jatrorrhizine, Tinosporidine, Cordifol, Cordifelone, N-trans-feruloyl tyramine as diacetate, Giloin, Giloinin, Tinosporic acid).

Medicinal Properties and Uses: The stem of *Tinospora cordifolia* is one of the constituents of several *Ayurvedic* preparations used in general debility, dyspepsia, fever and urinary diseases. The stem is bitter, stomachic, diuretic, stimulates bile secretion, causes constipation, allays thirst, burning sensation, vomiting, enriches the blood and cures jaundice. Dry barks of *T. cordifolia* have anti-spasmodic, antipyretic, anti-allergic, anti-inflammatory and anti-leprotic properties. *Tinospora cordifolia* is widely used in Indian *Ayurvedic* medicine for treating diabetes mellitus. It is reported to benefit the immune system in a variety of ways. The anti-stress and tonic property of the plant was clinically tested and it was found that it brought about good response in children with moderate degree of behaviour disorders and mental deficit. It has also significantly improved the Intelligent Quiescent levels. In a clinical evaluation, a compound preparation *Rumalaya* containing *T. cordifolia* was reported to reduce the pain in patients significantly. The *Ayurveda* literature reports that it can cause constipation if taken regularly in high doses. It has no side effect and toxicity.

Indigenous Knowledge: The extract of its stem is used in skin diseases. The root and stem of *T. Cordifolia* are prescribed in combination with other drugs as an antidote to snakebite and scorpion sting.

Common name: Indian squill

Botanical name: *Urgenia indica* (Roxb) Kunth.

Family: Liliaceae

Vernacular Names: Sanskrit- Vanapalandam; Hindi- Jangli piyaz, Kanda; English- Indian squill and Bengali- Jangli piyaz

Distribution: Found in drier sub-Himalayan tracts of western Himalayas; also in Bihar, Konkan and on the Coromandel Coast.

Plant details: It is a perennial plant with fibrous roots proceeding from base of a large, tunicated, nearly globular bulb, 4 to 6 inches long, the outer scales of which are thin and papery, red or orange-brown in colour. The bulb, which is usually only half immersed in the sand, sends forth several long, lanceolate, pointed, somewhat undulated, shining, dark-green leaves, when fully grown 2 feet long. From middle of the leaves, a round, smooth, succulent flower-stem rises, from 1 to 3 feet high, terminating in a long, close spike of whitish flowers, which stand on purplish peduncles, at the base of each of which is a narrow, twisted, deciduous floral leaf or bract. The flowers are in bloom in April and May and are followed by oblong capsules.

Chemical Composition: The chemical constituents of Squill are imperfectly known. Merck in 1879 separated the three bitter glucosidal substances, *i.e.*, scillitoxin, scillipicrin and scillin. The first two are amorphous and act upon the heart, the former being the more active. Scillin is crystalline and causes numbness and vomiting. Other constituents are mucilaginous and saccharine matter, including a peculiar mucilaginous carbohydrate named Sinistrin, an Inulin-like substance, which yields laevulose on being boiled with dilute acid. The name Sinistrin (in 1834, first proposed by Macquart for Inulin) has also been applied to a mucilaginous matter extracted from barley but it remains to be proved that the latter is identical with the Sinistrin of Squill. Calcium oxalate is also present, in bundles of long, acicular crystals, which easily penetrate the skin when the bulbs are handled, and it causes intense irritation, sometimes eruption, if a piece of fresh Squill is rubbed on the skin

Medicinal Properties and Uses: In small doses, it is expectorant, cardiac, stimulant, diuretic, deobstruent and emmenagogue. In large doses, it is emetic and cathartic. Physiologically, the drug slows heartbeat and increases the flow of urine. It is excreted by the bronchial genitourinary and gastrointestinal secretions. In excessive doses, it is a narcotic acrid poison causing nausea, stranguary and bloody urine. In the form of tincture, it is given in cardiac and renal dropsy, ascities, also in asthma, rheumatism, calculous and paralytic affections, leprosy and skin diseases.

Indigenous Knowledge: Bulb is rubbed on the soles of the feet to relieve burning sensation in chest. In small doses, it acts like an expectorant. Bulb mixed with some other ingredients is given in the form of syrup in acute bronchitis, where the sputa was tenacious and scanty, in chronic bronchitis associated with emphysema and in spasmodic croup. A powder of it is locally applied to remove warts.

Common name: Vetiver

Botanical name: *Vetiveria zizanioides* (L.) Nash.

Family: Poaceae

Vernacular Names: English- Vetiver; Hindi- Khus; Sanskrit- Abhaya; Tamil-Ilamichamver; Malyalam- Ramacham; Telugu- Avurugaddi veru and Kannada- Lavancha.

Distribution: The wild vetiver is found growing in plains throughout India. The best quality khus is available in Uttar Pradesh and Rajasthan.

Plant details: A perennial aromatic grass grows up to 2 meter in height. The leaves are narrow, linear and erect, spikelets grey in a panicle of numerous racemes and fruits oblong grains. It grows well in wastelands-sandy, marginal land where conventional cropping is uneconomic. Vetiver is extremely hardy and can tolerate submergence for a considerable period as well as withstand drought situation.

Chemical Composition: Chemical Composition of vetiver oil is extremely complex, containing over 100 sesquiterpene-type compounds and their derivatives, belonging to 11 structural classes. The main constituents of vetiver oil comprise of sesquiterpene hydrocarbons (γ-cadenene, clovene, α-amorphine, aromadendrine, junipene and their alcohol derivatives), vetiverols (khusimol, epiglobulol, spathulenol, khusinol; carbonyl derivatives), vetivones (α-vetivone, β-vetivone, khusimone; and ester derivative). Among them, three carbonyl compounds, *i.e.*, α-vetivone, β-vetivone and khusimone, are considered the primary odour influencing components. Beta-vetivone has the better odour, and it is considered the most important, while its major isomer nordihydro β-vetivone has a strong, rich, woody-peppery note. All these components individually and collectively contribute to the characteristic odour of the vetiver. Therefore, the α-vetivone, β-vetivone and khusimol can be considered as the *finger print* of vetiver oil.

Medicinal Properties and Uses: The main action of vetiver oil is on the nervous system and it is both sedating and strengthening in effect. It is excellent in the treatment of depression, nervous tension, debility, insomnia and many stress-related diseases and acts as an aphrodisiac where there is a clear connection between impotence or frigidity and stress. It may be used in massage blends and the bath. It has a rather powerful smell but is very pleasant when diluted. It stimulates the circulatory system and makes useful massage oil for elderly or debilitated people with poor circulation. It also helps to stimulate the production of red blood cells and is thus beneficial for anaemia. It makes a useful warming and pain-relieving rubbing oil, suitable for deep massage of muscular aches and pains, sprains, stiffness, rheumatism and arthritis. It may be added to sports oil blends and massaged into muscles before and after sports. In skin care, it helps to balance the secretion of sebum. It is used in lotions, compresses and baths for the treatment of oily skin, acne and weeping sores. Vetiver oil revitalizes the body by fortifying the red blood corpuscles crucial in transporting oxygen to all parts of the system. Increased blood flow could alleviate muscular aches and pains and said to be useful in cases of rheumatism and arthritis. Vetiver oil is particularly useful for jet lag, and for grounding and clarity while travelling.

Indigenous Knowledge: Traditionally the vetiver roots are used in sharbat preparations and also as pads for cooling houses.

Common name: Winter cherry

Botanical name: *Withania somnifera* (Stocks) Dunal.

Family: Solanoceae

Vernacular Names: Sanskrit- Ashvagandha; Hindi- Asgandh; English- Winter cherry; Bengali- Ashwagandha; Nepalese- Aasoganda; Sinhalese- Amukkara and Tamil-Amurkkuralckizhangu

Distribution: The plant is distributed throughout drier regions of India especially in wastelands ascending to an altitude of 5500 ft in the Himalayas. It is also cultivated.

Plant details: An erect branching shrub grows up to 150 cm in height. Leaves are ovate and alternate, up to 10 cm long. Flowers are greenish or greenish yellow arises from leaf axils. Fruits are globose berries, become reddish orange when mature and closed by the persistent calyx. Roots are fleshy, cylindrical and tapering towards the end with whitish grey colour and pure white inside.

Chemical Composition: Extraction with 45% alcohol yields the highest percentage of alkaloids. Isolation of nicotine, somniferience, somniferinine, withanine, withananine and pseudowithanine has been reported. Isolation of sucrose, B-sitosterol an acid, and a neutral compound, as also the occurrence of isopelletierine in the roots, have been reported. Roots contain tropine, pseudotropine, 3-tigloyloxytropane, choline, cuscohygrin, dl-isopelletierine and new alkaloids, *i.e.*, anaferine and anhygrin. Leaf contains withanone (empirical formula $C_{24}H_{32}O_5$ and the berries have amino acids.)

Medicinal Properties and Uses: It is tonic, alternative, astringent, aphrodisiac and nervine sedative. Seeds possess the property of coagulating milk. Leaves and root are narcotic. Root is also diuretic and deobstruent, tonic, alterative and aphrodisiac. Roots and eaves of *W.*

somnifera are used as a hypnotic in alcoholism and emphysematous dysphonea. Leaves are used as an anthelmintic and as an application to carbuncles. Root is used as an application in obstinate ulcers and rheumatic swelling. It is given in doses of about 2 gr in consumption, emaciation of children, senile debility, rheumatism, in all cases of general debility, nervous exhaustion, brain fag, loss of memory, loss of muscular energy and spermatorrhea. It infuses fresh energy and vigour in a system worm out owing to any constitutional disease like syphilis and in rheumatic fever.

Indigenous Knowledge: As nutrient and health restorative to the pregnant and old people, a decoction of the root is recommended. For scrofulous and other glandular swelling, fresh green root of *W. somnifera* reduced to paste with cow's urine or with heated water is applied to the affected parts. As a galactagogue, the decoction of the roots of *W. somnifera*, *Batatas paniculata* and licorice is recommended to be given in cow's milk. For improving sight, a mixture of *W. somnifera* powder, licorice powder, and juice of emblic myrobalans is recommended. Powdered root is very useful with equal parts of ghee and honey for impotence or seminal debility.

■■■

19 Seabuckthorn : A Therapeutic Food

M.K. Rana

Professor, Department of Vegetable Science, CCS Haryana Agricultural University, Hisar- 125004 (Haryana), India, E-mail: <mkrlotus@gmail.com>

Seabuckthorn is a winter hardy deciduous shrub with yellow or orange berries, and it is known by different names throughout the world like Siberian pineapple, sea berry, sand-thorn, or swallow-thorn. It is also known as *Tsermang* in Ladakh, *Chharma* in Himanchal Pradesh and *Chuk* or *Badriphal* in Uttaranchal. It is a unique and valuable plant currently cultivated in various parts of the world from the seashores in Europe to high altitudes and cold deserts of India, China, Mongolia, Russia, Afghanistan, Nepaland Pakistan. It has been growing wild in Ladakh region of Jammu and Kashmir, Lahaul and Spiti in Himanchal Pradesh, Chamoli and Badrinath in Uttaranchal in India besides limited patches in Arunanchal Pradesh and Sikkim in Northeast Region. According to a Chinese seabuckthorn expert, Professor Lu Rongsen, a total of about 3000 hectares seabuckthorn population is available in Pakistan, promising a harvest of 1200-2500 tonnes fruits annually. In China, it is famous as *magic plant* where more than 150 factories are producing 200 food preparations, medicinal and cosmetic products on commercial scale. The annual turnover of seabuckthorn fruit industry is more than six crores.

Historical Background

The berries of seabuckthorn are so rich in minerals and vitamins that it has been even speculated that the plant must have been cultivated by some ancient plant breeder. Legend tells how the ancient Greeks used seabuckthorn leaf in a diet for racehorses. Hence, its botanical name is Hippophae meaning shiny horse. According to another legend, seabuckthorn leaves were the preferable food of flying horse- Pegasus and allegedly were helpful to get Pegasus air born. It is native from northwestern Europe through central Asia to the Altai Mountains to western and northern China and the northern Himalayas. *Hippophae rhamnoides* L. ssp. turkestanica is found in Chitral and northern areas of Pakistan. Normally, it is spread throughout the Karakorum and Himalayan ranges at the altitudes of 1500-3500 m. The references to medicinal use of seabuckthorn were found in the ancient Greek texts attributing to Theophrastus and Dioskorid and in classic Tibetan medicinal texts dated to the times of Tang Dynasty (618-907 AD).

Scope of Production

Keeping in view its importance, seabuckthorn has high scope to be utilized as natural resource into value added products to eradicate poverty in the country. Development of seabuckthorn jam, squash, juice, candy, powder and oil on commercial scale is certainly gaining importance and could have a potential to play a viable role in giving boost to the economy and will lead to transfer of technology in the country. The seabuckthorn growers may fetch more money and the consumers may find hygienic and healthy products for consumption. It is expected that the out come of this article will bring awareness regarding the nutritional and economic importance of this fruit among the growers, food industries, pharmaceutical companies, government departments and private investors in the relevant areas.

Botany

Seabuckthorn *(Hippophae rhamnoides* L.) is a non-leguminous nodule bearing dioecious shrub of order Daphnales and family *Eleagnaceae*. Three Species of seabuckthorn namely, *Hippophae rhamnoides* L., *Hippophae salicifolia* and *Hippophae tibetana* have been reported from India. Two main types of variety, *i.e.*, red and yellow are grown in Pakistan. Seabuckthorn is a multi-branched thorny shrub reaching 2 to 4 m in height with stout branches forming a round, often-symmetrical head. It has brown or black rough bark and a thick grayish-green crown. Staminate and pistillate flowers are inconspicuous, appearing before the emergence of leaves, which are alternate, narrow, 4 to 6 cm long and lanceolate with a silver-grey color on the upper side. Flower buds are formed mostly on 2 years old woody branches differentiated during the previous growing season. Fruit is sub-globose, 6 to 10 mm long and 4 to 6 mm in diameter, turning yellow to orange when mature in mid September. The root system is characterized by nitrogen fixing nodules.

Composition

Seabuckthorn fruit is acidic in taste and rich in nutrients such as carbohydrates, proteins, organic acids, amino acids, globulins and albumins. The berries are also appeared to be a natural source of vitamin A, B (thiamine and riboflavin), C, E, K and P, carotenes and flavonoids. About 27 mineral elements are present in seabuckthorn, among these, the aluminum, calcium, iron, magnesium, phosphorus, manganese, tin, silica and barium contents are found in high quantity. Quercetin content in seabuckthorn was found to be 62 mg/kg and vitamin C content is 3-6 times higher than that of kiwi fruit. A greater reduction in the content of phenolics was observed in leaves dried at higher temperature for longer period, whereas, the carotenoids and chlorophyll concentrations in the leaves decreased with the increasing temperature, nonetheless, dried leaves were of a high nutraceutical quality. A study that D-fructose, D-glucose and ethyl-D-glucose are the major sugars, and malic acid, quinic acid and ascorbic acid reveals are the major acids present in seabuckthorn fruits. For the first time in seabuckthorn and evidently in any berry, the presence of ethyl β-D-glucopyranoside is reported. The yield of press juice of ssp. rhamnoides cv. Luchistaya, Prozrachnaya and ssp. mongolica cv. Indian Summer was 68, 69 and 66%, respectively and pectin methyl esterase (PME) was present in all seabuckthorn juices. The activity of PME was least at pH 3 to 5 and the highest activity was at pH 8, however, the heat treatment reduced the PME activity in juices. The nutritional composition of seabuckthorn is given below in Table 1.

Table 1 : Nutritional composition of seabuckthorn (values are in 100 g edible portion)

Constituents		Contents	Constituents		Contents
Fatty acids (%)	:	6-11	Vitamin C (mg)	:	266
Vitamin A (mg)	:	11	Vitamin E (mg)	:	162-255
Thiamine (mg)	:	0.04	Carotenoids (mg)	:	30-40
Riboflavin (mg)	:	0.56	Flavonoids (%)	:	0.1-1.0
Folic acid (mg)	:	80	Organic acids (%)	:	2-4

A stable isotope dilution assay for quantification of pantothenic acid in seabuckthorn berries, juice and concentrate showed that storage of seabuckthorn juice for 7 days at ambient temperature already resulted in a significant decrease in pantothenic acid of about 18%. The processing effects of juice production and subsequent concentration reveal a decrease of about 6-7% in the juice and of 32% in the juice concentrate. A new technology involving resin purification of proanthocyanidins extracted from seabuckthorn bark has been studied and by investigating single factor effects, the main factors affecting proanthocyanidins extraction were identified as extraction temperature, pH and ethanol concentration. After using this resin to purify the crude extract, a proanthocyanidin extract was obtained whose yield and purity were above 80% and greater than 95%, respectively. Analysis of leaves of male plants conducted by Field Research Laboratory (DRDO), Leh has revealed 18-22% crude protein and 4-5% fat and other micronutrients like phosphorous (1.52-2.23%), calcium (10.02 mg/kg), copper (23.52 mg/kg) and zinc (0.18-0.94%).

Oil Profile

Although the oil content of seabuckthorn is not much as compared to most oil crops, but it contains a lot of fat-soluble bioactive substance. The berries of ssp. rhamnoides contained higher proportion of oil in seeds, berries and seedless parts than the berries of ssp. sinensis. The mature seeds contain 8-20% oil, whereas, dried fruit pulp yields about 18-26% oil. Pomace, the residue after pulping of fruit, contains 15-22% oil. Out of total fatty acids of the oil, more than 80% are saturated fatty acids. Linoleic and α-linolenic acid comprised about 70% of seed oil fatty acids. Palmitoleic acid practically absent in seed oil, and more linoleic acid and less alpha-linolenic acid were found in the seed oil of ssp. sinensis than in the seed oil of ssp. rhamnoides.

Supercritical fluid extraction method employing carbon dioxide is found to be feasible method for extracting sufficient quantities of oils from both seeds and pulp. Linoleic acid (34%), α-linolenic acid (25%) and oleic acid (19%) were the major fatty acids in the seed oil, whereas, palmitic acid (33%), oleic acid (26%) and palmitoleic acid (25%) were the major fatty acids in the pulp oil. An efficient pilot scale process was developed to produce nutraceutical products from fresh seabuckthorn berries. Fresh berries were subjected to high pressure dewatering using a continuous screw press. The separated liquid phase containing 80-90% of pulp oil was clarified at 80°C and centrifuged to obtain pulp oil, clear juice and sludge. The pulp oil was remarkably rich in carotenoids, tocopherols and sterols with a characteristic fresh berry flavour. The clear juice obtained was free from oil and contained high amounts of vitamin C and phytochemicals such as polyphenols and flavonoids. Isohermnetin was the major flavonoid in the juice along with quercitin and quinic acid is the major acid in the juice. Seeds in the pressed cake were separated and extracted for oil. The seed is a source of seed oil, which is much unsaturated and shows promise because of its light absorption and emollient properties, as an ingredient in cosmetics, phytopharmaceuticals or ultra-violet skin-protectant preparations. It may be prepared by conventional extraction techniques or by supercritical carbon dioxide extraction. Phytosterols present in seabuckthorn seed oil are campesterol, clerosterol, avenasterol, lanosterol, sitosterol, sitostanol, lupeol, gramisterol etc. Avenasterol and sitosterol are quantitatively most important phytosterol. The oil profile of seabuckthorn is given below in Table 2.

Table 2 : Oil profile of seabuckthorn

Ingredient	Oil Content in		
	Seed	Pulp	Fruit residues
Unsaturated fatty acids (%)	87	67	70
Saturated fatty acids (%)	13	33	30

Food Value

The fruit is the main component of value, although the leaves are often made into seabuckthorn tea. The two main sources of valuable products are derived from berries, juice from the fleshy tissues and seed as a single from each berry. Fruit drinks were among the earliest seabuckthorn products developed in China. Seabuckthorn based juice is also popular in Germany and Scandinavian countries.

The juice provides a nutritious beverage, high in suspended solids and very high in vitamin C and carotenes. The juice may contain an oil phase trapped within the suspended solids, or the oil may be removed as pulp oil and provided separately. The pulp remaining after juice removal provides for extraction of *seabuckthorn yellow*, a pigment that has potential use as a food coloring material. Seabuckthorn mixed fruit jelly was prepared by blending seabuckthorn juice with papaya, watermelon or grapes in varying proportions maintaining a constant level of total soluble solids and acidity in the final product. The study on the development of value added seabuckthorn jam blended with prominent local fruits reveals that combination of apple or apricot with seabuckthorn at a ratio of 90:10 was highly appreciated among other combinations by the taste panelist and retailers. Pure seabuckthorn jam is observed very bitter, however, accepted as remedial use. The physico-chemical properties of the seabuckthorn berries jelly such as total soluble solids, reducing sugars, acidity and browning increased, while the total sugars, vitamin C, vitamin E, total phenols, total carotenoids and total anthocyanins decreased significantly during storage at 37°C temperature. The seabuckthorn berries jelly remained acceptable up to 6 months of storage under ambient temperature conditions when stored in PET bottles. Seabuckthorn extract can be used as a dietary supplement, especially by people who smoke, in order to prevent nicotine-induced oxidative stress. Fruits are used to make various food products like, squash, RTS, jam, jelly, puree, chayavanprash, pickles, *etc.* Seabuckthorn leaves are nutritive fodder for cattle, yak, sheep, goat, double humped camel, *etc.*

The yield of press juice for *Hippophae rhamnoides* ssp. sinensis and ssp. mongolica cv. Indian Summer were estimated 69.25 and 62.8%, respectively, whereas, the particulate of each juice was characterized microscopically and the particulate of cv. Indian Summer was much larger than that of ssp. sinensis. Much more oil was microscopically apparent in the cv. Indian Summer as expected from the above-mentioned yield data. Seabuckthorn juice is a product that can be derived from seabuckthorn berries. Fresh pressed juice separates into three phases when allowed to stand overnight in the refrigerator: an upper cream phase, juicy middle portion and sediment at the bottom. Enzymatic hydrolysis with commercial, broad-spectrum carbohydrate hydrolyzing enzyme preparations reduced the juice viscosity, assisted juice separation, and provided an opalescent juice. Soluble solids averaged 10.2° Brix, pH averaged 3.13, ascorbic acid averaged 174.2 mg/100 ml, and titrable acidity averaged 1.97% as determined on centrifuged juice. Washing berries by dipping reduced the soluble solids in juice suggesting uptake of water.

Technology for the preparation of herbal beverage, jam, sauce, *etc.* developed by Field Research Laboratory, Leh has been patented and utilized for commercial production of beverage, jam, sauce, tea, *etc.*, which are readily available in Indian market. Now a day, seabuckthorn beverage is available in the market under various brand names *viz.*, Leh berry, Sindhu berry, Ladakh berry, Siachen berry, Kargil berry, *etc.*

Medicinal Uses

According to scientific research, its leaves and fruits contain more than 200 bio-chemical substances, which are fundamental for human health. The super oxide dismutase (SOD) content of seabuckthorn fruit is even higher than that of *Ginseng*. For this reason, seabuckthorn is called as *miracle plant* and has high potential for its use in medicine and health care. Both the flavonoids and the oils from seabuckthorn have several potential applications. Seabuckthorn oil has much more nutritive and medicinal values and plays an important role in the inhibition of cancer.

There are five areas of research that have been focal points for their use as an aid to patients undergoing cancer therapy, a long-term therapy for reduction of cardiovascular risk factors, treatment of gastrointestinal ulcers, internal and topical therapy for a variety of skin disorders and as a liver protective agent, especially for chemical toxins and a remedy for liver cirrhosis. Oil has medicinal properties because of having vitamin E, antioxidants and essential fatty acids and have anti-radiation and anti-aging effect, thus, it is widely used in making UV-protective creams and beauty products. The seed oil has been found to enhance non-specific immunity and to provide anti-tumor effects in preliminary laboratory studies. An ingredient of the oil, palmitoleic acid, is a component of skin. It is considered a valuable topical agent in treating burns and healing wounds. This fatty acid can also nourish the skin when an adequate quantity of its oil is taken orally, and this is a useful method for treating systemic skin diseases such as atopic dermatitis. The only other major plant source of palmitoleic acid is macadamia nuts, and the oil is used to nourish the skin. Seabuckthorn oil is already widely used alone or in various preparations topically applied for burns, scalds, ulcerations and infections. It is an ingredient in sunblock-hippophae oil having UV-blocking activity as well as emollient properties, and it is an aid in promoting regeneration of tissues. The fruit may also be used for benefiting the hair, and hence, it refers to the good hair coat developed by the horses feeding off the plant.

In a double blind clinical trial in China, 128 patients with ischemic heart disease were given total flavonoids of seabuckthorn at 10 mg thrice a day for 6 weeks. The patients had a decrease in cholesterol level and improved cardiac function, and also they had fewer anginas than those receiving the control drug. No harmful effect of seabuckthorn flavonoids was noted in renal and hepatic functions. The mechanism of action may include reduced stress of cardiac muscle tissue by regulation of inflammatory mediators. In a laboratory animal study, the flavonoids of seabuckthorn were shown to reduce the production of pathogenic thromboses. A study of investigating effect of antioxidant rich juice of seabuckthorn on risk factors for coronary heart disease in humans revealed that although not significant, a 20 and 17% increase in plasma HDL cholesterol and triacylglycerol (TAG) concentrations were observed and seabuckthorn juice supplemented also resulted in a moderate decrease in the susceptibility of LDL to oxidation. Seabuckthorn has been shown to have a potent antioxidant activity, mainly attributed to its flavonoids and vitamin C content.

Seabuckthorn oil encapsulated with sodium octenyl succinate starch (SOSS) found to increase the glycemic response. A clinical trial demonstrated that seabuckthorn extracts helped normalize liver enzymes, serum bile acids and immune system markers involved in liver inflammation and degeneration. In addition, seabuckthorn oil protects the liver from damaging effects of toxic chemicals, as revealed in laboratory studies. Hippophae is traditionally used in the treatment of gastric ulcers, and the laboratory studies confirm the efficacy of seed oil for this application. Its functions may be to normalize output of gastric acid and reduce inflammation by controlling pro-inflammatory mediators. Seabuckthorn procyanidins has remarkable preventive effects on gastric ulcer induced by water immersion and restraint stress in rats. In China, a study was done to demonstrate faster recovery of the hemopoietic system after high dose chemotherapy (with 5-FU) in mice fed with seabuckthorn oil.

Seed and pulp oils have traditionally been used for treating skin diseases in China and Russia, but are not widely used in other countries. The effect of these on the fatty acid composition of skin glycerophospholipids of patients with atopic dermatitis was studied, and it was found that the seed oil slightly increased the proportion of docosapentaenoic acid and decreased the proportion of palmitic acid in skin glycerophospholipids. The results show that the fatty acid composition of skin glycerophospholipids is well buffered against short-term dietary modification. Topical application of seabuckthorn oil has been reported for skin therapy including sun, heat, chemical and radiation burns, eczema and poorly healing wounds. Russian cosmonauts used seabuckthorn cream for protection from cosmic radiation. The protective and curative effects of seabuckthorn oils were found on diseases involving the mucosa of stomach, urogenital tract and mouth. Oral administration of seabuckthorn oil is a promising alternative for treating chronic vaginal inflammation. Plant leaves, fruits and bark are used as medicine for the treatment of various ailments like cold, cough, malnutrition and skin diseases by local traditional doctors in Leh, Ladakh.

Other Uses

Seabuckthorn is used as fuel, fodder and for fencing fields as well as houses since time immemorial. Its extensive root system has excellent soil binding ability, which effectively controls soil erosion. Chinese reports say that it controls the soil erosion by wind up to 96.6% and can reduce runoff by 99.6%. It has an excellent ability to increase soil organic matter. Soil nitrogen has been found to be enhanced due to the presence of a symbiotic actinomycete, i.e., *Frankia* in its root nodules. Besides its nutritive value, it has potential use as a forest plant for greening of cold deserts through large-scale plantations, especially on waste and degraded lands, which can provide food, fodder, fence, improving soil fertility, checking erosion and can help in conservation of flora and fauna. Employment generation is another aspect, which has further improved seabuckthorn in wasteland and the economic condition of the local farmers. Thus, seabuckthorn, a wild thorny bush of Himalayas, has proved to be a golden bush in cold deserts of India.

Cultivation

Soil and Climate

It can withstand extreme temperatures from - 43 to + 40°C, and it is considered drought resistant. However, irrigation is needed in regions receiving less than 400 mm of rainfall annually. Seabuckthorn is found growing from 2550 to 4050 m above mean sea level under varied soil conditions ranging from rocky, sandy and marshy to moist soils and even in saline soils. It grows well on sand dunes and steep slopes of mountains and also on the banks of rivers. Seabuckthorn grows poorly on infertile sands containing no humus, and fails to grow on Aeolian quartz sands containing little moisture. It can be grown in the steppe and semi-desert zones on mantle sands rich in clay and colloidal particles with some humus and fresh groundwater accessible at a depth of 0.5-1.0 m and also on meadow-bog sandy loam soils buried by sand.

Propagation

Propagation of seabuckthorn from seed is difficult, and hence, it is mainly propagated through cuttings. Besides propagating from root suckers, 10-12 cm long shoot cuttings collected in June are used as planting material. The cut end is dipped in solution of naphthaline acetic acid, indole acetic acid, or indole butyric acetic acid, inserted at a depth of 2-3 cm in soil disinfected with 2% captan and shaded from direct sunlight with plastic netting; relative humidity should not fall below 80% and air temperature exceed 40°C. Shading is reduced from late July and removed in late August. The cuttings taken before the bud burst following spring are planted at a spacing of 2 m in lines 4 m apart in 30×30×25 cm holes receiving 1-2 kg organic manure mixed with phosphorus fertilizer 50-60 g. The optimum plant spacing for seabuckthorn is 4×3 and 4×2 m.

Seabuckthorn seeds are dusted with fungicide (TMTD) at 3 g/kg of seeds or fungicide is soil applied at 40 g/m^2 before sowing to avoid diseases in early growing periods. Satisfactory plant growth was observed in the second year when soil applied TMTD was mixed with NPK + peat + lime.

Soft woodcuttings are widely used propagation method for seabuckthorn. Cuttings are taken in June and rooted under plastic film shelters equipped with mist sprays and shaded by three layers of plastic net. The technique of raising plants from softwood cutting can be followed, where stock plants should be spaced at 2.0×0.5 m and cuttings are taken three times from the same plants within an interval of 30-40 days. Cuttings 15-20 cm long grown in the nursery for 1 year produced the best salable plants.

Nutrition

A single spring application of nitrogen 100 kg, phosphorus 200 kg and potassium 100 kg/ha is found to increase fruit size and number of fruits markedly at ripening. Top dressing generally had a favourable effect on plants, increasing shoot growth. Significant changes in fruit composition and fatty oil and vitamin C contents of several cultivars increased and titrable acidity decreased with the application of nitrogen, phosphorus and potassium.

Seabuckthorn fixes the nitrogen in the soil with the help of actinomycete *Frankia.* The application of phosphorus stimulated the nitrogen fixation by having direct effects on nodule dry matter and nodule function rather than indirectly *via* plant growth, and the plant uses both nitrogen fixation and combined N as nitrogen sources. The number of endophyte vesicles per unit cell area increased with age in first year seedlings and varied with mean air temperature in older plants. Molybdenum deficiency in sand culture led to a rapid decrease in vesicle formation and a decline in nitrogen fixing activity. Maximum N fixing activity and biomass of nodules occurred at only moderate levels of molybdenum.

Harvesting

Seabuckthorn plant reaches full bearing in 3 years of planting. Manual harvesting is difficult due to the number of thorns on the shrub, thus, manual cutter is used. Mechanical harvesting can be used in seabuckthorn but not all cultivars are suitable for mechanical harvesting. A tractor driven MOU-1 tree shaker has been developed, which is suitable for use on single stemmed trees in orchards with spacing of more than 5 m between rows and 2.0-2.5 m between plants. Defoliants may also be used to weaken the attachment of the berries to the branches. Berries can also be removed from cut branches by threshing, by dipping in liquid nitrogen, or by freezing at -16°C. Three female cultivars, *i.e.*, Frugana, Hergo and Leikora that make it possible to harvest continuously from early August to mid October are available in Germany. Good results were obtained with mechanical harvesting. Harvesting began when the fruits reached 80% of their maximum acidity content, and continued for 10-14 days for each cultivar.

■■■

20 Tulsi: A Cosmopolitan Herb of Religious and Medicinal Excellence

S.K. Gupta

Assistant General Manager, National Seeds Corporation Limited, Beej Bhawan, Pusa Complex, New Delhi- 110012,

E-mail: <skguptanscplan@yahoo.co.in>, <skguptaji@gmail.com>

Common English name: Sacred basil/holy basil, basil, common basil and sweet basil

Botanical or Latin Name: *Ocimum sanctum* Linn. Synonym *O. tenuiflorum. O. basilicum, O. basilicum* var. purpurascens (dark opal basil), *O. basilicum* var. citriodorum (lemon basil), *O. basilicum* var. minimum (Greek or bush basil)

Family: Lamiaceae/labiateae (mint family)

History

Basil has a rich and fanciful history bearing several name derivations and beliefs associated with it. Because of its popularity, it is often referred to as the *king of the herbs*. It is *Herb Royale* to the French. The common name basil may be derived from the Greek word basileus meaning *king*, or basilikon meaning *royal* or Latin word basiliscus referring to *basilisk* a mythical *fire-breathing dragon*, which was so repulsive that it could kill with just a glance. According to Roman legend, basil is the antidote to the venom of the basilisk. The botanical name *Ocimum* is derived from the Greek word *to be fragrant*. In the 1600's, the English used basil in flavouring their foods and also as an insecticide by hanging in doorways to ward off flies and other unwanted pests (evil spirits). Italians used basil as the sign of love. A pot of basil placed on the balcony meant that a woman was ready for her suitor to arrive, and if he brought a sprig of basil, she would fall in love with him. In America, basil has been grown for over 200 years. It was air-dried or preserved in layers of salt and kept in earthenware crocks. American colonists considered basil the essential ingredient in a snuff used to ease headaches. The first century A.D. Roman naturalist Pliny reported that basil relieves flatulence, which had been subsequently proven true. In the Far East, the herb had been used as a cough medicine, and in Africa, it has been used to expel worms. In the Christian tradition, it is said that it is grown around the place of Crucifixion. Basil (*tulsi*) is also mentioned in Shiite writings. Basil is one of the two most sacred plants in India and grown near houses. It is believed to possess spiritual powers, to give Divine protection (Vedas) and to purify air by discharging negative ions. It opens heart and mind, gives love, devotion, faith, compassion, clarity and cleanses the aura. Hindus believe that if a leaf of it is put in their mouth at time of death or buried with them, it would get them into heaven that is the level of popularity of sacredness in Hindu mythology.

Common Indian Name: *Krishnamul, Manjari, Ajaka, Brinda, Parnasa, Patra pushp, Suvasa and Krishna Tulsi* in Sanskrit, *Tulsi* in Hindi, Bengali, Gujarati and Punjabi, *Tulasi* in Oriya, *Tulashi* in Assamese *Tulshi* in Marathi, *Thulsi* in Tamil, *Trittavu* in Malayalam, *Oddhi* in Telegu, *Sree Tulsi* and *Vishnu Tulsi* in Kannada also called as Ram Tulsi, Kali Tulasi, Chajadha, *etc*.

Other Names: Albahaca, Albahacar/Alboharcar, Albácar (Spanish), Warung (Chuukese), Aring (Kosraen, Kadiring (Pohnpeian), Lamar (Yapese), Kha-kal-tun (Maya), Kemangen and St. Josephwort

Basil has 150 varieties. Each variety differs in height, colour of foliage and taste. Of the six common varieties of basil found in the United States, sweet basil and dwarf basil are the most popular. The description of some of the common basils is given below:

Common name	Latin name	Description
Sweet basil	*Ocimum basilicum*	It is most common type basil, flowers white, leaves bright green, 2 to 3 inch long, growth habit erect and clove like scent.
Genovese basil	*Ocimum basilicum 'Genovese'*	An Italian strain, regarded as the best variety for pesto and garlic dishes, leaves dark green up to 2 inches long, slow to bolt and growth habit erect.
Bush or Greek basil	*Ocimum basilicum* var. minimum	It is a dwarf variety with very small, leaves less than ½ inch long and pungent and flowers white. Plants are excellent for edging or containers. Flavour is preferred by many chefs. Varieties like Fine Green, Green Bouquet and Spicy Globe are widely available.
Purple basil	*Ocimum basilicum* var. purpurascens	It is grown for their ornamental foliage as well as their culinary uses. Flowers are soft lavender, leaf same in shape and size as sweet basil. Opal, Purple Ruffles and Red Rubin are excellent selections.
Lettuce-leaf basil	*Ocimum basilicum* var. crispum	Leaves are large and wide. Flavour is less pronounced than other green basils, sometimes, preferred for salads or sauces. Common varieties include Mammoth, Napoletano and Green Ruffles.
Scented basil	*Ocimum basilicum* var. odoratum	These basils possess flavours reminiscent of other plants. Cinnamon, lemon and licorice or anise basils all fit in this category. They can be used in recipes where a touch of a different flavour is desired. It is often used in fruit preserves or in custards and sorbets.

Common name	Latin name	Description
Holy basil	*Ocimum canum syn. Ocimum sanctum*	Leaves are small and fuzzy with a sweet, clove-like fragrance. Flowers are violet or white. It is used in religious ceremonies, not suited for culinary uses.
Camphor basil	*Ocimum kilimandscharicum*	It has a strong medicinal scent. Foliage is gray-green. It is not used for culinary purposes.
Peruvian basil	*Ocimum micranthemum*	It has a somewhat medicinal sweet flavour though it can be used in cooking. Flowering is sparse.
Thrysiflora basil	*Ocimum thrysiflora*	It is grown for its ornamental seed head. It forms a triangular shaped plant with a strong V-shaped branching habit. The seed head is a mound of purple flowers.

Habitat

Basil is one of the most popular annual herbs grown around the world for their distinctive flavour, essential oil and ayurvedic importance. It can easily be cultivated in a wide range of climates. The natural habitat varies from sea level to an altitude of 2000 m and in moist soil nearly all over the globe. It is native to Asia (India, Pakistan, Afghanistan, Iran, Thailand and so on) and commonly growing wild in tropical and subtropical regions of the world. It reached Western Europe only in the 16th century.

Plant Description

Basil grows up to 0.5-1.0 m high, depending on growing conditions. The low bushy plant has erect, much branched, fragrant, herbaceous, soft hairy stems and toothed, shiny, long/oval, green often-purplish leaves up to 5 cm long. It possesses many essential oil glands, which emit a spicy scent/aroma when bruised.

It has long purple flowers and smooth green foliage with a spicy fragrance. The flowers in closely whorled racemes in small compact clusters on cylindrical spikes vary in colour from white to red, sometimes with a tinge of purple or crimson and appear from June to September. The fruits are small and the seeds yellow to reddish in colour. Indian variety with green leaves is called *Shri* or *Rama Tulsi* (very common) and one with reddish leaves is called *Krishna* or *Shyama Tulsi*.

Holy basil is slightly hairy, whereas, Thai basil is smooth and hairless. Holy basil does not have the strong aniseed or licorice smell as in Thai basil. Holy basil has a hot spicy flavour, sometime, compared with cloves and its aroma is distinctively different to its close cousin. Sometimes, the Thai basil wrongly called holy basil in shops can be distinguished by their aroma and flavour.

A Plant of Tulsi

A Twig of Tulsi

Tulsi Flowers

Tulsi Leaf

Propagation

Plant sources are seeds and young plants. Good seeds are sown in moderate heat, 18-20ºC, during rainy season compost with added one-third extra sharp grit for best results. The seeds are covered very thinly with sieved compost or vermiculite. If plants are very well established, and growing in a greenhouse, the seeds may be saved for next year, which may be easier to germinate than bought seeds. Spent flowers are removed to prolong the life of plant.

Cultivation

As the seedlings are large enough to handle, pot up until they are in a 30 cm pot. If it is grown outside, it should be planted in rainy season in a sunny open place, perhaps near a south-facing wall. The potted plants are fortnightly supplied with a general liquid fertilizer and watered well, allowing the plants to dry between two irrigations. *Tulsi* is a perennial plant in tropics but is treated as a half-hardy annual in western countries. In north Europe, light levels are insufficient during winter to keep it growing, however, usually, it is best under cover on a sunny windowsill or a greenhouse, even during the summer but it can be grown in a sheltered place outdoors in mild areas. There are no safety concerns for this herb.

Like India in Pakistan too, it has immense medicinal value and use. In order to evaluate the genetic diversity among different genotypes of *Ocimum* for the development of better varieties, plant characteristics including phenotypic and genotypic variations as well as biochemical analysis were conducted in four genotypes of different localities of district Poonch, Azad Kashmir. The morphological studies based on leaf area, plant height, the number of racemes per plant, number of flowers per raceme, test weight (1000 seeds weight) and days to maturity indicated large diversity among genotypes. Total seed proteins in SDS-PAGE also produced diverse banding pattern among the genotypes compared. Due to the larger genetic diversity in germplasm of *O. sanctum* and its suitability for commercial cultivation, it was suggested to investigate on large scale about its genetic as well as biochemical properties for the production of commercial varieties and exploitation of plant for economic benefits of the local communities having small land holdings.

Diseases

Basil wilt occasionally causes plants to suddenly collapse and turn brown overnight. Do not try to save them but destroy them fully and wash hands thoroughly before touching other plants. This disease can be transmitted by infected seed, compost, or sap regurgitated from another infected plant by aphids, so wash off if any greenfly is seen on the plants.

Harvesting

Harvest young shoots in early morning for maximum flavour and scent. The leaves and flowering tops are gathered when the plant comes to flower. The best foliage flavour develops before flowering and it changes after opening the flowers. The terminal growth is removed leaving four sets of true leaves on the plant, which encourages bushier growth and increases yield. The herb is collected at four stages of vegetation, *i.e.*, from pre-flowering in July, flowering in August, end of flowering in September to fruit bearing in October. It is either used immediately or stored in perforated plastic bags in the refrigerator. For drying, the leaves are harvested early in the day but after drying of the dew, thereafter, the leaves are spread on screens or bundled loosely and air-dried. Warm air circulation <55°C aids colour retention. Sun dried leaves tend to be brownish in colour.

Constituents at various harvesting stages

To analyze the essential oil constituents in the air-dried aerial parts of *Ocimum sanctum* (syn. *Ocimum tenuiflorum*), the leaves are harvested in four stages of vegetation and the oil is isolated by hydrodistillation. The oil, thus, extracted from plants grown in Cuba, Brazil, India, Germany and Thailand contains eugenol as a main constituent (34, 17-79, 53, 24 and 55%, respectively), frequently together with a significant amount of beta-caryophyllene or α- and β-bisabolenes. Methyl eugenol is the main constituent of some oils of *Ocimum sanctum* from India (25%) and Thailand (38-52%), whereas, the oil from plants grown in Australia is found to contain mainly methyl chavicol (87%). The oil yield obtained at different growth levels is 0.68% at pre-flowering stage, 0.59% at flowering stage, 0.83% at end of flowering stage and 0.69% at fruit-bearing stage. The stage of growth for maximum oil yield (0.83%) is found to occur at the end of flowering. Fifty-seven constituents representing 90-94% of the oils have been identified. Quantitative composition of the oils varies significantly with the development of plant but content of the main group of constituents varies less than the content of individual compounds. Beta-bisabolene (13-20%), methyl chavicol (3-19%), 1,8-cineole (9-33%), eugenol (4-9%), (E)-α-bisabolene (4-7%) and α-terpineol (1.7-7%) are the main constituents of the oil. At the flowering stage, the content of oxygenated compounds is highest and of β-bisabolene- lowest. Chromatographical analysis shows at budding and early flowering stages, basil oil is rich in monoterpenes and the quantity of sesquiterpenes and phenylpropane derivatives increases only at later stages. Zinc, manganese and sodium are found significant, whereas, stearic, palmitic, oleic, linoleic and linolenic acids have been identified by gas liquid chromatography in the fixed oil of *Ocimum sanctum.* About 30 monoterpenoids, sesquiterpenoids and phenylpropanoids have been identified, especially linalool, methyl chavicol, eugenol, methyl eugenol and geraniol in *Ocimum basilicum.* The content of methyl chavicol and eugenol declines with drying.

Tulsi in Indian Tradition

The *Puranas*, *Vedas* and other scriptures point out the importance of Tulsi in religious worship. It is regarded as a goddess and a consort of Lord Vishnu and Krishna. A garland of Tulsi leaves is the first offering to Him as part of the daily ritual. Vaishnavites revere the Tulsi leaf because it pleases Vishnu the most, and thus, is as an inherent part of offerings of *naivedya*. Tulsi is ceremonially married to Vishnu annually on the eleventh bright day *Kaartik Shukla Dwadashi* of the *Kaartika* month in the lunisolar calendar. The ritual called *Tulsi Vivah* inaugurates the annual marriage season in India. This festival continues for five days and concludes on the full moon day, which falls in October or November. As with Diwali celebrations, there are usually clay lamps lit around the Tulsi plant and the house. They also wear *beaded necklaces* made of Tulsi stems. The manufacture of these necklaces is a *cottage industry* in places of pilgrimage. Tulsi means the incomparable one and it has been traditionally used by Hindus, and now by others for its diverse healing properties. It has been revered in India by the rishis (sages) for thousands of years as a prime herb in *Ayurvedic* treatment. They were keen to instill in people's hearts appreciation for the healing virtues to be derived from plants. Hindus grow Tulsi as a religious plant in the courtyard, temples and their farms. The surrounding atmosphere is filled with a type of charged aroma, an elixir of the finest quality, regarded as a kind of *elixir of life* and believed to promote longevity. When one begins to investigate its therapeutic and medicinal properties in indigenous systems of medicine, it is easy to see why the plant is considered so special. Among the many botanical treasures with which the country was blessed they discovered several that had a wide range of applications for curing common illnesses or preventing diseases altogether. Tulsi is mentioned in the *Rigveda* and *Charak*

Samhita, the central teaching of *Ayurvedic* medicine written at least four and two thousand years ago, respectively.

Herb Properties and Parts used

Basil is bitter and pungent in taste. Its leaves and oil have same medicinal properties as other mint species, demonstrating as cardio tonic, appetizer, balances *kapha* and *vata dosha*, antimicrobial, antiyeast and pesticide. The parts used are leaves, flowering tops, seeds, roots and essential oil.

Characteristics and Chemical Composition/Active constituents

Basil is a good source of beta-carotene, vitamin C, calcium and volatile/essential oil (about 1% including estragol, linalool, eugenol, methyl chavicol and small quantities of methyl cinnamate, cineole, caryophyllene, carvacrol, limatrol and other terpenes), tannins, basil camphor, flavonoids (apigenin and luteolin) and triterpene: urolic acid. Leaves are abundant in tannins like gallic acid, chlorogenic acid, *etc.* and they also contain alkaloids, glycosides and saponins along with bright yellow volatile oil. The major active constituents of leaves include urosolic acid. The seeds contain oil composed of fatty acids and sitosterol. The mucilage is compared of sugars-xylose and polysaccharides.

Volatile oil (0.4-0.8%) contains chiefly eugenol approximately 21% (reaches maximum in spring and minimum in autumn), β-caryophyllene 37% and a number of sesquiterpenes and monoterpenes, *viz.* bornyl acetate, β-elemene, methyleugenol, neral, β-pinene, camphene, a-pinene, *etc.*, ursolic acid, campesterol, cholesterol, stigmasterol, β-sitosterol and methyl esters of common fatty acids.

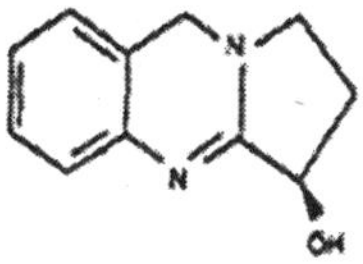

Vasicine

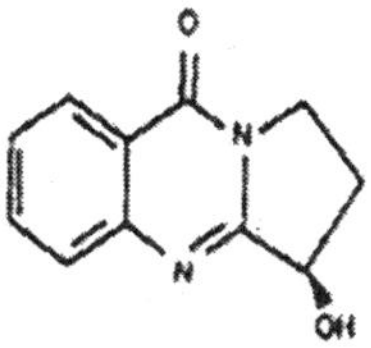

Vasicinone 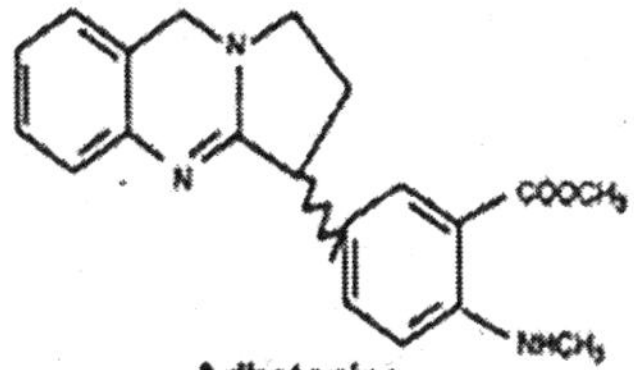 **Adhatonine**

Traditional Uses

Out of many uses, the most common is its culinary use in the western world, where its medicinal value is not widely appreciated. The fresh herb is used to flavour jelly, honey, tea and liquor and also foodstuffs such as vegetables, poultry, fish and pesto. Basil is commonly preserved in vinegar or in olive oil, which adds a delightful flavour for salad dressings. The flowers are also edible and can be an attractive addition to salads and other dishes.

As an aromatic herb, it is often used in potpourri and sachets. For centuries, the dried leaves have been mixed with stored grains to repel insects. The cosmetic industry uses basil oil in lotion, shampoos, perfumes and soaps. Essential oil extracted from *Karpoora Tulsi* is mostly used in herbal toiletry. Attractive foliage and flowers are used as an ornamental in the garden and as air purifier. The beaded necklaces are made up of its stems.

Chewing a couple of leaves before a meal helps to stimulate the appetite. Its tea taken after meal promotes digestion by increasing the flow of gastric juices, reducing gas and bloating, relaxing and promoting sleep. In *Ayurveda* basically, it is used in cold and coughs, fever for chills and intermittent fever like malaria. When inhaled in steam, it relieves nasal congestion. It has also been used to treat anxiety, stress and tension, bronchitis, asthma, skin diseases, canker sores, cuts and abrasions, flatulence, headaches and migraines, insect stings and snake bites, muscle tension, nerve pain, nervousness, sinusitis, tiredness and lethargy, abdominal distention, gastric disorders, absorption, anorexia, sore throats, constipation, diarrhea, dysentery, colic (air excess), urinary problems, memory, nerve tissue strengthening, heart tonic, rheumatoid arthritis, headaches, earaches, in lowering blood sugar levels and mouth ulcers. Aphrodisiac virtue has been attributed to it, where powdered Tulsi root with clarified butter/ghee is prescribed.

In Chinese medicine, basil is used for disturbances in renal function, gum ulcers and as a hemostyptic both before and after birth, in Belize to bring on delayed menstruation and to relieve painful periods, as an anti-inflammatory for stomachaches and earaches and to treat intestinal parasites, in Mexico to treat *susto*, or gastrointestinal blockage, in Guinea Bissau to ward off evil spirits, as a cleansing agent and mosquito repellent plants and in Kenya and Guatemala for ethno medicinal use. Chuukese mix it (Warung) with fish in coconut milk soup, flowers in *mwarmwars* (head garlands) and are also added to coconut oil that is used on the skin. For steam treatment in fever and in fever with running nose, the leaves are put in a cloth, pounded, and then, the juice is squeezed directly into the nose. It is also used in love magic. Kosrae treat head and chest colds with its steam under a layering of bed sheets with the pot of hot water having basil leaves and breathing the vapours produced by it. Pohnpei uses *Kadiring* in many ways. The leaves are used for cooking and also for curing headaches and colds by *umwulap* (steam bath). Suriname uses the juice of the leaves against fever, abdominal pains, in lowering the blood sugar level and as an antidote for snake and scorpion bites. Nigerians use *O. gratissimum* as native medicine.

Safety: Generally, it is recognized safe in United Kingdom and United States of America and is put on general sales list. In France, it is used in traditional medicine, whereas, in Canada, it has been approved as an over-the-counter drug, while in Germany, it is not yet approved by Commission E.

Basil Products

Basil Oil: Indian basil oil is of watery viscosity and is pale greenish-yellow in colour that is good in taste and more powerful than the plant itself. It has more uses in the kitchen and food productions. It refreshes mind and skin and it is widely used as massage oil.

Basil Powder: Its leaves are dried and ground into a powder form. People in India use this with honey in throat problems like sore throat or hoarseness of voice. However, many Indian pharmaceutical companies prepare herbal capsules out of it, which are useful for diseases like diabetes and blood pressure.

Basil Extract: The flowery stems are the medicinal part used for extraction process. It is used in India for reducing the negative effects of stress by lowering cortisol production in the adrenals.

Medicinal Actions and Uses

Key medicinal actions of basil include anti-catarrhal, stimulant/ antidepressant, antiseptic, Antimicrobial (antibacterial/antifungal/antiviral), insecticidal, anthelmintic, antispasmodic, anti-inflammatory analgesic, antipyretic particularly in malarial fevers, mosquito repellent, adaptogenic, carminative, febrifuge, diaphoretic, antiperiodic, nervine, expectorant, deodorant, cardiotonic, antidiabetic, blood purifier and lowers blood pressure.

The antimicrobial and insecticidal properties are due to the eugenol, present in the leaves, which inhibits the *in vitro* growth of *Mycobacterium tuberculosis* and *Micrococcus pyogenes* var. aureus and other bacteria and fungi. It has marked insecticidal and repellent activity against mosquitoes although it is not comparable to that of pyrethrum. Basil oil has been used as a potent anti-malarial drug. The oil from the green type is active against *Salmonella typhosa*. Ether and alcohol extracts of leaves are active against *Escherichia coli*. Basil leaves have properties similar to the currently available anti-tuberculosis drugs like Streptomycin and Isoniazide. The linalool and methyl chavicol, the constituents of oil, inhibit Gram-positive and Gram-negative bacteria, yeasts and moulds. Washing fresh lettuce with 0.1 or 1% (v/v) chavicol on survival of microbial flora was comparable with that affected by 125-ppm chlorine. The juice of fresh leaves, flower tops and slender roots is a very good antidote for snake and scorpion bite.

The ursolic acid has anti-allergic properties and can play a role in the management of immunological disorders such as allergic rhinitis, catarrh, allergies and asthma, which is commonly used with black pepper in bronchial asthma. Protective action against histamine-induced bronchospasm has been shown in animals. There is report that the solid extract of basil, in a dose of 500 mg x 3 for one week, significantly relieved the breathlessness in patients with tropical eosinophilia, however, there was no reduction in the eosinophil counts in peripheral blood.

Indian researches have shown its anti-inflammatory, analgesic and fever-reducing properties. The triglyceride fraction is anti-inflammatory in animals. It inhibits inflammatory causing enzymes in human body that contribute to rheumatic and arthritis pain and other kinds of inflammation. Recent studies suggest that this may be a COX-2 inhibitor, like many modern painkillers, due to significant amount of eugenol (1-hydroxy-2-methoxy-4-allylbenzene). For earache, 4 to 5 drops of basil extract in the ear in every 3 to 5 hours can minimize the earache and also take care of any mild infection that can be prevailing in the external ear canal. In case of irritation or redness of eyes, a little bit of honey is added to basil juice and 2 to 3 drops of this mixture can be administered to the effected eye to every 3 to 4 hours. It works well in such conditions.

Basil has anti-fertility effect by reducing the estrogen hormone levels in females and decreasing the sperm counts in males. In animals, the large dose of extract has shown antispermatogenic activity. The seeds are mucilaginous and demulcent and are given in disorders of genitourinary system/difficult urination. It prevents the accumulation of fat in the body (especially for women after menopause).

Since basil has the ability to lower down the blood pressure, it is thought to have an affinity for the heart as well as helping the body to adapt to new demands and stresses. Research into its ability to reduce blood sugar levels has gone on for several decades and found effective and useful in diabetes and also in significant reduction of total cholesterol levels. Leaf powder has significant effect on blood lipoproteins, glycated proteins and total amino acids in patients with non-insulin dependent diabetes mellitus. Further study showed that this beneficial effect on blood is due to its antioxidant properties.

The decoction of leaves can be used to bring down all types of fever, cough, cold, flu and respiratory tract infection. Only 5 ml juice of basil leaves with honey taken twice a day can take care of cough. Similarly, decoction of the root is given as a diaphoretic in malarial fevers. An infusion or decoction of its fresh leaves can be an alternative in chronic malarial fevers because of its action on the liver. A classic recipe advocates mixing basil, black pepper, ginger and honey to prevent infections like cough, cold, high fevers, mild indigestion, diminished appetite, nausea, gas pains, dysentery and malaise. Basil is an important constituent of many *Ayurvedic* cough syrups and expectorants. It helps to mobilize mucous in bronchitis and asthma. *Ocimum tenuiflorum* is used in *Surasadi Phanta* for cold and influenza.

Basil enhances adrenal function by stimulating the adrenal cortex and lowering cortisol levels. This results in reducing the negative effects of stress. Eugenol and methyleugenol showed a positive adaptogenic (antistress) activity. Adaptogens are a key class of botanical compounds that elicit a general tonifying effect on the body, generally enhancing one's resistance against physical and psychological stressors. These are ideal as part of a preventive health programme since they protect against many of the stress induced challenges of modern life, such as pollution, noise, hectic pace of life and environmental chemicals. The most noted adaptogens in modern times are the various species of ginseng- American and Asian- as well as Siberian eleuthero. *Ocimum sanctum* acts as an antioxidant and reduces cortisol production. Specifically, it has been shown to lower blood sugar. It frees ozone from sun's rays and oxygenates the body, cleanses and clears the brain and nerves, builds the immune system, relieves depression and the affects of poisons (first stages of many cancers). It increases the physical endurance and prevents stress-induced ulcers. In general-pharmacology, the aqueous extract has been shown to have hypotensive activity, to inhibit the smooth muscle contraction induced by acetylcholine, carbachol and histamine and to potentate the hexobarbitone sleeping time.

The ursolic acid and a host of flavonoids such as apigenin, polyphenols and anthocyanins elicit powerful free-radical squelching activity, supporting the traditional use of this botanical as a longevity and general health-promoting tonic. The leaves are used as a nerve tonic and as an agent to sharpen memory. The seeds are used as a general and heart tonic. The decoction of leaves along with tubers *Orchis latifolia* is considered a nourishing drink for old and weak people. Its oil is rich in vitamin C, carotene, calcium and phosphorus. Aphrodisiac virtue has been attributed to the root where powdered root with clarified butter/ghee is prescribed. It raises the human body immunity by increasing the antibody production. Basil contains trace mineral copper (organic form), needed to absorb iron.

Numerous other activities have been reported such as immune supporting, increased endurance, and improved cognitive function. It increases the glutathione, the body's primary antioxidant defense system. It shows some promise in protection from radiation poisoning and cataracts. The leaves juice is given for skin diseases internally as well as externally. Externally applied on chronic non-healing ulcers, skin disorders like (soothes) itching and ringworm. In urticaria, any kind of allergic skin rash, the paste of basil leaves can be applied over the affected part. This treatment significantly takes care of the burning sensation in the rash and the itching if any. Its powder is used for mouth ulcers. An infusion of the leaves is used as a stomachic in gastric and in vomiting, nausea, viral hepatitis and hepatic disorders of children. Seed's paste with milk is considered a harmless preparation for the stomach troubles of infants and small children.

Side Effect: The only side effect reported is constipation when the 5-27 g powdered leaves per day are taken.

Cautions in using basil

- The essential oil should not be used internally or externally during pregnancy or while breastfeeding the infants.
- Medicinally, it should not be given to young children or infants.
- Diabetics should cautiously use the herb medicinally.

Pharmacology

Medically, the plant provides a pharmacopoeia for the entire household. An infusion of basil leaves and ginger is India's most popular remedy for stomachaches in children. Its essential oil is an antiseptic and insect repellent, while its root paste soothes bites and stings, an antidote to snake venom and scorpion bites. It has numerous pharmacological activities like antioxidant, hypoglycaemic immunomodulatory or immunostimulant, antistress, analgesic, antipyretic, anti-inflammatory, wound healing, antiulcerogenic, antidiabetic, antihypertensive, CNS depressant, radioprotective, antitumor, anticancer, antibacterial, antithyroidic, antiHIV, *etc.*

Therapeutic Uses: It is used in cases of minor food poisoning, bronchial asthma, hiccups, cough, halitosis, abdominal worms, especially round worms and various kind of skin hypo pigmentation.

■■■

21 Blue Green Algae: Food for the Future

G.S. Unnikrishnan Nair

Assistant Director of Agriculture, Farm Information Bureau, Department of Agriculture; Government of Kerala, E-mail: <vanchiyurunni@gmail.com>

Blue green algae are gaining more importance as nutritious and medicinal food. The most prominent among blue green algae are Astaxanthin and Spirulina.

Astaxanthin

Astaxanthin is a dark-red pigment that occurs naturally in a wide variety of living organisms. Now, this wonderful pigment is produced commercially from micro-algae and it is finding application in a variety of fields from ornamental fish culture to human health. Salmon, red peppers, *etc* also contain the antioxidant astaxanthin, which can help in preventing heart disease, Alzheimer's and serve as a natural macular degeneration treatment.

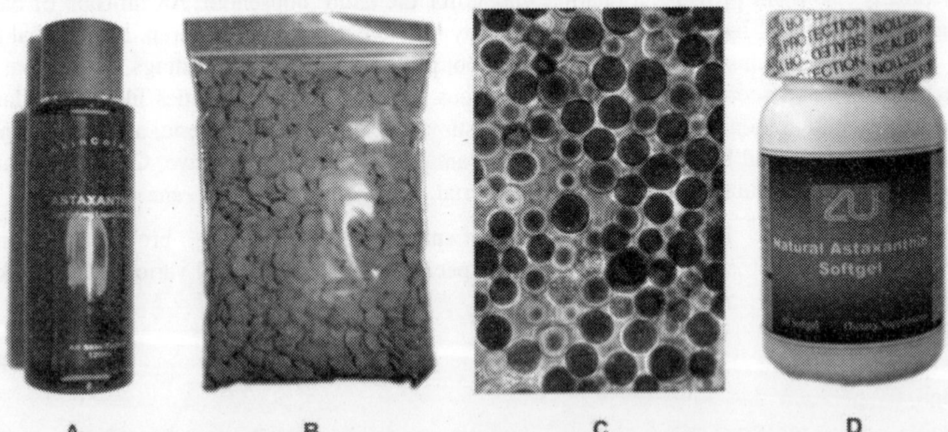

Figure 1 : A. Astaxanthin gel, B. Astaxanthin tablets, C. Astaxanthin and D. Natural Astaxanthin soft gel

A carotenoid

Astaxanthin belongs to the natural pigment group carotenoids, which are responsible for orange and red colours in plants and algae and for the wide range of blue, purple and reddish colour in aquatic animals. Carotenoids are lipid-soluble pigments and antioxidants, which participate as accessory pigments in the light-absorption process of photosynthetic organisms. Over 600 natural carotenoids have been recognized. Animals, including humans consume carotenoids as part of their diet.

Astaxanthin is present in many types of seafood, including salmon, trout, red sea bream, shrimp and lobster as well as in birds such as flamingo and quail. It is found in the marine algae and aquatic animals. Most crustaceans including crawfish, shrimp, crabs and lobster are tinted red by accumulated astaxanthin.

Astaxanthin unlike some carotenoids does not convert to vitamin A in the human body, and enhanced activity of astaxanthin is due to its molecular structure. It belongs to the oxygenated carotenoids or the xanthophyll group of carotenoids. After ingestion, astaxanthin is absorbed by the duodenal mucosa and transported to the liver where it binds with a lipoprotein for transport through body to the cells. The polar end groups attach to the outer and inner side of the cell membrane, where free radical attack first occurs. The hydroxy and carbonyl groups in astaxanthin help to anchor this molecule to the cell membrane, thus, strengthening it. Astaxanthin is an oxygenated carotenoid called a xanthophyll. Because it is oxygenated, a xanthophyll has greater antioxidant capacity than many other carotenoids. Within the xanthophylls, the astaxanthin molecule contains the longest conjugated double-bond polyene chain along with both hydroxy and carbonyl groups at each end. This configuration supports the greatest antioxidant capacity, its peroxyl radical chain-breaking abilities, its incorporation of free radicals into its polyene chain (thereby more effectively trapping them) and its enhancement of vitamin C as an antioxidant. While other carotenoids and antioxidants may perform one or two of these functions, astaxanthin does them all, and in most cases, does them better.

In addition to entrapping free radicals, astaxanthin also de-charges singlet and triplet oxygen and inhibits reactive oxygen, giving it anti-inflammatory properties. Astaxanthin may also increase gap junctional communication between cells (A gap junction or nexus is a specialized intercellular connection between a multitudes of animal cell types. It directly connects the cytoplasm of two cells, which allows various molecules and ions to pass freely between cells), which is a mechanism in the body thought to inhibit the growth of cancer cells.

Production

Currently, the primary natural source for astaxanthin is a marine microalge *Haematococcus pluvialis*. Astaxanthin biosynthesis in *H. pluvialis* is associated with the formation of cyst cells as the resting stage and is enhanced by environmental stresses such as intense light, drought, high temperature and high salt. This unicellular microscopic green alga accumulates the highest levels of astaxanthin in nature, commercially more than 40 g of astaxanthin per kilogram of dry biomass under suitable growth conditions. This is its natural mechanism to protect itself against environmental stress. Further, this algae's population double every week. It is observed that *H. pluvialis* cells accumulate 15,000 ppm of astaxanthin before harvesting.

For efficient production of astaxanthin from *H. pluvialis*, a two steps cultivation system is applied. In green stage of production, the optimal conditions for cell division are maintained to achieve the maximum cell number. In red stage of production, the optimal inductive conditions are provided for astaxanthin accumulation.

Once harvested the whole cells of the algae are spray dried, and then, carefully cracked to ensure bioavailability, thereafter, it is packaged in absence of oxygen to ensure stability. To process it to oleoresin and powder form, the broken cells are extracted in non-petrochemical solvents and distilled to obtain astaxanthin oleoresin with < 100 ppm of solvent residues in the final product. The oleoresin is then converted into powder form. In India, Parrys Neutraceuticals produce and market astaxanthin in powder, capsule and tablet form. Their algal facility is located at Oonaiyur in Tamil Nadu, twenty miles away from the historic town of Pudukottai.

Other than *Haematococcus pluvialis*, *Xanthophyllomyces dendrorhous* (yeast), *Euphausia superba* (krill) and *Pandalus borealis* (shrimp) are also sources of astaxanthin. Micro algae contain 15000-20000 ppm astaxanthin, while yeast contains up to 800 ppm crustaceans up to 400 ppm and salmon 3-40 ppm.

Astaxanthin occurs in certain plants, especially certain species belonging to the genus *Adonis*. One such species is *Adonis aestivalis*, where it occurs predominantly in the petals of the bright red flowers. In United States, a patent has been granted for production of astaxanthin from this flower. Now, the demand for natural astaxanthin has risen in such a way that it is emerging as a multi-billion dollar nutraceutical market. In addition to production from natural sources, astaxanthin is also chemically synthesized.

Valuable Antioxidant

The capacity of astaxanthin as an antioxidant is more powerful than vitamins C and E. In fact, it has 100-500 times the antioxidant capacity of vitamin E and 10 times the antioxidant capacity of beta-carotene and other carotenoids such as lycopene, lutein and zeaxanthin among others. Oxygen is necessary for the metabolic production of energy in our bodies. Mitochondria through electron-transport chain use oxygen to oxidize certain molecules and generate energy in the form of ATP. During this process, oxygen is reduced to water, producing several oxygen-derived free radicals or reactive oxygen species (ROS), which play an important role in various diseases. Oxygen free radicals are normally neutralized by natural antioxidants such as vitamin E, or enzymes such as superoxide dismutase (SOD). However, ROS become a problem when a decrease in either their removal or their overproduction occurs, resulting in oxidative stress. This stress and the resultant damage have been implicated in many diseases and a wealth of preventative drugs and treatments are currently being studied.

In a recent study researchers decided that carotenoids such as astaxanthin should be referred to as *antiradicals* rather than as free radical scavengers because they perform a double action against free radicals, either donating or accepting electrons. Typically, an antioxidant becomes unstable after quenching a free radical, and in this process, it becomes a free radical itself. However, astaxanthin due to its unique molecular structure that allows it to span the cell's lipid bi-layer and also because of its long double-bonded chain structure can stabilize free radicals by adding them to its structure rather than sacrificing an atom or electron. It stabilizes the cell membrane like a bridge because its polar end groups span the cell membrane, thus, increasing its rigidity and mechanical strength.

Oxidative stress leading to mitochondrial dysfunction cause many diseases such as cancer, neurodegenerative and lifestyle-related diseases. The efficacy of astaxanthin in limiting the damage produced by ROS-induced oxidative stress and improving health parameters in the tissues, and the body was demonstrated in a series of *in-vitro* experiments, in pre-clinical studies and in human models.

Scientists in Japan recently found that astaxanthin decreased the induced oxidative stress and protected cultured cells from free radicals. Moreover, astaxanthin has been found to provide many essential biological functions, including protection against lipid-membrane peroxidation of essential polyunsaturated fatty acids and proteins, DNA damage and ultra violet light effects. It also plays an important role in immunological defence. There is increasing testimonial evidence that astaxanthin may be effective in enhancing general well-being, improving the quality of life and enhancing the immune system. Recent studies have shown enhanced immune response and decreased DNA damage in human subjects following astaxanthin administration. Astaxanthin is capable of crossing the blood-brain barrier in mammals, a unique and important property in the realm of antioxidants. This characteristic allows astaxanthin to extend its superior antioxidant activity to the central nervous system, which being rich in unsaturated fatty acids is highly susceptible to oxidative damage by ROS.

Against Alzheimer's and Glaucoma

Over the last 10 years, astaxanthin has been shown to have beneficial effects for Alzheimer and Parkinson's, cholesterol, inflammatory, chronic viral and autoimmune diseases, dyspepsia, *etc.* Astaxanthin's low molecular weight and unique structure combine in allowing it to cross the *Blood Brain Barrier* and the *Blood Retina Barrier*, making it available to the eye, brain and central nervous system. This helps to ease oxidative stress that has been shown to contribute to ocular and neurodegenerative diseases such as Glaucoma and Alzheimer's.

Astaxanthin has been studied to have very powerful anti-inflammatory properties, hence, enhancing its function in eye, heart and joint health. Recent research from Taipei Medical University in Taiwan found that astaxanthin protected against induced cataracts in rats. The scientists backed up their finding by *in vitro* test, which supported their conclusion that astaxanthan was protective against cataracts that are usually caused by environmental insult. Certain carotenoids have been shown to help in protecting retina of the eye from oxidative damage. The eye lens focuses incoming light onto the photosensitive retina, which transmits visual signals to the brain. In central area of the retina lies the macula, which has the highest density of photoreceptors that provide visual acuity. Oxidation, as from sunlight exposure, degrades membranes and likely leads to damage or destruction of photoreceptor cells.

A recent United States rat study indicates astaxanthin is able to exert antioxidant effects to ameliorate retinal injury by staving off light-induced oxidation and protecting photoreceptors from degeneration. The carotenoid pigments lutein and zeaxanthin, which concentrate in the macula, absorb blue light and quench singlet oxygen radicals. As previously mentioned, astaxanthin similarly has been shown to be an excellent quencher of singlet oxygen radicals. The photoreceptor cells are a constituent of outer neuronal layer of the retina, which is a component of central nervous system. These study results suggest that astaxanthin may prevent and treat neuronal damage associated with age-related macular degeneration and may also treat ischemic reperfusion injury, Alzheimer's disease, Parkinson's disease, spinal cord injuries and other types of central nervous system injuries.

Good for Heart

Oxidative stress and inflammation are implicated in several different aspects of cardiovascular disease. Free radicals activate transcriptional messengers that contribute to endothelial dysfunction, start and succession of atherosclerosis, irreversible damage following stoke and heart arrhythmia. Researchers at the University of Hawaii found that due to its ability to extinguish free radicals, axtaxanthin is an appropriate candidate for development as a therapeutic agent for cardiovascular oxidative stress and inflammation.

Cholesterol is a primary indicator of cardiovascular health. It is now established that the gauge of health comes not so much from total cholesterol levels as from the ratio of high-density lipoproteins (HDL or *good* cholesterol) to low-density lipoproteins (LDL or *bad* cholesterol). A study at the University of Panama measured cholesterol levels of rats fed diets containing 1,000 parts per million beta-carotene, canthaxanthin and astaxanthin, respectively for 30 days. Those on astaxanthin and canthaxanthin (a natural carotenoid found in many plants as a natural colouring pigment of apples and many other fruits, vegetables and flowers) showed significant increase in HDL.

Astaxanthin may enhance the burning of fat during exercise and lead to improved muscle endurance, suggesting a new study with mice that needs to be repeated in humans. Mice supplemented with Astaxanthin were found to have accelerated body fat reduction or fat burning when combined with exercise compared to exercise just alone. According to researchers from Kyoto Prefectural University of Medicine, University of Shizuoka, University of Hyogo and University of Nagoya, astaxanthin could elevate fat utilization as energy substrate during exercise, at least partly, due to the antioxidant effect, leading to improvement of endurance in prolong exercise and efficient reduction of adipose tissue with sustained exercise. The study adds to a small but growing body of research linking the carotenoid to potential weight management benefits. Research shows that an antioxidant supplement called astaxanthin can improve energy and endurance levels in physically active people and even influence fat accumulation. A Swedish double blind placebo-controlled study of healthy men explored exercise performance after taking astaxanthin. The men took either 4 mg astaxanthin or placebo daily for up to six months. After three months, the men who took astaxanthin were able to perform more knee bends than those in the placebo group. At six months, the improvement was significant.

In another study, researchers found that a daily dose of 6 mg astaxanthin given to meter track athletes for four weeks caused them to accumulate lower levels of lactic acid, a substance that causes muscle fatigue and weakness.

The results of two animal studies were also promising. In one study, the scientists found that astaxanthin improved swimming endurance in mice while also significantly reducing the accumulation of fat. In another study conducted in 2008, it was revealed that astaxanthin promoted lipid metabolism rather than the use of glucose during physical activity. This led to an improvement in endurance and loss of fat (weight loss).

Skin Protection

Often referred to as nature's sunscreen, astaxanthin can block ultra violet damage of the skin. It improves skin health and beauty, remove wrinkles, increase skin's elasticity and moisture content while reducing inflammation and cell damage that leads to premature ageing. Study revealed the potential whitening effect of astaxanthin related to other whitening agents. Products are appearing that feature astaxanthin for its benefits to skin through topical application. However, studies have shown that oral supplementation works best.

Scientists in Rome, Italy compared the abilities of various carotenoids to modulate ultra violet A related injury. Fibroblasts from skin were exposed to moderate doses of ultra violet A, which stimulated cell death, increased levels of free radicals and other reactive substances, decreased antioxidant enzymes activities and promoted membrane disturbance and enzyme denaturing of hemoglobin. The fibroblasts were treated with carotenoids including astaxanthin 24 hours before exposure to ultra violet A. Of the carotenoids tested, only astaxanthin exhibited a pronounced photo protective effect and countered all of the ultra violet A induced alterations to a significant extent.

Astaxanthin dose-dependently reduced melanin deposition. According to the study by Biochemistry and Nutrition Department, Central Food Technological Research Institute, astaxanthin protects gastric mucosal injury that leads to gastric ulcer, by antioxidative mechanism.

Anti-cancer

Recent research has revealed astaxanthin's ability to increase cellular energy, provide protection against cancer, protection against damage to brain and nervous system and halt the progression of complications from diabetes. The ability of astaxanthin to inhibit the growth of colon cancer cells was studied at Catholic University School of Medicine in Rome, Italy. Astaxanthin inhibited cell growth in a dose and time dependent manner by arresting cell cycle progression and by promoting death of cancerous cells. These results document axtaxanthin as highly effective against colon cancer.

Several other studies have shown the effectiveness of astaxanthin as a preventative and treatment against cancer. In an experiment during 1994, astaxanthin protected mice from urinary bladder cancer. In 1995, the same Japanese research group showed that astaxanthin prevents oral cancer in rats and inhibits cancer in a manner more pronounced than that of beta-carotene. A further study by this group explored the effect of astaxanthin on colon cancer in male rats. They found a significant decrease in the incidence of colon cancer in the animals that were given astaxanthan.

Breast cancer was the subject of researchers in China. Looking into the effects carotenoids has on proliferation, cell cycle, apoptosis and the expression of cancer promoting *bcl-2* gene. They found that astaxanthan, lycopene, zeaxanthin and β-carotene all inhibited cell proliferation. The carotenoids also down regulated *bcl-2* gene expression.

According to researchers at the University of Minnesota, astaxanthin's ability to enhance immune responses in mice has a corollary effect of exhibiting antitumour activity. They fed the mice astaxanthin 40-mcg per kilogram body weight per day. Those mice had significantly smaller tumours than controls when supplementation was started one and three weeks before tumour inoculation.

Another mouse study at Washington State University in Pullman showed that astaxanthin inhibits the growth of mammary tumours in mice by modulating tumour latency and exerting antioxidant effects. Mice were fed a diet containing 0, 0.1, or 0.4% astaxanthin, beta-carotene, or cantaxanthin. Three weeks later, all mice were inoculated with tumour cells.

After 45 additional days, plasma concentrations of the carotenoids were undetectable in unsupplemented mice but pronounced in the supplemented group. In general, all three carotenoids decreased mammary tumour volume. Lipid peroxidation activity in tumours was lower in mice fed astaxanthin, which dose-dependently inhibited mammary tumor cell growth and was most effective of the three carotenoids.

Many carcinogens undergo detoxification by xenobiotic-metabolizing enzymes, which enhance the diversion of toxic byproducts toward detoxification pathways. Xenobiotics are chemicals that are foreign to the biological system and include naturally occurring compounds and drugs. Xenobiotic metabolism is the physical and chemical change that affects foreign substances in living organisms from uptake to excretion. Astaxanthin has been shown to induce rat lung and kidney xenobiotic-metabolizing enzymes.

A recent animal study reports on the benefits of astaxanthin against *Helicobacter pylori*, which in humans is associated with chronic type B gastritis, peptic ulcer disease, and more importantly, gastric carcinoma. In this study, conducted at the University of Lund in Sweden, mice infected with *H. pylori* were given daily treatments for 10 days of either algal meal rich in astaxanthin of various potencies (10, 50 and 100 mg/kg body weight), 400 mg/kg vitamin C, or control meals. After 1 and 10 days post-treatment, both the astaxanthin and vitamin C groups showed significantly lower *H. pylori* colonization levels and lower inflammation scores than controls, with astaxanthin showing a dose-dependent response.

Researchers in Japan examined the protective effects of axtaxanthin on neuroblastoma cells. Pre-treatment of the cells with astaxanthin preserved cellular integrity in a dose dependent manner, and inhibited mitochondrial dysfunctions, including lowered membrane potential and gene repair.

Reduce Diabetes

Flavangenol is an extract from French maritime pine bark that has become popular in Japan for its antioxidant potential and its ability to increase blood circulation. Scientists at Japan Women's University studied the effect of astaxanthin and flavangenol as well as a combination of both, in countering oxidative stress in rats with induced diabetes. These findings indicate that the astaxanthin and flavangenol combination was highly effective at preventing the progression of diabetic complications. Korean researchers found that under diabetic conditions, many damaged molecules including DNA, lipid peroxidation products and advanced glycation end-products (AGEs) eventually lead to pathological diabetic complications including nephropathy. These cells were treated in a high-glucose mix, and then, exposed to astaxanthin. According to the findings, the antioxidant effectively suppressed lipid peroxidation, total reactive species, superoxide and nitric oxide levels. Markers of inflammation, including inducible nitric oxide synthase (INOS) and cyclooxygeanase-2 (COX-2) were also suppressed.

Astaxanthin Patents

Some of the patents given to astaxanthin are *United States Nutraceuticals* for the use of astaxanthin for retarding and ameliorating central nervous system and eye damage; *AstaCarotene AB* (Gustavsberg, Sweden) for the use of astaxanthin for the treatment of autoimmune diseases, chronic viral and intracellular bacterial infections, immunoglobulin-rich milk production and use thereof, treatment of dyspepsia, method of increasing the production and improving the quality of semen, method of the prophylactic treatment of mastitis, oral preparation for the prophylactic and therapeutic treatment of *helicobacter* species infection, medicament for the improvement of duration of muscle function, or treatment of muscle disorders or diseases, agent for increasing the production of/in breeding and production mammals, method for increasing the production of/in breeding and of production animals in the poultry industry; *Cyanotech Corp* (*United States*) for the method for retarding and preventing sunburn by ultra violet light, method for retarding and ameliorating fever blisters and canker sores and method of retarding and ameliorating carpal tunnel syndrome; *Suntory Limited* (Japan) for the compositions normalizing circadian rhythm; *Itakura Hiroshige* (Inventor-Japan) for medicinal compositions having effects of ameliorating eye diseases and holding eye functions; *Aanesen berit annie* (Inventor-Oslo, Norway) for the use of di-esters of astaxanthin for enhancing the growth of farmed fish; *Fuji Chemical Industry Company Limited* (Japan) for the agents for relieving eye controlling function error; *Pacific Cooperative Limited* (Japan) for healthy nutrition composition containing chitosan, oligosaccharide and astaxanthin; *Lycored India* (New Delhi) for the use of carotenoids as anti-hypertension agents; *Suntory and Itano Limited* (Japan) for the use of astaxanthin in pharmaceuticals for treating stress, astaxanthin-containing food or drink; *Triarco Limited* (United States) for method of inhibiting 5 α-reductase with astaxanthin and *Anderson and Patterson* (Inventors- United States) for the method of inhibiting the expression of inflammatory cytokines and chemokines .

In Aquaculture

The primary use of synthetic astaxanthin today is as an animal feed additive to impart colouration, which includes farm-raised salmon and egg yolks. Salmon and trout like other animals cannot synthesize astaxanthin themselves and must obtain it from their diet. In fact, synthetic carotenoid, *i.e.*, coloured yellow, red, or orange pigments represent about 15-25% of the cost of production of commercial salmon feed. Until recently, many Aquaculture Companies used synthetic astaxanthin because of cost constraints but they are now turning to the natural source because of the excellent health benefits it offers. The feed trials have shown that apart from imparting an attractive colour, astaxanthin has been shown to prevent the oxidation of fats in rainbow trout during frozen storage, thus, preventing rancidity.

Astaxanthin is also useful in inflammatory, chronic viral and autoimmune diseases, dyspepsia, semen fertility improvement, muscle function, normalization of cardiac rhythm, stress management, benign prostatic hyperplasia (BPH), stroke (repairs damage caused by lack of oxygen anti-hypertension agent) and as an anti-hypertension agent.

Toxicity of natural astaxanthin is not believed to be a problem. The toxicity results on *H. pluralis* were negative. A Japanese 13 week's oral repeated dose rat study found *H. pluvalis* administration corresponding to 0.5, 1.5 and 5% astaxanthin, which showed no toxicological changes.

The demand for natural astaxanthin is now emerging in the multi-billion dollar nutraceutical market, and increasingly, medical researchers believe that astaxanthin may have significant pharmaceutical applications, while only a negligible part of today's market, the demand for such applications is expected to grow significantly in the near term as a result of numerous medical studies performed during the last 5 years in the area of astaxanthin applications. More and more research supports the conviction that a daily dose of ~5 mg of astaxanthin is of tremendous importance for health management. Thus, natural astaxanthin has fast become a widely preferred product in varied applications from health supplements to beverages and cosmetics and it may find wider applications in days to come

Spirulina

Spirulina is believed to be the first form of plant life on earth, nearly 3.5 billion years old. Through history, this remarkable aquatic plant has nourished people in Africa and America. Many essential nutrients recommended by experts to help protect our bodies are concentrated in Spirulina. It contains the most powerful combination of nutrients ever known in any grain, herb, or food.

Since its rediscovery in the 1960's, scientists around the world have extensively tested spirulina. Spirulina is single celled blue-green algae and is about 0.5 millimeter in length. It possesses a wonderful ability to thrive in situation much too harsh for other algae. Its adaptation to heat also assures that spirulina retains its nutritional value when subject to high temperature during processing and shelf storage, unlike many plant foodstuffs that quickly deteriorate at high temperature. This ability of spirulina to grow in hot and alkaline environments ensures its hygienic status, as no other organisms.

Researchers at NASA found it to be an excellent, compact space food for astronauts (1 kilogram of spirulina is equivalent to 1000 kg of assorted vegetables). International Health Organizations have hailed spirulina as one of the *greatest super foods on earth.* World Population Congress and World Food Conference have declared it as *best food for tomorrow.* The World Health Organization has found spirulina to be an excellent food for human consumption.

Super Nutrient

Today, health conscious people all over the world consume spirulina. An ideal food supplement, spirulina provides the body with entire essential nutrients required daily. For those moving away from processed food and chemical vitamins to natural foods, spirulina- nature's longest living food, will be a boon. People use spirulina in their own self-care strategy for more energy, nutritional insurance, weight control and cleansing. Used in between or before meals, it nurtures without the heaviness of bulky food. Athletes and joggers discover more endurance and strength. Seniors find better nutrients absorption. It is ideal and safe for children, pregnant and nursing mothers. In Japan and United States, business executives take spirulina tablets to combat stress. In past 20 years, it has been marketed and consumed as safe human food and has been approved as a food for human consumption by many governments, health agencies and associations of over 70 countries. Based on 30 years of safety and quality research, many countries and organizations have established Spirulina Quality and Safety Standards. In fact, spirulina is one of the cleanest, mainly naturally sterile foods found in nature.

Spirulina has many credits to its name- World's richest natural source of protein, minerals particularly iron, vitamins and world's richest natural source of pro-vitamin A, best natural source of anti oxidants, world's richest natural source of rare essential fatty acid GLA (gamma-linolenic acid), only vegetable source of vitamin B_{12}, *etc.* It has all of the essential amino acids in balance, which means it is a complete protein. It has more protein than the beef does. Spirulina has a protein content of 65%, while beef is having only 22% protein. Even better is that it is 90% digestible, which means that it is an easy complex protein to digest.

Scientists are discovering the benefits of polysaccharides, sulfolipids and glycolipids and the rainbow of natural pigments that give spirulina a deep green colour. Its deep green colour comes from its rainbow of natural pigments- chlorophyll (green), phycocyanin (blue) and carotenoids (orange) - that harvest the sun's energy. Chlorophyll is a natural cleanser and is often referred to as nature's green magic. Chlorophyll is anti-cancerous antioxidant and immune enhansing. Ten grams of spirulina contains 115 mg of chlorophyll. Spirulina beta-carotene is ten times more concentrated than carrots. Spirulina is the highest protein food with all the essential amino acids and has energy only a few calories. Protein content is 3 times that of meat. It is 60% easy to digest vegetable protein without the fat and cholesterol of meat.

A rare essential fatty acid is a key to health. Gamma-linolenic acid in mother's milk helps in developing healthy babies. Studies show nutritional deficiencies can block GLA production in your body, so a good dietary source of GLA can be important. Spirulina is the highest whole food with GLA after milk and evening prime rose oil. Iron is essential to build a strong system yet is the most common mineral deficiency. Spirulina is rich in iron, magnesium and trace minerals, and is easier to absorb than iron supplements. Spirulina has 50 times more iron than spinach and 10 times calcium than milk. It is high in vitamin B_{12} and B complex. Spirulina is the highest source of B_{12}, essential for healthy nerves and tissue, especially for vegetarians.

Spirulina does not have hard cellulose membranes similar to plant cells or a well-defined nucleus, yet its metabolic system is based on photosynthesis, and in fact, it has a photosynthetic conversion rate of 8 to 10% as compared to only 3% in such land-growing plants as soybeans.

Medicine par excellence

Spirulina helps in losing weight, protects eyesight, helps to stay young, reduce cholesterol, benefits the heart, calms the nerves and fights stress. Spirulina builds immunity against infections. Currently, there are 244 patents associated with this wonderful food.

A number of studies show spirulina or its extracts can prevent or inhibit cancers in humans and animals. Scientists at the Osaka Institute of Public Health in Japan have found that Spirulina promoted the activation of natural cancer-fighting substances in the body. The study was reported at the 30th Annual Meeting of the Japanese Society for Immunology held in November 2000. In this clinical study, the volunteers over 40 years of age were given 50 ml of Lina Green 21 (a spirulina extract) and the level of natural cancer-fighting substances were measured in their blood. The results showed that spirulina significantly increased the tumour killing ability of natural killer cells and interferon gamma. The interferon gamma and natural killer cell activity was increased 1-2 weeks after administration of spirulina and the activity continued for a surprising 12-24 weeks even after stopping the administration of spirulina. Many other similar studies by researchers at University of California, Davis and North Carolina State University point to the potential of spirulina for use in immuno therapy. In an earlier study reported at the 59th Annual Meeting of the Japanese Cancer Association (October, 2000), the Osaka group also found that co-administration of spirulina extract with a bacterial cell wall product significantly increased the regression of implanted tumor cells in mice According to the researchers, this new treatment approach holds great promise.

Some familiar forms of cancer are thought to be an outcome of damaged cell DNA running amok, causing uncontrolled cell growth. Cellular biologists have defined a system of special enzymes known as Endonuclease, which fixes the damaged DNA to keep the cells alive and healthy. When radiation or toxins deactivate these enzymes, errors in DNA go unrepaired, and cancer may develop. *In vitro* studies propose that the exclusive polysaccharides of spirulina improve cell-nucleus enzyme activity and DNA repair synthesis. This may be why a number of scientific studies, observing human tobacco users and experimental cancers in animals, report high levels of suppression of a number of important types of cancer. The subjects were fed either whole spirulina or treated with its water.

In April 1996, scientists from the Laboratory of Viral Pathogenesis, Dana-Farber Cancer Institute and Harvard Medical School, Boston, Massachusetts and Earthrise Farms, Calipatria, California, announced on-going research, saying that water extract of spirulina inhibits HIV-1 replication in human derived T-cell lines and in human peripheral blood mononuclear cells. A concentration of 5-10 mcg/ml was found to decrease viral production.

The United States National Cancer Institute has found that the sulfolipids in spirulina were active against the AIDS virus *in vitro*. Small amounts of spirulina reduced the viral replication, while larger amounts actually stopped the replication of the virus. Other tests found that a hot water extract of spirulina was found to inhibit the replication of HIV-1, mumps, herpes simplex-1, measles, human cytomegalovirus and influenza type A. Other *in vitro* studies show that spirulina could inhibit cancers and repair DNA damaged by cancer. Studies in India have shown that 1-gram of Spirulina per day caused a total regression of mouth cancer lesions in 44% of males. It also finds application as adjunct to cancer patients undergoing chemotherapy.

Spirulina has been found to have significant positive effects on people suffering from type 2 diabetes mellitus. It is reported that spirulina has the ability to reduce the fasting blood sugar levels in the body after 6 to 8 weeks of intake.

Spirulina today is most popular as a therapy for weight loss. It is mostly marketed now as the most vitamin rich appetite suppressant. Spirulina is actually one of the non-meat products, which contain the highest protein content, along with all the essential nutrients and amino acids, without the unnecessary carbohydrates or fats. Your waistline will surely be kept where you want it.

Based on the same concept as above, spirulina is also promising in terms of lowering blood cholesterol and triglyceride levels. Also, it is vital to promoting heart health in general. A double-blind placebo controlled study of the elderly in South Korea found that spirulina had a cholesterol-lowering effect resulting in a significant reduction in total cholesterol. The study also found spirulina boosted the immune system and had a favourable effect on antioxidant capacity of those in the study. A small study in Taiwan with 16 students supplementing with spirulina showed it a preventative of the skeletal muscle damage, which led to a longer time the students were able to exercise. The time to exhaustion during the all-out exercise was delayed in the students taking spirulina.

Spirulina is approved in Russia as a *medicine food* for treating radiation sickness. The children of Chernobyl suffer radiation poisoning from intake food grown on radioactive soil. Their bone marrow is injured, rendering them immuno-deficient. Radiation injured bone marrow cannot create normal red or white blood cells. The children are anemic and undergo from horrible allergic reactions. Children fed just five grams of spirulina in tablets form every day make impressive recoveries within 6 weeks. Spirulina does a great job of cleansing the kidneys of toxins of the mercury and toxins from some prescription drugs. Research found that it eliminated seven times the amount of dioxin from the body when compared to a controlled diet. After the Chernobyl accident in the Ukraine, a Russian patent was issued in 1994 for spirulina as a medical food for reducing the allergic reactions to radiation sickness.

Spirulina also provides a glowing skin, improves hemoglobin level in blood, fights fatigue and stress, helps in lowering blood pressure and correcting water imbalance, helps to relieve Chronic Fatigue Syndrome (CFS) and keeps you active and energetic.

It is good to overcome vitamin A deficiency and malnutrition, glaucoma and myopia. In formulations with other natural products, it is used as a general health supplement, liver corrective for liver disorder, burns therapy, skin grafting, control obesity and lactating agent for mothers. Spirulina is used in pimple lotions, facial masks, hair oil, shampoo, mineral bath, skin cleaner, *etc*. It has applications in pisciculture and sericulture as feed and in animal husbandry as calf-starter feed. This wonder gift of nature will find more applications in days to come.

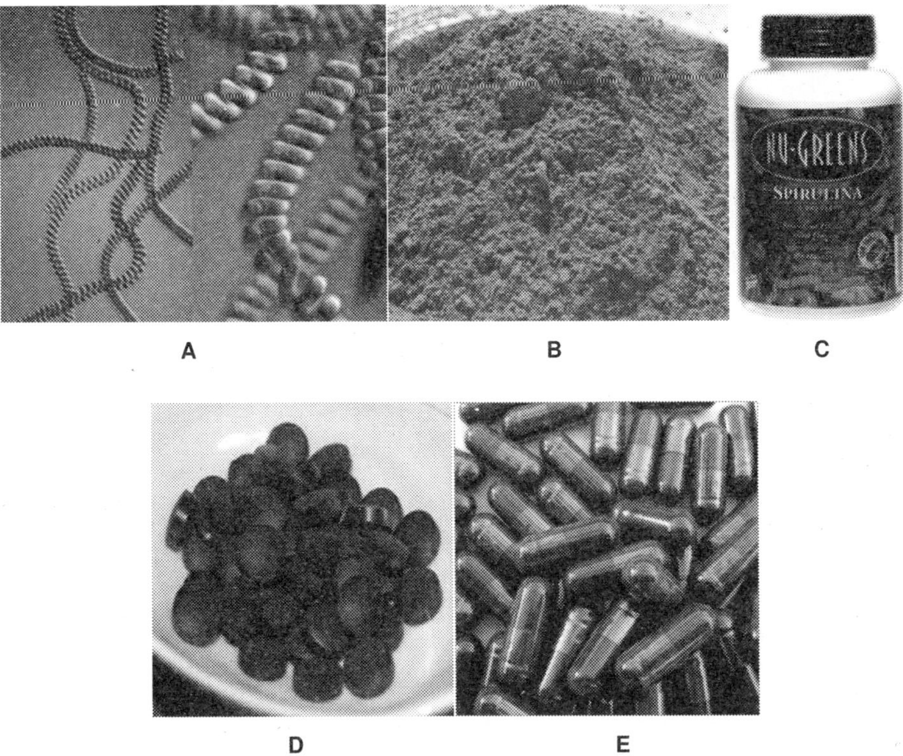

Figure 1 : A. Spirulina, B. Spirulina powder, C. Spirulina tablets, D. Spirulina tablet and E. Spirulina Capsule

Chlorella

Chlorella (*Chlorella vulgaris)* is single-celled algae produced as a source of protein for populations that could not afford protein of animal food. However, Chlorella offers not just protein but other nutrients also like high levels of chlorophyll, iron, vitamins C and B_{12} and beta-carotene. 60% of Chlorella is protein, which is in the form of amino acids. Chlorella contains 18 of 22 known amino acids and is considered a complete protein, meaning that it has the essential 8 amino acids. It is a far superior protein source than meat and fish. There is no other natural source of protein available that provides such a complete and balanced array of amino acids than Chlorella and Spirulina.

Amazing stories have surfaced of people getting noticeable improvement in health and well-being when consuming chlorella. Chlorella contains carotenoids, magnesium, and the super detoxifier chlorophyll, the health benefits of which have been well documented. Chlorella may be the perfect antidote to our refined foods, nutrients-poor diets and toxic environment. This alga is processed to make its tablets, powder and liquid extract. A person taking chlorella 5-7 grams per day will notice significant changes in digestion, energy and overall health.

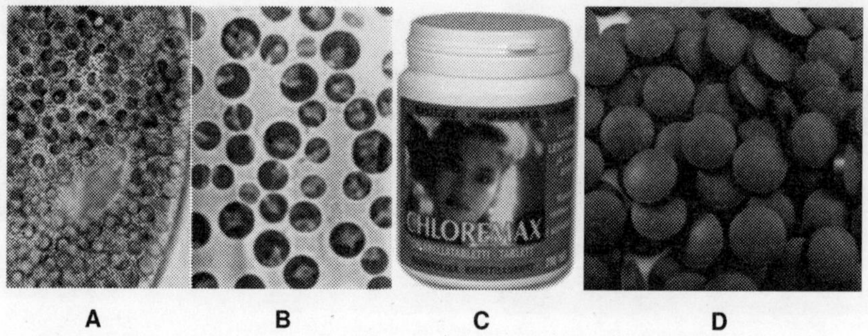

Figure 2 : A. Chlorella, B. Chlorella Cells and C & D. Chlorella Tablet

Highest percentage of chlorophyll

Chlorella contains high levels of chlorophyll, protein, iron, vitamins C and B_{12}, β-carotene and 19 amino acids. It is one of the most widely used food supplements in some Asian countries like Japan. One of the greatest food substances for cleansing the bowel and other elimination systems, the liver and the blood, is chlorophyll as found in all green vegetables, especially the leafy green vegetables. Chlorella contains the highest percentage of chlorophyll (1.7 to 7%) in the known plant world. It has over five times more chlorophyll than wheat grass and over ten times more than barley grass. The chlorophyll molecule is very similar to the hemoglobin molecule in blood, and it acts as a wonderful cleanser in the bowel, kidneys, liver and bloodstream. Chlorophyll or its derivatives stimulate the formation of erythrocytes in the blood and affect nutrition, synthesis of vitamins in plants, hormone action, tumors and number of diseases like anemia, cardiac hypertension, arteriosclerosis and others.

Much research, which has proven that chlorophyll or its derivatives influence bacterial and animal growth, metabolism and respiration, has been done. Since the chemical structure of chlorophyll is similar to the structure of hemoglobin, several scientists have suggested the use of chlorophyll as a medical therapy for anemia. Scientists of Liverpool University performed several studies on laboratory animals. In one study, the induced anemic rabbits given varying doses of chlorophyll (refined or fresh) were able to convert chlorophyll into hemoglobin, thus, correcting the anemia. Although the reaction occurred with both types of chlorophyll but the percentage of conversion was higher when the fresh chlorophyll was used.

Chlorella Growth Factor (CGF) is an amalgam of RNA, DNA, amino acids, vitamins and minerals whose exact properties have not yet been identified. In children, it promotes rapid growth, and in adults, it builds their immune systems while promoting rapid tissue healing, both internally and externally. It also strengthens ability of body to recover quickly from strenuous exercise and disease.

For detoxification of body

A clean bloodstream, with an abundance of red blood cells to carry oxygen, is necessary to a strong natural defense system. The cleansing action of chlorella on bowel and other elimination channels as well as its action on protection of the liver helps in keeping the blood clean. Clean blood insures that metabolic wastes are efficiently carried away from the tissues.

Japan, with the earliest and most catastrophic experience with nuclear byproducts and growing problem of industrial pollution, as exemplified in the outbreak of mercury-caused Minimata disease, has a special concern with environmental contamination and paid immediate attention to early medical journal reports of Chlorella's effect on pollutants. Chlorella is found to be effective in removing heavy metals like lead, mercury, *etc.* from the body. The detoxification capability of Chlorella is due to its unique cell wall and the material associated with it. The cell walls of Chlorella have been shown to have three layers of which the thicker middle layer contains cellulose microfibrils.

Chlorella is said to have healing powers and is used for reducing the problems caused by radiation exposure. It is also believed to be beneficial in protecting the body from ultraviolet radiation. A test in the United States involved the administration of the poisonous hydrocarbon chlordecone to animals, which were then fed Chlorella. The Chlorella interrupted the recirculation of the chlordecone and eliminated it from the animal's bodies. The researchers concluded that Chlorella was potentially effective as a detoxifier for not only chlordecone but for compounds with similar properties, such as dioxin and PCBs. As early as 1950, a paper in *Experientia* reported that guinea pigs on a diet rich in chlorophyll showed increased resistance to lethal X-rays. An experiment by the United States Army showed that the animals fed with chlorophyll-rich greens survived twice as long as the other animals when all were exposed to fatal levels of radiation.

A Japanese study of heavy metal poisoning with cadmium revealed that when 8 grams of Chlorella were administered to test animals daily, cadmium excretion increased threefold in the stool and sevenfold in the urine. According to Dr. Liang-Pin Lin of National Taiwan University, Chlorella absorbs heavy metals from water. Therefore, if Chlorella is used as a nutritional supplement, it will remove heavy metals such as lead, mercury and cadmium from the body. Chlorella strongly binds the cadmium, and thus, will not give it to the body.

Dr. T. Nagano at Shizuoka College of Pharmacy in Japan did a study in which rats were given Chlorella that contained cadmium to determine whether the cadmium would be absorbed from the Chlorella into the rats. In rats given only cadmium (without Chlorella), growth retardation was noted, while no problem with growth was seen in those given Chlorella containing cadmium. Blood levels of cadmium were determined and demonstrated that the cadmium that was bound to the Chlorella was not absorbed into the rat's bodies.

Chlorella has been used to detoxify people suffering from polychlorobiphenyl (P.C.B.) exposure. Dr. Ueda of the Kitakyushu City Institute for Environmental Pollution Research gave 30 patients who suffered from P.C.B. exposure daily doses of 4-6 grams of Chlorella for one year. Almost all improved being less tired, with better digestion and normal bowel movements. Chlordecone (kepone), another very harmful chlorinated hydrocarbon insecticide, was removed from the body more than twice as fast when Chlorella is taken by mouth.

Dr. S. Pore of the School of Medicine, West Virginia University did a study in which Chlorella given to rats speeded up the detoxification of this toxin, decreasing the half-life of the toxin from 40 days to 19 days.

Dr. D. Lorch and A. Weber reported that Chlorella walls absorb and hang onto lead. In 1979, a study carried out by Dr. Horikoshi showed that the cell components extracted from Chlorella even bind uranium. Chlorella is alkaline and helps balance the pH of the body.

Another experiment by scientists at National Taiwan University showed that Chlorella helped in protecting the rats from liver damage when they were fed a toxic chemical named ethionine. High chlorophyll content in chlorella is beneficial in treating bad odor, bad breath and foul smelling stools.

Preventive food

In traditional Japanese medicine, it is used for treating ulcers, hypertension, asthma, depression, diabetes, constipation, *etc*. The algae, which are a popular food supplement in Asia and has been used as energy-producing food for centuries, is often used to prevent or curb the spread of cancer, enhance immunity, promote a good balance of bacteria in the gut and lower down the blood cholesterol. Studies have shown that chlorella normalizes blood sugar in cases of hypoglycemia.

High blood pressure is one of the major risk factors in heart attack and stroke, which account for more fatalities. Laboratory experiments have shown that regular use of chlorella reduces high blood pressure and prevents strokes in rats.

Chlorella is now used as an adjunct supplement during radiation treatment for cancer. Research conducted in Japan suggests that chlorella may have antitumour activity against breast cancer. However, its main use in cancer therapy is to help in removing radioactive particles from the body after radiation treatment. Studies in mice have shown that Chlorella can protect against the development and spread of cancer, and other rodent studies have shown that it helps organisms get rid of toxic chemicals, such as dioxins. It also helps in carrying oxygen around the body and to the brain, that is why, chlorella is often called a *brain food*.

Good for Skin and Teeth

In 1941, Dr. S.L. Goldberg reported on treatment of 300 patients suffering from pyorrhea (bleeding gums and loosening teeth), noting significant improvement. Vincent's stomatitis is gum infection that occurs most often in persons suffering from stress with an underlying cause of a severe vitamin C deficiency. It was also treated by the doctor with chlorophyll. Goldberg noted that the use of chlorophyll resulted in a tightening of the teeth, the cessation of bleeding from the gums and the growth of new tissue to replace damaged tissue.

Hair loss can be prevented and reversed by taking chlorella. Weight loss is also possible with this alga. It is also used in the treatment of some skin conditions like skin ulcers. Thus, chlorella surely is a super food and a super-cleansing agent.

■■■

22 Mushroom–The Healthy and Healing Food

Moni Gupta[1], Sachin Gupta[2], Harsimrat K. Bons[3] and Neha Bhat[2]

[1]Division of Biochemistry and Plant Physiology, - Chatha, Sher-e-Kashmir University of Agricultural Sciences and Technology, Jammu

E-mail: <moniguptaskuast@rediffmail.com>

[2]Division of Plant Pathology, Faculty of Agriculture, Chatha, Sher-e-Kashmir University of Agricultural Sciences and Technology, Jammu

E-mail: <sachinmoni@rediffmail.com>, <nehabhat129@gmail.com>

[3]Krishi Vigyan Kendra, Samrala, Punjab Agricultural University, Ludhiana, Punjab

E-mail: <harsimratpau@gmail.com>

Mushroom is a macrofungus with the distinctive fruiting body, which can be either epigeous (above ground) or hypogeous (underground) and the large enough to be seen with naked eyes and to be picked by hand. The structure that we call a mushroom is in reality only the fruiting body on the fungus. Of the 14,000 known mushroom species, only 3,000 are regarded as prime edible mushrooms, of them, 200 are experimentally grown, 100 economically cultivated, 60 commercially cultivated and about 10 species have reached an industrial scale of production in many countries. Further more about, 2000 species are medicinal mushrooms along with a variety of health attributes. In the rural areas of developing world, the problems of malnutrition and poverty are alarming. Mushroom provides an ideal opportunity for elimination of protein malnutrition of people having cereals as staple food. Mushrooms supplement and complement the nutritional deficiencies, and are regarded as the highest producers of protein per unit area and time. Food and Agriculture Organization (FAO) has recommended mushrooms as a food item contributing significantly to the protein nutrition of the developing countries like India, which depend heavily on the cereal diets, however, the per capita consumption of mushroom in India is dismal 20-25 g as compared to 2-3 kg in United States of America and Europe.

Historically, most medicinal mushroom species were relatively scarce and were collected from the forests where they grew on dead or living trees and forest litter. They are predominantly lignocelluloses degraders. For medicinal purposes, they were almost always prepared either as hot water extracts, concentrates or in powdered form. Now a day, almost all of the important medicinal mushrooms have been subjected to large-scale artificial cultivation by solid substrate or low moisture fermentation, thus, removing the historical scarcity factor and allowing large commercial operations to develop. Mushroom cultivation is the only large-scale biotechnological process that creatively utilizes lignocelluloses.

The desirability of a food product does not necessarily bear any correlation to its nutritional value. Instead, its appearance, taste and aroma, sometimes, it can stimulate one's appetite (preference). In addition to nutritional value, mushrooms have some unique colour, taste, aroma and texture characteristics, which attract their consumption by humans. A large number of mushroom species are not only edible and nutritious but also possess tonic and medicinal qualities.

The public has become increasingly concerned about health and nutrition matters in recent years, and this has sparked the commercialization of natural food items consumed as dietary supplements. Mushrooms can be considered a functional food (medicinal or nutritional food) in this way. Such functional or medicinal food articles should not claim to cure disease but there are an increasing number of scientific studies that strongly support some functional food items, such as mushrooms as having a role in disease prevention and in some cases of bringing about a suppression or remission of a diseased state.

Supplementation of mushroom recipe in Indian diet will bridge protein gap and improve the general health of socio-economically backward communities. Earlier, mushrooms were considered as an expensive vegetable and were preferred by affluent peoples for culinary purposes. Currently, common populace also considers mushroom as a quality food due to its health benefits, and therefore, mushrooms constitute a most rapidly growing new food category to which the current health-oriented public is increasingly enjoying. The increasing awareness about high nutritional value accompanied by medicinal properties means that mushrooms are going to be important food item in coming days and at places may emerge as a substitute to non-vegetarian foodstuffs.

Nutritional Attributes of Mushrooms

Protein and Essential Amino Acids: Mushroom is considered a complete health food and suitable for all age groups, child to aged people. The greatest difficulty in feeding man is to supply a sufficient quantity of bodybuilding material, *i.e.*, protein. In terms of the amount of crude protein, mushrooms rank below animal meats but well above most other foodstuffs, including milk, which is an animal product. They are rich in proteins as compared to cereals, pulses, fruits and vegetables on dry weight basis.

Edible mushrooms provide high quality of protein, which can be produced with greater biological efficiency than animal protein. The crude protein content of mushrooms varies from 12 to 35% depending upon the species. Quantitative data related to the nutritive value of mushrooms are sparse. In the absence of feeding trials, alternative methods have been used to determine or predict the nutritional value of food items based on their content of essential amino acids. The proteins of commonly cultivated mushrooms contain all nine amino acids essential for humans as given in Table 1.

Table 1 : Essential amino acid composition of edible mushrooms

Amino Acids	Grams of amino acids per 100 grams of corrected crude protein							
	Agaricus bisporus	*Lentinula edodes*	*Pleurotus florida*	*Pleurotus ostreatus*	*Pleurotus sajor-caju*	*Volvariella diplasia*	*Volvariella volvacea*	*Hen's egg*[a]
Leucine	7.5	7.9	7.5	6.8	7.0	5.0	4.5	8.8
Isoleucine	4.5	4.9	5.2	4.2	4.4	7.8	3.4	6.6
Valine	2.5	3.7	6.9	5.1	5.3	9.7	5.4	7.3
Tryptophan	2.0	nd	1.1	1.3	1.2	1.5	1.5	1.6
Lysine	9.1	3.9	9.9	4.5	5.7	6.1	7.1	6.4
Threonine	5.5	5.9	6.1	4.6	5.0	6.0	3.5	5.1
Phenyalanine	4.2	5.9	3.5	3.7	5.0	7.0	2.6	5.8
Methionine	0.9	1.9	3.0	1.5	1.8	1.2	1.1	3.1
Histidine	2.7	1.9	2.8	1.7	2.2	4.2	3.8	2.4
Total essential amino acids[b]	38.9	36.0	46.0	33.4	37.6	48.5	32.9	47.1

Note: nd = not determined; [a] For comparison; [b] Excluding arginine and cystine

Sources: Bano, Z. and Rajarathnam, S., 1982. *Pleurotus* mushroom as a nutritious food. In: *Tropical Mushrooms – Biological Nature and Cultivation Methods.* Chinese University Press, Hong Kong, pp. 363-380; Li, G.S.F. and Chang, S.T., 1982. Nutritive value of *Volvariella volvaceae. Tropical Mushrooms-Biological Nature and Cultivation Methods.* Chinese University Press, Hong Kong, pp. 199-219.

For all mushrooms listed in Table 1, the most abundant essential amino acid is lysine. Mushroom protein is distinctly superior to that of vegetable protein on account of its essential amino acids, *i.e.*, alanine, proline, tryptophan, glutamic acid, aspartic acid, *etc.* content. The free amino acids composition differs widely. In general, they are rich in theronine and valine but deficient in sulphur containing amino acids, *i.e.*, ethionine and cysteine. The essential amino acid indices, amino acids scores and nutritional indices for mushrooms and other food articles are shown in Table 2.

Table 2 : Comparison of nutritive value of mushroom with various foods

Essential amino acid indices		Amino acid scores		Nutritional indexes	
100	Pork, chicken and beef	100	Pork	59	Chicken
99	Milk	98	Chicken; beef	43	Beef
98	Mushrooms (high)	91	Milk	35	Pork
91	Potatoes and kidney beans	89	Mushrooms (high)	31	Soybeans
88	Corn	63	Cabbage	28	Mushrooms (high)
86	Cucumbers	59	Potatoes	26	Spinach
79	Peanuts	53	Peanuts	25	Milk

Essential amino acid indices		Amino acid scores		Nutritional indexes	
76	Spinach and soybeans	50	Corn	21	Kidney beans
72	Mushrooms (low)	46	Kidney beans	10	Turnips
69	Turnips	32	Mushrooms (low)	8	Tomatoes
53	Carrots	23	Soybeans	6	Carrots
44	Tomatoes	18	Tomatoes	5	Mushrooms (low)

Note: Ranking based on essential amino acid indexes, amino acid scores, and nutritional indexes as calculated against the FAO reference protein pattern; biological values correlate closely with the EAA indexes. Values for mushrooms represent the mean of the three highest values (high) and three lowest values (low) as determined.

Source: Crisan, E.V. and Sands, A., 1978. Nutritional value. *The Biology and Cultivation of Edible Mushrooms*. Academic Press, New York, USA.

In terms of essential amino acid indices and amino acid scores, the most nutritive mushrooms (highest values) rank alongside meat and milk in potential nutritive value and considerably higher than most legumes and vegetables. The least nutritive mushrooms rank appreciably lower but are still comparable to some common vegetables. The values for protein content of 4 popular edible mushrooms, *i.e.*, *Agaricus bisporus*, *Lentinula edodes*, *Pluerotus sajor-caju* and *Volvariella volvacea*, which are commercially cultivated in various countries, range from 1.75 to 3.63% of their fresh weight. The value can be as high as 5.9%. However, an average value of 3.5 to 4% would appear to be more representative. On a dry weight basis, mushrooms normally contain 19 to 35% protein (Table 3).

Table 3 : Proximate composition of cultivated species of edible mushroom

Species	Moisture	Crude protein (N× 4.38)	Crude Fat	Carbohydrates		Crude Fiber	Ash	Energy Value
				Total	N-free			
Agaricus bisporus	78.3-90.5	23.9-34.8	1.7-8.0	51.3-62.5	44.0-53.5	8.0-10.4	7.7-12.0	328-368
Agaricus campestris	89.7	33.2	1.9	56.9	48.8	8.1	8.0	354
Auricularia sp.	89.1	4.2	8.3	82.8	63.0	19.8	4.7	351
Boletus edulis	87.3	29.7	3.1	59.7	51.7	8.0	7.5	362
Flammulina velutipes	89.2	17.6	1.9	73.1	69.4	3.7	7.4	378
Lentinula edodes	90.0-91.8	13.4-17.5	4.9-8.0	67.5-78.0	59.5-70.7	7.3-8.0	3.7-7.0	387-392
Pleurotus	92.2	25.0	1.1	59.2	–	12.0	9.1	261

Species	Moisture	Crude protein (N× 4.38)	Crude Fat	Carbohydrates		Crude Fiber	Ash	Energy Value
				Total	N-free			
eous								
Pleurotus florida	91.5	27.0	1.6	58.0	–	11.5	9.3	265
Pleurotus ostreatus	73.7-90.8	10.5-30.4	1.6-2.2	57.6-81.8	48.9-74.3	7.5-8.7	6.1-9.8	345-367
Pleurotus sajor-caju	90.1	26.6	2.0	50.7	–	13.3	6.5	300
Volvariella diplasia	90.4	28.5	2.6	57.4	40.0	17.4	11.5	304
Volvariella volvacea	89.1	25.9	2.4	–	45.3	9.3	8.8	276

Therefore, in amount of crude protein, mushrooms rank below most animal meats but well above most other food items, including milk, which is an animal product. The digestibility of mushroom protein is as high as 72 to 83% in comparison to proteins present in other foodstuffs and mushroom proteins are as good as animal proteins, thus, mushrooms are rightly called *vegetable meat.*

Fat

Mushrooms contain a high proportion of unsaturated fat and have low crude fat content, with a high proportion of polyunsaturated fatty acids (72 to 85%) relative to total content of fatty acids, which is a significant factor in regarding mushrooms as a health food. The fat content in different species of mushrooms ranges from 1.1 to 8.3% on a dry weight basis, with an average content of 4.0% (Table 3). In general, the crude fat of mushrooms has representatives of all classes of lipid compounds including free fatty acids, monoglycerides, diglycerides, triglycerides, sterols, sterol esters and phospholipids. Mushrooms do not have cholesterol. Instead, they have ergosterol that acts as a precursor for vitamin D synthesis in human body.

Vitamins

Edible mushrooms are a good source of several vitamins including thiamine (vitamin B_1), riboflavin (vitamin B_2), niacin (vitamin B_3), biotin (vitamin B_4) and ascorbic acid (vitamin C). The thiamine content ranges from 0.35 mg in *Volveriella volvacea* to 1.14 mg in *Agaricus bisporus*, 1.16 to 4.80 mg in *Pleurotus* spp. and 7.8 mg in *Lentinula edodes*. The niacin content varies from 54.9 mg in *Lentinula edodes*, 55.7 mg in *Agaricus bisporus*, 64.88 mg in *Volveriella volvacea* and 46.0 to 108.7 mg in *Pleurotus* spp. The riboflavin content was higher in *Agaricus bisporus* (5.0 mg) and *Lentinula edodes* (4.9 mg) than in *Volveriella volvacea* (1.63 to 2.98 mg). Folic acid (vitamin B_9) and cyanocobalamine (vitamin B_{12}), which are normally absent in vegetable food articles are present in mushrooms.

Carbohydrates and Fiber

The total carbohydrates content in mushroom vary from 26 to 82% on dry weight basis in different mushrooms. The digestible carbohydrates profile of mushroom includes starches, pentoses, methylpentoses, hexoses, disaccharides, amino sugars, sugar alcohols and sugar acids. *Pleurotus* species contain carbohydrates ranging from 46.6 to 81.8% as compared to 60% in *Agaricus bisporus* on a dry weight basis. Mushrooms contain high fiber content and they are useful food for diabetic patients with low carbohydrates and zero fat content. Mushrooms comprise about 80 to 90% of water and 8 to 10% of fiber. The crude fiber composition of the mushroom consists of partially digestible polysaccharides and chitin. The fiber content ranges from 7.4 to 27.6% in *Pleurotus* species as compared to 10.4% in *Agaricus bisporus* and to 4 to 20% in *Volveriella volvacea.*

Minerals

Mushrooms are a good source of minerals. The minerals present in the substrate are taken up by the growing fungal mycelium of mushrooms and are translocated to the sporophores (fruiting bodies). As in higher plants, the mineral of highest content is potassium, followed by phosphorus, sodium, calcium and magnesium. These are considered the major mineral constituents and copper, zinc, iron, manganese, molybdenum and cadmium make up the minor mineral elements. Essential minerals like potassium, phosphorous, calcium, sodium along with iron, copper, *etc.* are essential for the formation of bones and teeth of human beings. Copper is a mineral that has cardio protective properties, while potassium helps to lower elevated blood pressure and reduces the risk of stroke. Besides this, mushrooms are a rich source of selenium, which is an antioxidant that works with vitamin E to reduce the risk of prostate cancer. It is calculated that the concentration of potassium, phosphorus, sodium, calcium and magnesium constitute about 56 to 70% of the total ash content. Potassium is particularly abundant and accounts for nearly 45% of the total ash content. Sodium and calcium are present in approximately equal concentrations in all the mushrooms except for *Lentinula edodes* in which calcium is present especially in large amount. Since mushrooms possess low calorific value, high protein, high fiber content, high potassium to sodium ratio and low fat (rich in linoleic acid and devoid of cholesterol) make mushroom the choice of dietician for those suffering with or prone to heart diseases, diabetes, obesity, hypertension, *etc.* These properties are major contributing factors to the traditional recognition of mushroom as *healthy food.*

Nucleic Acids

The practical safe limit of nucleic acid intake of most adult populations suggested by the Protein Advisory Group of the United Nations System is a daily maximum of 4 g and only half the amount should be from a microbial source. This is because the humans lack the urate oxidase enzyme for the oxidation of uric acid, which is the sparingly soluble metabolic product of the purine bases guanine and adenine. As given in Table 4, *Pleurotus sajor-caju* contains the highest amount of nucleic acids (4.06% on a dry weight basis) among the four edible mushrooms studied. This is equivalent to 0.51% on a wet weight basis. Accordingly, it is safe to consume as much as 392.5 g of fresh *Pleurotus sajor-caju* daily. The limit could be even higher in the case of the other mushrooms with lower nucleic acid content, or around 20%

more after being cooked. Therefore, the content of nucleic acids in edible mushrooms should not be a limit to their use as a daily vegetable.

Table 4 : Nucleic acid content of four edible mushrooms

Mushrooms	Percent Moisture	Percent nucleic acid content (dry weight basis)		
		DNA	RNA	Total
Agaricus bisporus	89.07	0.17	2.49	2.66
Pleurotus cystidiosus	92.63	0.37	2.56	2.93
Pleurotus sajor-caju	87.45	0.21	3.85	4.06
Volveriella volvacea	89.21	0.29	3.59	3.88

Source: Li, G.S.F. and Chang, S.T., 1982. The nucleic acid content of some edible mushrooms. *European Journal of Applied Microbiology and Biotechnology*, 15: 237-240.

Medicinal Value of Mushrooms

Since thousands of years, mushrooms have been considered to have medicinal value and extensively used in folk medicine. The early herbalists were more interested in the medicinal properties of mushrooms than in their basic value as a source of food. Of the 14,000 to 15,000 species of mushrooms in the world, around 700 species have known medicinal properties. However, it has been estimated that there are about 1800 species of mushrooms those have potential medicinal attributes. Thus, mushrooms have vast prospects as source of medicines. Studies suggest that the probiotic quality of mushrooms helps to keep our body healthy and ward off diseases.

Most of the knowledge about the medicinal properties of mushrooms comes from the literature from the Far East, where mushrooms such as *Ganoderma lucidium*, *Lentinula edodes*, *Coriolus versicolor* and *Tremella fuciformis* have been collected, cultivated and used for the thousands of years. The recent upsurge of interest in traditional remedies for various physiological disorders and the recognition of numerous biological response modifiers in mushrooms have led to the coining of the term *mushroom nutriceuticals*. A mushroom nutriceutical is a refined and partially defined extract from either the mycelium or the fruiting body of a mushroom, which is consumed in the form of capsules or tablets as a dietary supplement (not in the form of food) and which has potential therapeutic application. In 1994 worldwide sales of medicinal mushrooms, mushroom extracts and various derived products were estimated at US $3.8 billion. With the renewed interest in traditional medicine, the nutraceutical aspects of mushrooms are being valued more and the exploitation of this potential is limitless. Specific biochemical compounds in mushrooms are responsible for improving human health in many ways. These bioactive compounds include polysaccharides, tri-terpenoids, low molecular weight proteins, glycoproteins and immunomodulating compounds. Hence, mushrooms have been shown to promote immune function, boost health, lower the risk of cancer; inhibit tumour growth, help in balancing blood sugar, ward off viruses, bacteria and fungi, reduce inflammation and support the body's detoxification mechanisms. Increasing recognition of mushrooms in complementing conventional medicines is also well known for fighting many diseases. Many traditionally used mushrooms belonging to genera *Auricularia, Flammulina, Ganoderma, Grifola, Lentinus, Trametes* (*Coriolus*) and *Tremella* have been demonstrated to possess significant medicinal properties. Mushroom also contains natural antibiotics, which inhibit microbial and other fungal infections. It also helps in healing ulcers and wounds caused by ulcers.

Many bioactive substances with immune-modulating effects have been isolated recently from mushrooms. These include polysaccharides, high-molecular weight polysaccharides and low molecular weight protein bound polysachharides, glycoproteins (lectins), triterpenoids and fungal immunomodulatory proteins. Though bioactive polysaccharides are widespread among mushrooms, different species can produce polysaccharides with different types of polysaccharide having different properties. Biologically active polysaccharides are best known mushroom derived substances, which are particularly effective in retarding the progress of various cancers and other diseases through immune stimulation rather than direct cytocidal effects and in alleviating the side effects of chemotherapy and radiation treatment through cell level regenerative effects.

Many, if not all mushrooms contain biologically active polysaccharides, which have antitumour and immune-stimulating properties. Three different anticancer drugs extracted from mushrooms have been approved by the Japanese Health and Welfare Ministry. They are Lentinan derived from *Lentinula edodes*, Schizophyllan derived from Suchirotake and polysaccharide kureha (PSK), also known as krestin from Kawaratake. Polysaccharide kureha sold in Europe and Japan is the best selling cancer drug in the world. In Japan, only 30% of the cancer treatment includes radiation, chemotherapy and surgery. There are only two common medicinal mushrooms, *Ganoderma lucidium* and *Agaricus blazei*, which have been demonstrated to have bioactive polysaccharides derived from these three sources, *i.e.*, fruiting body, mycelium and cultural filtrate (Table 5).

Table 5 : Bioactive polysaccharides isolated from mushrooms having antitumour and anticancer effects

Species	Tumor or Cancer Treated	Polysaccharide origin
Coriolous versicolor	Cancer of digestive organs, lung and breast	Culture Mycelium
Lentinula edodes	Cancer of stomach	Fruit body
Schizophyllum commune	Cervical cancer	Culture broth
Ganoderma lucidium	Antitumour	Fruit body and culture mycelium
Agaricus blazei	Antitumour	Fruit body and culture mycelium
Grifola frondosa	Antitumour	Fruit body and culture mycelium

The first three biologically active compounds developed from medicinal mushrooms were polysaccharides and all of them are beta-glucans. These are krestin from *Coriolous versicolor* mycelia containing (1→4), (1→3) or (1→4), (1→6)- β-D-glucan, lentinan, a high molecular weight (1→3)- β-D-glucan from *Lentinula edodes* fruiting bodies and Schizophyllan, a high molecular (1→3), (1→6)- β-D-glucan prepared from *Schizophyllum commune* cultural filtrates. Another popular edible mushroom Flammulina velutipes also has high antitumour activity. Thus, the intake of mushrooms and their extractable bioactive compounds appear to be effective in cancer prevention and growth inhibition. An important fact is the certainty that mushroom extracts, compared with other drugs show a very low toxicity when regularly consumed, even in high dosages. The current intense industrial interest worldwide in medicinal mushroom materials has resulted in a huge increase in a number of products reaching the retail market. However, a major problem associated with mushroom-based dietary supplements is their wide variability and the current lack of standard production and testing protocols necessary to guarantee product quality.

There are a number of direct and indirect benefits provided by medicinal mushrooms. Some of them are described hereunder:

Antitumour effects: The antitumour components of mushrooms vary in their chemical nature and include polysaccharides, proteins, glycoproteins and triterpenoids. Inclusion of cultivated mushrooms like *Lentinula edodes*, *Grifola frondosa*, *Agaricus blazei* and *Pleurotus* spp. in the diet are likely to provide some protection against some diseases, particularly some tumours. The main sources for antitumour polysaccharides are from mushroom cell walls that consist of chitin and cellulose. However, mushroom chitin does not have any antitumour activity. Antitumour polysaccharides found in the fruiting bodies, cultured mycelia and culture broth of mushrooms are described in Table 5.

Compounds restricting tumour activities are found in some mushrooms but only a limited number have undergone clinical trials. All forms of edible mushrooms, and white button mushrooms in particular, can prevent prostate and breast cancer. Fresh mushrooms are capable of arresting the action of 5-α-reductase and aromatase chemicals responsible for growth of cancerous tumours. The drug known as polysaccharide-K (Kresin) is isolated from *Trametes versicolor* (*Coriolus versicolor*), which is used as a leading cancer drug. Some mushroom-derived polysaccharides have ability to reduce the side effects of radiotherapy and chemotherapy too. Such effects have been clinically validated in mushrooms like *Lentinula edodes*, *Tramtes versicolor*, *Agaricus bisporous* and others. Mushrooms are very effective in preventing breast and prostate cancer due to anti-carcinogenic effects of β-glucans and linoleic acid present in mushrooms. The linoleic acid is helpful in suppressing the effects of estrogen, which causes breast cancer in women after menopause. The beta-glucans inhibit the growth of cancerous cells in case of prostate cancer. The presence of higher amount of selenium in mushrooms is very effective in inhibiting cancerous growth causing prostate cancer. Researchers have suggested that white button mushrooms can reduce the risk of breast and prostate cancer.

Antioxidant activity: Many edible mushrooms consist of a valuable source of biologically active compounds with potential for protecting cellular DNA from oxidative damage. Such protective compounds have possible commercial value as dietary supplements for offsetting adverse biological effects associated with coronary heart disease, cancer and age related neurodegenerative diseases. They might also facilitate the development of treatments for the repair of indiscriminate cellular DNA damage that occurs during certain forms of chemotherapy and radiotherapy. The polysaccharides from mushrooms are potent scavengers of super oxide free radicals. These antioxidants prevent the action of free radicals in the body, consequently reducing the ageing process. Mushroom contains a powerful antioxidant named Ergothioneine, found in *Flammulina velutipes* and *Agaricus bisporus*, which are very effective in protecting cells from damaging effects of free radicals and is necessary for healthy eyes, kidney, bone marrow, liver and skin and provides immunity to body.

Strengthens immunity: Medicinal mushrooms such as *Ganoderma lucidium*, *Lentinula edodes* and *Grifola frondosa* enhance immune function by stimulating cell mediated immunity. Such mushrooms seem to turn on cells in the immune system called T-cells that appear to have significant cancer fighting properties. Excellent presence of vitamin A, B-complex and C in mushrooms helps in strengthening of immune system. The products of medicinal mushrooms have been demonstrated to enhance the immune system and promote the natural defense system. They are also good for patients who have received treatments with radiotherapy or chemotherapy as they may help to reduce the side effects from such therapies, which include (i) increasing the number of leucocytes in the blood and enhancing the immune functions (ii) increasing and improving appetite, (iii) reducing pain, (iv) anti-emetic properties, (v) stopping hair loss, (vi) inducing tumour regression, (vii) potential antioxidant and genoprotective properties and (viii) general health improving effects. A regular intake may enhance the immune responses of the human body, thereby, increasing resistance to diseases, and in some cases, cause regression of a disease state. Mushrooms are capable of strengthening the immune system.

A diverse collection of polysaccharides (β-glucans) and minerals isolated from mushroom is responsible for upregulating the immune system. These compounds potentiate the host's innate (non-specific) and acquired (specific) immune responses and activate all kinds of immune cells. These medicinal polysaccharides are primarily modifiers of biological response where these polymers interact with the immune system to up-regulate or down-regulate specific aspects of the response of the host and this may result in various therapeutic effects. Their ability to enhance or suppress immune responses can depend on a number of factors including dosage, route of administration, timing and frequency of administration, mechanism of action or the site of activity. Several of these compounds have been shown to potentiate the host's innate (non-specific) and acquired (specific) immune responses and to activate many kinds of immune cells that are important for the maintenance of homeostasis, *e.g.*, host cells (such as cytotoxic macrophages, monocytes, neutrophils, natural killer cells, dendritic cells) and chemical messengers (cytokines such as interleukins, interferon and colony stimulating factors) that trigger complement and acute phase responses

Antiviral effects: Cochran was the first to report that antiviral substances were present in mushrooms. It has been found that an aqueous extract of *Lentinula edodes* fruiting body as well as the spores contained antiviral activity against A/SW 15 virus infection in mice. Anti-human immunodeficiency virus (HIV) activity, an inhibitory effect on HIV replication *in vitro*, was reported from an extract of the culture medium of *Lentinula edodes* mycelia. Sulfated lentinan completely prevented the HIV induced cytopathic effects. Anti-human immunodeficiency virus activities have been reported in water soluble extracts of *Ganoderma lucidium*. It has been reported that sulfated *Grifola frondosa* extract is able to prevent 97% HIV-infected T-helper lymphocytes from being destroyed *in vitro*.

Cardiovascular and renal effects: As already discussed, mushrooms are a good source of potassium, which reduces the risk of heart attack by lowering blood pressure. The edible mushrooms have little fat with higher proportion of unsaturated fatty acids and absence of cholesterol, and consequently, it is the relevant choice for heart patients and treating cardiovascular diseases. Minimal sodium with rich potassium in mushroom enhances salt balance and maintaining blood circulation in human. Hence, mushrooms are suitable for people suffering from high blood pressure. Regular consumption of mushrooms like *Lentinula edodes* and *Pleurotus* spp. decreases cholesterol levels. Mushrooms also contain insulin and enzymes, which help in breaking down the sugar or starch of the food. The natural antibiotics present in mushrooms can help in protecting diabetic patients from various infections. The lean proteins present in mushrooms help to burn cholesterol in the body. Thus, it is most preferable food for people striving to shed their extra weight.

Hematological effects: Lectins are proteins or glycoproteins with specific binding sites for sugars. They are not antibodies or enzymes but have a specific affinity toward glycosylated materials and they have become useful reagents in studies of cell surface structures. Some lectins have been shown to have antitumor and immunomodulatory activities. The interaction between the fungal lectin and the surface glycoproteins of red blood cells is an example of the hematological activities of edible mushrooms, and some fungal lectins have been characterized and purified from various mushrooms, *viz.* Volvatoxin from *Volvariella volvacea,* Pleurotolysin from *Pleurotus ostreatus* and from *Agaricus campestris* and *Flammunila velutipes.*

Regulates digestive system: The fermentable fiber as well as oligosaccharide from mushrooms acts as a prebiotics in intestine, and therefore, they anchor useful bacteria in the colon. This dietary fibre assists the digestion process and healthy functioning of bowel system.

Types of Medicinal Mushroom

Some traditional important medicinal mushrooms whose medicinal properties are well known are discussed below:

Reishi mushroom (*Ganoderma lucidium*): Reishi mushroom has been considered as a symbol of happy augury, good fortune, good health, longevity and even immortality. It is pharmacologically as well as commercially the most important medicinal mushroom in the world with current global trade of about 2 billion dollars. *Ganoderma lucidum* and related species have the longest historical usage for medicinal properties dating back at least four thousand years. In Japan it is called Reishi and in China and Korea, it is variously called Ling Chu and Ling Zhi (Mushroom of immortality). Reishi meets all the qualifications of being both a tonic and an adaptogen. A tonic strengthens invigorates organs and an adaptogen helps the body adapt to stress by readjusting to altered body conditions. Traditionally, it has been used widely in the treatment of hepatopathy, chronic hepatitis, nephritis, hypertension, arthritis, insomnia, bronchitis, asthma and gastric ulcer. It is used as a daily tonic to maintain and improve good health, increasing longevity, in the treatment of cancer and resistance to and recovery from diseases. Scientific studies have confirmed that the substances extracted from the mushrooms can reduce blood pressure, blood cholesterol and blood sugar level as well as inhibit platelet aggregation.

Ganoderma species are famous tonic in Chinese medicines. They are widely distributed in India on tree trunks. For the 4000 years, *Ganoderma lucidum* has been used as a part of Chinese and Japanese medicine especially for the treatment of most of the human ailments including chronic hepatitis, nephritis, hepatopathy, neurasthenia, arthritis, bronchitis, asthma, gastric ulcer, *etc.* It is known for managing cancer in combination with conventional therapy and also for its anti-HIV effect. It has been highly valued as a cancer cure due to the presence of various polysaccharides, organic germanium, adesnosine, triterpenoids, ganodermic essence and a number of other active elements having specific medicinal properties. It improves mucous flow and reduces coughing, lowers blood pressure and serum cholesterol, protects the liver and protects against ionizing radiation. It has been effective in the treatment of hypertension, diabetes, myotonia dystrophica, arthritis, dizziness nervous problems and many other disorders. It has also been reported to have curative effect on health conditions like neurasthenia, insomnia, chronic hepatitis, kidney disease, coronary heart disease, chronic bronchitis, bronchial asthma and gastric and duodenal ulcers.

Shiitake mushroom (*Lentinula edodes*): It is a mushroom with high nutritive value and used as a medicine for a remarkable array of health problems. It has an anti-tumour, antiviral and cholesterol lowering properties. Lentinan found in Shiitake has been licensed as an anti-cancer drug by the Japanese Food and Drug Administration. Lentinan has shown some effect on bowel cancer, liver cancer, stomach cancer, ovarian cancer and lung cancer and stimulates the production of T-lymphocytes and natural killer cells and can potentiate the effect of azidothymidine (AZT) in the anti-viral treatment of acquired immunodeficiency syndrome (AIDS). It is a food that activates the blood and is indicated in traditional medical treatments of colds, measles in children, bronchial inflammation stomachache, headache, dizziness, dropsy (fluid accumulation in tissues), small pox and even mushroom poisoning. It is also valuable in stimulating the immune system to increase the body ability to ward off cancerous tumours, viral infections and chronic syndrome. Shiitake mushroom helps in preventing high blood pressure, diabetes, atherosclerosis, kidney ailments, cataract, neuralgia, gallstones, numbness of the hand and feet, hemorrhoids and also improves sexual powers.

Maitake mushroom (*Grifola frondosa*): Maitake is a mushroom highly regarded in Japan for its medicinal and culinary properties. The studies show that the extracts from Maitake protect against hepatitis, tumours and are effective in regulating blood pressure, cholesterol and obesity. A polysaccharide known as beta-glucan, which activates immune cells, is believed to be largely responsible for Maitake anti-tumour properties. Maitake is rich in minerals such as potassium, calcium and magnesium and various vitamins such as riboflavin, niacin and vitamin D_2, fibers and amino acids. It has been frequently used for improving spleen and stomach ailments, calming nerves and mind and treating hemorrhoids. It may be effective for treating breast and colorectal cancers as well as therapeutic active against acquired immunodeficiency syndrome (AIDS). Researchers have also indicated that whole maitake has the ability to regulate blood pressure, glucose, insulin and both serum and liver lipids, such as cholesterol, triglycerides and phospholipids and may be useful for weight loss. Maitake has a hypoglycemic effect and may be beneficial for the management of diabetes as it contains alpha-glucosidase inhibitor

Caterpillar fungus (*Cordyceps sinensis*): Caterpillar fungus is considered a medicinal mushroom in oriental medicine. In India, it is found in the upper reaches of Uttarakhand forests. Cordycepin is a compound isolated from this mushroom. These are unique fungi that feed on insects and are parasitic on the larva of Lepidoptera. It is often used as a tonic and tranquilizing medicine and for treating general debility after illness, weakness, splitting of blood caused by tuberculosis, chronic coughing and asthma caused by senility, night sweating, spontaneous sweating, anemia and malignant. *Cordyceps* is effective for the treatment of high cholesterol, poor libido/impotence, arrhythmia, lung cancer and chronic kidney failure. *Cordyceps* also causes smooth muscle relaxation and is helpful in treating bronchitis and asthma. Caterpillar fungus might protect the liver from damage, and it has a hypoglycemic effect and may be beneficial for people with insulin resistance.

Mushrooms, akin to plants, have a great potential for the production of quality food. These are the source of bioactive metabolites and are a prolific resource for drugs. Knowledge advancement in biochemistry, biotechnology and molecular biology boosts application of mushrooms in medical sciences. From a holistic consideration, the edible mushrooms and its byproducts may offer highly palatable, nutritious and healthy food besides its pharmacological benefits. Still, there are enough challenges ahead. The protection against cancer afforded by *Ganoderma* products could be attributed to several compounds present in *Ganoderma*, *i.e.*, the tetro-glucans, lectins, terpenoids, steroids, nucleic acids and immune-modulatory proteins.

Until now, how these products works is elusive and vast number of potential wild mushrooms are not explored. Knowledge on dose requirement, route and timing of administration, mechanism of action and site of activity is also lacking. Numerous studies have shown that regular consumption of certain mushroom species either a regular food or as extracted compounds (nutriceuticals) is effective in both preventing and treating specific diseases, mainly through immunopotentation and antioxidant activity.

■■■

23 The Marine Plants and Human Health

Uche Fidelia Ijeoma[1] and Mgbahurike Amaka Augustina[2]

[1]Dept. of Pharmacognosy and Phytotherapy, Faculty of Pharmaceutical Sciences, University of Port Harcourt, Rivers state, Nigeria
E-mail: <uchefideliaijeoma@yahoo.com>
[2]Dept. of Clinical Pharmacy and Management, Faculty of Pharmaceutical Sciences, University of Port Harcourt, Rivers state, Nigeria
E-mail: <amaka_mgbahurike@yahoo.com>

Marine plants or sea vegetables are among the oldest and most powerful nutritious plants on earth. Marine plants, fresh water algae or sea vegetable or seaweed are marine algae, which are abundant all over the world. These macro-algae sometimes have root-like rhizoids, stem like stripes and leaf like structures.

Seaweeds or algae are marine plants that form the basis of a complex food web involving all the organisms living within a marine ecosystem. Seaweeds are eaten by fish, molluscs and crustaceans, which in turn are eaten by other animals, usually larger animals. In addition, seaweeds convert sunlight and dissolved nutrients into energy-rich organic compounds that the other organisms can eat. They also produce oxygen that all animals need to breathe and provide shelter and protection for many organisms.

Seaweeds are referred to as sea plants and as sea vegetables when they are edible. Sea plants are used by humans for food, medicine, body care products, animal feed and fertilizer for agriculture. There are three main groups of sea vegetables, *i.e.*, red, green and brown. The red ones are the trickiest to identify. Growing in a sunny place, they can appear green or yellow, but growing in darker places, they look purple, brown, or even black. Brown sea plants are always brown, greenish brown, or yellowish brown. Green sea plants are always green unless bleached white by the sun.

Seaweeds are divided into a number of groups or phyla, depending on pigmentation linked to a suite of other characters. Green algae (*Phylum chlorophyta*) are characteristically bright green since the chlorophyll pigment is not masked by other secondary pigments. Brown algae (*Phylum phaeophyta*) are characteristically brown in structure due to the presence of quantity of pigments masking green color of the chlorophyll. Red algae (*Phylum rhodophyta*) are characteristically red since the green chlorophyll pigment is masked by a combination of red pigment, phycoerythrin and the blue pigment, phycocyanin. Often under high light, the individuals become bleached to a brown or straw colour. There are also the microalgae, which include yellow green algae (*Phylum xanthophyta*), that often form a velvety covering over large areas of soft sediment and blue green algae (*Phylum cyanophyta*). Diatoms (*Phylum bacillariphyta*) are microscopic and characteristically have silica walled pillbox shaped cells divided into two parts, each halve known as valve.

Sea grasses

These are so called because many have ribbon like leaves containing veins. They are not true grasses but belong to Magnoliophyta. Sea grasses commonly form extensive beds or largely submarine meadows that assist in stabilizing unconsolidated surfaces and are important nursery grounds as well as a major food source for marine animals.

Seaweeds

Seaweeds are not a real weed but they are the large marine algae that grow in the coastal waters of many countries. They include thousands of species from microscopic plants called *phytoplankton* to huge floating/anchored plants commonly seen washed up on seashore.

Seaweed has been part of the traditional diet of all coastal cultures including the people of Japan, Korea, China, Iceland, Sweden, Wales, Scotland, Hawaii and South pacific islands. Many Asian cultures have attributed their long life spans and overall good health to their daily intake of sea vegetables.

A number of researches have shown the effectiveness of sea plants and algae on human and animal health. Marine and soft water algae are shown to have curative powers for tuberculosis, arthritis, cold, influenza and worm infestations. Some sea plants are effective vermifugal agent (kainic acid). It is claimed that the electrolytic magnetic action of the sea plants releases excess body fluid from congested cells and eliminates fatty wastes through the skin and replenishes depleted minerals particularly iodine and phosphorous.

Obesity is rare among *Polynesians* and other races that incorporate seaweed as a regular part of their daily diet.

Sea vegetables have been acknowledged as a detoxifier, balanced nourishment and a miraculous healing plant. Sea algae are believed to be the richest natural source of minerals, trace elements and rare earth elements. It is recommended that ocean or sea algae are the richest natural source of minerals, trace minerals and rare earth elements.

They are protective against radiation and environmental pollutant. They prevent assimilation of different radio nucleotides, heavy metals and other environmental toxins. Sea vegetables are virtually fat free, low calorie and of the richest sources of minerals in the vegetable kingdom, since they have ready access to the abundant mineral found in the ocean.

Sea vegetables contain high amount of calcium and phosphorous, magnesium, iron, iodine and sodium. They also contain vitamin A, B, C and E as well as protein and carbohydrates. When compared to plants that grow on land, sea vegetables are 10 to 20 times higher in vitamins and amino acids. Compared to herbs, sea vegetables are high in vitamins and minerals.

Yamamoto and his co-workers reported that several varieties of *Kombu mojaban*, a sea vegetable eaten in Asia and traditionally used as a decoction for cancer in Chinese herbal medicine, were effective in the treatment of tumour in laboratory animals.

Many types of seaweed are eaten by people because of their high vitamin and iodine content. Asian culture use seaweeds just as Americans use green beans and carrots.

One of the most prominent health benefits of seaweed is its ability to remove radioactive strontium and other heavy metals from the body. Seaweeds also have beneficial effects on animals like dogs, cats and horses.

Plankton

Planktons are tiny open water plants, animals, or bacteria. These organisms range in size from microscope bacteria and plants to large animal such as jellyfish. Generally, planktons have no swimming ability and these are transported through water by current and tide. Based on their size, plankton can be divided into three major groups, *i.e.*, phytoplankton, microscopic plants and bacteria. Zooplanktons are the microscopic animals. Macro-zooplanktons are the larger fish eggs and larvae, and pelagic invertebrates.

Planktons are used as indicators of environmental and aquatic health because of their sensitivity to environmental changes and short life span. Phytoplankton is useful indicators of high nutrient conditions due to their propensity to multiply rapidly in the right conditions. Zooplanktons are useful indicators of future fisheries health because they are food source for organisms at high trophic levels such as finfish. The best growth occurs in so-called *spring bloom* when many species of phytoplankton take advantage of the enhanced conditions provided at that time of the year.

Planktons have an alkaline pH, which is important, given the acidity to our diets high in refined sugars, soda pop and farmed large animals. The high density of nutrients found in algae is extremely important for many reasons. Perhaps, the most important is that these nutrients maintain human cell membranes in structure and function. This is vital for cell detoxification and for overall metabolism of human cells. In fact, the causes of diseases have been simplified to very specific mechanisms, all of which center on cell membrane function and structure. Inflammation, oxidation, toxicity and mitochondrial dysfunction keep the cell membranes from doing their job effectively. Algae contain high levels of antioxidants and anti-inflammatory micronutrients to fuel metabolism and detoxification. Also, they stoke the fires of the mitochondria, where cells make energy required to carry out their function.

Planktons supply about 90% of the oxygen on earth and are the basis of all of the life forms on earth. These microscopic plants from the ocean are the best source of rare minerals and elements, and they provide electrolytes, amino acids, glyconutrients, essential fatty acids omega-3 and omega-6, micro-nutrients, vitamins A, B_1, B_2, B_3, B_5, B_{12}, C, D, K, vitamin E production, potassium and iodine.

Marine Phytoplankton

Phytoplanktons are the plants that live in the ocean. Phytoplanktons are the floating primarily microscopic plant planktons, which obtain energy by photosynthesis. There are many species of phytoplankton, each of which has a characteristic shape. Collectively, phytoplanktons grow abundantly in the oceans around the world and are foundation of the marine food chain. Small fish and some species of whales eat them as food. Larger fish then eat the smaller fish. Humans catch and eat many of these larger fish. Since phytoplanktons depend upon certain conditions for growth, they are a good indicator of change in their environment. For these reasons and because they also exert a global-scale influence on climate, phytoplanktons are of primary interest to oceanographers and Earth scientists around the world. Phyto means plant and plankton means floating/suspended. Scientists at NASA theorize that some three and a half billion years ago, the world was changed forever. The appearance of tiny organisms with the ability to convert sunlight, warmth, water and minerals into protein, carbohydrates, vitamins and amino acids marked the beginning of life. Phytoplanktons, the single-cell plants, are the basis of all other life forms on planet earth, and they are the *vegetation* of the ocean.

Marine phytoplankton contains an abundance of nutrients, including the amino acids, vitamin C and the B group vitamins, which unlike other nutrients must be acquired each day through foods because they are water-soluble, and thus, they are not stored in the body as are the fat-soluble nutrients. Phytoplankton provides the most essential nutrients of any one-food source on earth. It can heal and prevent disease through its nutritive and detoxifying properties.

Commercially grown produce is devoid of many nutrients that are necessary for optimal human health. Phytoplankton is a most complete source of nutrients, which can maintain superior health and can help to bring about healing of serious diseases. Phytoplankton is a health necessity for anyone whose food intake is not primarily organically grown unprocessed raw foods. Even people with the best dietary habits possibly can find great health benefits from phytoplankton.

Phytoplankton can help detoxification, improve natural immunity and eliminate moodiness and even depression. The chlorophyll in sea vegetables also greatly aids assimilation of amino acids. Of the 20 amino acids, the human body can produce only ten and the other ten must be obtained through foods. If even one of the 20 amino acids are lacking, proteins and muscle will be adversely affected in the body's attempt to find the missing amino acid.

Eating huge amounts of all types of common food grown on the soil would still be impossible to find the varied and complex nutrition found in phytoplankton, which is the most complete health food known to humanity and a true super-food. With consumption of mainly processed foods, diseases resulting from malnutrition occur, which could be offset with the use of phytoplankton and addition of some natural foods to the diet.

Most natural raw foods have a high alkalinity and most cooked or highly processed foods make the body pH over-acidic. Phytoplanktons have an alkaline pH, which helps the humans to be healthier in another way because foods with high acidity levels contribute greatly to serious diseases.

Most people living in modern societies have a high acidity level, which greatly predisposes to all types of disease. A pH level from 7 to 14 is alkaline, 7 is neutral and 0-7 is acidic. Body pH levels need to be between 7.35 and 7.45, the perfect human blood pH is 7.4, and at levels above 7.8 or a bit below 6.8, the human body cannot survive.

Glyconutrients

Marine phytoplankton is a rich source of glyconutrients, which play an important role in good cell-to-cell communication. They do this through a system of *messages*, which are well-known molecules like hormones, neurotransmitters, enzymes, *ect*. They form a vast network of communication, which should have never been separated into different components. The Psycho-Neuro-Immune-Endocrine system of cell communication to coordinate our metabolism is now considered to be the cornerstone of health and function in cutting-edge research.

Healing, Detoxification and Phytoplankton

According to Dr. Jerry Tennant, M.D., marine phytoplankton has shown excellent result in healing process. In addition to providing excellent nutrition, phytoplankton is the best food source to provide protection against various heavy metals and other toxic chemicals, which are so prevalent in the environment, foods and water. Marine phytoplankton improves functioning of both cell metabolism and cell detoxification, which are vital to good health.

The ability of phytoplankton to assist with human detoxification and also assure superior nutrition is similar to the role of many naturopathic physicians who use preventative medicine and natural treatments to eliminate diseases. Toxicity and lack of proper nutrition, two of the main causes of diseases in modern society, are both greatly reduced and can be eliminated by the use of dietary phytoplankton.

Natural immunity, which comes from natural diet and lifestyle and elimination of toxins, becomes impaired immunity and diseases due to toxins, stress and insufficient sleep and nutrients. Other sources of health problems include hormones and unclean fodder added to dairy and meat animals, which then get tuberculosis and many other diseases.

Since phytoplankton not only help the humans with detoxification but also contains the most nutrients of anyone single food on earth, and sufficient nutrients and elimination of toxicity are the two primary needs to elimination of diseases, the use of phytoplankton is an invaluable habit for healing and maintaining good health.

Similar to land plants, phytoplankton synthesizes six-carbon compound through photosynthesis, making it available for higher trophic levels. The major environmental factors influencing phytoplankton growth are temperature, light and nutrients availability. Phytoplankton growth is usually limited to the photic zone, or the dept to which sunlight penetrates the water. Major groups of phytoplankton include the followings:

- Diatoms (phylum Bacillariophyta)
- Golden brown algae (Chrysophyta)
- Green algae (Chlorophyta)
- Blue green algae (Cyanophyta)
- Dinoflagellates (Pyrrophycophyta)
- Cryptomonads (Cryptophyta)
- Microflagellates (Prasinophyta, Euglenophyta and Protozoa)

Algae are also known as Prokariotes, or unicellular organisms without nucleus. An example is blue green algae like Spirulina. Another type of algae is Eukariotes, or unicellular organisms with a nucleus, such as green and red algae. Chlorella is a type of green algae. Larger algae are known as seaweed, and kelp is the best known of them.

The high density of nutrients found in algae is extremely important for many reasons. Perhaps the most important is that these nutrients maintain human cell membrane structure and function. This is vital for cell detoxification and for overall metabolism of human cells. The causes of diseases have been simplified to very specific mechanism, all of which centers on cell membrane function and structure. Inflammation, oxidation, toxicity and mitochondrial dysfunction keep the cell membranes from doing their jobs.

Algae contain high level of antioxidants and anti-inflammatory micronutrients to fuel metabolism and detoxification. They also stoke the fires of the mitochondrial where cells make energy required to carry out their function. Photosynthesis is the mechanism whereby plants in general and algae in particular harness life-sustaining solar energy. Thus, the very good evidence is that the algae are highly beneficial. The enrichment of cell membrane function through nutrients and prebiotic function of algae are the most important contributions to health from these organisms. Prebiotics are rich fibers that feed intestinal flora. It is precisely in the intestines that most of the immunologic, neurologic and hormonal systems are found, the very systems that the cells use to communicate through their cell membrane.

One of the most researched items on nutrition is the role of iodine in all aspects of cell functions. Its relative absence in the Mountain populations is generally thought to be at the root of many health problems particularly in thyroid function. Thyroid hormone is indispensable for practically all cell functions especially in the brain. Thus, it is known that populations living closer to oceans are generally healthier and live longer. Algae and fish in general and phytoplankton in particular are very high in iodine content.

Marine phytoplanktons (microscopic plants/organisms that live in the ocean) contain micronutrients and electrolytes that are exactly what human cell membranes need to carry out their metabolism. Marine phytoplankton contains almost everything needed for life and the rebuilding of healthy life. It contains 9 amino acids that the body cannot synthesize. The essential fatty acids (omega-3 and omega-6) are also present. Vitamin A (β-carotene), B_1 (thiamine), B_2 (riboflavin), B_3 (niacin), B_5 (pantothenic acid), B_6 (pyridoxine), B_{12} (cobalamine), C, D and major minerals and trace elements are all present.

Marine phytoplankton can reduce depression, harmful effects of stress and mood swing. It is rich in polymeric and basic healthy sugars such as polysaccharides. Phytoplankton also have alkaline pH which is important considering the acidity of diets that are high in refined sugars soda and farmed large animals.

Kelp (*Ascophylum nodosum*)

Kelp is unique seaweed that grows exclusively on North Atlantic coast of North America and Europe. It is found particularly in abundance on North Atlantic coast of Canada where it is cold, and clean growing environment is continuously replenished by the world largest tide. Kelp commonly known as rockweed or seaweed is recognized to be richest marine plant for animal feed and agricultural use. Kelp has long been a favorite vegetable source among ancient maritime culture, such as Japanese, Chinese, South Sea Islands and the Greenland Eskimos.

Kelp was first used medicinally to treat enlarged thyroid glands. Physicians did not know why kelp was effective until it was discovered that it was exceptionally rich in iodine and that enlarged thyroid was due to iodine deficiency. Because iodine works by stimulating the thyroid gland that controls the metabolism, it was noted that those who took iodine lost weight more easily. Based on these observations, kelp was used to assist weight loss. It has also been suggested that kelp positive effect in assisting metabolism may help in lowering the cholesterol.

This versatile seaweed, kelp, is also used to maintain healthy skin and hair. Its most dramatic application is its ability to neutralize heavy metal pollution and radiation in the body. Brown seaweed such as kelp contains alginic acid, which binds toxin in the intestine rendering them indigestible and eliminates them out of the system. The active ingredient in kelp, sodium alginate, can prevent the absorption of newly ingested heavy metals but have little effect on past exposures.

Kelp contains over 60 mineral elements, 21 amino acids, simple and complex carbohydrate and several plant growth factors. Being rich in all these amino acid, vitamin and trace elements, it is known as great promoter of glandular health, especially for pituitary, adrenal and thyroid glands. Since thyroid and pituitary glands regulate certain functions of digestion, kelp is beneficial in balancing out human and animal total system. Vitamin K, a fat-soluble vitamin in this seaweed, aids adrenalin regulation.

Kelp is useful in prevention of birth defects, dropsy, edema, fatigue, goiter, low blood pressure, schizophrenia, senility, intestinal ulceration, varicose vein, venereal diseases and weight loss.

Spirulina (*Cyanophyta*)

Spirulina, blue green algae, is an excellent source of chlorophyll and disease fighting compounds called phytochemicals. Phytochemicals are compounds in plants and fruits that imbue them with colour and protect them against cellular damage caused by free radicals.

Mouth cancer is one disorder that may be treated or prevented with phytochemicals especially beta-carotene, which is found in abundance in spirulina. Spirulina has been shown to be effective in the treatment of leukoplakia, a precancerous lesion in the mouth.

Spirulina has 62% amino acid, 20 times more protein than Soya protein and 200 times more than beef. It is also the richest source of vitamin B_{12} and it contains high levels of minerals like zinc. Spirulina has 10 times more carotenoids than carrots and it is rich in xanthophyl pigment like chlorophyll. It is also rich in oils, containing more omega-3 than fish oil. Plankton is also rich in polymeric and basic sugars, such as polysaccharides.

Spirulina is known as *wonder food* and it was part of the staple diet of ancient Aztec civilization. It contains phytonutrients, antioxidants and functional nutrients with proven health benefits. It has also been shown that spirulina has the ability to support immune system and contribute to overall health.

Chlorella Algae

Chlorella is a type of green algae. It has been found to be effective for mercury elimination. It is noted that some of chlorella beneficial effect stem from high concentration of chlorophyll and beta-carotene as well as from its unusual cell wall. Its cell wall binds effectively with heavy metals and pesticides, and they are eliminated safely out of the body system.

Chlorella also contains growth factor, chlorella growth factor, which is believed to strengthen immunity by increasing the activity of T- and B-cells with its defense against viruses. It also improves the activity of macrophages, which destroy viruses and other foreign bodies.

Chlorella growth factor isolated in the 1950s is an indicator of nucleic acid of chlorella. Nucleic acids (RNA and DNA) in the body direct cell growth and repair but decline with age and they are depleted by stress, pollution and poor diet. Chlorella, sometimes called green blood tonic and cleanser, has been reported to be beneficial in cases of anemia and has the ability to stimulate regeneration of tissue, accelerate the healing of wounds and inhibit the growth of some tumors. It is also thought to help in preventing degenerative diseases and slow down the aging process. Also, it has potentials in the treatment of cancer, Acquired Immunodeficiency Syndrome (AIDS), Epstein Barr disease and other equally chronic viral illnesses. In addition, chlorella is found to have potential radio-protective effect against Gamma rays. There are studies on chlorella detoxifying effect on cadmium, heavy metals and chlordecone, a harmful insecticide.

Table 1 : Algae nutritional properties and nutritional benefits known to man

Macro-elements
Calcium: Strengthens bones, calms nerves and lowers cholesterol
Potassium: Reduces hypertension and helps in preventing high blood pressure
Magnesium: Promotes tranquility, moderates mood swings and reduces migraine headaches
Sodium: Prevents sunstroke
Minor-elements
Boron: Strengthens bones, enhances mental clarity and unclogs arteries
Chromium: Moderates existent diabetes and prevents adult-onset diabetes
Cobalt: Repairs nerve cells and helps produce red blood cells
Copper: Eases arthritis and helps to produce red blood cells
Fluorine: Fights tooth decay and hardens bone
Iodine: Regulates body weight
Iron: Combats fatigue, lessens depression and decreases anemia
Manganese: Assists joint mobility
Molybdenum: Increases longevity
Nickel: Promotes cellular growth and reproduction nucleic

Selenium: Boosts the immune system and relieves anxiety
Silicon: Tightens the skin
Vanadium: Balances blood sugar
Zinc: ases Memory access, lessens acne outbreaks, boosts the immune system, reduces symptoms of the common cold and helps in avoiding prostate problems
Vitamins
Thiamine (vitamin B_1): Reduces fatigue, improves mental attitude and relieves tension
Riboflavin (vitamin B_2): Defends against free radical attacks provides physical energy and alleviates eye fatigue
Niacin (vitamin B_3): Reduces stress lowers cholesterol and helps to reverse atherosclerosis
Biotin (vitamin B_4): Produces healthy-looking hair
Pantohenic Acid (vitamin B_5): Reduces morning arthritis pain and alcohol toxicity
Pyridoxine (vitamin B_6): Boosts the immune system and relieves premenstrual tension
Folic acid (vitamin B_9): Improves mental function and prevents anemia
Vitamin B_{12}: Energizes the body, improves memory and repairs the nervous system
Vitamin C: Reduces duration of the common cold, reduces cancer risk and deters gum bleeding
Vitamin E: Boosts the immune system and keeps nerve tissue healthy
Amino acids
Alanie: Activates muscles and boosts the immune system
Arginine: Builds new muscles, detoxifies the liver and boosts the immune system
Asparagine: Gives energy to the brain and builds the immune system
Aspartic Acid: Helps in making healthy DNA and builds the immune system
Cysteine: Detoxifies carcinogens, essential fatty acids and reduces cardiovascular disease
Glutamic Acid: Reduces alcohol/sugar cravings
Glutathione: Defends against free radical attacks
Glutamine: Improves concentration
Histidine: Enhances nutrient absorption and removes toxic metals
Isoleucine: Builds muscle helps in repairing the liver
Lecithin: Dissolves cholesterol deposits and improves short term memory
Leucine: Reduces hypoglycemic symptoms
Lysine: Helps in preventing osteoporosis
Methionine: Enhances memory, elevates mood and removes heavy metals
Phenylalanine: Increases mental alertness and reduces sugar cravings

Proline: Raises learning ability and helps in repairing torn cartilage
Serine: Beautifies the skin
Threonin: Boosts the immune system and ameliorates skin tone
Tyrosine: Acts as an anti-depressant, promotes mental alertness and strengthens memory
Valine: Builds muscle tissue
Essential fatty acides
Gamma-Linolenic acid (GLA): Lowers cholesterol
Linoleic acid: Combats viral infections
Omega-3 fatty acids: Increases cell membrane flexibility, dissolves cholesterol deposits and reduces cardiovascular diseases
Omega-6 fatty acids: Reduces arthritis symptoms, improves skin tone and reduces cardiovascular diseases
Pigments
Beta-carotene: Improves vision, protects the cornea, helps in digestion, defends against free radical attacks and boosts the immune system
Bioflavinoids: Removes toxins from skin cells and builds the immune system
Chlorophyll: Promotes bowel regularity and cleanses interstitial tissues
Other substances
Electrolytes: Helps in kidney function
Fiber: Eliminates toxic wastes
Germanium: Helps control Epstein-Barr virus
Glycine: Calms nervous system
Glycogen: Increases physical stamina
IS: Defends against free radical attacks
RNA: Enhances brain function
Substance P: Sharpens the mind
Superoxide dismutase (SOD): Defends against free radical attacks

Sea vegetables can prevent assimilation of different radio-nucleotides, heavy metals such as cadmium and other environmental toxins. An experiment conducted by J.F. Stara at the Environmental Protection Agency showed that sodium alginate significantly reduced the amount of radioactive strontium in the bones of cats. Stara observed that radioactive strontium in the bones is re-secreted into the intestines where it is bound by alginate, neutralized, and then, excreted in the stools.

Aphanizomenon Flos-Aquae (AFA)

Afanizomenon flos-aquae are edible fresh water blue green algae. The algae are non-flowering plant that grows in almost every drop of sunlit water and every gram of soil, on tree barks and on the surface of rocks. Algae are the highest known source of chlorophyll, which is vital for rapid assimilation of amino acids. High-level chlorophyll foods help in the production of vitamin E in the body. It is observed that AFA helps to restore overall biochemical balance by nourishing the body at cellular level.

Thalassotherapy (hot seaweed bath)

Thalassotherapy, a Greek word thalsso or sea, is a treatment system using freshly heated seawater, which induces an exchange of minerals toxins between the blood and water.

Thalassotherapy benefits those with certain medical conditions such as arthritis and even those without arthritis. Ocean therapy is popular as a tool for healing deep emotional scars. Ocean therapy also includes the bathing hydrotherapy that has application to skin nutrition. Thalassotherapy is an ageless health restorative technique. Much of the population of the ancient Greek and Roman world soaked in sea water hot tubs and heated seaweed baths, drank and inhaled sea water, got sea water massage, aromatherapy, or algae wrap, had sea green facials and body wraps and use sea water pools for hydrotherapy and detoxification.

Coral (Seashell)

Corals are seashells from marine organisms. They contain every mineral needed by the body, and they are known for their synergistic relation with the body. Coral is used for orthopedic grafting and it contains many essential minerals that are chemically similar to the elements found in human skeleton, body fluid and blood. Other natural health enhancers and detoxifiers of benefit include:

- **Healing clay (healing earth):** Clay is renowned to have many uses in promoting health of plants, animals and humans. Bentonites, Pascalite as well as other types of healing clays have been used by indigenous cultures since ages. Benefits reported by people using liquid clay for a period of 2 to 4 weeks include improved intestinal regularity, relief from chronic constipation, diarrhea, whiter eyes, enhanced alertness, emotional uplift, improved tissue and gum repair and increased resistance to infections. Clay works on the entire organism.

  The eating of clay has been observed in native people and animals in all parts of the world since ages. Earlier description of widespread use of *kanwa* (earth from lick near Lake Chad) in Nigeria for animals and humans had been made.

- **Fulvic acid and Shilajit:** Fulvic acid and Shilajit are involved in complex biochemical reactions in living plants and animals, which directly influence the very metabolic process at cellular level. Humic substances are found in humus soil in trace amounts and are considered as nature's best medicine for plants, animals, humans and earth itself. Fulvic acid assists every stage of cellular metabolism and is the most powerful natural electrolyte known. It restores electrical balance to damaged cells, neutralizes toxins and can eliminate food poisoning within minutes

Shilajit comes from rocks in lower Himalayas and is the most important natural remedy of *Ayurvedic* medicine. The active principle in *Shilajit* is fulvic acid. Traditionally, it is considered a panacea and a strong kidney tonic. It increases the core energy power responsible for sexual and spiritual powers, which are depleted by stress and anxiety. *Shilajit* is used by the indigenous system of medicine in India, Hakims and Vaids and traditional healers in a number of diseases including genitourinary diseases, diabetes, chronic bronchitis, asthma, gall stones, jaundice, painful and bleeding piles, epilepsy, enlarged liver and spleen, dyspepsia, digestive disorder worms, renal and bladder calculi, nervous debility, sexual neurasthenia, hysteria anemia and in bone fracture.

Conclusion

The high density of nutrients found in algae is extremely important for many reasons. The most important is that these nutrients maintain human cell membranes in structure and function. This is vital for cell detoxification and overall metabolism of human cells. In fact, the causes of diseases have been simplified to very specific mechanisms, all of which center on cell membrane function and structure.

The micronutrients and electrolytes in plankton are exactly what human cell membranes need to carry out their metabolism. The composition of human plasma or fluid surrounding cell membranes is similar to that of the seawater.

■■■

24 Aquatic Food and Human Health

Anita Bhatnagar

Department of Zoology, Kurukshetra University, Kurukshetra-136119, Haryana India, E-mail: <anitabhatnagar@gmail.com>

There are people in the world so hungry that God cannot appear to them except in the form of bread (Smt. Indira Gandhi). A safe and nutritionally adequate diet is a basic individual right and an essential condition for sustainable development, especially in developing countries where the situation is grave. Over one billion people experience the hardship that hunger imposes, a figure that continues to rise even amidst the riches of the 21st century. Engulfed within a vortex of population growth, economic instability and climate change, food security has become an urgent challenge for national and global governance.

At global level, the increase in the average per capita supply and consumption of most plant foods such as grains, cereals, legumes, roots and tubers has already reached to a plateau. Very few avian (about four) and mammalian (about five) species account for the vast majority of the worldwide food production. Therefore, dairy animals and poultry, perhaps, will not be able to meet all the food requirements of human beings. On the other hand, there are many species of finfishes (teleosts), shellfishes (crustaceans and mollusks) and macroalgae that are grown across the earth. Therefore, fish and other aquatic products are considered most important and necessary to contribute to the food security of growing world population. Moreover, the dietary habits of people all over the world are creating a spurt in demand for aquatic foods, particularly because of the health conscious consumers.

Aquatic/seafood

Aquatic food refers to any aquatic animal or plant that is served as food and eaten by human beings. Aquatic foods include water animals, such as fish and shellfish including molluscs, crustaceans and echinoderms and also some plants such as macroalgae (Figure 1). More precisely, *all sea as well as freshwater edible aquatic products refer to Aquatic or seafood.* Edible aquatic weeds are also seafood and are widely eaten around the world. The harvesting of wild aquatic food is known as fishing and the cultivation and farming of aquatic food organisms is known as aquaculture, mariculture, or pisciculture. Aquatic/seafood is often distinguished from meat, although it is still animal and is included in a vegetarian diet. Aquatic/seafood is an important source of protein all around the world.

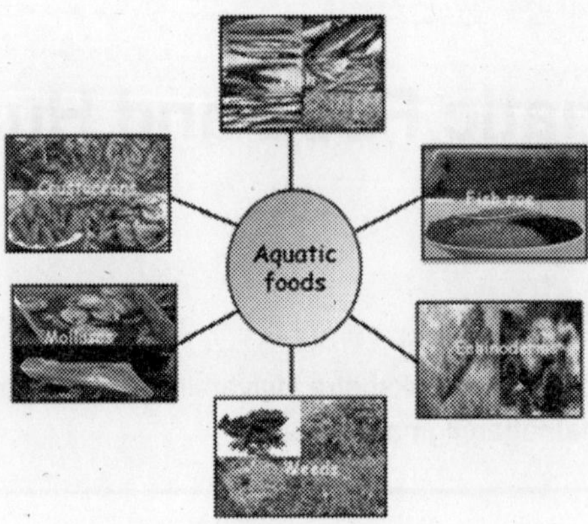

Figure 1. Important aquatic/see food

Aquatic/seafood is an important part of protein rich nutritive diet. They contain high-quality protein and other essential nutrients, low in saturated fat and contain omega-3 fatty acids. Amongst all aquatic foods, fish and shellfish are regarded as well-balanced diet, which can contribute to heart health, foetal development and children's growth and development. A statement *Look to the Seas* is the recent thinking for harnessing the aquatic wealth. This is a necessity to sustain over 16% of the global population, with 2.4% of land and 4.2% of water of the whole world.

About 20,000 species of fish, 200,000 species of invertebrates and 14,000 species of aquatic plants are available out of which about 500 of these species are now cultivated commercially, which include 314 fin fish, 74 crustaceans, 69 molluscs, 43 algae, 13 angiosperms, 12 sponges, 9 amphibians, 4 reptiles, 3 rotifers, 2 annelids, 2 mammals and 1 echinoderms. Seafood mainly includes fish, shellfish, roe and sea plants known as seaweeds.

Fish

Fish is a food consumed by many species, including humans. The word *fish* refers to both the animal and to the food prepared from it. Fish has been an important source of protein for humans throughout recorded history. There are over 20,000 species of fish, making them the most diverse group of vertebrates. However, only a small number of the total species are considered food fish and are commonly eaten. The principal food fish species groups are shown in Table 1.

Roe

Roe is the fully ripe egg masses of fish and certain marine invertebrates, such as sea urchins. As a seafood it is used both as a cooked ingredient in many dishes and as a raw ingredient.

Shellfish

Shellfish is a culinary and fisheries term for exoskeleton-bearing aquatic invertebrates used as food, including various species of molluscs, crustaceans and echinoderms. Although most kinds of shellfish are harvested from saltwater environments but some kinds are found only in freshwater.

- **Crustaceans:** Crustaceans include the members of family Crustacea a very large group of phylum Arthropoda, which include familiar animals such as crabs, lobsters, crayfish, shrimp, prawns, krills barnacles, *etc*. Some crustaceans commonly eaten are shrimp, prawn, lobster, crayfish and crabs. Over 60% by weight of all crustaceans caught for consumption are shrimp and prawns, and nearly 80% is produced in Asia.

    Prawn and scampi are the terms generally used interchangeably with shrimp. Prawn refers to freshwater shrimp or large saltwater shrimp and scampi is often used by restaurateurs to describe shrimp cooked in butter and garlic. There are many species of shrimps from cold-water, warm-water and freshwater. They are found in many colours, including reddish to light brown, pink, deep red, grayish-white, yellow, grey-green and dark green and category of sizes, *i.e.*, colossal, jumbo, extra-large, large, medium, small and miniature. Shrimps are available year-round, and as a rule, the larger the shrimp, the more expensive they are.

- **Molluscs:** Molluscs include the members of invertebrate phylum Mollusca. Molluscs enjoyed as a food source by humans include many species of clams, mussels, oysters, winkles and scallops.

- **Echinoderms:** Echinoderms are members of phylum Echinodermata and include marine organisms. The echinoderms are not as frequently harvested for food as the molluscs and crustaceans. However, sea urchin roe is quite popular in many parts of the world. Echinoderms sometimes pose a health threat to humans. The fine structure of the spines of certain species of sea urchins means that if the spine pierces the flesh, it may break off when an attempt is made to remove it and require the assistance of a physician to remove the remaining piece of spine fully.

    Echinoderms are also elements of many cuisines. Around 50,000 tonnes of sea urchins are captured every year, the gonads of which are consumed particularly in Japan, Peru and in France. The taste is described as soft and melting, like a mix of seafood and fruit. The quality depends on the colour, which can range from light yellow to bright orange.

    Sea cucumbers are also considered a delicacy in some countries of south east Asia and particularly the popular ones are the (Pineapple) roller, *Thelenota ananas* (*susuhan*) and the red *Halodeima edulis*. They are well known as *Trepang* in China and Indonesia. The sea cucumbers are dried and the potentially poisonous entrails removed. The strong poisons of the sea cucumbers are often psychoactive but their effects are not well studied. It does appear that some sea cucumber toxins restrain the growth rate of tumour cells, which has sparked interest from cancer researchers.

- **Sea plants:** Plants extracted from aquatic system and consumed by humans are called as seaweeds. Weeds are something we do not want. Seaweed implies something negative about the product. When trying to convince others to eat it, *sea plants* or *sea vegetables* may be more appropriate words to describe it, hence, in this account the word sea plants has been used.

Sea plants (seaweeds) include macroscopic, multicellular and benthic marine algae. Members of the red, brown and green algae are included under this caterory. Sea plants can also be classified by use as food, medicine, fertilizer, industrial, *etc.* There appearance somewhat resembles non-arboreal terrestrial plants. Sea plants are consumed by coastal people, particularly in East Asia, *e.g.*, Japan, China, Korea, Taiwan and Thailand. With the current trend for consumers to embrace organically grown foods and natural foods from clean environment, sea plants are receiving an increasing acceptance.

Important sea plants (seaweeds) used as food include sheets of dried *Porphyra* used in soups, *Chondrus crispus* (commonly known as Irish moss or carrageenan moss) used in producing various food additives, *Dulce* is mixed with milk, nutmeg, cinnamon and vanilla to make a common beverage. Sea plants are also harvested or cultivated for the extraction of alginate, agar and carrageenan, gelatinous substances collectively known as hydocolloids or phycocolloids. Hydrocolloids have attained commercial significance as food additives. The food industry exploits their gelling, water-retention, emulsifying and other physical properties. Agar is used in foods such as confectionery, meat and poultry products, desserts and beverages and moulded foods. Carrageenan is used in salad dressings and sauces, dietetic foods and as a preservative in meat and fish products, dairy items and baked goods.

- In China and some parts of Indonesia, packing of different sea plants is available, *viz.* yaki nori, roasted seaweed, frozen seaweed salad, roasted laver seaweed, rice ball seaweed, dried seaweed, seaweed hijikia, dried laminaria, wakame stem, *etc.*

Table 1 : Some examples of Fish, Roe, Shell fishes and Sea plants consumed as food

Fish	Roe	Shell fish			Sea plants
		Crustaceans	Molluscs	Echinoderms	
Ancovy carp	Caviar (sturgeon roe)	Crabs (mud crab, sand crab, king crab, dungeness crab and snow crab)	Abalone	Sea cucumber	*Porphyra*
Cod catfish eel	Ikura (salmon roe)	Lobster (American lobster, rock lobster and spiny lobster)	Clam	Sea urchin as Asian cuisines	*Chondrus crispus*
Haddock halibut	Kazunoko (herring roe and lumpfish roe)	Craw fish	Cockle		*Caulerpa*
Salmon	Masago (capelin roe)	Shrimp	Conch		*Fucus*
Herring	Shad roe	Prawms	Cuttlefish		*Gracilaria*
Mackerel	Tobiko (Flying-fish roe)		Mussel		*Laminaria*
Snapper			Octopus		*Macrocystis*
Scad			Oyster		*Monostroma*
Tilapia			Periwinkle		*Enteromorpha*
Sardine			Snail		*Undaria*
Trout			Squid		*Hizikia*
Tuna			Scallop		*Cladosiphon*
					Palmaria

Nutritional Benefits of seafoods

Nutrition is the net effect of the process by which an organism ingests and uses food for growth and maintenance of the body.

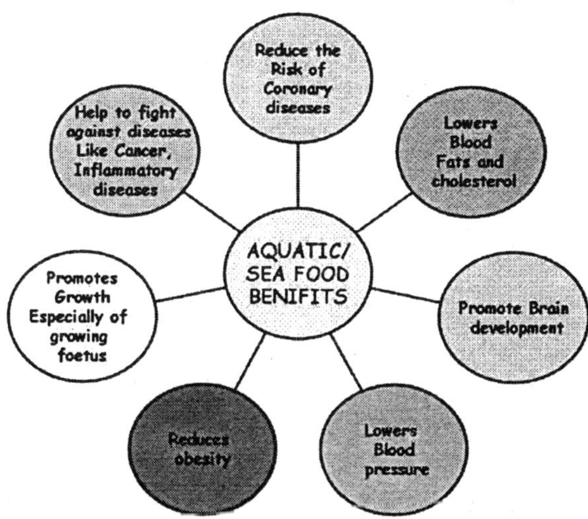

Figure 2 : Benefits of aquatic/see foods

Seafood is a nutrient-rich food source that is widely available (Table 3). It is a good source of high-quality easily digestible protein, contains all the essential amino acids, is low in saturated fat and is rich in many micronutrients. Seafood is often also a rich source of preformed long-chain polyunsaturated omega-3 fatty acids, *i.e.*, eicosapentaenoic acid (EPA) and docosahexaenoic acid (DHA), which are synthesized in limited amounts by the human body from α-linolenic acid (ALA), a fatty acid found in several vegetable, nut and seed oils, *e.g.*, walnut and flaxseed oils. In past several years, research has implicated seafood, particularly its contribution of EPA and DHA, in various health benefits identified for the developing fetus and infants and also for adults including those at risk for cardiovascular diseases (Figure 2).

As far as fish specifically is concerned, it is one of the main sources of protein and fat but it also has other qualities that make it very valuable as a nutrient.

- The fat in fish contains a larger portion of omega 3 and other unsaturated fatty acids that are healthier than the saturated fat of red meat. Large intakes of the saturated fat in red meat is one of the main causes of cardiovascular diseases, while the unsaturated fat of fish and vegetables does not have this effect. In fact, it may have a curative effect.
- In addition, the types and proportions of dietary fats are generally more *heart healthy* than the fats found in other protein foods. This is due to its relatively high share of unsaturated fat. Saturated fat is more thick or viscous than unsaturated fat. This is probably why fish, which is cold blooded, has higher contents of unsaturated fat. Fish that live in cold areas have the highest content of polyunsaturated fatty acids.

Excessive dietary intake of saturated fats can significantly raise the blood cholesterol level in humans. This increases the risk of cardiovascular diseases. Approximately 50% of the fatty acids in lean fish, *e.g.*, walleye and yellow perch and 25% in fattier fish, *e.g.*, channel catfish, herring, salmon and rainbow trout, are polyunsaturated fatty acids. In contrast, the polyunsaturated and saturated fatty acids in beef are 4-10 and 40-45%, respectively of the total fatty acids present.

- Fish is also valued as a source of omega-3 (n-3) fatty acids, very long chain polyunsaturated fatty acids which is critical for the development of the brain and retina and which may be protective of some chronic diseases. Eicosapentanoic acid (EPA) and docosahexanoic acid (DHA) are absent or present in much lower amounts in other foods. The omega-3 fatty acids are essential fatty acids, which means that the body can not build them from other building materials
- Fish is an excellent source of B vitamin niacin, which assists in the functioning of digestive system, skin and nerves. It is also important for the conversion of food to energy) and vitamin B_{12} is critical to buildup DNA and RNA, the maintenance of nervous system, fatty acid synthesis, homocysteine metabolism and energy production.
- Fish in general is a better source of vitamins D and A than beef, pork or chicken. The vitamin D is vital for the formation of bones.
- Fish can also contribute appreciable amounts of dietary calcium, heme-iron and zinc-the nutrients that tend to be low in people's diets.
- Fish is among the best sources of dietary selenium

Sea plants are a good source of iodine, which is necessary for normal functioning of thyroid glands and preventing goiter. Excess intake of sea plants may raise iodine to toxic levels. Sea plants may have curative properties for tuberculosis, arthritis, colds and influenza, worm infestations and even tumors. Sea plant extract is used in some diet pills. Other sea plant pills exploit the same effect as gastric banding, expanding in the stomach to make the body feel full.

One preserved packed sea plant called as salted sliced Wakame has anti-clotting, anti-tumor and anti-HIV effects.

Table 2 : Approximate Nutrition facts for some sea foods

Sea food	Serving size	Protein (g)	Fat (g)	Carbohydrates (g)	Calories
Fishes					
Cod	84 g	20	1	0	90
Catfish	84 g	17	6	0	130
Halibut	84 g	23	2	0	120
Trout	84 g	20	6	0	140
Haddock	84 g	21	1	0	100
Salmon	84 g	22	4	0	130

Sea food	Serving size	Protein (g)	Fat (g)	Carbohydrates (g)	Calories
Perch	84 g	21	2	0	110
Tilapia	84 g	22	2.5	0	110
Shellfishes					
Blue crab	84 g	20	1	0	100
Clams	About 12 small	17	1.5	6	110
Oysters	About 12 medium	10	4	6	100
Lobster	84 g	17	0.5	1	80
Shrimp	84 g	21	1.5	0	100
Scallops	About 6 large or 14 small	27	1	5	140

Source: US Food and Drug Administration, January 1, 2008

Risk associated with seafood consumption

Seafoods are actually and remarkably safe to eat. It accounts for not more than 10% of all the reported food-borne diseases. Nevertheless, seafood causes a significant number of food-borne diseases and deaths worldwide, hence, the people should be wary of them. The important points to be taken into consideration are given as under:

- Most health problems associated with seafood are due to either contaminant that is present in the environment where the seafood, *i.e.*, shellfish or fish, is grown or handled improperly.
- Many shellfish like oysters, mussels, clams, *etc.* hardly move, hence, they need to filter a lot of water to obtain food. In the process, disease causing bacteria, viruses and other microorganism that are present in their immediate environment tend to accumulate and concentrate in them.
- Raw or undercooked shellfish such as oysters, clams and mussels may be contaminated with bacteria, such as *Salmonella*, *Staphylococcus aureus* and *Vibrio* species, thus, it is risky to eat raw or half-cooked shellfish that are grown and harvested in polluted water because of the possibility of food poisoning.
- Pathogenic bacteria can also grow in a stale and poorly handled fish but fresh and properly prepared deep-sea fish can be eaten raw.
- Another health risk that shellfish carry on a seasonal basis is red tide. Red tide, which occurs at certain times of the year, is caused by the proliferation of dinoflagellates, a type of phytoplanktons that exude a lethal toxin, which is not destroyed by cooking.
- In fish, removal of the entrails or abdominal organs before cooking makes the fish ready for human consumption but there is no way to remove dinoflagellate toxin from

shellfish, hence, when the red tide warning is up, people should refrain from eating shellfish.

- The dumping of chemical pollutants into the sea is an additional growing concern associated with seafoods because their chemicals find their way into the food chain and accumulate in the tissues of sea creatures.
- The high iodine content of sea plants can produce iodine toxicity if large amounts of seaweed are consumed.
- Rotting sea plant is a potent source of hydrogen sulfide, a highly toxic gas and has been implicated in some incidents of apparent hydrogen sulfide poisoning. It can cause vomiting and diarrhea.

Ways to Prevent Seafood Based Illness

- Most seafood-borne diseases can be traced to pollution in area where the sea creatures are harvested. Seafood-borne illnesses are best prevented by proper monitoring of fish industry by the concerned government agencies.
- At personal level, as consumer, people can help to prevent these diseases by observing the following precautions:
  - Eat only shellfish and fish that are cultured or harvested from known safe areas.
  - Remove the entrails of fish before storing them in the freezer.
  - Preferably, shellfish and fish should be cooked thoroughly before they are consumed. Eat seafood raw only if they are fresh and handled properly. Seafoods that are safe to eat are deep-sea fish. Fresh water fish has to be cooked because some of the parasites they carry can also cause illness in humans.
  - Keep raw or live seafood separate from cooked seafood.
  - Wash knives, chopping boards and cooking utensils with soap and water before using them to prepare seafood.
  - Observe proper personal hygiene before handling seafood.

Tips for Purchasing Seafood

When buying seafood and fish, it is possible to find them fresh or frozen. Frozen seafood is as good as fresh. With the technological advances, as today, it is possible to process and freeze fresh catches immediately, usually right on the fishing vessel. The process of flash freezing is employed, in which the water contained inside the food items is subjected to very low temperature (below its melting/freezing point), causing water inside the food to freeze in a very short period of time. Therefore, frozen seafood can even be superior in quality to fresh seafood products.

Knowing how to choose fresh fish or seafood is a vital skill for a seafood cook. Unless you catch the fish yourself, you really have no way of knowing exactly how fresh it is but buying fresh fish is easy if you know what to look for fresh fish.

Tips for Buying Fresh Seafood

The first aspect you should consider when buying fresh fish or seafood has to do with the seafood seller/store. Pay attention to the cleanliness, appearance and smell (strong fishy odour is a sign of spoilage) of the local. As general rule, fresh fish and seafood should be storage at 0°C, which can be kept by refrigeration or ice.

Besides of that when selecting fresh fish or seafood pay attention to the following characteristics that indicates the freshness of the product, in order to safely enjoy their fine taste and good nutrition.

Tips to Identify Fresh Fish

Whole Fish

- Fish should smell fresh and mild, not fishy or ammonia-like.
- The eyes of a fresh fish should be clear, bright and a little protruding. Cloudy, pink and sunken eyes indicate that the fish has lost its freshness.
- The gills of the fish should be bright red or pink and free from slime.
- Fresh fish should have firm and elastic flesh. Test it by pressing with the finger the flesh delicately. It should spring back.
- Fish with shiny skin and scales that adhere tightly is another indicator of freshness.

Fish Fillets

- Fish fillets should have firm and elastic flesh, a fresh-cut, moist appearance, displaying no darkening or drying around the edges. It should be little evidence of bruising or reddening of the flesh from retention of blood. In addition, prepackaged fish fillets should contain a minimum of liquid because when seafood is stored in liquid deteriorates quickly.
- Look for vibrant flesh. All fish fade as they age. If the fillet still has skin, that skin should look as pristine as the skin on an equally good whole fish, *i.e.*, shiny and metallic.
- Smell it since the smell test is especially important with fillets. They should have no pungent aromas.
- Look for liquid on the fish, if so, that liquid should be clear, not milky. Milky liquid on a fillet is the first stage of rot.
- Press the meat with your finger. It should be resilient enough so your indentation disappears. If your fingerprint remains, do not purchase. This means fresh fish should have elastic flesh

Fresh shellfish

- Fresh mussels, oysters and clams are sold live because when they die, their internal organs start to deteriorate. This way, the fresh shellfish should have unbroken shells and should be closed. If the shells gape slightly, tap them: the shells should close up. If they do not close up, throw them away.
- The shells of fresh live shellfish should be shiny and moist. It should smell sweet and mild, like a fresh sea breeze.

- Fresh lobster and crabs are also sold alive because they spoil quickly when they die. Thus, when selecting them they should show leg movement. They will not be very active if they have been refrigerated but they should move at least a little bit.
- Shrimps are found shelled or unshelled, raw or cooked and fresh or frozen. When buying fresh shrimps, they should smell of the sea, not like ammonia.
- Raw shrimp meat should be firm and should smell mild. The shells should not have blackened edges, black spots or other dark areas because this indicates loss of freshness.
- While buying fresh squid, select those with clear and full eyes, the skin should be cream-coloured with reddish brown spots (when they start losing freshness, the skin turns pinkish) and the meat should be very firm.
- Fresh-shucked oysters should have a fresh smell. A clear slightly milky or light grey liquid should surround freshly shucked oysters.
- And last but not least, when shopping buy your fresh fish or seafood last, and then, take them home straight away to storage them appropriately.
- Take care of shelf life of seafood (Table 3).

Table 3 : Shelf life of important seafood

Sr. No.	Seafood	Shelf life
1.	Squid and freshly shucked clams	1-2 days
2.	Shrimp and scallops	2-3 days
3.	Freshly shucked oysters	5-7 seven days
4.	Mussels and clams in the shell	2-3 days
5.	Oysters in the shell	7-10 days
6.	Sea weed (dried)	6 months-2 years

Tips for Buying Frozen Seafood

- Frozen fish and shellfish should be packaged in a close-fitting, moisture-proof package. Do not buy frozen seafood if its package is open, torn, or crushed on the edges.
- Select packages from below the load line of freezer case in the store's freezer.
- Avoid the package defrosting during transportation. You can transport seafood in an ice-packed box from the store to home.
- Handling Tips after Buying
- After buying your seafood, take it home as quickly as possible, in order to put it in the refrigerator/freezer.
- Store your seafood in the refrigerator if you intend to use it within two days after purchase. In case, you do not use your seafood within two days after purchase, wrap it in moisture-proof paper or plastic wrap, place in a heavy plastic bag, or store in an airtight, rigid container and store it in freezer. Keep the refrigerator temperature between 34°and 40°F, and of your freezer at 0°F or colder than as close to -20°F as possible.

- Once your seafood is unfrozen, do not refreeze it.
- Store live shellfish in containers covered with clean, damp towels, or paper towels. Do not put live shellfish in an airtight container where they could asphyxiate and die.
- Shrimp, squid and shucked shellfish should be stored in leak proof bags, plastic containers, or covered jar.
- Pasteurized crabmeat heated in cans at a temperature high enough to kill bacteria but lower than that used in canning should be stored unopened in the refrigerator for up to 6 months and used within 4 days of opening.

Cooking Tips for General fish, *e.g.*, Tilapia

Bring fish either fresh or frozen. If it is Tilapia, it has a mild, sweet flavour that goes well with many different cooking styles. You can use it in any recipe calling for white fish. Whole fish also grills well.

Thawing

Thaw frozen seafood gradually by placing it in the refrigerator overnight. If you have to thaw seafood quickly, either seal it in a plastic bag and immerse it in cold water, or if the food is to be cooked immediately thereafter, microwave it on the *defrost* setting and stop the dcfrost cycle while thc fish is still icy but pliable.

When you are preparing fresh or thawed seafood, it is important to prevent bacteria from the raw seafood from spreading to ready-to-eat food. Consider the following steps important to avoid cross-contamination between raw and cooked foods:

- Wash hands thoroughly with soap and warm water before and after handling any raw food.
- Wash the cutting board with soap and hot water to remove food particles and juices after using it for raw foods such as seafood and before using the board for cooked or ready-to-eat food or preparing another food item.
- As an added precaution, sanitize cutting boards by rinsing them in a solution made of one teaspoon of chlorine bleach in one quart of water or running the plastic board through the wash cycle in your automatic dishwasher or consider using one cutting board only for raw food and another only for ready-to-eat foods such as bread, fresh fruit and vegetables and cooked fish.
- As a thumb rule, avoid using cutting boards that are made of soft porous materials, instead, choose those made of hard maple or plastic and make sure that they are free of cracks and crevices. Smooth surfaces can be cleaned more easily and thoroughly.

Apply the Heat

Grilling and broiling fish are similar. Both use high heat to cook the fish. Grilled fish can be whole or you can marinate fillets to add flavour. It will take about 5 minutes per side on the grill. Under the broiler, season your fish first. You can spread a sauce made of mustard or mayonnaise on top to brown after the fish has cooked for a few minutes. Both methods make a meal grand tasting.

Add Water

Poaching and steaming are the two excellent ways to cook fish. For poaching, fill your pan with a few inches of water, broth, or milk, add your fish and let it completely submerge. Cook for 3 to 6 minutes, depending on the size of fillet. Poaching creates a very delicate texture that complements many different dishes. Steaming is another popular method of cooking fish. Steaming can be done with water while you apply seasoning to the fish or you can mix up a nice flavourful marinade, which is used as the steaming liquid. If you use a marinade as your liquid and you are steaming whole fish, pour a little bit inside the body cavity of fish to add more flavours. Whole fish will need to steam about 20 minutes. If you are making fillets, 6 to 10 minutes will be sufficient.

Dunk it in Breading

Dip your fish in flour, egg wash and the crunchy coating of your choice to create a nice crispy coating. You can use seasoned breadcrumbs, panko, crushed cereal, instant potato flakes, or anything else that appeals to you. Once your fish is breaded, you can oven fry it or put it in a deep fryer. Both methods give you a nice crispy crunch when you bite into the fillet. If you are deep-frying them, you could cut up the fillets into smaller pieces and serve them as nuggets. This makes the dish appealing for the younger set, especially if you provide dips.

Important Tips for Cooking

- Some seafood may contain traces of toxic chemicals caused by polluted waters. Bake or broil fish so that the fats can drip away during cooking process, because many pollutants are stored in fatty part of the fish.
- If you plan to grill seafood on barbeque, it might dry out a little. Hence, keep a watch on the seafood.
- Most seafood should be cooked to an internal temperature of 63°C, but if you do not have a food thermometer, there are other ways to determine whether seafood cooked properly or not. Follow the below given tips:
  - **Fish:** Slip the point of a sharp knife into the flesh and pull it aside. The flesh should be opaque and separated easily. If you cooked the fish in the microwave, check it in more than one spot to ensure that it is completely cooked throughout.
  - **Shrimp and Lobster:** The flesh becomes pearly-opaque.
  - **Scallops:** The flesh turns milky white or opaque and firm.
  - **Clams, Mussels and Oysters:** Watch for the point at which their shells open, meaning they are done. Throw out the ones they do not open.

Care While Serving Seafood

Do not cross-contaminate since cross-contamination can happen once your seafood is cooked too. To keep your seafood safe while serving, the followings are the simple ways:

- Place the cooked seafood on a clean plate for serving. If the cooked food is placed on an unwashed plate that previously held raw seafood, bacteria from the raw food could contaminate the cooked seafood.
- Use clean utensils to serve food, not those used in preparation of the raw food.

Therapeutic Uses of Aquatic Foods

All aquatic foods can contribute significantly to human health if consumed properly. Apart from control of cardiovascular diseases, malnutrition and several other benefits, some species or their products have some specific therapeutic significance. Fish and other products are used as medicine, ground into vitamins, or processed into cosmetics and perfumes, lubricants, varnishes, soap and margarine.

Fish and Its Products as Medicine

- Flesh of some fishes such as *Heteropneustus fossilis* can help to fight anemia.
- Protein in fish can help to fight malnutrition related diseases.
- Fatty acids in fish can help to protect from cardiovascular disorders.

Fish Oil a Medical Miracle

- Omega-3 fatty acids in the form of high-dose pharmaceutical-grade fish oil are as close to a medical miracle as it will be seen in the 21st century. This advanced form of fish oil, which provides a very pure form of the fatty acids EPA and DHA, is believed by researchers around the world to offer extraordinary health benefits in treating chronic diseases, improving athletic performance and enhancing emotional well-being. It apparently does so by helping to reduce inflammation and increase blood flow throughout the brain and body.
- Medical researchers and practitioners from around the world are using high-dose fish oil (from 2 to 25 grams or more a day) not only for health maintenance but also to help in treating a variety of health conditions, including asthma, bronchitis, emphysema, high blood pressure, heart disease, arthritis, depression, hot flashes, Crohn's disease, ulcerative colitis, Alzheimer's, attention deficit disorder (ADD), multiple sclerosis (MS), Parkinson's disease, dyslexia and many more.
- Fish oil is used in the treatment of patients with immunoglobulin A (IgA) nephropathy, which is the most common glomerular disease worldwide. It has a variable course and leads to end-stage renal disease in a substantial number of cases. Trial conducted provided convincing evidence for protection against progressive renal disease after daily treatment for 2 years with fish oil providing 1.8 g eicosapentaenoic acid and 1.2 g docosahexaenoic acid-the two major omega-3 polyunsaturated fatty acids in fish oil.
- Cod liver oil is widely taken to ease the pain and joint stiffness associated with arthritis and has also been clinically proven to have a positive effect on heart, bone, as well as helping to nourish the skin hair and nails.
- Use of cod liver oil during pregnancy is associated with lower risk of Type I diabetes in the offspring. This effect was found only in mothers taking cod liver oil, not in mothers taking multivitamin supplements. Cod liver oil taken by nursing mothers improves the breast milk

by increasing the amount of fatty acids, which promote brain development, and the amount of vitamin A, which helps in preventing infections but the level of vitamin D is unchanged.

Shellfishes Control Cardiovascular Disease

One reason is that crustaceans such as lobsters and shrimp and mollusks such as oysters, clams and mussels, contain omega-3 fatty acids, which may control blood pressure and cholesterol levels. Among shellfish, oysters, mussels and crabs contain the most. Shellfish is also rich in protein and bone-friendly calcium, as well as zinc (linked with prostate health), iron and B vitamins, and it is relatively low in sodium and calories.

Molluscs and Medicine

- Sea-snail toxin could relieve chronic pain. Tests on rats hint that the chemical could be 10,000 times more potent than morphine, non-addictive and not cause side effects. Cone-shell snails (*Conus* spp.) dwell in coral reefs. They use their mouthparts to harpoon small fish and shoot conotoxins into them. These toxins paralyze prey by preventing nerve communication, and in small doses, the chemicals block pain. This toxin can be successfully used as medicine to cure pain (Nature Science update, 2002).
- Joan Ruderman and Tim Hunt researchers from Marine Biological Laboratory (MBL), Woods Hole, discovered a previously unknown protein in fertilized clam eggs. Interestingly, the protein accumulated, and then, disappeared as cells prepared to divide, and then, divided. The protein, dubbed cyclin, turned out to be a regulator of cell division and it is supposed that it can be used for treating cancer.
- The conopeptides derived from the venom of *Conus s*pecies of predatory sea snail have ultra high specificity, and they can be used to develop novel, highly efficacious drugs with minimal side effects for the treatment of central and peripheral nervous system disorders, and cardiovascular disorders.
- It has been reported that after removal of shell and mantle of *Pila*, *Bellamya*, *Lamellidens* and *Parreysia*, the flesh of these molluscs in various forms can be consumed for the cure of some aliments such as rheumatism, calcium metabolism, heart diseases, conjunctivitis, various gastrointestinal disorders giddiness, nervousness, and dehydration and also provide essential animal proteins, steroids, vitamins and minerals to the people of the region.
- The flesh of *Pila* and *Viviparous* sp. is used as medicine to cure asthma and swelling of joints.
- The foot portion of these edible shellfishes is eaten in the form of curry as well as in roasted form at regular intervals. The soup of eggs of these species are also used as medicine and given to children suffering from rickets.
- Soup prepared from the foot of *Bellamya* species is used to cure asthma, arthritis, joint pain and rheumatism.
- To get cured from conjunctivitis, *Bellamya bengalensis* are collected from pond and are kept in clean fresh water in a earthen pot for night and the water is used like eye drop.
- Soup prepared from the eggs of *Pila* is used to cure rickets in children.

- Soup prepared from the foot of *Lamellidens* sp. and *Parreysia* sp. is used to cure the cardiac ailments and to control blood pressure.
- The shell powder of *Lamellidens* sp. with honey is used for the remedy of giddiness and dehydration.
- Curry of the foot of *Bellamya* sp. is eaten regularly by original people of Kosi region to cure night blindness and for better eyesight.
- Soup and curry prepared from shellfishes (*Macrobra chium* sp. and *Paratelphua* sp.) are helpful in curing anemia, vitamin deficiencies, paralysis and arthritis and promoting strength.

Sea Plants as Medicines

Seaweeds offer a wide range of therapeutic possibilities both internally and externally. They are the best dietary sources of essential minerals.

- It is recommended that patients anticipating surgery eat 3-5 grams brown seaweed cooked as a vegetable broth daily for a week or two prior to surgery. Fucoidan in the pre-surgical patient diet seems to reduce the intensity of blood loss and vascular bed collapse shock during and after surgery. The mechanism for this positive effect is unclear.
- Patients undergoing radiation or chemotherapy seem to benefit from regular fucoidan consumption *via* brown seaweed broth before, during and after treatment. They report fewer and less intense adverse reactions, better recovery and sense of well-being.
- Age-related dementia and perhaps Alzheimer's can be prevented or suppressed by regular consumption of algin-rich brown seaweeds to slow the bioaccumulation of neurotoxic metals.
- Carrageenan eaten as red seaweeds can provide sugar molecules for glycoproteins secreted by mucous membranes and for cell surface aminoglycan labeling.
- Red algae containing carrageenan have been used for the treatments of respiratory ailments, especially intractable sinus infections and lingering pneumonias.
- Strong antiviral activity has been observed in a variety of heavily modified carrageenans and research continues on how to use this in commercial medications.
- People with low thyroid function can consume *Fucus* spp. of brown seaweeds.
- Dietary seaweeds effectively inhibit reabsorption of Dioxin mixture congeners excreted into the gut and promote their fecal excretion.
- Those dieting for weight loss release adipose-sequestered polychlorinated biphenyl (PCBs), dioxins and polybromated diethyl ethers (PBDEs) into their blood and lymph. Ingesting brown seaweeds may mitigate the negative consequences caused by weight loss toxin releases.

In this way, aquatic foods have great therapeutic significance. Our body with the help of right nutrition has the natural ability to defend, repair and restore health. Therefore, *if we eat wrongly, no doctor can cure us* and *if we eat rightly, no doctor is needed* (Victor G. Rocine Circa, 1930).

Balancing of Benefits and Risks of Seafood Consumption

Contamination of aquatic food sources whether by naturally occurring or introduced toxicants is a concern for consumers because of adverse health effects that have been associated with exposure to such compounds. Methyl mercury can accumulate in the lean tissue of seafood, particularly large predatory species such as swordfish, certain shark, tilefish and king mackerel. Lipophilic compounds such as dioxins and polychlorinated biphenyls (PCBs) can be found in the fatty tissue of some fish. High levels of particular microbial pathogens may be present during certain seasons in various geographic areas, which can compromise the safety of products commonly eaten raw, such as oysters. Additionally, some population groups have been identified as being at greater risk from exposure to certain contaminants in seafood.

Aquatic food products have been shown to contain varying amounts of heavy metals, particularly mercury and fat-soluble pollutants from water pollution, and for this, it is recommended by important Food and Drug Administration Agencies not to eat shark, king mackerel, swordfish and tile fish since they contain high levels of mercury. Pregnant women, older adults and people with weakened immune systems have an increased chance of getting a food borne illness called listeriosis (a food borne bacterial infection caused by gram-positive bacteria causing nausea and diarrhea and even lead to meningitis). These groups to reduce the chance of listeriosis disease from seafood should avoid refrigerated type of seafood. Refrigerated smoked seafood, such as salmon, trout, whitefish, cod, tuna, or mackerel, can be found in the refrigerated section of grocery stores and delicatessens. They should be avoided. There is no danger of getting listeriosis from canned or shelf-stable smoked seafood.

The scientific assessment and balancing of the benefits and risks associated with seafood consumption is a complex task. Diverse evidence, of varying levels of completeness and uncertainty, on different types of benefits and risks must be combined to carry out the assessment required as a first step in designing consumer guidance. In light of uncertainty in the available scientific information associated with both nutrients intake and contaminant exposure from seafood, it is required that seafood/aquatic food consumers should be educated properly. It then balanced the benefits and risks to come to specific guidance for healthy consumption for the population as a whole and as an appropriate for specific target population.

Image of Seafoods amongst consumers

Fish and other seafoods have been repeatedly described as a health promoting food category. However, in many countries, consumers do not meet the recommended daily intake level of consuming two servings of fish per week as recommended by Scientific Advisory Committee on Nutrition in the year 2004. This behaviour may be associated partly with the image of fish. Though the image of seafood is generally positive, the differences have been reported between the images of farmed versus wild fish. Consumers perceive farmed fish as being of lower quality as compared to fish captured in the wild. It was also suggested that despite its possibly preferable sensory properties, quality consistency and guarantees, the image of fish from aquaculture could influence the perception of fish products negatively. The image of farmed fish by means of perceived quality and sensory characteristics has been suggested to be less positive than the image of wild fish. Even though the overall quality image of farmed fish may still be positive, in general, the perceived image related to the safety, healthiness and sustainability is a valuable field for an expansion of the existing knowledge. A less positive image of farmed fish when compared to wild fish could be a barrier for the development and acceptance of products originating from aquaculture.

From perspective of the aquaculture sector and farmed fish industry, the major benefit pertains to the possibility of building trust following transparent communication. Transparency on the subject of production methods likely to increase consumer trust will not harm the image of fish from aquaculture. Increased trust in the production system and a positive image of the product, *i.e.*, farmed fish, are preconditions for a possible increase in the consumption of fish, particularly farmed fish. Finally, the positive associations between subjective knowledge and the image of farmed seafood showed that increasing consumers' knowledge could not harm but only improve the image of seafood from aquaculture.

Consumer Approach in Development of New Seafood Product Concepts

The health benefits of seafood consumption are well known, especially with respect to lowering the risk of coronary heart disease. For many other diseases, such as diabetes type II, cancer, cognitive decline, or development, more efforts are needed to demonstrate the health effects of eating seafood. Public health organizations in various countries recommend that fish should be consumed at least two times per week. However, the average fish consumption is considerably less frequent than recommended.

However, the demand for fish is expected to exceed all the available supplies in the near future owing to the revolutionary changes taking place in the dietary habit of people all over the world and the consumer's look towards the fishery products as healthy food.

Barriers to the Consumption of Seafood

- Product quality
- Consumer attitudes toward choosing fish for a meal
- Involvement with seafood
- Consumer food choice habits
- Beliefs about risks and benefits related to health
- Convenience

Consumer Values Regarding Seafood

- Healthiness
- Satiation
- Convenience
- Visibility and trust
- Freedom of choice
- Successful preparation
- Image improvement
- Availability

- Price

Indian Fisheries

Fisheries is a sunrise sector in Indian agriculture, with high potentials for diversification of farming practices, rural and livelihood development, domestic nutritional security, employment generation, export earnings as well as tourism. The possibilities extend from vast seas to high mountains with valued coldwater species. Untapped potentials exist in island systems to ornamental fishes to value added products. With third position in fisheries and second in aquaculture, India has high potentials in the sector for rural development, domestic nutritional security, employment generation, gender mainstreaming as well as export earnings.

Sustainable Sea Food

Sustainable seafood is food from either fished or farmed sources that can maintain or increase production in the future without deteriorating the ecosystems from which it was acquired. The sustainable seafood movement has gained momentum as more people become aware about both over fishing and environmentally destructive fishing methods.

In general, slow-growing fish that reproduce late in life is quite vulnerable to over fishing, whereas, seafood species that grow quickly and breed young, such as anchovies and sardines, are much more resistant to over fishing. Therefore, sustainability in seafood production can be achieved by:

- Obtaining sea food species from farmed sources
- Increasing consumption of fast growing species
- Monitoring abundance of native stocks
- By using non-destructive fishing methods
- By protecting breeding grounds of important sea food animals
- By containing pollution of water bodies

From the foregoing account, it is clear that the connection between seafood and health is undeniable, yet information available to consumers is confusing. Seafood can be a delicious and heart healthy part of our diets. In fact, a well-balanced diet that includes a variety of fish and shellfish can contribute to heart health and children's growth and development, but as with any type of food, it is important to handle seafood properly in order to reduce the risk of food borne illness. Following the basic food safety tips for buying, preparing and storing fish, the shellfish and other seafood products all can safely enjoy the fine taste and good nutrition of seafood. Majority of population does not consume sufficient seafood especially fatty fish to obtain health benefits, thus, should be encouraged to increase consumption.

■■■

25 Herbal Significance of Flax Seed

S.K. Gupta

Assistant General Manager, National Seeds Corporation Ltd., Beej Bhawan, Pusa, New Delhi-110012, E-mail: <skguptanscplan@yahoo.co.in>, <skguptaji@gmail.com>

The researchers of Duke University Medical Centre, North Carolina reported in June 2007 that the growth rate of prostrate cancer decreased or slowed down in men (scheduled for surgery), whose food was supplemented by Flax seed. Whether it can prevent, the prostrate cancer is to be ascertained. Such findings trigger the curiosity to peep into the past about this botanical herb.

History

Flax seed has a long and extensive history. Originating in Mesopotamia, the flax has been known since the Stone Ages, the culinary use is from times of ancient Greece, and health benefits were widely praised in ancient Rome. Flax seed has been used by various civilizations around the world for more than 10,000 years. The production of linen goes back 5000 years and of cloth in northern Europe dates back to Neolithic times. Pictures on tombs and temple walls at Thebes depict flowering flax plants. In North America, flax was introduced by the Puritans. In fact, rural India has been advocating the use of Flax seed and Flax seed oil for quite a long time. In modern times, its nutritional benefits are being rediscovered.

Introduction

Flax is native to the region extending from the eastern Mediterranean to India and was probably first domesticated in the Fertile Crescent. It was extensively cultivated in ancient Egypt. The New Zealand flax is unrelated. Flax is a member of the genus *Linum* of Linaceae family. The species name is *usitatissimum*, meaning the most useful that would definitely describe the versatility and nutritional value of this tiny seed. It is also known as common flax or linseed in English, *San*, *Alsi* in Hindi, Gujarati and Punjabi, *Ali Vidai* in Tamil, *Jawas* and *Alashi* in Marathi, *Tishi* in Bengali, *Pesi* in Oriya, *Agasi* in Karnataka, *Avise ginzalu* in Telugu and *Cheruchana vithu* in Malyali.

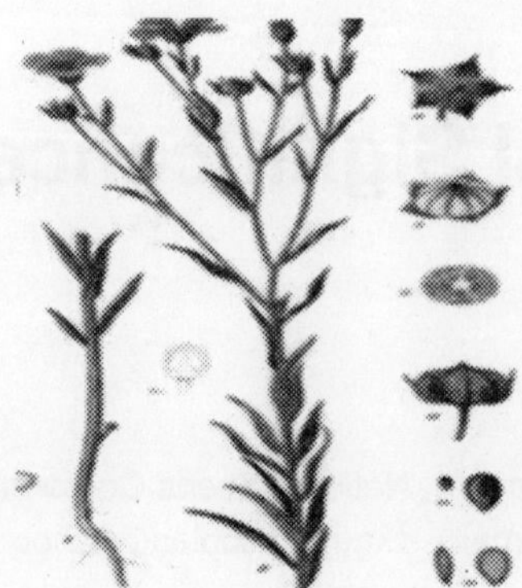

Plant parts of Flax, Courtesy: Wikipedia

Description

Flax is an erect annual, non-edible, self-pollinating plant growing to 120 cm tall with slender stems. The leaves are glaucous green, slender lanceolate, 2-4 cm long and 3 mm broad. The flowers are pure pale blue and 1.5 to 2.5 cm in diameter with 5 petals. The fruit is a round dry capsule with 5-9 mm diameter, containing several glossy reddish brown seeds with flat and oval shape, crisp chewy texture and pointed tip that measures around 2.5x5.0x1.5 mm. The flax seeds come in two basic varieties, *i.e.*, brown and yellow (golden) in colour.

Flax Flower

Yellow Flax seed

Brown Flax seed

(Courtesy: Wikipedia)

Though brown flax can be consumed and has been for thousands of years, it is better known as an ingredient in paints, fibre and cattle feed. Both types have the similar nutritional values and equal amounts of short-chain omega-3 fatty acids. The exception is a type of yellow flax called solin, which is very low in omega-3 fatty acids and has a completely different oil profile. North Dakota State University has created a new variety of the yellow flax called omega as a source of food. It has a nutty-buttery flavour, with a level of the beneficial omega-3 fatty acids comparable to brown flax.

Habitat and Cultivation

It is a winter (*Rabi*) season crop grown under diverse agronomic conditions for its fibre (fibre flax), seed oil (linseed) or both fibre and seed oil (dual-purpose linseed). The most suitable soils besides the alluvial are deep friable loams containing a large proportion of organic matter. Heavy clay soils are out rightly unsuitable. Flax is harvested for fibre production after approximately 100 days, when the base of the plant begins to turn yellow. If the plant is still green, the seed will not be useful and the fibre will be underdeveloped. The fibre degrades once the plant is brown. It is pulled up with the roots (not cut) to maximize the fibre length, thereafter, the flax is allowed to dry, the seeds are removed and is then retted.

Varietal Improvement

Fifty years of breeding efforts under All India Coordinated Research Project on Oilseeds (Linseed) have produced more than 51 improved varieties and made significant impact on Indian linseed production and productivity. A few of the latest notified varieties are SUYOG (JLS 27) in 2004, Baner (KL 224), Indira Alsi 32 (RLC 81) and Kartika (IA 18) in 2005, Sharda (LMS-4-27) in 2006, Deepika (RLC 78) 20/07/2007 and RLU 6 in 2007, LC 2063, Azad Alsi 1 (LMS9-2K), Himani (KL 214) and RLC 92 (IC 555926) in 2008.

The genetic relatedness among these cultivars and genetic erosion that might have occurred during the breeding process is not known. It is well known that information on the extent and distribution of genetic variation among the linseed germplasm are essential for target specific breeding programmes. Traditionally, morphological characters have been used to establish the description of genotypes based on distinctness, uniformity and stability but these characters are often influenced by environment. Recently, molecular markers such as Restriction Fragment Length Polymorphism (RFLP), Amplified Fragment Length Polymorphism (AFLP), Random Amplified Polymorphic DNA (RAPD) and Simple Sequence Repeat (SSR) have been developed with distinct superiority over morphological markers and allozymes in numbers and polymorphisms. Polymerase Chain Reaction-based molecular markers like RAPD are relatively quick, need low amount of DNA and detect good number of polymorphisms without prior sequence information of target genome.

Uses of Flax

Flax is grown for its seeds and fibres both. Various parts of the plant have been used to make fabric, dye, paper, medicines, meal, oil, fishing nets and soap. It is also grown as an ornamental plant in gardens as flax is one of the few plant species capable of producing truly blue flowers with shades of purple, although not all flax varieties produce blue flowers.

Flax fibre

Flax fibres are amongst the oldest fibre crops in the world. Flax may refer to the unspun fibres of flax plant. Flax fibre is soft, lustrous and flexible. It is stronger than cotton fibre but less elastic. The best grades are used for linen fabrics such as damasks, lace and sheeting. Coarser grades are used in manufacturing of twine and rope. It is also a raw material for the high-quality paper industry for the use of printed banknotes and rolling paper for cigarettes. The major flax fibre producing countries are the former USSR, Poland, France, Belgium, Ireland and the Czech Republic.

Flax seed

It can be easily and successfully incorporated and accepted in wide range of recipes. The seed is also available whole or milled for use in breads. It is generally used with wheat flour, as it would not give its distinctive taste. Flax seeds can replace the oil or shortening in a recipe because of its high oil content. If a recipe calls for one-third cup of oil, use of 1 cup of milled flax may replace the oil- a 3:1 substitution ratio. If ground is to be used just, 1 or 2 tablespoons may be added in water, milk, tea, coffee, or yoghurt or sprinkled on salads, cooked vegetables, sandwich, or hot or cold cereals like oatmeal or any other food item where a warm, earthy and subtly nutty flavour is appropriate. For instance, one tablespoon of ground flax seeds and three tablespoons of water may serve as a replacement for one egg in baking by binding the other ingredients together in breads, muffins, pancakes and cookies. It adds two appealing textural features to foodstuffs, *i.e.*, tenderness due to fat content and crunchiness from the seed coat. The ground one is better because it simply passes through the body. The fresh grinded preserves best flavour and nutrition. Flax seed sprouts with a slightly spicy flavour are edible. In rural India, for a very effective nourishment, a sweet (*Laddoo*), prepared by roasted and grounded flax seeds mixed with sugar or jaggery and other supplements like gum and dry fruits is served to women especially after delivery and during menstrual period.

Though there is no set recommendation for daily intake for omega-3 fatty acids, it has established adequate intake amounts between 1.1 and 1.6 gram per day for adults. One tablespoon of ground flax seed provides 1.6 gram of omega-3 fatty acids. Tips for including flax seed in daily diet are:

- Add a tablespoon of ground flax seed to your hot or cold breakfast cereal.
- Add a teaspoon of ground flax seed to mayonnaise or mustard when making a sandwich.
- Mix a tablespoon of ground flax seed into an 8-ounce container of yogurt.
- Bake ground flax seed into cookies, muffins, breads and other baked goods.

Flax seed oil

Flax seeds produce vegetable oil, which is one of the oldest solvent-processed commercial oils and has been used for centuries as a drying oil in painting and varnishing. The oil is used in cooking and is ingested in the capsule form for various other disorders. The flax seed oil (2-3 g) is good although no side effects have been detected with people who consume more. Flax seeds oil is most commonly consumed in salad dressing instead of olive oil or in capsules. For best absorption, the flax seeds oil is taken with food. It is easily mixed with juices and other drinks, and its nutty and buttery flavour complements cottage cheese, yoghurt and many other foodstuffs. However, cooking is not done with it, as this will deplete the oil's nutrients content. Flax seeds oil can be applied externally to soften the dry skin.

Nutritional Profile of Flax seed

Flax seed is an excellent source of nutrients, including carbohydrates, sugar, essential fatty acids (EFA), especially omega 3 fatty acids, amino acids, soluble and insoluble fibre, minerals and vitamins. Flax seeds meal and flour provide a very good source of fibre that can lower cholesterol levels, reduce the cancer-causing chemicals, help relieve constipation and stabilize blood sugar levels in diabetic patients. It is also a good source of folate and vitamin B_6 as well as minerals such as manganese, magnesium, phosphorous, sodium and copper.

Table: Nutritional composition of Flax seed (per 100 g edible portion)

Constituents	Contents	Constituents	Contents
Dietary fiber (g)	27.30	Iron (mg)	5.73
Protein (g)	18.29	Zinc (mg)	4.34
Fat (g)	42.16	Thiamine- vitamin B_1 (mg)	1.644
Carbohydrate (g)	28.88	Riboflavin- vitamin B_2 (mg)	0.161
Sugars (g)	1.55	Niacin- vitamin B_3 (mg)	3.08
Calcium (mg)	255	Pantothenic acid- vitamin B_5 (mg)	0.985
Magnesium (mg)	392	Folate- vitamin B_9 (µg)	0
Phosphorus (mg)	642	Ascorbic acid- vitamin C (mg)	0.6
Potassium (mg)	813	Energy (kcal)	530 (2230 kJ)

Source: USDA Nutrient database

Magnesium helps to reduce the severity of asthma, hypertension, heart attack and stroke, prevents the blood vessel spasm that leads to migraine attacks, and generally, promotes relaxation and restores normal sleep patterns (in *Archives of Internal Medicine* journal). In addition, it is concentrated in lignan phytonutrients- a phytoestrogen (antioxidant) as well as fibre. Its metabolites can bind to estrogen receptors, hence, inhibiting the onset of estrogen-stimulated breast cancer.

Flax Seed Oil and Omega 3 and Omega 6 fatty acids

Many people cut out all oils, thinking they are all bad. This is not the case and the body needs essential fatty acids, *i.e.*, omega 3 and 6 fatty acids (polyunsaturated fats), which are considered good since they regulate a large number of mechanisms including increasing the fluidity of cell membranes and improving their gate-keeping abilities. These mechanisms help to keep toxins out and bring nutrients into cells. Just like other essential vitamins and minerals, essential fatty acids influence the activation of cell genes, acting as second messengers and critical for the synthesis of good eicosanoids- a family of hormone-like substances that help in cell maintenance on a minute-to-minute basis and also reduce inflammation in the body, help blood from clotting and keep blood vessels dilated. It has abundance (50-60%) of alpha linolenic acid (ALA), which is an essential omega 3 fats, "essential" meaning that alpha linolenic acid must be included in diet because the human body cannot synthesize it. The foodstuffs high in alpha linolenic acid content are hempseeds, walnuts, pumpkin seeds, Brazil nuts, sesame seeds, avocados, some dark leafy green vegetables (kale, spinach, purslane, mustard greens, collards, *etc.*), canola (cold-pressed and unrefined), soybean, wheat germ oil, salmon, mackerel, sardines, anchovies and albacore tuna.

The omega 3 fatty acids help to absorb the iodine and this is very useful in treating conditions where this element is present in small amounts. Flax seeds and fish oils are the richest sources of omega 3 fatty acids, that is why, flax seeds oil is compared with fish oil. Alpha linolenic acid shows a lot of promise in fighting diseases. Being high in high-density lipoprotein (good fat), it also helps in reducing cholesterol in the body. Technically, it effects the production of prostaglandins (hormone-like substances synthesized in the body from fatty acids) from the hormones, which in turn affect other areas of the body such as a positive effect on body immunity, the skin's ability to fight inflammation, heart circulation and many others. Flax seeds are rich in alpha linolenic acid (ALA), an omega-3 fat that is a precursor to the form of omega-3 found in fish oils called eicosapentaenoic acid (EPA). Since the fats are found in their isolated form in flax seed oil, it is a more concentrated source of ALA than the seeds themselves. Alpha linolenic acid, in addition to providing several beneficial effects of its own, can be converted in the body to EPA, thus, providing EPA's beneficial effects also. For this conversion to take place readily, however, depends on the presence and activity of an enzyme called delta-6-destaurase, which, in some individuals, is less available or less active than in others. In addition, delta-6-desaturase function is inhibited in diabetes and by the consumption of saturated fat and alcohol. For these reasons, higher amounts of ALA-rich flax seeds or its oil must be consumed to provide the same benefits as the omega-3 fats found in the oil of cold-water fish. Yet research indicates that for those who do not eat fish or wish to take fish oil supplements, flax seed oil does provide a good alternative. A study published in the Journal of Nutrition found that flax seed oil capsules providing 3 grams of alpha-linolenic acid daily for 12 weeks- an amount that would be provided by 3 tablespoons of flax seed oil a day- increased blood levels of EPA by 60% in a predominantly African-American population with chronic illness.

Alpha linolenic acid, eicosapentaenoic acid and docosahexaenoic acid (DHA) are alike in keeping the body's cell membranes flexible and elastic to help cells work properly and in blocking the actions of some compounds that cause inflammation. By such blocking, omega 3 fatty acids help to reduce the risk of chronic diseases like cardiovascular, stroke, diabetes, cancer and arthritis that are marked by inflammation. Docosahexaenoic acid, other omega 3 fatty acid, helps the eye, brain and nervous system to develop properly, therefore, infants specially have need for this and even the aging adults also. One tablespoon per day of flax seeds oil should provide the recommended alpha linolenic acid daily to an adult. However, time-released effects of consuming nuts and other linolenic-rich foodstuffs are considered more beneficial than a once-daily oil intake.

These two omega 3 and 6 essential fatty acids have very important but different roles in the body. One may gain significant health benefits by increasing the level of omega 3 fatty acids in diet. However, it is important to have a balance of both. Good health requires the right ratio of omega 3 fatty acids to omega 6 fatty acids that is around 1:2. Excessive intake of either type of fat can cause health problems. American diets are typically high in omega 6 fatty acids but low in omega-3 fatty acids, *i.e.*, around 1:2 to 1:5. Taking such diets may cause to produce substances that cause inflammation and negatively affect body's response to diseases making more susceptible to heart disease, inflammatory conditions such as arthritis and psoriasis and infections and can lower the body immunity. The ratio present in flax seed oil is about 4:1. Therefore, flax seed oil is a good source of omega 3. One should consume every day small portions of flax seed or flax seed oil. However, if flax seed oil is used in the diet for long time, without other oils, it may cause omega 6 fatty acids deficiency symptoms. Therefore, the best way is to blend flax seed oil with other oils that contain more omega 6 in order to get the right balance such as corn, safflower, cottonseed, sesame and sunflower.

Health benefits of Flax seed

Health benefits of flax seeds in connection with heart disease and anti-inflammatory agent are extensive. Flax seeds are used throughout the world for food, medicine and many other consumer good products. Omega 3 fatty acids, often in short supply in populations with low-fish diets, promote heart health by reducing cholesterol, blood pressure and plaque formation in arteries. Soluble fibres along with omega 3 fatty acids help in healthy blood lipid production making them an increasingly popular to the diets of many health conscious consumers. The natural therapists have often seen the beneficial results of ground up flax seeds and cold pressed flax seeds oil into human diet. Lignans benefit the heart and possess anti-cancer properties, particularly the kind of tumour found in human, post-menopausal breast cancer and prostate cancer. It may also lessen the severity of diabetes by stabilizing blood-sugar levels.

Protection from Diseases

Omega 3 fatty acids are responsible for forming a group of hormones like compounds called prostaglandins, which help in fighting many diseases. They make the cell membrane permeable, which helps in protection of cells against invading bacteria toxins and viruses.

Omega 3 rich Flax seeds Protect Bone Health

It improves the absorption of calcium. Alpha linolenic acid in flax seeds and walnuts promotes bone health by preventing excessive bone turnover when consumption of foodstuffs rich in omega 3 fat results in a lower ratio of omega 6 to omega 3 fats in the diet. Similarly, the diets rich in omega 3 fatty acids from fish (DHA and EPA) naturally result in a lowered ratio of omega 6 to omega 3 fats and reduce bone loss. Researchers think this is because omega 6 fats are converted into pro-inflammatory prostaglandins, while omega 3 fats are metabolized into anti-inflammatory prostaglandins.

Anti-Inflammatory Benefits

Omega 3 fatty acids are used by the body to produce Series 1 and 3 prostaglandins, which are anti-inflammatory hormone-like molecules, in contrast to the Series 2 prostaglandins, which are pro-inflammatory molecules produced from other fats, notably the omega 6 fatty acids, which are found in high amounts in animal fats, margarine and many vegetable oils including corn, safflower, sunflower, palm and peanut. Omega 3 fatty acids help to reduce the inflammation in asthma, osteoarthritis, rhcumatoid arthritis, migraine headaches, gout, lupus, joints and kidneys and osteoporosis. It reduces the swellings and intensity of joint pain and also heals sprains and bruises faster. The gel of flax seeds has been used as a poultice on injured areas in many Indian homes for quite a long time. It shortens recovery time for fatigued muscles after exertion and increases the body's production of energy and also stamina.

Flax seed provides cholesterol-lowering benefits

Flax seeds meal and flour stimulate brown fat cells and increase the metabolic rate making it easier to burn off fat, which lower body mass index, total cholesterol (HDL and LDL), triglycerides, platelet reactivity and neutrophil activity. In a study involving 40 patients with high cholesterol (greater than 240 mg/dL), daily consumption of 20 g of ground flax seeds was compared to taking a statin drug. After 60 days, significant reductions were seen in total cholesterol, low-density lipoprotein (bad cholesterol), triglycerides and the ratio of total to high-density lipoprotein (good cholesterol) in both groups. In those eating flax seeds, significant reductions were seen in total cholesterol (-17.2%), LDL-cholesterol (-3.9%), triglycerides (-36.3%) and the ratio of total cholesterol/HDL-cholesterol (-33.5%) were observed in the diet + flax group compared to baseline. Similar reductions were seen in those taking the statin. Benefits did not significantly differ between the two groups.

Protect Diabetes

Obesity is wide spreading nutritional disorder and Diabetes Mellitus type 2 is a gathering storm. High doses of alpha linolenic acid may reduce blood cholesterol and triglyceride levels, which have shown to stabilize blood sugar and also aids in circulation which is an important factor in diabetes. However, some people with T-2 diabetes cannot properly metabolize alpha linolenic acid so flax seeds oil may be of no benefit to them. Lastly, lignan-rich fiber has also been shown to decrease insulin resistance, which, in turn, reduces bio-available estrogen, which also lessens breast cancer risk. And, as insulin resistance is an early warning sign for type 2 diabetes, flax seed may also provide protection against this disease. Dieticians recommend it during pregnancy for having gestational diabetes.

Protect Cancer

Flax seeds as part of a low-fat diet reduce the exposure of colon cells to cancer-causing chemicals and may slow growth of tumours in men with prostate-cancer and women with breast-cancer. According to a study published in the journal Urology, 25 men with prostate cancer supplemented with 3 tablespoons of ground flax seeds daily for an average of one month had a greater rate of prostate tumour cell death than who did not follow this diet.

Protect Heart Disease

Alpha-linolenic acid produces substances that reduce the formation of blood clots- sticky platelets-, which can reduce the risk of angina, heart attack and stroke in patients with atherosclerosis or diabetic heart disease by improving the function and integrity of arteries that carry blood throughout the body and to the brain.

Lowers High Blood Pressure (Hypertension)

Omega 3 fatty acids lower the both systolic and diastolic blood pressure, prevent the blood vessel spasm, which leads to migraine attacks, and generally promotes relaxation and restores normal sleep patterns. One tablespoon of flax seeds oil per day appears to be effective in lowering blood pressure.

Rich in Fibre Beneficial in Digestion

Flax seed has a potential to increase laxation because it provides dietary fibres, which absorb water, also help in water retention, increasing intestinal bulk that helps to ease constipation and bowel movements, therefore, is useful in controlling constipation, hemorrhoids, biventricular disorders. Since it combats inflammation, it is useful in repairing any intestinal tract damage and in keeping the gallstones at bay, and sometimes, it dissolves existing stones. It can often improve the function of the liver.

Protect from Allergy and Asthma

Being a good source of magnesium, flax seeds help to protect from allergic responses and reduce the severity of asthma by keeping airways relaxed and open.

Special Protection for Women's Health

Flax seed meal and flour have been studied quite a bit lately for their beneficial protective effects on women's health. Flax seed is particularly rich in lignans, special compounds also found in other seeds, grains and legumes that are converted by beneficial gut flora into two hormone-like substances called enterolactone and enterodiol. These hormone-like agents demonstrate a number of protective effects against breast cancer and are believed to be one reason a vegetarian diet is associated with a lower risk for breast cancer. Studies show that women with breast cancer and women who are omnivores typically excrete much lower levels of lignans in their urine than vegetarian women without breast cancer. In animal studies conducted to evaluate lignans beneficial effect, supplementing a high-fat diet with flax seed flour reduced early markers for mammary (breast) cancer in laboratory animals by more than 55%.

In a study published in the American Journal of Clinical Nutrition, when postmenopausal women ate a daily muffin containing 25 grams of soy protein, 25 grams of ground flax seed or a placebo muffin containing neither for 16 weeks, the estrogen metabolism of those eating flax seed but not soy or placebo was altered in several important protective ways:

- Levels of 2-hydroxyestrone, a less biologically active estrogen metabolite thought to be protective against breast cancer, increased significantly.
- The ratio of 2-hydroxyestrone (the protective estrogen metabolite) to 16 alpha-hydroxyestrone (an estrogen metabolite thought to promote cancer) increased.
- Blood levels of the estrogen fractions (estradiol, estrone and estrone sulfate) did not change significantly, which is important since estradiol is involved in maintaining bone mass.

So eating about 25 grams of ground flax seed each day will affect the way estrogen handled in postmenopausal women in such a way that offers protection against breast cancer but will not interfere with estrogen's role in normal bone maintenance.

In addition to lessening a woman's risk of developing cancer, the lignans rich flax seed can promote normal ovulation and extend the second, progesterone- dominant half of the cycle. The benefits of these effects are manifold. For women trying to become pregnant, consistent ovulation significantly improves their chances of conception. For women between the ages of 35 and 55 who are experiencing peri-menopausal symptoms such as irregular menstrual cycles, breast cysts, headaches, sleep difficulties, fluid retention, anxiety, irritability, mood swings, weight gain, lowered sex drive, brain fog, fibroid tumors and heavy bleeding, a probable cause of all these problems is estrogen dominance. Typically, during the 10 years preceding the cessation of periods at midlife, estrogen levels fluctuate while progesterone levels steadily decline. Flax seed, by promoting normal ovulation and lengthening the second half of the menstrual cycle, in which progesterone is the dominant hormone, helps to restore hormonal balance, which can relieve some cases of Premenstrual Syndrome (PMS). In addition, flax seeds oil is often recommended as a galactagogue.

Preliminary research also suggests that flax seeds may serve a role in protecting post-menopausal woman from cardiovascular disease. In a recent double-blind randomized study, flax seeds reduced total cholesterol levels in the blood of postmenopausal women who were not on hormone replacement therapy by an average of 6%.

Fend Off Dry Eyes

Dry eye syndrome afflicts more than 10 million Americans. Artificial tears offer only temporary relief. Expensive prescription drugs promise help but at the cost of potentially serious side effects. A research published in the American Journal of Clinical Nutrition involving nearly 40,000 female health professionals aged 45-84 found that essential fatty acids- the omega-3 fats (found in high amounts in cold water fish and flax seeds) and the omega-6 fats (found in red meat, safflower, sunflower, soy and corn oils)- play an important role. Women whose diets provided the highest amounts of omega-3 fatty acids had a 17% lower risk of dry eye syndrome compared with those consuming the least of these beneficial fats. In contrast, a diet high in omega-6 fats but low in omega-3s significantly increased dry eye syndrome risk. Women whose diets supplied a high ratio of omega-6 to omega-3 fatty acids had a 2.5 folds higher risk of dry eye syndrome compared to those with a more balanced intake of fatty acids.

Researchers specifically looked at eating tuna fish- a main source of omega-3 fatty acids in the American diet- compared with women eating less than one 4-ounce serving of tuna a week:

- Women who ate 2 to 4 servings of tuna per week had a 19% lower risk of dry eye syndrome.
- Women eating 5 to 6 servings of tuna per week had a 68% lower risk of dry eye syndrome.

These findings suggest that increasing dietary intake of omega-3 fatty acids may reduce the risk of dry eye syndrome, an important and prevalent cause of ocular complaints. In addition to tuna fish, omega-3 fatty acids are richly supplied by other fatty fish (such as salmon, mackerel, halibut, sardines, and herring), and flax seeds/flax seed oil. A variety of cold-water fish and flax seeds can improve eyesight and perception of colours.

Helps in Skin Disorders

Flax seeds oil is well known to have a beneficial effect on the skin as it assists with conditions such as acne, eczema, psoriasis, dandruff, sunburn, rosacea and irritant skin disorders like itching, swelling and redness. It can help in the treatment of multiple sclerosis. The essential fatty acids target the sites of inflammation and bring about an overall soothing, also help to revitalize and soften dry skin, ensure healthy hair and nails, prevent nails from cracking and breaking and help to combat the effects of aging.

Protect the Nervous System and Brain

Flax seeds oil helps to reduce the severity of nerve damage and also aids in the triggering of nerve impulses. Since it nourishes the nerve, it may possibly be of some use in the treatment of Parkinson's disease. It helps to combat the effects of aging and another important area where it is of great help is the brain. Doctors generally believe that the best way to treat mood, thought and memory disorders is with pharmaceutical medicines that directly influence levels of brain chemicals such as serotonin, dopamine and acetylcholine. Although drugs have very important clinical uses, they are only part of the solution. A comprehensive approach to treating cognitive disorders should include foods and supplements that benefit the overall health of brain cells. These include omega-3 fatty acids found in flax seed and fish. One way to influence brain health through diet is to consume the right fats and oils. About 60% of the brain consists of lipids (fats), which make up the lining, or cell membrane of every brain cell. The types of fat present in the brain influence its structure and function. The omega 3 fatty acids help to retain emotional health of a person, helping to tackle depression and possibly Alzheimer's disease.

Protect Autoimmune Disorders

The disorders such as rheumatoid arthritis and ulcerative colitis cause immune cells to attack healthy tissues in the body. Flax seeds favourably affect immunity with the help of two major components *eicosanoids* and *cytokines* converting to a substance in the body that can inhibit this autoimmune reaction. Alpha linolenic acid and lignans modulate the immune response and may play a beneficial role in clinical management of autoimmune diseases.

Flax seed Oil with Cottage Cheese

In mid 1950's, Dr. Johanna Budwig Germany's premier biochemist began her extensive research on the benefits of flax seed because of its rich essential fatty acid content. Dr. Budwig assisted many seriously ill helping them to regain their health through a simple regimen of flax seed oil blended with cottage cheese.

She supplemented her patient's diets with 2 tablespoons of flax seed oil combined with one-forth cup of cottage cheese and found that a yellowish green substance in their blood was replaced with a healthy red pigment, hemoglobin. She also found that the tumors receded and disappeared, anemia was alleviated and vital energy was regained. Here is the blend:

- One cup organic cottage cheese
- Two to five tablespoons of flax seed oil
- One to three tablespoons of freshly ground up flax seed

- Enough water to make it soft
- A little cayenne pepper (red hot chilly pepper)

Why cottage cheese should be consumed? The essential fats bind to the proteins in the cottage cheese making them more effective because it is easily absorbed by body.

Safety

While flax seeds contain cyanogenic glycosides compounds at normal levels and without protein malnutrition, researchers currently maintain that this is not of concern and should cause no adverse effects (50 g to be a safe amount for most people). The heat employed by cooking has been found to eliminate the presence of these compounds. Some people have gastrointestinal symptoms, such as flatulence and bloating, when they first begin to incorporate flax seeds into their diet. It is suggested to start with a small amount such as one teaspoon and slowly build up to intake goal. When increasing fibre intake in the diet, it is also a good idea to include fluid (water) intake as well. Though the Institute of Medicine has not set a recommended daily intake for omega 3 fatty acids, it has established adequate intake amounts of between 1.1 and 1.6 g per day for adults and 1 tablespoon of ground flax seeds provides 1.6 g of omega 3 fatty acids. Presence of antioxidants provide a long self-life and easy safe storage, however, flax seeds oil used for dietary supplementation should be kept in the refrigerator or freezer.

■■■

26 Therapeutic Significance of Oilseeds

Anureet Kaur[1], G.P.S. Dhaliwal[2] and Harsimrat K. Bons[3]

[1]Assistant Agronomist, P.A.U. Regional Station, Bathinda, Ph.D. Scholar, Department of Plant Breeding, CCS H.A.U., Hisar, E-mail: <anureet_1@yahoo.com>
[2]Assistant Professor (Horticulture), KVK, Samrala (Ludhiana), Punjab Agricultural University, Ludhiana, Punjab, India

The role of dietary fats and oils in human nutrition has created wide spread interest among consumers, researchers and food producers. The essential fats play important role in growth, reproduction, vision, skin health and metabolism of cholesterol and can help in preventing heart diseases and anti-immune diseases such as rheumatoid arthritis. The oilseeds, like castor, sunflower, sesame, flaxseed, peanuts, mustard, *etc.*, when consumed one or other way help in keeping oneself fit. The natural benefits and curative properties of these oilseeds have been given as under:

Common name: Castor (Arind)

Botanical name: *Ricinus communis*

Family: Euphorbiaceae

The average composition of castor seed oil/fatty acid chains is given below:

Constituents	Contents	Constituents	Contents
Ricinoleic acid (%)	85-95	Stearic acid (%)	1-0.5
Oleic acid (%)	6-2	Palmitic acid (%)	1-0.5
Linoleic acid (%)	5-1	Dihydroxysterric acid (%)	0.5-0.3
Linolenic acid (%)	1-0.5	Others (%)	0.5-0.2

Castor oil is obtained from the beans or the seeds of the castor plant. It is colourless or pale yellow coloured oil with a mild or no odour and taste. This oil has been in use for the centuries in many parts of the world like India, China and Egypt for alleviating a number of health conditions. It is a triglyceride of fatty acids, out of which ricinoleic acid, oleic acid and linoleic acid are the main components, which account for most of the medicinal properties of this oil. Here is a brief account of what for the castor oil is used as well as the health benefits of castor oil.

One of the well-known and most important castor oil benefits is that it can relieve constipation. Castor oil is a well-known laxative, and hence, can be effective in alleviating constipation and bleeding piles. It can promote regular bowel movement, and due to its anti-inflammatory properties, it can provide relief in inflammatory bowel disorders.

Along with relieving constipation, it is used for skin burns, sunburns, minor cuts or abrasions and skin disorders. Ricinoleic acid found in castor oil is a very powerful agent and can inhibit the growth of bacteria, yeast and virus. It can relieve the pain and inflammation associated with infections. The fatty acids, *i.e.*, ricinoleic acid, oleic and linoleic acid found in castor oil, can provide relief in rheumatism, arthritis and gout. It is also effective in the treatment of warts. Massaging castor oil daily on the affected area for few minutes can help in removing warts. The oil is also suitable for hair care, especially in case of excessive hair loss and protects the scalp and hair from fungal and microbial infections. This smooth oil is readily absorbed by the skin, and hence, can alleviate dry skin conditions and remove blemishes and pigmentation. It can be used for skin ulcers, finger and toenail infections. For women, the oil has additional health benefits, as it can be used for menstrual disorders like delayed and painful menstruation and is also used for inducing labour in pregnant women.

Castor oil and its derivatives are essential ingredients in many drugs like miconazole, paclitaxel, sandimmune, tacrolimus, saperconazole and nelfinavir mesylate. Miconazole is used as an anti-fungal agent, while tacrolimus and sandimmune are used for suppressing the immune system. On the other hand, paclitaxel is mainly used in chemotherapy and nelfinavir mesylate is used as Human Immunodeficiency Virus (HIV) protease inhibitor.

Common name: Sunflower (Suraj mukhi)

Botanical name: *Helianthus annuus*

Family: Asteraceae

The nutritional value of sunflower per 100 grams is given below:

Constituents	Contents	Constituents	Contents
Energy (KJ)	3699	Vitamin K (μg)	5.4 (5%)
Carbohydrates (g)	0	Vitamin E (mg)	41.08 (274%)
Fat (g)	100	Palmitic acid	4-9%
Saturated	9.748	Stearic acid	1-7%
Monosaturated	83.594	Oleic acid	14-10%
Polysaturated	3.798	Linoleic acid	48-74%

Sunflower oil is non-volatile oil extracted from sunflower seeds. It is commonly used in food as frying oil and in cosmetic formulations as an emollient. The sunflower oil benefits are manifold, especially imparted by its health-enhancing nutrients content can choose from two types of sunflower oils, *viz.* linoleic and high oleic, that are available in the market. Sunflower oil containing high oleic is known to be rich in monounsaturated fatty acids but it is the linoleic type that contains polyunsaturated fat, which is commonly known and used as sunflower oil.

Sunflower oil benefits, especially when they are considered in the light of heart diseases, give it an edge over the commonly used cooking oils as follows:

Diet and cardiovascular benefits: Sunflower oil is high in the essential vitamin E and low in saturated fat. Therefore, sunflower oil of any kind has been shown to have cardiovascular benefits. This benefit of sunflower oil is further enhanced by the phytochemicals, such as choline, phenolic acid, *etc.*, present in it. These chemicals are known to prevent heart diseases.

Skin care: It is found that sunflower oil is useful in lowering the risk of infections in the low birth weight (pre-term infants). In these babies, there is always a risk of infection because of their underdeveloped organs, including skin. Sunflower oil works as a protective barrier, when applied to the skin of these babies.

Fights free radicals: One of the benefits of the sunflower oil is that it acts as an antioxidant due to the presence of vitamin E in it. As an antioxidant, it neutralizes cancer-causing free radicals. Vitamin E, in the sunflower oil, also acts as moisturizer, by helping the body cells retain water inside them. For this property, sunflower is often used in cosmetic products. The vitamin E present in oil prevents rheumatoid arthritis, asthma and cancer. It also contains vitamin B, Tocopherols, carotenoids, selenium, copper, iron, zinc, calcium, omega-6 fatty acids. The folate or folic acid, present in sunflower oil, helps the body in the manufacture of new cells. Magnesium helps in preventing muscle cramps, while tryptophan helps in relaxing the brain and promoting sleep by helping the body in the production of neurotransmitter, *i.e.*, serotonin.

Common name: Linseed (flaxseed)

Botanical name: *Linum usitatissimum*

Family: Linaceae

The nutritional value of linseed 100 grams is given below:

Constituents	Contents	Constituents	Contents
Total dietary fiber (g)	14	Manganese (mg)	traces
Mucilage	12	Zinc (mg)	0.0057
Protein (g)	26	Copper	traces
Oil (g)	35	Sulfur	0.21
Calcium (mg)	0.21	Chlorine	0.043
Magnesium (mg)	0.38	α-Linolenic acid- omega-3 (g)	20.7
phosphorus	0.70	Oleic Acid (g)	6.7
Potassium (mg)	0.74	Linoleic acid- omega-6 (g)	4.9
Sodium (mg)	0.046	Stearic acid (g)	1.4
Iron (mg)	.0077	Palmitoleic acid (g)	1.8

Linseed is the richest source of omega-3 fatty acids (50-60% omega-3s). It contains almost twice as much of the omega-3s as fish oil, which may be beneficial for reducing inflammation leading to atherosclerosis, preventing heart disease and arrhythmia and is required for normal infant development. The poultice of flaxseed is prepared for curing boils and skin abscesses. The poultice is prepared by rolling 3 tablespoons of flaxseeds in a 6 inch square of clean white cloth. Both ends of the cloth are tightened and the middle portion is dipped in the boiling water. The infected area is covered with poultice followed by a dry cloth. The flaxseed retains the heat, providing relief to the infected area. The poultice is removed when it has cooled. The inclusion of linseed in our diet can help in following ways:

Heart disease: Omega-3s lowers the high blood pressure by lowering the high blood cholesterol and triglyceride levels by as much as 25 and 65%, respectively. Omega-3s decrease the probability of a clot blocking the artery in the brain (stroke), heart (heart attack), lungs (pulmonary embolism), or the organ (peripheral vascular disease, gangrene).

Cancers: Omega-3s are known to dissolve tumor. Germany has over 1000 documented cases of successful cancer treatment using linseed oil along with additional support. The recent research shows that omega-3s kill human cancer cells on the same culture. Breast, lung and prostrate cancer cell lines were studied.

Diabetes: According to Scientists, diabetes has its origin in deficiency of omega-3s as well as omega-6 fatty acids, and it is made worse by current lack of vitamins and minerals.

Arthritis: Omega-3s have been found to be effective in the successful treatment and prevention of arthritis. Both fish and linseed oils have been used. More recently, research using combinations of the omega-3 and -6 fatty acids found that 60% of rheumatoid arthritics were able to discontinue completely their non-steroidal anti-inflammatory drugs (NSAIDS) and another 20% were able to reduce their dosages of NSAIDS in half.

Asthma, Allergy and Inflammatory Tissue Conditions: Linseed oil can relieve asthma noticeably, sometimes, within a few days of starting to take the oil. It also helps to decrease the allergic response. Since the body must be rebuilt, a longer time is needed before allergies are alleviated, thus, total nutritional support is required. The meningitis, bursitis, tendinitis, tonsillitis, gastritis, ileitis, colitis, arthritis, phlebitis, prostatitis, nephritis, splenitis, hepatitis, pancreatitis, otitis, *etc.* as well as psoriasis and lupus, all of these inflammatory conditions may be helped by omega-3s.

Skin conditions: Linseed oil is famous for its ability to make the skin smooth, soft and velvety. It will also alleviate those skin conditions whose origin is the lack of the omega-3s in the diet.

General health and vitality: Linseed oil helps the kidneys in removing sodium and water. Water retention (edema) is involved in swollen ankles, some forms of overweight, premenstrual syndrome and late stages of cancer and cardiovascular disease. One of the most noticeable signs of improved health from the use of linseed oil is increased vitality, *i.e.*, more energy. Athletes notice that their fatigued muscles recover from exercise more quickly. Omega-3s increase stamina and calmness under stress.

Other conditions: Linseed oil can also be helpful in multiple sclerosis (in places where essential fatty acid consumption is high, multiple sclerosis is very rare). Omega-3s are necessary for visual function (retina), adrenal function (stress) and sperm formation, cystic fibrosis (oils containing omega-3 will loosen the viscous mucous secretions and relieve breathing difficulties), some cases of sterility and miscarriage, some glandular malfunctions, some behavioural problems (schizophrenia, depression, manic-depressive disorder, *etc.*), addictions (to drugs or alcohol) and pathologically deviant behaviours.

Medicinal applications: For relieving headache, linseeds are ground in cold water and applied on painful area, *i.e.*, forehead. This is also useful in case of wounds on head. In the conditions of insomnia, equal quantities of linseed and castor oil are mixed, beaten well in bronze plate and applied the poultice on the eyes in order to get sound sleep. The reddishness in the eyes may also be reduced by applying grinded seeds (viscous pulp) around the eye. The pain in the ear can be overcome by boiling linseed in onion juice and pouring it in the ears to reduce swelling. Roasted linseeds can be taken along with honey for curing cough and liver enlargement. A home remedy prepared from linseeds is very effective for a tuberculosis patient. The linseeds (25 g) are soaked overnight, sieved, and then, heated with water and consumed with lemon juice. For dry cough and respiratory problems, about 3 g linseeds are grinded, soaked in 250 g of boiling water, kept for one hour, and then, consumed with unrefined sugar. It is also helpful for easy urination.

Joint pain can be relieved by applying a mixture of isabgol and grinded linseeds on the painful area. Hip and back pain can be calmed by massaging the area with dried ginger (*Zanziber officinale*) powder mixed with hot linseed oil. The poultice of linseed oil is very beneficial in case of swelling in arthritis. The lep of grinded linseeds on the infected area can heal insect bite. The mixture of linseed oil and limewater (lime should sediment in bottom) makes a white ointment known as carron oil. The application of this ointment can bring immediate relief on burnt area.

Common name: Peanut or Groundnut (mungfali)

Botanical name: *Arachis hypogaea*

Family: Fabaceae

The nutritional value of peanuts per 100 grams is given below:

Constituents	Contents	Constituents	Contents
Total dietary fiber (g)	9.0	Potassium (mg)	332 (7%)
Fat (g)	48	Iron (mg)	2 (16%)
Saturated fat (g)	7	Zinc (mg)	3.3 (33%)
Monounsaturated fat (g)	24	**Vitamins**	
Polyunsaturated fat (g)	16	Thiamine (mg)	0.6 (46%)
Carbohydrates (g)	21	Riboflavin (mg)	0.3 (20%)
Protein (g)	25	Niacin (mg)	12.9 (86%)
Minerals		Pentothenic acid (mg)	1.8 (36%)
Calcium (mg)	62 (6%)	Pyridoxine (mg)	0.3 (23%)
Magnesium (mg)	184 (50%)	Total folate (µg)	246 (62%)
Phosphorus	336 (48%)		

Constituents	Contents	Constituents	Contents
Amino acids		Valine (g)	1.052
Tryptophan (g)	0.244	Histidine (g)	0.634
Threonine (g)	0.859	Arginine (g)	3.001
Isoleucine (g)	0.882	Alanine (g)	0.997
Leucine (g)	1.627	Aspartic acid (g)	3.060
Lysine (g)	0.901	Glutamic acid (g)	5.243
Methionine (g)	0.308	Glycine (g)	1.512
Cystine (g)	0.322	Proline (g)	1.107
Phenylalanine (g)	1.300	Serine (g)	1.236
Tyrosine (g)	1.020	Energy (kcal)	570 kcal

Peanuts, also commonly referred to as groundnuts, are legumes like black beans or lentils. One pod of peanut comprises two to three kernels, which are loaded with loads of nutrients. The nutritional facts of peanut are given as under:

Protein provider: Peanuts are high in high quality protein content. They contain more protein than most nuts or legumes. The protein, which is required for body metabolism and builds up of body cells, is the building block of the body. Vegans and children need high protein in their diet, thus, for such people peanuts are good source of protein. It is recommended that growing children, expecting women and nursing mothers consume roasted groundnuts with jaggery and goat's milk. It is said to provide resistance and immunity against dangerous infections like Hepatitis and also Tuberculosis

Mineral provider: Peanuts are loaded with several essential nutrients like copper, magnesium, zinc and phosphorous. The body needs these minerals for various purposes in our body. In fact, some of these minerals even help in protecting against heart disease.

Vitamins provider: Vitamins are important for growth and development of cells and tissues of the body. They assist in smooth functioning of the body and also regulate body metabolism. They also play different roles in fighting infections and in the formation of tissues and bones. Since peanuts are high in folate, pregnant women consuming them will reduce the risk of birth defects and other anemia related medical conditions. Groundnut and its products are very beneficial in the treatment of hemophilia and other such inherited blood disorders. People suffering from nose bleeding also benefit from eating groundnuts and it is also helpful in reducing excessive menstruation bleeding in women.

Lowers blood cholesterol: Peanuts comprise beneficial monounsaturated and polyunsaturated fats, which are known to lower the cholesterol levels in blood. Low levels of cholesterol in the blood will reduce the risk of heart disease and heart attacks due to arterial blockage.

Antioxidants provider: Peanuts are known to comprise high concentrations of polyphenolic antioxidants, which are associated with reducing stomach cancer risks by decreasing the production of harmful carcinogenic nitrosamines. Another polyphenol antioxidant named resveratrol, which is known to have cancer-fighting properties, is also present in peanuts. It also has the ability to fight degenerative nerve diseases, viral and fungal infections and Alzheimer's disease. The vitamins E in peanuts also have wonderful antioxidant properties, which is useful to destroy the free radicals that harm cell membranes.

Common name: Sesame (til)

Botanical name: *Sisamum indicum*

Family: Pedaliaceae

The nutritional value of sesame seed kernels dried (decorticated) per 100 grams is given below:

Constituents	Contents	Constituents	Contents
Total dietary fiber (g)	11.6	Leucine (g)	1.500
Fat (g)	61.21	Lysine (g)	0.650
Sugars (g)	0.48	Methionine (g)	0.880
Carbohydrates (g)	11.73	Cystine (g)	0.440
Protein (g)	20.45	Phenylalanine (g)	0.940
Minerals		Tyrosine (g)	0.790
Calcium (mg)	975	Vallne (g)	0.980
Magnesium (mg)	345 (93%)	Histidine (g)	0.550
Phosphorus	667 (95%)	Arginine (g)	3.250
Potassium (mg)	370 (8%)	Alanine (g)	0.990
Iron (mg)	14.5 (116 %)	Aspartic acid (g)	2.070
Sodium (mg)	47 (2%)	Glutamic acid (g)	4.600
Amino acids		Glycine (g)	1.090
Tryptophan (g)	0.330	Proline (g)	1.040
Threonine (g)	0.730	Serine (g)	1.200
Isoleucine (g)	0.750	Energy (kcal)	630

The sesame seeds are exceptionally rich in minerals and contain vitamin B_1 (thiamine) and vitamin E (tocopherol). They contain lignans, including unique content of sesamin, which are phytoestrogens with antioxidant and anti-cancer properties. Among edible oils from six plants, sesame oil had the highest antioxidant content. Sesame seeds also contain phytosterols associated with reduced level of blood cholesterol. The nutrients of sesame seeds are better absorbed if they are ground or pulverized before consumption, as in *tahini*. Women of ancient Babylon consumed halva, a mixture of honey and sesame seeds to prolong youth and beauty, while Roman soldiers ate the mixture for strength and energy.

Bone health: It is said that a handful of sesame seeds contains more calcium than what a glass of milk contains. The calcium that these seeds contain help maintain bones and prevent disorders related week bones and teeth. It is especially beneficial in preventing osteoporosis in middle aged and older women. This is one of the important health benefits of sesame seeds.

Anti-cancer: The sesamin and sesamolin that sesame seeds contain are known to impart anti-cancer properties to sesame seeds. Their consumption is known to prevent the development and growth of cancerous cells and aid cancer treatment.

Migraine relief: Magnesium contained in sesame seeds helps prevent trigeminal blood vessels spasms that trigger migraine attacks. Regular consumption of sesame seeds is also known to eliminate the migraine attacks and related symptoms like headache and dizziness.

Strengthens nervous system: The vitamin E that sesame seeds contain helps in strengthening the nervous system and prevents disorders and ailments of the nervous system. This is one of the most important sesame seeds nutritional benefits.

Skin care: Sesame seeds and the oil obtained from them are very beneficial for skin and hair. The oil is known to have properties that can soothe skin burns and prevent various skin related disorders. The consumption of sesame seeds can benefit all the skin types.

Intestine and digestive health: Sesame seeds help in preventing intestinal worm infection and also eliminate worms from the intestines. The dietary fiber that they contain helps in maintaining good digestive health, and prevents and cures constipation and other related disorders.

Cardiovascular health: The sesamolin and sesamin present in seeds help in lowering and preventing high blood pressure and cholesterol. Thus, they protect against the development of cardiovascular diseases and disorders.

Asthma prevention: The magnesium present in sesame seeds helps in preventing asthma attacks and various other respiratory system disorders by preventing airway spasms.

Menopausal relief: Consuming sesame seeds have known to bring a great amount of relief in menopausal women as it helps in dealing with various menopausal symptoms like hot flushes and insomnia. This is one of the essential black sesame seeds, which provide health benefits. Other sesame seeds health benefits include rheumatoid arthritis relief, regulation of body metabolism, faster wound healing, *etc.*

Common name: Safflower (Taramira)

Botanical name: *Carthamus tinctorius*

Family: Asteraceae

Safflower oil has the highest source of polyunsaturated fats than that found in any other type of vegetable oil, *i.e.*, almost 79% polyunsaturated fatty acids (linoleic acid). It also contains 13% of monounsaturated fatty acids (oleic acid) and only 8% of saturated fatty acids. The other essential nutrients present in safflower oil are omega-6 fatty acids, cis-linoleic acid, vitamin E, *etc.* Due to the presence of these ingredients in safflower oil, this oil is said to have various benefits to the health of individuals. The safflower oil benefits to overall health are as follows:

Weight loss: As safflower oil contains high amounts of polyunsaturated fats, it helps in creation of prostaglandins in the individual's body. These hormone like compounds aid in strengthening the cell membranes and also in proper functioning of the body. One of the best health benefits of safflower oil is that it is an effective way of eliminating the excess fat accumulated in the body, thus, helps in losing weight. Brown fat is present in several areas of the body, which is a major factor that aids in burning calories. The cis-linoleic acid present in safflower oil is converted into gamma linolenic acid (GLA) that triggers the brown fat to burn calories.

Reduction in cholesterol levels: Another benefit of using safflower oil is that it helps in the reduction of cholesterol level in blood, which reduces the risk of heart diseases in the individual. Apart from this, safflower oil is also said to strengthen immune system of the body, which increases its ability to resist the appearance of certain infections as well as diseases.

Diabetes and premenstrual syndrome: Many doctors and dietitians recommend the use of safflower oil to obese people suffering from diabetes. Research has been conducted to find the benefits of safflower oil in older diabetic women who are overweight. These women were asked to take safflower oil supplements for a period of sixteen weeks, and the results showed loss of excess fat, accompanied with a significant decline in their blood sugar levels. Apart from diabetic patients, using safflower oil is an effective method of curing premenstrual syndrome (PMS). Consuming safflower oil instead of hormonal drugs helps in regularizing menstrual cycles, thus, getting rid of the problem of irregular periods.

Healthy skin and hair growth: Safflower oil also promotes hair growth and a healthy skin as it contains vitamin E. People who have dry, permed and coloured hairs are recommended to get a regular safflower oil massage for healthy looking hair. This odourless oil is mild, has a very light texture, and so easily absorbs into the scalp. People who are suffering from excess hair fall can find effective hair regrowth with the application of safflower oil on the hair. Apart from benefiting the hair, there are also many safflower oilskin benefits. People who have dry skin should use safflower oil, which helps in moisturizing the skin, especially during winter, thus, making it smooth and healthy. People are also recommended the use of safflower oil for acne treatment. Application of safflower oil on the skin helps in getting rid of other skin problems like scars, wrinkles, *etc.* along with acne.

Common name: Mustard (Sarson)

Botanical name: *Brassica* sps.

Family: Brassicaceae

There are number of *Brassica* species, which have many beneficial properties related to health.

Rapeseed (*Brassica napus*): Canola is actually a cultivar of rapeseed plant, which is a member of the mustard family. The plant was originally developed by cross- breeding the rapeseed plant, mainly to produce edible oil from its seeds. The oil derived from the seeds of such canola or genetically engineered rapeseed plant has been found to contain much lower level of erucic acid, and hence, considered suitable for human consumption. Eventually, several studies have revealed that canola oil contain some important nutrients, required for human health.

Canola oil has been found to contain some essential fatty acids, which can have beneficial effects on the heart and cardiovascular system. About 218 g of canola oil contains 16 g of saturated fats, 61.2 g of polyunsaturated fats, 137.9 g of monounsaturated fats and 0.9 g of trans-fats. The same amount of canola oil can provide about 1927 calories. Apart from these, omega-3 essential fatty acids and omega-6 fatty acids are other important nutrients. Much of the known benefits of canola oil can be attributed to the presence of high levels of monounsaturated and polyunsaturated fatty acids. Unsaturated fatty acids are known for their beneficial effects on human heart. These fatty acids are known to reduce the risk of heart and cardiovascular diseases. Again, canola oil can provide significant amount of omega-3 and omega-6 fatty acids, which may increase the level of good high-density lipoprotein (good cholesterol) in the body. These essential oils may also improve immune functions and promote brain health.

Canola oil does contain a small quantity of saturated fats, which is much lower than that is found in many other cooking oils. This is the reason why canola oil is considered as superior oil than many other commonly used cooking oils.

Seeds of rapeseeds when powdered with salt is said to be a folk remedy for cancer. Oil is used in massage and oil baths, believed to strengthen the skin and keep it cool and healthy. It is applied with camphor for rheumatism and stiff joints. Roots are used in chronic coughs and bronchial catarrh.

Indian mustard (*Brassica juncea*): Whole plant with flowers possesses bitter, aperients and tonic. The oil is used as an external stimulant application in chest infections especially of children.

Black mustard (*Brassica nigra*): Leaves have been extensively used as a ready, simple and safe rubefacient. Seed is acrid, bitter and powerful stimulant. Powdered seed combined with white mustard in the form of flour used as a simple vesicant and rubefacient, thus, used in paralysis. The mustard oil when dissolved in rectified spirit or spirit of camphor is a very useful application for chilblains and chronic rheumatism.

White mustard (*Brassica alba*): Seeds are powerful acrid, pungent, laxative and are used as a remedy in dyspepsia and other complaints accompanied with torpidity of the bowels. Flour made into paste with water is applied to the skin as a stimulant cataplasm or sinapism.

Common name: Olive (Jaitoon)

Botanical name: *Olea europaea*

Family: Oleaceae

Olive oil is a natural juice, which preserves the taste, aroma, vitamins and properties of the olive fruit. Olive oil is the only vegetable oil that can be consumed as it is, freshly pressed from the fruit.

Olive oil is composed mainly of the mixed triglyceride esters of oleic acid and palmitic acid and of other fatty acids, along with traces of squalene (up to 0.7%) and sterols (about 0.2% phytosterols and tocosterols). The composition varies with altitude, cultivar, region, time of harvest and extraction process. Olive oil contains a group of related natural products with potent antioxidant properties that give the extra-virgin unprocessed olive oil its bitter and pungent taste and are esters of tyrosol and hydroxytyrosol, including oleocanthal and oleuropein.

The beneficial health effects of olive oil are due to both its high content of monounsaturated fatty acids and its high content of antioxidative substances. Studies have shown that olive oil offers protection against heart disease by controlling low-density lipoprotein (*bad* cholesterol) cholesterol levels while raising high-density lipoprotein (the *good* cholesterol) levels. No other naturally produced oil has as large an amount of monounsaturated as olive oil, mainly oleic acid. Olive oil is very well tolerated by the stomach. In fact, olive oil's protective function has a beneficial effect on ulcers and gastritis. Olive oil activates the secretion of bile and pancreatic hormones much more naturally than prescribed drugs. Consequently, it lowers the incidence of gallstone formation.

Olive oil and heart disease: Studies have shown that people who consumed 25 ml (about 2 tablespoons) of virgin olive oil daily for 1 week showed less oxidation of low-density lipoprotein and higher levels of antioxidant compounds, particularly phenols, in the blood. While all types of olive oil are sources of monounsaturated fat, extra virgin olive oil, from the first pressing of the olives, contains higher levels of antioxidants, particularly vitamin E and phenols, because it is less processed. Olive oil is clearly one of the good oils, one of the healing fats. Most people do quite well with it since it does not upset the critical omega 6 to omega 3 ratio, and most of the fatty acids in olive oil are actually omega-9 oil, which is monounsaturated.

Olive oil and colon cancer: Spanish researchers suggest that including olive oil in diet may also offer benefits in term of colon cancer prevention. Their study results showed rats fed diet supplemented with olive oil had a lower risk of colon cancer. In fact, the rats that received olive oil had colon cancer rates almost as low as those fed fish oil, which several studies have already linked to a reduction in colon cancer risk.

Skin benefits: Olive oil is rich in vitamin A, E and K. The other ingredients are polyphenol, oleocanthol squalene, triterpenes, *etc.* All these micronutrients are beneficial for hair growth and improvement of texture of skin. Olive oil for hair loss is mainly used because it prevents formation of dihydrotestosterone, a hormone, which is highly detrimental for scalp. Olive oil being rich in omega 9 oil, a monounsaturated fatty acid, does not pose any health hazard. Olive oil is the preferred grade for moisturizing the skin, especially when used in the oil cleansing method (OCM), which is a method of cleansing and moisturizing the face with a mixture of extra virgin olive oil, castor oil (or another suitable carrier oil) and a select blend of essential oils. Massaging olive oil over the skin keeps it soft and supple. It helps to get rid of sunburn. Rub olive oil all over your face for a week to get rid of the tan. Not only the tan disappears but the quality of the skin also improves. This is one of the best essential oils, used widely in aromatherapy. A body massage with olive oil is extremely helpful to alleviate muscle fatigue and stress. Olive oil is an effective remedy for chapped lips. Massage your lips everyday with olive oil to get the best results. They also help to prevent lips from drying.

It is extremely helpful for people who are suffering from a snoring disorder. Just have a teaspoon of olive before going to bed in order to lubricate the throat muscles.

Olive leaves

Olive leaf is used in various forms like fresh olive leaf and dried olive leaf (especially used for making olive leaf tea) and olive leaf extract, which is utilized in liquid concentrate and capsule forms and also in lotions and soaps. Olive leaf-tea benefits are observed in the treatment of common cold, fever, *etc*. Due to the abounding olive leaf-extract benefits, it is marketed as one of the greatest herbal essences.

Olive leaf health benefits: Olive leaf is good for the overall health of a person. Due to its abundant antioxidant properties, olive leaf benefits are seen in its preventing the movement of free radicals that damage the cells in the body. This makes the olive leaf a better weapon against life-threatening diseases like cancer. Antioxidants in olive leaf increase energy levels and are therefore beneficial in the treatment of fatigue. Olive leaf benefits are also observed in reducing the blood sugar levels and as a result, it is very helpful in controlling and preventing the onset of diabetes. One of the olive leaf benefits is that it helps maintain the overall heart health. It prevents oxidation of low-density lipoprotein, which leads to various cardiovascular diseases and high blood pressure levels. Olive leaf is rich with many anti-inflammatory, antiviral and antibacterial properties and hence, it supports the immune system well too. The healing properties of olive leaf provide a better prevention from many diseases of the immune system, like human immunodeficiency virus (HIV)/acquired immune deficiency syndrome (AIDS). Olive leaf is also used as an herbal remedy for arthritis, headache, asthma, body ache, yeast infection, muscle pain, diarrhea, herpes, *etc*. It is also known to improve digestive health. Along with improving the nutrient absorption process, it reduces the risk of ulcers, gallstones and colon cancer.

Common name: Neem

Botanical name: *Azadirachta indica*

Family: Meliaceae

Neem is useful as a tonic and astringent that promotes wound healing. All parts of the plant like leaves, bark and oil possess a lot of significant pharmacological properties and are very effective in curing various diseases. It has detoxifying benefits that help in maintaining healthy circulatory, digestive, respiratory and urinary systems. The neem extract possesses anti-diabetic, anti-periodic, anti-bacterial and anti-viral properties. It is also beneficial in treating malarial fever and cutaneous diseases. It is used as a poultice in boils and is antiseptic, demulcent, a tonic in catarrhal affections, stomachic and stimulant. It is useful in snake bite, scorpion sting, hypoglycemic, in rheumatism, as an analgesic, antipyretic, sedative, antibacterial, antiprotozoal, antiviral and other skin diseases.

Neem oil comprises of mainly triglycerides and large amount of triterpenoid (azadirachtin) compounds, which are responsible for bitter taste. *Neem* oil also contains steroids like campesterol, beta-sitosterol and stigmasterol. The composition of various fatty acids present in *neem* oil are linoleic acid 6-16%, oleic acid 25-54%, hexadecanoic acid 16-33%, octadecanoic acid 9-24%, fair amounts of alpha-linolenic acid and 9-hexadecenoic acid. The various health benefits of using *neem* products are as follows:

Psoriasis: In clinical studies, *neem* extracts and oil were found to be as effective as coal tar and cortisone in treating psoriasis. However, there was none of the usual side effects accompanying the use of *neem* as there was with coal tar and cortisone. The application of *neem* extracts and oil on the skin reduces redness and itching problems.

Eczema: In case of eczema, clinical studies demonstrate that even the application of weaker *neem* leaf extracts effectively cured acute conditions of eczema. The usage of soap or shampoo containing *neem* oil can easily relieve the itching and redness of eczema. For specific areas of intensity, the application of *neem* cream on skin after bathing or shampooing with *neem* oil products solves the problem.

Acne: *Neem* effectively kills the bacteria that cause acne and reduces inflammation produced by acne.

Other skin related problems: Dry skin, wrinkles, dandruff, itchy scalp, skin ulcers and warts are other conditions that can be effectively resolved by the use of soaps, lotions and creams, containing *neem* leaf extracts and oil.

Periodontal disease: *Neem* twigs and leaves are used to brush teeth, and this helps to keep gums free of disease and infection. The ancient ayurvedic practice of using *neem* to heal and rejuvenate gum tissue and to prevent cavities and gum disease is verified in modern clinical studies. The various studies have shown that *neem* bark is more active than the leaves against certain bacteria and that *neem* based toothpastes and mouthwashes containing *neem* oil significantly improved Pyorrhea. In addition to this, studies have identified a number of compounds in the *neem* tree that effectively regulate immune system functions.

Diabetes: *Neem* has also been identified effective in treating and preventing diabetes. Drinking a cup of tea out of the leaves or taking extracts of *neem* leaf, significantly and consistently reduced insulin requirements, insulin fast and insulin sensitive forms of diabetes.

Arthritis: Traditionally, *Ayurveda* has recommended the use of *neem* leaf, seed, bark and oil for reducing arthritic pain and inflammation and for halting the progression of disease. The pain, inflammation and swelling of the joints in arthritis can be greatly reduced by different compounds in *neem*. In numerous clinical studies, the polysaccharides and other compounds of *neem* leaf extract have produced a reduction in inflammation caused by arthritis. One of the major components in *neem* seed, which has demonstrated a significant effect against arthritis, is nimbidin. Other components that possess anti-inflammatory properties are limonoids and catechin.

In addition to its anti-inflammatory effects, there are several leaf compounds like the above-mentioned polysaccharides, catechin and limonoids, which are clinically proven to be more potent than aspirin (acetylsalicylic acid) in decreasing prostaglandin synthetase and the pain it causes.

Rheumatism: Rheumatism or rheumatoid arthritis is an autoimmune condition that can cause serious pain and inflammation. Modern medicine uses strong drugs like phenyl butazone and cortisone to relive pain and inflammation of rheumatism, however, these drugs can cause many uncomfortable side effects. Clinical studies have conclusively demonstrated that the anti-inflammatory activities of various compounds in *neem* leaves are equally as effective as phenyl butazone and cortisone.

In addition to relieving pain and inflammation, the active constituents in *neem* leaves are clinically proven to inhibit the prostaglandin mechanism that causes pain, more effectively than aspirin. The consumption of *neem* leaf capsules and the topical application of slightly warmed *neem* cream can help to alleviate pain and inflammation of this condition.

Cancer: *Ayurvedic* practitioners and herbalists have successfully used *neem* to reduce tumors. Clinical research using the polysaccharides and limonoids found in *neem* leaves, bark and seeds demonstrated some remarkable effects against cancers. These compounds not only reduced tumors and cancers but also found effective against lymphocytic leukemia. An amazing reduction in tumor size was reported when *neem* extracts were injected around the tumors.

Others: Heart diseases including high blood pressure, blood clots, cholesterol and arrhythmia or rapid heartbeat can be effectively cured by using *neem*. The blood disorders including poor circulation, blood poisoning and kidney problems and digestive disorders, including heartburn or indigestion, peptic or duodenal ulcers, gastritis and hemorrhoids are few more problems that can be solved by using *neem* products. The nervous disorders like anxiety, epilepsy, hives and common diseases like tuberculosis, bronchitis, conjunctivitis, allergies, bad breath, hangover, stress, insomnia and smoking are effectively cured by the use of *neem*.

■■■

27 Organic Farming: Revival of Ancient Practice

S.K. Gupta

Assistant General Manager, National Seeds Corporation Ltd., Beej Bhawan, Pusa, New Delhi-110012, E-mail: <skguptanscplan@yahoo.co.in>, <skguptaji@gmail.com>

Since inception of agriculture, organic farming by default has been in practice, which is gradually being modified according to our requirements for quantity and quality. By using agricultural chemicals such as fertilizers, pesticides, herbicides and so on, no doubt we solved our short-term goal but left a dangerous legacy for future generations. There are continuous deterioration of environment due to unavoidable pollution of the soil, water and air, causing detrimental effects to health of the plants and animals that grow in this environment and are prone to disease. It is we to decide to do whatever we can, may be in a very small way to help restore the balance. Among all farming systems, organic farming is gaining wide attention among farmers, entrepreneurs, policy makers and agricultural scientists for varied reasons, *i.e.*, it minimizes the dependence on chemical inputs, thus, safeguards/improves quality of resources and environment. Being labour intensive, it provides an opportunity to increase rural employment.

Earlier days, the consumers interested in organic farming would look for chemical-free fresh or minimally processed food. They mostly had to buy directly from growers: *Know your farmer, know your food* was the motto. Personal definitions of what constituted *organic* were developed through first hand experience by talking to farmers, seeing farm conditions and farming activities. In a world increasingly concerned with climate change, managing the carbon footprints and reducing the impact on the environment, organic is the way forward. So, adopt a healthy lifestyle by eating natural organic foodstuffs.

Organic Farming

It is a form of Agriculture Production System, which avoids or largely excludes the synthetic fertilizers, pesticides, herbicides, plant growth regulators, livestock additives, genetically modified organisms and follows natural way of farming based on the ancient teachings. Possibly organic farmers rely on crop rotation, crop residues, animal manures and mechanical cultivation to maintain soil productivity and tilth to supply plant nutrients and to control weeds, insects and other pests.

Codex Alimentarius Commission, a joint body of Food and Agriculture Organization and World Health Organization, defines organic agriculture as holistic food production management systems, which promotes and enhances agro-ecosystem health, including biodiversity, biological cycles and soil biological activity. This is accomplished by using, possibly in agronomic, biological and mechanical methods as opposed to using synthetic materials to fulfil any specific function within the system.

According to the International Organic Farming Organization (IFOAM), the role of organic agriculture in the operations of farming, processing, distribution, or consumption, is to sustain and enhance the health of ecosystems and organisms from the smallest in the soil to human beings.

- Organic agriculture should sustain and enhance health of the soil, plant, animal, human and planet- the earth as one and indivisible.
- Organic agriculture should be based on living ecological systems and cycles, work with them, emulate them and help to sustain them.
- Organic agriculture should build on relationships that ensure fairness with regard to the common environment and life opportunities.
- Organic agriculture should be managed in a precautionary and responsible manner to protect the health and well being of the current and future generations and the environment.

Organic farming is also associated with support for principles beyond agricultural practices, such as fair trade and environmental stewardship although this does not apply to all organic farms and farmers.

History

Long ago, mankind lived a simple life in harmony with nature. Modern lifestyles and the population explosion snatched that peace from us and left us trying to increase production and disfiguring the nature. We invented new ways and came upon advances in chemical agriculture, biochemistry, plant breeding and genetic engineering in the early 20th century led to profound changes in agriculture. Fields grew in size and cropping became specialized to make efficient use of machinery, irrigation, fertilization, pesticides and herbicides. The British botanist Sir Albert Howard, the father of modern organic agriculture, studied traditional farming practices in West Bengal, India and regarded them as superior to modern one. The first use of the term organic farming is credited to Lord Northbourne in 1940. In 1972, the International Organic Farming Organization was founded in Versailles, France to diffuse the information on the principles and practices of organic agriculture, while small independent producers and consumers initially drove to organic farming, meanwhile the volume and variety of organic crop grow to a large-scale. In 1980s, various farming and consumer groups worldwide began pressing for government regulation of organic production, led to enact the legislation and certification standards in 1990s. Since then, the retail market for organic farming in developed economies has grown about 20% annually due to consumer demand.

Methods

Organic farming is often understood as a form of agriculture that use only organic inputs in the supply of nutrients and management of pests and diseases rather it refers to the concept of farm as an organism. It is a specialized form of diversified agriculture, wherein farming is managed using local resources alone. In traditional rain fed agriculture or in drought conditions (with low external inputs), organic agriculture has shown the significant potentials to increase and sustain the yield. Organic farms that raise livestock and poultry for meat, dairy and eggs provide animals with natural living conditions and feed. Ample, free-range outdoor acces, for grazing and exercise is a distinctive feature and the overcrowding in their habitation is avoided. Feed is also organically grown and the addition of drugs including antibiotics is prohibited, thus, animal health and food quality are pursued in a holistic fresh air, exercise, and good food approach.

- **Organic Manures and Fertilizers:** Enhancing soil health by biological process is the part of organic farming. The natural production of nutrients in the soil, driven by microorganisms, is referred to as feeding the soil to feed the plant. Variety of methods are employed, including careful crop selection, crop rotation, mechanical controls such as row covers, traps, green manure, cover cropping and application of compost, mulching, processed natural fertilizers such as plant and animal residues and various mineral powders, *i.e.*, rock phosphate and greens- a natural source of potash. Nitrogen availability from organic resources often limits to realize full yield potential of cereal crops. Farmyard manure usually contains nitrogen 0.5-0.8%, phosphorus 0.2-0.4% and potassium 0.8-1.0% with no mention of quality of organic matter or alternative methods of efficient use. This nutrient rich farmyard manure helps to raise crop productivity even at lower application rates. Development of several compost production technologies like vermi-composting, microbe-mediated phospho composting, nitrogen-enriched phospho-composting, *etc.* improves the quality of composts through enrichment with nutrient-bearing minerals and other additives. These manures have the capacity to fulfil nutrients demand of crops adequately and promote the activity of beneficial macro- and micro-flora in the soil. In addition, use of liquid manures prepared through fermentation of green leafy materials, cattle urine and other locally available resources are common. The basic need in organic farming is to increase input use efficiency at each step that is achieved partly through reducing losses and adoption of new technologies for enrichment of nutrients content in manure. Livestock keeping at farms is an age-old practice, as they play a major role as the intermediary between the utilization of crop residues or fodder produced at the farm and the return of nutrients as manure. Dairying in particular has helped number of small and marginal farmers to improve their income. Animal dung, crop residues, green manure, biofertilizers and bio-solids from agro-industries and food processing wastes are some of the potential sources of nutrients. While animal dung has competitive uses as fuel, it is extensively used in farmyard manure.
- **Organic Pesticides:** Pesticides create a hazardous work environment such as chemical accidents and effects of long-term exposure, airborne after spraying and in the groundwater after wash-away. These are well-known risks and the matter of concern. Synthetic fertilizers and pesticides increase crop susceptibility to pests, while organic crops have more tolerance as well as resistance against pests. Organic rice has thicker cell walls and lower levels of free amino acids than conventional ones. Soil-borne root diseases are generally less severe on organic than conventional ones,

however, no consistent difference is observed in foliar diseases between the two systems. Effective organic pest control involves the cumulative effect of many techniques, allowing for an acceptable level of pest damage, encouraging or introducing beneficial organisms, which provide benefits in addition to pests control, soil protection and improvement, fertilization, pollination, water conservation, season extension, *etc.* and these benefits are both complementary and cumulative in overall farm health. A thorough understanding of pest life cycles and interactions requires taking right decisions at right time in pest control, *i.e.*, growing crops naturally resistant to diseases and pests or choosing sowing times to prevent pest and disease outbreaks, careful management in both time and space of planting not only prevents pests but also increases population of natural predators having natural capability to control insects, diseases and weeds. The other methods are clean cultivation, improving soil health to resist soil pathogens and promote plant growth, rotating crops, encouraging natural bio-agents such as pheromone attractants to trap the pests, using physical barriers to protect from insects, birds and animals and modifying habitat to encourage pollinators and natural enemies of pests. Organic farming standards do not allow the use of synthetic pesticides but they do allow the use of specific pesticides derived from plants. The most common organic pesticides accepted for restricted use by most organic standards include Bt, pyrethrum and rotenone. Some of them, such as *rotenone*, have high toxicity to fish and aquatic creatures with some toxicity to mammals including humans.

Choosing a diverse range of plants can help to regulate pests in a variety of ways, including (i) masking the plants from pests, depending on proximity of the companion or intercrop and (ii) producing olfactory inhibitors, odours that confuse and deter pests. The plants often used in vegetable gardens to deter insects are *basil* for flies and mosquitoes, *catnip* for flea beetle, garlic for Japanese beetle, *horseradish* for potato bugs, *marigold* for workhorse of pest deterrents, Mexican bean beetles, nematodes and others, *mint* for white cabbage moth, *nasturium* for aphids, squash bugs and striped pumpkin beetles, *pot marigold* for asparagus beetles, tomato worm and general garden pests, *peppermint* for white cabbage butterfly, *rosemary* for cabbage moth, bean beetles and carrot fly, *sage* for cabbage moth and carrot fly, *southern wood* for cabbage moth, *summer savory* for bean beetles, *tansy* for flying insects, Japanese beetles, striped cucumber beetles, squash bugs and ants, *thyme* for cabbage worm, *wormwood* for animals from garden and *neem* for general use.

The biological pest control is economically successful, and even when it has been less successful, it still produces a benefit to cost ratio.

- **Organic Herbicides:** Weeds are so far controlled mechanically, thermically and with mulches. They are expensive and they may not be affordable for commercial production and are less effective than synthetic one but of course do not inject unnatural chemicals into the environment. These include:
- **Spices:** The extract of natural spices is now effectively used in patented herbicides.

- **Vinegar:** 5-20% acetic acid solutions to higher concentrations are effective but destroy mainly surface growth, thus, to treat regrowth respraying is needed. Resistant plants generally succumb when weakened by respraying.
- **Steam:** It is applied commercially but now it is considered uneconomic and inadequate. Steam kills only surface growth, not the underground, and hence, respraying to treat regrowth of perennials is needed.

Productivity

One study found a 20% lower yield from organic farms using 50% less fertilizers and 97% less pesticides. However, studies comparing yield have mixed results. Supporters claim that organically managed soil has higher quality and higher water retention. This may help to increase yield in drought years. One study of two organic farming systems and one conventional found that in one year's severe crop season drought, the organic soybean yield was 52 and 96% higher and organic maize yield was 37% higher in one-system but 62% lower in the other. Some studies are also consistent in showing that organic farms are more energy efficient. However, alternative views hold that this may be deceptive based on fossil fuel usage but ignoring energy costs of ploughing and other laborious practices used to maintain yield. Furthermore, Michael Pollan, author of *The Omnivore's Dilemma,* notes that in the whole chain of food production and distribution, only one-fifth of the energy is used on the farm. Yet a report published by the Department of Environment, Food and Rural Affairs (DEFRA), Britain's environment and farming ministry, concluded that shifts toward a local food production and distribution system, as advocated by many organic farming proponents, would actually increase the amount of energy being invested in food due to higher level of small-scale transport systems, which suffer from inefficiencies compared to standard large-scale supermarket systems.

In a study, the Danish Environmental Protection Agency found that organic farms of potatoes, sugar beet and seed grass produce as little as half the output of conventional. Such findings and the dependence of organic farming on manure from low-yield cattle have prompted criticism from many scientists that organic farming is environmentally unsound and incapable of feeding the world population. Among these critics are Norman Borlaug, father of the *green revolution* and winner of the1970 Nobel Peace Prize, who asserts that organic farming practices can at most feed 4 billion people, after expanding cropland dramatically and destroying ecosystems in the process.

A 2006 study suggests that converted organic farms have lower pre-harvest yield than the yield of their conventional counterparts in developed countries (92%) and that have higher pre-harvest yield than the yield of their low-intensity counterparts in developing countries (132%). The researcher attributes this to a relative lack of expensive fertilizers and pesticides in the developing world compared to the intensive, subsidy-driven farming of the developed world. Nonetheless, the researcher purposely avoids making the claim that organic methods routinely outperform green-revolution (conventional) methods. This study incorporated a 1990 review of 205 crop comparisons, which found that organic crops had 91% of conventional yield. A major United States survey published in 2001, analyzed results from 150 growing seasons for various crops and concluded that organic yields were 95-100% of conventional yields.

The repeated studies in 2003 have found that organic farms withstand severe weather conditions better than conventional farms, sometimes yielding 70-90% more than conventional farms during droughts. A 22 years farm trial study by Cornell University concluded in 2005 that organic farming produces the same corn and soybean yield as the conventional ones but consumes less energy and contains no pesticide residues. The results were attributed to lower yield in general but higher yield during drought years. A study of 1,804 organic farms in Central America hit by Hurricane Mitch in 1998 found that the organic farms sustained the damage much better, retaining 20 to 40% more topsoil and smaller economic losses at highly significant levels than their neighbors did.

However, a 21 years Swiss study found an average of 20% lower organic yield, along with 50% lower expenditure on fertilizers and energy and 97% less pesticides. A long-term study by United States Department of Agriculture and Agricultural Research Service (ARS) scientists concluded that contrary to widespread belief, organic farming could build up soil organic matter better than conventional no-till farming, which suggests long-term yield benefits from organic farming. An 18 years study of organic methods on nutrient-depleted soil concluded that conventional methods were superior for soil fertility and yield in a cold-temperate climate, arguing that much of the benefits from organic farming are derived from imported materials, which could not be regarded as *self-sustaining*.

In 2008, the United Nation Environmental Programme (UNEP) and United Nation Conference on Trade and Development (UNCTAD) issued a report, which stated that organic agriculture could be more conducive to food security in Africa than most conventional production systems and that it is more likely to be sustainable in the long-term. The report assessed 114 projects of over two hundred crop comparisons in 24 African countries, finding that yield had been more than doubled where organic or near-organic practices had been used and that soil fertility and drought resistance improved. The difference in yield between organic and non-organic methods was small, with non-organic methods resulting in slightly higher yield in developed areas and organic methods resulting in slightly higher yield in developing areas.

Genetically Modified Organisms (GMOs)

Organic farming wholesomely rejects genetically modified products, plants and animals. Therefore, genetically modified organisms-free is a popular marketing point for organic food. Proponents argue that genetic engineering could play an important role with a rapidly expanding global population in ending world hunger by creating higher volumes of produce, without additional land. It could also help to create healthier food and ensuring proper nourishment, to make farming more profitable. It is a technique, not a characteristic of the organisms. Moreover, humans have used selective breeding to modify crops and livestock for tens of thousands of years. Opponents insist for negative impact on food quality, plant, or animal health. The genetically modified plants can unintentionally cross-pollinate with other plants on nearby ecosystems including organic farms is one of the greatest concerns. Contamination may lead to products being incorrectly labelled as organic or genetically modified organisms-free, reducing the value of crop and losses to the farmer, as it cannot be sold as organic. At least two-thirds of conventional corn, soybeans and canola in the United States contain traces of from genetically modified varieties.

The Environment

In conventional agriculture, peasants apply only necessary fertilizers in order to minimize expenditures and waste pollutants. Organic farmers must use fertilizers such as manure, which contain fixed amounts of various elements. Interestingly, many growers rely on manure that is not organic (comes from animals not fed and raised organically) to continue fertilization. Technically, this does not violate the traditional definitions of organic produce because there are no inorganic components added to the manure. Conventional agriculture is rapidly depleting natural resources, particularly fossil fuels and fresh water and seriously polluting soil, water and air. Some critics, most notably the Nobel Laureate Dr. Norman Borlaug, contend that adopting organic farming on a global scale would be more detrimental to the environment. He asserts that if organic farming is to feed the globe, it will require a dramatic increase in cropland area, which is not possible.

Environmental impact

Inorganic farming depletes soil, crop, animal and human heath. The soil sprinkled with fertilizers, pesticides and weedicides is choked, and their molecules, which reduce the microbial content of the soil, are blocked. Based on various surveys and studies to compare conventional and organic systems of farming, the general consensus is that organic farming is less damaging due to the following reasons:

- Organic farms do not consume or release synthetic chemicals into the environment some of them have the potential to harm soil, water and local terrestrial and aquatic wildlife.
- Organic farms are better than conventional at sustaining diverse ecosystems, *i.e.*, populations of plants and insects, as well as animals.

When calculated either per unit area or per unit of yield, organic farms use less energy and produce less waste such as packaging materials for chemicals.

Sustainability

It is common to equate organic farming with sustainable agriculture but the two are not synonymous. Sustainability is a broad concept considering many levels such as environmental health, economic profitability and social and economic equity. One goal of sustainability would be a balance between what is taken out of the soil with what is returned to it, without relying on outside inputs. Large organic farms that rely on machinery and automation and purchased inputs will have sustainability issues similar to the large conventional farms.

Rural Infrastructure

Proponents condemn conventional practices for putting small, independent farmers out of business, destroying rural communities in the process, while small-scale organic farming encourages local economies and provides social and employment alternatives, thus, improving the quality of life.

Organic Food

Organic food is produced according to legally regulated standards in crop production and food processing (without ionizing radiation or food additives and genetically modified organisms). For animals, they were reared without the routine use of antibiotics and growth hormones. Historically, organic farms have been small family-run farms and organic food comprised mainly fresh vegetables. Early consumers would look for chemical-free fresh or minimally processed food. Mostly had to buy directly from growers or small stores, and now, these are becoming more widely available. United States have 17 to 20% growth for the past few years, while sales of conventional food, though still larger in size, have the growth rate only 2 to 3% a year. This large growth is predicted to continue and many companies are jumping into the market.

Food Quality: Organic foods have lower levels of pesticides, medicinal and hormonal residues and nitrate contents (generally associated with intensive use of nitrogenous fertilizers). In common sense, the food grown in unnatural, sheltered, chemically assisted ways is not as good for people as the naturally grown food. By current standards, the counter argument is that there has been no demonstration of a functional difference between organically and conventionally produced food, and that assisted food is actually healthier, and thus, more nutritious. Critics claim that organic food could be less safe than non-organic food, by the risk of exposure to biological contaminants and food-borne diseases, *e.g.*, the use of manure, well known for carrying human pathogens and presence of mycotoxins from molds.

Natural Food Diet: It is the avoidance of all unnatural and refined/processed ingredients like refined sugars, flours, milled grains, hydrogenated oils, artificial sweeteners, food colours and flavouring and other similar ingredients.

Taste: In 2001, study by researchers at Washington State University concluded under the judgment by a panel of tasters that organic apples were sweeter. Along with taste and sweetness, the texture as well as firmness of the apples was also rated higher than those were grown conventionally. These differences are attributed to the greater soil quality resulting from organic farming techniques.

Photo by René Piamonte IFOAM Training Manual on Seed Saving (Courtesy: Wikipedia)

Nutritional Value: Organic foodstuffs contain loads of minerals and vitamin C, which are required by the body. In fact, it has been analyzed by health experts that organic farming is more nutritious in the sense that it has 27% more vitamin C, 29% more magnesium, 21% more iron and 14% more phosphorous, when compared with conventional foodstuffs. Some studies have shown higher nutrients levels in organic fruits and vegetables. However, due to the difficulty with designing such experiments, the evidence is not considered conclusive. In 2002, a review of all past studies on the subject found no proof that organic farming offers greater nutritional values, more consumer safety or any distinguishable difference in taste.

Eating Raw Organic Food: Cooking food change, the very composition of fat, protein and carbohydrates, also tends to interfere with the bioelectrical energy, necessary for sustaining life that often deteriorates the nutritional value of food due to a loss of vital nutrients and enzymes. One of the major benefits of eating raw organic food is that not only it tastes better than conventional but also contains a far greater nutritive quality, serves as perfect health improving remedy and makes to feel more energized. This ensures the nutritional value of the food, which remains enzymatically and nutritionally intact and also aids in helping to lose weight faster. Organic foodstuffs are easily absorbed in the body, thus, helping in ensuring smooth bowel movements. Consumption of raw organic food goes a long way in cleansing the bloodstream and eliminating the waste products that are responsible for the formation of toxins in the body so intake of raw organic food in daily meals may improve the health fitness.

Health Benefits: Since organic food is natural and fresh, it boasts off its rich taste. Proponents claim that organic foodstuffs contain phenolic compounds that protect from developing heart disease and cancer. So, move towards leading a healthy lifestyle by eating organic food articles.

The Seattle P-I turned the study on its head, shedding more heat than light on the subject with an article *Harmful pesticides found in everyday food products*. To improve public health for all, a more useful focus would be the primary source of toxins, more than 150 chemicals found in the typical American food. The regular and significant exposure to toxic chemicals are the laundry detergent, plastics, cleaners, tobacco, cosmetics, shampoos, lotions and even children's toys.

Cost Analysis

Organic products typically cost 10 to 40% more than to their conventional counterparts. Similarly, the processed organic foodstuffs also vary in price. Prices may be higher because organic produce is produced on a smaller scale and may need to be milled or processed separately. Furthermore, there is an increase in shipping costs from more centralized production in otherwise regional markets. In case of dairy and eggs, the animal's requirements such as the number of animals that can be raised per acre, or the breed of animal and its feed conversion ratio affect the cost.

Pesticides and Farmers

There have been many studies on the health effects of pesticide exposure to farm workers. Even when pesticides are used correctly, they still end up in the air and bodies of farm workers. Amongst these, organophosphate pesticides are associated with acute health problems such as abdominal pain, dizziness, headaches, nausea, vomiting, as well as skin and eye problems. In addition, there have been many other studies that are associated with more severe health problems such as respiratory, memory disorders, dermatologic conditions, cancer, depression, neurological deficits, miscarriages and birth defects.

Pesticide residue: A study published in 2002 showed that organically grown foodstuffs consistently had about one-third as many residues as conventionally grown. Monitoring of pesticide residues in the United States is carried out by the Pesticide Data Program, created in 1990 as a part of United States Department of Agriculture (USDA). It has since tested over 60 different types of food for over 400 different types of pesticide with samples collected closer to the point of consumption. Their results in 2005 indicate that 29.5% of all samples tested contained no detectable pesticides, 30% contained single pesticide and slightly over 40% samples contained pesticides more than one. Several studies corroborate this finding that while 77% of conventional food carries synthetic pesticides residues, only about 25% of organic food does.

A study published in 1993 by the United States National Research Council determined that for infants and children, the major source of exposure to pesticides is through diet. Another study in 2006 measured the levels of organophosphorus pesticide exposure in 23 schoolchildren before and after replacing their diet with organic food, wherein it was found that levels of organophosphate pesticide exposure dropped dramatically and immediately when the children switched to an organic diet. Food residue limits established by law are set specifically with children in mind and consider a child's lifetime ingestion of each pesticide. There is controversial data on the health implications of certain pesticides. Certain genetic modifications can allow some non-organic farming to avoid the usage of pesticides without any loss in yield. However, genetically modified crops are the subject of controversy on their own.

The United States Environmental Protection Agency and State Agencies periodically review the licensing of suspect pesticides but the process of delisting is slow. For example, the pesticide Dichlorvos or 2,2-dichlorovinyl dimethyl phosphate (DDVP)- a household pesticide related to World War II-era nerve agents, which the Environmental Protection Agency has almost banned on several occasions since 1970s but it never did so despite considerable evidence that this pesticide is not only carcinogenic but dangerous to the human nervous system, especially in children. The Environmental Protection Agency proposed continued its sale during the year 2006.

Herbicide: The herbicide Atrazine, *e.g.*, has been shown in some experiments to be a teratogen, even at concentrations as low as 0.1 parts per billion to emasculate male frogs by causing their gonads to produce eggs effectively turning males into hermaphrodites.

Facts and Statistics of Organic Farming and Food

While organic farming accounts for 1-2% of total food sales worldwide, the organic farming market is growing rapidly, far ahead of the rest of the food industry, in both developed and developing nations. The infrastructure for handling, logistics and supply-chain management will also come into existence as quickly as the consumption for organic farming is increasing to help in the growth of organic agri-businesses in different nations:

United States: Organic farming sales have grown by 17 to 20% a year for the past few years, while sales of conventional food have grown at only about 2 to 3% a year. In 2003, organic products were available in nearly 20,000 natural food stores and 73% of conventional grocery stores. In 2005, organic food accounts for 2.6% of total food sales. Two thirds of organic milk and cream and half of organic cheese and yogurt are sold through conventional supermarkets.

Europe: In the European Union (EU 25), 3.9% of the total utilized agricultural area is used for organic production. The countries with the highest proportion of organic land are Austria (11%) and Italy (8.4) followed by Czech Republic and Greece (both 7.2%). The lowest figures are shown for Malta (0.1%), Poland (0.6%) and Ireland (0.8%).

Austria: In 2007, 11.6% of all farmers produced organically. The government has created incentives to increase the figure to 20% by 2010. The 4.9% of all food products sold in Austrian supermarkets (including discount stores) in 2006 was organic and 8000 different organic products were available in the same year.

Germany: The baby food is almost organic and over 30% of bread baked in *Munich*.

Italy: the existing legislation- school lunches to be organic since 2005.

Poland: Since 2004, the products of certified organic farms are allowed to carry the European Union organic farming label. In 2005, 1.68 lakh hectares were organic and the organic market value is estimated at Eur 50 million (2006).

United Kingdom: By January 2005, half of the land- 6.86 lakh hectares- was managed to organic standards. Organic farming sale increased from £100 million in 1993-94 to £1.21 billion in 2004 (an 11% increase on 2003).

Cuba: In Cuba, after the collapse of the Soviet Union in 1990, the government converted the entire country to organic agriculture. However, they encourage genetically modified crops, particularly those with enhanced nutritional content and vaccines.

Economics of Organic Farming

The replacement of external inputs by farm-derived resources normally leads to a reduction in variable input costs. The interest in organic agriculture in developing countries is growing because it places more reliance on the natural and human resources available, requires less financial input and provides safe food while conserving the environment. Organic methods often require more labour, providing rural jobs but increasing costs to urban consumers. The prices for organic food have been and continue to be higher because farmers have to meet stricter quality standards and more labour. The benefits of organic farming are relevant both to developed nations (environmental protection, biodiversity enhancement and reduced energy use and carbon dioxide emission) and to developing countries like India (sustainable resource use, increased crop yield without over-reliance on costly external inputs, environment, biodiversity protection, *etc.*).

Issues related to Organic Food and Frequently Asked Questions

All aspects of organic farming and organic food are under debate. Environmentalists, food safety advocates, small independent farmers, consumer protection, social justice and labour groups and a growing number of food consumers are ranged against agribusiness and current government agricultural policies. As public awareness increases, the number of obstacles also increases. The controversy centers on the overall value and safety of chemical agriculture with organic farming popularity. The most often debated issues are:

- Can organic farming produce enough food for everybody?
- Is it possible to meet the nutrients requirements of crops entirely from organic sources?
- Is there any significant environmental benefit of organic farming?
- Is the food produced by organic farming superior in quality?
- Is organic agriculture economically feasible?
- Is it possible to manage insect-pests and diseases in organic farming?

Future Prospects

In recent years, organic agriculture has grown tremendously. Considering this rapid growth, it is within the nature of organic farming to keep it from becoming a large scale industrial business as conventional farming has become. Despite the growth in the organic food market over the last decade, future of the small, independent farmer, organic or otherwise, is as much in jeopardy. The current food distribution system favours high-volume production and large farming operations. The popularity of organic farming is gradually increasing and now organic agriculture is practiced in approximately 130 countries of the world, mostly in Australia, Argentina, Italy, Canada and United States. Some countries have reached a substantial proportion (close to or more than 10%). These include Sweden, Austria, Switzerland, Finland and Italy. Presently, the total organically managed area is more than 24 million hectares worldwide.

Although production of organic crops is increasing across the globe, sales are concentrated in the industrialized parts of the world. The American market amounted to US $1 billion in 1994, $13 billion in 2003 and $ 16 billion in 2006. The demand for organic food is steadily increasing both in developed and developing countries with annual average growth rate of 20-25%, as in 2001, the estimated total world organic food market value of certified organic products was estimated to be $20 billion. In 2002, this was jumped to $23 billion, in 2006, $40 billion and by 2007, more than $46 billion according to Organic Monitor (Willer/Kilcher 2009). In recent years, both Europe (2007: 7.8 million hectares) and United States (2007: 2.2 million hectares) have experienced strong growth in organic farmland. However, this growth has occurred under different conditions. While the European Union has shifted agricultural subsidies to organic farmers in recognition of its environmental benefits, the United States has taken a free market approach. As a result, as of 2007, four percent of the European Union's farmland was organically managed compared to just 0.6% in United States (Willer/Kilcher 2009). IFOAM's most recent edition of *The World of Organic Agriculture: Statistics and Emerging Trends 2009* lists the countries which had the most areas in 2007, wherein, the country with the most organic farm, *i.e.*, more than 12 million hectares, is Australia followed by Argentina, Brazil and the United States. In total, 32.2 million hectares were under organic management in 2007 compared to 11 million hectares grown during 1999 (Willer/Kilcher 2009).

A growing consumer market is naturally one of the main factors encouraging farmers to convert to organic agriculture. Increased consumer awareness of food safety and environmental concerns has contributed to the growth in organic farming over the last few years. The beginning of the 21st century saw multinational food corporations taking major stakes in the organic market, and this has dramatically increased the variety, availability and falling cost of processed organic food. The government, private sector and producer associations each have a necessary role to play in promoting and facilitating marketing of organic produce. How the various pieces fit together in order to increase value and marketability of farmers' produce is a challenge and will require additional work to make the system function properly.

Controversies

A. Organic food not so healthy

Critics claim that organic food could be less safe than non- organic food by the risk of exposure to biological contaminants and food-borne diseases, *e.g.*, the use of manure, well known for carrying human pathogens and presence of mycotoxins from molds. Infinitesimal levels of all kinds of chemical exist in all human bodies. Detectable levels of radiation exist all around us but none of those represents a threat to human. According to the Mayo Clinic, no medical or scientific data demonstrate organic produce to be safer or more nutritious than conventionally grown. A number of European organizations, including the Food Standards Agency (FSA), the French Food Safety Agency and the Swedish National Food Administration, support that assessment that there is no difference in food safety and nutrition between the both.

Tom Sanders, the professor of nutrition and dietetics at Kings College London, claimed that there is little evidence that an organic diet provides greater nutritional benefits over the non- organic food or those non-organic crops are more harmful than their organic competitors. It is a fantastic opportunity (for organic farming industry) to make money out of people because of charging more for the organic farming. The Food Standard Agency, which issues advice to consumers, also states that organic food is not significantly different from food produced conventionally.

- **Fruits and vegetables in general nutritious organic or otherwise:** There is an established proverb that an apple a day actually does help keeps the doctor away. Diets rich in fruits and vegetables help prevent stroke, heart attack and cancer. People who eat apples, cherries and pears instead of high-fat snacks weigh less and have more energy. All apples are full of vitamins that build blood cells, keep the teeth and gums healthy and aid in the growth and repair of all body tissues. Fruits and vegetables are safe and healthy food choices for all, regardless of whether they are grown conventionally or organically. No credible scientific study has demonstrated anything else (Jim Hazen, executive director of the Washington State Horticultural Association February 28, 2008, PT). A recent study by Chensheng Lu once again confirmed that organic produce is no more healthful than conventionally grown fruits and vegetables. Lu's study demonstrates fruits and vegetables are safe for people of all ages to consume, no matter how produced. Some researchers claim that health risks associated with conventional fruits and vegetables is zero. People thirsting for better health should keep buying and consuming fruits and vegetables. The point is that if you truly want to eat healthfully, organic or conventional does not matter, they claim. All are for marketing linking with good health.

- **Mystery milk:** Rule change would keep consumers in the dark. Most Utah consumers want to know what is in their food and how it is produced. Some prefer fruits and vegetables that are grown without pesticides, and milk and meat that are produced with no growth hormones. Food grown under those conditions and other federally mandated standards can be labeled certified organic with a description of the required production standards (Tribune Editorial: February 28, 2008). That is not to say that non-organic food is unsafe or inferior. The two are simply produced differently and consumers have a fundamental right to make a choice. To do that, they have to have the information contained on food labels but a Utah House member who is also a non-organic dairy farmer wants to restrict labeling for organic food, keeping important information out of the hands of consumers. Kerry Gibson, R-Ogden, says that labeling milk as having been produced without the use of bovine growth hormone rBST is misleading and implying that non-organic products are not as safe as the labeled organic. Gibson and others, including Monsanto, the company that produces rBST and makes a lot of money selling it to farmers, are proposing a Utah Agriculture Department rule change. The new rule would prohibit milk being labeled *rBST-free* or any food labels or ads that make any claim, true or otherwise that can not be proven by an analytical test. A sentence on a soup can reading, *just the right herbs and spices in a flavourful broth* would have to be verified by an analytical test. The rule would allow milk to be labeled *rBST-free* as long as the label also stated that there is no significant difference between milk derived from cows treated with artificial growth hormones and those not treated.
- **Organic chickens might have a better life but not the taste:** There is no difference in the taste between free range and factory raised chickens, according to a taste test by Australian food experts (Times of India 4th March 2008 basically reported by Reuters in Sydney). Consumer advocate *Choice* had a panel of food experts to taste eight different roast chickens- 2 organic, 2 free-range, one corn-fed and 3 regular factory farmed but despite the higher cost of free range, corn-fed and organic chickens, the expert could not tell the significant difference in taste. All agreed that chickens do not have as much taste as they used to. The narrow range of fast maturing breeding stock now used in almost all bird's production made any difference in different methods effectively meaningless. Organic or free-range chickens have their own importance such as better life but it does not necessary mean a better taste. To find a chicken with real flavour was to find a smaller producer who uses traditional breeds and lets the chickens live longer or buy the largest chicken you can find.

B. Organic foods – safe and healthy

- **Garden Organic (UK) disputes claims against organic benefits:** The United Kingdom's leading authority on organic growing, *Garden Organic*, is disputing claims made on 26 February 2008 on BBC 2's Horizon programme, *Professor Regan's Supermarket Secrets* that there is little scientific evidence of organically produced food having any nutritional differences or benefits compared to non-organic food. The Midlands based charity, celebrating 50 years at the forefront of organic horticulture this year, also criticizes the programme for disregarding the massive environmental benefits that organic methods bring.

Director of Research of Garden Organic, Dr. Margi Lennartsson, said that despite the programme's assertions, there is a growing body of evidence, demonstrating that organic food contains desirable components such as vitamins and minerals and fewer harmful ones such as pesticides and additives. Later on new evidence emerged that comparing organic and industrially produced crops indicates higher vitamin C levels and polyphenols in kiwis, higher levels of phenols, flavonoids and vitamin C in apple puree, higher content of vitamin C, vitamin B, carotene and flavonoids in tomatoes and higher polyphenol content in peaches.

- **Mothers on organic garming produce nutritious milk:** Milk from breastfeeding mothers whose diets contain a large proportion of organic food has also been found to contain significantly higher level of conjugated linoleic acids (an antioxidant) than those eating conventional food. A study also showed that United Kingdom completely organic milk has on an average 68% higher level of the essential omega-3 fatty acids and a healthier omega-3: omega-6 profile than non-organic milk.
- **Organic fruit and vegetables have better health benefits:** Early results of the extensive £12 million four year European Union Quality Low-Input Food study published in October 2007 have indicated that organic fruits and vegetables have health benefits and in fact contain 40% more antioxidants as compared to non-organic fruits and vegetables. The results also showed higher levels of other beneficial minerals such as iron and zinc. The United States life expectancy has been climbing for some time in part because of the increasing affordability and availability of fruits and vegetables. So, one should keep in mind that the life expectancy curve precedes the aggressive marketing of organic produce. In addition, United States cancer rates are leveling off or declining across the board that also predates the wildly successful marketing of organic products.
- **Environmental safety:** Most concerning thought is that the programme failed to pick up on the environmental benefits of organic horticulture and agriculture that in turn will have an impact on us all. The sustainable approach that comes with organic growing not only protects us from a cocktail of chemicals but also ensures our land and soil remain healthy for generations to come and the natural balance of wildlife and biodiversity is preserved.

Regarding Climate Change, Anthony Meleca argues that organic agriculture- with its emphasis on closed nutrients cycles, biodiversity and effective soil management- has the capacity to mitigate and even reverse the effects of climate change. According to the Rodale Institute, which has been comparing organic farming systems and conventional systems since 1981, organic agriculture can also be used to mitigate global warming by decreasing fossil fuel emissions and sequestering carbon in the soil. The elimination of synthetic nitrogen in organic systems decreases fossil fuel consumption by 33 percent (LaSalle) and carbon sequestration takes carbon dioxide out of the atmosphere by putting it in the soil in the form of organic matter, which is often lost in conventionally managed soils. Carbon sequestration occurs at especially high levels in organic no-till managed soil according to the Rodale Institute. Organic agriculture can reduce the level of negative externalities from (conventional) agriculture. Whether this is seen as private or public benefits depends upon the initial specification of property rights. However, it is clear that agriculture has been undervalued and underestimated as a means to combat global climate change.

- **Benefit to farmers:** While organic farms have lower yield, organic methods require no synthetic fertilizers and pesticides. The decreased cost on those inputs, along with the premiums to which the consumers pay for organic produce, creates higher profits for organic farmers. Organic farms have been consistently found to be as or more profitable than conventional farms with premiums included, but without premiums, profitability is mixed (Lotter 2003:11). In 2009, a review concluded that organic production was more profitable in Wisconsin, when including price premiums.

Organic Food Regulations

Small farms grew vegetables and raised livestock using organic farming practices with or without certification and the individual consumer was monitored. As demand for organic food continues to increase, high volume sales through mass outlets such as supermarkets are rapidly replacing the direct farmer connection. However, for supermarket consumers, food production is not easily observable and product labeling like *certified organic* is relied on. Government regulations and third-party inspectors are looked to for assurance, thus, organic farming production is legally regulated. Currently, the United States, the European Union, Japan and many other countries require producers to obtain certification in order to market food as organic.

Various alternative organic standards, generally bypassing formal certification, are emerging with their own definitions. That can be expensive and cumbersome. One such the *Authentic Food Standard* proposed by leading United States organic farmer Eliot Coleman, includes criteria that are incompatible with current agribusiness:

- Fresh fruits and vegetables, milk, eggs and meat products are produced within a 50 miles radius of place of their final sale.
- The seed and storage crops (grains, beans, nuts, potatoes, *etc.*) are produced within a 300 miles radius of their final sale.
- Only traditional processed foods such as cheese, wine, bread and lacto-fermented products may claim, *made with authentic ingredients*.

Standards and Certification

Organic farming is defined by formal standards regulating production methods, and in some cases, the final output for organic agriculture. Two types of standard exist, *i.e.*, voluntary and legislated. As early as the 1970s, private associations created standards against which organic producers could voluntarily have themselves certified. An international framework for organic farming is provided by the International Federation of Organic Agriculture Movements (IFOAM), the international democratic umbrella organization. For IFOAM members, organic agriculture is based upon the Principles of Organic Agriculture and the IFOAM Norms. In

1980s, the governments began to introduce organic production guidelines. In 1990s, a trend toward legislation of standards began, most notably the EU-Eco-regulation developed in the European Union. Quality Assurance International, an organization established in 1989, is currently the largest private certifier of organic food systems. In 1991, the European Commission formulated the first government system to regulate organic labelling. In one go, the European Ecoregulation (EEC) 2092/91 set the rules in 12 countries, creating a huge market. Legislated standards vary from country to country. Europe had become the most prominent market and an increasing number of suppliers all over the world accepted this niche as a new challenge and a rewarding option to export high quality and high priced specialty products. Consequently, many countries like Japan, India (2001) and the United States (2002) have also legislated. Non-governmental national and international associations also have their own production standards, which must be accredited by the government where production is regulated.

Internationally, equivalency negotiations are underway, and some agreements are already in place, to harmonize certification between countries, facilitating international trade. There are also international certification bodies, including members of the International Federation of Organic Agriculture Movements (IFOAM), the Organic Crop Improvement Association (OCIA) and Ecocert. Where formal agreements do not exist between countries, organic product for export is often certified by agencies from the importing countries, who may establish permanent foreign offices for this purpose. Ecocert is the world's largest organic certification organization with offices in 20 countries and operating in over 85 countries. They certify over 40,000 farms and companies worldwide.

- **Organic Certification:** In general, any business directly involved in food production can be certified, *e.g.*, seeds, farming, food processing, retailers and restaurants, involving a set of production standards for growing, storage, processing, packaging and shipping, which vary from country to country. Organic certificate, particularly where mandated by law, as in the United States and European Union, is increasingly being seen by individual organic farmers and consumers as a contentious issue. In United States, serious complaints have been lodged with the United States Department of Agriculture against the largest United States Certifying Agency, and the United States Department of Agriculture itself has been taken to court. In a number of countries, including the United States, Bulgaria, Iceland, Norway, Romania, Switzerland, Turkey, Australia, India, Japan, the Philippines, Korea, Taiwan, Thailand, Argentina, Costa Rica, Tunisia and in the European Union, organic farming is defined by law, hence, the commercial use of the term *organic* to describe farming and food products is regulated by the government and certification is available to farms for a fee. It is usually illegal for a non-certified farm to call itself or its products *organic*. Elsewhere, for example, in Canada, the voluntary certification is available, while legislation may be pending.

The Australian Quarantine and Inspection Service (AQIS) is the controlling body for organic certification because there are no domestic standards for organic produce within Australia. Currently, the government only becomes involved with organic certification at export, meaning AQIS is the default certification agency. The largest importer of Australia's organic produce (by weight) is Japan (33.59%), followed by the United Kingdom (17.51%), France (10.51%) and New Zealand (10.21%). The largest certifier of organic products is

Australian Certified Organic, which is a subsidiary of Biological Farmers of Australia, the largest organic farmer's collective in the country.

In the United Stetes, federal organic legislation defines three levels of organics. Products made entirely with certified organic ingredients and methods can be labeled *100% organic.* Products with at least 95% organic ingredients can use the word *organic.* Both of these categories may also display the USDA organic seal. A third category, containing a minimum of 70% organic ingredients, can be labeled *made with organic ingredients.* In addition, the products may also display the logo of the certification body that approved them. Products made with less than 70% organic ingredients can not advertise this information to consumers and can only mention this fact in the product's ingredient statement. Similar percentages and labels apply in the Europe.

In Canada, certification was implemented at the federal level on 30th June 2009. Mandatory certification is required for organic products in import, export and inter-provincial trade, or that bear the federal organic logo.

European countries acquired comprehensive organic legislation with the implementation of the Europe-Eco-regulation 1992. Supervision of certification bodies is handled on the national level. In March 2002, the European Commission issued a Europeanwide label for organic food. However, for most of the countries, it was not able to replace existing national product labels. It was relaunched in 2009 with a design competition for a new logo to be used throughout the Europe from July 2010. The final logo will be chosen through a public vote online till 31st of January 2010.

In the United Kingdom, organic certification is handled by a number of organizations, regulated by the department of Environment, Food and Rural Affairs of which the largest are the Soil Association and Organic Farmers and Growers. In Sweden, organic certification is handled by the organization with members such as farmers, processors, trade and also consumer, environmental and animal welfare interests. In Ireland, organic certification is available from the Irish Organic Farmers and Growers Association. In Greece, organic certification is available from eight organizations approved by Europe. In France, organic certification was introduced in 1985. It has established a green-white logo of *AB-agriculture biologique.* The certification for the *AB label* fulfills the Europe regulations for organic food. In Belgium, a similar process as in France is being used where certification of the *Biogarantie* label is overseen by the *Public Association Sans but Lucratif* (ASBL) administration. In Germany, the national label was introduced in September 2001. The national *Bio-label* in its hexagon green-black-white shape has gained wide popularity. In 2007, there were 2431 companies having certified 41708 products. The popularity of the label is extending to neighbouring countries like Austria, Switzerland and France.

In India, the Agricultural and Processed Food Products Export Development Authority (APEDA) regulates the certification of organic products as per National Standards for Organic Production. The National Programme on Organic Production (NPOP) standards for production and accreditation system has been recognized by European Commission and Switzerland as equivalent to their country standards. Similarly, United States Department of Agriculture has recognized NPOP conformity assessment procedures of accreditation as equivalent to that of United States. With these recognitions, Indian organic products duly certified by the accredited certification bodies of India are accepted by the importing countries. In China, the China Green

Food Development Center awards two Standards, *i.e.*, A and AA, while the former standard does permit some use of synthetic agricultural chemicals and the latter is more stringent.

- **Legal Definition:** To be certified organic, the products must be grown and manufactured in a manner that adheres to standards set by the country they are sold. In Australia: *NASA Organic Standard*, in United Kingdom, *Organic Farmers and Growers Standards*, in United States, *the National Organic Program (NOP) Standards*, in European Union, *the EU-Eco-Rgulation*, in Japan, *the Japanese Agricultural Standards (JAS)*. In India, the standards for organic agriculture were announced in 2001 and the *National Programme on Organic Production (NPOP)* is administered under the Ministry of Commerce.

The *Societe Generale de Surveillance (SGS) Organic Production Standard* is a worldwide programme for organic production and processing. Organic agricultural and horticultural systems promote the sustainable production of food of optimum quality and quantity. The programme has been developed with reference to International Federation of Organic Agriculture Movements, European Community Regulation (EEC) No 2092/91, United States Department of Agriculture Rulings, Japanese Agricultural Standards (JAS) requirements and the Codex Alimentarius Guidelines. SGS has been Guide 65 (EN 45 011) accredited by the RvA, Japanese Ministry of Agriculture, Forestry and Fisheries and the United States Department of Agriculture to carry out certifications against the SGS Organic Production Standard. In addition, SGS's local organic programme is accredited by authorities in Austria.

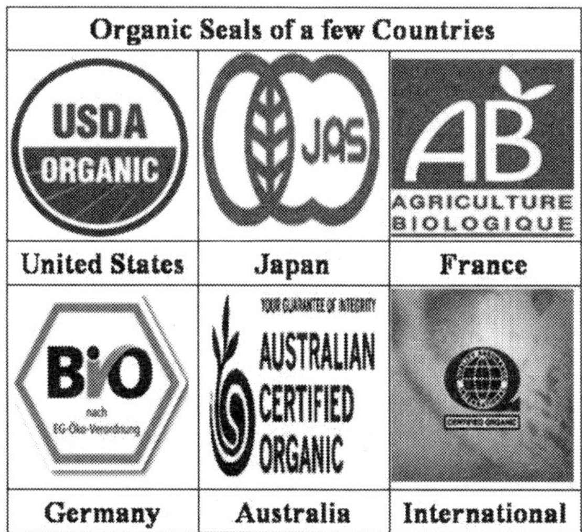

Organic Farming Association of India (OFAI)

The association registered under the Indian Societies Registration Act was formed four years ago with the objective to produce organic food for Indian consumers and to maintain the living fertility of Indian soils. Its memorandum of association was written and approved after a wide consultation with organic farmers. Indian agriculture continues to remain a source of livelihood for mostly small farmers and peasants. It is the organization of the country having grassroots organic farmers, which includes the active involvement of women farmers in the decision-making structures of the association. Such involvement is mandatory and reflected in the association's byelaws. The primary objective of the association is to promote organic farming within the country and to take all such means that are available to achieve this purpose. OFAI's labeling scheme is meant to provide an assurance of guarantee of organically grown produce exclusively for domestic consumers and not for the export purposes. Unlike other organic farm certification systems, OFAI certification is done through the agency of trained organic farmers themselves, and it does not accept farm inspectors who do not themselves practice organic agriculture. Organic Farming Association of India firmly opposes to the introduction of genetically modified organisms in agriculture and campaigns against it.

National Innovations Foundation (NIF)

National Innovations Foundation team is promoting traditional knowledge by collecting information on food habits from people, researching on their knowledge and then adding value to these products. They are, thus, disseminating information about the nutritional value and health benefits of organic farming and for conservation of forgotten, diverse organic farming habits that has eight times more fibre than wheat and maize. The National Innovations Foundation study also reveals that an herbal tea promises to protect from cold, acts as a blood purifier and relieves stress, similarly an herbal juice from the flower of cactus helps in blood circulation and increases haemoglobin in body, a daily cup of Nero (palm tree extract) juice is an antidote for arthritis, cardiac soup made of bottle gourd, carrots, onion, tomato and lime reduces cholesterol and other fatty acids in the body. Residents of a cosmopolitan city inhale hazardous intoxicants, which can be ameliorated with a cup of Tulsi sudha (basil leaves, ginger and raw sugar) or Kashmiri kava. Motivated by the National Innovations Foundation multiple for the health benefits of organic farming, Indian Institute of Management, Ahmadabad has planned to open an organic food outlet in their campus, giving students a chance to earn the wealth of health as well.

The concept of organic agriculture in India

Organic farming is not new to Indian agriculture. Some indigenous cultures had been farming organically for centuries. Several forms of organic farming are being successfully practiced in diverse climate particularly, in rain fed, tribal, mountains and hill areas of the country. The organic agriculture began as a reaction of insiders (agricultural scientists and farmers) to develop consciously in Central Europe and India in the early 20th century in response to industrialization. Much of the forests produce of economic importance like herbs, medicinal plants, *etc.* by default come under this category. According to the various certification bodies, the area of globally certified wild harvested plants is at least 10.7 million hectares.

Presently, only 35% of total cultivable area, where irrigation facilities are available, is covered with fertilizers and in the remaining 65% of arable land, mainly rain-fed, negligible amount of fertilizers are being used. Farmers in these areas often use organic manure that is readily available either in their own farm or in their locality. The north-eastern region provides considerable scope and opportunity for organic farming due to least utilization of chemical inputs. Such land is estimated about 18 million hectares being exploited for organic. With the sizable acreage under naturally organic/default organic cultivation, India has tremendous potential to grow crops organically and emerge as a major supplier of organic products in the world. Need is for putting up a clear strategy on organic farming and its link with the markets. According to a conservative estimate, around 600 to 700 million tonnes of agricultural waste is available in the country every year but most of it is not being used properly. India produces about 1800 million tonnes of animal dung per annum. Even if two thirds of the dung is used for biogas generation, it is expected to yield about 440 million tonnes per year of manure, which is equivalent to 2.90 million tonnes nitrogen, 2.75 million tonnes phosphorus and 1.89 million tonnes potassium. We must convert waste into wealth by converting this biomass into bio-energy, nutrients to starved soil and fuel to farmers.

The Task Force on Organic Farming appointed by the Government of India observed that in vast areas of the country, where limited amount of chemicals are used and have low; productivity could be exploited as potential areas for organic agriculture. Arresting the decline of soil organic matter is the most potent weapon in fighting against unabated soil degradation and imperilled sustainability of agriculture in tropical regions, particularly those under the influence of arid, semiarid and subhumid climate. Future of sustainable development of agriculture, next to water, depends on arresting fall in organic matter in soils by the application of organic manure to improve the soil organic carbon for sustenance quality and future agricultural productivity.

■■■

28 Plant Pigments and Human Health

S. Sheokand[1], S. Madan[2] and M.K. Rana[3]

[1]Department of Plant Physiology, E-mail: <sunita_sheokand@hau.ernet.in>

[2]Department of Biochemistry, E-mail: <smadan_2008@rediffmail.com>

[3]Department of Vegetable Science, CCS Haryana Agricultural University, Hisar-125004, Haryana, India, E-mail: <mkrlotus@gmail.com>

Plants are a rich source of natural pigments. The varied colours of fruits and vegetables are due to the presence of pigments in them. The green colour of leaves is because of the chlorophylls, the orange or yellow colour of fruits is due to carotenoids and the red, blue and purple colours of flowers, fruits and vegetables can be attributed to the presence of anthocyanins. The natural plant pigments are considered to have bioactive (or functional food) value since unlike the traditional classes of nutrients (proteins, vitamins, fat and minerals), they are not generally considered vital to survival, but instead in a wide range of ways, they have the properties that can promote optimal human health. The nutraceutical value of natural plant pigments such as carotenoids, anthocyanins, chlorophylls and betalains is based on their recognized biologically active properties. These pigments have been implicated in regimes to maintain human health, to protect against chronic disease incidence, or to restore wellness by repairing tissues after disease has been established. Plant pigments appear to serve three main functions, *i.e.*, act as antioxidants, function like chemo-preventors and regulate hormone levels. Hippocrates recommended as far as 2500 years ago, *let the food be your medicine.*

Figure 1 : Different colours of fruits and vegetables due to plant pigments

Anthocyanins

Anthocyanin (in Greek, anthos means flower and kyanos means blue) pigments are responsible for the red, purple and blue colours of many fruits, vegetables, cereal grains, and flowers. In plants, their roles as pollination attractants and phytoprotective agents are well known. Today, interest in anthocyanin pigments has intensified because of their possible health benefits as dietary antioxidants. Over 300 structurally distinct anthocyanins have been identified in nature. Anthocyanins are one class of flavonoid compounds, which are widely distributed plant polyphenols. The flavonoids, perhaps the most important single group of phenolics in food articles, comprise a group of over 4000 C_{15} aromatic plant compounds with multiple substitution patterns. The primary players in this group include the anthocyanins (cyanidin, pelargonidin and petunidin), the flavonols (quercetin and kaempferol), flavones (luteolin and apigenin), flavanones (myricetin, naringin, hesperetin and naringenin), flavan-3-ols (catechin, epicatechin and gallocatechin), and although sometimes classified separately, the isoflavones (genistein and daidzein). Phytochemicals in this class are frequently referred to as bioflavonoids due to their multifaceted roles in human health maintenance, and anthocyanins in food are typically ingested as components of complex mixtures of flavonoid components. There is considerable anecdotal and epidemiological evidence that dietary anthocyanin pigments and polyphenolics may have preventive and therapeutic roles in a number of human diseases. Anthocyanins are glycosides of polyhydroxy and polymethoxy derivatives of 2-phenylbenzopyrylium or flavylium salts (Figure 2). The differences between individual anthocyanins relate to the number of hydroxyl groups, the nature and number of sugars attached to the molecule, the position of this attachment, and the nature and number of aliphatic or aromatic acids attached to sugars in the molecule.

Table 1 : Naturally occurring anthocyanidins

Name	Abbrev	Colour	Name	Abbrev	Colour
Apigeninidin	Ap	Orange	Luteolinidin	Lt	Orange
Aurantindin	Au	Orange	Malvidin	My	Bluish-red
Capensinidin	Cp	Bluish-red	Methylcyanidin	5-MCy	Orange-red
Cyanidin	Cy	Orange-red	Pelargonidin	Pg	Orange
Delphinidin	Dp	Bluish-red	Peonidin	Pn	Orange-red
Europinidin	Eu	Bluish-red	Petunidin	Pt	Bluish-red
Hirsutidin	Hs	Bluish-red	Pulchellidin	Pl	Bluish-red
6-Hydroxycyanidin	6OHCy	Red	Rosinidin	Rs	Red
Luteolinidin	Lt	Orange	Tricetinidin	Tr	Red

To date, there are 18 known naturally occurring anthocyanidins or aglycones, which are listed in Table 1. Only six anthocyanidins, *i.e.*, pelargonidin (Pg), peonidin (Pn), cyanidin (Cy), malvidin (Mv), petunidin (Pt) and delphinidin (Dp), are common in higher plants. The glycosides of the three non-methylated anthocyanidins (cyanidin, delphinidin and pelargonidin) are the most widespread in nature, being present in 80% of pigmented leaves, 69% of fruits and

50% of flowers. The distribution of the six most common anthocyanidins in edible parts of the plants is cyanidin (50%), pelargonidin (12%), peonidin (12%), delphinidin (12%), petunidin (7%) and malvidin (7%).

Figure 2 : The flavylium cation: R1 and R2 are H, OH, or OCH_3; R3 is a glycosyl or H, and R4 is OH or a glycosyl

Figure 3 : Blueberries a rich source of anthocyanins

The free radical scavenging and antioxidant capacities of anthocyanin pigments are the most highly publicized of the modus operandi used by these pigments to intervene with human therapeutic targets, but in fact, research clearly suggests that other mechanisms of action are also responsible for observed health benefits. The roles of anthocyanin pigments as medicinal agents have been well-accepted dogma in folk medicine throughout the world, and in fact, these pigments are linked to an amazingly broad-based range of health benefits, *e.g.*, anthocyanins from *Hibiscus* sp. have historically been used in remedies for liver disfunction and hypertension, and bilberry (*Vaccinium*) anthocyanins have an anecdotal history of use for vision disorders, microbial infections, diarrhea and diverse other health disorders. Oral intake of anthocyanosides from black currants resulted in significantly improved night vision adaptation in human subjects and similar benefits were gained after administration of anthocyanins from bilberries. Three anthocyanins from black currant stimulated regeneration of rhodopsin (a G-protein coupled receptor localized in retina of the eye) and formation of regeneration intermediate was accelerated by cyanidin 3-rutinoside. These studies strongly suggest that enhancement of rhodopsin regeneration is at least one mechanism by which anthocyanins enhance visual acuity.

Anthocyanins have demonstrated marked ability to reduce cancer cell proliferation and to inhibit tumor formation. The capacity of anthocyanin pigments to interfere with the process of carcinogenesis seems to be linked to multiple potential mechanisms of action including inhibition of cyclooxygenase enzymes and potent antioxidant potential. Fruit extracts with significant anthocyanin concentrations proved to be effective against various stages of carcinogenesis.

Table 2 : Anthocyanin content (mg per 100 g food) in some edible plants

Foodstuff	Anthocyanin	Foodstuff	Anthocyanin
Aubergine (eggplant)	750	Orange	~200
Black currant	130-400	Radish	11-60
Black berry	83-326	Rasp berry	10-60
Blue berry	25-497	Red currant	80-420
Cherry	350-400	Red grape	30-750
Choke berry	200-1000	Red onions	7-21
Crane berry	60-200	Red wine	24-35
Elder berry	450	Straw berry	15-35

Source: Lauro, G.J. and Francis, F.J., 2000. Natural food colours. In: *Science and Technology.* IFT Basic Symposium Series 14, Marcel Dekker

The role of anthocyanins in cardiovascular disease protection is strongly linked to oxidative stress protection. Protection from heart attacks through administration of grape juice or wine was strongly tied to the ability of anthocyanin rich products to reduce inflammation and enhance capillary strength and permeability and to inhibit platelet formation and enhance nitric oxide (NO) release. Ingestion of these extracts, which contained mixtures of delphinidin and cyanidin, modulate cognitive and motor function, to enhance memory and have a role in preventing age-related declines in neural function. It has been reported that administration of isolated semi-purified anthocyanins from purple sweet potato enhanced cognitive performance as assessed by passive avoidance tests in ethanol-treated mice and also effectively inhibited lipid peroxidation in rat brain tissues. By administering blueberry extracts with significant anthocyanin content (but not purified pigments), it was noted that the blueberry-supplemented diets led to effective reversal of age-related deficits in various neural and behavioral parameters (memory and motor functions).

Anthocyanins extracted from purple corn, when provided to mice in tandem with a high-fat diet, effectively inhibited both body weight and adipose tissue increases. Typical symptoms of hyperglycemia, hyperinsulinemia and hyperleptinemia provoked by a high-fat diet did not occur when mice also ingested isolated anthocyanins. The experiments suggest that anthocyanins, as a functional food component, can aid in the prevention of obesity and diabetes.

Carotenoids

Carotenoids are widely distributed natural fat-soluble plant pigments in both the plant and animal kingdoms but are exclusively of plant origin. They are isoprene derivatives with a high degree of unsaturation. Because of the presence of many conjugated double bonds, they give red, orange, or yellow colour to fruits and vegetables. As the number of double bonds increases, the wavelength of the absorbed light increases, giving the compound an increasingly red appearance as an example, the pigment of tomato (lycopene) and that of carrot (carotene) are red. There are more than 600 known natural carotenoids, and all of them are synthesized only in plants. Carotenoids may be classified as:

- Hydrocarbon carotenes or un-oxidized carotenoids such as α-carotene, β-carotene and lycopene
- Xanthophylls, which are oxygenated derivatives of carotenes, or in which some of the double bonds have been oxidized. The examples of xanthophylls include lutein, astaxanthin, canthaxanthin, zeaxanthin and capsorubin.

Table 3 : The common carotenoids and their sources

Carotenoids	Source
Carotenes	
Alpha-carotene	Carrot and water melon
Beta-carotene	Banana, jack fruit, maize, mango, papaya, pumpkin, water melon, red pepper, carrot, spinach and choke berry
Lycopene	Tomato, blood oranges, red grapefruit and water melon
Xanthophylls	
Capsanthin	Red pepper
Violaxanthin	Orange juice and red pepper
Lutein	Spinach

Carotenoids are the compounds of great dietary importance. The epidemiological studies strongly suggest that consumption of carotenoid-rich foodstuffs reduces the incidence of several diseases such as cancers, cardiovascular diseases and age-related macular degeneration. In addition to being antioxidants, carotenoids also help to rejuvenate the body by promoting the growth of healthy cells and inhibiting the growth of unhealthy ones. The beneficial actions of carotenoids are optimal when a variety of carotenoids is consumed together rather than alone. Health benefits of carotenoids are derived from fruits and vegetables in the diet, particularly from cooked products containing oil, or from supplements of their extracts, such as tomato sauce, dried tomatoes, or those suspended in oil. In general, the greater the intensity of colour, the higher is the level of carotenoids.

Xanthophylls, such as lutein, are abundant in a number of yellow or orange fruits and vegetables such as peaches, mango, papaya, prunes, acorn squash and oranges. Food sources of lutein and zeaxanthin include corn, egg yolks, green vegetables such as broccoli, green beans, green peas, Brussels sprouts, cabbage, kale, collard greens, spinach and lettuce and fruits such as kiwi and honeydew melon. Lutein and zeaxanthin are also found in nettles, algae and the petals of many yellow flowers. They provide better adsorption when consumed in food items or as a supplement.

Table 4 : Composition of some carotenoids in fruits and vegetables

Foodstuffs	Contents of major carotenoids in mole percent				
	Lutein	Zeaxanthin	Lycopene	α-carotene	β-carotene
Maize (corn)	60	25	0	0	0
Cucumber	38	4	0	0	4
Apple	20	19	23	13	5
Green grapes	31	25	0	0	7
Kiwi	54	0	5	3	16
Mango	2	16	6	0	20
Orange	7	15	11	8	11
Orange juice	15	20	0	3	8
Peach	13	5	0	10	50
Broccoli	22	0	0	0	27
Carrot	2	2	0	43	55
Green lettuce	15	0	0	16	0
Green pepper	39	36	20	0	0
Pea	41	0	0	0	5
Pumpkin	49	0	0	0	21
Red pepper	7	0	8	24	3
Spinach	47	0	4	0	16
Tomato fruit	6	0	82	0	12
Tomato juice	13	11	57	12	16
Yellow pepper	12	0	0	1	0

Source: Sommerburg and Coworkers, 1998 Fruits and vegetables that are sources for lutein and zeaxanthin: the macular pigment in human eyes. *Br. J Ophthalmol*, **82**: 907-910.

Major carotenoids

Beta-carotene

Beta-carotene is one of the orange dyes found in most green leaves and in carrots. Carotene is a terpene, an orange photosynthetic pigment, important for photosynthesis. Carotenes are hydrocarbons and most carotenes are found in yellow, orange and red vegetables. Beta-carotene is made up of eight isoprene units, which are cyclized at each end. Beta-carotene is the most potent precursor to vitamin A, which is considered a conditionally essential nutrient. Beta-carotene becomes an essential nutrient when the dietary intake of retinol (vitamin A) is inadequate. The richest sources of β-carotene are yellow, orange and green fruits and vegetables, such as carrots, spinach, lettuce, tomatoes, sweet potatoes, broccoli, cantaloupe and winter squash). Beta-carotene is the best studied, and in most countries, this is the most prevalent carotenoid in diet.

Figure 1 : Structure of β-carotene

Figure 2 : Carrots showing the red colour of β-carotene

Carotenoids are converted to vitamin A mainly in the intestine and liver. About 10% of dietary carotenoids are converted to vitamin A in the body, and they contribute 25% of our total vitamin A requirements. The most well known carotenoids of this group are beta- and alpha-carotene.

Lycopene

Lycopene is a carotenoid pigment responsible for red colour of the tomato, water melon, papaya, pink guava and pink grapefruit. It is a natural fat-soluble pigment found in certain plants and microorganisms. It serves as an accessory light gathering pigment and protects the organisms against the toxic effects of oxygen and light. Its colour is due to many conjugated double bonds, more than any other carotenoid. This configuration is responsible for lycopene's special ability to neutralize the free radicals. It protects against breast, skin and prostate cancer. It is a very potent anti-oxidant and detoxifier, and it helps to inhibit the production of cholesterol. Lycopene is more than twice as powerful as β-carotene at quenching free radicals.

Figure 3 : Structure of lycopene

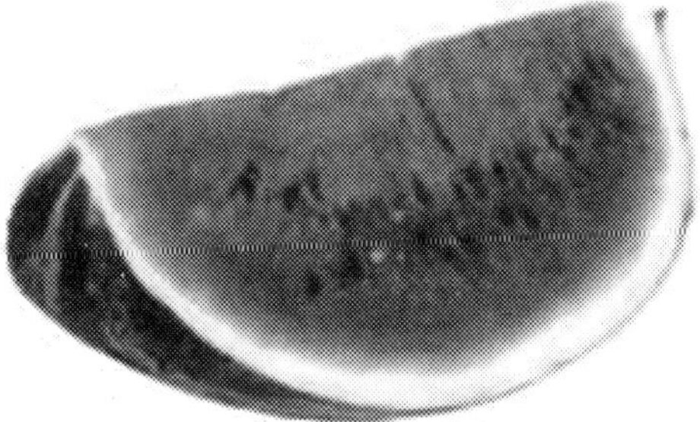

Figure 4 : Watermelon showing the red colour of lycopene

Lutein and zeaxanthin

Lutein is a carotenoid pigment found in many green vegetables, and it is usually derived from marigold flower petals. Lutein and zeaxanthin are the two major components of the human macula and they are often referred as macular pigments of the retina. These macular pigments are not evenly distributed throughout the human retina. They have been shown to protect against blindness caused by macular degeneration. These pigments filter out potentially phototoxic blue light and near-ultraviolet radiation from the macula.

Figure 5 : Structure of zeaxanthin

Figure 6 : Maize showing the yellow colour of xanthophylls

Many carotenoids are antioxidants that protect cells against free radicals by neutralizing them before they cause oxidative damage. As antioxidants, carotenoids protect cells from damage by unstable oxygen molecules called free radicals. Hydrogen peroxide, singlet oxygen, nitrogen oxides, superoxide anion and other reactive oxygen species insulting the body from either endogenous or exogenous routes can be inactivated by carotenoids. In fact, carotenoids are considered the most potent biological quenchers of singlet oxygen system. In majority of these reactions, the carotenoids break down to biologically active degradation products like β-carotene. Chemically, the polyene chain length of carotenoid molecules is correlated with efficiency in quenching singlet oxygen, with lycopene being the most efficient type of carotenoid. Carotenoids can also protect cells from oxidative damage. Beta-carotene may terminate lipid oxidation as a chain-breaking antioxidant. Both, beta-carotene and lutein protect cells from lipid peroxidation and membrane damage as measured by decreased cellular release of lactate dehydrogenase. An advantageous association of carotenoid intake and protection from type 2 diabetes was recently investigated with glucose metabolism in men at high risk. Dietary carotenoids were inversely associated with fasting plasma glucose concentrations and plasma. The β-carotene concentrations were inversely associated with insulin resistance.

Also like flavonoids, the carotenoids provide protection to vision and eye function, and against macular degeneration and cataracts. Two carotenoids present within the entire retina and macula, lutein and zeaxanthin (classified together as xanthophylls), account for the yellowish cast in macular region of the eye. Carotenoids are able to absorb light in visible range of the spectrum. By filtering out the sun's harmful ultraviolet (UV) rays and by keeping free radicals from damaging the retina, the carotenoids may help to protect against macular degeneration, which caused by blue light-mediated free radical damage to the retina, the leading cause of irreversible blindness in aging adults over the age of 65. Lutein has been loosely referred to as *the eye-protective nutrient*, and its intake has been shown in epidemiological trials to be inversely associated with ocular diseases such as age related macular degeneration and cataracts.

All carotenoids help to prevent heart disease and heart attack by inhibiting the formation of harmful low-density lipoprotein (bad cholesterol). Since they are very soluble in fat and insoluble in water, carotenoids circulate in lipoproteins along with cholesterol and other fats. It has been reported that a 2 weeks dosage of tomato juice, rich in carotenoids, effectively lowers the low-density lipoprotein in male subjects, which may be due to the inhibition of cholesterol synthesis.

A higher dietary intake of carotenoids offers protection against developing certain cancers, *e.g.*, lung, skin, uterine, cervix, gastrointestinal tract. Carotenoids may inhibit the growth of cancer cells. Epidemiological studies have shown that people with high beta-carotene intake and high plasma levels of beta-carotene have a significantly reduced risk of lung cancer. Lycopene and beta-carotene, taken along with vitamins C and E, may help to protect the body against the effects of chemotherapy or radiation. Carotenoids are associated with the inhibition of several types of cancers including cervical, esophageal, pancreatic, lung, prostate, colourectal and stomach. It has been observed that very low carotenoid intakes may be associated with increased risk of breast cancer. In particular, the carotenoid lycopene has been extensively researched for its role in cancer inhibition. Lycopene's role in colon carcinogenesis has been documented in rats. Lutein and zeaxanthin intake has been linked to reduced risk of prostate cancer.

Chlorophylls

Chlorophyll is the green pigment found in most plants, algae and cyanobacteria. Its name is derived from the Greek word *chloros* meaning green and *phyllon* meaning leaf. Chlorophyll absorbs light most strongly in blue and red but poorly in the green portions of the electromagnetic spectrum. Hence, the leaves appear green in colour because of chlorophyll present in them. It is one of the greatest miracles of Nature that chlorophyll and its companions, in the presence of sunlight, produce sugars (carbohydrates). The process is called *photosynthesis*, which beyond doubt is the most important of all biochemical reactions and one that helps plants to grow, flower, produce seed and sustain life on Earth.

Chlorophyll is often described as *blood of the plants* since its molecular structure is absolutely identical to hemoglobin except for the center atom. In hemoglobin, this is iron, whereas, in chlorophyll, it is magnesium (Figure 3) and chlorophyll-rich food artcles have been linked with increasing hemoglobin content. Most plant pigments comprise only a small percentage of common foodstuffs, thus, it may require ingestion of concentrated supplements to achieve biologically active levels for some human health therapies. Chlorophylls, in contrast, are typically consumed in much higher doses in a diet that incorporates green and leafy vegetables. While all green plants contain chlorophyll '*a*', and most vegetables that we eat contain both chlorophyll '*a*' and chlorophyll '*b*', some of the vegetables contain particularly high amounts of total chlorophyll. Best studied of all the vegetables is spinach (*Spinach oleracea*), which contains about 300-600 milligrams per ounce. Adequate amounts of chlorophyll can be found in regular consumption of fresh green vegetables like broccoli, spinach, kale, lettuce, beet greens, Asian greens, peas, green beans, asparagus, leeks and basically anything green and edible.

Chlorophylls in green foods are credited with many roles in human health maintenance including reduction of blood pressure, blood sugar control, brain activation, antimutagenic and anticlastogenic effects and more. Chlorophyll has anti-inflammatory, antioxidant and wound-healing properties. Chlorophyll has the power to regenerate our bodies at the molecular and cellular level and is known to help in cleansing the body, fight infection, help in healing wounds and promote health of the circulatory, digestive, immune and detoxification systems. Chlorophyll and chlorophyllin (a water-soluble, food-grade, stable sodium and copper salt of chlorophyll) are able to form tight molecular complexes with certain chemicals known or suspected to cause cancer, including polyaromatic hydrocarbons found in tobacco smoke. It is known to protect against carcinogens found in fungus-laden foods such as nuts and grains, the toxins from cooked meats. It blocks the metabolism in body of harmful chemicals known as procarcinogens that damage DNA. Research supporting chlorophyll's roles in cancer chemoprevention indicates that these pigments are capable of deactivating potential carcinogens in the digestive tract, *e.g.*, heterocyclic amines that provoke colorectal cancer and preventing absorption. Chlorophyll may reduce the binding of carcinogens to DNA in the liver and other organs. Chlorophyll and chlorophyllin may have some antimutagenic and anticarcinogenic potential, may help protect against some toxins and may ameliorate some drug side effects. Chlorophyll is a good source of antioxidant nutrients. Antioxidant nutrients such as vitamins A, C and E help to neutralize harmful molecules (free radicals) in the body that can cause damage to healthy cells. Chlorophyll is an efficient deliverer of magnesium and helps in carrying blood oxygen to the cells and tissue. Chlorophyll has chelation ability, which means it can bind heavy metal in the body and expel it. This is important for people who are, or have been exposed to heavy metal poisoning, especially with lead and mercury.

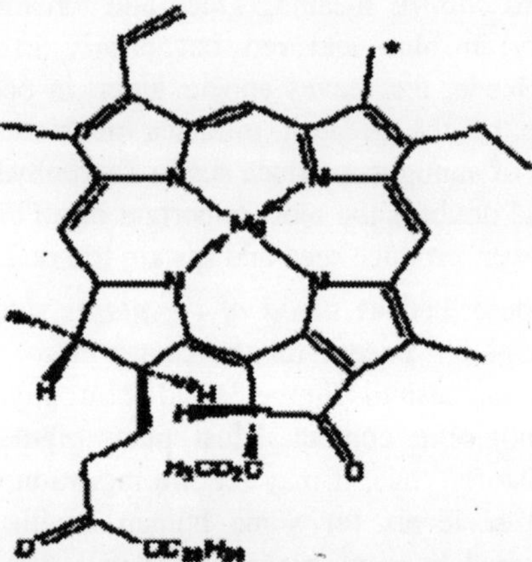

Figure 7 : Structure of chlorophyll '*a*'

Figure 8 : The green coloured chlorophyll in leaves

Chlorophyll has been studied for its potential in stimulating tissue growth and in stimulating red blood cells in connection with oxygen supply. Chlorophyll also removes carbon dioxide and carbon monoxide, and has been found to reduce fecal, urinary and body odour. Chlorophyll may be beneficial in the treatment of calcium oxalate stone disease and that they may have some anti-atherogenic activity. Chlorophyll has been used traditionally to improve bad breath as well as to reduce the odour of infected wounds. Chlorophyll also contains protein, folic acid, iron, calcium, vitamin K and C, which are all essential in building and repairing red blood cells and boosting our immune system.

Betalains and Other Pigments

The name *betalain* comes from the Latin name of common beet (*Beta vulgaris*), from which betalains were first extracted. The deep red colour of beets, bougainvillea, amaranth and many cacti results from the presence of betalain pigments. Betalains are a class of red and yellow indole-derived pigments found in plants of the Caryophyllales, where they replace anthocyanin pigments as well as in some higher order fungi. Both betalains and anthocyanins are water-soluble pigments found in vacuoles of the plant cells. Betalains are aromatic indole derivatives synthesized from tyrosine (Figure 4). They are not related chemically to the anthocyanins and are not even flavonoids. Each betalain is a glycoside and consists of a sugar and a coloured portion. They are most often noticeable in petals of the flowers but may colour the fruits, leaves, stems and roots of plants that contain them. They include powerful antioxidant pigments. There are two categories of betalains given below:

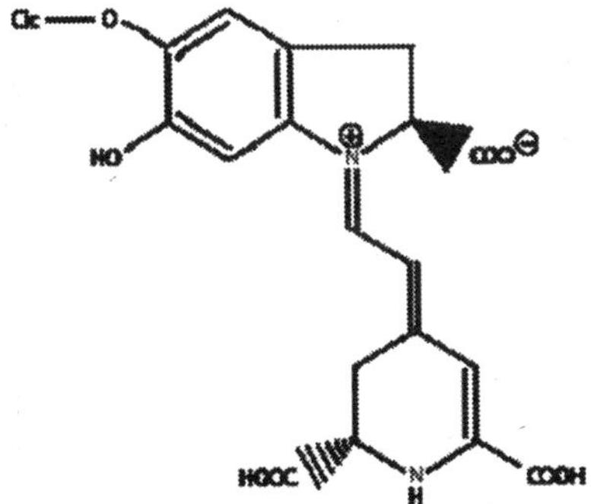

Figure 9 : Structure of betanin- a betalain

Figure 10 : Beetroots showing betalain colour

- Betacyanins include the reddish to violet betalain pigments.
- Betaxanthins are those betalain pigments, which appear yellow to orange. The betaxanthins present in plants include vulgaxanthin, miraxanthin and portulaxanthin and indicaxanthin.

The very potent antiradical scavenging activity of betalains has been demonstrated repeatedly in a wide range of assays. Betalains shows a cationized structure with positive nitrogen in a polyene system. Their cyclic amine, which is similar to that of antioxidant ethoxyquine, has reasonably been considered the reactive group conferring to this class of molecules reducing properties. Therefore, these phytochemicals (betacyanins and betaxanthins) carry a phenolic and an acyclic amine group, which are excellent electron donors and are able to stabilize radicals. In a study of different beetroot phenotypes, it was the most highly pigmented (β-cyanin-rich) beets, which were most efficient at free radical scavenging and inducing quinine reductase (a phase II enzyme important as a marker of initiation stage of carcinogenesis) in murine hepatoma cells *in vitro*. The betaxanthin indicaxanthin showed *in vitro* positive actions against *b*-thalassemia by incorporation into the redox machinery of *b*-thalassemic red blood cells to defend them from oxidation. Recently, it has been found that betalain enrichment of human low-density lipoproteins effectively increased their resistance to oxidation. Preliminary cancer-chemopreventive studies with crude beetroot extract have shown that the protection against skin and lung tumors in animal (mice) models is slightly higher than that afforded by anthocyanin pigments from red onion skin or carotenoids from red bell peppers. In particular, one pigment in the betalain group, betanin, appears to be a potent anticancer compound. Beetroot extract rich in betalain pigments has demonstrated chemopreventive properties against both lung and skin cancers. The antimalarial activities of *Amaranthus spinosus* L. and *Boerhaavia erecta* L. containing betalains have also been reported. Damaged dead cells must be removed otherwise they will become the focus of infection and illness. Betalains assist macrophage to engulf the dead cell and eliminate it from the body. Then, toxins are drained away from the cell and the protein matrix of Betalains permeates and reinforces the cell, returning it to health.

Another pigment having widespread health benefits is curcumin (Figure 5), the yellow pigment found in the curry spice turmeric. Turmeric (*Curcuma longa* L. family Zingiberaceae) has been used historically as a component of Indian *Ayurvedic* medicine since 1900 BC to treat a wide variety of ailments. Research in the latter half of the 20th century has identified curcumin as responsible for most of the biological activity of turmeric.

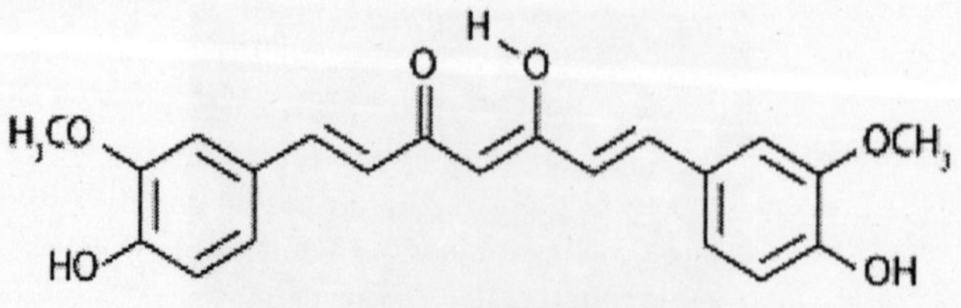

H
O O
H_3CO OCH_3
HO OH

Figure 11 : Structure of curcumin

Figure 10 : Turmeric showing curcumin

Previous researches found that curcuminoids have antioxidant, antibacterial and anti-inflammatory effects upon human body. In addition, the turmeric intake helps in supporting the health of liver, prevents high doses of cholesterol in organism and is being studied by the scientists for its tumor-blocking ability. Turmeric as an anti-cancer agent has been found to block the uncontrollable growth of melanoma tumor cells, the cells that lead to breast and colon cancer, leukemia, *etc.* Curcumin was shown to induce apoptosis among B lymphoma cells and inhibits the multiplication of leukemia cells in laboratory studies. It has been shown to possess potent antiproliferative and proapoptotic effects in melanoma cells and neuroblastoma cell lines by suppressing the IkB kinase and NF-kB activity. The efficacy of curcumin to offer a dose- and time-dependent decrease in cell viability, cell cycle arrest and induction of apoptosis is encouraging to the development of a natural drug with known NFk-B inhibitory activity and little toxicity in humans, unlike the existing inhibitors. Various studies have reported that curcumin inhibits angiogenesis, the formation of new blood vessels from the host vasculature to feed the growing tumor in non-malignant and malignant cells, and reduces the number and size of existing tumors and decreases the incidence of new tumor formation. It can also affect proteins related to cell-cell adhesion and inhibits the production of cytokines relevant in tumor growth. Furthermore, curcumin can enhance cancer cells' sensitivity to certain drugs commonly used to combat cancer and can potentially improve the effectiveness of radiation treatment.

Cucurmin's positive effects on cardiovascular health have been widely reported. Oxidation of low-density lipoproteins (LDL) is implied to play a vital role in atherosclerosis, deposition of cholesterol and lipids on arterial walls to form damaging plaques to block the blood flow. In this regard, many *in vivo* animal studies have clearly shown curcumin's ability as an antioxidant to prevent the free radical mediated oxidation of cellular and sub-cellular membranes that are associated with atherosclerosis. Moreover, it helps to lower the total cholesterol levels and increase high-density lipoproteins (HDL) level. In addition to cholesterol, excess levels of homocysteine are generally regarded as an independent risk factor for cardiovascular disease. Researchers suggest that elevated cholesterol alone may not cause heart disease. However, it is quite possible to have a cardiac arrest even with normal cholesterol levels but with elevated homocysteine levels. Curcumin has been found to prevent the disfunction of the endothelial cells, caused by the elevated homocysteine and may provide a role in the treatment of patients with high homocysteine. Similarly, curcumin was also shown to inhibit completely the effect of dangerous marker protein called C-reactive protein on endothelial cells.

The anti-inflammatory action of curcumin includes lowering histamine levels and increasing the production of natural cortisone by adrenal glands. Among arthritis patients, it significantly improved morning stiffness, walking time, joint swelling, *etc.* when a daily dosage of 1200 mg of curcumin for two weeks was given. In a diabetes mellitus rat model study, administration of curcumin was found to offer a reduction of blood sugar, Hb and glycosylated Hb levels significantly. In case of cataract, the rats treated with naphthalene to induce cataract and kept on a diet supplemented with only 0.005% (w/w) curcumin were found to have significantly less opacification of lenses as compared to controls. Being an efficient singlet oxygen quencher at physiological or pharmacological concentration, it may be useful pharmacologic agent diseases such as erythropoietic protoporphyria, pellagra and cataractogenesis. Curcumin can also chelate toxic metals like cadmium and lead and potentially reduce their neurotoxicity and tissue damage. More interestingly, a possibility has also been reported as inhibitors of HIV-1 and HIV-2 proteases against acquired immunodeficiency syndrome (AIDS).

Conclusions

Plant pigments have multiple roles in human health protection, maintenance and therapy through a diverse array of biological activities, evidenced through *in vitro*, *in vivo* and clinical trials, as well as through extensive epidemiological evidence. As a body, the combined evidence of numerous research trials indicates that variety of different pigments in a well-balanced diet is key means to prevention and therapy for a range of chronic human disease conditions.

■■■

29 Polyunsaturated Fatty Acids (PUFA)

Meenakshi Virmani

Assistant Endocrinologist, Department of Animal Production Physiology, College of Veterinary and Animal Sciences, Lala Lajpat Rai University of Veterinary and Animal Sciences, Hisar (Haryana) India, E-mail: <virmanim2003@yahoo.com>

Fatty acids are components of fat mainly derived from triglycerides or phospholipids. These are mostly bound or attached to other molecules. When they are not attached to other molecules, they are known as *free* fatty acids. Fatty acids are not highly soluble in water. They are found in oils and other fats that make up different foods and perform a number of functions in the body. Metabolism of fatty acids in the body yields large quantities of ATP, thus, they act as source of energy for many cells in the body. They help in carrying oxygen to all parts of the body through the bloodstream. They aid in cell membrane development and function and are necessary for strong organs and tissue.

Fatty acids are organic, that is, they contain both carbon and hydrogen molecules. They are linear hydrocarbon chains with a methyl (CH_3-) and a carboxyl (-COOH) end. The most naturally occurring fatty acids have a chain of an even number of carbon atoms from 4 to 28. Carbon atoms in fatty acids are identified by Greek letters based on their distance from the carboxylic acid. The carbon atom closest to the carboxylic acid is the *alpha* carbon, the next adjacent carbon is the *beta* carbon and the chain is further named so on. In a long-chain fatty acid, the carbon atom in the methyl group is called the *omega* carbon since *omega* is the last letter of the Greek alphabet. Fatty acids that vary in their number of carbon atoms and double bonds are classified as:

- Saturated fatty acids (SFA, no double bonds)
- Monounsaturated fatty acids (MUFA, one double bond)
- Polyunsaturated fatty acids (PUFA, two or more double bonds)

Saturated fatty acids (SFA, no double bonds): Saturated fatty acids have all bonding positions between carbon atoms occupied by hydrogen atoms and have no double bonds between the individual carbon atoms of the fatty acid chain. The chain of carbon atoms is thus fully *saturated* with hydrogen atoms. There are many kinds of naturally occurring saturated fatty acids, which mainly differ in number of carbon atoms, from 3 carbons (propionic acid) to 36 (hexatriacontanoic acid). The sources of saturated fat include animal fats such as cream, cheese, butter, ghee, suet, tallow, lard and fatty meats as well as certain vegetable products such as coconut oil, cottonseed oil, palm kernel oil and chocolate.

Monounsaturated fatty acids (MUFA, one double bond): Monounsaturated fatty acids are the fatty acids that have a one double bond in the fatty acid chain and all of the other carbon atoms in the chain have single bond. The common monounsaturated fatty acids are Palmitoleic acid (16:1 *n*-7), cis-vaccenic acid (18:1 *n*-7) and oleic acid (18:1 *n*-9). Palmitoleic acid has 16 carbon atoms with the first double bond occurring 7 carbon atoms away from the methyl group. Cis-vaccenic acid has 18 carbon atoms with the first double bond occurring 7 carbon atoms away from the methyl group while oleic acid has 18 carbon atoms with the first double bond occurring 9 carbon atoms away from the carboxylic acid group. Monounsaturated fats are found along with saturated fat in natural foods such as red meat, whole milk products, nuts and high fat fruits such as olives and avocadoes.

Polyunsaturated fatty acids (PUFA, two or more double bonds): Polyunsaturated fatty acids are the fatty acids in which more than one double bond exists within the chain. They are determined by the position of these double bonds in relation to the end of the molecule. Since they have multiple double bonds, polyunsaturated fats are chemically unstable and prone to oxidation. Polyunsaturated fatty acids are present in dietary fats and oils and are known as essential fatty acids. They cannot be synthesized in the body and must be obtained *via* dietary sources. These fats are the *good fats*, which are a necessary part of the human diet.

In this chapter, we will be discussing about the polyunsaturated fatty acids in detail along with their uses.

Chemistry of Polyunsaturated Fatty Acids

Polyunsaturated fatty acids have at least two double bonds. Their structural, physical and chemical properties depend upon the number of double bonds in the chain. They can assume a *cis*- or *trans*-conformation depending on the geometry of double bond. In *cis*- conformation, hydrogen atoms are on the same side of the double bond, whereas, in the trans-conformation, they are on opposite sides. The lack of the extra hydrogen atoms on the molecule's surface typically reduces the strength of compound's intermolecular forces, thus, significantly lowering the melting point of compound. This property can be observed in unsaturated vegetable oils, which remain liquid even at relatively low temperature, while butter is much more saturated and is mainly solid at room temperature. *Trans*-fats are more similar to saturated fat than are the *cis*-fats and they solidify at a higher temperature than *cis*-fats.

Depending on the position of first double bond, the polyunsaturated fatty acids (PUFA) are further sub-classified as:

- Omega-3 PUFA
- Omega-6 PUFA
- Omega-9 PUFA

Omega-3 PUFA: The ω-3 PUFA or *n*-3 PUFA are a family of unsaturated fatty acids that have in common a final carbon-carbon double bond in the *n*-3 position, *i.e.*, the third bond from the methyl end (*n* or *ω* end) of the fatty acid. This is named from *n*-end because the *n* end is more stable energetically and is never changed during physiologic transformations in the human body but from the carbonyl end or from double bonds in the middle of the chain, other carbohydrates compounds can be synthesized.

$$CH_3CH_2CH{=}CHCH_2CH{=}CHCH_2CH{=}CHCH_2CH_2CH_2CH_2CH_2CH_2CH_2COOH$$

or

Figure 1 : Chemical structure of alpha-linolenic acid (ALA)

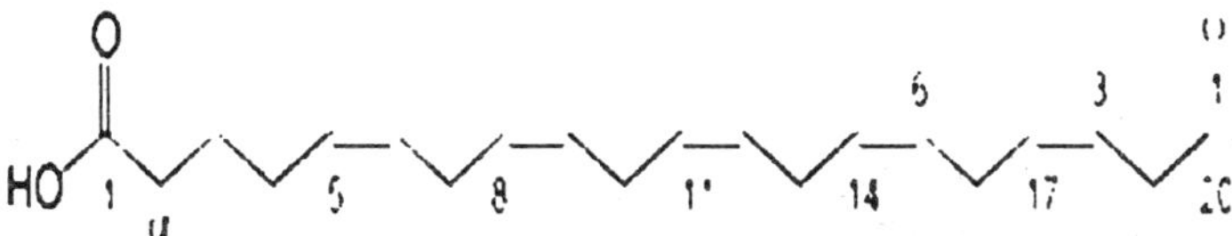

Figure 2 : Chemical structure of eicosapentaenoic acid (EPA)

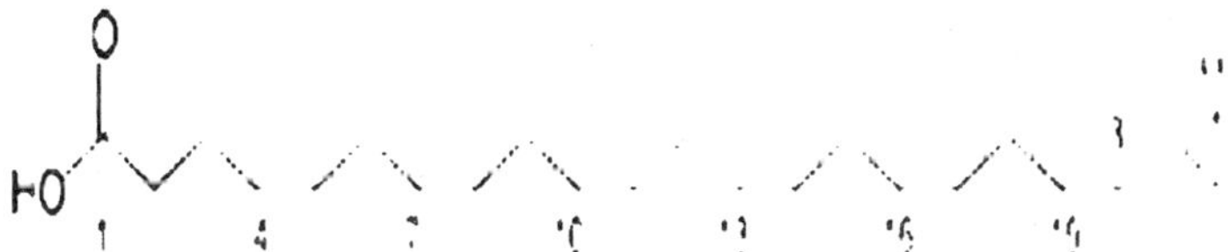

Figure 3 : Chemical structure of docosahexaenoic acid (DHA)

Nutritionally important *n*-3 fatty acids include α-linolenic acid (18:3, *n*-3; ALA), eicosapentaenoic acid (20:5, *n*-3; EPA) and docosahexaenoic acid (22:6, *n*-3; DHA). These three polyunsaturated fatty acids have either 3, 5, or 6 double bonds in a carbon chain of 18, 20, or 22 carbon atoms, respectively. All double bonds are in the *cis*-configuration. In other words, the two hydrogen atoms are on same side of the double bond. The human body cannot synthesize *n*-3 fatty acids *de novo* but it can form *long chain* 20-carbon unsaturated *n*-3 fatty acids (like EPA) and 22-carbon unsaturated *n*-3 fatty acids (like DHA) from the *short chain* eighteen-carbon *n*-3 fatty acid α-linolenic acid. The efficiency of conversion of the short chain *n*-3 fatty acids to long chain forms (EPA and DHA) is much more in women as compared to men. The major omega-3 polyunsaturated fatty acids in the diet are given below:

Sr. No.	Common name	Lipid name	Chemical name
1.	Alpha-linolenic acid (ALA)	18:3 (*n*-3)	*all-cis*-9,12,15-octadecatrienoic acid
2.	Stearidonic acid (SDA)	18:4 (*n*-3)	*all-cis*-6,9,12,15,-octadecatetraenoic acid
3.	Eicosatrienoic acid (ETE)	20:3 (*n*-3)	*all-cis*-11,14,17-eicosatrienoic acid
4.	Eicosatetraenoic acid (ETA)	20:4 (*n*-3)	*all-cis*-8,11,14,17-eicosatetraenoic acid
5.	Eicosapentaenoic acid (EPA) (Timnodonic acid)	20:5 (*n*-3)	*all-cis*-5,8,11,14,17-eicosapentaenoic acid
6.	Docosapentaenoic acid (DPA)	22:5 (*n*-3)	*all-cis*-7,10,13,16,19-docosapentaenoic acid

Sr. No.	Common name	Lipid name	Chemical name
7.	Docosahexaenoic acid (DHA) (Cervonic acid)	22:6 (*n*-3)	*all-cis*-4,7,10,13,16,19-docosahexaenoic acid
8.	Tetracosapentaenoic acid	24:5 (*n*-3)	*all-cis*-9,12,15,18,21-tetracosapentaenoic acid
9.	Tetracosahexaenoic acid (Nisinic acid)	24:6 (*n*-3)	*all-cis*-6,9,12,15,18,21-tetracosahexaenoic acid

Omega-6 PUFA: The *n-6* fatty acids (popularly referred to as ω-6 fatty acids or omega-6 fatty acids) are a family of unsaturated fatty acids that have in common a final carbon-carbon double bond in the *n*-6 position, *i.e.*, the sixth bond, counting from the end opposite the carboxyl group.

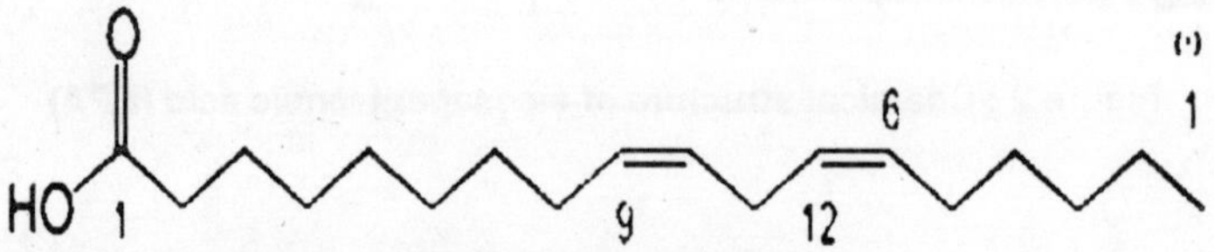

Figure 4 : Linoleic acid

or

$$CH_3CH_2CH_2CH_2CH_2CH{=}CHCH_2CH{=}CHCH_2CH_2CH_2CH_2CH_2CH_2CH_2COOH$$

9, 12-octadecadienoic acid (Linoleic Acid)

The biological effects of *n*-6 fatty acids are largely mediated by their conversion to *n*-6 eicosanoids that bind to diverse receptors found in every tissue of the body. The conversion of tissue arachidonic acid (*n*-6 PUFA) to *n*-6 prostaglandin and *n*-6 leukotriene hormones helps in drug development and treatment of different disease conditions. Competitive interactions of *n*-6 fatty acids with *n*-3 fatty acids affect the relative storage, mobilization, conversion and action of *n*-3 and *n*-6 eicosanoid precursors. The major omega-6 PUFA in the diet are as under:

Sr. No.	Common name	Lipid name	Chemical name
1.	Linoleic acid	18:2 (*n*-6)	*all-cis*-9,12-octadecadienoic acid
2.	Gamma-linolenic acid (GLA)	18:3 (*n*-6)	*all-cis*-6,9,12-octadecatrienoic acid
3.	Eicosadienoic acid	20:2 (*n*-6)	*all-cis*-11,14-eicosadienoic acid
4.	Dihomo-gamma-linolenic acid (DGLA)	20:3 (*n*-6)	*all-cis*-8,11,14-eicosatrienoic acid
5.	Arachidonic acid (AA)	20:4 (*n*-6)	*all-cis*-5,8,11,14-eicosatetraenoic acid
6.	Docosadienoic acid	22:2 (*n*-6)	*all-cis*-13,16-docosadienoic acid
7.	Adrenic acid	22:4 (*n*-6)	*all-cis*-7,10,13,16-docosatetraenoic acid
8.	Docosapentaenoic acid (Osbond acid)	22:5 (*n*-6)	*all-cis*-4,7,10,13,16-docosapentaenoic acid

Omega-9 PUFA: The *n*-9 fatty acids (popularly referred to as ω-9 fatty acids or omega-9 fatty acids) are a family of unsaturated fatty acids, which in common have a final carbon-carbon double bond in *n*-9 position, *i.e.*, the ninth bond from the methyl-end of fatty acid. Omega-9 is the most abundant fatty acid in nature and is not in short supply in our diets. When the body does not have enough omega-3 or omega-6, it tries to compensate by producing omega-9 fatty acids to take their place. The major omega-9 fatty acids are given as under:

Sr. No.	Common name	Lipid name	Chemical name
1.	Oleic acid	18:1 (*n*-9)	9-octadecenoic acid
2.	Eicosenoic acid	20:1 (*n*-9)	11-eicosenoic acid
3.	Mead acid	20:3 (*n*-9)	5,8,11-eicosatrienoic acid
4.	Erucic acid	22:1 (*n*-9)	13-docosenoic acid
5.	Nervonic acid	24:1 (*n*-9)	15-tetracosenoic acid

Dietary Sources of PUFA

Most fatty acids can be synthesized in the body but human beings lack the enzymes required to produce two fatty acids *viz.*, α-linolenic acid and linoleic acid. These are called the essential fatty acids and must be acquired from the diet. Although the long chain omega-3 polyunsaturated fatty acids *viz.*, eicosapentaenoic acid and docosahexaenoic acid can be synthesized by the human body but the rate of synthesis may not be sufficient to meet requirements and it is, therefore, recommended that good sources of these fatty acids, namely oil-rich fish, are also included in the diet.

Omega-3 PUFA

Vegetables oils are the major sources of alpha-linolenic acid (ALA). In particular, ALA is found in the chloroplast of green leafy vegetables, such as purslane and spinach and in seeds of flax/linseed, walnuts, *etc*. The richest source of ALA is Purslane (*Portulaca olearacea*), a vegetable used in soups and salads along the Mediterranean region and in Middle East and is known to be a source of eicosapentaenoic acid (C20:5 *n*-3, EPA). The other sources include canola oil (*Brassica napus*), nuts, seeds, vegetables, some fruits, egg yolk, poultry and meat. Fish is the main source of EPA and docosahexaenoic acid (C22:6 *n*-3, DHA).

Omega-6 PUFA

Food oils that are the major source of *n*-6 fatty acids include palm oil, soybean oil, rapeseed oil and sunflower oil. The other dietary sources of *n*-6 fatty acids include poultry, eggs, avocado, nuts, cereals, whole-grain breads, most vegetable oils, evening primrose oil, borage oil, blackcurrant seed oil, flax/linseed oil, hemp oil, cottonseed oil, corn oil, safflower oil, pumpkin seeds, acai berry and cashews.

Omega-9 PUFA

Olive oil is the best source of omega-9 fatty acids. The *n*-9 fatty acids are present in olives, avocados, almonds, peanuts, sesame oil, pecans, pistachio nuts, cashews, hazelnuts, macadamia nuts, rapeseed, wallflower seed and mustard seed.

Unlike *n*-3 and *n*-6 fatty acids, the *n*-9 fatty acids are not classified as essential fatty acids (EFA). This is because they can be created by the human body from unsaturated fat and are therefore not essential in the diet. Also due to the lack of an *n*-6 double bond, these do not participate in the reactions that form the eicosanoids.

Potential Health Benefits

Omega-3 PUFA

The *n*-3 fatty acids reduce blood triglyceride levels and regular intake may reduce the risk of secondary and primary heart attack. Among *n*-3 fatty acids, Eicosapentaenoic acid (EPA) and Docosahexaenoic acid (DHA) have been shown to have much more cardiovascular benefits as compared to α-linolenic acid. People with certain circulatory problems, such as varicose veins, may benefit from supplements containing EPA and DHA, which stimulate blood circulation, increase the breakdown of fibrin, which is involved in clot and scar formation and additionally reduces the blood pressure. There are many products in the market, which claim to contain health-promoting omega 3 but contain only α-linolenic acid (ALA), not the EPA or DHA. These products contain mainly higher plant oils, which must be converted by the body to create DHA, and thus, they are considered less efficient.

The *n*-3 fatty acids supplementation might be helpful in cases of depression and anxiety alone and in conjunction with medication. Some potential benefits have been reported in conditions such as rheumatoid arthritis and cardiac arrhythmias. In addition, *n*-3 fatty acids intake may reduce the risk of ischemic and thrombotic stroke. However, large amounts of *n*-3 fatty acids may actually increase the risk of hemorrhagic stroke.

Cardiovascular disease: Supplementation of *n*-3 fatty acids reduces death due to cardiovascular reasons. Very low-density lipoprotein (VLDL) is a bad type of cholesterol and elevated triglycerides can be deleterious for cardiovascular health. Administration of purified EPA improves the thickness of carotid arteries along with improving blood flow in patients with unhealthy blood sugar levels. Eicosapentaenoic acid (EPA) is a promising treatment for the prevention of major coronary events, especially non-fatal coronary events. Consumption of high amounts of *n*-3 fatty acids increases high-density lipoprotein (HDL) cholesterol and decreases triglycerides (fatty material that circulates in the blood) and less heart disease. Oxidative modification of low-density lipoprotein (LDL) is thought to play an important role in the development of atherosclerosis. Serum level of eicosapentaenoic acid is inversely related to the levels of anti-oxidized LDL antibodies.

Antiarrhythmic effects: Dietary *n*-3 PUFA at low concentration reduces the severity of cardiac arrhythmia and coronary heart disease (CHD). At high doses, dietary *n*-3 PUFA has several beneficial properties given as under:

- Act favourably on blood characteristics by reducing platelet aggregation and blood viscosity
- Are hypotriglyceridemic
- Exhibit antithrombotic and fibrinolytic activities
- Exhibit antiinflammatory action
- Reduce ischemia/reperfusion-induced cellular damage. This effect is apparently due to the incorporation of eicosapentaenoic acid in membrane phospholipids.

Hypolipidemic effects: The hypolipidemic effects of *n*-3 fatty acids are similar to those of *n*-6 fatty acids, if they replace saturated fats in the diet. The *n*-3 PUFA have the added benefit of consistently lowering serum triacylglycerol concentrations, while the n-6 fatty acids do not and may even increase them.

Antithrombotic effects: The antithrombotic effects of *n*-3 PUFA are due to decrease in platelet aggregation, decrease in whole blood viscosity and an increase in bleeding time, while the diet rich in *n*-6 fatty acids shifts the physiologic state to one, which is prothrombotic and proaggregatory, with increases in blood viscosity, vasospasm and vasoconstriction, and decreases in bleeding time.

Hypotensive effects: Supplementation of diet with high doses of *n*-3 PUFA reduces the blood pressure. The most effective *n*-3 PUFA is DHA rather than EPA.

Diabetes: The type 2 diabetes is a multigenic, multifactorial disorder characterized by hyperglycemia in the presence of insulin resistance, hypertriglyceridemia and the development of vascular complications. Men and women with type 2 diabetes have 3 and 5 folds higher cardiovascular mortality, respectively than the non-diabetic population, and this higher risk is also carried by non-diabetic first-degree relatives of type 2 diabetes subjects. The *n*-3 fatty acid intake along with oral therapy for diabetes can lower the triacylglycerol concentrations with no adverse effects on glycemic control. It is also known that the concentration of serum leptin hormone in patients with type 1 diabetes mellitus is influenced by the type of fat in the diet. It has been reported that *n*-3 fatty acids decrease leptin gene expression both *in vivo* and *in vitro*.

Anti-Inflammatory Effects: The beneficial effects of *n*-3 fatty acids are derived in part from their effect on the immune system. Metabolism of EPA and ALA (*n*-3 fatty acid) leads to the generation of eicosanoids such as prostaglandins and leukotrienes that are less potent inducers of inflammation.

Bowel Disease: Consumption of long chain omega-3 fatty acids had shown positive effect less without failing in Crohn's disease, but in ulcerative colitis (inflammatory bowel diseases), the patients reported significant improvement. Alpha-linolenic acid has been proved better to decrease bowel disease than docosahexaenoic and eicosapentaenoic acids. However, consumption of fish oil may sometimes cause flatulence and diarrhea.

Arthritis: Supplementation with *n*-3 fatty acids can modulate the expression and activity of degradative and inflammatory factors that cause cartilage destruction during arthritis. The *n*-3 fatty acid supplementation can specifically affect regulatory mechanisms involved in chondrocyte gene transcription, and thus, alleviates several physiological parameters that cause and propagate arthritic disease. Incorporation of *n*-3 fatty acids into articular cartilage chondrocyte membranes results in a dose-dependent reduction in:

- Expression and activity of proteoglycan degrading enzymes (aggrecanases)
- Expression of inflammation-inducible cytokines (IL-1 alpha and TNF-alpha) and cyclooxygenase-2

Psoriasis: Intravenous *n*-3-fatty acid administration causes reduction of psoriasis, which may be due to changes in inflammatory eicosanoid generation.

Cancer: Daily consumption of eicosapentaenoic acid and docosahexaenoic acid (omega-3) along with low fat diet and supplementation of vitamin C, vitamin E, β-carotene and selenium has been proved not only to slow but even checks the progression of cancers in early stages of disease (particularly breast, colon and prostate cancer). Oral *n*-3 fatty acid supplementation benefits cancer patients, improving appetite, weight and quality of life. Omega-3 fatty acids reduce prostate tumour growth, slow down the histopathological progression and increase the survival. High levels of docosahexaenoic acid, however, the most abundant *n*-3 PUFA in erythrocyte membranes, are associated with a reduced risk of breast cancer. Omega-3 fatty acids reduce cancer risk by markedly inhibiting activator protein 1 (AP-1), which is a transcription factor that promotes cancerous proliferation and metastasis.

Immune Function: Supplementation of PUFA causes improvement in immune function maturation.

Brain: Long-chain *n*-3 fatty acids may help prevent psychotic disorders in children and adolescents. The *n*-3 fatty acid known as Ethyl eicosapentaenoic acid (E-EPA) may prevent memory impairment and speed up recovery from major depression that occurs due to accumulation of amyloid and tau proteins in the brain. The *n*-3 supplementation in children with learning and behavioural problems leads to improvement in children's behaviours as well as permanent central nervous system regeneration.

Stress: It is reported that intake of decosahexaenoic acid (found in fish oil), an omega-3 polyunsaturated fatty acid, helps to reduce stress and decrease the aggressive tendencies of young adults, perhaps by modulating stress.

Depression: There is relationship between the drop in omega-3 fatty acid consumption and the risk of depression. Treatment with omega-3 fatty acid could be useful under certain specific conditions.

Autism: Autism remains one of the few conditions classified as a syndrome, defined only in terms of observable symptoms. Sufferers from autism often have a range of physiological problems and alterations in their fatty acid metabolism, particularly those related to eicosanoid production. The plasma phospholipids of autic children contain 23% less DHA than those of normal children, while the total omega-3 acid condition is 20% lower. It has been reported that supplementation of PUFA in fish oil is beneficial in treating autism.

Omega-6 PUFA

Omega-6 fatty acids may be useful for the following health conditions:

Diabetic neuropathy: Gamma linolenic acid (GLA) administration for a longer period may reduce symptoms of nerve pain in people with diabetic neuropathy. People who have good blood sugar control may find GLA more effective than those with poor blood sugar control.

Rheumatoid arthritis: Supplementation of *n*-6 fatty acid may reduce pain, swelling and morning stiffness in patients of arthritis. If GLA is supplemented for arthritis, it may take 1-3 months for benefits to appear.

Allergies: Omega-6 fatty acids from food or supplements, such as GLA, have a longstanding history of folk use for allergies. Women and children who are prone to allergies appear to have lower levels of GLA in breast milk and blood. However, there is no good scientific evidence that taking GLA helps in reducing allergy symptoms.

Attention deficit/hyperactivity disorder (ADHD): Clinical studies suggest that children with ADHD have lower levels of EFAs, both omega-6s and omega-3s. Essential fatty acids (EFAs) are important for normal brain and behavioral function.

Breast cancer: Women with breast cancer show a better response to drugs like tamoxifen (a drug used to treat estrogen-sensitive breast cancer) when taken along with *n*-6 PUFA.

High blood pressure (Hypertension): Gamma linolenic acid may help in reducing high blood pressure, either alone or in combination with omega-3 fatty acids- eicosapentaenoic acid (EPA) and docosahexaenoic acid (DHA). In people with intermittent pain in the legs while walking caused by blockages in the blood vessels, GLA when given in combination with EPA reduces systolic blood pressure.

Mastalgia: The *n*-6 PUFA may reduce breast pain and tenderness in people with cyclic mastalgia. It may also help in reducing the symptoms to a lesser extent in people with non-cyclic mastalgia.

Osteoporosis: People who do not get enough of some essential fatty acids (particularly EPA and GLA) are more prone to bone loss than those with normal levels of these fatty acids. In women over the age of 65 years with osteoporosis, the supplementation of EPA and GLA for a long period causes less bone loss as well as an increase in bone density.

Omega-9 PUFA

Omega-9 fatty acids are not linked as much with health benefits, but still, they should be a part of a healthy diet. These should be present in the body in healthy ratios as compared to omega-3 and omega-6. Omega-9 fatty acids may also aid in the production of prostaglandins, which have many great health benefits.

Omega-9 plays a role in lowering the cholesterol levels, and thereby, reduces the risk of cardiovascular disease. The other health benefits of omega-9 fatty acids are given as under:

- Reduce atherosclerosis (hardening of the arteries)
- Reduce insulin resistance, thereby, improving glucose (blood sugar) maintenance
- Promote healthy inflammation responses
- Improve immune function
- Provide protection against certain types of cancer

Cancer prevention as well as heart protection is still possible with omega 9 fatty acids but these should be a much smaller part of the diet since they can also include saturated fats, which are harmful to the heart in larger quantities.

Negative Health effects

Omega-3 PUFA

Non-cardiac health risks: Suspected risks of EPA and DHA (*n*-3 fatty acids) consumed in excess of 3 grams per day may include the possibility of:

- Increased incidence of bleeding
- Hemorrhagic stroke
- Oxidation of omega-3 fatty acids forming biologically active oxidation products
- Increased levels of low-density lipoproteins (LDL) cholesterol or apoproteins associated with LDL cholesterol among diabetics and hyperlipidemics
- Reduced glycemic control among diabetics

Cardiac risks: Persons with congestive heart failure, chronic recurrent angina pectoris, or evidence that their heart is receiving insufficient blood flow are advised to talk to their doctor before taking *n*-3 fatty acids. In congestive heart failure, cells that are only barely receiving enough blood flow become electrically hyperexcitable, which leads to increased risk of irregular heartbeats, and in turn, causes sudden cardiac death. Normally, *n*-3 fatty acids seem to stabilize the rhythm of the heart by effectively preventing these hyperexcitable cells from functioning, thereby, reducing the likelihood of irregular heartbeats and sudden cardiac death. For most people, this is obviously beneficial and it would account for most of the large reduction in the likelihood of sudden cardiac death. Nevertheless, for people with congestive heart failure, the heart is barely pumping blood well enough to keep them alive. In these patients, *n*-3 fatty acids may eliminate enough of few pumping cells that the heart would no longer be able to pump sufficient blood to live, causing an increased risk of cardiac death.

Omega-6 PUFA

Supplementation of excessive levels of *n*-6 fatty acids, relative to *n*-3 fatty acids, may increase the probability of a number of diseases and depression. Excess *n*-6 fats interfere with the health benefits of *n*-3 fats, in part because they compete for the same rate-limiting enzymes. A high proportion of *n*-6 to *n*-3 fat in the diet shifts the physiological state in tissues towards the pathogenesis of many diseases such as prothrombotic, proinflammatory and proconstrictive.

The chronic excessive production of *n*-6 eicosanoids from *n*-6 fatty acids is associated with heart attacks, thrombotic stroke, arrhythmia, arthritis, osteoporosis, inflammation, mood disorders, obesity and cancer. In order to treat and manage these conditions, cyclooxygenase (COX)-1 and COX-2 inhibitor medications are given to treat inflammation and pain, work by preventing the COX enzymes from turning arachidonic acid into inflammatory compounds, thereby, blocking the effects of arachidonic acid. A high consumption of omega-6 polyunsaturated fatty acids (PUFAs), which are found in most types of vegetable oil, may increase the likelihood that postmenopausal women will develop breast cancer. Similar effect was observed on prostate cancer in men.

Omega-9 PUFA

There are no recorded side effects from the use of omega-9 fatty acids. As *n*-9 fatty acids are typically used in ratio with omega-3 and omega-6, caution should be taken while using with anticoagulants or antiplatelet drugs due to the increased risk of bleeding. Oleic acid and monounsaturated fatty acids also have been positively associated with breast cancer risk.

Biological Significance of PUFA

The biological effects of *n*-3 fatty acids are largely mediated by their interactions with the *n*-6 fatty acids. The ratio of omega-6 to omega-3 in modern diets is approximately 15:1, whereas, ratios of 2:1 to 4:1 have been associated with reduced mortality from cardiovascular disease, suppressed inflammation in patients with rheumatoid arthritis, and decreased risk of breast cancer.

Importance of PUFA in brain function: Polyunsaturated fatty acids, especially AA and DHA are acylated into membrane phospholipids. Polyunsaturated fatty acids account for 21-36% of the fatty acids in the cell membranes but the proportion of fatty acids with 20 or 22 carbon atoms varies considerably between tissues. The nervous system is the organ with the second largest concentration of lipids after adipose tissue. The adult brain contains approximately 50-60% of its dry weight as lipid, and approximately 35% of the lipids are PUFA, most of which are long-chain PUFA (EPA and DHA). Neuronal tissues, such as the brain, retina and synaptic membranes are especially high in DHA. This relative distribution gives some indication of a possible important role in the membranes of these tissues *viz.*, high concentration of DHA in synaptic membranes correlates with development of the synapses. The *n*-3 PUFA in neuronal membranes affect the activity of ion pumps and channels, and the precise fatty acid composition of the neuronal membranes affects the tertiary and quaternary structures of membrane-bound receptors and associated neurotransmitter functioning.

Importance of PUFA in bone growth and repair: Bone is a highly active metabolic tissue, continually changing throughout life. Bone growth is regulated by complex interactions between an individual's genetic potential, environmental influence and nutrition. Bone metabolism is under the control of many hormones and growth factors, including activated vitamin D, estrogen, growth hormone, insulin, insulin-like growth factor, parathyroid hormone and various eicosanoids that are produced from PUFA. Among these eicosanoids, PGE2 plays a major role. At low levels, PGE2 apparently stimulates bone formation. The mechanism for this may be that PGE2 increases the production of insulin-like growth factor (IGF), a potent growth stimulator for bone, cartilage and muscle, found in abnormally low levels in women with osteoporosis. High or excessive levels of PGE2 swamp this effect, bone formation is reduced and resorption is increased. Furthermore, PGE2 has been shown to mediate the effects of 1, 25-$(OH)_2$ vitamin D_3, TNF-alfa and growth factors, thus, enhancing bone resorption.

Importance of PUFA during pregnancy and fetal development: Polyunsaturated fatty acids are vitally important structural elements of cell membranes, and therefore, essential for the formation of new tissues, as occurs during pregnancy and fetal development. As the central nervous system is particularly rich in AA and DHA, and the brain has its growth spurt in third trimester of the pregnancy and during early childhood, thus, an appropriate pre- and post-natal supply of PUFA is essential for normal fetal and neonatal growth, neurologic development and function, the activity of retinal photoreceptors and learning and behaviour. Maternal concentrations of PUFA, especially DHA is associated with sleep and wake states of newborns, which helps in assessing the functional integrity of the central nervous system. Higher maternal plasma DHA during pregnancy is associated with more mature neonatal sleep-state patterning. Docosahexaenoic acid and AA have been shown to be associated with increased prenatal and neonatal growth. The fetal PUFA status may promote fetal and early neonatal development, thereby, improving the *starting condition* and general prognosis of infants born pre-term. Moreover, the DHA content of maternal plasma phospholipids is significantly lower in multiparous than in primiparous women. The infants born to multiparous women have significantly less DHA in their umbilical tissue phospholipids than the infants born to primiparous women. This has important consequences on growth and function of the central nervous system, and consequently, on neurologic and cognitive development of the infants.

Assessment of PUFA Status

The polyunsaturated fatty acids status of an individual can be assessed by the followings:

- **Total amount of EFA and PUFA in plasma or erythrocyte phospholipids:** If insufficient essential polyunsaturated fatty acids are available to meet PUFA requirements, the body starts to synthesize certain fatty acids that are normally not present. Therefore, these fatty acids can be essential PUFA status markers. Among these, the best-known marker is Mead acid (C 20:3, *n*-9). The synthesis of this fatty acid is promoted if there are insufficient concentrations of LA and LNA to meet the need for the synthesis of long-chain PUFA. Eicosapentaenoic acid and DHA inhibit Mead acid synthesis. The presence of Mead acid indicates a general shortage of all essential PUFA.
- **Essential PUFA status index:** The ratio between all essential PUFA (the sum of all *n*-3 and *n*-6 fatty acids) and all nonessential unsaturated fatty acids (the sum of all *n*-7 and *n*-9 fatty acids) determines the PUFA status index. The higher the essential PUFA status index, the better is the essential PUFA status.
- **Ratio between DHA and osbond acid:** It is a reliable indicator of the functional DHA status under steady state conditions. When there is a functional shortage of DHA, the body starts to synthesize the most comparable long-chain PUFA of the *n*-6 family, osbond acid (22:5 *n*-6), thus, the increase in osbond acid concentration refers to the decline in DHA levels in the body.

■■■

30 Green Portals for Vibrant Health

S. Bhanumati[1], S. Sheokand[2] and Rajyasri Narayan[3]

[1]Chemistry Department, Gargi College (Delhi University), Siri Fort Road, New Delhi-110049, E-mail: <sbhanumati@gmail.com>

[2]Department of Plant Physiology, CCS Haryana Agricultural University, Hisar-125004, Haryana, India, E-mail: <sunita_sheokand@hau.ernet.in>

[3]English Language Teaching Specialist and Trainer, E-mail: <rajyasri21@yahoo.co.in>

Dieticians the world over unanimously agree that we are what we eat. Our body is the product of what we ate yesterday, the day before and the countless days even before our birth in this world since the nourishment provided by our mothers while we rested in the womb as passive in takers. This is the reason why there is so much emphasis on nutritive value of the food for the prospective mothers.

Go green, eat green, drink green, be green and live green. This is the anthem of a new generation of conscientious people who desire to live healthy and in harmony with nature. Living green means living closer to nature and the abundance it provides. The term *green* in relation to food encompasses eating green vegetables, which are high in nutritive and therapeutic value. All greens are low calorie and low sugar food. The power of chlorophyll is present all around us. Look at the heartiness of pine trees and the intimidating nourishing power of the great Redwood, and finally yet importantly, look at the grass that surrounds us all. It is the most abundant life force on earth, and the reason why this planet can support life and by incorporating chlorophyll back into our diet, we can combat many of the ills that we suffer from today.

Figure 1 : Green leaves showing the green colored pigment chlorophyll

Green food is the power plants for humans because it is through the green pigment chlorophyll, which absorbs energy from the sunlight and transmits to our bodies. Chlorophyll is the pigment that gives plants, algae and cyanobacteria their green colour. It is the molecule, which absorbs sunlight and uses its energy to synthesize carbohydrates from carbon dioxide and water. This process is known as photosynthesis and is the basis for sustaining the life processes of all plants. Since animals and humans obtain their food supply by eating plants, photosynthesis can be said to be the source of our life also. Chlorophyll absorbs light energy and is activated (shown with*) to reduce atmospheric carbon dioxide in the presence of water to synthesize plant food. During this process, carbohydrates are synthesized and oxygen is liberated. The process of photosynthesis is as follows:

$$\text{Chlorophyll} \xrightarrow{\text{Light}} \text{Chlorophyll*}$$

$$6CO_2 + 6H_2O + \text{Chlorophyll*} \xrightarrow{\text{Light}} C_6H_{12}O_6 + 6O_2$$

Since oxygen is given out at the expense of carbon dioxide, a carbon dioxide to oxygen balance and equilibrium is maintained in the atmosphere. When the greens are extracted with ethyl alcohol, crystalline chlorophyll is obtained. The natural pigment is a dark green waxy solid. The chlorophyll '*a*' (melting point 117-120°C) is bluish black, while the chlorophyll '*b*' (melting point 120-130°C) is dark green in colour.

Chlorophyll contains magnesium metal in a complex form. Hence, it is a source of magnesium for the body. Nature has bestowed the human body with a similar pigment, haemoglobin, which performs a vital function as oxygen carrier, besides carrying the vital nutrients to various constituent parts of the physical body. Haemoglobin, the red pigment of our blood, contains iron (Fe^{++}) in a complex form. The basic structure of chlorophyll is a porphyrin ring similar to that of heme in hemoglobin, although the central atom in chlorophyll is magnesium instead of iron. The long hydrocarbon (phytol) tail attached to the porphyrin ring makes chlorophyll fat-soluble and insoluble in water. Mainly two different types of chlorophyll (chlorophyll '*a*' and chlorophyll '*b*') are found in plants. The small difference in one of the side chains allows each type of chlorophyll to absorb light at slightly different wavelengths (Figure 1 and 2). The distribution of various chlorophylls and bacterio-chlorophylls is shown in Table 1. The restricted distribution of other chlorophylls may have an evolutionary significance. Chemical structure of chlorophyll is a chlorin pigment, which is structurally similar to and produced through the same metabolic pathway as other porphyrin pigments such as heme. At the center of the chlorin ring is a magnesium ion. The chlorin ring can have several different side chains, usually including a long phytol chain.

The most widely distributed form in terrestrial plants is chlorophyll '*a*'. It is the only chlorophyll found in all oxygen evolving photosynthetic organisms and is also the only chlorophyll found in the core complex of photosystems (P_{680} and P_{700}), *i.e.*, the chlorophyll '*a*' protein complexes that absorb light at 680 and 700 nm, respectively. The long and uninterrupted chain of carbon and hydrogen atoms carries on from the core of the chlorophyll molecule and links it to the internal casing of the chloroplast, which is the cell organelle where the process of photosynthesis occurs. When a chlorophyll molecule soaks up a photon of light, electrons present in the molecule become active and move to higher energy planes. In turn, this kicks off a series of complex chemical reactions in the chloroplast enabling the energy obtained from the sunlight to be stored in chemical bonds, *i.e.*, ATP and NADPH.

Table 1 : The distribution of chlorophylls and bacterio-chlorophylls

Pigments	Organism
Chlorophyll '*a*'	All plants algae and the photosynthetic bacteria, *i.e.*, Cyanophyta and Prochlorophyta
Chlorophyll '*b*'	All plants, the algal Chlorophyta and Euglenophyta and bacterial Prochlorophyta
Chlorophyll '*c*'	Algal Phaeophyta (brown algae), Pyrrophyta (dinoflagellates, Bacillariophyta (diatoms), Chrysophyta, Prasinophyta and Crypophyta
Chlorophyll '*d*'	Some Rhodophyta (red algae) and Chrysophyta
Chlorophyll '*e*'	Reported from Xanthophyta
Bacteriochlrophyll '*a*' and '*b*'	Purple bacteria including Chromatiaceae and Rhodospirillaceae
Bacteriochlrophyll '*c*', '*d*' and '*e*'	Green and brown sulphur bacteria, *i.e.*, Chlorobiaceae and Chloropflexaceae

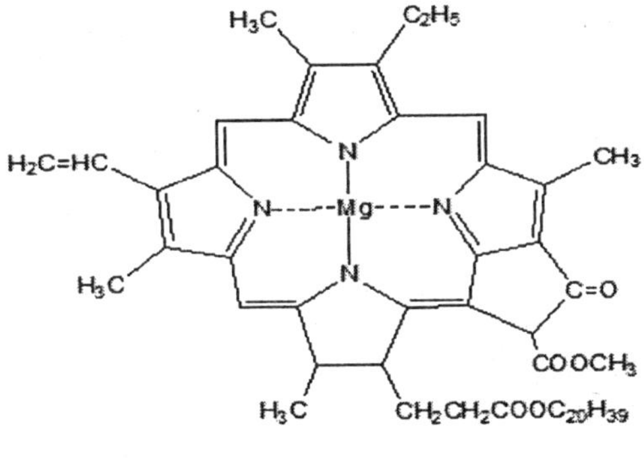

Figure 1 : Structure of Chlorophyll '*a*'

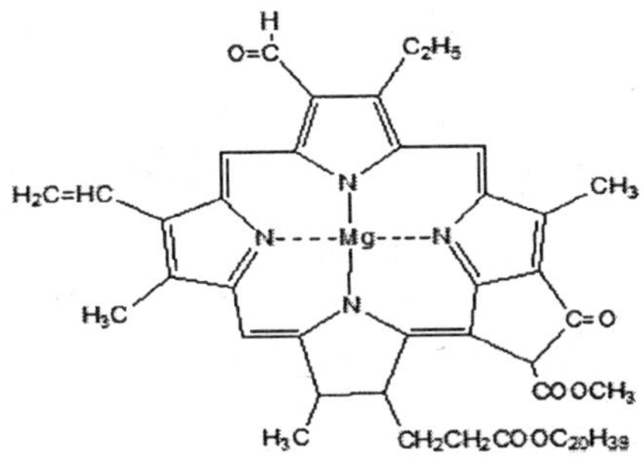

Figure 2 : Structure of Chlorophyll '*b*'

The different side groups in the 2 chlorophylls *tune* the absorption spectrum to slightly different wavelengths, so that the light, which is not significantly absorbed by chlorophyll '*a*' at 460 nm, will instead be captured by chlorophyll '*b*', which absorbs strongly at that wavelength. Thus, these two kinds of chlorophyll complement each other in absorbing sunlight. Plants can obtain all their energy requirements from blue and red parts of the spectrum. However, there is still a large spectral region between 500 and 600 nm, where very little light is absorbed. This light is in the green region of the spectrum, and since it is reflected, this is the reason plants appear green. Chlorophyll absorbs so strongly that it can mask other less intense colours. Some of these more delicate colours (from molecules such as carotene and quercetin) are revealed when the chlorophyll molecule decays in the autumn, and the woodlands turn red, orange and golden brown. Chlorophyll can also be damaged when vegetation is cooked since the central magnesium atom is replaced by hydrogen ions. This affects the energy levels within the molecule, causing its absorbance spectrum to alter. Thus, cooked leaves change colour- often becoming a paler, insipid yellowy green.

As a phytochemist, I have observed that in peak summer, the edible leaves develop stronger taste than in winters and rainy season. Similarly, it is also noticed that in a particular soil, some plants develop a particular hue and taste. During a study on pigeon pea plant (*Cajanus cajan*), a number of newer chemical components were found present in plant roots collected from the Rajasthan state, which were almost negligible in plant roots collected from the district of Godavari. Similarly, the latter had some chemical components, which were missing in the former. Hence, it is difficult to accommodate the chemistry and the chemicals present in these varieties. In this narration, it can safely be said that certain soil and climatic variations can trigger the synthesis of certain chemicals, which can be perceived as taste, colour, or appearance by common man. It is advantageous to learn about more varieties of the same species of plants for adequate use in all places.

Health Benefits of Chlorophyll

The various health benefits of chlorophyll are listed below:

- The researches on animals have shown that chlorophyll derivatives such as chlorophyllin exhibit antioxidant activity as good as vitamin C. The function of chlorophyll in animals is suggested to be inhibition of lipid peroxidation and protection of mitochondria from oxidative damage induced by various free radicals and other reactive oxygen species.
- Chlorophyll has an antibacterial and an anti-inflammatory effect. It is used to promote wound healing. Chlorophyll reduces inflammation in conditions ranging from arthritis and sciatica to stomach ulcers and acid reflux. In addition to skin conditions like eczema and psoriasis, it hastens skin grafting, cures chronic sinusitis, overcomes chronic inner-ear inflammation and infection, reduces varicose veins, heals leg ulcers and rectal sores, successfully treats inflammation of the uterine cervix, gets rid of parasitic vaginal infections, reduces typhoid fever and cures advanced pyorrhea in many cases.
- It has been seen to help in the growth and repair of tissues.

- Chlorophyll helps in neutralizing the pollution that we breathe in and intake. It forms complexes with polyaromatic hydrocarbons found in tobacco smoke everyday, hence, it is a good supplement for smokers.
- Chlorophyll may be beneficial in the treatment of calcium oxalate stone disease and that they may have some anti-atherogenic activity. It also breaks down calcium-oxalate stones, which are created by the body to neutralize acid for elimination. Calcium oxalate stones are better known as kidney stones. It is also found to be useful in chelating heavy minerals.
- It had been seen to have a good potential in stimulating red blood cells to improve oxygen supply. Research has shown that chlorophyll and its derivatives can increase red blood cell production.
- Chlorophyll also removes carbon dioxide and carbon monoxide and is an effective deodorizer to reduce bad breath, urine, fecal waste and body odour.
- Chlorophyll is known to have a cleansing effect on the intestine. Clinical uses of chlorophyll are to clear up foul smelling odours, neutralize strep infections and heal wounds. It may reduce the ability of carcinogens to bind with the DNA in different major organs in the body. Research indicates that chlorophyll reduces carcinogen binding to DNA in the target organ by inhibition of carcinogen activation enzyme or degradation of ultimate carcinogens with the target cells. Inhibiting the activity of the enzyme that attacks the DNA sequence significantly reduces the incidence of mutagenicity and the onset of cancer. Noncompetitive inhibition of carcinogenic enzymes by chlorophyll has been reported. Dietary inhibitors of mutagenesis and carcinogenesis are of interest since they may be used for human cancer prevention and treatment. Chlorophyll has demonstrated to be an inhibitor of carcinorgenesis in the gut. Research shows that there is sufficient systemic chlorophyll distribution for these inhibitory mechanisms to be operative beyond the gut and in the whole animal. Oral administration of chlorophyll has been studied for the last 50 years. High levels of oral ingestion of chlorophyll are required for the chemotherapeutic effects of chlorophyll to be effective. Another important property of chlorophyll is its potential to reduce metastasis. Chlorophyll has been shown to take an active part in controlling the enzymes that are involved in mitosis. By controlling this rapid cell division chemically, the rapid onset of tumors can be reduced significantly, potentially giving the cell the ability to repair itself. In addition to being locally efficacious, this reduces the chances of tumor cells being transferred to other parts of the body to be effective.
- Chlorophyll is rich in vitamin K necessary for blood clotting. Naturopathic physicians use chlorophyll for women with heavy menstrual bleeding and anaemia. Vitamin K helps in forming a compound in urine that inhibits growth of calcium oxalate crystals, so chlorophyll helps with kidney stones. Vitamin K also enhances adrenal activity, so chlorophyll rich food items help in maintaining steroid balance for a more youthful body.
- Magnesium deficiency is a common day epidemic due to diets high in sugar and low in green food items. Magnesium is essential for healthy brain, muscle and nervous system function. Magnesium ion containing proteins are associated with several enzymatic reactions. Chlorophyll efficiently delivers magnesium and helps the blood

in carrying the oxygen much needed to all cells and tissues. Magnesium ions play critical roles in many aspects of cellular metabolism. It stabilizes structures of proteins, nucleic acids and cell membranes by binding to the macromolecule's surface and promoting specific structural or catalytic activities of proteins, enzymes, or ribozymes and has a critical role in cell division. It has been suggested that magnesium is necessary for the maintenance of an adequate supply of nucleotides for the synthesis of RNA and DNA. It is necessary for both the action of insulin and the synthesis of insulin. Magnesium helps the muscles to relax. Without sufficient magnesium, the body accumulates toxins and acid residues, degenerates rapidly and ages prematurely. Green food articles also contain other major nutrients that are essential for a healthy bloodstream and nervous system such as vitamins A, C, K, folic acid, B_6, copper, potassium and calcium.

Sources of Chlorophyll

Chlorophyll is present in all plants even though the content may vary depending on species and plant parts growth stage. The sources of chlorophyll generally used for therapeutic purposes are listed below:

Cereal Grasses (wheat, oat, or barley grass juice): Cereal grasses are a nutrient dense high chlorophyll food. Though wheat and barley grass have similar nutritional qualities, barley grass is milder and more bitter than the slightly sweet wheatgrass. People who are allergic to wheat or gluten are usually not allergic to wheat in its grass stage. Barley grass is a great alternative for individuals who cannot tolerate wheatgrass. Cereal grasses are an excellent source of many nutrients and enzymes.

Wheat grass: The solid matter in wheatgrass juice contains 70% crude chlorophyll and many vital nutrients and minerals that can theoretically support human life for weeks or even months. Wheatgrass juice is also a good source of key vitamins like folic acid and minerals like iron, potassium and copper. Wheat grass juice is specially known for multitude health benefits. It has an ability to increase hemoglobin and iron within the blood, has a positive effect on liver, makes the lymph system clean, restores balance in the body, removes toxic metals from the cells and restores vitality. The juice of wheat grass acts as a detergent in the body and it is used as a body deodorant.

Barley grass: Barley grass contains an astounding amount of vitamins and minerals, including five of the B group vitamins (the often hard to obtain vitamin B_{12} being among them). It also contains folic acid, pantothenic acid, beta-carotene, vitamin C and E. Recent laboratory analysis on green barley grass has reported traces of more than 70 minerals, among them calcium, iron, magnesium and phosphorus (which is determined, in part, by soil conditions). Barley grass also contains 18 amino acids, which are the building blocks of proteins. Green barley grass is also said to have a highly alkalizing effect since it contains buffering minerals such as sodium, potassium, calcium and magnesium.

Research done on patients with ulcerative colitis showed that those given barley grass showed significant improvements in their symptoms, including fewer episodes of standard anti-inflammatory medication including steroids. Ulcerative colitis is linked with low levels of friendly bacteria in the gut and an accumulation of toxins in the bowel. Barley grass helps in reversing this by stimulating the growth of friendly bacteria, which in turn help in reducing inflammation and improves symptoms. Barley grass helps in combating ulcerative colitis by lowering the amount of inflammatory chemicals in the bowel, including one called epithelial NF-k and by balancing the water content of the bowel contents. Studies also show that barley grass increases the amount of friendly bacteria in the patients' bowels. Other studies done in Japan have shown that barley grass tea significantly improves the flow of blood as well as lowers the amount of low-density lipoprotein (LDL) cholesterol. Barley grass juice improves stamina and increases energy and is also known to improve texture of the skin. Barley juice is especially beneficial for individuals with chronic diseases and health problems associated mainly with ageing. It prevents oxidation of LDL in patients with type II diabetes as people with type II diabetes have an increased risk of atherosclerosis.

Oat grass: As part of the cereal grass family, oat grass is an excellent source of beta-carotene, vitamin K, folic acid, calcium, iron, protein, fiber as well as vitamin C and B group vitamins. Oat grass (*Avena sativa*) has a relaxing and stimulating action that nourishes and strengthens the nervous system. It may be valuable for arthritis, rheumatism, stress, depression, debility, exhaustion, nerve tremors, epilepsy, convalescence, palpitations, nervous headache, nervous stomach, nervous breakdown, cholesterol lowering, herpes, and menopause symptoms. It is also used for thyroid and estrogen deficiencies, degenerative diseases such as MS and for colds, especially recurrent or persistent.

All these cereal grasses can easily be grown on roofs in small earthen pots filled with a mixture of 1 part garden soil, 1 part sand and 1 part compost. About 3-4 cups of hard winter wheat grass seeds (preferably organically grown) are soaked for 3 to 8 hours in water. A couple of drops of organic vinegar are added to acidify the water, which keeps the fungus or mold from forming. The seedlings are then transferred to earthen pots or trays filled with a mixture of soil, sand and compost. The seedlings harvested 8-14 days after sowing. Growing wheat grass to about seven inches tall is optimum for its health benefits. Their juice can be extracted by grinding in mixer after attaining a height of about 12 to 15 cm and can be taken after adding few drops of lemon juice.

Algae: It will not be an exaggeration to state that algae are responsible for almost 90% of the total photosynthesis taken place on our planet. The phytoplanktons and other algae are consumed by the fish population, which in turn return to man as seafood with rich nutrients. Spirulina, chlorella and wild blue-green algae contain more chlorophyll than any other food on the planet. In addition to chlorophyll, they are also a great source of complete protein, beta-carotene and nucleic acids (RNA and DNA). Spirulina is an especially rich source of protein, beta-carotene and the beneficial fatty acid gamma-linolenic acid (GLA). It also contains a pigment called phycocyanin, which has antioxidant and anti-inflammatory properties. The cell wall of spirulina contains muco-polysaccharides, which are composed of amino acids, sugars and protein. The nutrients in muco-polysaccharides are completely digestible unlike the nutrients within the cell walls of other micro-algae. Although chlorella has a similar nutrient composition to spirulina but chlorella contains a little less protein, much less beta-carotene and much more chlorophyll and nucleic acids than spirulina. Chlorella has more chlorophyll than any other food and it contains a higher amount of fatty acids when compared to other micro-algae. Another health benefit of chlorella is that it has a tough outer cell wall that binds with heavy metals, pesticides and other carcinogens. It then carries the toxins safely out of the body.

Wild blue-green algae offer similar nutritional benefits as spirulina and chlorella. Laminaria and Gracilaria are the food algae widely used in China since ancient times. In Japan, the edible seaweed, Porphyra alga, very rich in vitamins B and C, was popularly being eaten with rice, which was cultivated widely in later years for its food value. Hundreds of algal species and seaweeds are already in use the world over as food plants due to their nutrients content.

Green Leafy Vegetables: Dark leafy greens are some of the most nutrients-dense foods on the planet. They are richer than carrots in usable carotene, which is the building block for vitamin A. Folic acid is also found in abundance in the leaves of bright green vegetables. The word folic comes from the Latin word folium, meaning leaf. Low levels of folic acid have been related to poor fetal development, mental deterioration and anemia. We also want to be sure that we get enough of the antioxidant vitamin C in our diets. Besides citrus fruits, green leafy foods are packed with this essential nutrient. Chlorophyll is abundant in leafy vegetables, and generally, to a lesser extent in fruits, *e.g.*, in spinach, chlorophyll content can be as high as 1% on dry weight basis. Not only the green leafy vegetables are a rich source of chlorophyll but they are also an excellent source of many other nutrients including vitamins A and C, calcium, potassium and fiber. The amount of each nutrient varies with each green leafy vegetable. Several varieties of green leafy vegetables including chenopod, collard greens, fenugreek, kale, mustard greens, kohlrabi, spinach and Swiss chard are available in most grocery stores and farmer's market stands. They all contain their own unique composition of nutrients. The green leaves and fruits of drumsticks or horseradish (*Morunga olifera*) have excellent cures for eye problems.

Table 2 : Chlorophyll content of some raw vegetables

Food	Serving	Chlorophyll (mg)
Spinach	1 cup	23.7
Parsley	½ cup	19.0
Cress, garden	1 cup	15.6
Green beans	1 cup	8.3
Arugula	1 cup	8.2
Leeks	1 cup	7.7
Endive	1 cup	5.2
Sugar peas	1 cup	4.8
Chinese cabbage	1 cup	4.1

All over the world to derive health benefits, people have used the greens of *agasthi*, amaranth, bamboo sprouts, beet leaf, broccoli, Brussels sprouts, cabbage, celery, chenopod (*bathua*), dill, drumstick, kohlrabi, lettuce, mustard leaves, parsley, radish leaves, spinach, spring onion greens, trifolium (tipatia) and Indian spinach (poi). Gram greens, pea leaves, white bottle gourd and pumpkin leaves also find use in culinary preparations. The well-known curry and *bael* leaves, sacred basil, mint, coriander leaves, papaya, *giloi* and *jad Amla* leaves are of great therapeutic value.

There are several greens in your kitchen or herbal gardens, which possess remarkable medicinal properties. The following plants are very common but have enormous proportion of uncommon curatives properties in them.

Horseradish or drumstick (*Moringa oleifera*): Horseradish is rich in enzyme peroxidase. This enzyme catalyzes the oxidation of a variety of organic and inorganic substrates using hydrogen peroxide (H_2O_2) as an oxidant. Horseradish therefore is an effective scavenger when consumed in food.

Coriander (*Coriander sativum*): *Dhania* well known in Hindi is a memory tonic. The green leaves work as carminative and a stimulant with a cooling and smoothing effect on the digestive system. It is an antispasmodic, antibilious, antiflatulant and is a diuretic. It can be used for salad dressing, as a paste with a little salt in sandwich, in various preparations for its aromatic flavour and also with turmeric for healing wounds.

Fenugreek (*Trigonella foenum*): The green leaves are wonderful for keeping and maintaining the sugar level low in blood. For joint pains and arthritis, the bitter leaf extract is very good, besides the seeds, which serve to lubricate the stiff joints. The clean and washed leaves can be finely chopped and made into dough with coarse wheat flour. Rolled into loaf and eaten in the principal meal with suitable side dishes is excellent for managing a host of physical ailments. A combination of fenugreek curry with potatoes is a delicacy in north Indian cuisine. Mildly boiled (with the lid on the pot) and later eaten with bland curd is another special dish with nutrition and taste.

Onion springs (*Allium cepa*): The springs are excellent stimulants, diuretic, expectorant, rubefacient and emmenagogue, similar to its bulbs. The juice is used to treat headache, hysteric fits, insect bites and skin rashes. It is useful in treating anemia, cholera, toothache, piles and also in heart problems.

Bermuda grass (*Panicum dactylan*): It is called *arugampul* in Tamil and *durba ghas* in Hindi. It is an antidote to body toxins and the raw juice is best as medicine in stomach problems such as loose motions, vomiting, *etc.*

Basil (*Ocimum sanctum*): The basil leaves are expectorants, antipyretics (lower body temperature in fever) anti-allergic and bronchodialators.

Ghekuar (*Aloe vira*): There are varieties of cacti of which the succulent leaves are filled with a jelly. A spoonful of the jelly takes with water in empty stomach as the first thing in the morning has multiple benefits. It is an excellent coolant, and hence, is very handy in thermal and radiation burns for external application.

Designing Dietary Medicinal Foods from the Greens

Tender green leaves are so positive in their therapeutic value that a course of medicinal diet can be designed and prescribed to patients in various cases. A host of grass and leaf eating animals like horses, cows, bulls, antelopes, deer, the elephant, goats and even the camels can boast on their strength and agility. The North American bison owes its exceptional strength to grass of the *prairie*. The leaf is the source of natural immunity. It is widely considered that the immunity of goats to tuberculosis is due to the leaf intake by goats in large quantities. It is therefore reasonable to design food for the consumptive with plenty of greens and leaves. Even an extract of tender green edible leaves will help a consumptive to recuperate fast. Normal blood contains alkaline salts, which neutralize the acid wastes generated in the system, and thus, facilitate their elimination. On the contrary, if the acid wastes are excessive, food medicines as greens come to our rescue and create the vital influx.

Some Tips to Take Chlorophyll

The plant is a real biochemical link between the elements, Sun, Air, Water, Earth and the organic life of animals including the human beings. Animals do not have the potential to create a life substance from the inorganic matters alone. Even when the necessary inorganic material is carefully supplied, the animal still cannot flourish. The animal body is adjusted to live upon plant tissues or upon animals containing countless other substances besides proteins, carbohydrates and fats.

In human body, the minerals are organically associated with the vitamins and in plants with chlorophyll. This is nature's arrangement for proper absorption and non-irritating elimination of the excess. A good lot of roughage is provided by different greens in the form of dietary fibers through vegetable peels, leaves and other edible parts. This helps in preventing constipation and helps the bowel movement. Diet experts often recommend cooking greens after proper washing without peeling. It is a common observation that the plants from which leaves have been plucked do not grow to their normal size.

The greens help in the body's detoxification system. However, eating too much of the greens alone is not too good for any individual since the body has to spend the reserve energy for removing the harmful excess.

Recipes to Consume Green Veggies for Taste and Nutrition

According to the food pyramid guidelines, an average adult needs 3 cups of leafy greens per week. Greens are generally taken in the form of salads, soups, vegetable juice and cooked vegetables. Given below are some easy tips for making the greens more palatable and nourishing after washing with liberal water. Cucumber and green sweet peppers all make nutritious green salads. Some simple items designed for health benefits are given below:

Soups: Green soups are incredible for health and in taste as well. It provides an easily digestible way to include many vitamin-rich green vegetables in your daily menu. The most common green soup is the spinach soup in which boiled and ground spinach may be mixed with a number of other vegetables (peas, tomatoes, onions, ginger, *etc.*) to increase the health benefits. The mushroom-spinach soup recipe is given below:

The ingredients of golden mushroom-spinach-basil soup are 1 teaspoon extra-virgin olive oil, chopped onions (about 2 large or 3 medium onions), ½ teaspoon sea salt, diced potatoes (about 2 medium potatoes), chopped celery (about 2 stalks), 1 large carrot diced, 3 to 4 cups water, 2 tablespoons cashew butter, 2 tablespoons tamari or shoyu, sliced mushrooms, sliced ½ cup fresh basil leaves, finely chopped ½ bunch fresh spinach, sea salt and fresh ground pepper. Heat oil in a pot and add onions and salt. Cover the pot and simmer on low heat stirring occasionally, until onions cook down to a nice mush (15-20 minutes). Add potatoes, celery, carrot and water to the onion mush and cover and simmer until potatoes are soft (15-20 minutes). Put soup mixture into bender with cashew butter and tamari and blend until smooth.

Run soup through a strainer to remove celery strings and return soup to pot. Heat 2 teaspoons oil in a small skillet and add mushrooms and sauté until soft, and thereafter, add chopped basil and spinach. Stir for a few minutes, and then, add salt and pepper to taste and serve.

Salads: Salad greens are usually eaten raw. In general, the darker the color, the more is the nutritious. Iceberg lettuce, for example, is extremely low in nutrients compared with its more colorful relatives, *i.e.*, romaine lettuce has 8 times the vitamin A and 6 times the vitamin C as iceberg lettuce. When you have a choice, a variety of greens is always best, as each has its own constellation of nutrients. Salads should be washed properly as they could be a source of bacterial infection. Preferably, the organically grown vegetables should be used for salads.

Green salad: Mix chopped lettuce leaves and minced spring onions in proportion to suit your taste, garnish with grated cheese, and then, add lemon, black pepper and salt according to taste.

Ring salad: Cut cucumber, capsicum, tomatoes, carrots and radish as slices/rings/discs, garnish with chopped green coriander, and then, add salt and lemon juice according to taste.

Cucumber *Raita*: Add curd to grated cucumber, garnish with green coriander finely chopped and the, add salt and a little cumin powder.

Greens with *Daal*: The right combination of the pulses with the greens is given below:

- Spinach with chana/moong daal
- Chcnopod (bathua) with moong daal
- Mustard, radish saag and kohlrabi without any daal
- Fenugreek leaves with pigeon pea daal
- French/broad/lima/cluster beans with moong daal
- Cabbage with moong/chana daal
- Bottle/bitter/ash/snake/ribbed gourd with moong/chana daal
- Cowpea pods, Mungra phalli, guar phalli with moong daal

The greens have to be cooked with *daals* in pressure cooker. After adding turmeric, salt and coconut, it is served hot.

Industrial Uses of Chlorophyll

In the food industry, chlorophyll is used as a natural pigment ingredient in processed foods. Because of its strong green colour and consumers, growing preference for natural food's chlorophyll is gaining importance as food additive. The large number of health benefits of chlorophyll is encouraging food processors to switch from artificial pigments to chlorophyll based natural colouring. Chlorophyll is also being used in cosmetic industry (chlorophyll '*a*', known as natural green). It is used in soaps and cosmetic products.

Conclusion

This chapter has been written with only the relevant technical, chemical and biochemical information so that a reader even with a little general knowledge can make use of the contents and benefit health wise. However, for inquisitive readers, consulting some books on diet and biochemistry and of course the books on Naturopathy will be more enriching.

The word SWASTHYA, which is an equivalent to the English word health, has been derived from the two words SWA meaning oneself and STHYA meaning stabilizing. The meaning implied is that an individual needs to stabilize oneself by ones own efforts. An unsystematic, erratic life style, negligence to ones duties, casually handling responsibilities, consuming anti health commodities are all within ones control and with determination can be rectified. Diet and brain are co-related. *Hurry, burry and curry are the roots of self-invited worries*. Stress is the out come of these and clinically it is proved that the flow of the harmful biochemical catecholamine increases under stressful conditions. Therefore, it is sensible to observe a principled life style, which incorporates physical exercise, power naps and a positive attitude besides the nutritive eats and drinks.

Let your food be your medicine and let your medicine be your food - Hippocrates, the Father of Medicine.

■■■

31 Exploring Biocolours for Cosmetic and Textile Industries

Ram Singh[1], Geetanjali[2] and Richa Srivastava[1]

[1]Department of Applied Chemistry, Delhi Technological University, Bawana Road, Delhi-110 042, India, E-mail: <singh_dr_ram@yahoo.com>

[2]Department of Chemistry, Kirori Mal College, University of Delhi, Delhi- 110 007, India

The colours have been part of our life for the centuries. The use of colour by humankind extends back to 150,000 to 200,000 years when the ice age man buried his dead in red ochre or painted their bones red and believed that colour red itself was life giving. Turmeric, henna, capsicum red, extracts of various flowers and leaves, monascus pigments from moldy red rice and indigo are some of the natural colours used from centuries for colouring foods, cosmetics, textiles and other products. In Egypt, malachite green was used as pigments in cosmetics before the era of dynasties, *i.e.*, as early as 5000 BC. Egyptian women were known to have used henna to dye their hair, carmine to redden their lips and kohl or kajal to blacken eyebrows, lids and lashes. For thousands of years, it has been a common practice in India to tint the faces yellow with saffron and to dye the feet red with henna. Chinese women used vegetable extracts to dye their feet, cheeks and tips of the tongues. Romans used white lead and chalk on their faces and blue and gold dyes on their hair and beards.

The main purpose of using colours in textiles and cosmetics has been to increase appeal, to mask processing loses and to aid in adulteration. In drugs, they were mainly used to avoid confusion among the various drugs. The sources of natural colourants were plants, animals and minerals. In 1856, Sir William Henry Perkin discovered the first synthetic organic dye stuff, mauve and soon a host of new synthetic colours was added to the market. The use of some of these synthetic colours in foods began in Europe almost immediately and it was soon extended to drugs and cosmetics. Of particular concern was the fact that substances known to be poisonous were often incorporated into cosmetics and that colours were frequently used to hide poor quality cosmetic and textile material and to add weight or bulk to certain cosmetics but during the last one and half decades, there has been much concern at international level about the use of safe colouring agents. This was mainly due to the ill effects of synthetic dyes on human beings. Germany was the first country to take initiative to put ban on numerous azo dyes. Later, the countries like India, Netherlands and others followed the ban on selected synthetic dyes. Because of continuous reports of health hazards due to synthetic dyes, people have started looking for alternatives in the form of biocolours.

Definition of Biocolour

Natural colours or biocolours are synonyms and has the same meaning. They are the coloured chemical compounds, either single compound or in combination, isolated from nature, which includes the sources like agricultural, biological and minerals. Hence, one can define biocolours as the coloured compound(s) obtained from nature. How can be forgotten the pleasing colours of beetroot, turmeric, peas, red cabbage, red capsicum, roses and many others? Nature has imparted biocolour to all of these and many others. The history of biocolour starts simultaneously from the evolution of humankind. Even the Egyptian mummies have been found wrapped in cloths dyed from the madder plant. Alexander the Great is supposed to have deceived the Persians into thinking that his army was wounded by sprinkling his soldiers with a red dye, probably madder juice, which contains the dye alizarin, and hence the use of *Rubia* species as a source of biocolour is known to be as old as mankind and is being used now a day also. Woad, also known as Dyer's Woad (*Isatis tinctoria*) was cultivated as the source of a blue dyestuff for over 2000 years in Europe and was only superseded around 50 years ago by indigo. This dyestuff was first extracted from the subtropical *Indigofera* species.

It could be very well understood that historically, biocolours were used to colour clothing or other textiles, and by the mid-1800's, the chemists began producing synthetic substitutes for them. By the early part of this century, only a small percentage of textile dyes was extracted from plants, but again now, we are looking towards nature for safe colours.

Chemical Aspects of Biocolour

Colour plays an important role in acceptability of materials by the consumers because the quality of product is often judged by its colour. It does not only lead to rapid identification of the product but also responsible for its ultimate acceptance, therefore, it becomes a vital constituent in the industries. The major class of biocolour includes flavonoids: (1), chalcones (2), aurones (3), anthraquinones (4), carotenoids (5), flavins (6), betacyanins (7), anthocyanins (8), cytochromes (9), chlorophylls (10) and indigo (11). The suitable chemical structures are responsible for imparting the colours of different shades. Hence, it is always recommended to understand the basic structure of colour imparting molecules since not all chemical compounds show colour. To impart colour, a molecule must be associated with the presence of certain group called chromophores, *e.g.*, unsaturated groups such as the nitro ($-NO_2$), nitroso (N=O), azo (N≡N), ethylene ($CH_2=CH_2$), carbonyl groups (-C=O), *etc*. Usually, a single chromophoric group may not be sufficient to produce colour but if two or three such groups are present in a conjugated system, colour usually appears in a compound. Nature has its own way to introduce these groups in any molecule. The chemical structures of major class of biocolours are given below:

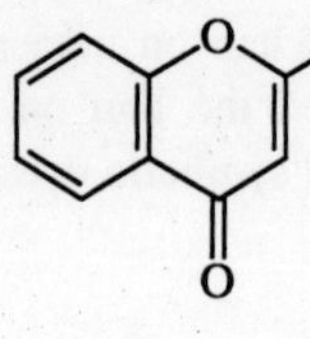

O O 1 OH O 2 O O 3

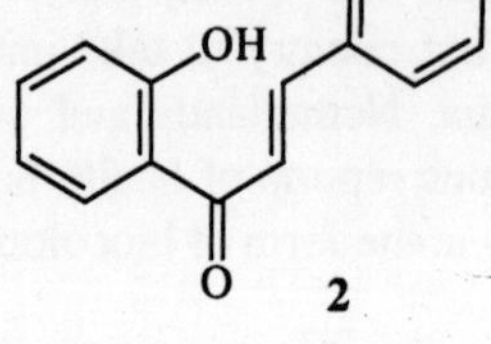

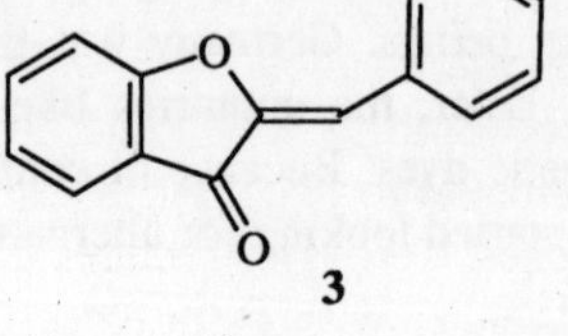

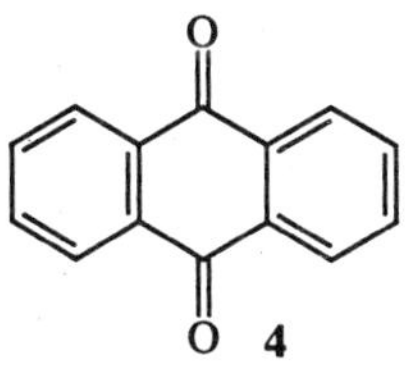

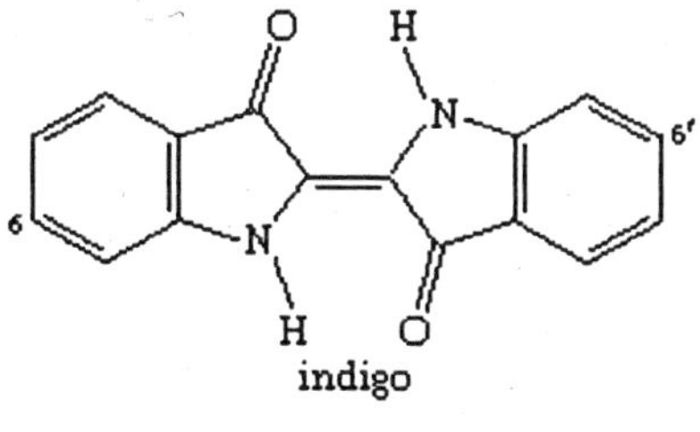

indigo

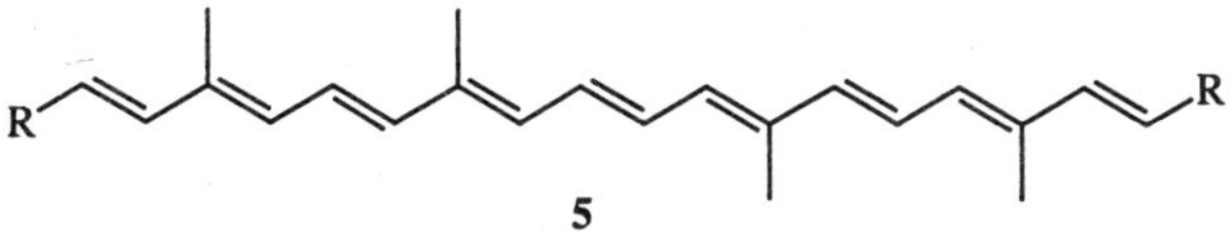

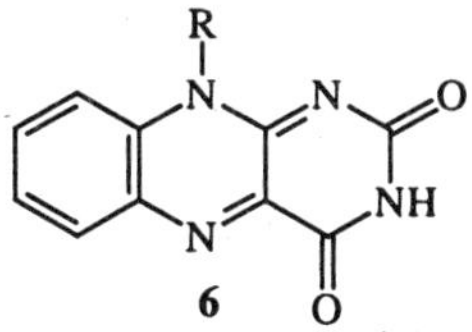

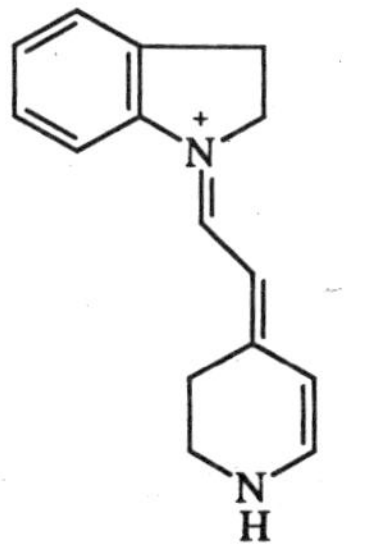

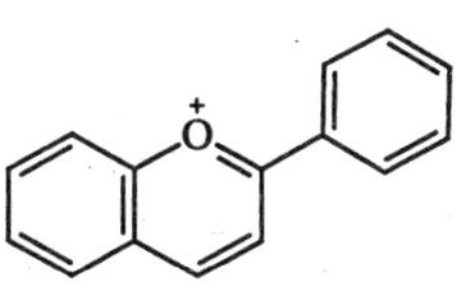

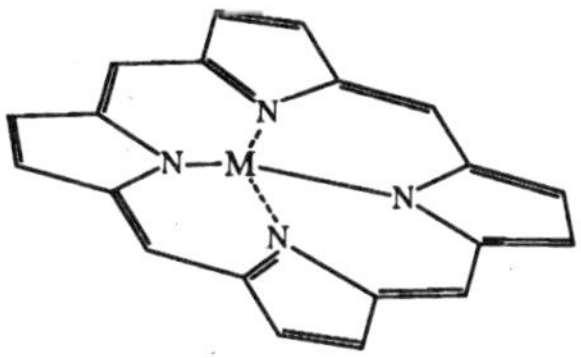

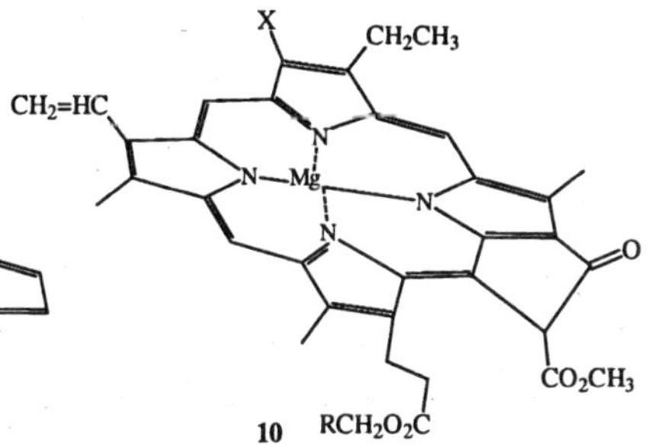

11

The chemical structures are sensitive to their chemical and physical environment. For example, anthocyanins (12), a major group of biocolour compounds, are pH dependent colour sensitive and give different colour under different pH. The anthocyanin equilibrium structure is the flavylium cation, from which secondary structures are derived either by proton transfer to produce bluish quinonoidal bases and corresponding ionized quinonoidal bases, or by hydration, to form colourless carbinol pseudobases that can be tautomerised to colourless retro-chalcones. Tertiary structures can result from molecular association of primary or secondary forms, either with themselves (self-association) or with another molecule in solution (intermolecular copigmentation). Quaternary structures, which result from medium or solvent effects on the colour of any of the preceding structures, are likely to be of significance only in strongly alcoholic solutions. Apart from pH, the colour produced by anthocyanins is dependent on many other factors. This is not like synthetic colourants, which generally obey the Beer- Lambert law and yield 100% of their colour irrespective of pH. The complete anthocyanin colour is developed only in strongly acidic solutions (low pH). In isolation, they have little colour above pH 3.5, but in natural media, they become much more coloured by copigmentation with other components present in plants, which may themselves be colourless. In acid solution (pH less than 2), anthocyanins display a negative deviation from Beer's law (colour increases less than linearly with increase in concentration). However, at higher pH values, a positive deviation from Beer's law is observed. The reason for this greater than linear growth of colour results from self- association and co-pigmentation.

Health Hazards associated with Synthetic Colours

One of the most important and technically most difficult parts in the cosmetic and textile industries is the usage of colours. There are always very high expectations of consumers in this regard. Consumer always wants eye-catching shades with lesser health hazards. This section of the chapter is going to highlight some of the important health hazards related with synthetic dyes used in the above industries. The hazardous effect is associated at all levels starting from their manufacturing until applications as given below:

- **Chemical Pollution:** The production of synthetic colours requires strong acids, alkalis, solvents, high temperatures and heavy metal catalysts. The improper discharge of the chemicals causes lot of problems to the inhabitants around.
- **Energy Pollution:** Petroleum is the starting material for most of the synthetic dyes. This leads to direct link with the price of dyes to the price of petroleum. In addition, since synthesis is energy intensive, the process is sensitive to energy prices and it generates greenhouse gases.
- **Hazardous Waste Generation:** Since synthetic production of dyes needs very toxic and hazardous chemicals, it also generates a hazardous waste, the disposal of which is a major environmental and economic challenge.
- **Toxic and Allergic Effects:** There is occupational safety issues involved since production processes use the toxic and hazardous materials. People working there are susceptible to the various health hazards.

Cosmetic Industries

The use of cosmetics is not a modern phenomenon. Cleopatra is said to have bathed in milk to whiten and soften her skin. More than 3000 years ago, Greek women used poisonous lead carbonate to achieve a pale complexion, costing some wearers their lives. As per the definition of FDA, the cosmetics are intended to apply to the human body for cleansing, beautifying, promoting attractiveness, or altering the appearance without affecting the body's structure or functions. In general, cosmetics are the substances used to enhance the appearance or odour of the human body. Cosmetics include very wide range of products, *i.e.*, skin-care creams, lotions, powders, perfumes, lipsticks, fingernail, toe nail polish, eye and facial makeup, hair colours, hair sprays and gels, deodorants, hand sanitizer, baby products, bath oils, bubble baths, *etc*. Most cosmetics contain a combination of at least some of the following ingredients such as water, emulsifier, preservative, thickener, colour, fragrance and pH stabilizers.

For the colouration of cosmetic products, the solubility plays vital role. Usually three types of colours, *i.e.*, water-soluble, fat-soluble and water dispersible pigments, are available in the market. The types of colour present in selected cosmetics are tabulated below:

S.No.	Cosmetic Product	Colour
1.	Shower Gel	Water soluble colourant
2.	Shampoo, Foam Bath	Water soluble colourant
3.	Soap	Water soluble colourant and water dispersible pigments
4.	Eye makeup	Pigments and paint lacquers
5.	Shading Cream	Pigments
6.	Lipstick	Organic pigments, eosin and paint lacquers

Unfortunately, the ingredients in cosmetics can have unintended side effects, *e.g.*, skin allergies to specific ingredients are a common problem. Allergies to cosmetic products can be due to chemicals such as added synthetic colour, fragrances and preservatives. This can lead to a skin rash where the product is applied. Some people, while not allergic to a specific ingredient, may nevertheless find that a product irritates their skin because it damages the outer layers, which is a condition known as irritant dermatitis. Sometimes, over-washing of hair or skin with soaps and detergents can strip the skin's natural protective oily layer, resulting in dry and scaly skin. Alternatively, excessive use of make-up or oily moisturizers can block pores and aggravate acne. The adverse effect of colour additives in selected cosmetics is discussed in this section.

Sindoor is applied by most of the Hindu women as a traditional symbol of auspiciousness. Most of the sindoor available in the market are made from synthetic dyes, filler materials, lime and salts of lead and mercury to provide fast and brilliant red shade. Some commercially prepared sindoor contain toxic, low-grade red lead oxide (Pb_3O_4). Prolonged use leads to health hazards like lead poisoning. In addition, due to the lack of strict regulations in local market, it is not surprising that the red colour might be obtained by using Rhodamine B dye, which is known to induce hereditary disorders. Red colour may also be obtained by using mercury sulphate, which can cause skin cancer. In addition, the prolonged use of these chemicals leads to dermatological disorders.

Another traditional cosmetic, kajal or surma, has also been reported as a lead-hazardous eye make up. The main components are lead sulphide (PbS) along with lead oxide, iron oxide, amorphous carbon, zinc oxide (ZnO), calcium carbonate ($CaCO_3$) and boric acid (H_3BO_3). These chemicals have their adverse effect on health.

Lipstick is a well-known cosmetic item and is very popular among women. Some of the lipstick uses lead compounds for colouration, which is toxic. Experts have found that many famous and expensive lipstick brands contain unacceptably high and dangerous to the health levels of lead. Scientists believe that lead causes numerous health problems, including reproductive dysfunction. Frankfurt, a test of lipsticks commissioned by a German consumer protection magazine, has found more than half of them to be potential health hazards. In August 2009 issue, Oeko-Test (Eco-Test) reported that of the 306 lipsticks tested, 21 different brands and various shades of colour, the 121 lipsticks had received a failing grade and 44 a grade of unsatisfactory. Sometimes, certain shades contained potentially harmful dyes, while others did not. Also, the coal tars, which are the basic ingredients from which synthetic dyes are formed can cause allergy, nausea, dermatitis and drying of lips, even if, present in very less amount. Ninety eight percent of the chemicals used in lipstick have not been tested for human safety.

The nail cosmetics are also not free from health hazards. Some dyes are harmful to both the applicants and the one who applies to the customers. The nail cosmetics include the application of nail polish, artificial nails, nail elongators and/or nail treatments that are usually dictated by current fashion trends. With the increasing volume of clients, cosmetologists are being put at risk of increased and prolonged exposures to various hazards in the workplace.

The overall use of hair dyes made from synthetic colour is also not free from serious health hazards. Hair dye products are divided into three categories, *i.e.*, permanent, semi-permanent and temporary hair colours. Permanent hair colours are the most popular hair dye products. They are further divided into oxidation hair dyes and progressive hair dyes. The products of oxidation hair dye consist of a solution of dye intermediates, *e.g.*, *p*-phenylenediamine, which form hair dyes on chemical reaction and preformed dyes, *e.g.*, 2-nitro-*p*-phenylenediamine, which already are dyes and are added to achieve the intended shades. Another type of oxidation hair dye contains a solution of hydrogen peroxide (H_2O_2), usually 6%, in water or a cream lotion.

The progressive hair dye products contain lead acetate as the active ingredient. Lead acetate is approved as a colour additive for colouring hair on the scalp at a concentration not exceeding 0.6% w/v, calculated as metallic lead. Bismuth citrate, the other approved colour additive, is used to a much lesser extent. Progressive hair dyes change the colour of hair gradually from light straw colour to black by reacting with the sulfur of hair keratin as well as oxidizing on the hair surface but the increase in permissible amount of these chemicals leads to many types of health problem. A large survey by the American Cancer Society found that dye users had a slightly lower rate of fatal cancers than people who did not colour their hair.

Semi-permanent and temporary hair colouring products are solutions of various coal tars. The studies on hair dye products say that 71 hair dye products evaluated were found to contain ingredients derived from coal tar. Several studies have linked long-time coal-tar hair dye use to bladder cancer, non-Hodgkin's lymphoma and multiple myeloma. A research conducted in 2001 by the USC School of Medicine found that women using permanent hair dye at least once a month more than doubled their risk of bladder cancer. A link between hair dye and non-Hodgkin's lymphoma was established in 1992 when a study conducted by the National Cancer Institute found that 20% of the cases of non-Hodgkin's lymphoma might be linked to hair dye use.

A synthetic colouring compound tartrazine (13) is a coal-tar derivative, which is used to colour cosmetics and related products. This is a synthetic lemon yellow azo dye. This was one of the first synthetic pigments patented in 1884. Tartrazine is now known to be a catalyst in hyperactivity, other behavioural problems, urticaria, angioedema, paresthesias, vomiting, migraine, rhinorrhea and nasal obstruction, coughing, asthma attacks and purpura. Behavioural changes like irritability, restlessness and sleep disturbance are associated with the ingestion of tartrazine in some children.

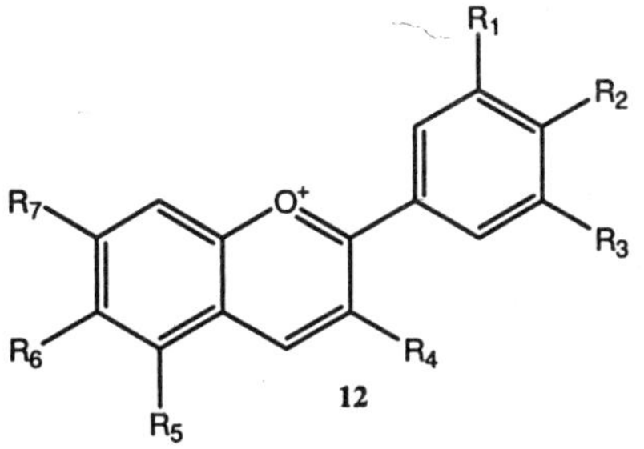

12

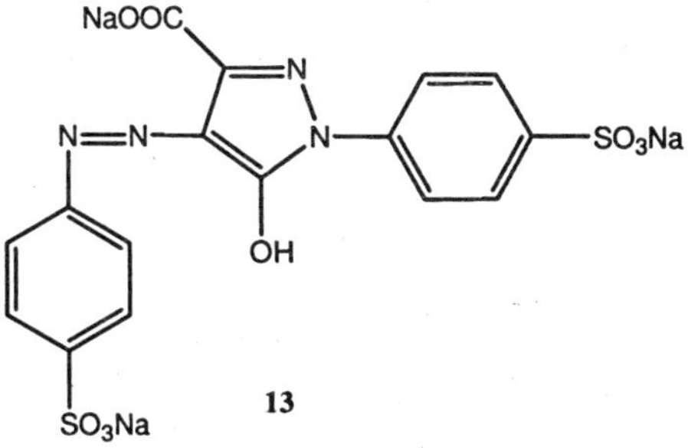

13

Textile Industries

The textile production industry is one of the oldest and most technologically complex of all industries. The industry produces and uses approximately 1.3 million tonnes of dyes, pigments and dye precursors, valued at around $23 billion, almost all of which are manufactured synthetically. Textile dyes form a large group of textile chemicals and comprise over 8,000 different compounds with almost 40,000 commercial names. The first synthetic dye was discovered in 1865 and given the name mauve. After that, the synthetic dyestuff industry developed rapidly with the introduction of another very important dye, *i.e.*, magenta. The subsequent development of water-soluble azo dyes represented a landmark in the synthetic dye industry. The most important dye structures include triphenylmethane compounds, indigoid and azo structures, azines, thiazines, anthraquinone derivatives and phthalocyanines. Today, the chemicals used to produce dyes are often highly toxic, carcinogenic, or even explosive. The chemical aniline, the basis for a popular group of azo dyes are considered deadly poisons giving off carcinogenic amines and dangerous to work with, also being highly flammable. In addition, some other harmful chemicals used in the dying process include dioxin (a carcinogen and possible hormone disrupter), toxic heavy metals such as chrome, copper, and zinc (the known carcinogens) and formaldehyde (a suspected carcinogen).

The huge utilization leads to many health hazards that can broadly be categorized into four groups:

- Their production process requires hazardous chemicals, creating worker safety concerns.
- They may generate hazardous wastes.
- These dyes are not environment friendly.
- Some fiber dyes with synthetic colour are not suitable for consumers.

Since clothing comes into prolonged contact with one's skin, toxic chemicals are often absorbed into the skin especially when one's body is warm and skin pores have opened to allow perspiration. Though adults are more resistant, one can see the symptoms in children with red cheeks and ears, dark circles under eyes and also sometimes behaviour or learning problems.

Most synthetic fabrics, from towels to dress shirts to bed linens, are treated with chemicals during and after processing. These chemicals not only leach into the environment, leaving an impact on groundwater, wildlife, air and soil but also they may be absorbed or inhaled directly. The dyeing and printing plants near residential area are a major concern for health. Contamination of wastewater by unfixed dyes is a potential health hazards. In ordinary dyeing, fixation of over 90% of the dyestuff can be achieved but fixation level of only 60% or less is common in printing with reactive dyes. Hence, during washing off the printed fabric, more than one-third goes to wastewater. This wastewater is toxic, mutagenic and carcinogenic. In addition, the colouration in wastewater is easily detectable and has inhibitory effect on the process of photosynthesis and aquatic life and reach to human health through food chain.

In mid 1950s, a crucial technological advance occurred with the discovery of reactive dyes that react directly with the fiber itself. Reactive dyes produce textiles with specific colourfastness that had not been possible with the earlier water-soluble dyes. The reactive dyes such as Reactive Blue 19 (RB-19), also known as Remazol brilliant blue have very low fixation efficiency (75-80%), mainly due to the competition between the formation of reactive form and the hydrolysis reactions. These dyes are extensively used in textile industries as their reactive group binds on textile fibers by covalent bonds formation. They are mainly used for dyeing cellulosic fibers like cotton and rayon and also used for silk, wool, nylon and leather.

Some reactive dyes are even recognized as respiratory sensitizers. Breathing in respiratory sensitizers can cause occupational asthma. Once a person is sensitized, re-exposure to even very small amounts of the same dye may result in allergic symptoms such as a runny or stuffy nose, watery or prickly eyes, wheezing, chest tightness and breathlessness. Some dyes can cause similar allergic skin reactions. A small number of dyes, based on the chemical benzidine, are thought possibly to cause cancer. The list of adverse effect is enormous, and hence, there is an urgent need to search for safe and reliable dyes in the form of biocolours.

Role of Biocolour

This is well known fact that the dyes of the ancient world were derived from plant and earth pigments. However, for most people, the colours available to them were limited because of geographical isolation, expense, or rarity of certain plant or mineral substances and due to the lack of proper standardization and reproducibility of shades. One important reason for less-reproducibility is the varied amount of secondary metabolites present at varied places. The practices and techniques of dyeing have remained mostly unchanged until the mid 19th century. Usually, the dyes were mixed natural resins, animal fats, or drying oils to produce paints, boiled with, or painted onto cloth as a dye. Later, the biocolours lost the race with synthetic dyes. Within a very short time, these synthetic dyes taken over by the centuries-old traditions of natural dyeing, but with passing of time, public became aware about the health hazards and environmental issues related with synthetic dyes. The consumer once again looked towards the phytochemists and related researchers to have a re-look at the potential of biocolour.

The choice of biocolour instead of synthetic ones is certainly due to health concerns, though real safety and healthiness of natural substances is not scientifically established yet. Still, the finding out and exploitation of naturally occurring dyes have strong motivations that are equally important for both cosmetics and textile sectors. Therefore, the interest in all plants able to provide pigments to be used by industry is increasing. Some of these plants are already known, *e.g.*, Isatis, rubia, safflower, *etc.* before the worldwide. They are already being used to this purpose.

To take few examples, people have been using madder (*R. tinctorum*) and other *Rubia* species for dyeing silk, cotton, jute and other textiles along with cosmetic additives. The ancient Indian dyers used turmeric for yellow colours, madder from plants other than *R. tinctorum* and indigo. They had the knowledge of dyeing plant fibers and of reducing the vat dye before dyeing with indigo. A high level of textile printing was achieved. However, all preservation techniques, except for mordant preservation, were of Chinese rather than Indian origin. Silk was dyed with a colourant extracted from *R. tinctorum* with highest dyeing capacity at pH 3.5 and 40°C for 120 minutes, similar to results with another biocolour molecule purpurin. Purpurin is the major colourant present in the roots of *R. cordifolia* for dyeing nylon. A change in hue from yellowish brown to red was observed in dyeing with aluminium mordanting. A concentration of 6 g dye from *R. tinctorum* extracts per 100 ml provided the maximum optical density in the dye liquor. The best results for dyeing silk with *R. tinctorum* extracts have been determined with the mordant concentration for alum, copper sulfate, stannous chloride, ferrous sulfate and potassium dichromate (8 g/100 g silk). Post-mordanting gave the best results with ferrous sulfate, copper sulfate and chrome. Alum was best with pre-mordanting and stannous chloride with simultaneous mordanting.

Several studies have been done for the optimization of procedure for dyeing of silk and the effects of pigments obtained from various *Rubia* species. The aqueous extract of *R. iberica* was used for dyeing wool in the carpet industry. Wool dyed with these natural dyes has good fastness to alkaline agents or sunlight. Extraction of rhizome of the madder *R. tinctorum* is used at various pH values of the media and in the presence of different mordants for dyeing wool yarn with various colours. The madder-dyed wool carpets had a high light fastness and resistance to soap solutions. Bleached jute fibers were dyed with *R. cordifolia* using aluminium sulfate as mordant. The red colour of candies is obtained with water or alcohol extract of *R. cordifolia* root, cultured cells of *R. cordifolia* and/or *R. cordifolia* root fermentation products and applied with alum, organic acid salts and carbonic acid salts. Due to its importance in the dye industry, experiments are being conducted to make industrial extraction unit.

Most people are using biocolour as hair dyes. The most widely used is *henna*. Natural to many Muslims and native to the East Indies and North Africa, the henna plant (*Lawsonia Inerinus*) is the mother of hair dyes. It varies in effect, depending on where it grows. Moroccan henna is the lightest in colour, while Iranian henna is the richest and probably the most expensive. The leaves are dried then crushed into a fine powder and mixed with water to form paste that can be used to dye hair.

A well-known biocolour lipstick is shikonin. This high-tech Japanese product is a big hit on the cosmetics market. It was reported that within a few days, two million items were sold in Japan in 1985, in spite of having high cost of 3500 yen. The red colourant in this lipstick is a purly biological product, a naphthaquinone derivative obtained from *Lithospermum erythrorhizon*. The herbal lipstick is in high demand.

The list of selected plants, which can be used to isolate biocolours is given in Table.

S.No.	Family	Species (common name)	Plant part	Biocolour[#]
1.	Mimosaceae	*Acacia catechu* (Cutch)	heartwood	Brown/red
2.	Mimosaceae	*Acacia nilotica* (Suntberry)	Seed pods	Blue
3.	Mimosaceae	*Adenanthera pavonina*	Wood	Red
4.	Acanthaceae	*Adhatoda zeylanica* (Adusa)	Leaves	Green
5.	Rutaceae	*Aegle marmelos*	Rind of fruit	Light red
6.	Boraginaceae	*Arnebia benthamii*	Root	Red
7.	Boraginaceae	*Arnebia nobilis*	Root	Red
8.	Boraginaceae	*Arnebia hispidissima*	Root	Red
9.	Caesalpiniaceae	*Bauhinia purpuria*	Bark	Purple
10.	Bixaceae	*Bixa orellana* (Annatto)	Seeds	Orange-yellow
11.	Chenopodiaceae	*Beta vulgaris* (Beetroot)	Roots	Red
12.	Fagaceae	*Butea monosperma*	Flower	Bright yellow
13.	Solanaceae	*Capsicum annuum* (Paprika/lal mirch)	Ripe fruits	Red
14.	Asteraceae	*Carthamus tinctorious* (safflower)	Flower	red
15.	Caesalpiniaceae	*Cassia tora*	Seeds	Blue
16.	Zingiberaceae	*Curcuma longa*	Roots	Bright yellow
17.	Fabaceae	*Indigofera tinctoria* (Indigo)	Leaves	Blue
18.	Brassicaceae	*Isatis tinctoria* (Waod)	Leaves	Blue
19.	Boraginaceae	*Lithospermum erythrorhizon*	Root	Red
20.	Lythraceae	*Lawsonia alba*	Leaves	Brown
21.	Sapotaceae	*Madhuca indica*	Bark	Reddish yellow
22.	Euphorbiaceae	*Mallotus philippensis*	Fruits	Red
23.	Fagaceae	*Quercus robur* (English oak)	Bark	Yellow
24.	Fagaceae	*Quercus tinctoria*	Bark	Yellow
25.	Rubiaceae	*Rubia tinctorum* (European madder)	Roots	Red to orange
26.	Rubiaceae	*Rubia cordifolia* (Indian madder)	Roots	Red to orange
27.	Rubiaceae	*Rubia sikkimensis* (Naga madder)	Roots	Red to orange
28.	Polygonaceae	*Rumex hastatus* (churki)	Leaves	Green
29.	Polygonaceae	*Rumex nepalensis*	Leaves	Green
30.	Asteraceae	*Solidago species* (Golden rod)	Flower	Yellow
31.	Combretaceae	*Terminalia arjuna* (Arjun)	Bark	Light brown
32.	Verbenaceae	*Tectona grandis*	Leaves	Yellow
33.	Asteraceae	*Tagetes species* (Marigold)	Flower	Yellow
34.	Lythraceae	*Woodfordia fruiticosa*	Leaves and flowers	Pink/red

[#]colour changes with different mordents

The method used for the extraction of biocolour depends upon the plant parts under utilization. In most of the cases, the root or rhizome materials are cut into pieces, soaked in water for 2-3 days. This is further made paste that squeezed and filtered. The resulting extract is boiled and concentrated. However, the floral parts and leaves, when used for biocolour, are crushed and dissolved in water for 12-24 hours. Further, this is heated to get the required concentration. The biocolour so obtained can be differentiated into different shades by mixing with different mordents. The usual mordents being used for biocolours includes dilute solution of organic manure, cow dung or urine, curd water, rock salt, lemon juice, lime water, *etc.*

Major Limitations Associated with Biocolours

- Availability and reproducibility of desired colour shades
- Colour yield and process of extraction
- Change in content of dyes with environmental impact
- Complexity in dyeing process
- Limitations in the number of suitable biocolours
- Limitations in dyeing the types of textiles
- Mainly allow wool, natural silk, linen and cotton to be coloured
- Great affinity in blending dyes
- Lack of standardization and inadequate degree of fixation
- The major difference in fastness properties
- The utility cost is high because they can not be mass produce in limited space and require about large acres of land per kilogram of biocolour produced.

Major Advantages Associated with Biocolours

- In most of the cases, they can be obtaining from renewable sources.
- Less health hazards are associated.
- Sometimes, they act dual purpose, as ciocolour and as health cure.
- Practically, very mild chemical reactions are involved.
- Usually, the waste products are not harmful, and hence, less disposal problems.
- They require innovations as lot of creativity is used to use these dyes judiciously.
- They are eco-friendly and have better biodegradability.
- They are soft in colour, cool to eyes and less hazardous.
- They help to provide employment for the rural people and preserve our traditional craftsmanship.
- They save energy because the raw materials are not from petroleum products.
- Waste material is used for the manufacture of compost.

The future of biocolour is increasing day by day but to make the biocolour more market compatible, some of the steps are highly recommended:

- A complete linkage must be developed between the cultivation, collection, processing and utilization.
- A proper documentation of their traditional use and scientific support for the traditional use is necessary.
- Suppliers of raw materials and their national and international market must be well identified.
- Exploration and research on new sources must be continued.
- In case of plants, the research and development must focus on the use of aerial parts rather than the roots.
- Improvement in biocolour yield should be done by biotechnological processes.
- The technical ratings, low cost, durability and reproducibility for textile market must be increased.
- A global market strategy must be developed to make the people aware about the benefits of using biocolours.

Conclusion and Future Prospects

India's vast plant resources represented by 20,000 species inhabiting wide range of ecological niches have enormous economic potential. The rich biodiversity of India ecosystem has not been assessed for their colour yielding potential. The commerce of biocolour can be considered as a measure of economic prosperity of a nation, especially the country like India whose rich biodiversity is an asset. One can very easily say that biocolours are one of the important uses of plants as this is associated with our cultural practices, rituals, arts and crafts and over all personal satisfaction. No doubt, the problems are associated with the isolation, extraction, reproducibility, *etc.* with the biocolours but considering the increasing health hazards of synthetic dyes, there is a constant need to harness the nature. At present, biocolours are being used by selected hobby groups, designers, traditional dyers and printers, non-government organization, museums, academic institute and research association/laboratories and very few industries.

The field studies on the agronomy of biocolour-crops have indicated that they can be grown successfully, as crops, using modern agricultural systems. The cultivation process can be made as simple as cash crops. Seed can be drilled directly and modern husbandry can be applied using chemical weed control, fertilizer application and mechanical harvesting. The commercial production of biocolour will require further investigation, although, work so far has indicated that the production of biocolour on a commercial-scale may be possible in near future.

■■■

32 Free radical Scavenging activity of Some Medicinal Plants

Gayatri Nahak and R.K. Sahu

B.J.B. Autonomous College, Bhubaneswar, Orissa, India,
E-mail: <gayatri_bioteq@yahoo.co.in>, <sahurajani.sahu@gmail.com>

In recent years, there is an upsurge in the areas related to newer developments, especially the role of free radicals and antioxidants, in the prevention of disease. Thus, it will be pertinent to examine the possible role of free radicals in disease and antioxidants in its prevention, especially the status of subject matter and future prospects, in this present chapter.

Antioxidants

Antioxidants are the substances that neutralize free radicals or their actions. Nature has endowed each cell with adequate protective mechanisms against any harmful effects of free radicals, *i.e.*, super oxide dismutase (SOD), glutathione peroxidase, glutathione reductase, thioredoxin, thiols and disulfide bonding, are buffering systems in every cell. Alpha-tocopherol (vitamin E) is an essential nutrient, which functions as a chain-breaking antioxidant that prevents the propagation of free radical reactions in all cell membranes in the human body. Ascorbic acid (vitamin C) is also part of the normal protecting mechanism. Other non-enzymatic antioxidants include carotenoids, flavonoids and related polyphenols, α-lipoic acid, glutathione, *etc*.

Levels of Antioxidant Action

Antioxidants, capable of neutralizing free radicals or their actions, act at different stages. They act at the levels of prevention, interception and repair. Preventive antioxidants attempt to stop the formation of reactive oxygen species (ROS), which include super oxide dismutase (SOD) that catalyses the dismutation of super oxide to hydrogen peroxide (H_2O_2) and catalase that breaks it down to water. Interception of free radicals is mainly by radical scavenging, while at the secondary level, the scavenging of peroxyl radicals is affected. The effectors include various antioxidants like vitamins C and E, glutathione, other thiol compounds, carotenoids, flavonoids, *etc*. At the repair and reconstitution level, mainly repair enzymes are involved.

Behaviour of Antioxidants

The free radicals play a very important role in human health and beneficial in combating against several diseases. Before we understand their working, it is necessary to have a brief idea about free radicals. During a chemical reaction (oxidation), one reactant loses an electron, which is called oxidant or free radical, while the other gains an electron. In living organisms, oxygen in unstable form is the most common free radical. This is called reactive oxygen species (ROS) and is generated during various metabolic activities. Contaminants in the environment as well as normal metabolism of a cell can change molecule into a free radical. The examples of reactive oxygen species are OH, O_2, H_2O_2, O_3, HOCI, RO_2 and RO. Any molecule can become a free radical by either losing or gaining an electron. Once initiated, these free radicals get involved in chain reaction with stable types. The compounds thus formed have longer stability in body and increase the potential for cellular damage. Free radicals damage the cell at the site of their operation causing serious disorders. Plaque may accumulate in arteries on oxidation. Low-density lipoprotein (bad cholesterol) functions as free radical and damages the free artery lining. It hampers the blood circulation, which may lead to heart attack.

Stages of free radicals

- Initiation
- Propagation
- Termination

The initiation and propagation stages have already been discussed earlier. However, termination or neutralization of free radicals is achieved by antioxidants or enzymatic mechanism. It is necessary to balance the fee radical and antioxidant activity.

Effects of Excess Free Radicals

- Free radicals damage DNA, RNA, proteins and enzymes.
- Lead to the formation of tumors and cause cancers.
- Cardiovascular diseases
- Nervous disorders
- Premature ageing
- Parkinson's and Alzheimer's diseases
- Rheumatic and pulmonary disorders

Concept of Oxidative Stress

The relation between free radicals and disease can be explained by the concept of oxidative stress. In a normal healthy human body, the generation of pro-oxidants in the form of reactive oxygen species (ROS) and reactive nitrogen species (RNS) are effectively kept in check by the various levels of antioxidant defense. However, when it is exposed to adverse physicochemical, environmental, or pathological agents such as atmospheric pollutants, cigarette smoking, ultraviolet rays, radiation, toxic chemicals, over nutrition and advanced glycation end products (AGEs) in diabetes, this delicately maintained balance is shifted in favour of pro-oxidants resulting in oxidative stress. It has been implicated in the etiology of several (>100) human diseases and in the process of ageing.

Requirement of Antioxidants in the Body

A number of processes are taking place in our body like breathing, breaking up of protein in the body or exposure of body to air pollution or ultraviolet radiation leads to the formation of free radicals that aid in the process of oxidation, which further leads to several health problems. If the body is healthy, it is capable enough to break down these free radicals before they become harmful. In case, the formation of free radicals exceeds in the body, it can damage the cells and tissues. This damage can be counteracted by antioxidants, which prevent the free radical formation, and if the formation occurs, it helps in their removal from blood stream. Thus, antioxidants play potentially beneficial role in the risk reduction and treatment of diseases.

Molecular Damage Induced by Free Radicals

All the biological molecules present in our body are at risk of being attacked by free radicals. Such damaged molecules can impair cell functions and even lead to cell death eventually resulting in diseased states.

Lipids and Lipid Peroxidation

Membrane lipids present in sub-cellular organelles are highly susceptible to free radical damage. Lipids when reacted with free radicals can undergo the highly damaging chain reaction of lipid peroxidation (LP) leading to both direct and indirect effects. During lipid peroxidation, a large number of toxic byproducts are also formed that can have effects at a site away from the area of generation, behaving as second messengers. The damage caused by lipid peroxidation is highly detrimental to the functioning of cell. Lipid peroxidation is a free radical mediated process. Initiation of peroxidative sequence is due to the attack by any species, which can abstract a hydrogen atom from a methylene group (CH_2), leaving behind an unpaired electron on the carbon atom (•CH). The resultant carbon radical is stabilized by molecular rearrangement to produce a conjugated diene, which then can react with an oxygen molecule to give a lipid peroxyl radical (LOO•). These radicals can further abstract hydrogen atoms from other lipid molecules to form lipid hydroperoxides (LOOH) and at the same time propagate lipid peroxidation further. The peroxidation reaction can be terminated by a number of reactions. The major one involves the reaction of LOO• or lipid radical (L•) with a molecule of antioxidant such as vitamin E or α-tocopherol (α-TOH) forming more stable tocopherol phenoxyl radical, which is not involved in further chain reactions. This can be recycled by other cellular antioxidants such as vitamin C or glutathione (GSH).

$$LH + {\bullet}OH \rightarrow L{\bullet} + H_2O$$

$$LOO{\bullet} + LH \rightarrow L{\bullet} + LOOH$$

$$LOO{\bullet} + \alpha\text{-}TOH \rightarrow LOOH + \alpha\text{-}TO{\bullet}$$

The process of lipid peroxidation gives rise to many products of toxicological interest like malondialdehyde (MDA), 4-hydroxynonenal (4-HNE) and various 2-alkenals. Isoprostanes are unique products of lipid peroxidation of arachidonic acid and recently tests such as mass spectrometry and ELISA-assay kits are available to detect isoprostanes.

Reaction with Carbohydrates

Free radicals such as •OH react with carbohydrates by randomly abstracting a hydrogen atom from one of the carbon atoms, producing a carbon-centered radical. This leads to chain breaks in important molecules like hyaluronic acid. In synovial fluid surrounding joints, an accumulation and activation of neutrophils during inflammation produces significant amounts of oxyradicals, which is also being implicated in rheumatoid arthritis.

Damage to DNA

Oxidative damage to DNA is a result of interaction of DNA with ROS or RNS. Free radicals such as •OH, eaq- and H• react with DNA by addition to bases or abstractions of hydrogen atoms from the sugar moiety. The C4-C5 double bond of pyrimidine is particularly sensitive to attack by •OH, generating a spectrum of oxidative pyrimidine damage products, including thymine glycol, uracil glycol, urea residue, 5-hydroxydeoxyuridine, 5-hydroxydeoxycytidine, hydantoin and others. Similarly, interaction of •OH with purines will generate 8-hydroxydeoxyguanosine (8-OHdG), 8- hydroxydeoxyadenosine, formamidopyrimidines and other less characterized purine oxidative products. Several repair pathways repair DNA damage. 8-OHdG has been implicated in carcinogenesis and is considered a reliable marker for oxidative DNA damage.

Damage to Proteins

Oxidation of proteins by ROS/RNS can generate a range of stable as well as reactive products such as protein hydroperoxides that can generate additional radicals particularly upon interaction with transition metal ions. Although most oxidized proteins that are functionally inactive are rapidly removed but some can gradually accumulate with time and thereby contribute to the damage associated with ageing as well as various diseases. Lipofuscin, an aggregate of peroxidized lipids and proteins accumulates in lysosomes of aged and brain cells of patients with Alzheimer's disease.

Significance of Antioxidants in Relation to Disease

Antioxidants may prevent and/or improve different diseased states. Zinc is an essential trace element, being a co-factor for about 200 human enzymes, including the cytoplasmic antioxidant Cu-Zn SOD, isoenzyme of SOD mainly present in cytosol. Selenium is also an essential trace element and a co-factor for glutathione peroxidase. Vitamin E and tocotrienols (such as those from palm oil) are efficient lipid soluble antioxidants that function as a *chain breaker* during lipid peroxidation in cell membranes and various lipid particles including low-density lipoprotein. Vitamin E is considered as a *standard antioxidant* to which other compounds with antioxidant activities are compared, especially in terms of its biological activity and clinical relevance. The daily dietary allowance varies between 400 to 800 IU. Vitamin C (ascorbic acid) is a water-soluble free radical scavenger. The daily recommended dietary allowance is 60 mg. Apart from these carotenoids such as β-carotene, lycopene and lutein, the other carotenoids function as important antioxidants and they quench $^1O_2$ and ROO•. Flavonoids, mainly present as colouring pigments in plants also function as potent antioxidants at various levels.

Antioxidants and Protection against Human Disease

A number of epidemiological studies have shown inverse correlation between the levels of established antioxidants/phytonutrients present in tissue/blood samples and occurrence of cardiovascular disease, cancer, or mortality due to these diseases. However, some recent meta-analysis shows that supplementation with mainly single antioxidant may not be that effective, a view that contrasts with those of preclinical and epidemiological studies on consumption of antioxidant-rich food articles. Based on the majority of epidemiological and case control studies, the recommendations were made for daily dietary intake of some established antioxidants like vitamin E and C as well as others. Requirement for antioxidants in Indian conditions differ from that of industrialized western countries due to the nutritional differences. There are also a number of dietary supplements rich in antioxidants tested for their efficacy. Many laboratories from India are working on the antioxidant effect of plant compounds, mainly derived from natural sources that are capable of protecting against such damage. Such studies show that compounds with potent antioxidant activity include carotenoids, curcumin from turmeric, flavonoids, caffeine present in coffee, tea, *etc*., orientin, vicenin, glabridin, glycyrrhizin, emblicanin, punigluconin, pedunculagin, 2-hydroxy-4-methoxy benzoic acid, dehydrozingerone, picroliv, withaferin, yakuchinone, gingerol, chlorogenic acid, vanillin (food flavouring agent) and chlorophyllin (a water-soluble analogue of chlorophyll).

Newer Therapeutic Approaches Using Antioxidants

Antioxidant-based drugs/formulations for the prevention and treatment of complex diseases like atherosclerosis, stroke, diabetes, Alzheimer's disease (AD), Parkinson's disease, cancer, *etc*. appeared over the past three decades. Free radical theory has greatly stimulated interest in the role of dietary antioxidants in preventing many human diseases, including cancer, atherosclerosis, stroke, rheumatoid arthritis, neuro-degeneration and diabetes. Dietary antioxidants may have promising therapeutic potential in delaying the onset as well as in preventing the ageing population with Alzheimer's disease and its related complications. Two neuro-protective clinical trials, *i.e.*, deprenyl and tocopherol antioxidant therapy of Parkinson's study, are available with antioxidants. By fusing ancient wisdom and modern science, India can create excellent products. Therefore, it has embarked on a fast track programme to discover new drugs by building on traditional medicines and screening the diverse plants and microbial sources of the country. In terms of its size, diversity and access to talent and resources, this programme is not only the world's largest project of its kind but is also unique.

Ayurveda, Antioxidants and Therapeutics

Employing a unique holistic approach, Ayurvedic medicines are usually customized to an individual constitution. Ayurvedic Indian and traditional Chinese systems are living great traditions and have important roles in bio-prospecting of new medicines from medicinal plants, which are also rich sources of antioxidants. Current estimate indicates that about 80% of people in developing countries still rely on traditional medicine-based largely on various species of plants and animals for their primary healthcare. *Ayurveda* remains one of the most ancient and yet living traditions practiced widely in India.

Sources of Antioxidants, Phytonutrients and Functional Foods

Natural compounds, especially derived from dietary sources provide a large number of antioxidants. Some beverages such as tea are also rich sources of antioxidants. A growing body of evidence suggests that moderate consumption of tea may protect against several forms of cancer, cardiovascular diseases and the formation of kidney stones, bacterial infections and dental cavities. Tea is particularly rich in catechins, of which, epigallocatechin gallate (EGCG) is the most abundant.

Medicinal Plants as Sources of Antioxidants

Since time immemorial, man has been using plant extracts to protect himself against several diseases and also to improve his health and life-style. No doubt, plants are serving several purposes whether health, nutrition, beauty, or medicinal. With the development in techniques and recent researches, it has been proved that certain non-nutritive chemicals in plants such as terpenoids and flavonoids, which were earlier thought to be of no importance to human diet, possess antioxidant properties. The plants are susceptible to damage caused by active oxygen and thus develop numerous antioxidant defense systems resulting in formation of numerous potent antioxidants. Many aromatic, medicinal and spice plants contain compounds that possess confirmed strong antioxidant components. The essential oils derived from aromatic plants not only serve as fragrance and flavour agents but also as dietary antioxidant expected to prevent several diseases caused by free radicals. Apart from the dietary sources, Indian medicinal plants also provide antioxidants. Currently, a large and ever-expanding global population base prefers the use of natural products in treating and preventing medical problems. This has influenced many of pharmaceutical companies to produce new antimicrobial formulations extracted from plants or herbs. Herbs are from natural plants, therefore, they are often considered harmless compared with western medicines. Plant species still serve as a rich source of many novel biologically active compounds. However, very few plant species have been thoroughly investigated for their medicinal properties. Thus, there is renewing interest in phyto-medicine during last decade, and now a day, many medicinal plant species are being screened for pharmacological activities. At present, plant and herb resources are unlimited. Natural products from plants have provided the pharmaceutical industry with one of its most important sources of lead compounds and up to 40% of modern drugs are derived from natural sources, using either the natural substance or a synthesized version.

Why plants are huge source of antioxidants?

Plants are potential sources of natural antioxidants. They absorb the sun's radiation and generate high levels of oxygen as secondary metabolites of photosynthesis. On the other hand, oxygen is easily activated by ultraviolet radiation and heat from the sunlight to produce toxic reactive oxygen species, which are highly reactive because they can interact with a number of cellular molecules and metabolites thereby leading to a number of destructive processes causing cellular damage. Plants produce various antioxidant (enzymes and non-enzymes) compounds to counteract and detoxify these reactive oxygen species in order to survive. Hence, naturally occurring phyto-compounds possessing antioxidant and anti-inflammatory properties appear to contribute to their chemo-preventive or chemo-protective activity, which in turn, by the alternative medicine, has been used to the benefit of human beings.

Plant's Most Abundant Antioxidant Substances

Phenolic substances, which are known to possess high antioxidant activity, are actually common phytochemicals in fruits and leafy vegetables. Plants containing phenolic compounds have been reported to possess strong antioxidant properties. Most of these phenolics are classified into two principal groups of phenol, *i.e.*, carboxylic acids and flavonoids, the latter being the most significant. These phenolic compounds are found abundantly in all parts of the plant, such as wood, bark, stems, leaves, fruit, root, flowers, pollen and seeds. The antioxidant activity of phenolic compounds is based on their ability to donate hydrogen atoms to free radicals. Many phenolic compounds, particularly flavonoids, exhibit a wide range of biological effects, including antioxidant activity, antibacterial, antiviral, anti-inflammatory, antiallergic, anti-thrombotic, vasdilatory actions and the ability to lower the risk of coronary heart diseases. The protective effects of diets high in fruits and vegetables have been attributed to the presence of these compounds. A study on the plant extracts of *Bidens pilosa* L. var. radiata (Asteraceae) revealed that it could serve as a good source of caffeoylquinic acid derivatives and flavonoid glycosides which are attributed to *Bidens pilosa* significant antioxidant activity and inhibitory effect on nitric oxide (NO) production (an inflammatory mediator) in macrophages. Therefore, the study proposed the potential dietary value and benefits of *Bidens pilosa* extract based on phytochemical characteristics and the observed bioactive properties. Another example of the antioxidant effect of flavoniods is *Scutellaria baicalensis* Georgi (Lamiaceae), which is a widely used herb in traditional medical systems of China and Japan. The major constituents of *S. baicalensis* are flavonoids (baicalein, baicalin and wogonin), which have been associated with various properties, *e.g.*, antioxidant, anti-inflammatory, antithrombotic, antibacterial and antiviral.

Antioxidants and Free Radicals Scavenging Bioactivity

The term antioxidant is used to describe a dietary component that can function to decrease tissue damage by reactive oxygen. A good example of potent plant's antioxidant is anthocyanins, which are water-soluble pigments present in plants. They are known to be powerful antioxidants, which help to protect the plants from ultraviolet damage, particularly, ultraviolet-B radiation that proves to be one of the most relevant environmental risks because of its hazardous effects to human being, such as premature skin aging and especially skin photocarcinogenesis.

Most living organisms possess enzymatic and non-enzymatic defense systems against excessive production of reactive oxygen species. However, different external factors (smoke, diet, alcohol and some drugs) and aging decrease the efficiency of such protecting systems, resulting in disturbances of the redox equilibrium established under healthy conditions. The homeostatic balance between reactive oxygen species and endogenous antioxidants is important in maintaining healthy tissues. Excessive reactive oxygen species states are important in diseases such as acute respiratory distress syndrome and idiopathic pulmonary fibrosis. Redox imbalance increases the breakdown of extra-cellular matrix component hyaluronan into lower molecular weight fragments that in turn activate innate immune responses and perpetuate tissue injury. Thus, antioxidants that scavenge reactive oxygen species may be of great value in preventing the onset and propagation of oxidative diseases like autoimmune, cardiovascular and neurovascular diseases. Recently, more attention has been paid to the role of natural antioxidants, mainly phenolic compounds, which may have higher antioxidant activities than those of conventional vitamins C, E and α-carotene. Therefore, antioxidants, which can neutralize free radicals, may be of central importance in the prevention of carcinogenicity, cardiovascular and neurodegenerative changes associated with ageing. Epidemiological studies showed that the consumption of fruits and vegetables could protect humans against oxidative damage by inhibiting or quenching free radicals and reactive oxygen species.

Fruits and vegetables are rich sources of many food factors including minerals, vitamins and phytochemicals that may act as antioxidants. The antioxidant activity of fruits and vegetables is often assumed to be of greatest importance in combating a number of degenerative diseases since free radical-related damage has been implicated in causing many of these conditions. Thus, a daily consumption of five or more servings of antioxidant-rich food provides the body with the essential antioxidants needed to prevent degenerative diseases, premature ageing symptoms, chronic fatigue and general disability. Doctors and nutritionist have long known that antioxidants are needed by the human body for optimal well being, especially for maintaining a healthy body system and defense mechanism against cell damage. Therefore, the more the discovered antioxidant, the more the benefit to human beings has developed. The daily consumption of *Piper sarmentosum* (kadok of Piperaceae family) and *Morinda elliptica* Ridley (*Mengkudu jantanm* of Rubiaceae family) leaves, edible medicinal plants commonly used in Malaysian traditional medicine, can help in maintaining energy, general ability and fitness even during ageing. Naringenin, a naturally occurring super oxide scavenger, was found in the methanolic leave extracts of *P. sarmentosum* and *M. elliptica.* Thus, these plants could be considered as potent antioxidant food. Therefore, if these plants are consumed daily, they could scavenge access free radicals in the human biological system and could prevent oxidative related diseases. Antioxidant food supplies the body with the essential antioxidant nutrients needed to enhance the immune system, to eliminate excess free radicals and to keep the oxidative stress state in balance. In a similar study, it was found that the consumption of *Morinda citrifolia* L. (Rubiaceae family), locally known as (mengkudu), may have potential health effects. All the examined fractions, demonstrated high antioxidant activity. The potency of some of these compounds could provide a scientific basis for the health benefits claimed for *M. citrifolia* in folk medicine. Some of the Korean medicinal plants and their isolated polyphenols have exhibited effective free radicals scavenging activity and may act as promising agents for scavenging free radicals and treating diseases associated with excess free radicals, *e.g.*, the Rosaceae family plants, *i.e.*, *Rosa rugosa* and *Rosa davurica*, showed strong radical-scavenging activity. The most effective medicinal plant from families other than Rosaceae was *Cedrela sinensis* (Meliaceae family), followed in order by *Nelumbo nucifera* of Nelumbonaceae family, *Eucommia ulmoides* of Eucommiaceae family, *Zanthoxylum piperitum* of Rutaceae family, *Cudrania tricuspidata* of Moraceae family and *Houttuynia cordata* of Saururaceae family.

Anti-Lipid Peroxidation and Immunomodulatory Effects of Antioxidants

An oxidant is any substance when present at low concentration significantly delays or prevents oxidation of cell content like proteins, lipids, carbohydrates and DNA. Antioxidants can be classified into three main types, *i.e.*, first line defense antioxidants, second line defense antioxidants and third line defence antioxidants. (SOD), Super oxide dismutase (SOD), catalase (CAT), Ascorbate peroxidase (APX), Glutathione Peroxidases (GPX), glutathione reductase (GR), Dehydroascorbate reductase (DHAR), Monodehydroascorbate reductase (MDHAR) and some minerals like selenium, manganese, copper and zinc come under first line defence antioxidants. Super oxide dismutase mainly acts by quenching of super oxide (O_2^-) and catalase hydrogen peroxide (H_2O_2) to water and oxygen. Glutathione peroxide is selenium containing enzyme, which catalyses the reduction of hydrogen peroxide and lipid hydroperoxide, generated using reduced glutathione as substrate. Selenium and vitamin E both appear to be necessary for efficient scavenging of peroxides from cytosol and cell membrane, respectively. Copper exerts its antioxidant activity through cytosolic super oxide dismutase.

Zinc is an element essential for normal growth, reproduction and other different functions of the body. It is a component of several enzymes like cytosolic superoxide dismutase, alcohol dehydrogenase, alkaline phosphatase, carbonic anhydrase, *etc.* Glutathione (GSH), vitamin C, uric acid, albumin, bilirubin, vitamin E (mainly α- tocopherol), carotenoids, flavonoid, *etc.*, come under second line defence antioxidants. Beta-carotene is an excellent scavenger of single oxygen. Vitamin C interacts directly with radicals like O_2^- and HO (hydroxyl). Glutathione is a good scavenger of many free radicals like O_2^-, $\bar{HO}$ and various lipid hydroperoxides and many help to detoxify many inhaled oxidizing air pollutants like ozone, NO_2 and free radicals in cigarette smoke in the respiratory tract. Vitamin E scavenges peroxyl radical intermediates in lipid peroxidation and is responsible for protecting polyunsaturated fatty acid (PUFA) present in cell membrane and low-density lipoprotein (LDL), against lipid peroxidation. Flavonoids are phenolic compounds, present in several plants, which inhibit lipid peroxidation and lipoxygenases.

The most important chain breaking antioxidant is α-tocopherol, present in human membranes. Vitamin C and α-tocopherol both help to minimize the consequences of lipid peroxidation in membranes. A major antioxidant defense of human body is to prevent O_2 and H_2O_2 from reacting to form dangerous species such as hydroxyl ions, by binding transition metal ions in forms that will not stimulate free radical reactions. Thus, safe sequestration of iron and copper ions into forms that will not catalyze free radical reactions is an important antioxidant strategy in the human body. Third line antioxidants are a complex group of enzymes for the repair of damaged DNA, damaged protein, oxidized lipids and peroxides and peroxyl lipid radical. These enzymes repair the damage to biomolecules and reconstitute the damaged cell membrane, *e.g.*, lipase, protease, DNA repair enzymes, transferase, methionine sulphoxide reductase, *etc.*

In *Ayurveda*, formulation of some rasayanas with well-defined antioxidant properties has been done. Rasayanas are a group of non-toxic polyherbal drug preparation, which are immuno-stimulatory, and thereby, they prevent the causation of diseases and promote health and longevity. It is reported that when the balance between reactive oxygen species production and antioxidant defenses is lost, the oxidative stress results, which through a series of events deregulates the cellular functions and leads to various pathological condition, *viz.* acquired immunodeficiency syndrome (AIDS), ageing, arthritis, asthma, atherosclerosis, cardiovascular dysfunction, cataract, diabetes, gastro-duodenal pathogenesis, genetic disorders, inflammatory diseases, ischemic reperfusion injury, liver disorder, muscular dystrophy, neurodegenerative diseases, Parkinson's dementia, Alzheimer's disease, amyotrophic lateral sclerosis, pulmonary fibrosis, radiation damage, retinopathy, rheumatism, skin disease porphyries and senile dementia stroke.

As an example for the indirect antioxidant effect of plants, the antioxidant potential in herbal barks extract of five therapeutically important medicinal plants native to India was investigated. They are *Crataeva nurvala* of Capparidaceae family, *Buchanania lanzan* (Anacardium) and *A. marmelos* of Rutaceae family, *Dalbergia sissoo* of Fabaceae family and *Cedrela toona* of Meliaceae family. All of them showed an excellent lipid peroxidation inhibitory potential and a comparative high NO quenching capacity, which was found to be the highest in *Crataeva nurvala*. Hence, the bark of *Crataeva nurvula* has the highest antioxidant capacity. Spices and herbs, which are abundant in the daily meals of Asian countries, have been proved to possess high antioxidant activity that can be applied for preservation of lipids and reduce lipid peroxidation in biological systems. The antioxidant activity of spice extracts are retained even after boiling for 30 minute at 100°C, indicating that the spice constituents are resistant to thermal denaturation. The antioxidant activity of these dietary spices suggests that in addition to imparting flavour to the food, they possess potential health benefits by inhibiting the lipid peroxidation. Some of the dietary constituents commonly used in Indian food articles such as cloves (*Syzygium aromaticum* of Myrtaceae family), licorice (*Glycyrrhiza glabra* of Fabaceae family), mace (*Myristica fragrans* of Myristicaceae family) and greater cardamom (*Amomum subulatum* of Zingiberaceae family) have significant ability to inhibit lipid peroxidation due to their polyphenol content, strong reducing power and super oxide radical scavenging activity. Cloves showed the highest antioxidant activity probably due to the higher polyphenol content as compared to other spices.

The water and alcoholic extract (1:1) of commonly used spices (garlic, ginger, onion, mint, cloves, cinnamon and pepper) inhibits the oxidation of fatty acid and linoleic acid in the presence of soybean lipoxygenase. Among the spices tested, cloves exhibited the highest effect, while onion showed the least antioxidant activity. The relative antioxidant activities decreased in the order of cloves, cinnamon, pepper, ginger, garlic, mint and onion. Spice mix namely ginger, onion and garlic, onion and ginger and ginger and garlic showed cumulative inhibition of lipid peroxidation, thus, exhibiting their synergistic antioxidant activity. Currently, research interest has focused on various herbs that possess hypolipidemic, antiplatelet, antitumor, or immune-stimulating properties that may be useful adjuncts in helping to reduce the risk of cardiovascular disease and cancer. In different herbs, a wide variety of active phytochemicals, including flavonoids, terpenoids, lignans, sulfides, polyphenolics, carotenoids, coumarins, saponins, plant sterols, curcumins and phthalides have been identified. In addition to delivering antioxidant and other properties, herbs and spices can be used in recipes to partially or wholly replace less desirable ingredients such as salt, sugar and added saturated fat.

Since several metabolic diseases and age-related degenerative disorders are closely associated with oxidative processes in the body, the use of herbs and spices as a source of antioxidants to combat oxidation warrants further attention. In functioning processes of the immune system, such as phagocytosis, the reactive oxygen and nitrogen species are generated. The generation of reactive oxygen species by phagocytes is one of the irreplaceable microbicidal tools of innate immunity. If they are left unchecked, they can affect the components of immune system by inducing oxidative damage. This is more so in the elderly or during inflammation where there is excess generation of these reactive species than can be taken care of by the defenses in the form of antioxidants. Dietary supplementation with antioxidants may greatly help in such conditions. There are some indications of possible benefits of antioxidant supplementation. Polyphenols have well-known antioxidant-based immunomodulatory activities, which can be as therapeutic agents in the inflammation-driven damaging oxidant load.

A Brief Description of Common Antioxidant Plants

A long list of antioxidant plants has been given in Table 1. Screening of plants is done by measuring the antioxidant activity through various *in vitro* models like DPPH (2,2-Diphenyl-1-Picryl hydrazyl) method, nitric oxide method, ABTS method, ORAC method, TBARS assay, *etc.* and *via* various *in vivo* method using rats or mice.

Common name: Neem

Botanical name: *Azadirachta indica* A. Juss

Family: Meliaceae

Bark is bitter, astringent, acrid, refrigerant, depurative, antiperiodic, vulnerary, demulcent, insecticidal, antiseptic, ophthalmic, stomachic, anathematic, and tonic. It is useful in colic, dyspepsia, skin diseases, eczema, lecoderma, intermittent fever, ulcers, burning sensation of tumour, earache, headache, malaria, filarial, intestinal worm and blood sugar lowering action. It is also used in leprocy, skin disease, blood pressure, tooth problems. Few drops of neem seed oil is applied on genitals before sexual intercourse for contraception. The *Azadirachta indica* extract of obtained from ethanol showed 50.48±0.03% at 200 (μg/ml) with showing IC_{50} value of 0.080 μg/ml. The values are also comparable with commercial antioxidant L-ascorbic acid (90.0%).

Common name: Chinese ginger

Botanical name: *Boesenbergia rotunda*

Family: Zingiberaceae

Boesenbergia rotunda exhibits antibacterial, antifungal, anti-inflammatory, analgesic, antipyretic, antispasmodic, antitumor and insecticidal activities. The rhizome of *B. rotunda* is generally used as a culinary spice in Thailand and also has been used for the treatment of oral diseases (dry mouth), stomach discomfort, stomach pain, leucorrhoea, diuretic, dysentery and inflammation. The rhizomes are used in traditional medicine as antiseptic and for the treatment of stomachache, diarrhea, dermatitis, dry cough and mouth ulcers gastrointestinal disorders and post-natal treatment. DPPH free radical scavenging activity in *Boesenbergia rotunda* rhizome was 4.29±0.55, which expressed as mg ascorbic acid equivalent antioxidant capacity (AEAC) in 1 g dry sample.

Common name: Golden shower, Indian laburnum, golden rain

Botanical name: *Cassia fistula* L.

Family: Fabaceae

It is useful in skin diseases, leprosy, tuberculosis glands, syphilis, cures burning sensation, snake bite, cures feeling of heat in body, washing septic wounds, blindness and flatulence, skin diseases rheumatism, laxative, antiperiodic, heals ulcers, erysipelas a menorrhagia, allay swelling due to cold and cows in intestinal problem. It is also used in swelling of neck of cattle, ripening banana, tuberculosis, cough, flatulence, common cold laxative, scabies, swelling of testis, lecoderma, symptom, common cold cough. The DPPH assay showed the scavenging effects of all extracts of *Cassia fistula* (75.15% at 1.0 mg/ml for the flower and 70.37% for the pod, at the same concentration). The RSA values were also remarkably good for stem (61.27% at 1.0 mg/ml) and leaves (62.69% at 1.0 mg/ml), at 1.0 mg/ml of extract tested.

Common name: Wild turmeric

Botanical name: *Curcuma aromatica* Salisb

Family: Zingiberaceae

Curcuma aromatica, also called Kasturi haldi in Oriya and Hindi, has long been used as a medicinal plant for relieving pain and inflammation and for the treatment of cancer and lung diseases in India and as a spice for contributing flavor in India. The rhizome of *C. aromatica* has better beneficial physiologic functions but it is used as an ointment or a powder. In addition, the oily components of *C. aromatica* have shown antifungal activity, anti-inflammatory activity and antitumor and pharmacologic effects. The hydro ethanolic extract of *curcuma aromatica* was found to have highest reducing power (3.717±0.065 µg/ml) when compared to standard ascorbic acid (3.5±0.0007 µg/ml).

Common name: Crown Flower, Blue Madar, or Swallow wort

Botanical name: *Calotropis gigantean* (Linn) R.Br.

Family: Apocynaceae

It is cardiac stimulant, blood purifier, febrifuge, alternative, antiperiodic, antispasmodic, diaphoretic, digestive, emetic, expectorant, sedative, stomachic, suppurative, sharpness, hearing, removes filarial, elephantiasis, abdominal pain, leprosy, anasarca, asthma, cold cough, dysentery, eczema, ear sore, enlargement of spleen, eye trouble, piles, toothache and worms. It is used for expelling thorn, toothache, epilepsy, toothache, pneumonia, nasal sore, rheumatism, hernia, prisonous, insect bite. *Calotropis gigantea* shows moderate antioxidant activity when compared with standard antioxidant L-ascorbic acid whose antioxidant activity at different concentrations like 400, 600 and 800 μg were 84, 88 and 94%, respectively and the antioxidant activity of L-ascorbic acid at 800 μL (positive control).

Common name: Veldt Grape or Devil's Backbone

Botanical name: *Cissus quadrangularis L.*

Family: Vitaceae

The stem pulps of *Cissus quadrangularis*, a commonly consumed diet in India, have been used for fracture healing, eye diseases, chronic ulcer, tumors, asthma and piles. At higher dose of CQE (1000 mg/ml) the percentage of inhibition on DPPH radical and super oxide radicals was increased by 75.6 and 75.3%, while catechin showed an inhibition of 79.3 and 77.1%, respectively.

Common name: Kolinji in Tamil, Vempali in Telugu

Botanical name: *Galega purpurea*

Family: Papilionaceae

Galega purpurea plant thrives in Southern parts of India and grows on hard stony ground too difficult to be rooted. Various parts of the plant are widely used in the folk medicine for the treatment of cough, asthma, bilious febrile attacks, arthritis and rheumatism. Decoction of the root useful in then management of enlargement and obstruction of the liver, spleen and kidney. The root is useful in the treatment of dyspepsia, chronic diarrhea and ulcers. The methanolic extract of *Galega purpurea* root exhibited 71.32% at 125 μg/mL where standard drug α-tocopherol at the same concentration exhibited 89.17% inhibition.

Common name: Anantamula

Botanical name: *Hemidesmus indicus* R.Br.

Family: Asclepiadaceae

Hemidesmus indicus is useful in constitutional disorder, diarrhoea, gonorrhoea, skin diseases, agalactia, dysentery, fever, malaria, vomiting easier delivery, excess menstruation and itching skin. It is useful in general debility, syphilis, stomatitis, wound healing and piles. The methanolic extract of *H. indicus* at 100 μg/ml showed strong radical scavenging activity with percentage decrease of 77.0%. The values are also comparable with commercial antioxidant L-ascorbic acid (90.0%).

Common name: Physic nut

Botanical name: *Jatropha gossypifolia* Linn.

Family: Euphorbiaceae

It is an emmenagogue, emetic purgative useful in boils, carbuncles, itches, eczema, soreson tongue, anaemia, gonorrhoea, fixing loose teeth, burns, cuts, dysentery, sores on feet, snakebite, jaundice, wounds, as a haemostatic and forms a protecting film. It is used in rheumatism, gonorrhoea and wounds caused by burns. The antioxidant activity of ethyl acetate fraction of *Jatropha gossypifolia* showed 15.6±0.9%. Results are comparable with the standards (gallic acid, BHT and tocopherol).

Common name: Henna or Neutral henna

Botanical name: *Lawsonia inermis* L.

Family: Lythraceae

It is antifertile and cures burning sensation, headache, leucorrhoea, jaundice, sprain, rheumatism, menorrhogia, ulcers of mouth, spermatorrhoea and weakness. It is used in eye infection, eczema, white leprosy, dhobis itch, abortion, constipation of cattle. The leaf extract showed 67.67±5.48% of antioxidant activity.

Common name: Harsinghar

Botanical name: *Nyctanthus arbortistis* Linn

Family: Nyctanthaceae

It is useful in obstinate fever, rheumatism, eye pain, worm infection, expelling of chick, phlegm, inflammation of boils, tonic for hair, breast tumour, anthelmintic, vermifuge, in veterinary disease, blood purification, malaria, skin eruption caused by spider urine, cures baldness and is used as tooth brush. The scavenging effect of plant extracts and standard (ascorbic acid and BHT) on the DPPH radical decreases in the following order: ascorbic acid > Butanol > Ethyl acetate > BHT > Petroleum ether and ascorbic acid was found to be 93.88% at concentration of 10 mg. BHT, Butanol, Ethyl acetate and Pet ether was found to be 97.42, 95.22 84.63 and 82.04% at concentration of 100 mg, respectively. In the present study, different extracts of *Nyctanthes arbor-tristis* leaves showed concentration dependent free radical scavenging activity.

Common name: Wild leadwort

Botanical name: *Plumbago zeylianica* L.

Family: Plumbaginaceae

It is abortificient, antiseptic, appetizer, antifertility and diuretic, useful in diarrhoea, dysentery, dyspepsia, eczema, fever, headache, indigestion, leprosy, leucoderma, malaria, muscular pain, paralysis, piles, skin diseases, spleen complaint and wombs. It is used as abortificient and wound healing. The methanolic extract of *P. zeylanica* at 100 µg/ml showed strong radical scavenging activity with 73.41%. The values are also comparable with commercial antioxidant L-ascorbic acid (90.0%).

Common name: Casteor plant or Palma christi

Botanical name: *Ricinus communis* Linn.

Family: Euphorbiaceae

It is heating, carminative, useful in inflammations, pains, ascites, fever, glands, asthma, bronchitis, leprosy, rectum diseases, burns and headache, night blindness, earache, increases biliousness, rheumatism, cures wound, used in boils, used as a febrifuge, stomachache, lever diseases, abdominal pains, headache, sunstroke. It is also used in eczemas, bone fracture, sciatica, foot crack, pimples, headache, obesity and leprosy. The antioxidant activity of ethyl acetate fraction of *Ricinus communis* showed 93.2±0.7%. Results are comparable with the standards (gallic acid, BHT and tocopherol).

Common name: Katuk, Star gooseberry or Sweet leaf

Botanical name: *Sauropus androgynus* L. Merr.

Family: Meliaceae

A decoction from the roots is said to relieve urinary disorders and treat fever. Leaves are given as vegetables to nursing mothers to stimulate breast milk production. Leaves are also known to help the womb recover after giving birth. The roots and leaves are sometimes used as medicine. The leaves and roots are used to relieve fever and treat urinary problems. The juice from its leaves is dropped into the ear as a remedy for earache. The ethanolic fraction of *Sauropus androgynus* leaves showed 62.90±0.03% of antioxidant with IC_{50} value 0.016 µg/ml. The values are also comparable with commercial antioxidant L-ascorbic acid (90.0%).

Common name: Beleric, Bastard myrobalan

Botanical name: *Terminalia bellerica* (Gaertn) Roxb.

Family: Combretaceae

It is useful in diarrhoea, eye diseases, scorpion bites, apathae, ascitis, asthma, bleeding of gum, flatulence, stomach complain, grey hair and insanity. It is astringent, antidropsical, antibilious, antileprotic, anti-inflammatory, stomachic, antiasthmatic, strangury and sore throat. The fruit extract of *Terminalia bellerica* showed 76.29±3.06% of antioxidant activity in DPPH.

Common name: Black Myrobalan or Chebulic Myrobalan

Botanical name: *Terminalia chebula* Retz.

Family: Combretaceae

It is astringent, antiparalytic, purgative, tonic, carminative, alterative, stomachic, febrifuge, laxative, antiasthmatic and antidysnteric. It strengthens brain and enriches blood. It is useful in piles, cold, opthalmia, sore throat, dental carries, bleeding, ulceration of gums, burns, scalds, spleen diseases, diuretic and cardiotonic. It is used in swelling of testicles, dhobis itch, pimples, swelling of dental roots and inflammation of children due to heavy dysentery. The fruits of *Terminalia chebula* showed 85.36±2.44% of antioxidant activity in DPPH assay.

Common name: Veldt-Grape, Winged Treebine

Botanical name: *Vitis quadrangularis*

Family: Vitaceae

The herb has two anti-carcinogenic compounds such as vincristine and vinblastine, which have been found to be effective in curing various types of cancer. *Vinca* relieves anxiety, thus, it has tranquilizing properties. It contains various alkaloids, which are sedative, hypotensive, tranquilizer and anti-cancerous. It is also effective in the treatment of diabetes as it has the property to lower down the blood sugar. *Vinca* lowers the blood pressure in high blood pressure patients. The herb has calming properties and also relieves from the depression like symptoms. It was clear that all the parts of *V. rosea* showed the highest antioxidant activity and followed the trend flower (97.44%) > stem (93.80) = root (93.84) > leaves (83.72) > seed (80.28), and the antioxidant activity of flower was higher than L-ascorbic acid at 800 μg.

Conclusion

The antioxidant components of the natural products constitute the major source of human health promotion and maintenance. Plant antioxidants are natural plant products that inhibit the adverse effects of the reactive oxygen species produced in plants, thus, antioxidants enable the plants to survive. Day by day, many researches have shown the potential of phytochemical antioxidants as health benefactors because of their ability to neutralize free radicals, reactive oxygen species, or oxidants responsible for the onset of cell damage. Taken together, it was concluded that antioxidant components of the natural products constitute the major source of human health promotion and maintenance. Moreover, antioxidant activity of the natural products includes detoxifying toxins, scavenging free radicals, wiping out excess ROS and anti-lipids peroxidation. Indirectly, antioxidant components are useful as anticancer, anti-inflammatory, antimicrobial, antilipids, antiallergic and antidiabetics. The main conclusion is that the nature is still the perfect source for health promotion and for the supplementation of safe drugs. Moreover, natural products and their antioxidant activity might represent the best solution to prevent the side effects of many commercially available drugs and to counteract the rapid emergence of multi-drug resistant microbes against the commonly used antimicrobial agents. Great attention is mandatory from all the scientists around the world to explore many underestimated plants with highly effective antioxidant activity. It is worth mentioning that synthetic antioxidants were found to be harmful to health, while most of the natural antioxidants from plant source proved to be safer for health and possess better antioxidant activity.

■■■

33 Fruits and Vegetables: A Potent Source of Phytochemicals

M. Masroor A. Khan and Tariq Aftab

Plant Physiology Section, Department of Botany, Aligarh Muslim University, Aligarh-202 002, Uttar Pradesh, India, E-mail: <masruur@gmail.com>

Today, science has a better understanding of why fruits and vegetables should be part of a healthy eating plan. One probably enjoys them for their wonderful flavour and bright colours but fruits and vegetables are good for health too. An eating pattern packed with plenty of fruits and vegetables includes 2 to 3 servings of fruits and 3 to 5 servings of vegetables each day. Fruits and vegetables provide important nutrients including antioxidants such as vitamin C, vitamin E, beta-carotene, lycopene, folic acid, ellagic acid and many more. Antioxidants have disease-fighting properties that protect cells from damage by free radicals. Antioxidants work by neutralizing free radicals that are formed when body cells burn oxygen for energy. They also help in keeping the immune system healthy and reduce the risk for cancer and other diseases. Antioxidant supplements may interfere with the killing of cancer cells if taken during chemotherapy or radiation treatment. Studies in humans are still needed to find out more about this. However, eating fruits and vegetables high in antioxidants is still considered safe during cancer treatment. There are a number of studies suggesting that the antioxidants in fruits and vegetables are the main factors for the observed efficacy of these food articles in reducing the incidence of chronic diseases including heart diseases and cancers. The free radical scavenging activity of antioxidants in foodstuffs has been substantially investigated and reported below:

Researches in the field of nutrition and chronic disease causation during last few decades have brought out exciting significant progress in providing an understanding of specific risk factors and chemo-preventive agents. The major health problems considered are cardiovascular diseases and the nutritionally linked cancers, including those in the stomach, colon, breast, prostate and ovary. The vegetables and fruits have been shown to inhibit the complex processes of initiation and development of these diseases. One aspect involved in initiation and development of both cardiovascular diseases and the cancers noted are abnormal oxidative processes leading to the generation of hydroxy radicals and peroxy compounds. In part, the protective role of vegetables and fruits is to provide antioxidant vitamins and specific polyphenols that display a powerful inhibition in oxidative reactions. Epidemiological studies as well as laboratory experimentation have produced sound evidence in support of the fact that specific antioxidants are accounted mechanistically for inhibition.

Among a variety of dietary antioxidants, polyphenolic compounds, vitamin E, vitamin C and carotenoids are believed to be effective nutrients in the prevention of oxidative stress-related diseases. Thus, antioxidants have become a topic of increasing interest recently. Previous epidemiological studies have consistently shown that consumption of fruits and vegetables has been associated with reduced risk of chronic diseases, such as cardiovascular diseases and cancers and neurodegenerative diseases, including Parkinson's and Alzheimer's diseases as well as inflammation and problems caused by cell and cutaneous aging. Most young people lead a sedentary life and make common nutritional errors, among which we highlight an insufficient consumption of fruits and vegetables. To encourage vegetable consumption through nutritional counseling among the young and elderly, it is important to bear in mind that not all vegetables have the same phenolic composition and not all phenolics have the same antioxidant capacity. Therefore, it is important to review the list of vegetables and fruits, which have the highest antioxidant capacity to recognize and introduce them regularly into the diet and to know how much antioxidant capacity the vegetables and fruits do have.

Nutraceuticals Found in Fruits and Vegetables

Although fruits and vegetables are the obligatory supplements of the diet but many people do not know what they are taking with their food. Here, we highlight different main active constituents, which provide with natural antioxidants, anti-ageing and anticancerous compounds present in fruits and vegetables.

Lycopene

Lycopene being a potent antioxidant has received a lot of attention from researchers in recent years. Lycopene is a pigment that gives vegetables and fruits, such as tomato, pink grapefruit and watermelon, their red colour. Lycopene is an open-chain unsaturated carotenoid that also appears to have strong antioxidant capabilities. Several studies suggest that consumption of food rich in lycopene is associated with a lower risk of prostate cancer and cardiovascular disease. The protective effect of antioxidants on heart disease has been well documented. Researchers have observed that the level of lycopene in body fat is an indicator of lycopene content in the diet. Lycopene is not produced in the body, thus, one can only obtain its benefits by eating foodstuffs rich in lycopene. Other fruits and vegetables such as watermelon and pink grapefruit also provide lycopene. According to research repots, watermelon contains 40% higher lycopene content than tomato.

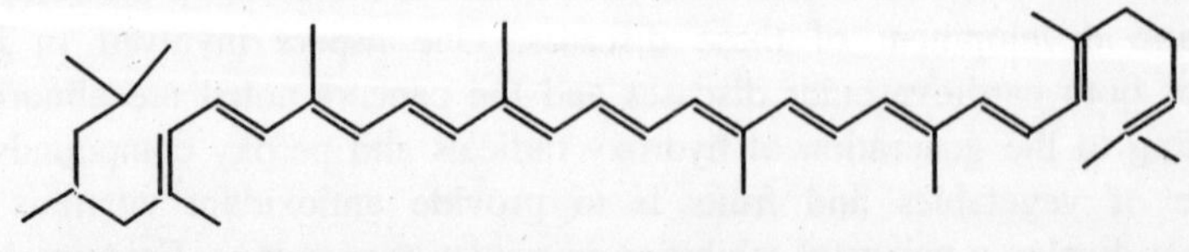

Several varieties of tomato were screened to find out which one is having greater amount of lycopene than the other varieties. Furthermore, lycopene content has been found to increase by the foliar application of gibberellic acid. Since tomato also contains vitamins, potassium, carotenoids and other antioxidants, other compounds in tomato, acting either alone or with lycopene, may be responsible for some of the protective effects attributed to lycopene.

Beta-carotene

Beta-carotene is a member of the carotenoid family, a group of powerful antioxidants that also includes α-carotene, lycopene, zeaxanthin and lutein. However, among the carotenoids, only α- and β-carotenes are converted to significant amount of vitamin A in the body, and β-carotene is by far the most plentiful carotenoid found in fruits and vegetables. β-carotene also contains flavonoids, which are antioxidant substances giving colour and flavour to orange and red-coloured fruits and vegetables. Food articles such as carrot, tomato, sweet potato, cantaloupe, watermelon, apricot and mango are good source of beta-carotene. Romaine lettuce, spinach, kale and broccoli are also rich in carotenoids. The epidemiological evidence of β-carotene appeared particularly promising, which was supported by experimental evidence and possible biological mechanisms, including its antioxidant activity.

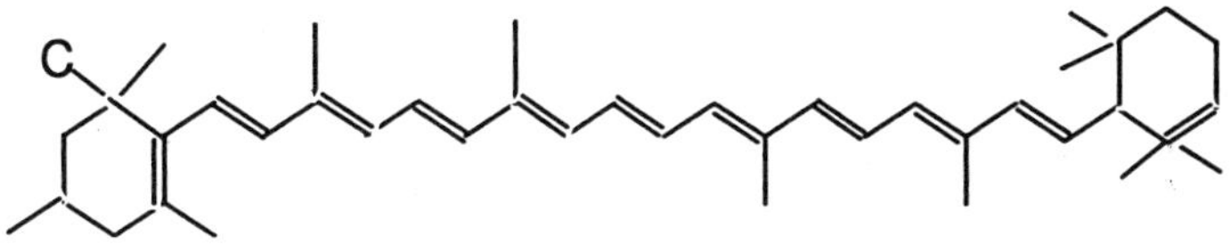

C

As far as oxygen radical absorbance capacity of β-carotene is concerned, it is a potent anti-aging phytonutrient of the highest order, standing firmly alongside vitamin C, vitamin E and mineral selenium in its ability to reduce illness, improve overall health and increase longevity. Studies conducted over several years indicate that beta-carotene plays a significant role in reducing the effects of certain illnesses such as heart disease, decreased immune function, cataracts and forms of cancer and in certain cases. Seven cultivars of carrot have been screened for their β-carotene content in which the cultivar Tropical Red was having more β-carotene content than other varieties.

Ascorbic Acid

Ascorbic acid is white to light-yellow crystals or powder and is water-soluble. One form of ascorbic acid is commonly known as vitamin C. The name ascorbic acid is derived from a- meaning no and scorbuticus meaning scurvy, which is a disease caused by the deficiency of vitamin C. Ascorbic acid, the accepted name for vitamin C, is available in reduced form (*L*-ascorbic acid) and oxidized form (*L*-dehydroascorbic acid). It is found in citrus fruits, green peppers, red peppers, strawberries, tomatoes, broccoli, Brussels sprouts, turnip and other leafy vegetables. Vitamin C is declined gradually as fruits tend to maturity.

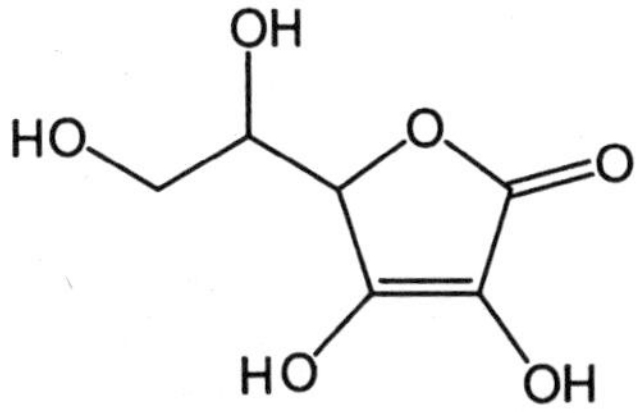

OH
HO O O
HO OH

Ascorbate acts as an antioxidant by being available for energetically favourable oxidation. Many oxidants (typically, reactive oxygen species), such as the hydroxyl radical (formed from hydrogen peroxide), contain an unpaired electron, and thus, are highly reactive and damaging to humans and plants at molecular level. This is due to their interaction with nucleic acid, proteins and lipids. Reactive oxygen species oxidize ascorbate first to monodehydroascorbate and then dehydroascorbate. The reactive oxygen species are reduced to water, while the oxidized forms of ascorbate are relatively stable and unreactive, and does not cause cellular damage.

Vitamin E

Vitamin E is a fat-soluble vitamin that exists in eight different forms. Each form has its own biological activity, which is the measure of potency or functional use in the body. It is a collective name for a set of related α-, β-, γ- and δ-tocopherols and the corresponding four tocotrienols, which are rich in antioxidant properties. Of these, α-tocopherol has been most studied since it has the highest bioavailability.

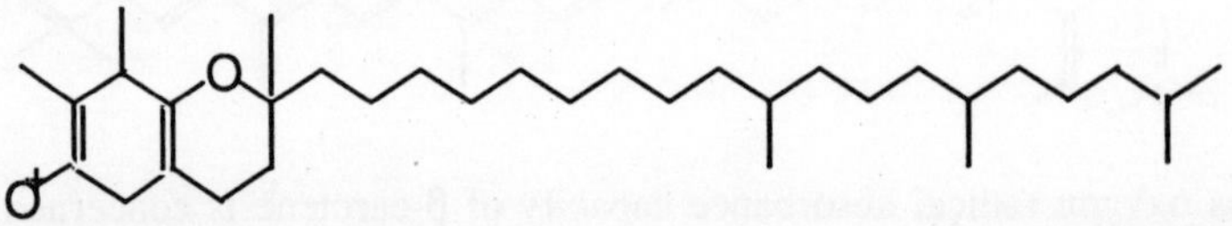

Like vitamin C, vitamin E plays a significant role as an antioxidant, thereby protecting body tissues from the damage of oxidation. It is important in the formation of red blood cells and the use of vitamin K. Many people also use it to help minimize the appearance of wrinkles, as it is valued for its ability to soothe and heal broken or stressed skin tissues. It is mostly found in blackberries, bananas, apples, almonds and peanuts.

Ellagic Acid

Ellagic acid is a polyphenol antioxidant found in numerous vegetables and fruits including raspberries, strawberries, cranberries, walnuts, pomegranates and other food articles of plant origin. The antiproliferative and antioxidant properties of ellagic acid have spurred preliminary research into the potential health benefits of ellagic acid consumption. Eating berries or other natural sources of ellagic acid is generally considered safe. These food items should be the part of a balanced diet that includes several servings of fruits and vegetables each day.

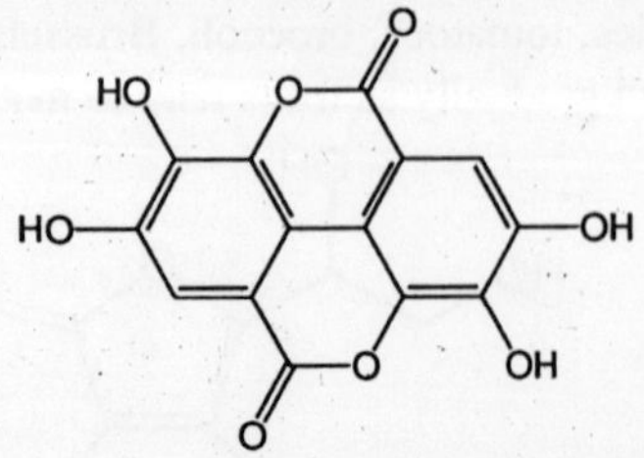

Ellagic acid has antiviral and antibacterial properties. Recent studies have indicated that ellagic acid may have anticarcinogenic effects against liver, esophageal, prostate and colorectal cancer cell lines. Ellagic acid obtained from dietary sources appears to be safe but its long-term toxicity effects have not been evaluated.

Folic Acid

Folic acid (also known as vitamin B_9 or folacin) and folate (the naturally occurring form) are forms of the water-soluble vitamin B_9. Vitamin B_9 (folic acid and folate inclusive) is essential for numerous bodily functions ranging from nucleotide synthesis to the remethylation of homocysteine. It is especially important during the periods of rapid cell division and growth. Both children and adults require folic acid to produce healthy red blood cells and prevent anemia.

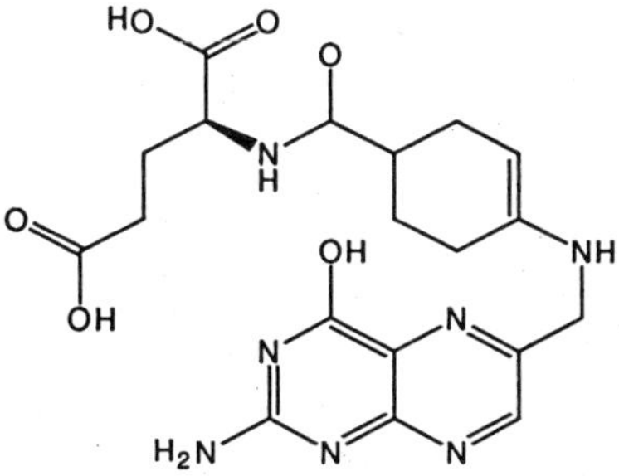

Leafy green vegetables are a good source of folic acid. Other folate rich vegetables are peas, corn, cauliflower, green pepper, beetroot, mustard greens and ladyfinger. This is especially important during periods of rapid cell division and growth such as infancy and pregnancy. Folate is needed to make DNA and RNA, which are the building blocks of cells. It also helps in preventing changes to DNA that may lead to cancer. Both adults and children need folate to make normal red blood cells and prevent anemia. Folate is also essential for the metabolism of homocysteine and help to maintain normal levels of this amino acid.

Flavonoids

Flavonoids are polyphenolic compounds that are ubiquitous in nature and are categorized, according to chemical structure, into flavonols, flavones, flavanones, isoflavones, catechins, anthocyanidins and chalcones. Over 4,000 flavonoids have been identified, many of which occur in fruits, vegetables and beverages (tea, coffee and fruit drinks). The flavonoids have aroused considerable interest recently because of their potential beneficial effects on human health. They have been reported to have antiviral, anti-allergic, antiplatelet, anti-inflammatory, antitumor and antioxidant activities. Like alpha-tocopherol (vitamin E), flavonoids contain chemical structural elements, which may be responsible for their antioxidant activities.

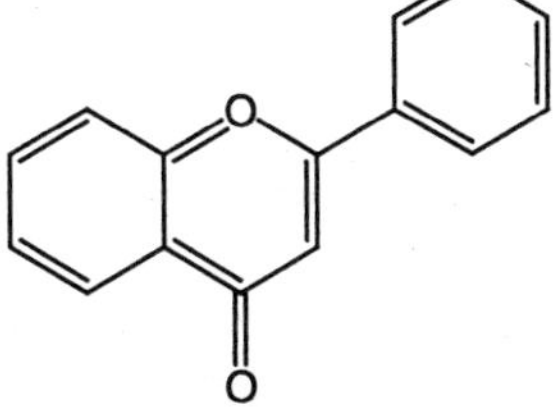

Flavonoids may help to provide protection against diseases by contributing, along with antioxidant vitamins and enzymes, to total antioxidant defense system of the human body.

Epidemiological studies have shown that flavonoid intake is inversely related to mortality from coronary heart disease and to the incidence of heart attacks. The capacity of flavonoids to act as antioxidants depends upon their molecular structure. The position of hydroxyl groups and other features in the chemical structure of flavonoids are important for their antioxidant and free radical scavenging activities. Quercetin, the most abundant dietary flavonol, is a potent antioxidant because it has all the right structural features for free radical scavenging activity.

Quercetin

Quercetin is a bioflavonoid (or flavonoid), which is a type of pigment found in almost all herbs, fruits and vegetables. Bioflavonoids provide the body with anti-inflammatory and antioxidant protection, and quercetin is one of the most powerful and effective herbal anti-inflammatory and antioxidant supplements in the market today. Indeed, quercetin has been shown to help in preventing the development of a variety of conditions related to inflammation and free-radical damage, including arthritis, allergies, macular degeneration, heart disease, gout and various forms of cancer.

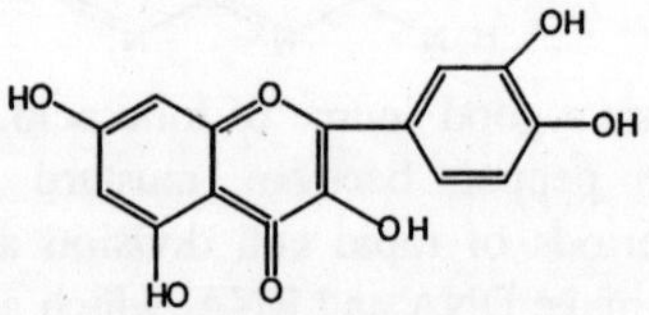

Quercetin occurs in food as an aglycone (attached to a sugar molecule). Only a small percentage of the ingested quercetin will get absorbed in the blood. It is found in many common foodstuffs including apple, tea, onion, nuts, berries, cauliflower and cabbage.

Anthocyanins

Anthocyanins (from the Greek word anthos meaning flower + kyanos meaning blue) are water-soluble vacuolar pigments that may appear red, purple, or blue according to pH. They belong to a parent class of molecules called flavonoids synthesized *via* the phenylpropanoid pathway. Anthocyanins occur in all tissues of higher plants, including leaves, stems, roots, flowers and fruits. Anthoxanthins are their clear white to yellow counterparts occurring in plants. Anthocyanins are derivatives of anthocyanidins, which include pendant sugars.

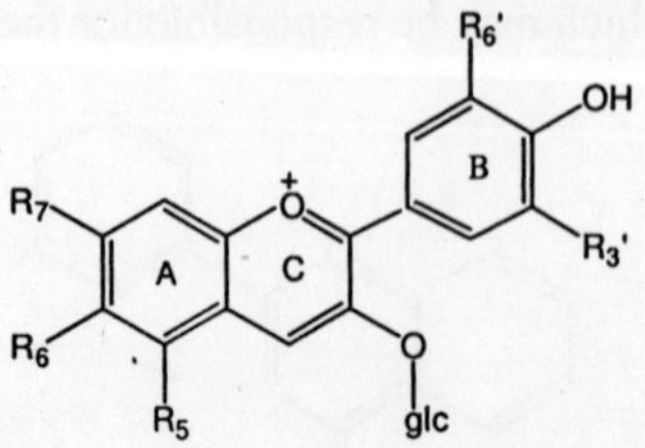

Anthocyanins are one of the largest and most important groups of water-soluble pigments in most species in the plant kingdom. They are accumulated in cell vacuoles and are largely responsible for diverse pigmentation from orange to red, purple and blue in flowers and fruits, such as blackberry, red and black raspberries, blueberries, bilberries, cherries, currants, blood orange, elderberries, grapes and vegetables including red onion, radish, red cabbage, red lettuce, egg-plant, red-skinned potato and purple sweet potato.

Allicin

Allicin is an organic compound obtained from garlic. It is also obtainable from onions, and other species in the family Alliaceae. This colourless liquid has a distinctively pungent smell due to the presence of SO group. This compound exhibits antibacterial and anti-fungal properties. Allicin is garlic's defence mechanism against attacks by pests. Allicin is synthesized from alliin when garlic is crushed or bruised. The reaction is catalyzed by an enzyme, alliinase and as a result quite complex. Allicin is a very reactive compound. It is slightly soluble in water and completely soluble in alcohol, which is what would be expected of an organic molecule containing OH. It decomposes on heating, and in the body, it oxidizes the haemoglobin in the blood to methemoglobin.

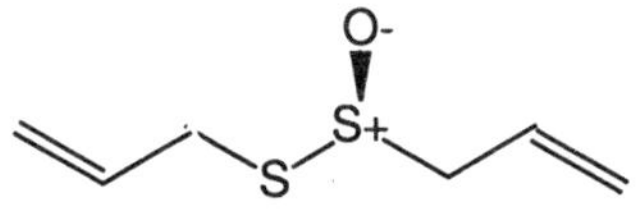

O-
S+
S

Curcumin

Turmeric has long been used in the Chinese and Ayurvedic systems of medicine as an anti-inflammatory agent among other things. Much of the observed activities of *Curcuma longa* seem to be due to curcumin. It is known to relieve pain and inflammation and is safe when given orally. Several studies have reported the antioxidant property of curcumin, augmenting endogenous antioxidant levels. It also down regulates nitric oxide formation. Curcumin is the principal curcuminoid of the popular Indian curry spice turmeric, which is a member of ginger family Zingiberaceae. The other two curcuminoids are desmethoxycurcumin and bis-desmethoxycurcumin. The curcuminoids are polyphenols and are responsible for the yellow colour of turmeric.

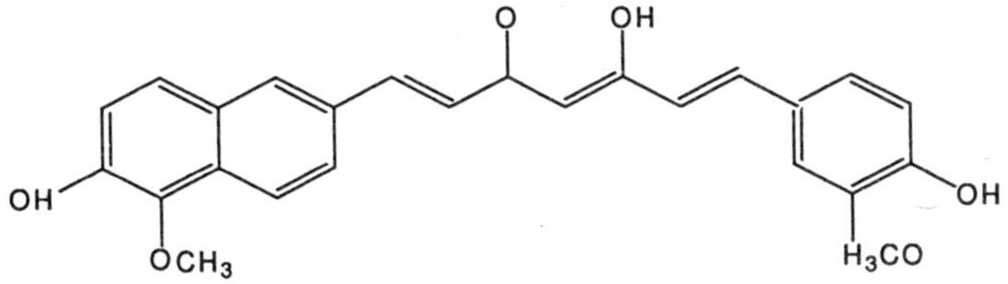

O OH
OH OH
OCH_3 H_3CO

Curcumin can exist in at least two tautomeric forms, *i.e.*, keto and enol. The enol form is more energetically stable in the solid phase and in solution. Curcumin incorporates several functional groups. The aromatic ring systems, which are polyphenols, are connected by two α,β-unsaturated carbonyl groups. The two carbonyl groups form a diketone. The diketone form stable enols or is easily deprotonated and form enolates, while the α, β-unsaturated carbonyl is a good Michael acceptor and undergoes nucleophilic addition. Agronomical studies also demonstrated that pre-harvest combined application of triacontanol and kinetin can ameliorate the yield and curcumin content of turmeric.

Lutein

Lutein is a member of carotenoid family, a group of powerful antioxidants. Lutein is a xanthophyll, which is a type of carotenoid found in high concentrations in dark-green leafy vegetables. Like other well-known carotenoids α- and β-carotene, lutein is a powerful antioxidant that gives foods a red, orange, or yellow colour. Like β-carotene, lutein is an antioxidant that protects our cells against damage caused by dangerous, naturally occurring chemicals known as free radicals. However, unlike α- and β-carotene, lutein cannot be converted by the body to vitamin A. All dark-green leafy vegetables, especially kale, turnip greens, collard greens and spinach, are good sources of lutein. It is also found in fruits and vegetables, such as tomatoes, carrots, corn, and squash, which have red, orange, or yellow pigments. Consuming any of these foodstuffs can also provide with added antioxidant protection and health benefits, as they are generally high in other carotenoids as well.

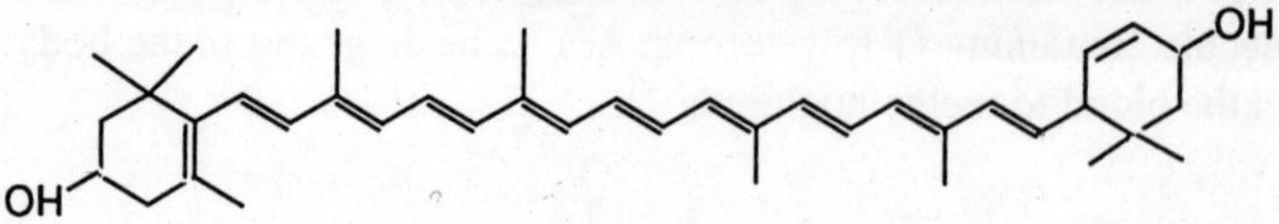

OH

OH

Limonene

Limonene is a hydrocarbon, classified as a cyclic terpene. It is a colourless liquid at room temperature with an extremely strong smell of orange. It takes its name from the lemon, as the rind of lemon like other citrus fruits contains considerable amount of this chemical compound, which is responsible for much of their smell. Limonene is found in essential oils of citrus fruits and many other plant species. Industrial limonene is produced by alkali extraction of citrus residues and steam distillation. It is reported that limonene has anti-cancerous effects since it increases the levels of liver enzymes involved in the detoxification of carcinogens. The glutathione S-transferase (GST) is a system, which eliminates carcinogens. Limonene seems to promote the GST system in the liver and small bowel, thereby, decreasing the damaging effects of carcinogens. Animal studies demonstrated that dietary limonene reduced mammary tumour growth.

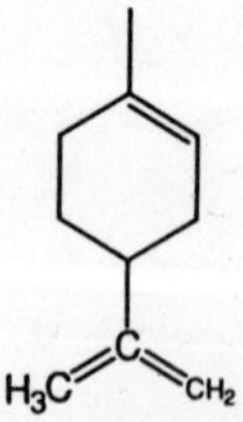

H_3C C CH_2

Daidzein

Daidzein belongs to the group of isoflavones and has the typical C_6-C_3-C_6 structure. Isoflavones are also called phytoestrogens because they have some estrogen activity. Daidzein is the aglycone (molecule without sugar) of glycoside daidzin. Other well-known isoflavones are genistein and glycitein. Daidzein has both weak estrogenic and weak anti-estrogenic effects. Daizein has also antioxidant activity. Epidemiological studies have long shown that people who consume a lot of soybean have reduces incidences of prostate cancers. This benefit of soybean could be explained by the anti-cancer and antioxidant activity of daidzein. Daidzein seems to reduce the risk of osteoporosis. Studies have shown that daidzein stimulates the formation of osteoblasts, which are the cells that produce bone mass.

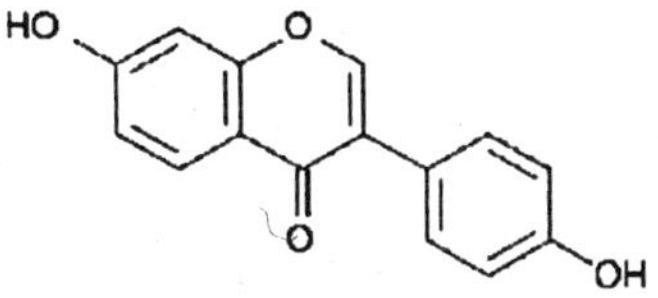

Genistein

Genistein is one of the most abundant isoflavones in soybean. Isoflavones belong to the group of flavonoids. Because of its similar structure to that of human estrogen, it is also called a phytoestrogen, together with daidzein. Genistein is derived from the hydrolysis of glycoside genistein. The main source of genistein is soybean. Other legumes, such as chickpeas, contain small amounts of genistein.

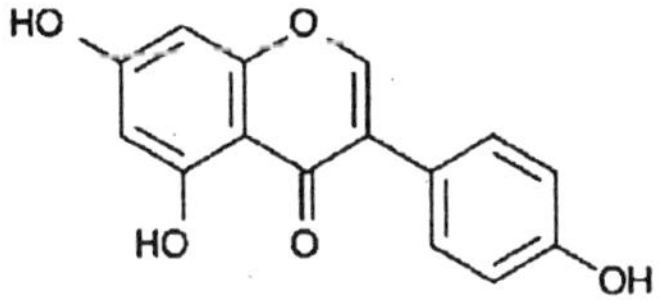

Genistein has many health benefits and works on several fronts. It acts as a phytoestrogens, antioxidant, anti-cancer agent and it could help the people with metabolic syndrome. The estrogenic activity of genistein has been confirmed in many studies. Among isoflavones, genistein has the strongest estrogenic activity. The estrogenic effect of genistein may explain its protective action against osteoporosis and its possible effect on body weight reductions. Genistein seems to reduce the risk of some hormone related cancers, principally breast and prostate cancer. Epidemiological studies show that consumption of isoflavones may protect against breast and prostate cancer.

Resveratrol

Resveratrol is a flavonol belonging to the group of flavonoids. It is produced by the plant as a defence against diseases. Resveratrol is present in many plants and fruits, including red grapes, eucalyptus, spruce, blueberries, mulberries, peanuts and giant knotweed. In addition, red wine contains a lot of it. The longer the grape juice is fermented with the grape skins, higher will be the resveratrol content. *In vitro* studies have shown that resveratrol inhibits the oxidative damage caused by heavy metal cadmium. The antioxidant activity of resveratrol reduces damage to endothelial cells exposed to nitrite radicals and protects skin cells against damage caused by ultraviolet radiation.

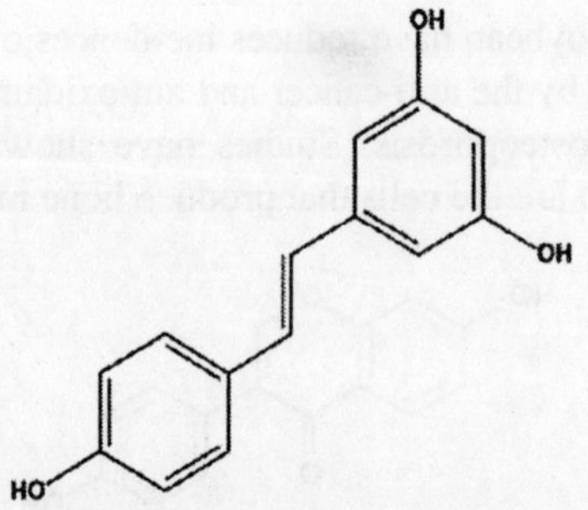

OH

OH

HO

The antioxidant action of resveratrol helps in preventing damage to DNA but it also influences the transcriptions of genes responsible for redox metabolism and inhibits proliferation of cancer cells. Resveratrol appears to decrease tumor promotion activity by inhibiting the enzyme cyclooxygenase-1, which converts arachidonic acid to substances that promote tumour growth. In vitro experiments provide support for resveratrol to serve as a candidate preventive agent against prostate cancer but *in vivo* effects of resveratrol and the mechanisms of action of resveratrol on prostate cancer prevention remain largely unknown.

Beta-sitosterol

Beta-sitosterol is a phytosterols or plant sterol. The structure of beta-sitosterol is similar to that of cholesterol. Beta-sitosterol differs from cholesterol by the presence of an extra ethyl group. There are many plant sources of beta-sitosterol but the most important are wheat germ, rice bran, flax seeds, peanuts, soybeans, pumpkin seeds and corn oil. Beta-sitosterol is mainly known and used for its cholesterol lowering property but studies have shown that the phytochemical may have other health benefits, *i.e.*, easing symptoms of benign prostatic enlargement, reducing risk of cancer and prevention of oxidative damage through its antioxidant activity. Epidemiological and experimental studies have suggested a protective role of beta-sitosterol in the development of some types of cancer such as breast, colon and prostate cancer. *In vivo* studies have shown that the phytochemical inhibited proliferation and induced apoptosis in human solid tumours such as colon and breast cancers.

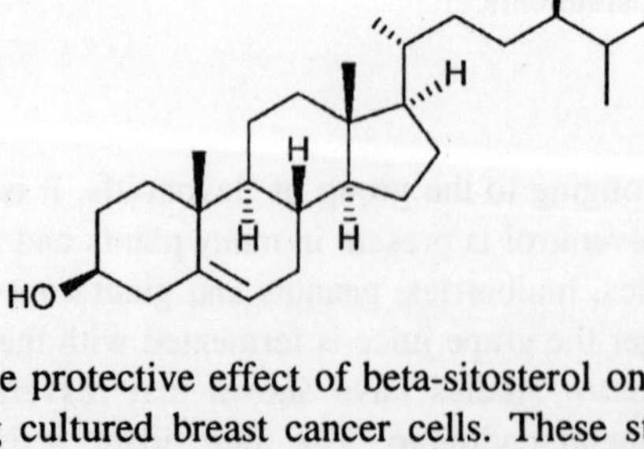

H

H

H H

HO

The studies about the protective effect of beta-sitosterol on breast cancer only involved *in vitro* experiments using cultured breast cancer cells. These studies clearly show that the phytochemicals kill breast cancer cells and are not toxic to normal cell. Clinical studies linking beta-sitosterol and breast cancer are still missing but some scientists suggest that it may improve the efficiency of tamoxifen, a drug often used to treat breast cancer.

Astaxanthin

Astaxanthin, which is similar in structure than beta-carotene, is a red carotenoid pigment. The small differences in structure cause large differences in the chemical properties. Astaxanthin is produced by microscopic small plants, *i.e.*, the micro-alga *Haematococcus pluvialis*. *Haematococcus* algae can contain up to 30 g of astaxanthin per kg of dried algae. Astaxanthin is a powerful antioxidant. The free radical scavenging activity of astaxanthin protects the lipids from peroxidation and reduces oxidative damage of LDL-cholesterol (thereby, reducing arterial plaque formation), cells, cell membranes and mitochondrial membranes. Astaxanthin increases strength and endurance. Astaxanthin seems to improve the immune system by increasing the number of antibody producing cells. Astaxanthin enhances antibody production by exerting actions on T-cells and T-helper cells. Astaxanthin is used to treat neurodegenerative conditions such as Alzheimer's and Parkinson's disease. Astaxanthin protects the eyes and skin from sun radiation damage by quenching singlet and triplet oxygen. Studies with rats show that astaxanthin reduces retinal injury. Studies have shown the anti-cancer effects of astaxanthin in rodents.

Conclusion

Fruits and vegetables provide us with a range of compounds, many of which have more than one role, being involved both with immediate good health and with protection against diseases, such as cancer, heart conditions, stroke and diabetes, that develop over a long period. They are very important to our health because they are whole foodstuffs, created by nature, that are rich in a large amount of nutrients. The processed food artiles that we so commonly eat, can never compare to the health benefits provided by strawberries or broccoli, which have fiber, vitamins and enzymes built right in. Fruits and vegetables help to protect us from illnesses that take time to develop. What we eat now will affect our health in twenty years time, the time taken for some illnesses such as heart disease and cancer to develop. Fruits and vegetables are sources of a range of antioxidants, the so-called ACE vitamins (vitamins A, C and E), which help to prevent DNA damage. Lycopene found in tomatoes and the compound lutein found in broccoli, spinach and red peppers are also potent antioxidants.

Other protective compounds from fruits and vegetables include carotenoids, flavonoids, isothiocyanates and other substances that are often grouped together as phytochemicals or nutraceuticals. They simply mean plant chemicals and contribute to the vivid, bright colours found in fruits and vegetables, which have a profound effect on our health. Researchers are currently studying hundreds of these phytochemicals. Antioxidant substances, in their unique natural location and natural combination with other nutrients, contained in fruits and vegetables, scavenge free radicals and protect cells from excessive DNA damage. They prevent the blue print of the body from being damaged.

By eating fruits and vegetables of a variety of different colours, one can get the best all-around health benefits. The phrase eating a rainbow of fruits and vegetables is a simple way of remembering to get as much colour variety in the diet as possible, so that one can maximize his intake of a broad range of nutrients. The colour of fruits and vegetables is a small clue as to what vitamins and nutrients are included. By getting a variety of different coloured fruits and vegetables, you are guaranteed a diverse amount of essential vitamins and minerals.

■■■

34 Role of Dietary Antioxidants in Human Health

Gayatri Nahak[1], Vandana Mathur[2] and R.K. Sahu[1]

[1]B.J.B. Autonomous College, Bhubaneswar, Orrisa, India
[2]54-Devnagar, Jodhpur-342008, Rajasthan, India
E-mail: <gayatri_bioteq@yahoo.co.in>, <vandanam22@yahoo.co.in>, <sahurajani.sahu@gmail.com>

Certain fruits such as apple and vegetables like brinjal, on cutting and exposure to air turn brown. This browning has been attributed to air oxidation of phenolic substances, present in most of the plant materials. The enzyme called polyphenoloxidase is present in plants, and on exposure to air, these are activated and initiate air oxidation and polymerization of phenolic compounds producing brown colour. Similarly, freshly extracted juice of certain fruits turns brown on exposure to air, developing an unpleasant taste. Oils and fats, particularly those rich in polyunsaturated fatty acids, on exposure to air undergo oxidative decomposition, becoming viscous and giving foul smell. This is referred as rancidity of fats, which makes the fat unfit for human consumption. In case of deep-fried foodstuffs, *e.g.*, potato chips and other *namkeens*, the air contact is increased due to large surface area. The storage life of these foodstuffs is poor and such foodstuffs develop staled odour and taste on exposure to air. Hence, these food articles must be stored in tightly closed containers and preferably be packed under nitrogen. To a certain extent, these processes can be minimized by the use of suitable antioxidant.

Normally, it is presumed that molecular oxygen (at ground energy level or in triplet stage), *i.e.*, the form in which it is present in the air, at a concentration of about 21% by weight, is harmless for most living things (man, animals and plants). This is largely true. Principally, an air oxidation of plant phenolics prevents harmful oxidation of more vital constituents of living organisms, *i.e.*, tissues. It is a common observation that those parts of plants, which are exposed to strong sunshine, are coated with thick and dark brown bark, containing oxidatively polymerized phenolics or lignins. This is a defense mechanism of plants against air oxidation of its more vital components.

Synthetic Antioxidants

Some synthetic organic compounds, *e.g.*, butylated hydroxytoluene (BHT), butylated hydroxyanisole (BHA), propyl gallate and other synthetic phenolic compounds, when added at a very low concentration to a fruit juice or edible oil, can stop their air oxidation, preventing the development of brown colour followed by unpleasant odour and rancidity, respectively.

Such substances are referred as *synthetic antioxidants*. There has been a controversy regarding the use of synthetic antioxidants in food since some food scientists and medical personals do not regard them as completely safe and non-toxic food additives. However, by long-term experiments, it is now amply proved that at permitted concentration (< 0.1%), these antioxidants are completely harmless to human beings and protect against harmful air oxidation of food. In fact, natural occurrence of phenolic antioxidants in plants like rosemary and gum guaiac (nordihydroguaiarctic acid or NDGA) has shown that these antioxidants act as natural-defense in plant system against air oxidation under adverse conditions.

In this article, we are only concerned with antioxidants, which are naturally present in food or synthetic antioxidants used as food additives. It may, however, be mentioned that apparently harmless air under ambient conditions can also cause oxidative damage to certain industrial commodities, *e.g.*, petroleum based fuels, lubricants and rubber goods, which are protected against deterioration by additives that are antioxidants.

Naturally Occurring Antioxidants

Vitamin C (ascorbic acid), vitamin A (retinal), vitamin E (tocopherol) and melatonin (hormone) occurring in plant or animal tissues are known to act as antioxidant for the plants in which these occur and for the animals or human being, using such plants for food. Antioxidant nature of vitamin C can be realized from the fact that apple juice turns brown in air, whereas, lemon and orange juice retains the original colour due to the presence of natural antioxidant, *i.e.*, vitamin C. Similarly, vegetable oils containing vitamin E do not easily become rancid. Sulfur compounds, *e.g.*, α-lipoic acid and those present in garlic and onions (thiol and mercaptans) are also antioxidants.

Most important group of naturally occurring antioxidants is of phenolic compounds. Thick and dark coloured bark of tropical trees, *e.g.*, *kher*, *babool* and *neem* have bioflavonoids and their polymerized products tannins in their bark, which are act as antioxidants, protect the plant against air oxidation (oxygen), which gets activated by ultraviolet-radiation present in bright and harsh sunlight. Brightly coloured flower and fruits, *e.g.*, red and black grapes, blackberries, blackcurrant and beetroot also contain anthocynins and flavones, which are phenolic substances with antioxidant properties. Seed coat of many plants, *e.g.*, wheat bran, oilseed and legume husk are rich in antioxidants (tocopherols as well as bioflaonoids), which protect them from air oxidation, when they need to be stored in contact with air, for a long period till used as food or germinated to grow into plant.

Human body has its own antioxidant enzymes, some of which have metal ions, *e.g.*, copper, zinc, manganese, selenium and iron in their prosthetic groups. These enzymes, acting as redox systems, mop up free radicals under normal conditions. Additionally, the dietary antioxidants are transported through blood to various tissues and organs of the body, where they act as antioxidants. Human food should contain both water-soluble and lipid soluble antioxidants, which are transported through blood and stored in appropriate tissues in the body.

Free Radicals as Reactive Intermediates in Oxidative Degradation

Respiratory oxygen generates free radicals in a metabolic process in various parts of body, where they are mopped up by antioxidant enzymes and with dietary antioxidant augmenting the action of antioxidant enzymes. Highly reactive type of atoms or fragments of organic molecules, which are called free radicals, are the most common intermediates in harmful air oxidation of plants or human tissues. Most of the normal molecules constituting human body, and our surroundings have electrons, constituting the chemical bonds, in paired form and in lower energy level. Unlike these, free radicals have at least one unpaired electron. A free radical is formed when an electron pair bond, in a molecule breaks in such a way that each of the fragments of the parent molecule carries an unpaired electron of original electron pair bond. Chemists call it *hemolytic fission* of a covalent bond. This is shown in the following equation:

$$A : B \rightarrow A\cdot + B\cdot$$

Free radicals have a tendency to seek an electron from an electron pair bond of another molecule. In this process, it creates a new free radical and starts a chain reaction. Termination of a free radical chain reaction can result by disproportion reaction or by mutual combination of two free radicals, and sometimes, due to the formation of a less reactive, or a non-reactive radical. Molecular oxygen is paramagnetic and it can be regarded as weak *bis* radical. On activation, it becomes an active free radical. Thus, molecular oxygen on reacting with a fat molecule at the site of unsaturation produces peroxide or a hydroperoxide, which on decomposition sets up a free radical chain reaction. An antioxidant can quickly terminate such chain reaction. The reactivity of an antioxidant is measured in terms of *oxygen radical absorbing capacity* (ORAC).

The mechanism of air oxidation of a lipid, at the site of unsaturartion and that of an antioxidant, disrupting a chain reaction are shown below:

$$A + O_2 \rightarrow AO_2 \text{ and}$$

$$AO_2 + B \rightarrow AO + BO \rightarrow A + B + O_2$$

Here, the substrate 'A', reacting with oxygen molecule, forms adduct ($AO_{2'}$), which is a peroxide or a hydroperoxide of a lipid. This is followed by reaction of the antioxidant 'B', which acting as a catalyst terminates the chain reaction and regenerates molecular oxygen.

In absence of an antioxidant, the homolytic fission of peroxide (-O:O-) generates free radicals and sets up a chain reaction. One antioxidant molecule is capable of neutralizing millions of free radicals, which explain their affectivity, even when present in traces.

Normal human metabolic processes produce oxygen radicals. It has been realized that limited amount of free radicals produced in the body, controls bacterial infection by destroying them in the human body. However, if produced in large excess, they can damage cells and tissues of the body and damage the molecules in a cell wall, making it permeable to toxic substances and pathogens. Damage of DNA in cell nucleus can result in cell mutation and disease such as cancer.

Oxidative Stress

The body is normally under a dynamic equilibrium between free radical generation and quenching. The physiological defense systems to counteract free radicals encompass endogenous enzyme systems, such as catalase, glutathione reductase and superoxide dismutase, as well as glutathione, urate and coenzyme Q, or exogenous factors (β-carotene, vitamin C, vitamin E and selenium). All these molecules have an antioxidant effect due to their ability to transform reactive oxygen species into stable and harmless compounds or by scavenging both reactive oxygen species and RNS with a redox-based mechanism. Very recently, a main role in the fight against oxidative stress has been assumed by enzymes such as heme oxygenase (HO) and biliverdin reductase (BVR). Heme oxygenase is a microsomal enzyme, which metabolizes heme into ferrous iron, carbon monoxide and biliverdin (BV). The latter is then reduced by biliverdin reductase into bilirubin (BR), a molecule endowed with strong antioxidant and antinitrosative activities. Interestingly, all these protective factors act in a concerted way, enhancing the antioxidant defense system of the cell. When the balance between ROS/RNS and antioxidants turns in favour of the former, oxidative or nitrosative stress occurs. Although oxidative stress is associated with most diseases but routine assay methods are not now a day available in the clinical practice. A strategy widely used to determine oxidative stress is measurement of malonyldialdehyde, F2-isoprostanes, or 8-hydroxydesoxyguanosine. Actually, these molecules are regarded as the most reliable markers available. A classic example of an oxidation product apparently leading to disease, is oxidized cholesterol in low-density lipoprotein (LDL), which displays a higher atherogenic potential than native LDL and mainly involved in the pathogenesis of atherosclerosis and coronary heart disease (CHD).

At cellular level, a large body of data clearly demonstrated that reactive oxygen species, when produced in low amounts and in a controlled manner, are physiological components of the signaling generated by cytokines, growth factors and neurotrophic peptides, although they may also activate apoptotic cell death. Extracellularly generated reactive oxygen species can diffuse through anion channels into the cytoplasm, and the resulting variation in the cell redox state leads to modulation of an array of transcription factors (*e.g.*, NF-kB, AP-1), protein kinases (*e.g.*, AKT, JNK, p38) and receptor activated MAP kinases involved in apoptosis. Moreover, the proapoptotic molecules Fas and Fas ligand (FasL) undergo positive transcriptional regulation after exposure to oxidants. Interestingly, Krammer and Colleagues demonstrated that *in vitro* administration of vitamin E suppresses FasL mRNA expression and protects T-cells of HIV-1 infected individuals from Fas mediated apoptosis. Moreover, it was demonstrated that administration of combinations of vitamin E and C to cultures of human umbilical vein endothelial cells (HUVEC) treated with lipopolysaccharide could prevent apoptosis by upregulation of *Bcl-2.*

Dietary Antioxidants

Dietary antioxidants are normally present in human food and these include vitamins A, C and E besides carotenoids, *e.g.*, β-carotene present in carrot, yellow fruits (pumpkin, mangoes and papaya) and green vegetables. Carotenoids, which are precursor of vitamin A, are even more effective as antioxidant than vitamin A. These are also present as the red pigment (lycopene) in tomatoes and red peppers. Vitamin C is present in citrus fruits and green vegetables. Plant pigments other than green and yellow are phenolic compound, which are collectively known as bioflavonoids, *i.e.*, other than chlorophyll and carotenoids, which are antioxidants, widely distributed in plants such as tea, coffee, cocoa, gambier leaves, marigold flowers and kutch. Herbal tea and herbal infusion have been used in traditional systems of medicine (*Ayuervedic* and *Yunani* decoction). Their healing power has been attributed to the presence of antioxidants, thus, providing a scientific basis for their action. Vegetable oils, *e.g.*, sunflower oil, peanut oil and soybean oil are rich in vitamin E. Trace metal ions, *e.g.*, selenium, copper, iron, manganese and zinc found in shellfish and nuts, also act as antioxidants and can neutralize harmful effect of free radicals.

Just how important is to supplement natural dietary antioxidants, is a subject of debate, because often the claims are exaggerated. However, one thing is certain that vegetarian food is a better source of antioxidants, and more and more persons in the Western Countries are now adapting to vegetarian food. Common sources of dietary antioxidant are shown in Table 1.

Table 1 : Antioxidants and their source

Antioxidant type	Source
Antioxidant Vitamins	
Vitamin C (ascorbic acid)	Citrus fruits, green vegetables, strawberries, black currants and guavas
Vitamin A (retinal)	Milk products, eggs, meat, fish and lever
Vitamin E	Vegetable oils (soybean, peanut, sunflower), nuts, peanuts, whole grains, legumes and almonds
Caratoniods (β-carotene and lycopene)	Yellow fruits (mangoes, papaya, pumpkin), carrots, red peppers and tomatoes
Bioflavanoid	Fruits, lemon, cherries, blackberries, blackcurrants, tea, coffee and cocoa
Antioxidant Minerals	
Selenium	Shellfish, fish, herring, cod fillet, dairy products and avocado
Manganese	Plant food
Iron	Pigmented fruits, meat, egg and sea food
Copper	Lever, shellfish, coca, nuts, mushrooms and vegetables
Zinc	Grains, meats, fish, oyster, eggs, diary products and vegetable

Functions of Food Preservatives and Antioxidants

Packaged food contains added preservatives (bisulphite, benzoic acid, sorbates, *etc.*) to increase their shelf life. These preservatives are effective against microbial spoilage of food, but when they come in contact of air, food is also spoiled due to air oxidation. The later is purely a chemical process. Toxicity of packaged food can result in a type of food poisoning known as botulism, and antioxidants are added to packaged food to guard against such infection. Some food processors prefer the use of natural antioxidants, which can be herbal extracts. Herbal extracts can be economical yet safer than synthetic antioxidants. Thus, the solvent extracted pigments from tamarind seed coat contain antioxidant suitable for preventing rancidity of oils and improving shelf life of fried food items. Bioflavnoids form marigold flower and catechu, *i.e.*, catchin present in *katha*, contains antioxidants suitable for use in food. Rice and corn bran oils are rich source of antioxidant, *i.e.*, vitamin E. It has been mentioned that ultraviolet-radiation can activate oxygen molecule (singlet stage), which can damage skin tissues and cause skin disease (cancer) on long exposure to sun. To prevent it, cosmetic companies use vitamin E and other antioxidants in skin-care products.

Dietary food fibers, which are water-soluble non-starch polysaccharides, also act as antioxidants. These are depolymerized by free radicals into harmless products (sugars and lower polysaccharides). Thus, they act as free radical scavengers or antioxidants. Water-soluble as well as fat-soluble antioxidants are known and appropriate choice of an antioxidant should be made, based on the nature of food.

The Different Types of Antioxidants

Most people have heard the term *antioxidant* but the likelihood is that a vast majority of these people do not actually know what the term truly refers to the fact that there are two different types of them. In a scientific sense, the difference between the two types of antioxidant is actually quite significant, but to an average person, the difference may not be that impressive.

Activity of the Water Soluble Antioxidants

Water-soluble antioxidants are referred to as hydrophilic antioxidants. Basically, they are able to assist the body in the process of cell cytosol and help out in the blood plasma. In other words, they take a hand on approach to ridding the body of harmful free radicals and pollutants. Following is a list of the most common water-soluble antioxidants:

- Ascorbi Acid
- Gluathione
- Lipoic Acid
- Uric Acid

As you can see, most of these are acids. Their concentration in the human serum and liver varies to a large extend, and some of these antioxidants are certainly more powerful. As you would expect, any food, which contains any four of the above antioxidants in large quantities, is definitely beneficial to help in eradicating the harmful oxidants from the human body.

Activity of the Lipid Soluble Antioxidants

Unlike water-soluble antioxidants, the lipid soluble versions do not actively go out seeking to destroy rogue cells in the human body. These are the antioxidants, which in fact have a much more passive role in keeping the human body healthy.

Antioxidant Nutrients

Antioxidants from our diet appear to be of great importance in controlling damage by free radicals. Each nutrient is unique in terms of its structure and antioxidant function.

Vitamin E (tocopherol): This is actually a generic term that refers to all entities (eight found so far) that exhibit biological activity of the isomer tocopherol (an isomer is one of two or more molecules that have the same chemical formula but different atomic arrangements). Alpha-tocopherol, the most widely available isomer, has the highest biopotency or strongest effect in the body. Because it is fat-soluble (can *only* dissolve in fats), α-tocopherol is in a unique position to safeguard cell membranes- largely composed of fatty acids- from damage by free radicals. Alpha-tocopherol also protects the fats in low-density lipoproteins (LDLs, or the *bad* cholesterol) from oxidation.

Vitamin C (ascorbic acid): This is a water-soluble vitamin. As such, it scavenges free radicals that are in an aqueous (watery) environment, such as inside cells. Vitamin C works synergistically with vitamin E to quench free radicals. Vitamin C also regenerates the reduced (stable) form of vitamin E.

Vitamin A (β-carotene): This fat-soluble vitamin is the most widely studied of the 600 carotenoids identified to date. It is thought to be the best quencher of singlet oxygen (an energized but uncharged form of oxygen that is toxic to cells). Beta-carotene is also especially excellent at scavenging free radicals in low oxygen concentration.

Selenium: This is a trace mineral-element, which is needed only in very small quantity but without which one cannot survive. This mineral forms the active site of several antioxidant enzymes including glutathione peroxidase.

Similar to selenium, the minerals manganese and zinc are trace elements that form an essential part of various antioxidant enzymes.

Antioxidant Enzymes

The antioxidant enzymes superoxide dismutase (SOD), catalase (CAT) and glutathione peroxidase (GPx) serve as your primary line of defense in destroying free radicals.

Superoxide dismutase first reduces (adds an electron to) the radical super oxide (O_2^-) to form hydrogen peroxide (H_2O_2) and oxygen (O_2).

$$2O_2^- + 2H - SOD -> H_2O_2 + O_2$$

Catalase and GPx then work simultaneously with the protein glutathione to reduce hydrogen peroxide and ultimately produce water (H_2O).

$$2H_2O_2 - CAT -> H_2O + O_2$$

$$H_2O_2 + 2\text{ glutathione} - GPx -> \text{oxidized glutathione} + 2H_2O$$

The oxidized glutathione is then reduced by another antioxidant enzyme, *i.e.*, glutathione reductase.

Together, they repair oxidized DNA, degrade oxidized protein and destroy the oxidized lipids (fat-like substances that are a constituent of cell membranes). Various other enzymes act as a secondary antioxidant defense mechanism to protect the body from further damage.

Other Antioxidants

In addition to enzymes, vitamins and minerals, there appear to be many other nutrients and compounds that have antioxidant properties. Among them is *coenzyme* Q_{10} (CoQ_{10}, or ubiquinone), which is essential to energy production and can protect the body from destructive free radicals. Besides, *uric acid*, a product of DNA metabolism, has become increasingly recognized as an important antioxidant. Additionally, the substances in plants called *phytochemicals* are being investigated for their antioxidant activity and health-promoting potential.

Infections and Diseases Caused by Free Radicals

Common diseases arising due to the harmful action of free radicals are emphysema, lungs inflammation, atherosclerosis, osteoarthritis, immune-deficiency, high blood pressure, joint problem, skin-brown patches and cancer. Factors responsible for excessive formation of free-radical are human metabolic processes, over exposure to high energy radiation, *e.g.*, radio-active rays, ultra-violet rays, X-rays, bright sun light, exposure to environment polluted due to smoke and smog, cigarette smoke, industrial gases and reactive industrial chemicals, industrial effluent and also due to poor nutrition and intense exercise. Blood cholesterol is oxidized by free radicals and this oxidized form of cholesterol is more damaging to arteries than the *native cholesterol.*

Certain cooking methods, *e.g.*, deep-frying and frequent reuse of cooking oil, excessive charring of meat products during roasting, grilling and barbecuing, produce potential precursor of free radicals and cancer. Left unchecked, these unstable and potentially harmful chemicals create conditions for infection and reduce immunity. Research has shown that a diet naturally rich or supplemented with antioxidants can provide protection against above-mentioned diseases and prevent premature aging.

How do Antioxidants Act to Neutralize Harmful Free Radical?

Some of the antioxidants act as sacrificial molecules, *i.e.*, they preferentially react with free radicals, thereby getting destroyed and producing harmless byproducts but protecting vital constituents of the body. Such molecules are themselves reducing agents, *e.g.*, vitamin C and plant pigments, *e.g.*, anthocynins and flavones or the bioflavonoids. In other cases, particularly certain metal ions acting as antioxidants are constituents of antioxidant metal-enzymes, which are naturally present in the body. These metal ions can exist in a variable oxidation state. After neutralizing free radicals by being oxidized to a higher oxidation state, these are reconverted, by redox systems of the body, into the original lower oxidation-state. Thus, these act as catalysts and react in a continuous cycle. Vitamin C is also regenerated by body redox system and acts as free radical scavenger.

Moderate amounts of tea, coffee, cocoa and chocolate, which contain antioxidant bioflavonoids, are good for health. Chewing of betel leaf can be harmful due to tobacco, but catechu (katha) used in it contains a bioflavonoid catechein, which is an effective antioxidant.

Effects of Antioxidants on Free Radicals

Fortunately, nature has an answer to free radicals in the form of antioxidants, which can stop the chain reaction of ever-increasing free radical destruction by neutralizing free radicals. They do this by donating one of their electrons to the free radical. This spares the DNA and other elements within cells from losing their electrons to free radicals, thus, it prevents cellular damage and mutations, which cause disease.

Because antioxidants themselves become oxidized in this process, the body needs a constant source of antioxidants.

Antioxidants also help in preventing the formation of oxidative chain reactions. Antioxidant enzymes like superoxide dismutase, catalase and glutathione peroxidase prevent oxidation by reducing the rate at which oxidative chain reactions are initiated. Thus, instead of waiting free radicals to create a long chain of free radicals, antioxidants scavenge initiating radicals and destroy them before oxidation is set in motion.

Whilst the body produces some antioxidants naturally, the actual levels of antioxidants we get depend on the diet. We should eat at least five servings of fruits and vegetables per day. Bright coloured produce like tomatoes, watermelon, red peppers, kale, broccoli, spinach, berries and citrus fruits are good source of antioxidants.

Antioxidants can also be found in coffee and tea, especially in green and white varieties. Dark chocolate is also a source of antioxidants but it should be consumed in moderate amount. Even red wine contains antioxidants but more than one glass of wine per day can negate the beneficial effects.

Antioxidants, the Immune System and Related Disorders

The protective function against external pathogens carried out by the immune system is by itself a source of reactive oxygen species, since activated neutrophils, produce free radicals to a significant extent. Moreover, during the inflammatory process, activation of phagocytes through the interaction of pro-inflammatory mediators, or bacterial products with specific receptors result in the assembly of multi-component flavor-protein NADPH oxidase, which catalyzes the production of large quantities of super oxide anion radical (O_2^-). In addition to classical reactive oxygen metabolites, activated neutrophils and monocytes release the hemoprotein myeloperoxidase (MPO) into the extracellular space, where it catalyzes the oxidation of Cl^- by H_2O_2 to yield hypochlorous acid (HClO), which is a non-specific oxidizing and chlorinating agent that reacts rapidly with a variety of biological compounds, such as sulphydryls, polyunsaturated fatty acids, DNA, pyridine nucleotides, aliphatic and aromatic amino acids and nitrogen-containing compounds. Moreover, apart from their direct toxic effects, neutrophil-derived oxidants may promote tissue injury indirectly by altering the protease/antiprotease equilibrium that normally exists within the intestinal interstitium. The oxidative inactivation of important protease inhibitors, coupled to the oxidant-mediated activation of latent proteases, creates a favourable environment for neutrophils that allows degradation of interstitial matrix through elastases, collagenases and gelatinases, as well as injury to epithelial cells. However, not only immune cell produce reactive oxygen species necessary for the microbicidal activity but they are also sensitive to external reactive oxygen species, due to their high polyunsaturated fatty acids (PUFA) content.

Immune cells are atypical, as compared with other somatic cells, in which they contain high levels of antioxidant vitamins, presumably providing protection against lipid peroxidation and immuno-suppression, both of which are well known risks posed by high polyunsaturated fatty acids content. The reactivity of immune cells to exogenous reactive oxygen species has been shown to be age-dependent. In fact, lymphocytes from elderly individuals appear to be more sensitive to exposure to hydrogen peroxide than lymphocytes from young adults. Moreover, it has been demonstrated that a micronutrients deficiency can be the cause of suppression of immune function affecting both innate T-cell-mediated immune response and adaptive antibody response, thus, altering the balanced host response. Therefore, an adequate intake of vitamins and antioxidant elements seems to be essential for an efficient function of immune system. Micronutrients deficiency occurs in various conditions, such as eating disorders, tobacco smokers, chronic diseases and aging. During aging, changes in immune system are frequent and associated with increased susceptibility to infections. Antioxidant vitamins and trace elements contribute to maintain an effective immune response, *e.g.*, administration of vitamin E supplement to healthy elderly patients produced an increased antibody titer to both hepatitis B and tetanus vaccine, and thus, enhancing T-cell mediated functions. In conclusion, maintaining adequate antioxidant status may provide a useful approach in attenuating cell injury and dysfunction observed in some inflammatory or autoimmune disorders.

Autoimmunity has been for decades considered the result of a breakdown in self-tolerance. At the present, it is known that autoimmunity is a physiological process. This phenomenon becomes pathological when the number of auto-reactive cells and particularly the avidity of their receptors for auto-antigens, increases. Triggering of the disease usually depends both on the increase in immunogenicity of the target cell, which may be secondary to a viral infection (Chediak-Higashi syndrome and Griscelli syndrome by EBV), and on the individual's own capacity to recognize the auto-antigens [HLA or T cell repertoire in Familial hemophagocytic lymphohistiocytosis (FHL)]. Moreover, apart from the genetic defects that may predispose to autoimmune diseases, one must take into account the environmental factors that are implicated in the development of such pathologies. Among them, an important role is played by xenobiotics such as chemicals, drugs and metals. Iron, aluminum and manganese readily cross the blood brain barrier *via* specific or non-specific carriers and contribute to nervous tissue damage. The toxic effects of metals are mediated through free radical formation or enzyme inhibition. In addition, metals may act as immuno-suppressants (cytostatically), or as immuno-adjuvants (through non-specific activation of immune response). Several mechanisms are proposed on how metals may act within the immune system to induce autoimmunity.

Patients suffering from scleroderma develop auto-antigens with metal-binding sites. After metal binding, free radical species are generated which fragment auto-antigens, thereby exposing cryptic epitopes, which may then trigger autoimmunity. Taken together, these findings underlie the importance of exogenous factors in the pathogenesis of autoimmunity. Nevertheless, not all these elements appear sufficient to provoke *chronic* autoimmune diseases such as multiple sclerosis (MS), myasthenia gravis, insulin dependent diabetes mellitus (IDDM), or Hashimoto's thyroiditis, and the passage to chronic disease is usually secondary to a defect in immuno-regulation.

Several classes of regulatory T-cells, such as Th2, $CD25^+$ and natural killer (NK) T-cells, are implied in autoimmune pathologies. In an animal model of a Th2-dominated autoimmune syndrome, the administration of antioxidant N-acetyl-cysteine (NAC) induced a decrease in mast-cell expression of both IgE and IL-4. Of major interest is the discovery of therapeutic potential of a new benzoquinone-containing product derived from wheat germ fermentation. This latter has been shown to have immune restorative properties since it affects the Th1/Th2 network by inhibiting the Th2 response. Another beneficial effect of this molecule is its anti-metastatic activity, shown in various human malignancies and Jurkat leukemia cell line.

Intriguingly, the combined treatment with wheat germ and vitamin C profoundly inhibited metastasis formation in various tumor models of different origin [Lewis lung carcinoma, B16 melanoma and human colon carcinoma xenografts (HCR25)]. On the contrary, wheat germ had no toxicity on peripheral blood leukocytes (PBLs) at doses that affected tumor cells. The crude powder extract of fermented wheat germ inhibits nucleic acid ribose synthesis primarily through the non-oxidative steps of the pentose cycle. Curiously, another quinone compound, carnosic acid quinone, like wheat germ, recovers potent antioxidant activity upon standing.

Keeping in mind the importance of oxidative stress in the regulation or deregulation of immune system, the use of antioxidants in such diseases has been reasonably proposed. Rheumatoid arthritis (RA) is a classic example of autoimmune disease. Joint inflammation in rheumatoid arthritis is characterized by invasion of T-cells in the synovial space and proliferation of activated macrophages and fibroblasts in the synovial intima. Therefore, in rheumatic joint, there is an increased activity of fibroblasts and leucocytes, which produce reactive oxygen species. Very recently, antioxidants have successfully been used as adjuvant therapy in rheumatoid arthritis. Although the results obtained with rheumatoid arthritis seemed to be very promising but the indiscriminate use of antioxidants in autoimmune disorders is not recommended. In fact, autoimmune lymphoproliferative syndrome (ALPS), multiple sclerosis, type 1 diabetes and multiple autoimmune syndrome, have been linked to decreased Fas functionality, and as discussed previously, antioxidants may up-regulate Fas and FasL *in vitro*. Increasing evidence provides support that oxidative stress and apoptosis are closely related physiological phenomena and are implicated in diseases including autoimmune diseases. Therefore, molecules that target both apoptosis-related signal transduction and oxidative stress, like antioxidants, are likely to result in the improvement of these pathologies.

A novel possible approach to modulate immune system, thus, preventing autoimmunity, or transplant rejection is the activation of cytoprotective and antioxidant enzymes such as HO^{-1}. Heme oxygenase-1, the inducible isoform of HO, is a key protein in the cell stress response and its up-regulation is a common event during pro-inflammatory conditions. Recent work clearly demonstrated that regulatory T-cells over express HO^{-1} and release CO under pro-oxidant conditions. Carbon monoxide may inhibit the proliferation of effect or T-cells, thus, reducing the immune response and prevent autoimmunity and/or graft reaction. Dietary antioxidants, in particular polyphenols, have been shown to increase HO^{-1} expression in different *in vitro* systems and the potential use of this natural substances to regulate immune response should be addressed carefully.

Antioxidants, Cancer and Neurodegenerative Disorders

It is well known that the dietary consumption of fruits, vegetables, herbs, or their phytochemical constituents aids in cancer prevention. It is believed that the antioxidant properties of such food articles protect the cells from reactive oxygen species-mediated DNA damage that can result in mutation and subsequent carcinogenesis. reactive oxygen species-induced DNA damage can take many forms, ranging from specifically oxidized purine and pyrimidine bases, to DNA lesions such as strand breaks, sister chromatid exchanges (SCEs) and the formation of micronuclei. However, the equation *antioxidant = benefit* is not always true. *In vivo* experiments demonstrated that retinol increases both the humoral and the cell-mediated immune response could enhance immune surveillance against tumorigenesis. Retinol may influence the immune response by quenching free radicals, which could lower the level of immuno-suppressing lipid peroxides, alter arachidonic acid metabolism, *etc.*

In last few years, many studies have been conducted to investigate the effects of vitamins on disease prevention. The first results have been encouraging and a wide number of people are taking antioxidant supplements with the aim to improve their health. These studies, initially, have shown that a high consumption of fruit and vegetables decreases risks of lung cancer in healthy individuals and a combination of β-carotene, vitamin E and selenium reduced stomach cancer mortality in China. Conversely, supplemental β-carotene alone or in combination with retinol or vitamin E did not have any effect on cancer risk, or increased the development of lung cancer in smokers. In the light of these first contrasting results and as a consequence of the wide antioxidant consumption in the general population, various systematic reviews to estimate the association between antioxidant use and disease prevention, in particular for primary cancer incidence and mortality, have been issued. These reviews share the opinion that antioxidant supplementation *per se* does not prevent cancer. On the contrary, some antioxidant elements seem to be harmful for health.

Recent studies have confirmed the relationship between beta-carotene and an increased incidence of cancer among smokers but not among non-smokers. Moreover, beta-carotene supplementation is associated with increased cancer-related mortality. Vitamin E treatment also appears to be associated with a slightly increased incidence of lung cancer. Other studies report that combination of vitamin A and other antioxidants, significantly increases mortality related to neoplastic diseases. According to these studies, selenium would be the only element displaying beneficial effects, as it has been shown that it reduces total cancer incidence, an apparently sex-related effect, as it is predominant among males rather than in females.

The reason, why β-carotene may exert dual activity, *i.e.*, antioxidant or pro-carcinogenic, has been debated for quite a long time. The first hypothesis is that at high concentrations, β-carotene stimulates free radical production, whereas, at lower concentration, β-carotene exerts antioxidant activity. Furthermore, in the presence of cigarette smoke-derived free radicals, β-carotene is cleaved into many derivatives, which are very unstable and may trigger further oxidation. A recent corollary to this theory is the evidence that β-carotene, either alone or in combination with cigarette smoke condensate, repressed HO^{-1} expression in both rat fibroblasts and human lung cancer cells. The reduced expression of HO^{-1} accounted for a reduced production of CO and bilirubin both of which have a marked antiproliferative effects. Vitamin E has also been shown to act at the immune system level. In fact, supplementation with this vitamin can increase production of antibodies and enhance cell-mediated immunity in both experimental animals and in humans.

Neurodegenerative diseases, such as Parkinson's disease (PD), Alzheimer's disease (AD) and amyotrophic lateral sclerosis (ALS), as well as multiple sclerosis (MS), are triggered, at least in part, by oxidative and nitrosative stress and also sustained by inflammatory cytokine production. Similarly, autoimmunity mainly contributes to the pathogenesis of multiple sclerosis, characterized by central and peripheral loss of nerve myelin. Though the specific sources of damaging ROS and the affected target structures differ between the neuronal pathologies, the following general features can be defined. Increased levels of oxidation-altered metabolites are found in post-mortem tissues in many of the neurodegenerative diseases listed before. An oxidative stress response and compensatory defense reactions can be seen in the affected neural cells. Further, disturbances of mitochondrial metabolism are observed, which may account for an increased leakage of reactive oxygen species originating from the respiratory chain. However, in addition to direct induction of oxidative stress, the metabolic disorders underlying every single disease can also indirectly generate an oxidative microenvironment, *e.g.*, *via* induction of a local immune response. On this basis, antioxidant and antinflammatory drugs, such as polyphenols and non-steroidal anti-inflammatory drugs (NSAIDs) have been proposed in the treatment of different neurodegenerative diseases. However, both polyphenols and NSAIDs gave rise to some problems when used in clinical setting. Due to their scarce bioavailability, only a negligible amount of polyphenols reaches brain tissue and the concentration achieved is much lower than that efficacious *in vitro*. As far as NSAIDs, *ad hoc* designed clinical trials with a large number of patients, clearly demonstrated that these drugs do not have any significant effect in slowing cognitive decline in patients suffering from mild-to-moderate AD. Similar disappointing results have been obtained in the treatment of amyotrophic lateral sclerosis, a systemic motor neuron disease that affects corticospinal and corticobulbar tracts, ventral horn motor neurons and motor cranial nerve nuclei.

Approximately 10% of the cases are familial and have been linked to point mutation in the gene encoding for copper or zinc super oxide dismutase (SOD). Mice transgenic for mutated SOD1 develop symptoms and pathologies similar to those in human amyotrophic lateral sclerosis. Mutant SOD1 toxicity is mediated by damage to mitochondria in motor neurons, and this may trigger the functional decline of motor neurons and the onset of amyotrophic lateral sclerosis in mice. Unfortunately, though the role played by free radical to the pathogenesis of amyotrophic lateral sclerosis has been demonstrated, antioxidants did not have any effect to prevent or slow down its progression. Desnuelle and his associates clearly demonstrated that alpha-tocopherol, given together with riluzole did not affect the survival and motor functions in amyotrophic lateral sclerosis patients in respect to the group treated with riluzole alone. Novel compound, such as AEOL-10150 (Aeolus), structurally related to the SOD catalytic site, is under phase-I clinical investigation but further clinical trials will be necessary to evaluate the real efficacy of this compound for the therapy of amyotrophic lateral sclerosis.

Cellular and Molecular Studies on the Potential Prevention Action in Heart Disease and Carcinogenesis

Reactive oxygen species (ROS) and the related oxidative damage have been implicated in the pathogenesis of various human chronic diseases. Lycopene is one of the most potent antioxidants and has been suggested to prevent carcinogenesis and atherogenesis by protecting critical biomolecules including lipids, low-density lipoproteins, proteins and DNA. Several studies have indicated that lycopene is an effective antioxidant and free radical scavenger. Lycopene, because of its high number of conjugated double bonds, exhibits higher singlet oxygen quenching ability compared to beta-carotene or alpha-tocopherol. In *in vitro* systems, lycopene was found to inactivate hydrogen peroxide and nitrogen dioxide. Using pulse radiolysis techniques, Mortesen and his associates demonstrated its ability to scavenge nitrogen dioxide ($NO_2^{\cdot}$), thiyl ($RS^{\cdot}$) and sulphonyl ($RSO_2^{\cdot}$) radicals.

Lycopene is highly lipophilic and is most commonly located within cell membranes and other lipid components. Therefore, it is expected that in the lipophylic environment, lycopene will have maximum reactive oxygen species scavenging effects. Lycopene was shown to be the most effective antioxidant in protecting the 2,2'-azo-bis(2,4-dimethylvaleronitrile) (AMVN)-induced lipid peroxidation of the liposomal membrane. Oxidative modification of low-density lipoprotein is hypothesized to be the key step in the atherogenic process, and low-density lipoprotein associated antioxidants provide protection against this oxidation. *In vitro* lycopene and other carotenoids are able to inhibit oxidation of low-density lipoprotein. Lycopene was also found to protect lymphocytes against $NO_2^{\cdot}$-induced membrane damage and cell death twice as efficiently as β-carotene. *In vivo* antioxidant effects of lycopene and its interaction with the host and other dietary antioxidants are now being investigated.

A number of studies have used tissue culture and *in vitro* systems to demonstrate potential disease preventive action of lycopene and provided a mechanistic hypothesis. Levy and his co-workers showed that lycopene inhibited the growth of human endometrial, mammary and lung cancer cells grown in cultures and was more effective than α- or β-carotene. Lycopene along with vitamin D_3 synergistically inhibited cell cycle progression and induced differentiation of the HL60 promyelocytic leukemia-cell line. In mouse embryo fibroblast-cells, lycopene-enrichment upregulated gap-junction communication by enhancing the expression of the connexin43 gene, which encodes a major gap junction protein, and thereby, acted as an anticarcinogenic agent. Lycopene was also shown to protect against microcystin CR-induced mouse hepatocarcinoma by suppressing the phosphorylation of regulatory proteins and arresting cells in the G0/G1 phase of cell cycle. Preliminary *in vitro* evidence indicates that lycopene reduces cellular proliferation induced by IGF-1 in various cancer cell lines. In a recent investigation, lycopene, together with alpha-tocopherol at physiological concentration, synergistically inhibited cell proliferation of an androgen insensitive prostate carcinoma cell line. In the J774A.1 macrophage cell line, lycopene was shown to act as a hypocholesterolemic agent by inhibiting the HMG-CoA reductase pathway.

Conclusion

The field of antioxidants is moving rapidly. About 20 years ago, the hypothesis that diet might have a substantial influence on the development of some pathology, such as cancer, has been raised by many scientists. In this light during the last decade, efforts have been made to analyze the effects of plant food and synthetic antioxidants on the development and prevention of chronic diseases. Now a day, antioxidants are used on a large scale to try to obtain and preserve optimal health. While there is no doubt that the correct balance between endogenous and exogenous antioxidant capacity is essential to life, the curative power of antioxidants has often been over estimated. In fact, according to the popular idea *if one is good two is better*, antioxidants are taken in excess too often and the risk to originate diseases instead of preventing them is quite high. It is noteworthy to underlie that as for all drugs, antioxidants may give important side effects if not correctly used or in combination with other drugs. Vitamin A, E and β-carotene for instance, have been shown to have pro-oxidant effects at higher doses or under certain conditions.

■■■

35 Health Benefits of Plant Enzymes, Juice and Detoxification

Uche Fidelia Ijeoma

Head of the Department, Department of Pharmacognosy and Phytotherapy, Faculty of Pharmaceutical Sciences, University of Port Harcourt, Rivers State, Nigeria, E-mail: <uchefideliaijeoma@yahoo.com>

The healing elements that are found in green vegetable juices are so powerful and synergistic. No combination of synthetic medicines or supplements could ever match their power. It is no secret that vegetables are good for you, hence, it stands to reason that vegetable juicing is one of the best things you can do for good health. Vegetable juice provides your body with a very concentrated source of nutrients and enzymes, particularly when organic vegetables are used. It also helps to detoxify your body and boost the immune system with high doses of minerals and vitamins that are easy to absorb. In order to get the most benefit from eating vegetables, it is best to eat them raw. Of course, it can be difficult to eat large quantities of raw vegetables but drinking several glasses of vegetable juice daily is relatively easy. Veggies are generally lower in sugar and higher in nutrients and enzymes than fruits, so vegetable juicing is an especially good way to nourish the body especially when you are sick. One of the most important results of giving your cells a healthy dose of vegetable juice is that it will help to maintain the pH balance of your body, *i.e.*, alkalinizing ability. Apart from alkalinizing ability, other main benefits of green vegetable juicing include source of plants enzymes, chlorophyll, trace minerals, vegetable protein and detoxification abilities.

Plant Enzymes

Vegetable juicing also provides a concentrated source of enzymes to aid digestion. Proper digestion is just as important as nutrients concentration, since undigested nutrients cannot be utilized by the body. Because the typical diet includes foodstuffs that are difficult for the body to digest, many people do not have enough digestive enzymes to break down foodstuffs and utilize nutrients, however, the good news is that with a few vegetable juice drinks a day, the digestive enzymes can easily be replenished.

Green foodstuffs or juices contain phytonutrients that are capable of initiating chemical reactions. These enzymes can act as catalyst at cellular level, initiating cleansing, detoxifying and rebuilding the body while enhancing the immune system.

Plant sources contain both antioxidant and digestive enzymes. The enzyme complexes that are found in green juice are identical to those enzyme complexes manufactured in the body. This is very important when we realize that our bodies have limited activity to manufacture enzymes, and when we run out, death may result. Cooking and processing of food eliminates the enzymes, while raw unprocessed foodstuffs contain these enzymes in abundant. Of all the raw food articles, green food items provide the largest source of both digestive and antioxidant enzymes.

Green foodstuffs contain antioxidant enzymes because those enzymes defend the plant from damage caused by sun, including its ultraviolet and radioactive rays. The protective enzymes are quite similar in function to those found in human body.

In fact, enzymes provide stimulus for every chemical reaction that occurs in the body. They make the life possible as stated by Burton Goldberg Group. According to Edward Howell M.D., no mineral, vitamin, or hormone can do any work without enzyme. They are manual workers that build the body from the proteins, carbohydrate and fats. The body may have the raw building materials but without the workers, it cannot begin.

Plant Enzyme Therapy

Plant derived enzymes are used in enzyme therapy. They are prescribed to enhance the body's vitality by strengthening the digestive system. When proper digestive system is strengthened or restored, many acute and chronic conditions may also be remedied.

As stated by Burton and Goldberg Group, the enzyme therapy can be an important first step in restoring health and well-being by helping to remedy digestive problems. Plant and pancreatic enzymes are complementary ways to improve digestion and absorption of essential nutrients. Treatment includes enzyme supplements, coupled with a healthy diet that features whole foods.

Digestion of food takes place in many stages, *i.e.*, beginning from the mouth, moving to stomach, and finally, through the small intestine. At each step, specific enzymes break down different types of food. An enzyme designed to digest protein, for instance, has no effect on starch, and an enzyme that works in the mouth will not be active in the stomach. This process is made possible by different degree of acidity along the digestive track that allows certain enzymes to function while deactivating others.

Digestion of food by enzymes starts in the mouth. The enzymes continue their works in the stomach, plant enzymes from juice or taken as supplement also join in and become active. The food then enters the upper part of small intestine where the pancreas (a digestive organ that feeds enzymes into the gut) provides particular enzymes to further break down the food. Final breakdown of remaining small molecules of food occurs in the lower small intestine. Ideally, these enzymes can work together, digesting food and delivering nutrients to cells to maintain their health: *Protocols in enzyme therapy are based on this sequence of events*, as stated by Burton Goldberg Group.

Plant enzymes are essential for proper digestion of food. They play an important role in promoting good health. This is the basis of treatment in plant enzyme therapy. According to Howard F. Loomis Jr., D.C. of Forsyth, Missouri, the ability to absorb nutrients from the food we eat is the foundation of good health. If we treat digestive disorders, other complaints often clear up as a result.

Enzyme therapy needs to be combined with good eating habits. Fresh fruits, vegetable juice, nuts and seeds can provide plentiful plant enzymes and plant enzymes are only meant to supplement those that naturally occur in food.

The four categories of plant enzymes have used in plant enzyme therapy. Protease digests protein, amylase digests carbohydrates, lipase digests fats and cellulose digests fiber. Plants are main and only source of cellulose as the human body is unable to produce it. Numerous plant enzyme formulations in the market combine these enzymes.

Plant enzymes function in the stomach, predigesting the food, and plant enzyme therapy uses this to its advantage. This phenomenon was explained by Dr. Howard in his statement that if the stomach is performing its proper role and we are eating our food uncooked, a large portion of the intake would be partially digested before reacting with the stronger digestive juice found there. Moreover, fewer of your body's internal digestive enzymes will be called upon to perform the digestive function. It is the easing of the body enzymes' workload, which is thought to contribute substantially to the healing effects of enzyme therapy. According to Dr. Howell, when the body receives plentiful supplies of enzymes, its internal enzyme supplies are preserved for the important work of maintaining metabolic harmony. As a result, many body systems are strengthened.

Enzymes in the stomach can digest 30 to 40% of the starch we eat. Hence, Dr. Lee stated that by eating raw food and taking good enzyme producing juice, 30% of the protein and 10% of the fat can be digested in the stomach in less than one hour.

Cooking food can destroy these important plant enzymes. They are more heat-sensitive than vitamins and are the first to be destroyed during cooking. They are destroyed by heating above 48°C, and according to Dr. Lee, are deactivated, or destroyed by pasteurizing, canning and microwaving.

However, while raw foodstuffs or juices are recommended, a 100% raw food diet is not necessary. Dr. Loomis explained that some people might have problems digesting uncooked food because of lack of cellulose. According to him, people who rarely eat raw food can have problems when they finally eat uncooked food articles or vegetables because they do not chew their food thoroughly. Chewing liberates the cellulose out of the food, however, when they eat raw food and do not chew properly, the cellulose is never released. Cellulase may also be lacking because of the way the food was handled or processed by the suppliers. Some supermarket vegetables loss their cellulose content be they sprayed them with sulfites, which can destroy these enzymes.

Plant Enzyme Deficiency

Eating a predominantly cooked-food diet could result to various inflammatory pancreatic hypertrophy, a toxic colon and allergies because of inflammation conditions such as bronchitis, sinusitis, cystitis, arthritis and rhinitis and may be accompanied by fever, redness, swelling and pain. Pancreatic hypertrophy results when a diet lacking in enzyme puts an extra strain on the enzyme production of the pancreas. If the pancreas fails behind in its work, the organ will become hypertrophy (enlarge).

Low levels of enzymes can also cause damage to the colon. The undigested food can remain in the intestine. The molecules will then be converted into toxins, which are transported by the blood to liver for detoxification. If the liver is overworked, it will be unable to detoxify the blood, and this becomes poison to the body system.

Insufficient plant enzymes can also lead to leukocytosis (an increased white blood cells count). A rise in white blood cells is a sign that immune system is being mobilized. It accompanies many pathological conditions including infections and poisoning but it can also immediately occur after eating breakfast, lunch, or dinner.

Dr. Lee stated that the digestive leukocytosis occurs a mere thirty minutes after eating cooked. This does not occur when a person eats raw food because of the presence of plant enzymes in the food; such a response puts additional stress on the immune system. This implies that when one eats cooked food, ones immune system is stimulated as if one has an infection.

According to Dr. Lee, when a patient's plant enzyme deficiencies are addressed, many other conditions can be resolved from digestive ailment and common sore throats to hay fever, ulcers and Candida.

Chlorophyll Power

Chlorophyll, a green pigment of plant, is among the most powerful elements that exist in the universe. Chlorophyll has found an application in the treatment of many diseases such as diabetes, wounds, ulcers, low iron and body odour. Chlorophyll has gained wide acclaim for its ability to prevent disease. The chlorophyll has similar structure with hemoglobin. The major difference in their structure is that magnesium is the center element in chlorophyll while iron is center element in hemoglobin. Hence, chlorophyll can replace hemoglobin functionally as stated by the scientist. It implies that chlorophyll can be converted to hemoglobin, which increases the flow of oxygen to all parts of the body. Therefore, chlorophyll can create an aerobic environment, thereby, preventing development of disease state. An aerobic environment is also beneficial to friendly bacteria and detrimental to bad bacteria. The good bacteria promote digestive health, while bad bacteria contribute to disease, body odour and bad breaths.

Trace Minerals

Green vegetables are the most recommended source of minerals. Magnesium and iron are two key minerals, both for men and women but chlorophyll creates an environment suitable for delivering and proper maintenance of mineral qualities in human.

When we ingest food rich in chlorophyll, magnesium is immediately delivered to the muscle tissues. Magnesium is the mineral, which is very essential for the maintenance of both cardiac and respirator systems. Magnesium is the mineral that relieves muscle cramping, allows our nerve endings to be soothed and allows us to breathe efficiently.

By drinking green juice (abundant in both chlorophyll and iron), we create the environment for efficient utilization of both iron and magnesium by the body and this strengthens the human blood. When we ingest minerals through plants, we can never suffer toxicity. The mineral from the green vegetable juice will either be utilized by the body or eliminated by the body. They do not accumulate in the body as to cause unwanted effects like synthetic minerals.

There are many other benefits of trace minerals. Potassium is important for digestive health, chromium helps in controlling blood sugar and selenium aids the metabolism. Calcium reverses osteoporosis and helps in building stronger bones and beautiful hair. While calcium is probably the most popularly consumed mineral supplement, the bioavailability of such products is a fraction of what one can get by consuming green vegetable juice.

Alkalizing Agent

An optimum pH level is essential for maintaining good health. The pH scale ranges from 1 to 14 with a measure of 1 being extremely acidic and a measure of 14 being extremely alkaline.

The reason of pH (or potential hydrogen) to be important is that too often our cells become too acidic. The more acidic the cells are, the less able they are to transport oxygen throughout the body. In this regard, acid is toxic to the body's cells. The ideal pH level is slightly alkaline between 7 and 7.5. Unfortunately, the environment tends to contribute too much acid, making it difficult to maintain a healthy pH balance without a concerted effort to include alkaline food items like vegetable juices in the diet

Acidity and alkalinity are measured on pH scale of 0 to 14, with seven being neutral. The scores above 7 are alkaline and scores below 7 are acidic. Some part of the body environment must be alkaline for the body in order to remain healthy. For instance, human blood has to be maintained at PH of 7.4. If the human blood does not have enough alkalinity, it will start to draw alkaline from other parts of the body. Other parts of the body that should be alkaline include mouth, the small intestine, and at time, the urinary tract. When the blood cannot get its required pH from the raw materials we consume, the first part of body it will rob its alkalinity is the mouth. This is important because when our mouth becomes more acidic, there are some common health problems that will manifest. These include cavities, plague, bleeding gums, periodontal disease and bad breath.

Green juice can reverse this dangerous process. When green juice product is taken, profound effects will manifest, such as strengthening of gums, a cessation of bleeding when brushing of flossing, a reduction of plague and an elimination of cavities.

The small intestine also needs to be alkaline in order to regulate the passage of glucose through the intestine. The benefits of regulated blood sugar are numerous, and this includes potential weight loss, increased energy and higher metabolism. Both, hypoglycemic and diabetics benefit significantly from green juice, which alkalinizes the small intestine, and this regulates the passage of glucose through the intestinal barrier. This results in energy, stamina and endurance.

Vegetable Protein

Lightweight vegetable proteins contain eight essential amino acids and are especially rich in lucine, Isolucine and valine. These peptides act as the building blocks for the immune system, enhance reproductive function and create lean connective tissue, which leads to dense and heavy permanent muscle on the body.

If proper quantity and quality of peptides are not provided for the body, the body will start to break down. The implication is that the body will start to draw upon the muscle tissue in an effort to get raw material to build the immune function. This will create much stress on the body and begins to break down the immune system. By taking green vegetable juice, therefore, our body will be protected from many health hazards or diseases that threaten humankind because of deficiency of some or all of the phytochemicals present in green vegetable juice.

Detoxification

Juicing with vegetables also helps to detoxify the body and enables the cells to absorb and utilize nutrients better. If cells of the body are loaded with toxins, adding more nutrients will be of little value, so the ability of vegetable juicing to detoxify cells is as important to overall digestion and health as the consumption of large amounts of nutrients and enzymes.

The best vegetable juices are made from organic vegetables that are available in the in season. Green vegetables should always be used in a healthy vegetable juice recipe. Some of the best vegetables for juicing are lettuce, kale, carrots, parsley, chard and broccoli. One can experiment with recipes to find the flavour combinations that one likes best.

Detoxification Therapy

Detoxification is the natural method of removing or neutralizing toxins by the body through the kidney, liver, feces, exhalation, urine and perspiration. The industrial revolution has caused the accumulation of toxins in the human system faster than they are eliminated. Other factors that cause the accumulation of toxins in the human body are pesticides, food additives, heavy metals, anesthetics, residue of pharmaceutical drugs, legal drugs (alcohol, tobacco and caffeine) and illegal drugs (heroin, cocaine and marijuana). Most drinking water contains several chemicals, including excessive levels of lead. Chemicals are added to food supply and some solvents like emulsifiers and preservatives are used in food processing and storage, which can remain in the body for a long period. When people consume these foods, especially seafood, meat and poultry, they ingest all the chemicals and pesticides that have remained as contaminants accumulating in the food chain, this is why, the people are exposed to chemicals today in far greater concentration than in previous generations.

Every product such as gasoline, paint, household cleaners, cosmetics, pesticides and dry cleaning fluid also pose a serious threat since the body cannot easily break them down. As the earth becomes more polluted, the body inadvertently becomes a filter that traps these pollutants.

According to Marshall Mandell, M.D., the current level of chemicals in the food and water supply and the indoor and outdoor environment has lowered the threshold of resistance to diseases and has altered our body's metabolism, causing enzyme dysfunction nutritional deficiencies and hormonal imbalances.

Leon Chaitow, N.D. also stated that a body with healthy immune system, efficient organs of eliminating and detoxification and a sound circulatory and nervous system can handle a great deal of toxicity but if they have been damaged from chronic exposure from environmental pollutants, restoring these functions, organs and systems can be accomplished only through detoxification therapies including fasting, chelating and nutritional, herbal and homeopathic methods that accelerate the body's own natural cleansing process.

Benefits of Detoxification

Detoxification process, juice and water fast together with cleansing diet prevent chronic diseases like ulcer, cardiovascular problems, arthritis, diabetics and obesity. The modern diet with excess animal proteins, fats, caffeine, alcohol and chemicals inhibits the optimum functions of our cells and tissues. Therefore, the cleansing of toxins and waste products will restore optimum function and vitality. The aim of detoxification is to cleanse the body system of toxins and waste products. These revitalize the body's optimum function and healing capacities. The most important and longest lasting effect of detoxification therapy is the reduction of stress on the immune system. Other benefits include vitality, reduced blood pressure and blood fats (cholesterol and triglyceride) and an improved assimilation of minerals and vitamins. Detoxification is instrumental for the maintenance of normal function and integrity of intestinal flora and skin disorders. Dr. Haas stated that most people will feel mentally and physically rejuvenated after detoxification therapy, with a corresponding reduction in symptoms and disease.

Indications of Detoxification

Everyone has a specific level of tolerance that cannot be exceeded if good health is to be maintained. If the amount of toxins within the body stays below that level, the body can usually adapt and rid itself of these but when the system is overwhelmed, the body's defense mechanism malfunctions, and symptoms such as fatigue, confusion, aggression, or mental disorder may occur.

Some indications that the body may need detoxification are headaches, joint pain, recurrent respiratory problems, back pain, allergy symptoms, insomnia, mood changes, and food allergies. Conditions like arthritics, hemorrhoids, sinus congestion, constipation, ulcers, psoriasis and acne can indicate the need for detoxification.

Laboratory test can also be used to check the need for detoxification. The test can involve stool, urine, blood, or liver function as well as hair analysis.

Methods of Detoxification

There are different methods of detoxification available. These include fasting and specific diets, colon therapy, vitamin therapy, chelation therapy, hydrotherapy and hyper thermia. It may be necessary to consult with professionals in selecting a detoxification programme.

Fasting and Diets

Fasting programs and specific dietary regimens are some of the easiest and most inexpensive and effective methods of detoxification. They involve only purchase and consumption of healthy food. The result depends upon both the patient and length of the fast. Fasting is often combined with enemas and bowel stimulation for ridding of the body bacteria. Also, while a detoxification program is not basically concerned with weight loss by applying a more efficient and healthy way to, weight loss can be the result. Before embarking on any kind of fast or diet, one should always consult with a qualified health professional.

There are two types of fast namely water fast and juice fast as being described below in detail:

Juice Fasting

Evarts G. Loomis prescribes vegetable juice (equal parts of carrot diluted with water 1:1 and celery) for detoxification. If desired, the green vegetables such as green beans, zucchini, watercress and parsley can be added. Dr. Loomis also prescribes a detoxifying cocktail that combines garlic, lemon juice, grape fruit juice and olive oil. This should be given at bedtime because the liver, a major detoxification organ is most active between 11.00 p.m. and 1.00 a.m. according to traditional Chinese medicine.

Alkaline-Detoxification Diet

There is need to detoxify the body tissues of protein and waste that can create both inflammatory and degenerative changes when interacting in the body. Dr. Haas stated that when foodstuffs oxidize in the body, they leave a residue or ash. If the foodstuffs are acidic, they leave an acidic residue (acid ash, sulphur, phosphorus, chlorine, or uncombusted acid radicals), which can be harmful. Success with this method is achieved by elimination of acid, forming foodstuffs from your diet.

Essentially this diet consists of fruits and vegetables plus fresh sprouts and millets. The diet also focuses on particular eating principles, including exceptional chewing (30 to 35 times per mouthful), drinking quality water plus the steamed vegetable water and eating all kinds of food prior nightfall (advisedly before 6:30 p.m.).

Colon Therapy

Colonic irrigation is one of the most effective ways to cleanse the large intestine of accumulated toxins and waste products. Its function is to draw toxins from the blood and lymph back into the colon for excretion. In this program, purified water (which may contain vitamins, herbs, friendly bacteria, or oxygen) is introduced directly in the rectum. Colon therapy can be combined with massage, nutritional program and special diets to help cleanse the bowels and aid in treatment of parasitic infections.

Vitamin C Therapy

Vitamin C is very important in healing and maintaining good health. The vitamin C as detoxification agent combines with certain toxins in the body and destroys them. Roberts Cathcart stated that the vitamin C functions as free-radicals scavenger, neutralizing the immunosuppressive toxins produced by infectious diseases.

Chelation Therapy

In chelation therapy, a synthetic amino acid known as EDTA (ethylenediaminetetraecetic acid) is administered intravenously and binds to various toxic metals in the blood, such as lead, cadmium and aluminum. The toxins are the flushed from the body through the kidney. It has been found that besides using EDTA to treat lead poisoning, EDTA can also remove calcium and plaque present in the walls of arteries in atherosclerosis, the major cause of heart disease.

Hyperthermia– Heat Stress Detoxification

The only detoxification program that has proven successful in removing fat-stored toxin from the body is hyperthermia or heat stress detoxification. According to Zane Gard, heat stress can also remove calcium deposits from the blood vessels and break down scar tissues from their walls. He went further to say that hyperthermia can affect numerous body systems including cardiovascular, endocrine, neurological, neuromuscular, bronchopulmonary, blood, skin and immune.

Hyperthermia can remove chemicals such as Bis-(*p*-chlorophenyl) acetic acid] (DDA) [a metabolite of DDT], polychlorinated biphenyls (PCB's) and dioxin from fat cells.

Biotoxic Reduction Program

Biotoxic reduction program (BTR) was established by Gard and Brown. This is a medically managed detoxification program designed to reduce the body's internal toxic load of chemicals, heavy metals, medical or street drugs and alcohol. The following health evaluation should be carried out by the Physicians before embarking on Bio-Toxic Reduction (BTR) program. These include a physical examination, health survey, immunologic profile, antibody level testing, toxic panel and heavy metal analysis as well as the usual complete blood count (CBC), blood chemistry, electrocardiogram readings and personality and perception testing.

Hydrotherapy

Hydrotherapy has been used to treat diseases and injury by many people of different cultural origin. According to Burton Goldberg Group, hydrotherapy is the use of water, ice, or steam at hot or cold temperature to maintain and restore health. The treatments include full body immersion, steam baths, Sitz baths, colonic irrigation and the application of hot and /or cold compresses. Hydrotherapy is effective for treating a wide range of conditions and can easily be used in the home as part of self-care program.

Hot or cold water administered externally or internally could be effective in the treatment of conditions ranging from stress to the many toxins, bacteria and viruses causing disease.

Mechanism of Hydrotherapy

According to Douglas Lewis N.D., hot water produces a response that stimulates the immune system and causes white cells to migrate out of the blood vessels and into the tissues where they clean up toxins and assist the body in eliminating wastes. Therapeutically, hot water soothes and relaxes the body, and through the reflex action of the nerves, it can affect almost every system of the body.

Dr. Lewis also stated that cold water discourages the inflammation by means of vasoconstriction of the blood vessels. Lewis added that short cold-water treatment may actually increase fever and only long cold-water treatment pulls heat from the body for fever reduction.

Contrast therapies are those that alternate between hot and cold water in the same treatment. They can stimulate the adrenal and endocrine glands, reduce congestion, alleviate inflammation and activate the system.

Hydrotherapy techniques can be performed in the home or clinical set up. Some of the hydrotherapy techniques that can be done in our homes include ice and contrast baths and showers, foot and hand baths, cold water treading, steam and compresses and packs. Hydrotherapy techniques in clinical set up include hyperthermia, whirlpool baths and neutral baths.

Clinical Application of Hydrotherapy

Different forms of hydrotherapy are used by the naturopathic Physicians, alternative Practitioners and Physical therapists at clinics, hospital and health sectors to treat a wide range of conditions.

Hyperthermia (as fever-induction therapy) is a form of hydrotherapy that deliberately induces fever in the patient who is unable to mount a natural fever response to pathogens (disease-causing organism and toxins). Fever is often regarded as an undesirable symptom of illness but holistic practitioners see it as the body's defense against invading organisms. Fever stimulates the immune system by increasing the production of antibodies and interferon (a group of proteins released by white blood cells that combat a virus).

According to Burton Goldberg group, laboratory research has pro-shown Human Immunodeficiency Virus (HIV) is temperature sensitive and suffers greater inactivation per unit time at progressively higher temperature above the normal body temperature (37°C). Dr. Lewis also stated that a hot immersion bath, if done without raising body temperature and heat rate too quickly could be used for adjunctive treatment for a diverse number of diseases from upper respiratory infection to sexually transmitted diseases and from cancer to AIDS. Hyperthermia also helps in the treatment of chronic fatigue syndrome.

Whirlpool Bath

This can rehabilitate injured muscle and joints and alleviate the stresses in everyday life. Whirlpool is also effective in healing skin sores, infected wounds, edema (swelling) and minor frostbite pains. Some physical therapists use it to sooth burns in patients and improve the circulation in paraplegic and polio victims.

Neutral Baths

According to Bolton Goldberg group, neutral baths are full immersion therapy that submerges the body up to the neck in water from 33 to 37°C. The soothing nature of the neutral bath calms the nervous system and is effective for treating emotional and mental disturbances and insomnia. Dr. Chitowu stated that clinical studies reveal that two hours in a neutral bath is effective in reducing the excessive fluids retention in patients suffering from mild heart conditions and cirrhosis of the liver. It is also beneficial to the swollen joints of rheumatoid patients. Natural baths help to promote detoxification from alcohol and drugs abuse since the neutral baths help the body rids itself of large amount of toxin-laden fluids.

Hydrotherapy at Home

Ice and contrast are effective therapies for trauma relief. Any injury likes sprains and strains or acute inflammation, like tendinitis calls for an immediate application of cold. The ice should be applied as often as twenty minutes every hour for the first twenty four to thirty six hours post-trauma.

Trauma or chronic conditions also respond well to contrast therapy. Alternating hot and cold increases the circulation to bring the vital nutrients to the area and to move the waste products out. Simply apply alternating hot and cold packs to the affected area begging with hot for 3 minutes, and then, cold for thirty to sixty seconds. These should be repeated 3 times in one sitting, always finishing with cold, one to three times per day depending upon severity of the condition.

Baths and showers are soothing both mentally and physically. Not only do they relieve general pains, and they can also ease internal congestion and digestive ills.

More so, water temperature varies according to individual needs. Hot baths and showers are relaxing and stimulating to immune system. By inducing perspiration, they facilitate the detoxification process. Cold baths and showers can tonify muscles and reduce inflammation and fatigue. Long cold baths are useful to reducing fevers but are not recommended for home use and should only be used under direction of a physician.

Cold showers also can increase the nerve force, stimulate the glandular system and have a positive effect on the central nervous system. Alternating hot and cold showers can be an excellent way to increase circulation and stimulate organ function. It is always necessary to start with hot and to end with cold.

Sitz Baths

These are traditionally European folk remedy in which the pelvis is immersed in hot or cold water. A hot sitz is particularly helpful for problems involving pelvic region, including uterine cramps, painful ovaries or testicles and hemorrhoids. A cold sitz bath taken for two minutes or less can be used for inflammation, constipation, virginal discharge and impotence but should not be used for urinary tract infections.

Foot and hand baths are excellent for drawing blood away from inflamed parts of the body, or drawing congestion away from the organ. They can help to relieve insomnia, sore throats, colds, menstrual cramps and pain from gout, neuralgia and headache.

Hot footbath is the ideal remedy for shivering, cold hands or feet, nausea and dizziness or faintness. When mustard powder is added to the footbath container, it can be used to relieve cold or sore throat.

Alternating hot and cold footbaths has profound effects on the nerve and reflex points of the feet. These baths also help to relieve toothaches, neuralgia, headaches (when used in conjunction with a cold compress on the head), ankle swelling, foot infections, blood poisoning and abdominal congestion. Hands also contain many reflex points that affect the entire body. A cold hand bath can stop a nosebleed and release sunstroke. A hot hand bath can relieve cramps in the hands from overuse in athletics, writing, or sewing, and even alleviate an asthma attacks.

Steam is an excellent cleanser and deep moisture treatment for the skin. It also helps to break up congestion from the common cold and flu.

Compresses and packs are particularly effective for applying heat or cold to specific parts of the body. They stimulate the immune system and increase the body's white blood cell count.

Cold water treading is one of the most important preventive water treatments. The feet are immersed in cold water, after shower or bath and marched in a place for five seconds to five minutes. Afterwards, the feet are rubbed vigorously with a towel, especially the soles. It is believed that by building tolerant to cold, it is possible to develop resistance to infectious disease. It is recommended for this practice to be a part of daily self-care routine for everyone, young and old.

■■■

36 Fermented Food and Human Health

Leela Wati[1], Kamla Malik[1] and Kushal Raj[2]

[1]Department of Microbiology
[2]Department of Plant Pathology
CCS Haryana Agricultural University, Hisar-125 004, Haryana, India,
E-mail: <lwkraj@hau.ernet.in>, <kamlamalik@rediffmail.com>

Fermentation is a form of energy-yielding microbial metabolism in which an organic substrate, usually a carbohydrate, is incompletely oxidized, and simultaneously, an organic carbohydrate acts as the electron acceptor. Fermentation is one of the classical methods to preserve food. Man has been eating fermented food for millennia. Different cultures around the world have one or the other fermented food.

Fermentation is one of the oldest and most economically methods for producing and preserving food articles. Each fermented food is associated with a unique group of microflora, which increases the level of protein, vitamins, essential amino acids and fatty acids. The deficiency of protein and vitamins is the major worldwide problem. In this regards, production of fermented food will become increasingly important to the world's diet.

In the Indian subcontinent, making use of fermented food and beverages using local food crops and other biological resources are very common but the nature of the products and the base material vary from region to region. Indigenous fermented food items such as fermented soybean, fish, milk, meat and a leafy vegetable contribute to a large proportion of the daily food intake in Northern India. Fermentation has been used for a long time as an effective and low cost means to preserve the quality and safety of the foodstuffs. The fermentation process causes enrichment and improvement of food through flavour, aroma and change in texture, preservation by producing organic acids, nutritional enrichment and reduction of endogenous toxins and in the duration of cooking, and thereby, fuel requirement.

All around the world, the fermented foodstuffs and beverages are an important part of human diet. These food items are made from plant and animal materials in which bacteria, yeast and molds play an important role by modifying the material physically, nutritionally and organoleptically. Globally, there are varieties in fermented foods and beverages and so do the ingredients and recipes for making them. In some places, they make up a minor 5% of daily intake, while in others their role can be as substantial as 40%. Fermented foods can be fried, boiled or candied, or consumed in curries, stew, side dishes, pickles, confectionery, salads, soups and desserts. They can be in the form of pastes, seasoning, condiments, masticators and even colourants. Fermented drinks can be either alcoholic such as beer and wine or non-alcoholic like buttermilk, certain teas, or things that contain vinegar. Though most fermented foods have health-promoting benefits, their global consumption is declining since the traditional food systems give way to the influence of a western diet and fast foods. Fermented foods are processed with enzymes or microbial mediated conversion of sugars into products such as alcohol and lactic acid under conditions with little or no oxygen present. When consumed by humans, the fermented foods often introduce microflora that inhabit the human body.

Lactic acid fermentation and alcohol fermentation produce familiar items like yogurt and wine. Hence, the fermented foods play an important role in providing food security, enhancing livelihoods and improving the nutrition and social well being of millions of people around the world, particularly the marginalized and vulnerable.

Food fermentation is practiced by human cultures all over the world. Preserved food is a major component of human survival in places where it is a necessity. Production of fermented food does not require knowledge of the biologically mediated nature of fermentation because the biota that carryout fermentation is present worldwide. Fermentation organisms are ambient in the environment, whether humans make use of them or not. Fermentation that unavoidably affects the human food supply worldwide is a natural process. Wild fermentation bacteria and yeast cover the continents and permeate ecosystems, in the air, soil, water and guts of animals. They are a natural resource available to people all over the world. Though fermentation is everywhere, it is rarely understood and used for thousands of years in human nutrition. Now a day, the health promoting functionality of fermented foods becomes more and more in the focus of scientific and public interest.

Types of Fermented Food

There are varieties of fermented foodstuffs available to supplement our daily diets. Fermented food articles based on the substrate are categorized into the following groups:

- **Beverages:** Wine, beer and *kombuch*, *schnaps*, *arak*, coffee and cocoa
- **Cereal products:** Bread, pancakes, tape *ketan*, *bhatura*, *kulcha*, *dhokla*, *kenkey*, *injera* and *ogi*
- **Dairy products:** Curd, yoghurt, *lassi*, kumis and cheese
- **Fish products:** Fish sauce
- **Fruit and vegetable products:** *Kimchi*, sauerkraut and pickles
- **Legumes:** Soymilk, soy sauce, tofu, *tempeh*, *natto*, *miso*, *dosa*, *kinema*, *soumbal* and *aidli*
- **Meat products:** Sausages, raw ham (bacon), fermented meat products, *e.g.*, salami
- **Starch crop products:** Poi, *kefir*, *ang-kak*, *gari*, tape *ketella* and *bwiru*
- **Miscellaneous products:** Poi, edible mushrooms (champignons), *rakfisk*, *pidan*, *nata*, *ragi* and asyn

Fermented Foodstuffs Popular in Different Parts of the World

For most post-agricultural people, some forms of fermented food were a standardized component of the traditional diet. *Magou*, *nham*, or *kimchi* are fermented food articles eaten around the world. Other fermented food items, which are more familiar to Westerners, are yogurt, sauerkraut and sourdough bread. In the United States, sauerkraut (fermented cabbage), *tempeh* (fermented whole soybean), kefir, *kombucha*, standard pickling, chocolates and yogurt (fermented milk) tend to be the most popular fermented food items. In East and Southeast Asia, *natto* (fermented soy), *kimchi* (fermented cabbage), soy sauce, fermented fish sauce, fermented shrimp paste are popular fermented food articles. In Central Asia, kumis (fermented mare milk), kefir and *shubat* (fermented camel milk) are famous, while in the Middle East, fermented pickles, various yogurts and *torshi* (mixed vegetables) are popular fermented foodstuffs. In Europe, sauerkraut, kefir, crème *fraiche* and *rakfisk* (salted, fermented trout) are the fermented food items of choice. In the Pacific region, poi (fermented mashed taro root) and kanga *pirau* or rotten corn are very common fermented food articles. In Africa, fermented cassava products (like *gari* and *fufu*) are a major component of the diet of more than 800 million people, and in some areas, these products constitute over 50% of the diet. India probably has the most diverse variety of fermented foodstuffs. Among them, *idli, dosa, dhokla, khaman, batura, kulcha* and traditional curd are the most popular Indian fermented food items.

Numerous fermented food items are consumed around the world. Each nation has its own types of fermented food, representing the staple diet and the raw ingredients available in that particular place. Though the products are well known to the individual, they may not be associated with fermentation. Indeed, it is likely that the methods of producing many of the world's fermented foods are unknown and they have come about by chance. Some of the more obvious fermented fruit and vegetable products, *i.e.*, beers and wines, are the alcoholic beverages. However, several more fermented fruit and vegetable products arise from lactic acid fermentation and are extremely important in meeting the nutritional requirements of a large proportion of the world's population. The examples of fermented fruit and vegetable products from around the world are given in Table 1.

Table 1 : Fermented foods from around the world

Name and region	Types of product
Indian subcontinent	
Acar, achar, tandal achar and *garam nimboo acharappam*	Pickled fruit and vegetables
Gundruk	Fermented dried vegetable
Lemon pickle, lime pickle and mango pickle	
Southeast Asia	
Asinan, burong mangga, dalok, jeruk, kiam-chai, kiam-cheyi, kong-chai, naw-mai-dong, pak-siam-dong, paw-tsay, phak-dong, phonlami-dong, sajur asin, sambal tempo-jak, santol, si-sek-chai, sunki, tang-chai, tempoyak, vanilla, *amazaka, atsara, belacan*, com *ruou, douchi, lambanog, kombucha, narezushi, miso*, sake and stinky tofu	Pickled fruit and vegetables
Bai-ming, leppet-so and *miang*	Fermented tea leaves
Nata de coco and *Nata de pina*	Fermented fruit juice

Name and region	Types of product
East Asia	
Bossam-kimchi, *chonggak-kimchi*, *dan moogi*, *dongchimi*, *kachdoo kigactuki*, *kakduggi*, *kimchi*, *mootsanji*, *muchung-kimchi*, *oigee*, *oiji*, *oiso baegi*, *tongbaechu-kimchi*, *tongkimchi* and *totkal kimchi*	Fermented in brine
Cha-ts'ai, *hiroshimana*, *jangagee*, *nara senkei*, *narazuke*, *nozawana*, *nukamiso-zuke*, *omizuke*, *pow tsai*, red in snow, *seokbakji*, *shiozuke*, *szechwan cabbage*, *takana*, *takuan*, *tai-tan tsoi*, *tsa tzai*, *tsu*, *umeboshi*, *wasabi-zuke* and *yen tsai*	Pickled fruit and vegetables
Hot pepper sauce	Sauce
Africa	
Fruit vinegar	Vinegar
Hot pepper sauce	Sauce
Lamoun makbouss, *mauoloh*, *msir*, *mslalla* and olive	Pickled fruit and vegetables
Oilseeds, *ogili*, *ogiri* and *Hibiscus* seed	Fermented fruit and vegetable seeds
Wines	Fermented fruits
Americas	
Cucumber pickles, dill pickles, olives, sauerkraut, chichi and *kombucha*	Pickled fruit and vegetables
Lupine seed and oilseeds	Pickled oilseed
Vanilla, Tabasco, *tibicos* and wines	Fermented fruits and vegetables
Middle East	
Kushuk	Fermented fruit and vegetables
Lamoun makbouss, *mekhalel*, olives, *torshi* and *tursu*	Pickled fruit and vegetables
Wines	Fermented fruits
Europe and World	
Mushrooms and yeast	Moulds
Olives, *sauerkohl* and *sauerruben*	Pickled fruits and vegetables
Grape vinegar and wine vinegar	Vinegar
Wines and citron	Fermented fruits
Quark, kefir, *filmjolk*, *crème fraiche*, *smetana* and *skyr*	Cultured milk

Source: Adapted from G. Campbell-Platt (1987)

History of Fermented Foodstuffs

Exactly how fermented food articles entered the human diet is unclear. Evidence of fermented foodstuffs dates back over 3000 years in several ancient civilizations, *i.e.*, Babylon, Egypt, Mexico and Sudan. The earliest sign of wine dates from about 8000 years ago in Georgia, and there is evidence that people were fermenting drinks in Babylon circa 5000 BC, Egypt circa 3150 BC, Mexico circa 2000 BC and Sudan circa 1500 BC. Fermented leavened bread was produced in Ancient Egypt circa 1500 BC and milk was fermented in early Babylon circa 3000 BC as well. There was also evidence of fermented meat products being produced for King Nebuchadnezzar of Babylon. China is thought to be the birthplace of fermented vegetables and the use of *Aspergillus* and *Rhizopus* moulds to make food. The book called *Shu-Ching* written in the Chou dynasty in China (1121-256 BC) refers to the use of *chu* a fermented grain product. The research efforts undertaken by the Danish Carlsberg scientists greatly accelerated the gain of knowledge about yeast and brewing. Knowledge about traditional fermentation technologies and products has been adapted over generations for centuries.

The history of fermented food is lost in antiquity. It may have been a mere accident when people first experienced the taste of fermented food. The first fermentation must have started with the storage of surplus milk, which resulted in a fermented product the next day. After drying, fermentation is the oldest food preservation method. Fermentation became popular with the dawn of civilization because it not only preserved food but also gave it a variety of tastes, forms and other sensory sensations. Slowly, people have realized the nutritional and therapeutic value of fermented foods and drinks, and this has made fermented foods even more popular.

Fermentation Process

Fermented foodstuffs are processed with enzymes or microorganisms that break down the carbohydrates content of food under conditions with little or no oxygen present. Fermentation organisms produce alcohol, lactic acid, acetic acid and all *bio-preservatives* that retain nutrients and prevent spoilage. Lactic acid fermentation and alcohol fermentation produce familiar items like yogurt and wine. The fermentation process breaks down the lactose, thus, mitigating a potentially problematic sugar and decreasing the carbohydrates content. Fermentation is a relatively efficient low energy preservation process, which increases the shelf life and decreases the need for refrigeration or other form of food preservation technology. It is therefore a highly appropriate technique for use in developing countries and remote areas where access to sophisticated equipment is limited.

Benefits of Fermented Foods

Many different health benefits that are now being associated with fermented foods are as follow:

- Enrichment of diet through the development of a diversity of flavours, aromas and textures in food substrates
- Preservation of substantial amounts of food through lactic acid, alcohol, acetic acid and alkaline fermentations

- Biological enrichment of food substrates with protein, essential amino acids, essential fatty acids, minerals, vitamins and even antibiotics
- Detoxification during fermentation and/or degradation of antinutritrive factors like protease inhibitors, phytic acid, amylase and other enzyme inhibitors, such as tannins, lectins and saponins
- A decrease in cooking times and fuel requirements
- Functional (probiotic) effects

Preservation

Fermentation is one of the classical methods to preserve food items. Preservation of foods by fermentation is a widely practiced and ancient technology. Fermentation ensures not only increased shelf life and microbiological safety of a food but also may also make some foods more digestible. Fermentation allowed people to preserve foodstuffs that would have spoiled for use during non-growing seasons and in times of famine. Vegetables, fruits, milk, fish and meat are highly perishable, and our ancestors used whatever techniques they could discover to store foods from seasons of plenty for later consumption. Before the modern era of hot water bath canning and vinegar brines, people used to preserve vegetables and fruits in cans and other airtight containers using lactic-acid fermentation. Fermentation increased the vitamin and mineral content of the food, provided a rich source of valuable digestive enzymes and preserved the food for months at a time.

There are several options for preserving fresh fruits and vegetables including drying, freezing, canning and pickling. There are examples from around the world of the fermented foods that have played role in preserving food to enhance food security and increasing the range of raw materials that can be used to produce edible food products and removing anti-nutritional factors to make the food safe to eat.

Improved Flavour

Humans have always appreciated the distinctive flavors resulting from the transformative power of microscopic bacteria and fungi. Fermented foods and drinks are quite literally alive with flavour and nutrition. The lactic acid causes the food to pleasantly sour. The flavours of fermented foodstuffs tend to be strong and pronounced as of stinky aged cheeses, tangy sauerkraut, rich earthy *miso* and smooth sublime wines. Cacao beans must first be fermented before they are turned into delicious rich dark chocolate. This deepens the colour and enriches the flavour, but most importantly, it destroys the astringent tannins present in raw cacao.

Nutritional Value

Fermented foods supply proteins, minerals and other nutrients. Fermentation can increase the concentration of vitamins and minerals in foods. The process also increases the digestibility of foods, improving nutrient absorption. The traditional societies soaked, sprouted, or fermented their grains prior to consuming them. Sprouting, fermenting and soaking grains can increase vitamins and minerals content availability by 300-500%.

Detoxification

Fermentation can render previously inedible or even dangerous food article edible and somewhat nutritious. The lectins, gluten and phytates in grains can be greatly reduced by fermentation. Fermentation also removes toxins from foods. Cassava is an enormous tuber native to the tropical regions of the Americas that has also become a staple food in equatorial regions of Africa and Asia. Certain varieties of cassava contain high levels of cyanide and are poisonous until they have undergone a soaking fermentation. The fermentation process eliminates cyanide, rendering the cassava edible and nutritious. Nitrites, prussic acid, oxalic acid, nitrosamines and glucosides are some other potentially toxic chemicals found in food articles that can be reduced or eliminated by fermentation. The fermentation process also destroys natural plant inhibitors such as goitrogens in soybeans, which inhibits the thyroid gland and increases the availability of isoflavones. All grains contain a compound called phytic acid, which can block absorption of zinc, calcium, iron, magnesium and other minerals and lead to mineral deficiencies. Fermenting grains by soaking them before cooking neutralizes phytic acid, rendering the grains more nutritious.

Many fruits and vegetables contain naturally occurring toxins and anti-nutritional compounds. These can be removed or detoxified by the action of microorganisms during fermentation. For instance, the fermentation process that produces the Sudanese product *Kawal* removes the toxins from the leaves of *Cassia obtusifolia* and fermentation is an important step in ensuring that cassava is safe to eat.

Effects on Weight Loss

Fermented food articles tend to be nutritionally dense and filling and some are low-calorie and possess qualities that support weight loss efforts. People who drink *kombucha* report benefits like enhanced wellbeing, increased energy and weight loss.

Fermented foods do not directly affect the success of your weight loss plan. A fermented food comes from their ability to help manage blood sugar. Many reports suggested that fermented fruits and vegetables cause a lower glycemic response than their fresh counterparts, meaning that a person with diabetes or hypoglycemia can eat the fermented versions of these foods and is maintaining blood sugar levels low, including in people without diabetes or hypoglycemia, which supports weight loss. Fermented foods possess qualities that support weight loss efforts but do not cause weight loss. They stop craving for sugar and carbohydrates in processed foods since cravings for sugar and carbohydrates are the number one cause of weight gain. Losing these cravings means losing weight fast

Health Benefits

Fermented foodstuffs are not only important sources of preserved nutrients but have also great potential in maintaining health and preventing diseases. Traditional fermented food articles can serve as functional food when eaten for specified health purposes, because of their content of one or more nutrients or non-nutrient substances, which can confer health benefits.

Since ancient times, fermented foods are considered extremely beneficial to one's health as they encourage the growth of bacteria required by the intestine, help digestion, elevate overall nutrition levels and strengthen the body's immune system. Fermented foods are also known to provide the body with an additional amount of omega-3 fatty acids, lactic acid, B-group vitamins and various other chemicals that help in fighting off cancer and harmful bacteria. There are various types of fermented food commercially available now a day such as *kimchi, nham, balao-balao* and others. *Kimchi* is a popular fermented food consumed in Korea. It is a spicy fermented cabbage, which is known to have medicinal properties along with even anti aging benefits. The United States alternatives of available fermented foods would include sauerkraut, which is fermented cabbage, kefir, fermented milk, or yogurt. Fermented foods are also known to help detoxification and preservation of foods. Additionally, fermented milk has *Lactobacillus acidophilus*, which in turn produces good bacteria like *Lactocidin, Acidophillin, etc.*, which prevent the occurrence or growth of harmful microorganisms in the body. Overall, the fermented foods are an excellent source of minerals, amino acids and vitamins, and hence, should constitute an important part of our diet. The B-group vitamins that increase the energy levels are found in ferment milk, whereas, vitamin C is obtained from fermented vegetables, which are also easier to digest. Various cereals or grains typically contain harmful phytates, gluten and lectins, which can be considerably reduced by fermenting them. During fermentation of milk, lactose is broken down and reduces the carbohydrates content making it healthier and more nutritious.

Throughout ancient history, fermented foodstuffs have played a role in sustaining thriving civilizations, and many cultures continue to consume these health-promoting traditional food articles. Captain James Cook, the eighteenth-century English explorer who extended the far reaches of the British Empire, was recognized by the Royal Society for having conquered scurvy (vitamin C deficiency) among his crews by sailing with large quantities of sauerkraut. On his second round the world voyage in the 1770s, sixty barrels of kraut lasted for twenty-seven months and not single crewmember-developed scurvy, which previously had killed huge number of the crews of long sea voyages. Among the many lands Cook discovered and delivered into the Crown's realm were the Hawaiian Islands. The Polynesian people who crossed the Pacific Ocean and populated Hawaii more than a thousand years before Captain Cook also sustained themselves through the long voyage with fermented food, in this case poi, a thick starchy taro root porridge still popular in Hawaii and throughout the South Pacific. Roman soldiers often subsisted on long-fermented sourdough bread, which survived long treks well (conquering the known world on a diet of bread-fermentation must be effective stuff). The Inuit traditionally wrap whole seabird carcasses in seal pelts and bury them underground to ferment for months, and rotting fish is another feature of their traditional diet. Fermented dairy is a major aspect of the traditional *Masai* diet, as is clotted steer's blood. Fermented foodstuffs improve health in following ways:

Digestion: Fermented food articles provide nutritionally dense and easily digestible choices. This process not only preserves nutrients but also it breaks them down into more easily digestible forms. Soybean, which is extraordinarily protein-rich food, is largely indigestible without fermentation. Fermentation breaks down the soybean's complex protein into readily digestible amino acids, giving us traditional Asian foodstuffs such as *miso, tempeh* and *tamari* (soy sauce), which have become staples in contemporary Western vegetarian cuisine.

Milk is difficult for many people to digest. *Lactobacllli* (a type of bacteria present in fermented dairy products and many other types of ferments) transform lactose, the milk sugar that so many humans cannot tolerate, into easier to digest lactic acid. Likewise, wheat that has undergone fermentation is easier to digest than unfermented wheat. A study in the journal Nutritional Health compared unfermented and fermented versions of a mix of barley, lentils, milk powder and tomato pulp and found that starch digestibility almost doubled in the fermented mixture.

Cabbage, the starting material for sauerkraut, is very high in glutamine, an amino acid that is also very beneficial to the gut. It is usually easier to digest fermented cabbage than unfermented cabbage because it is somewhat predigested by the fermentation process.

Lactic acid fermented weaning foods are traditionally produced in developing countries, both to improve the safety of the food and to improve its digestibility. Starchy porridges are commonly fed to weaning infants in developing countries. By acidifying the porridge through lactic acid fermentation, starch is hydrolyzed into shorter chains of glucose and dextrose, which reduce the viscosity of porridge and increase its energy density. Thus, the child is more able to meet its energy requirements.

New nutrients: Fermentation also creates new nutrients. As microbial cultures go through their life cycles, they create B vitamins, including folic acid, riboflavin, niacin, thiamine, and biotin. According to Food and Agriculture Organization of the United Nations, which actively promotes fermentation as a critical source of nutrients worldwide, fermentation improves the bioavailability of minerals present in food. Bill Mollison, author of the Permaculture Book of Ferment and Human Nutrition, calls the action of fermenting foodstuffs- *a form of pre-digestion. Natto*, a Japanese form of fermented soybeans, is high in vitamin K_2 (MK-7), which is vital for bone, cardiovascular and dental health. Fermenting the foods produces certain nutrients like an increase in the B complex (folic acid and vitamin B_{12}) and vitamin K.

Healthy heart: Fermentation of milk is reported to produce certain peptides (chains of amino acids) that inhibit an enzyme, which tends to increase blood pressure. Studies on humans indicate that fermented milk could help lower blood pressure when used on a daily basis. Some studies have reported a modest reduction in cholesterol levels after consumption of fermented milk products for two to four weeks. According to research conducted by scientists in Japan, *natto* has the highest fibrinolytic activity among 200 food items available worldwide. The enzyme produced in the fermentation process, *nattokinase*, is a very powerful agent contained in the sticky part of *natto* that dissolves blood clots, which can lead to heart attacks, strokes and senility.

Other effects: There are tonnes of possible benefits of fermented foodstuffs to our body, including protection from rotavirus diarrhea, reduction in children's cavities and prevention of reoccurrences of inflammatory bowel disease.

Anticancerous: Some fermented food items have been shown to function as antioxidants, scavenging cancer precursors known as *free radicals* from the body. Many studies have shown that traditionally fermented soy aids in preventing and reducing a variety of diseases including certain forms of heart disease and cancers. One study on the culturing method involved in the production of the Japanese traditional food *Miso* concluded that the culturing process itself led to a lower number and growth rate of cancer cells. *Miso*, a fermented form of soybean is particularly rich in isoflavone, aglycones, genistein and daidzein, which are believed to be cancer preventatives.

Natto also contains vitamin K_2 and isophrabon, which help to prevent diseases such as osteoporosis and breast cancer and to slow down the aging process. According to World Health Organization, the Japanese, who consume large amounts of fermented soy foods like *natto* and *miso* along with green tea, ginger and ocean herbs, have the longest lifespan of any people in the world.

Probiotics: The fermented foodstuffs introduce helpful probiotics the *good* bacteria that our body is dependent absolutely on such as lactobacilli, which increase the quantity, availability, digestibility and assimilation of nutrients in our body to our guts. Another metabolic byproduct of probiotics is a variety of short chain fatty acids that are actually used as a source of fuel by intestinal cells. The combination of keeping the intestinal cells built up and functioning well, the increased nutrients and the crowding out of pathogenic bacteria help to boost our immune system. Some amino acids, growth factors, antioxidants and antimicrobial factors are also produced. Yogurt, kefir, buttermilk and some other fermented food articles include live strains of bacteria that support the production of good bacteria in the human intestine and may also support the immune system. Consuming foodstuffs that contain live lacto-bacteria has protective benefits against harmful, pathogenic microorganisms. Not all fermented foodstuffs are still alive when we eat them. Certain food items, by their nature, cannot contain live cultures. Breads, for instance, must be baked, thereby killing the organisms present in them. However, many fermented-food articles can be consumed live, especially those involving *Lactobacilli*, and alive is the most nutritious way to eat them.

Immunity: Fermented food items are not only a great source of variety in our diet but also they can be an important way to build our intestinal health and immune defenses. Cells of *Lactobacilli* create omega-3 fatty acids, essential for cell membrane and immune system function. According to some reports, the culturing process generates copious amounts of naturally occurring ingredients like super oxide dismustase, glucose tolerance factor (GTF) chromium, detoxifying compounds like glutathione, phospholipids, digestive enzymes and β-1,3-glucans.

Fermented foodstuffs and drinks provide so many benefits from improved digestion, energy and immunity to more youthful skin and hair.

One of the most healthful fermented foods is kefir. Kefir is an ancient cultured enzyme-rich food full of friendly microorganisms that balance your *inner ecosystem* and strengthen immunity.

Antimicrobial: The lactic acid-producing *Lactobacilli* in this product can favourably alter acidity of the intestine, helping in preventing the overgrowth of unfriendly putrefactive bacteria, moulds and yeasts such as *Candida Albicans*, giving you a better chance of achieving optimal health and preventing many chronic disorders. Probiotics produce certain antimicrobial compounds that kill some of the undesirable pathogens in the intestines, including yeasts, viruses and bacteria. The associated decrease in the production of potentially cancer-causing toxins by the unfavorable microorganisms may help in decreasing the incidence of colon cancer. Probiotics also make certain vitamins needed by the body including folic acid, vitamin B_6, niacin and vitamin K.

Kimchi, a fermented spicy cabbage, is very well studied for its medicinal, antimicrobial and anti-aging properties. Scientists found that chickens infected with avian flu started to recover after being fed a *kimchi* extract.

Fermenting food increases the vitamin content: Fermentation processes can increase levels of vitamins in the final product. In Southern Africa, sorghum beer contains relatively high levels of riboflavin and nicotinic acid, which are important for people consuming a high maize diet. Pellagra (a vitamin deficiency disease associated with high maize diets) is unusual in communities in which sorghum beer is consumed. Palm wine in West Africa is high in vitamin B_{12}, which is very important for people with low meat intake, and who subsists primarily on a vegetarian diet. *Pulque* is a traditional alcoholic drink made in Mexico. It is an important source of vitamins. Similarly, *idli* (a lactic acid bacteria fermented product consumed in India) is high in thiamine and riboflavin.

Side Effects of Fermentation

During food fermentation or decomposition, certain waste products are produced by the bacteria, which break down the food. The major byproducts are alcohol, ammonia, acids and vinegar. Many fermented foods, such as soy sauce, contain a significant amount of alcohol. Of course, the alcohol in fermented foods is usually a small quantity but even small amounts of alcohol affect the body cells. Another product of fermentation is ammonia. Fermented soy may be as much as 15% *ammonia*, which is dangerous to health. Vinegar, in the form of *acetic acid*, also results from food fermentation. This acid gives fermented foods sour or sharp taste. That sharp taste is a signal to the body that the food should not be eaten, as it is harmful. Vinegar prevents the digestion of foods, thus, a food filled with vinegar and other similar byproducts would seem to be indigestible. During fermentation, acid is produced in the form of *lactic acid*, which is a waste product. In exercised or worked harder might notice a stiffness or soreness in muscles. That stiffness results from a buildup of lactic acid in the muscles. Try to eat fermented foods that contain lactic acid, which may not make stiffness. Other acids are also present in fermented foods. Carbonic acid is found in fermented foods and soft drinks. All of these acids, which are feeding on the decomposing or *fermented* foods, are the wastes produced by the bacteria.

Many commercially processed fermented foods are not actually fermented at all but have some chemicals added to produce some of the qualities of traditionally fermented food artificially. Olives are treated with lye, and then, soaked in a brine solution with acetic and lactic acid, and the chemicals such as potassium sorbate and sodium benzoate. Similarly, pickles are not fermented but soaked in salt and vinegar. Sauerkraut often has vinegar added to create sour taste rather than allowing the natural sour flavour to develop as a result of fermentation. Most commercially fermented products are pasteurized. Pasteurization, which kills the beneficial bacteria and stops the fermentation process, is considered necessary in large-scale production of foods that will be bottled and transported because it extends the shelf life and halts the continued production of carbon dioxide by the live bacteria, which would cause dangerously high pressure to build up inside sealed jars and could lead to explosions.

Conclusion

Although already used for thousands of years in human nutrition, it is today that the health promoting functionality of fermented foodstuffs becomes more and more in the focus of scientific and public interest. Fermented foods enjoy worldwide popularity as attractive, wholesome and nutritional components of our diets. They are produced on an enormous scale empowering a huge variety of ingredients and manufacturing techniques. When traditional homemade foods or high-tech products derived from genetically modified organisms are used, the safety of the consumers remains important. Fermented foods have always been generally regarded as safe. Consuming fermented foods on a regular basis will not only provide better nutrition but will enhance your inner ecology, which will heighten overall health. Fermented foods may positively affect the level of absorption of various nutrients and may defend against infectious diseases that limit growth potential.

The benefits of fermented food are not stemming from some magical property inherent to them but rather the simple fact that introducing beneficial bacteria into our bodies restores the balance of intestinal flora. Despite our best attempts to recreate perfect primal environments through diet and exercise, we still live in an increasingly sterile world. Introducing fermented foodstuffs into diet can help in normalizing the things and get our guts in good shape. Sadly, modern culture has sacrificed many of the advantages of traditionally fermented healthy food articles for faster and cheaper methods of mass production. Increases in viruses, intestinal parasites and chronic health problems are warning signs that it is time to return to the use of traditionally fermented food items.

Kombucha makes for a great refreshing drink so just make sure you watch the sugar content. People have been eating fermented food items for thousands of years willingly and even longer accidentally. The evidence shows there is definitely something to it, and I think it can be a vital part of a healthy Primal diet. The intestinal world is a complex ecosystem. Some bacteria can ruin our health, while others will be very protective of our intestinal world. Bacterial secretions can directly affect the cytochrome 450 detoxification system by which the liver purifies the blood of harmful chemicals that are both inadvertently created by the body and ingested from the environment.

The fermenting the foods also produces certain nutrients like an increase in the B complex (folic acid and vitamin B_{12}) and vitamin K. Probiotics also produce short-chain fatty acids, an important energy source for the colon. Some amino acids, growth factors, antioxidants and antimicrobial factors are also produced. Fermentation also destroys certain anti-nutrients like phytates and oxalates that interfere with the absorption of minerals like iron and calcium.

People with lactose intolerance are unable to digest lactose, the milk sugar, leading to bloating, flatulence and abdominal pain and diarrhoea. Consumption of fermented dairy products such as curd and buttermilk helps in such cases because the bacteria produce the enzymes necessary to break down lactose. Probiotics also reduce intolerance symptoms such as diarrhea. Fermented food items, especially dairy foodstuffs, contain certain strains of bacteria that promote health of the gastrointestinal tract. Prebiotics have also been studied for their beneficial gastrointestinal effects.

The energy boost, which lacto-fermented power food articles delivers is due to the fermentation process, which gives better assimilation of proteins, carbohydrates, fats and micronutrients including minerals, vitamins, flavanoids, carotenoids and hundreds of other newly discovered small protective molecules called antioxidants.

An unique blend of fermented beans, grains, sprouts, seeds and vegetables, which uses 2 strains of special symbiotic yeast and 12 strains of friendly *Lactobacilli* bacteria, which have been toughened to withstand changes in acidity, alkalinity and changes in temperature. Some people underestimate the importance of fermented food articles and health benefits that you can experience include lower cholesterol, efficient digestive functioning and a stronger immune system.

People who start eating a diet high in raw and fermented foodstuffs reverse the course of cancer, stop diabetes in their tracks and notice an increased level of heart fitness. Fermented foodstuffs cannot only be a great source of variety in your diet but can be an important way to build your intestinal health and immune defences.

With benefits like improved digestion, better skin, stronger immunity and anti-aging effects on body ecology, the fermented food items and drinks are a cornerstone of healing your body and restoring your immunity. *Kombucha* is a fermented tea that has become popular in health food stores and tastes a bit like sparkling apple cider. Consumed for thousands of years, there is a great deal of mystery surrounding *kombucha* and its health benefits.

In addition to beneficial bacteria and yeast, *kombucha* contains gluconic acid, a small amount of alcohol and vinegar. Depending upon how the *kombucha* is made, it may also contain B vitamins. Research has shown that *kombucha* may have anti-microbial effects against harmful bacteria like *E. coli*, a range of vitamins and minerals and possibly anti-fungal properties as well. Still, there are some reasons that *kombucha* may not be the ideal probiotic drink, especially if you struggle with Candida. *Kombucha* tea is made from wild airborne yeast and bacteria that facilitate its fermentation process.

Thus, fermented foods offer tremendous potential for promoting health, improving nutrition and reducing the risk of various diseases worldwide. Advances in fermented foods production technology have led to a variety of products that suit diverse cultural tastes. Infants, children, adults and elderly can consume fermented foods for their good taste and their general nutritional value.

■■■

37 Honey and its Uses

S.K. Sharma[1], Yogesh Kumar[1] and M.K. Rana[2]

[1]Department of Entomology, E-mail: <sk_sharma@hau.ernet.in>
[2]Professor Department of Vegetable Sciences, CCS Haryana Agricultural University, Hisar, Haryana, India, E-mail: <mkrlotus@gmail.com>

Honey was the most important primary bee product used by the man in ancient time and the history of its use is parallel to the history of man. The evidences of its use as a food source and as a symbol employed in religious and therapeutic ceremonies are available in the literature. Honey is a natural sweet substance produced by honeybees from the nectar of blossoms or from the secretion of living plant parts or excretions of plant sap sucking insects on to which honeybees collect, transform and combine with specific substances of their own and store in the comb to mature and ripen.

Honey contains natural sugars (glucose and fructose), minerals (magnesium, potassium, calcium, sodium chlorine, sulphur, iron and phosphate) and vitamins (B_1, B_2, B_3, B_5, B_6 and C). Besides these, copper, iodine and zinc exist in it in small quantity. Several kinds of hormone are also present in it. It is also fat free, cholesterol free and sodium free.

The types of flower from which the bees collect nectar determine flavour and colour of the honey. Buckwheat honey is dark in colour with a distinct flavour, whereas, cotton honey is whitish in colour. The darker shades of honey are believed to have more antioxidants. Honey, natural syrup produced by bees, is similar to invert sugar with a small but variable excess of levulose (fructose). The composition and flavour of honey varies with the plant source of nectar, processing and storage but a typical composition is 41% fructose, 34% glucose, 18% water, and 2% sucrose with a pH of 3.8 to 4.2. Honey has a fuel value of about 3307 cal/kg.

Honeys are classified mainly as per the floral origin (litchi, *Jamun*, mustard honey, *etc.*), seasonal honey (spring, Greeshma, or *kartika* honey) or according to the place where honey is produced (Mahabaleshwar honey of Maharashtra, Coorg honey of Karnataka, Kashmir honey of Jammu and Kashmir and English honey of United Kingdom).

Honey due to its purity, natural versatility and varied taste provides health benefits. It not only stimulates taste buds and appetite but also offers incredible antiseptic, antioxidant and cleansing properties for our body and health from eye conjunctivitis to athlete foot. Despite high acid content, it is easily digestible because sugar molecules in honey can convert into other sugars, *e.g.*, fructose to glucose. It helps the kidneys and intestines to function better. It is cheap, making it potentially useful for treating wounds in earthquake-stricken and war-torn areas where running water is scarce and often contaminated.

Patterns of Honey Production

New patterns of honey production have been developed to bring awareness about the honey among the beekeepers and honey packers and to meet the increasing demand of consumers. These are given as under:

- Generally, the traditional honey hunters or tribal kill the bees by fire, and cut and squeeze the comb along with pollens, brood, wax, *etc*. Honey obtained by pressing or squeezing of combs is known as *pressed* or *squeezed honey*, which is highly contaminated and spoiled soon due to fermentation.
- From hives, the honey frames are removed and extracted with the help of centrifugal machine. Honey collected from the combs by centrifugal machine is termed as *apiary* or *extracted honey*. This can be done repeatedly without disturbing the colonies.
- Comb honey may be available in pieces of the original comb, cut and wrapped (cut comb) or pieces of comb in glass jar of liquid honey (chunk honey). Very small frames filled with honey by the bees or sealed honey in comb is termed as *section comb*.

Formation of Honey

Bees prepare honey mainly from nectar of the flowers but other plant saps and honeydew are also used for this purpose. Since each bee sucks the liquid up through its proboscis and collects it into the honey sac, a small amount of enzymes is added and water is evaporated. The worker bees collect nectars the sugar-rich flowers and return to the hive, which the bees use their *honey stomachs* to ingest and regurgitate the nectar a number of times until it is partially digested, and then, it is stored in the honeycomb. Nectar is high in both water content and natural yeasts, which if unchecked, would cause the sugars in the nectar to ferment. After the final regurgitation, the honeycomb is left unsealed. Bees inside the hive fan their wings, creating a strong draft across the honeycomb, which enhances evaporation of much of the water from the nectar. The reduction in water content, which raises the sugar concentration, prevents fermentation. The bees at this stage seal the cells by wax capping and honey at this stage is considered as *ripe honey*, which is removed from the hive by the beekeeper, has a long shelf life and will not ferment.

Physical Characteristics of Honey

Viscosity: The freshly extracted honey is a viscous liquid. Its viscosity depends on a large variety of substances, and therefore, varies with its composition and particularly with its water content. Viscosity is an important technical parameter during honey processing since it reduces honey flow during extraction, pumping, settling, filtration, mixing and bottling. Raising the temperature of honey, a phenomenon widely exploited during industrial honey processing, lowers its viscosity. However, concerning viscosity, some honeys show different characteristics, *i.e.*, heather (*Calluna vulgaris*), manuka (*Leptospermum scoparium*) and *Carvia callosa*, which are described as thixotrophic, meaning they are gel-like (extremely viscous) when standing still and turn liquid when agitated or stirred. By contrast, a number of *Eucalyptus* honeys show the opposite characteristics. Their viscosity increases with agitation.

Density: Another physical characteristic of practical importance is density. Honey density, expressed as specific gravity, is greater than water density but it also depends on water content of the honey. Because of the variation in density, it is sometimes possible to observe distinct stratification of honey in large storage tanks. The high water content (less dense) honey settles above the denser honey. Such inconvenient separation can be avoided by more thorough mixing.

Hygroscopicity: The strong hygroscopic character of honey is important both in processing and for final use. In end-products containing honey, this tendency to absorb and hold moisture is often a desired effect in pastries and breads. However, during processing or storage, the same hygroscopicity may become problematic, causing difficulties in preservation and storage due to excessive water content.

Surface tension: It is the low surface tension of honey, which makes it excellent humectants in cosmetic products. The surface tension varies with the origin of honey and is probably due to the presence of colloidal substances. Together with high viscosity, it is responsible for foaming characteristics of the honey.

Thermal properties: For designing a honey processing plant, its thermal properties have to be taken into account. According to its composition and state of crystallization, the heat absorbing capacity, *i.e.*, specific heat varies from 0.56 to 0.73 cal/g/°C and the thermal conductivity varies from 118 to 143x10 cal/cm^2/sec/°C. Therefore, one can calculate the amount of heat, cooling and mixing necessary to treat a certain amount of honey, *i.e.*, before and after filtration or pasteurization. The relatively low heat conductivity combined with high viscosity leads to rapid overheating from heat sources, and thus, there is need for careful stirring and for heating only in water baths.

Colour: Colour of liquid honey varies from clear and colourless (like water) to dark amber or black. Various honey colours are all nuances of yellow amber, like different concentrations of caramelized sugar, which has been used traditionally as a colour standard. Colour varies with botanical origin, age and storage conditions but transparency or clarity depends on the amount of suspended particles such as pollens. Less commonly occurring honey colours are bright yellow (sunflower), reddish undertones (chestnut), greyish (eucalyptus) and greenish (honeydew). Once crystallized, the honey turns lighter in colour since the glucose crystals are white. Some of the honeys reported as white as milk in some parts of East Africa are finely crystallized honeys, which are almost water white, *i.e.*, colourless in their liquid state.

The most important aspect of honey colour lies in its value for marketing and determination of its end use. Darker honeys are more often for industrial use, while lighter honeys are marketed for direct consumption.

Crystallization: Crystallization is an important characteristic for honey marketing, not for price determination. In temperate climate, most of the honeys crystallize at normal storage temperature due to the fact that honey is an over saturated sugar solution, *i.e.*, it contains more sugar than that remains in solution. Many consumers still think that if honey has crystallized, it has gone bad or adulterated with sugar, although crystallization results from the formation of monohydrate glucose crystals, which vary in number, shape, dimension and quality with the honey composition and storage conditions. The optimum temperature for fast crystallization is around 14°C. However, the presence of solid particles, *e.g.*, pollen grains and slow stirring, usually results in quicker crystallization. Slow crystallization produces bigger and more irregular crystals.

During crystallization, water is freed. Consequently, the water content in liquid phase increases, and with it, the risk of fermentation is increased. Thus, the partially crystallized honey may have preservation problems, which is why the controlled and complete crystallization is often induced deliberately.

Composition of Honey

The average composition of honeys, more or less representative of all honeys, is described below:

Major Constituents

Sugars: These sugars account for 95 to 99% of the honey dry matter, and majority of these are the simple sugars, *i.e.*, fructose and glucose, which represent 85-95% of the total sugars. Generally, fructose is more abundant than glucose. This predominance of simple sugars, particularly the high percentage of fructose, is responsible for most of the physical and nutritional characteristics of honey. Other sugars, such as disaccharides (sucrose, maltose and isomaltose) and a few trisaccharides and oligosaccharides are also present in small quantity. Though quantitatively these sugars are of minor importance, their presence can provide information about adulteration and botanical origin of the honey.

Water: Quantitatively, it is the second most important component of honey. Its content is critical since it affects the storage of honey. Only the honey with less than 18% water can be stored with little to no risk of fermentation. The final water content during production depends on a number of environmental factors such as weather and humidity inside the hive, nectar conditions and treatment of honey during extraction and storage. It can be reduced before or after extraction by special techniques.

Minor constituents

Organic acids: In honey, organic acids are most important, and of these, gluconic acid, which is a byproduct of enzymatic digestion of glucose, usually predominates. Organic acids are responsible for the acidity of honey and contribute largely to its characteristic taste.

Minerals: These are present in very small quantity, potassium being the most abundant. Dark honeys, particularly honeydew honeys, are richest in minerals.

Nitrogenous compounds: Among nitrogenous compounds, the enzymes originated from salivary secretions of the worker honeybees have an important role in the formation of honey. Their commercial importance is not related to human nutrition but related to their fragility and uniqueness. Thus, their reduction or absence in adulterated, overheated, or excessively stored honeys serves as an indicator of freshness. The main enzymes in honey are invertase (saccharase), diastase (amylase) and glucoseoxidase. Traces of other proteins, enzymes, or amino acids as well as water-soluble vitamins are thought to result from the presence of pollen grains in honey.

Virtually absent in newly produced honey, hydroxymethylfurfural (HMF) formed during storage or heating is a byproduct of fructose decay. Thus, its presence is considered the main indicator of honey deterioration.

Though some of the substances responsible for honey colour and flavour have been identified, the majority are still unknown. It is more than likely that honeys from different botanical origins contain different aromatic and other substances, which contribute to specific colour and flavour, and thus, allow distinguishing one honey from another.

Contamination during production

The location of colonies in industrial zones or other areas with considerable air pollution such as cities, can lead to considerable contamination to the various hive products with noxious or toxic chemicals. In Canada, United States of America, United Kingdom and Italy, honeybees were used to monitor environmental pollution since the accumulations of certain metals and other substances could be measured in hive products, mostly in pollens but also in honey. Agricultural use of toxic chemicals is another common and very likely source of contamination.

Radioactive contamination throughout Europe after the Chernobyl nuclear reactor incidence showed in nectars and honeys for a considerable time. Since most of the contamination was due to plant uptake of radioactive elements replacing normally occurring minerals, the overall content remained relatively low although closer to the accident scene, and immediately after the incidence, the safety limits were exceeded. This was mostly due to short-lived iodine isotopes, as for example in Austria.

The worldwide exchange and shipping of honeybee colonies and queens have led to the introduction of new honeybee diseases in many parts of the world. Unadapted bees cannot resist the new infections, and hence, help from the beekeeper is required. Such help usually involves chemical treatments. If unsafe chemicals are used or even if relatively safe chemicals are applied in exaggerated quantities or at inappropriate times, honey is contaminated. Problems with such contaminations have increased in recent years. Buyers are increasingly alert and test regularly for residues. Another source of contamination is the treatment of combs against wax moth during storage. All available chemical treatments leave residues in the wax and only abundant aeration (ventilation) for at least a couple of weeks can reduce the hazard. A well-ventilated storage without chemicals is preferred.

Contamination during Harvesting

Many harvesting methods are available to separate the bees from their honeycombs. Combs can be taken out one at a time, and shaking and brushing may remove the bees. Whole supers can be cleared of bees with a strong air blower. An inner cover or special board with a one-way bee escape can be placed below the honey super. Up to one deep or two shallow supers can thus be cleared in 24 hours, if enough space is available below. However, this method cannot be recommended, if the colonies are sitting unprotected in the sun, as it may cause melting of combs in unventilated supers. None of these three methods will contaminate the harvested honey.

The use of unpleasant smelling chemicals to drive the bees away is a technique preferred by many beekeepers because it is quick and easy. Some of the chemicals, *e.g.*, carboxylic acid, benzaldehyde, nitrobenzene and others, are illegal for use in many countries, leave unpleasant flavour and odour, are toxic and are absorbed by wax and honey. Careful use of butyric acid, marketed as *Bee-go* in the United States of America, has so far not been proven to produce any contamination, but in general, the use of chemicals during harvesting cannot be recommended.

Excessive use of smoke during harvesting will flavour the honey quickly. Microscopic contamination with soot can also be detected. No chemical should be included in the smoke. Though unavoidable with some bees, the heavy use of smoke can be reduced by selecting more favourable (perhaps more inconvenient) harvesting time (weather and time of day) and shorter and more frequent harvests.

Cleanliness

Honey in combs (be it in supers of frame hive beekeeping or in the broken combs from top bar or traditional fixed comb beekeeping) already needs to be regarded as a food product. From a microbiological point of view, mature honey is such a stable product that neither it alters nor it permits the multiplication of bacterial or fungal organisms. It can nevertheless be contaminated either by non-biological substances or by potential human pathogens. Therefore, every caution and care in hygiene should be taken to prevent any form of contamination.

This general requirement must be taken into account during all processing phases. Already in the comb, contrary to many beekeepers' beliefs, honey is exposed to the danger of contamination since the surface area of contact with the environment is very large. Contact with humid air (during days between harvesting and extraction), with soil (supers set on the ground, truck bed, honey house floor or combs and frames dropped on the ground), unprotected transportation on dirt roads or in dirty buckets without a lid during comb harvesting and exposure to insects and other animals, can adversely affect honey quality.

The extraction room needs to be exceedingly clean as well as the space where the honey supers or combs are stored prior to processing. If processed outside, processing should not be done during a windy or rainy day. All surfaces, hands and containers coming in contact with the honey need to be particularly cleaned. The need for clean water may influence the site of processing centres or the feasibility of beekeeping in certain areas. In many countries, there are explicit rules to which any honey producer has to adhere, as far as the minimum facilities and cleanliness in the extracting room are concerned. Containers and processing equipments need to be made of material compatible with this very acidic food. No utensil made of copper, iron and zinc should be used as these elements dissolve into the honey and may affect the colour and flavour and may reach to a toxic level. If further processed into other products, the chemical reactions of contaminants with other ingredients may cause strange discolouration and off-flavour. Instead, stainless steel, glass and food grade plastic can be recommended. Galvanized steel (zinc) may be used on surfaces, which come into contact of honey only for short periods, such as in extractors. The used containers need to be free of any odour since honey will absorb these very quickly. Storage containers made of improper material can be coated completely with beeswax or food grade plastic liners to avoid any direct contact. However, there is no adequate protection if the containers have been used previously for toxic chemicals.

Processing of Honey

Uncapping is the first real step of honey processing. It consists of removal of thin wax layer that seals the honey cells. The wax caps can be sliced off with a thin long sharp knife heated by steam or electricity. Large number of frames is more rapidly processed with partially or completely automated uncapping machines, which cut or chop the wax caps with blades, chains, or wires.

In comb harvesting, the equivalent step is the comb selection (eliminating pieces of comb with pollens or even brood something that should have already been done during harvesting), the removal of bees, *etc.* and subsequently thorough mashing of combs. Processing proceeds further by either letting this wax and honey mixture separate by dripping through a screen (strainer) or by pressing it in special honey presses Processing further proceeds by either letting this wax and honey mixture separate by dripping through a screen (strainer) or by pressing it in special honey press. Modified centrifugal extractors can also be used.

After uncapping, the honey-frame processing proceeds to centrifugal extraction. Extractors range in size from a manual two frames model to motorized units extracting more than 12 deep supers at a time. More commonly, 24 to 72 frames radial extractors are used for commercial enterprises. The smaller units for part-time beekeepers can be made out of recycled materials. Though honey can be extracted faster and more completely at higher temperature, the combs may become softer and break. Therefore, the extraction temperature should not exceed 30°C.

Honey after extraction from the hive should be processed as soon as possible. Processing of honey is a sticky operation, in which time and patience are required to achieve the best results. Processing is necessary to destroy sugar tolerant yeast cells to prevent fermentation and to retard granulation in honey. It is recommended that processing for 20 minutes at 60°C, for 10 minutes at 65°C and for 2.5 minutes at 70°C does not change its quality. It is important to remember that honey is a food. Thus, it must be handled hygienically, and all processing equipments must be cleaned properly. After processing, the honey is filled in wide mouth glass bottles and food grade plastic jars of different sizes.

Purification

The next step is the removal of any impurities such as wax particles, other debris and air bubbles incorporated during extraction. There are two practical techniques, *i.e.*, (i) settling and (ii) straining. The first simply consists of leaving the honey in a suitably large container, so that impurities, *i.e.*, air bubbles, wax particles, insect pieces and other organic debris floating to the surface, can separate according to their specific weight, while mineral and metallic particles drop to the bottom. The surface scum can be removed carefully, or honey can be drawn off near the bottom for bottling without disturbing either surface scum or bottom sediment. Settling velocity varies with particle size (the smallest particles settle the slowest), container size and honey viscosity, *i.e.*, moisture content and temperature.

At a temperature of 25-30°C, settling is rather quick and can be completed in few days. Tanks have to be well covered to avoid excessive contact with air. The process can be accelerated by letting thc honey flow through special buffer tanks, prior to filling into the settling tanks. In these buffer tanks, the honey is heated through a water jacket, similar to a water bath, and then, forced to flow up and down through several compartments in the process of which impurities remain at the surface such a device works well with medium quantity, and once heated like this, the honey can also be filtered more easily. Subsequent settling frees the honey of air and foam, and if the containers are big enough, it allows some mixing of extractions from various colonies, *i.e.*, blending to achieve a certain degree of uniformity of the end-product. The disadvantage is the cost of containers for the extra storage lasting several days, which in large operations requires several very large tanks and large amounts of extra space.

Straining can be used instead of, or in addition to, settling. It is more frequently used in larger processing plants, where many tonnes of honey are processed every day and where it is inconvenient and uneconomic to immobilize honey for as long as is required for settling. Strainers can be simple metallic screens, preferably covered with a fine nylon mesh (fine nylon stockings are the best) or a nylon sack filter submerged in a tall narrow tank. The sack-like filter can also be made of several layers of increasingly finer metal screens (perforated metal sheets). These filters have the advantage of a large filter surface, which can be submerged to avoid any further inclusion of air. The finest mesh size used commonly has holes of 0.1-0.2 mm diameter. The temperature, for this kind of straining, must be near 30°C.

Finer filtering is usually only done in association with pasteurization and heating of honey to 77-78°C. It serves the purpose of removing all fine materials, including pollens, in order to delay crystallization for as long as possible. Such filtration requires high-pressure filters with diatomaceous earth. Since it requires heating, and particularly, because it removes some natural ingredients such as pollens, this honey cannot be sold as table grade honey in European Economic commission (EEC). Consumers in some countries regard it as inferior in quality, while it is the preferred quality for supermarkets and other large marketing chains, which want a product with a long shelf life in a homogeneous liquid state.

All the above purification methods can only be applied to liquid honeys. Therefore, it is preferable to use them immediately after extraction, when honey is still naturally liquid at right temperature. However, in processing plants of large buyers, it is often necessary to purify honeys that have already crystallized. In this case, the honey has to be melted first without destroying any of its characteristics.

Even the small buyer sometimes has to clean purchased honey since most of the beekeepers do not process their honey to sufficient standards for inclusion in other products and often not even well enough for bottling for direct retail sale. Here too, it is important to proceed as soon as possible after purchase, before the crystallization commences. On a small to medium scale, settling is usually the least expensive and least labour-intensive method, particularly if the honey barrels can be stored for a few days at a temperature range of 30-35°C. Additional straining assures that the raw product offers at least a minimum standard of hygiene requirements.

Extracted, cleaned, or purified honey is ready to be consumed directly or to be included into other products but processing technology does not end here. Other techniques are employed to prepare a product of uniform, constant and agreeable appearance, or to prevent the only possible storage problem, *i.e.*, fermentation.

Moisture content

Moisture content of honey is practically the most important quality parameter since it affects storage life and processing characteristics. Though moisture can be removed after extraction, only completely ripe honey, *i.e.*, combs with more than 75% of the honey cells sealed should be harvested. To achieve such results prior to end of the nectar flow, the colony has to have sufficient super space for storing incoming and ripening honey. When the average atmospheric humidity is not much above 60%, moisture content below 18% may be expected in the honey. In humid climate, even the sealed cells can contain honey with more than 24% even 28% moisture content. Combs containing fresh nectar should never be harvested since it can dilute and spoil the whole harvest unless the purpose of harvesting the honey is making beer.

Post-harvest reduction of moisture content can be achieved by leaving honey supers in warm rooms at 30 to 35°C and circulating warm air through them. At this time, the surface area of the comb relative to the honey mass is still fairly large and does not require any extra equipment for efficient evaporation. In relatively cool climate, the circulation of air heated to 35°C can reduce moisture content in open honey cells by 1 to 3%. This is the easiest and cheapest of all post-harvest moisture controls. The relative humidity of the air at 35°C temperature has to be controlled. If it is more than 60%, aerial moisture will have to be removed by a dehumidifier. In tropical climate, the air temperature will have to be considerably higher (damaging to honey) or prior dehumidification of the air will be necessary. This requires a small sealed room and a dehumidifier.

Prevention of Fermentation

Fermentation is the only microbiological alteration to which honey is susceptible. Only osmophilic yeasts can grow in high sugar concentration but their presence is ubiquitous in honey, nectar, hive interiors, dust and soils, and their rate of multiplication increases proportionally up to a certain point with increasing water content. Below 18% moisture content, there is little probability of fermentation, but even at concentration below 17.1%, the risk of fermentation cannot be excluded completely. This aspect of fermentation depends on factors such as the quantity of yeast and other growing factors, *i.e.*, honey temperature and the distribution and availability of water following crystallization.

Appropriate and expensive cold storage, but above all, careful production techniques can prevent fermentation. After all precautions and care, if honey cannot be harvested at less than 18% water content, excessive moisture should be removed. One of the previously described methods if carefully either used and if honey has not yet fermented can prevent fermentation without degrading the honey.

Another method is based on pasteurization and the destruction of the yeasts. The osmophilic yeasts found in honey die after only a few minutes of exposure to temperature between 60 and 65°C. If the honey is heated and cooled quickly, enough with special heat exchangers feasible only on an industrial scale, very little damage occurs to the honey. Often these pasteurization treatments have two functions, *i.e.*, (i) the prevention of fermentation and (ii) the postponement of crystallization. Relatively small quantities of honey with high moisture content do not justify complicated and costly pasteurization or drying. They should be directed towards market with immediate consumption, for processing into other food items or for fermented drinks. Such honey should not be considered for shipping over long distances as containers might explode. Careful heating in a water bath to wax melting temperature (about 65°C) and subsequent cooling in a water bath with running water may prolong storage life. For small quantities, this is an acceptable compromise between spoilage by fermentation and some loss of quality by heating. However, under most circumstances, the water baths are overheated and honeys are not properly stirred and cooled down rapidly enough. Therefore, pasteurization on a small scale can only be recommended in emergencies and not as a routine procedure as it is used in many places. The pasteurized honey needs to be bottled hot in a clean environment in order to prevent reinfection with omnipresent yeasts.

Heating

Any application of heat has a negative effect on honey through the loss of thermolabile aromatic substances, which is proportional to the temperature and duration of heating. Therefore, the basic concept is to heat up the honey only to the lowest temperature and for the shortest period consistent with the desired technical objective.

Honey owes its distinctive characteristics not to the stable major compounds that can be found in any other sweet product, such as sugar, molasses, syrup and marmalade but to the multitude of minor components originating from the nectar and the bees themselves. Many of these substances, which give honey its specific aroma and flavour, and some of the biological activities are unstable over time and thermolabile, *i.e.*, they are destroyed by heat. This uniqueness and fragility afford honey its legal protection and consumer preference, at least in most of the Europe. All the following precautions concerning heating, storage and further processing are made in consideration of these fundamental quality characteristics of honey. With respect to the above-mentioned effect on its quality and its peculiar physical characteristics, honey needs to be heated always with particular care. Its low thermal conductivity makes uniform heating throughout a large body of honey very difficult and the use of high temperature heat sources like open flames or a boiling water bath may quickly lead to local overheating. This may cause significant alterations of the honey's characteristics, even caramelization. Heating in industrial plants is required in order to reduce the honey viscosity or to melt the crystals, especially large surface heat exchange systems are used with a heat source only a few degrees above the temperature to which the honey is to be heated.

For the melting of crystallized honey in large containers, thermo-regulated rooms or water baths are maintained at temperature between 35 and 50°C. The melting time for different size containers and temperature differs and depends on the type of honey. For heating smaller quantities, only indirect heating through water bath method should be used and the water should never be more than 5-10°C hotter than the desired temperature of honey. It should certainly never be boiled. Afterwards, an appropriate cooling, like running water through the water bath, has to be provided. In large containers, heating can be provided externally or internally with special heating coils through which hot water is circulated. Usually, some kind of mixing device like propellers, blades, or recirculating pumps are added in order to facilitate heat exchange. In fact, in large processing plants, smaller containers are preheated in warm rooms kept at 60-70°C. As soon as the honey softens at 35 to 40°C, it drips out of the inverted containers over grids or inclined surfaces kept at 35-45°C. Before reaching the high room temperature, it is then pumped out of the hot room into large melting vats where the melting of crystals is completed.

Storage

Storage containers for liquid or crystallized honey should be made of either glass or stainless steel or coated with food-approved plastic, paint, or beeswax. Nothing should be allowed to impart any odour to the honey. Particularly, if the used containers are recycled, care must be taken that they are clean and have no slightest residual odour. Honey readily absorbs odours of all kinds, *e.g.*, these can readily be absorbed by a beeswax coating and then passed into the honey. Containers previously used for toxic chemicals, oils, or petroleum products should never be used for storing any bee product, even after coating with paint, plastic, or beeswax.

Openings in wholesale containers have to be big enough to facilitate removal even of crystallized honey. To keep the moisture out, lids have to be airtight and all products should be kept away from heat and preferably light. In addition, most products containing honey should be protected from excessive moisture by special packaging, *i.e.*, baked products in moisture proof clear plastic bags, caramels in separate plastic or waxed paper wraps and single portions of liquid or pulverized product in laminated foil envelopes made of aluminum foil covered with plastic or plastic and paper envelopes. Storage rooms should have a temperature near 20°C and a relative humidity of less than 65%. Storage of honey at more than 25°C temperature causes increasing quality loss with time, due to progressive chemical and enzymatic changes.

Honey is considered a stable product, in the sense that the bacteria and fungi normally responsible for food spoilage do not spoil it. However, the products containing honey are the preferred targets for such organisms, and therefore, demand pasteurization (stabilization with heat) or chemical preservatives (according to product requirements) plus adequate storage and protection from recontamination after production. Proper storage and packaging together with quick marketing and consumption will reduce or eliminate the need for preservatives.

Fermentation remains the major threat to unprocessed honey, whether it is liquid or crystallized. Therefore, storage conditions have to prevent fermentation through either low temperature storage or by preventing further absorption of moisture.

Even honeys, which are not susceptible to deterioration by yeasts, can be subjected to other progressive alterations due to chemical and enzymatic action. These changes include organoleptic characteristics such as colour, taste and aroma, together with a loss of biologically active substances (inactivation of enzymatic and antimicrobial activity). Substantial changes may also occur in the sugar composition with an increase of disaccharides and other complex sugars and a corresponding decrease in simple sugars. Other transformations of the initial composition include an increase of acidity and HMF content. These changes occur in all honeys but at different rates according to their initial composition (more moisture and a lower pH result in faster changes) and storage temperature (higher temperature also leads to faster changes). Same changes take place even faster during and after the heat treatments of various processing technologies. Though the damaged honey is not dangerous to human health, it nevertheless loses some of its nutritional and organoleptic values. Therefore, in almost all countries, legal limits are set for the degree of *ageing* or deterioration of honey for food use. Heat and sunlight (mostly the ultraviolet spectrum) can destroy the quality of honey both in brief high exposure and in low-level exposure over a long period. Some decay is unavoidable but it should be kept to a minimum. Ultraviolet radiation destroys glucose oxidase and most of the antibacterial activity. Since it is difficult to give a precise preservation limit for honey due to the large variability of different factors, HMF and diastase are used as indicators of damaging treatment received by a honey during either processing or storage. Therefore, decreasing half-life, *i.e.*, faster disappearance of diastase can be equated with increasing damage to honey. However, initial diastase contents vary in different honeys and have to be known for fresh untreated material. Hydroxymethylfurfural is used more frequently as an indicator since its value is close to zero in very fresh honeys (other than a few tropical honeys) and its level increases with time and exposure to heat.

In warm climate, it is important to protect storage vessels from overheating and possibly cool them by special shading or ventilation. Processing, moving and selling honey have to be as fast as possible. Care also needs to be taken that the honey is not damaged by overheating during trucking (particularly during parking in direct sunlight) or while waiting for reloading in harbours or railroad yards. The same is true for small bottles of honey sold at roadsides or in market stands. They should never be left in the sun energy since only relatively low temperature is required.

Adulteration

Most simple adulterations of honey can be detected if certain characteristics exceed the legal quality standards, *e.g.*, by a high sucrose content (> 8%) if simple cane or beet sugars are added, or by high HMF values, if acid hydrolyzed corn syrup is used. The latter has fructose to glucose ratio similar to honey.

Simple field methods for detection of adulteration without laboratory equipment are based on taste, viscosity or its solubility in cold water. If a droplet of honey poured into cold water stays together without dissolving rapidly, it is most likely pure honey. This can be observed best against the light with a dark background. If edges of the droplet or the thread start dissolving during pouring, the honey is likely to have been adulterated or has very high water content. In any case, it should be kept separate from other honey until tests that are more precise can be carried out.

Major Effects Commonly Attributed to Honey

Nutritional benefits: Honey is said to facilitate better physical performance and resistance to fatigue, particularly for repeated effort. It also promotes higher mental efficiency. Therefore, it is used by both the healthy and the sick for any kind of weakness. Improved growth of non-breast-fed infants, improved calcium fixation in bones and curing anemia and anorexia may all be attributed to some nutritional benefit or stimulation from eating honey. Honey has long been used in medicines not only as a valuable item in the diet but also as a medicine remedy. In Indian medicines, it was considered that honey could be used both as a remedy and as an analeptic. The tonics prescribed for good health are mainly prepared from honey. Milk with honey is considered as very nutritious food and preserve youth. Use of honey in daily diet gives pleasure. In other words, honey consumption alone or with other natural products medicines has beneficial effects on the human body system.

Antibacterial property of honey: Experiments have shown that honey inhibits bacterial growth and kills bacteria. The ancient Egyptians and Greeks used honey to embalm the dead. The ancient Greeks and Romans also used honey to preserve meat, which kept well in it and retained its natural flavour. Some researchers advocated that these properties are due to the presence of high sugar content. Antibacterial activity is probably the most studied biological activity of honey even the diluted honey has shown antibacterial activity. In normal honey, it is attributed to high sugar concentration, an elusive substance generically termed *inhibin* and acidity (pH range 3.5-5.0). Much of this activity was later attributed to hydrogen peroxide (H_2O_2) - an enzymatic byproduct during the formation of gluconic acid from glucose. The responsible enzyme, glucoseoxidase, is inactive in concentrated normal honey. Thus, in honey solutions (diluted honey) with right pH, the antibacterial activity is largely due to the presence of hydrogen peroxide. The biological significance of such a mechanism arises from the requirement to protect immature honey (with high moisture content) inside the colony until higher sugar concentration is achieved. Antibiotic substances or inhibitors present in honey are of great value and are considered the product of secretory activity of worker bees. Honey prepared by *Express Method* has been found to inhibit the growth of *Streptococci*, *Staphylococci*, *Typhoid bacillus*, *Coliform* bacteria, *Paratyphoid* bacterium A and *Paratyphoid* bacterium B.

Both the mechanisms can partially explain the sterilizing effect of honey on wounds and some of its efficacy against cold infections but it does not explain its beneficial effect on burn wounds and faster wound healing with less scarred tissue. After storage in honey for up to 12 weeks, skin grafts acceptance has been achieved 100%. Antibacterial activity varies greatly between different types of honey. In addition to glucoseoxidase, honey seems to contain mostly other unknown substances with antibacterial effects, among which are polyphenols. These other factors have been identified in few cases, but as a whole, there are few scientific studies on various claims of the beneficial effects of honey. However, it has been well demonstrated that most of the antibacterial activities of honey are lost after heating or prolonged exposure to sunlight.

Antimycotic property of honey: Foodstuffs like flour, sugar, jams, unglazed sweets, fruits and beverages acquire unpleasant odour and flavour due to the development of mould, whereas, the honey possesses antimycotic property and does not become mouldy if stored properly. Natural and honeys obtained by *Express Method* have been reported to inhibit moulds, and thus, are antimycotic.

Honey and digestive system: Honey has a beneficial effect on digestion and is stomach's best friend. Honey is said to improve food assimilation and to be useful for chronic and infective intestinal problems such as constipation, duodenal ulcers and liver disturbances. According to clinical reports published in scientific journals, honey cures various gastrointestinal disorders.

- It is particularly good as a laxative and when taken regularly helps in keeping the gastrointestinal tract in normal working orders.
- Honey can be used as a remedy or as part of the diet in the treatment of several gastrointestinal ailments particularly in gastritis or gastric ulcer caused by hyperacidity.
- Eating honey by patients suffering from ulcers proves beneficial as honey brings acidity and the secretion of gastric juices back to normal. In this way, it relieves the

patients from heartburn, belching and cramps. Honey has dual effect in the treatment of ulcers, *i.e.*, (i) a local healing effect and (ii) general build up of health particularly nervous system. The gastric and duodenal ulcers develop when the receptors of these organs cease to function properly. For ulcers treatment, honey should be taken by dissolving it in warm or boiled water. In this way, honey is quickly assimilated without irritating the intestine and it dilutes the mucus and lowers acidity in stomach. On the other hand, honey in cold-water increases acidity, slows down the digestion stomach contents and irritates the intestine. Honey taken just before meals stimulates the secretion of gastric juices. Therefore, it must be taken one and a half to two hours before or three hours after meal.

- Honey has beneficial effect on liver due to its chemical composition particularly high glucose content. Glucose not only feeds the tissue cells of liver but also increases glycogen reserve and improves the process of tissue replacement. Glycogen also helps the liver in rendering bacterial poisons harmless more effectively.
- Food taken by any organism remains in stomach for sometimes (few hours) and it is subjected to the action of gastric juices, which increases hyperacidity. This hyperacidity is reduced if honey is taken.
- Honey mixed with curd, grounded and boiled wheat, boiled buckwheat, barley and apples is not only beneficial for the patients but also for healthy persons.
- Lemon Juice with honey and olive oil is an effective remedy for diseases of the liver and gall bladder.

Beneficial effect of honey on respiratory system: In temperate climate and places with considerable temperature fluctuation, honey is a well-known remedy for cold and mouth, throat or bronchial irritations and infections. The benefits, apart from antibacterial effects, are assumed to relate to the soothing and relaxing effect of fructose.

- Honey alone or mixed with other food items and medicines is considered as all-purpose remedy for cold. One tablespoon of honey mixed with one glass of milk or warm water improves the condition of people suffering from cold. Juice of one lemon in 100 g of honey also helps in curing cold. The above mixtures are more useful when taken at bedtime or the patient should not move in cold weather.
- Honey is especially good for the elderly suffering from lungs problems, they get rid of sputum and it soothes a cough.
- A mixture of honey and rose petals has proved beneficial during early stages of tuberculoses. Honey is an excellent remedy for pulmonary tuberculosis. In this disease, no specific curative properties can be attributed to it. It has merely been noted that honey is generally a restorative and it helps the organism to fight tuberculosis bacterium.
- Taking honey 100-150 g every day improves appetite, increases hemoglobin and weight and reduces cough and the amount of sputum.
- Honey inhalations in upper respiratory tract with steam using 10% solution of honey proved beneficial in patients suffering from dryness in pharynx and loss of voice, and inhalations for 7 days showed improvement in disappearance of scabs and hoarseness.

Inhaled honey not only has beneficial effect on mucous membrane of the nose and throat but also on the alveoli of lungs (through which it enters the blood).

- Some researchers have recommended honey and lemon juice for catarrh of the throat.

Honey for healing wounds: Honey is used in not only moisturizing and nourishing cosmetic creams but also in pharmaceutical preparations applied directly on open wounds, sores, bedsores, ulcers, varicose ulcers and burns. It helps against infections, promotes tissue regeneration and reduces scarring also in its pure unprocessed form. Honey can make a layer on wounds due to the reason that it contains about 80% sugars. Moreover, it can easily be washed from wounds, as honey is soluble in water. Honey increases the glutathion in wounds that results in fast recovery of wounds. If applied immediately, honey reduces blistering of burns and speeds up regeneration of new tissue. Glutathion plays important role in oxidation-reduction process in the organism and stimulates cell growth and division. Many case histories are reported in the literature for human as well as veterinary medicines used to treat sores and open wounds.

- Infected wounds and sores in mouth can be cured with a mixture of fish fat and honey. Honey and fish oil were used to treat extensive wounds in 10 to 12 days.
- Honey ointment containing pine tar has been used to heal wounds.
- Honey stimulates growth of tissue in slow healing wounds and it can be used to cure gunshot wounds. Slow healing wound and ulcers can be treated with an ointment containing fish oil 20 g, xeroform 3 g and honey 80 g. However, fish oil and xeroform may be allergic for some people.
- Patients have been treated with torpid wounds by means of honey electrophoresis. Honey electrophoresis treatment to patients with fractures due to gunshot wounds, complicated by inflammation of bone marrow included active development of granulation. Wounds after treatment became clean and healed.
- Placing a thin film of cooked flour and honey mixture on wounds and replacing it after 12 hours helps in curing wounds and results in new and healthy cells.
- A cream prepared from equal parts of honey, rye flour and olive oil is applied thrice a day on many sores and open wounds even on gangrenous wounds in horses.
- A mixture of honey and cod liver oil suspended in a simple non-reactive cream base has been successfully tested on open wounds in humans.

Honey for eye disorders: Honey has long been considered a most effective remedy for many eye diseases. It has not lost its importance today even when medicine has been enriched with antibiotics, *etc*. Clinical cases claim that honey reduces and cures eye cataracts, conjunctivitis and various afflictions of cornea if applied directly into the eye are known from Europe, Asia and Central America. This is said to be truer for Meliponid and Trigonid honeys from South and Central America and India. There are also case histories of ceratitis rosacea and corneal ulcers, healed with pure honey or 3% sulphidine ointment in which Vaseline was replaced by honey. Honey for direct use in eyes must be taken directly from honeybee combs. Extracted honey should never be used for putting in eyes because it may contain some impurities like dust, pollens, *etc*. Pollen grains sometimes cause injury to cornea of the eye.

- Honey mixed with onion juice, clover, or wheatgrass has been found good for eyes.

- Honey is a good remedy for burns especially those affecting the eyes and an excellent cure for inflammation of the eyes.
- Eucalyptus honey as an ointment treats swelling of the eyelid, conjunctiva and cornea sores on the corneal membrane.
- Honey mixed with 3% sulphapyridine ointment has been found highly effective on very slow healing sores.
- Honey has been extensively used to treat herpetic and ulcerous Keratitis and as a means of resorption in cases of opacity of the cornea or vitreous body, immature or initial cataract and burns affecting the eyes.
- Honey when smeared on the eyeball prevents people from going blind.

Effect of honey on heart related disorders: Honey is very beneficial and it should be taken in cases where cure depends on the activity of heart since it not only stimulates the heart, when given digitalis but also provides nutrition. The heart muscles work continuously and needs glucose in order to make up the energy. Honey is beneficial to heart because it contains much easily assimilated glucose. In various types of cardiac disease, the weakened muscles of the heart have been found strengthened by honey.

- It has been found by researchers that when 0.1% glucose is added to the physiological saline in which an isolated heart was immersed, the heart worked outside the body for four days.
- Circulation through the coronary arteries improves due to the expansion of veins by honey.
- Taking honey on an average 70 g/day regularly for one month improves the condition of patients with heart trouble. In patients with heart troubles, honey normalizes the blood composition, increases hemoglobin level and improves cardiovascular tonus.

Beneficial effects of honey on nervous system: The beneficial effect of honey on nervous system is well known and can be attributed to the fact that hypertonic solutions of glucose provide effect rapidly in the treatment of some diseases of nervous system.

- Honey reduces/cures headaches and improves eyesight.
- Taking honey helps in normal sleep and one feels stronger, less irritable and cheerful.
- Glass of water with 2 tablespoons of honey and juice of ½ lemon or two tablespoons of honey before going to bed cures the people suffering from exhaustion and helps in normal sleep.
- Bran soaked in water or vitamin B_1 mixed with honey is an excellent remedy for strengthening the nerves.
- An intravenous injection of 10 ml of 20-40% solution of honey frequently is given before the radiation treatment to prevent the radiation sickness that often develops in patients during radiation therapy.

Honey and skin care: As we age, the skin begins to lose its ability to absorb and retain moisture and tends to become dry and wrinkled. Honey plays a great role in the prophylactic means for preserving beauty of the skin. Healthy skin in turn protects the body against harmful external influences. Honey can be used as a cosmetic because of its permeability and the fact that it feeds the muscle layer of skin with glucose and other nutrients. Honey is hygroscopic and absorbs skin secretions and its microbicidal properties protect the skin and have an antiseptic effect. Honey can be used as a cosmetic in the form of face packs, lotions, ointments, baths, *etc.*

- Honey and cucumber lotion is a beneficial care for skin and can be prepared by mixing 100 ml juice of fresh cucumbers in 100 ml of 40% alcohol. This mixture should be kept in closed container at a cool and dark place for 8 days. Mixture prepared in this way is then mixed with honey as per requirement. The pack should be applied with the help of muslin cloth after proper cleaning with warm water.
- Honey mixed with water combats premature deterioration of skin. Washing face and neck with a mixture of 2 tablespoon of honey in 2-3 liters of boiled water is an effective care.
- Addition of honey 200-250 g to bathing water has a beneficial effect on human body. Honey alone or mixed with equal parts of egg yolk soften and strengthen the skin. Honey and flour face pack (honey 50 g + flour 30 ml + water 30 g) has been proved beneficial for dry skin.
- The most widely used face packs are (i) honey 100 g + alcohol (surgical spirit) 25 g + boiled water 125 ml and (ii) honey 50 g + wheat flour 30 g + water 30 g and (iii) honey 1 tablespoon + glycerin 1 tablespoon + yolk of one egg.

Honey and diabetes: Frequently, claims are voiced that honey is good for diabetics. This is unlikely to find confirmation because of its high sugar content. However, it is better than products made with cane sugar. It revealed that insulin levels were lower when compared to the uptake of equal caloric values of other foodstuffs but blood sugar level was equal or higher than in other compared products shortly after eating. In healthy individuals, the consumption of honey produced lower blood sugar readings than the consumption of same quantity of sucrose.

Scientific Evidence

According to scientific evidence, it would be better to consider honey as a food rather than a medicine. Most of the benefits described above, at least for internal use, can most likely be ascribed to nutritional effects of some kind. Benefits of honey mentioned above need confirmation from a medical specialist and it is suggested that honey for prevention and cure of diseases must be used after consultation with him.

Use of Honey as Food

Honey is most commonly consumed in its unprocessed state, *i.e.*, liquid, crystallized, or in the comb. In these forms, it is taken as medicine, eaten as food, or incorporated as an ingredient in various food recipes.

As food, honey is mainly composed of simple sugars, *i.e.*, glucose and fructose, which form the basis of almost all indications on how, when and why to use it. The main consideration is the fact that honey provides immediately available calories (quick access to energy without requiring lengthy or complicated digestive action) from which it derives its energy value for healthy and sick people. The same direct absorption also carries a risk of pathological sugar metabolism, such as diabetes and obesity.

Honey is often recommended because of its other nutrients like minerals and vitamins but their quantity is so low that it is unrealistic to think that they can provide any significant supplement in a deficient diet. Micronutrients in unprocessed honey can be assumed to be of the highest possible quality. Thus, from nutritional point of view, a synergistic balancing effect that unlocks the availability of other nutrients already present is one of the most plausible though it is untested hypothesis.

However, honey is considered a food only in few societies such as those of the industrialized countries in Europe and North America, Latin America, North Africa, the Near East and increasingly in Japan. In most parts of Africa, it is used for brewing honey beer and to a much lesser degree as medicine. In most of the Asia, it is generally regarded as a medicine or at most an occasional sweet. High per capita consumption in industrialized nations does not reflect the consumption of unprocessed honey per person but includes a very large quantity of honey used as a food ingredient in industrial food production.

In order to increase consumption and to make the various honeys more attractive, a large variety of semi-processed and pure honey products are marketed. One of the more appreciated forms of selling natural honey seems to be honey in its natural comb. Including pieces of comb honey in jars with liquid honey (chunk honey) is very attractive to many consumers and appears to disperse suspicions of adulteration. Creamed honey (soft, finely crystallized honey) is a very pleasant product, which is convenient in use because it does not drip. In some countries, the appearance of marketed honey, *i.e.*, liquid, crystallized, or semi-crystallized and with or without wax particles, *etc.* is not very important. Therefore, it can be bottled as it is. In other countries, consumers want not only clean honey but also prefer liquid honey. Consumer education may change this attitude, particularly where it is based on widespread but false belief that honey crystallizes because it is adulterated with sugar. However, many honeys require special processing to remain in liquid form.

Slow crystallizing honeys can be sold or used without further processing. Light coloured honeys are particularly suitable for sale as comb honey in special clear packages but any kind of honey can be sold as comb honey as long as the combs are evenly sealed and relatively new with white or light yellow wax. In blending different honeys, attention has to be paid to the final ratio of glucose to fructose and the possible need for additional heat treatment. Fast and slow crystallizing honeys low in moisture content can be processed to prolong their liquid state or can be forced to crystallize under controlled conditions to achieve a soft and uniform consistency. Uniformly crystallized honey is attractive both visually and for its convenience of use. It is also less likely to ferment than badly crystallized or semi-crystallized honeys. Among other factors, different storage temperatures in different climates influence the crystallization and speed of re-liquification of honeys. Most of the honeys stored above 25°C temperature remain liquid or reliquify slowly but lose much of their aroma in few months.

Honey as Food Ingredient

The traditional use of honey in food preparations has been substituted in most cases by sugar and more recently by various sugar syrups derived from starch. These exhibit similar composition and characteristics but at a much reduced cost. At the same time, as part of the increasing appreciation of more natural products in many countries, honey has been *rediscovered* as a valuable food, and therefore, confers also as an ingredient, an enhanced market value to the end product. Many honey containing industrial products which were developed in the last decades but which did not have the expected success, are currently being remarketed more successfully.

Outside of the thousands of *homemade* recipes in each cultural tradition, honey is largely used on a small scale as well as at an industrial level in baked products, confectionary, candy, marmalades, jams, spreads, breakfast cereals, beverages, milk products and many preserved products. In particular, the relatively new industry of natural health and biological products uses honey abundantly as a sweetener of first choice, together with non-refined sugars in substitution of refined sucrose (cane and beet sugar). In fact, honey can substitute all or part of the normal sugar in most products. Limitations are presented on one side by cost and handling characteristics and on the other side by the natural variations in honey characteristics, which change the end product, make it more variable and require more frequent adjustments in the industrial formulations.

Baked Product: Aside from the already mentioned above, honey confers several other advantages such as a particular soft, spongy (springy) consistency, which persists longer. Products that contain honey also dry out more slowly and have a lesser tendency to crack. These properties are due to the hygroscopicity of honey, a trait that the honey has in common with other sweeteners high in fructose, like acid-hydrolyzed corn syrup or other syrups made from starch and fruit juices. Another advantage consists of uniform baking with a more evenly browned crust at lower temperature. These characteristics too are mostly due to the fructose content. Yet another advantage is an improved aroma, conferred by relatively small percentage of honey (up to 6% by weight of the flour) in sweet cakes, biscuits, breads and similar products. Since most beneficial effects can be obtained with relatively small quantities, the baking industry prefers strong flavoured honeys, thus, maximizing flavour for the lowest possible cost.

Confectionery Products: Honey is still included in many traditional confectionery products, such as torrone from Italy, turon from Spain, nougat from France and halvah from Turkey and Greece, which are consumed locally in considerable quantities but are also exported. For the production of caramels (bonbons), honey is used only in very small quantity since its hygroscopicity presents a major disadvantage, *i.e.*, it reduces the preservation time and softens the caramels at the surface causing them to stick together. Some caramels made with special machinery have a liquid honey core. In gelatinous or gum products, honey can be used in the same way as the other flavouring agents (aroma or fruit pulp). The chocolate industry uses honey in only few products. One Swiss chocolate in particular, in which honey is included in the form of broken nougat, can be found worldwide.

Breakfast cereal industry: Honey is used in breakfast cereal either in its liquid or in dried and pulverized form both for better flavour and for increased consumer appeal. It can be mixed with cereal flakes and dried fruits or applied as a component of sweetening and flavouring film, which covers the flakes. The dryness or hardness of the cereal can be adjusted with honey content and with the degree of drying.

Candy Bars: Numerous *snack candy bars* are marketed in which honey constitutes the binding and sweetening agent. Other ingredients of the mixture can be dried fruits (like raisins, figs, apples, apricots, prunes, dates, pineapple, papaya, *etc.*), nuts (like hazelnuts, walnuts, almonds, brazil nuts, pistachios, ground nuts, cashew nuts, *etc.*) and seeds (sesame seeds, sunflower seeds, linseeds, coconut flakes, *etc.*), cereals of all kinds (rolled, as flakes, or in puffed form) and possibly other ingredients such as milk powder, pollens, cacao, carob and aromas. The ingredients are chopped to various sizes and mixed with the hot honey and sugar. Depending on the composition and the degree of heating of sugars (including honey), a more or less solid product is obtained after cooling. Some can be cooled in moulds, some are cut after cooling and others, which remain soft, have to be layered between wafers or biscuits and coated with chocolate.

Products of Honey Fermentation

In many regions, honey is or was the only or the most accessible source of fermentable sugars. In some parts of non-Islamic Africa, the traditional manufacture and consumption of honey beer is still very common. The base is crudely pressed or drained honey, often with added brood or pollens. An additional nutrient base is generally provided for the yeasts, which may add characteristic flavour as well. Occasionally, the other available sugars or sugar sources are added but always the beverage is consumed before fermentation is finished. Preparation by a skilled brewer (in East Africa most commonly women) can be as fast as 5 to 6 hours. Consumption most commonly by men is usually still faster.

In Europe, traditional fermented honey products, some similar to African honey beers and others more refined for longer storage, have largely been abandoned and replaced with grape wines and grain beers. The fundamental problem with mead, the honey wine drinkable only after some months or years of maturation, is that without precise control of the yeasts and other microorganisms growing in the must, the final flavour can often be very disappointing. The must of honey water by itself does not contain sufficient yeasts, or the right kind of yeasts or nutrients to allow rapid fermentation. The yeasts most commonly found in honey (Zygosaccharomyces) grow only in concentrated solutions with more than 50% sugar. Unlike honey beer production, even if sufficient yeast is added at the onset to produce rapid fermentation, the whole process lasts much longer, during which strong flavour derived from other microorganisms can develop. Probably also in order to cover unpleasant flavour, old mead recipes often prescribed the addition of fruits or aromatic herbs. The beverage is then referred to as metheglin.

New microbiological understanding of fermentation process leads to better controlled working conditions and more production, which is reliable. The result is better control of final flavour, and as a consequence, the production of meads is becoming more popular again.

Through refermentation by careful addition of honey or incomplete primary fermentation prior to bottling, sparkling mead can be produced. Refermentation with selected yeasts can also produce sherry mead. In Poland, meads with extremely high sugar contents are traditionally produced from musts using equal volumes of honey and water. This *Dwojniak* has to mature for very long period of at least 5 to 7 years but the primary fermentation is similar to that one mentioned earlier for medicinal syrups from plant extracts and can be conducted with the honey's own osmophilic yeasts. A must with a honey to water ratio of 1:2 requires only 3 years of ageing and is prepared with a special strain of wine yeast (Malaga type). This *Trojniak* is still sweet and is preferably made from cornflower (Centaurea) honey.

Honey vinegar can be produced from mead in the same way as the wine vinegar. Unless there is some very special appreciation of this unique flavour, the production is hardly economically feasible. Mead can also be used as a base from which distilled alcoholic beverages can be produced. Such production is usually for home consumption only.

Other Industrial Uses

The tobacco industry is estimated to use more than 2000 tonnes of honey annually to improve and preserve tobacco's aroma and humidity. Since tobaccos, at least in part, are valued according to the rate at which they dry, the importance of honey for the more valuable tobaccos can easily be understood.

Wax moth larval diets sometimes contain honey to improve survival rates. These larvae are raised for scientific experiments and fish bait and could be used as human food as well.

The cosmetic industry uses honey as a skin moisturizer, softener and restorer of the skin's own moisturizing factors in creams, soaps, shampoos and lipsticks. Because of its stickiness, it can only be employed in small quantities.

Precautions to be Taken

Honeys of some flowering species are reported in the literature as toxic because of their content of active ingredients from nectars or honeydews, which are noxious or toxic to humans. Although these honeys are not very common but they can be of particular importance in some localities. The plant families and genera from which a few species have been reported to produce toxic honeys include Ericaceae (*Rhododendron*, *Azalea*, *Arbutus*, *Andromeda* and *Kalmia*), Solanaceae (*Datura*, *Hyoscyamus* and *Atropa*) Compositae (*Senecio jacobaea* or ragwort), Lagnonaceae (*Gelseminum*), Ranunculaceae (*Aconitum*), some species of the genus *Euphorbia* in South Africa and the honeydew of *Coriaria arborea* from New Zealand. Bitter and off-flavoured honeys are produced from many more species.

The consumption of honey by infants less than one year of age is not recommended by the United States Food and Drug Administration. This recommendation is based on the correlation of some infant deaths with the ingestion of botulism spores (*Clostridium botulinum*) from honey. The spores were recognized in very few samples of Californian honey for the first time in 1976 and more recently in the United Kingdom and Italy. Other surveys in Italy, France and Norway have not found any *C. botulinum* spores in any honey samples. Though omnipresent in the environment and in many foodstuffs, the spores normally are not capable of developing in the intestines of adults or children. However, since very young infants often have less acidic digestive tracts and less competition from a bacterial flora yet little developed, *Clostridium* spores may develop in their intestines. The toxin produced by the bacteria binds irreversibly to motor nerve endings and can only be overcome by the growth of new nerve endings. Some of the typical symptoms are mild paralysis (failure to thrive), moderate to severe paralysis, which requires hospitalization and fulminate (sudden and intense) paralysis, which can lead to death without warning. Other milder symptoms can be constipation, listlessness, lethargy, diminished appetite or activity and lack of muscle control.

Spores in honey cannot grow or multiply unless they have a watery, anaerobic (no oxygen) medium with a more or less neutral pH. Therefore, even in the newborn infant, special conditions have to come together to allow the spores to germinate and produce their toxin, such as after changes in the intestinal flora due to antibiotics, anomalies of intestinal secretions, or others. The risk for a child less than one year old contracting this infection has been estimated to be about 1 in 12,000. Such a risk is likely to be higher during the first one or two months and lower during the second half of the year.

In India and many other parts of the world, honey is given to newborn babies during the first few days of life as a special tonic, particularly if they were born weak or prematurely. In countries with better infant nutrition and generally more hygienic conditions, honey may be eliminated from the diet of an unweaned child since it is not an essential food. Under less favourable nutritional conditions, the risk of ingesting *C. botulinum* spores through eating honey and the resulting possibility of death must be weighted against the benefit of strengthening the young organism against many other more common stresses and diseases.

Normally, *Clostridium botulinum* causes problems only in badly preserved products (contaminated or insufficiently boiled preserves or conserves, pH near 7, absence of oxygen and storage at room temperature). Cooking the product at 80°C for a few minutes will destroy the toxin and bacteria but spores will survive up to 130°C. The presence of *C. botulinum* in honey is not due to any carelessness or mistreatment of the honey, nor can it be eliminated with normal processing other than ultra-filtration. Conditions in properly stored honey will not allow the bacteria to grow and produce any toxin. Fortunately, it is not a risk for anybody with a healthy intestinal flora or above the age of six months or a year.

Products containing honey and intended for human consumption or cosmetics should be treated as carefully as any food item and with even greater care regarding storage if they are prepared with no preservatives.

■■■

38 Nutritional Aspects of Milk

R.S. Dabur and Vandana Dhaka

Department Animal Product Technology, College of Animal Science, CCS Haryana Agricultural University, Hisar, Haryana, India, E-mail: <dabur1958@gmail.com>

Milk production began 6000 or 8000 years ago or earlier. Practically everywhere on earth, the man started domesticating animals. As a rule, herbivorous multipurpose animals were chosen to satisfy his need of milk, meat, clothing, *etc*. Milk has been used for human consumption for thousands and thousands of years, as proven by cave drawings showing cows being milked. It is one of the most nutritionally complete foodstuffs available and is a natural source that contains a whole host of minerals and vitamins needed for maintaining good health.

Today, cow's milk is still one of the most popular (especially in western countries) animal milks consumed by humans, while in India cow's and buffalo's milk is commonly used as human food but milk from sheep, goat, yak, horse and camel is also being used. Around the world, people drink milk from many other animals including goat, camel, reindeer, sheep and water buffalo. The composition of milk from different mammalian species (per 100 g fresh milk) is as follows:

Mammalian species	Protein (g)	Fat (g)	Carbohydrates (g)	Energy (kcal)
Cow	3.2	3.7	4.6	66
Human	1.1	4.2	7.0	72
Water buffalo	4.1	9.0	4.8	118
Goat	2.9	3.8	4.7	67
Donkey	1.9	0.6	6.1	38
Elephant	4.0	5.0	5.3	85
Monkey, rhesus	1.6	4.0	7.0	73
Mouse	9.0	13.1	3.0	171
Whale	10.9	42.3	1.3	443
Seal	10.2	49.4	0.1	502

Source: Indian food industry 29(2), March-April 2010: 49

Milk is secreted by the mammary gland of mammals to feed their offspring and is first food we swallow since birth. It is particularly good for children and the elderly, for the thin and weak, for those who are convalescing from a fever/disease and is a food filled with nutrients such as carbohydrates, protein, fat, minerals and vitamins. It contains large amount of essential nutrients, and it has rightly been recognized as nature's single most complete food. Nutritionally, milk has been defined as the *most nearly perfect food.* Its popularity in food market is based on its multi-dimensional functions as a repository of nutrients. The quality and proportions of these nutrients gave milk the ubiquitous status in our dietary programme. As a food, milk serves the following broad purposes such as (i) growth, (ii) reproduction, (iii) supply of energy, (iv) maintenance and repair and (v) appetite satisfaction. The requirements of these categories vary with the individual, and in some instances, not all the stated functions of food need to be served, *e.g.*, adults no longer require food for growth as the infants require. The functions of a food are served specifically through various nutritionally important components, comprising proteins, carbohydrates, lipids, minerals, vitamins and water. The functions of macro- semi-micro- and micro- nutrients are given as under:

Constituents	Contents	Functions
Macronutrients	Carbohydrates and fat	Sources of energy
	Proteins	Building blocks
Semi-micronutrients	Essential fatty acids	Membranes enzymes
	Amino acids	Hormones
	Calcium, magnesium and phosphorus	Hard structures
Micronutrients	Minerals	Catalysis function
	Vitamins	Vision and metabolism regulation

The macronutrients, semi-micronutrients and micronutrients found in milk are given as under:

Minerals		Vitamins	
Cations	**Anions**	**Fat soluble**	**Water soluble**
Iron	Iodine	Vitamin A (retonic acid)	Vitamin B_1 (thiamine)
Zinc	Sulphur	Vitamin D (calciferol)	Vitamin B_2 (riboflavin)
Calcium	Boron	Vitamin E (tocopherol)	Vitamin B_3 (niacin)
Magnesium	Selenium	Vitamin K (phylloquinone)	Vitamin B_6 (pyridoxine)
Potassium	Vanadium	Beta-carotene	Vitamin B_9 (folate/folic acid)
Sodium	Chloride		Vitamin B_{12} (cynocobalamine)
	Fluoride		Vitamin C (ascorbic acid)

The group, milk and milk products, is one of the basic food groups, with a minimum of 2 cups of milk a day recommended for adults, 4 cups for teenagers and 3 cups for children. This requirement can be fulfilled by either drinking milk or eating dishes made of milk. It is an outstanding source of calcium and phosphorus for bones and teeth and contains riboflavin, vitamins B_6, A and B_1 in significant amounts. Milk helps to balance human diet by supplementing good quality protein, calcium and vitamins particularly, vitamin A, riboflavin, niacin and folic acid. In addition, milk contains several bio-protective molecules that ensure health security to human beings.

In Indian *Ayurvedic* literature, milk has been outlined as a symbol of purity, divinity, auspicious same as temple ingredient, unity in molecular diversity, a fundamental article of diet with good therapeutic properties, a appellant strengthening tranquilizing and an improver of the appetite. *Ayurvedic* uses human milk and milk from various animals such as cows, goats, sheep, camel, water buffaloes, mares and elephants. It contains the essence, the juice of many medical plants and is therefore a life giver. Type of feeding and the places where the animals graze influence the milk's consistency and nutritional richness. From the lush upland meadows, cows give the richest and heaviest milk; low-lying, waterlogged, or parched fields afford light and less creamy milk.

Ayurvedic literature also point out the effects of certain containers on the properties of milk. Milk out of copper vessel calms *Vata*, out of a gold vessel calms *Pitta*, out of silver bowl calms *Kapha* and out of a brass vessel helps to make blood. According to *Ayurveda*, the milk of different species has dissimilar characteristics, which are reported as under:

- **Cow milk:** It is sweet and at the same time astringent. It is light and rejuvenating health tonic since it strengthens the intellectual and vital power. According to old tact, it is an elixir of life. It has reducing action on all three *Doshas*. Milk from black cow reduces *Pitta* and *Vata* and from white cow agitates.
- **Goat milk:** It is much prized in the *Ayurvedic* texts. It is astringent and sweet and recommended for diarrhea, tuberculosis, bronchitis, fever and hemorrhages. Goat milk when boiled and allowed to cool reduces *Vata* and *Pitta* if consumed in the morning.
- **Sheep milk:** Sheep milk is highly nutritious, richer in calcium, phosphorus, potassium, and magnesium and vitamins A, B and E than the cow milk. It contains a higher portion of short and medium chain fatty acids, which have recognized health benefits, *e.g.*, short-chain fatty acids have little effect on cholesterol in humans and make the milk easier to digest. It is good for patients suffering from disturbed *Vata*, and it relieves bronchitis and gout.
- **Mare milk:** It has a lower content of fat and vitamin A but four times more vitamin C. It contains magnesium, potassium, nitrogen, calcium, phosphorous and sodium. Mare's milk is easier to digest than cow's milk. Its texture is indeed thinner than cow's milk, and it is sweeter and very light on the stomach. Its protein composition is closer to human milk than to cow's milk. Mare's milk can be an alternative for adults suffering from metabolic, gastrointestinal and liver problems and also for recovering after surgery and severe illness, cholesterol problems, allergic to cows' milk, stress, skin problems, stiff joints and just to keep fit and fine. This milk strengthens the body, boosts the immune system and increases a person's energy and vitality. In case of metabolic disorders, it stimulates internal cleansing.
- **Human milk:** It is used as a medicine only when it is cold. It is digestible and its fat has a strengthening effect and nourishes the seven *Doshas*.

According to modern chemist point of view, milk is *born* with a unique chemical structure involving myriad nutrients with the primary purpose of ushering neonates. Milk is a tapestry of major and minor chemical entities with the understanding of maintaining colloidal and hydrophobic members of milk, formulating an assembly with distinct function and purpose. There is a relationship between structures of these varieties of molecules in milk with its function. The simple way of describing milk composition would be that it has two major types of constituent, *i.e.*, one group, which remains as suspended materials, gives the milk white look and the other group is the soluble material. The suspended constituent comprises two major milk nutrients, *i.e.*, (i) casein micelles and (ii) milk fat. Milk is white due to these floating bodies. Chemically, this phenomenon is narrated as the colloidal system in milk. The other major nutrient, the milk fat having no love for water (hydrophobic), remains as an emulsion with the aqueous phase of milk. It comprises the chemical family known as triglycerides along with some small proportions of di- and mono-glycerides. It also has phospholipids and cholesterols as a minor relative. The soluble constituents, which are water loving (hydrophilic), are nutrients present in milk comprising, (i) whey proteins, (ii) carbohydrates and (iii) minor components. The protein group has α-lactalbumin (LA), β-lactoglobulin, proteose, peptone and enzymes. Lactose is the main member in the carbohydrate group and oligosaccharides are present as minor relative. All these members contribute to the solids present in milk, whereas, the major bulk of milk is water occupying about 82-88% of the space.

Apart from the above-mentioned major constituents of milk, which are the synthetic outcome of mammary gland, the other occupants of milk namely minerals, vitamins, pigments, milk enzymes and hormones appear in milk as a result of either accidental or inevitable leakage from the mammary gland along with the secretary constituents of milk.

Milk Fat

Milk fat is composed of mainly triglycerides (96-99%) with a small amount of diglycerides (0.3-1.0%), phospholipids (0.1-0.3%), cholesterol (0.2-0.4%), free fatty acids (0.1-0.4%) and traces of monoglycerides and others. The short chain fatty acids (up to C8) content in milk is relatively small. The major fatty acids in milk fat are palmitic acid (24-28%), oleic acid (23-28%), myristic acid (13-14%) and stearic acid (11-12%). Polyunsaturated fatty acids that occur in milk fat only at about 1% level are linolenic and linoleic acid. As mentioned above, milk fat has relatively high content of short-chain and medium-chain fatty acids, which are more easily absorbed than the long-chain ones.

In milk triglycerides, short chain fatty acids chiefly butyric acid and caproic acid are found in the outer position and the long chain fatty acids occupy middle position 2. The triglycerides containing short chain fatty acids are hydrolyzed more rapidly by lipase than triglycerides containing long chain fatty acids. It is interesting to note that milk lipase is specific for hydroxy position 1 and 3 in the milk triglycerides. Butyric acid is uniformly distributed in milk fat, and it is thought that a molecule of triglycerides always contains only one molecule of butyric acid, which makes the triglycerides more rapidly attacked by lipase.

Milk fat is a good source of vitamin A. The average total vitamin A content in cow's milk fat is 2000 IU and in buffalo's milk fat 900 IU per 100 g. Both, vitamin A and carotene contribute to the total vitamin A activity in milk fat. The carotene contributes on an average of about 30% of it in cow's milk fat. Vitamin D content varies from 20 to 60 IU per 100 g in cow's milk fat. The average vitamin E content varies from 2.6 mg per 100 g in buffalo's milk fat to 3.5 mg per 100 g in cow's milk fat. About 95% of the vitamin E consists of alpha-tocopherol, which has the highest vitamin E activity, while the remainder consists of gamma-tocopherol. Milk fat also contains lactones, which contribute to the flavour of milk fat to some extent. The compounds responsible for the typical aroma of milk fat are acetoin, diacetyl, aldehydes, ketones, lactones, alcohols, esters, dimethyl sulphide and free volatile fatty acids.

A 250 ml of cow milk contains 9-10 g of fat and an equal amount of buffalo milk contains 15-22 g fat, which has a saturated form of energy and useful for infants and children to meet their energy requirement. Milk fat also supplies essential fatty acids. For human health, the types and composition of fatty acids in dietary fat are very important, as saturated fatty acids (SFA) and mono-unsaturated fatty acids (MUFA) are mainly oxidized for cellular energy need and the excess is stored in the body, whereas, long-chain polyunsaturated fatty acids (PUFA) such as linolenic and linoleic acid are converted to integral components of cell membranes and essential for membrane functions (fluidity, receptor activity and membrane bound enzymes).

Minimum fat requirement of a person is about 20 g visible fat per day and excessive intake of fat may lead to a risk factor for obesity, cardiovascular diseases and other associated complications. Dietary fat should provide 15-40% of total energy depending on the age, *e.g.*, 40% of the total energy requirement of a child should come from fat, and out of which, 20% of its energy should be furnished by essential fatty acids. However, for normal person, 30% of its energy requirements should be met with fat. The fat present in cereals, pulses and vegetables (invisible fat) provides almost 7% energy, of which 28-29% is furnished from essential fatty acids.

In past few years, milk fat has received adverse publicity in terms of its effect on blood cholesterol level of the consumers. However, the nutritional and metabolic effects of milk fat cannot be judged solely based on its fatty acid components. Other important considerations that should be born in mind are (i) the proportion of total calorie derived from milk fat, (ii) the form in which the milk fat is consumed, *i.e.*, whether the fat is consumed pure or along with non-fat portion of milk, (iii) digestibility and other dietary value of milk fat and habitual diet with which it is consumed (composition of invisible fat of foodstuffs and (iv) other dietary factors which alter lipid metabolism, atherogenesis and thrombosis). In plasma, lipids are circulated in combination with proteins called lipoproteins, and these are as follows:

- Chylomicrons (a microscopic particle of triglycerides produced in the intestines during digestion)
- Very low-density lipoproteins (VLDL)
- Low-density lipoproteins (LDL)
- High-density lipoproteins (HDL)

Dietary lipids are largely carried by chylomicrones. Very low-density lipoproteins transport lipids are those that are synthesized in the body. Low-density lipoproteins carry cholesterol from liver to the various peripheral tissues including blood vessels. The plasma level of LDL is strongly correlated with atherogenesis (formation of atheromas on the walls of the arteries). Since high-density lipoproteins scavenge excess cholesterol from the tissues to liver for degradation, it is considered a protective molecule.

The perception of milk has changed over the past 25 years from one of being the ideal food to one of being nutritionally detrimental mainly due to the fatty acid composition of its fat component, however, now, it has been discovered that milk contains a number of compounds, which may have positive nutritional benefits. It also appears that the association between saturated fatty acids in milk and effects on cholesterol may have an over simplification. It is accepted that the hypercholesterolaemic saturated fatty acids in milk fat are confined to lauric (C12:0), myristic (C14:0) and palmitic acid (C16:0) with shorter chain saturated fatty acids and stearic acid having no cholesterol raising effect. Indeed, bovine milk fat contains two fatty acids, which may have important beneficial effects on human health, namely conjugated linoleic acid- CLA (cis-9, trans-11 linoleic acids- C18:2) and butyric acid (C4:0). Also monounsaturated fatty acids have been shown to be beneficial in altering the proportions of LDL and HDL cholesterol and it is possible to increase the concentration of principal monounsaturated fatty acid, oleic acid (C18:1) in milk by optimizing the diet of the milch animal.

Milk fat is most easily digestible of various fats and oils. The degree of digestibility of milk fat is 99%, while that of natural palm oil is 91%. Since milk fat is easily digested and absorbed, it puts relatively little strain on the body. Therefore, it is a valuable dietary constituent in diseases of stomach, intestinal tract, liver, kidney, gall bladder and disorders of fat digestion. The reason for the good digestibility of milk fat is found in the state of dispersion of milk fat globules and its fatty acid composition. Milk fat globule particles up to the size of 100 μm in diameter can pass through directly into the lymphatic ducts. In infants, the fat is absorbed into the circulation as particles *via* lymphatic system since the glands responsible for the digestion of fat are not fully developed. The fat from mother's milk is more easily absorbed than that of cow's milk. Mother's milk also contains lipase, which plays an important role in digestion of fat. The mother's milk lipase is highly active and is stimulated by bile salts. It seems that even large particles are able to penetrate the intestinal epithelium probably chiefly at the tips of the villi (a-minute hair like projection on mucous membrane).

On the other hand, other dietary fats have to be at least partially emulsified in the small intestine by the bile, pancreatic enzymes and intestinal lipases before they can pass through the intestinal wall either in the form of a fine emulsion, or mainly, in the form of degradation products. When milk fat emulsion is broken and dehydrated in the form of butter oil or ghee, the enzymatic breakdown takes place in the lumen and fine dispersion of a mixture of triacylglycerols, diacylglycerols and monoacylglycerols, and free fatty acids are formed before they are absorbed by the mucosa.

The short-chain fatty acid are taken up directly by the blood of portal vein in the form of free fatty acids (FFA), while long-chain fatty acids first form chylomicrones, and then, reach the blood circulation *via* lymph vessels. The digestibility of fat is also affected by the position of the individual fatty acids in the triacylglycerols (TG) because lipase attacks first of all the other fatty acids, producing 1,2 diacylglycerols (DG) followed by 2-mono acylglycerols (MG). The fat is absorbed in the form of 2-mono acylglycerols and free fatty acids.

Milk fat has a relatively low content of essential fatty acids (EFA). However, it must not be concluded that milk fat should therefore be replaced with another fat with higher linoleic acid content in order to prevent deficiencies. The essential fatty acids requirement is only 30% of the total calories. It can be assumed that the essential fatty acids are available in required amount in normal diet, since two-third of the essential fatty acids requirement is met from invisible fat present in dietary cereals, pulses and vegetables. Mother's milk fat has relatively higher essential fatty acids content than cow's or buffalo's milk fat. The babies fed purely on cow's milk or buffalo's milk do not receive sufficient amounts of essential fatty acids, even then, the risk of a deficiency of linoleic acid is not likely, provided that this diet is not given for too long. The deficiency of essential fatty acids is manifested as histological changes of skin, an infantile eczema and low growth rate. The fat content of mother's milk represents more than 50% of the energy, which satisfies the body's high-energy requirements. In baby, feeding a sufficient fat supply is essential for thriving babies, a rosy and smooth skin, a good subcutaneous fat deposit, as well as resistance to bacterial infections.

Milk Proteins

Milk is an excellent source of high quality protein. The proteins in milk have balanced amino acid profiles and good digestibility. In addition, they contain high levels of amino acids, which are deficient in vegetable proteins. Because of this, milk proteins are excellent *complementary* proteins. Milk proteins are synthesized in mammary gland of the animal. Milk normally contains about 0.5% of nitrogen of which 95% constitutes milk protein and 5% non-protein nitrogen. Fat content is closely followed by milk proteins at about 3.4-3.6%. Milk protein is a heterogenous mixture and can be separated based on solubility at specific pH values and specific salt solutions. Milk proteins are subdivided into casein, comprising approximately 76-80% of the total milk proteins, and the whey proteins comprising roughly 20-24%. The whey proteins are of higher nutritional value than casein.

Milk proteins are outstanding sources of essential amino acids. If casein is coagulated from milk, the remaining proteins in whey are known as whey protein. They are lactalbumins and lactoglobulin. Lactalbumins resemble the blood albumins but are not identical. They are soluble in water and their concentration in milk is around 0.30%. Lactalbumins are composed of three major fractions, *i.e.*, (i) β-lactoglobulin 50-60% of the total lactalbumins, (ii) α-lactalbumins 15-20% of the total lactalbumins and (iii) blood serum albumins 5-6% of the total lactalbumins. The remaining minor fraction comprises of a large variety of other proteins. The alpha-lactalbumin is quite rich in tryptophan. Lactoglobulins are identical to blood serum globulins and are soluble in water. In colostrums, they are present in large quantities and help in the transfer of passive immunity from mother to child. Their concentration in normal milk is around 0.1%, but in colostrums, it can be as high as 5-10% and so important in the development of immune system. It serves as the *first vaccine* to the new born. Colostrums globulins can be further identified as euglobulins (true) or pseudoglobulins. A number of enzymes (minor proteins) such as lactoferrin, lysozyme, and lactoperoxidase are also present in milk. Lactoferrin is an iron binding glycoprotein that occurs in cow milk at a level of 0.2 mg/ml. Mother's milk contains large amounts of lactoferrin (0.1-0.2 g/100 ml).

Milk proteins are rich in essential amino acids. Especially whey proteins are of high nutritional quality, containing 51% essential amino acids, compared to 45% in casein. The Protein Efficiency Ratio (3.1), Biological Value (91) and Net Protein Utilization (82) of milk protein are very close to that of egg proteins (3.8, 100 and 94, respectively). Lactalbumin is superior to cascin having Biological Value (BV), Net Protein Utilization (NPU) and Protein Efficiency Ratio (PER) 100, 92 and 3.6, respectively, however, the corresponding values for casein are 77, 76 and 2.5, respectively. Only 14.5 g of lactalbumin or 28.5 g milk proteins is sufficient to meet the daily requirement of essential amino acids for adult humans. The digestibility of milk proteins is rated higher (96%) than that of plant proteins (74-78%). Because of their high biological value, the milk proteins are useful in the diet of patients suffering from liver and gall bladder diseases, hyperlipidaemia and diabetes. Patients with impaired kidney functions rely on protein with high biological value for relieving strain on the excretory function of the kidney. The milk proteins are also used in slimming diets.

Many milk-derived peptides possess functional properties. Several peptides with opium like (sleep inducing) activity have been extracted from the degradation products of milk proteins. These include β-casomorphins from β-casein, exorphin from α_{S1} casein, β-lactostensin from lactoglobulin and serorphin from serum albumin. These opium-like peptides have been shown to prolong gastrointestinal transit time exerting anti-diarrhoeal effect. They also stimulate secretion of insulin and somatostatin. Glycomacro peptide (GMP) derived from K-casein induces production of cholecystokinin, a hormone associated with satiety. Glycomacro peptide and other fractions of K-casein digest inhibit the adhesion of actinomycetes and streptococci to erythrocytes and binding of cholera toxins to its receptor. Glycomacro peptide is free from aromatic amino acids, and therefore, a suitable protein substitute for those suffering from hereditary disorder of aromatic amino acid metabolism, such as phenyl ketonuria (PKU). Angiotensin-converting enzyme (ACE) located in different tissue splits two amino acids from C-terminal end of angiotensinogen I converting it into angiotensinogen II, which is a highly hypertensive octa-peptide. Peptides with anti-hypertensive activity that act through inhibition of ACE have been identified in the sequence of bovine and human β- and α_{S1}-casein. Recently, α-lactalbumin and β-lactglobulin fragments that inhibit angiotensin-converting enzyme have also been characterized.

Several immune-stimulatory peptides have been identified from both bovine and human casein and whey proteins. These peptides have been shown to stimulate the phagocytic activities of marine and human macrophages and enhance resistance against certain bacteria. Certain peptides from casein stimulate the production of immuno-globulins.

Breast-fed babies fall ill less often than the babies do on diets based on cow milk, indicating the presence of antimicrobial substances in mother's milk. The substances in mother's milk, which have an antimicrobial effect, are immunoglobulins, lactoferrin, lysozyme, lactoperoxidase and vitamin B_{12}-binding protein. The immunoglobulins, mainly 1gA are not broken down by the digestive enzymes. Thus, they not only act against the microorganisms in the intestine but also prevent the absorption of foreign proteins. Lactoferrin plays an important role in the resistance against intestinal infection, particularly *Escherichia coli*. Lactoferrin has both bacteriostatic and bactericidal properties. The bacteriostatic effect is of lactoferrin is due to its iron binding ability making iron unavailable for iron requiring bacteria. Mother's milk also contains a large amount of an unsaturated vitamin B_{12}-binding protein, which competes with bacteria that have a vitamin B_{12}.

A number of enzymes are also involved in the milk immune system. These are lactoperoxidase, xanthinoxidase and lysozyme. Lysozyme has a direct effect by breaking down the cell wall of gram-positive bacteria. The lactoperoxidase-thiocyanate-water system is an antibacterial system. Lactoperoxidase and thiocyanate are found in milk and other tissue secretions, and water is produced by lactic acid bacteria or by the action of xanthinoxidase. Thiocyanate is oxidized by water and lactoperoxidase to an intermediate product that destroys the microorganisms.

Milk proteins can be added to cereal based products to increase their lysine and threonine content. Whey proteins have even better supplementary value. A mixture of whey protein, wheat, rice, or maize has protein efficiency ratio even greater than that of whey proteins. Whey proteins can raise the biological value of soy proteins because of their high concentration of sulphur amino acids.

Milk Carbohydrates

Carbohydrates other than lactose are present in milk but they are in traces and of little importance in processing of milk. Lactose alone represents the milk carbohydrates in general discussions. Milk contains about 4.6-7.0% lactose depending upon the species. It is synthesized by the mammary glands and occurs in nature in milk only and so called as milk sugar. It is about one-fifth as sweet as sucrose. It is a disaccharide, and on hydrolysis, it yields one molecule of glucose and one of galactose. It is important for the development of cerebrocide tissues of brain. Galactose may be important because its derivatives, *e.g.*, galactosamine, are constituents of several glycoproteins and glycolipids, which are important constituents of cell membranes. Young mammals have a limited capacity to synthesize galactose. Some persons are intolerant to milk because of lactose intolerance of their body. The concentration of lactose is highest in human milk.

Lactose, the principal milk sugar, is slowly metabolized and has slight laxative effect. The blood glucose does not rise rapidly on lactose diet. Milk consumption, therefore, enables the diabetic person to obtain the biologically highly valuable milk proteins without running the risk of rise in blood glucose levels. Considerable portion of lactose passes into the large intestine where it promotes the growth of lactic acid producing bacteria and lowering the pH of intestine that increases the peristalsis. Lactic acid also creates a desirable condition that inhibits the growth of proteolytic and putrefying bacteria in the intestine and replaces them gradually with acidophilic bacteria, thus, lactose very effectively favours the growth of acidophilus bacteria. Microbial flora in the intestine of infant fed on mother's milk consists almost entirely of anaerobic lactobacilli, *e.g.*, *Lactobacillus bifidus*. Glycoproteins and oligosaccharides, present in mother's milk in high concentration, promote the growth of bifidus bacteria, and thus, provide the baby resistance against intestinal infection.

Several investigations have shown that lactose promotes the utilization of calcium, magnesium and phosphorus. The acidic condition created by fermentation of lactose by intestinal bacteria may increase the solubilization and absorption of calcium. Part of the effect may also be due to the ability of lactose to form soluble complex with calcium. The lactose effect on calcium absorption has been observed in human infant and rats but not in human adults. Lactose promotes the diffusion of calcium across the intestinal wall, which is the only route of calcium transport in human infant since the active transport system is not developed at the time of birth. In adults, the calcium transport is predominantly *via* calcium carrier protein, induced by vitamin D hormone, which is not influenced by lactose.

However, lactose has 2 major nutritionally undesirable consequences, *i.e.*, lactose intolerance and galactosaemia. Lactose intolerance is caused by an insufficiency of intestinal β-galactosidase. In absence of this enzyme either lactose is partially hydrolyzed or not hydrolyzed at all in the small intestine, and since disaccharides are not absorbed, it passes into the large intestine where it causes an influx of water, causing diarrhoea and is fermented by intestinal microorganisms, causing cramping and flatulence.

Milk Vitamins

Milk is a rich source of vitamins in terms of not only their contents but also their better bioavailability. A liter of cow milk approximately contains fat-soluble as well as water-soluble vitamins like B-complex group. Concentration of major vitamins in cow milk is as follows:

Constituents	Content (per litre)	Constituents	Content (per litre)
Vitamin A (IU)	1299.5	Biotin (mcg)	19.6
Vitamin B_1 (mg)	0.39	Vitamin C (mg)	9.69
Vitamin B_2 (mg)	1.67	Vitamin D (IU)	41.25
Vitamin B_3 (mg)	0.87	Vitamin E (IU)	1.54
Vitamin B_6 (mg)	0.43	Vitamin K (mcg)	41.25
Vitamin B_{12} (mg)	3.68	Folate (mcg)	61.57

Source: http://www.food-fortification.com

Vitamin A deficiency is a major cause of widespread blindness among children in India. A 250 ml serving of cow milk contains vitamin A sufficient to meet 75% daily vitamin A requirement of pre-school child. Vitamin A activity is present in milk as retinol, retinyl esters and as carotenes. Whole cow's milk contains an average of 52 μg retinol and 21 μg carotene per 100 g. Human milk and colostrums contain an average of 58 and 155 μg retinol per 100 g, respectively. In addition to their role as pro-vitamin A, the carotenoids in milk are responsible for the colour of milk fat. Vitamin A has a number of roles in the body, *i.e.*, it is involved in the vision process, in cell differentiation, in growth and bone remodeling and in the immune system. The body will tolerate a wide range of vitamin A intakes [500-15000 μg Retinal Equivalents (RE) per day] but insufficient or excessive intakes result in illness. Vitamin A deficiency (<500 μg RE per day) results in night blindness, xerophthalmia (progressive blindness caused by drying of eye cornea), keratinization (accumulation of keratin in digestive, respiratory and urogenital tract tissues) and finally exhaustion and death. The recommended dietary allowance for vitamin D is 10 and 5 μg per day for persons aged 19-24 and over 25 years, respectively. The RDA values for vitamin D are 10 μg per day for persons over the age of 65 years and for pregnant or lactating women. Consuming milk 1 liter per day will supply only 10-20% of the RDA.

Vitamin E is a very effective antioxidant. It can easily donate hydrogen from the phenolic -OH group on the chromanol ring to free radicals. The resulting vitamin E radical is quite unreactive as it is stabilized by delocalization of its unpaired electron into the aromatic ring. Thus, vitamin E protects the lipids (particularly polyunsaturated fatty acids) and membranes in the body against damage caused by free radicals. The role of vitamin E is of particular importance in the lungs where exposure of cells to oxygen is greatest. Vitamin E also exerts a protective effect on red and white blood cells. It has been suggested that the body has a system to regenerate active vitamin E (perhaps involving vitamin C) once it has acted as an antioxidant. The physiological role of vitamin K is in blood clotting and is essential for the synthesis of at least four of the proteins (including prothrombin) involved in this process.

Vitamin K also plays a role in the synthesis of a protein (osteocalcin) in bone. Vitamin K deficiency is rare but can result from impaired absorption of fat. Vitamin K levels in the body are also reduced if the intestinal flora is killed, *e.g.*, by the long time use of antibiotics. Vitamin K toxicity is rare but can be caused by excessive intake of vitamin K supplements. Symptoms include erythrocyte haemolysis, jaundice, and brain damage and reduced effectiveness of anticoagulants. The RDAs for vitamin K for people aged 19-24 years are 70 μg and 60 μg per day for men and women, respectively.

Milk is one of the richest natural sources of riboflavin (vitamin B_2). A 250 ml serving of cow milk contains riboflavin equivalent to 50% of the daily requirement of a pre-school child. Although milk contains only small amount of preformed niacin (vitamin B_3), nevertheless, it is a very good source of this vitamin. Niacin can be synthesized in the body from tryptophan, which is present in milk proteins in good amount (480 mg/liter). In body, 60 mg tryptophan is metabolized to give rise to 1 mg niacin. Indeed, milk is used as dietary ingredient for patient suffering from pellagra, a disease caused by niacin deficiency. For vegetarian, milk is a sole natural source of vitamin B_{12}, as this vitamin is present only in animal food. Milk is also a good source of folic acid, deficiency of which causes megaloblastic anemia mainly in pregnant women of low-income groups.

Milk Minerals

These are different organic and inorganic salts present in milk. They are represented fairly by ash percentage. The ash content of milk remains relatively constant at 0.7-0.8%. These salts are present either in the true solution phase or in colloidal phase, attached with the proteins. The major salts in milk are calcium, phosphates, citrates, sulphates, chlorides, copper, sodium and bicarbonates. All others are present in minutes or trace quantity. Studies of salts of milk are important because they are nutritionally important and more over influence the stability of milk protein. Concentration of major minerals in cow milk is as follows:

Constituents	Content (per litre)	Constituents	Content (per litre)
Calcium (mg)	1277.3	Manganese (mg)	0.04
Chlorine (mg)	1031.36	Molybdenum (mcg)	20.63
Copper (mg)	0.1	Phosphorus (mg)	963.28
Iodine (mcg)	237.21	Potassium (mg)	1567.66
Iron (mg)	0.52	Selenium (mcg)	15.47
Magnesium (mg)	138.2	Sodium	505.36

Source: http://www.food-fortification.com

Recent research has shown that poor nutritional status with respect to calcium is related to diseases like osteoporosis, hypertension and colon cancer. The hypertensive patients have shown significant reduction in blood pressure in response to increased calcium intake. The effect of calcium on blood pressure is mediated by: (i) an increase in urinary excretion of sodium, (ii) preventing the rise in vitamin D hormone, which increases blood vessel resistance, (iii) relaxing smooth muscle cells, which lines the blood vessels, (iv) suppressing the renin-angiotensin system and (v) increasing production of endothelial relaxing factors.

Introduction of increased dietary calcium through dairy products has been shown to reduce incidence of colon cancer and hyper-proliferation in colonic mucosa of rodents. Calcium produces these effects by neutralizing deconjugated bile acids and free acids, thereby, removing their mitogenic/toxic influence.

Milk and dairy products are the most important source of calcium in readily available form. In western countries, milk and dairy products provide up to 75% of the total calcium intake. A 250 ml serving of cow milk contains calcium equivalent to 60% of Indian Council Medical Research's RDA for adults. Equal amount of buffalo milk contains 95% calcium of recommended dietary allowance for adults. Calcium in all forms of milk is equally well utilized. Incorporation of milk in diet also improves the bioavailability of calcium from vegetable foodstuffs. The factors that contribute to better availability of calcium from milk include lactose, protein and phosphorus.

Milk has lysine that enhances diffusion of calcium across the intestine. A part of calcium in milk is associated with phosphorylated casein. The casein phosphopeptides, released in the gastrointestinal tract during digestion, form soluble complex with calcium phosphate salts and improve the diffusion of calcium across the intestine. Milk is rich in phosphorus that reduces urinary calcium excretion and counters balances. Milk and most dairy products, except some processed cheese, have near 1:1 calcium to phosphorus ratio considered ideal for retention of calcium in the body.

The calcium content of human milk is considerably lower than that of cow milk but calcium from human milk is better utilized by infants. This could be due to the nature of fatty acids, the higher calcium to phosphorus ratio and the higher concentration of lactose in human milk. Long chain fatty acids that can render calcium unavailable by forming insoluble complex are present in milk fat at 2 positions of triglyceride molecules and are absorbed as monoglycerides since pancreatic lipase hydrolyses fatty acids from 1 and 3 positions, which are esterified with short chain fatty acids. This certainly contributes to the high bioavailability of calcium from breast milk.

Milk Enzymes

More than 25 enzymes are identified in milk. The enzymes originate from the mammary tissues and enter the milk accidentally during the secretory process. Some enzymes are bound to the surface of fat globules, some present in the free colloidal systems and some in association with casein miscelles. Some of the important enzymes and their actions are as follows:

- **Lipase:** It is fat splitting enzyme and causes rancid flavours in milk. There are two types of lipases, *i.e.*, (i) plasma lipase, which is in the skim milk phase and is activated by some or other form of mechanical agitation and (ii) membrane lipase, which is adsorbed on the fat globule membrane.

- **Analase:** This enzyme causes splitting of starch.
- **Xanthine oxidase:** This enzyme is conjugated with flavin adenine dinucleotide (FAD), iron and molybdenum. When FAD is degraded by an enzyme, it produces flavin mononucleotide (FMN) and riboflavin.
- **Peroxidase and catalase:** These enzymes decompose hydrogen peroxide.
- **Phosphatase:** This enzyme can split certain phosphoric acid esters, which formed the basis of phosphatase test for checking the pasteurization efficiency.

Therapeutic Significance of Milk Constituents

Milk proteins contain a surplus of certain essential amino acids, *i.e.*, lysine and threonine, which can raise the biological value of vegetable proteins. Milk proteins can be added to cereal based products to increase their lysine and threonine content. Whey proteins have even better supplementary value and can raise the biological value of soybean proteins because of their high concentration of sulphur amino acids. Digestibility of milk proteins is rated higher (96%) than that of plant proteins (74-78%). Because of their high biological value, the milk proteins are useful in the diet of patients suffering from liver and gall bladder diseases, hyperlipidaemia and diabetes. Patients with impaired kidney functions rely on protein with high biological value for relieving strain on the excretory function of the kidney. The milk proteins are also used in slimming diets.

Milk proteins have high buffering power. Therefore, it is useful in the treatment of inflammation of mucous lining of stomach and of stomach ulcers, preventing hyperacidity. Milk and milk products are used as a source of proteins in hyperuricaemia and gout disease. In contrast to other foods, these do not contain purines, which are precursors in the synthesis of uric acid that causes gout when deposited in the joints or may lead to the formation of urinary calculi.

Many milk-derived peptides possess functional properties. Several peptides with opium like activity (sleep inducing) have been extracted from the degradation products of milk proteins. These include β-casomorphins from β-casein, exorphin from αS_1 casin, β-lactostensin from lactoglobulin) and serorphin from serum albumin. These opium-like peptides have been shown to prolong gastrointestinal transit time exerting anti-diarrhoeal effect. They also stimulate secretion of insulin and somatostatin.

Glycomacro peptide (GMP) derived from K-casein induces production of cholecystokinin, a hormone associated with satiety. Glycomacro peptide and other fractions of K-casein digest inhibit the adhesion of oral *Actinomycetes* and *Streptococci* to *Erythrocytes* and binding of cholera toxins to its receptor. Glycomacro peptide is free from aromatic amino acids, and therefore, a suitable protein substitute for those suffering from hereditary disorder of aromatic amino acid metabolism, such as phenyl ketonuria.

Angiotensin-converting enzyme (ACE) located in different tissue splits two amino acids from C-terminal end of angiotensinogen I converting it into angiotensinogen II, which is a highly hypertensive octa-peptide. Peptides with anti-hypertensive activity that act through inhibition of ACE have been identified in the sequence of bovine and human β- and αS1 casein. Recently, α-lactalbumin and β-lactglobulin fragments that inhibit ACE have also been characterized.

Several immune-stimulatory peptides have been identified from both bovine and human casein and whey proteins. These peptides have been shown to stimulate the phagocytic activities of murine and human macrophages and enhance resistance against certain bacteria. Certain peptides from casein stimulate the production of immuno-globulins.

Casecidin, a chymosin digest of casein *in vitro*, inhibits *Sarcina*, *Bacillus subtitis*, *Diplococcus pneumoniae* and *Streptococcus pyrogenes*. Similarly, fragments of human β-casein have a protective effect against *Klebsiella pneumoniae*. Iracidin, a αS_1-casein fragment, has therapeutic and prophylatic effect. Lactoferricin, an acid-pepsin digestion product of lactoferrin, has stronger bactericidal activity compared to the native molecule.

Short and medium-chain fatty acids with 4 to 12 carbon atoms, which occur in a relatively high concentration in milk fat, are reported to have antibacterial and fungicidal activity against gram-negative bacteria and certain moulds. Milk fat has a protective effect against human tooth decay. This effect has been ascribed in part to adsorption of milk fat onto the enamel surface and in part to anti-microbial effect of milk fatty acids.

Recently, much attention has been directed towards an unusual fatty acid, conjugated linoleic acid, which is naturally present in milk and dairy products. Since the first report of Pariza's research group of CLA anti-mutagenic activity, there has been an explosion of interest to discover several other biological activities in different experimental model. As on today, several bio-protective properties of CLA have been demonstrated such as anti-carcinogenic activity in mouse fore-stomach, mammary cancer prevention in rats, anti-carcinogenic activity in rat's colon, anti-carcinogenic activity in skin, anti-atherogenic activity in rabbits, reduction in body fat and immuno-modulation. CLA is a milk fat molecule considered good for everything.

Milk fat has several other bioprotective molecules, *e.g.*, sphingomyleins, butyric acid, myristic acid, β-caroteines and fat-soluble vitamins that have potential to inhibit the process of carcinogenesis and atherogenesis and stimulate immune systems. While butyric acid has anti-cancer properties, myristic acid helps the body to fight against infection by activating macrophages.

Several studies have shown that milk reduces serum cholesterol levels of the consumers. Experiments with volunteers have shown that cholesterol levels do not rise when as much as 2 liters of milk is consumed daily. On the contrary, the cholesterol level is reduced. Experiments with animals have shown that even buffalo milk, containing 7% fat, lowers the plasma cholesterol levels. Both decreased formation and increased breakdown of cholesterol are responsible for the cholesterol lowering effect of milk. Orotic acid and another nucleotide associated with proteose-peptone fraction of milk proteins and calcium are suggested to have cholesterol reducing properties. Milk slows down the biochemical processes leading to atherogenesis in rabbits fed on atherogenic diet. It has been suggested that the regular intake of milk keeps the blood vessels healthy.

Several investigations have shown that lactose promotes the utilization of calcium, magnesium and phosphorus. The acidic condition created due to the fermentation of lactose by intestinal bacteria may increase the solubilization and absorption of calcium. Part of the effect may also be due to the ability of lactose to form soluble complex with calcium. The lactose effect on calcium absorption has been observed in human infant and rats but not in human adults. Lactose promotes the diffusion of calcium across the intestinal wall, which is the only route of calcium transport in human infant since the active transport system is not developed at the time of birth. In adults, the calcium transport is predominantly *via* calcium carrier protein, induced by vitamin D hormone, which is not influenced by lactose.

Since lactose is slowly absorbed, it has slight laxative effect, which is due to lowering of the pH that increases the peristalsis of intestine. The blood glucose does not rise rapidly on lactose diet. Milk consumption, therefore, enables the diabetic person to obtain the biologically highly valuable milk proteins without the risk of rise in blood glucose levels.

■■■

39 Tea: A Nature's Boon to the Mankind

M.K. Rana

Department of Vegetable Science, CCS Haryana Agricultural University, Hisar- 125 004, Haryana, India, E-mail: <mkrlotus@gmail.com>

Tea was discovered first time in China where it has been consumed for its medicinal properties since 3000 BC. It is prepared from tender leaves of the plant named *Camellia sinensis* var. sinensis, and *Camellia sinensis* var. assamica. Based on different processing procedures, it can be classified into three principal groups, *viz.* green tea (unfermented), black tea (fully fermented), and oolong tea (semi-fermented). Today, about 75% of the tea produced worldwide is black type, about 23% is green type, and about 2% is oolong type. The tea manufacturers cautiously control weather to acquire a variety of taste profiles, and for how long the tealeaves are kept away from air, a process called fermentation.

Green tea is prepared directly from freshly harvested leaves by inactivating the enzymes by steaming them to avoid enzymatic oxidation of catechins, and the black tea is prepared by fermenting the fresh leaves for longer time; this fermentation process involving polyphenols oxidase enzyme oxidizes the simple polyphenols (catechins) present in fresh tealeaves and converts the catechins to a more complex condensed molecules, which give the typical black color and strong astringent flavor. The oolong tea, also known as semi-fermented tea, is prepared by ending the oxidation process by firing the leaves shortly after rolling, and then drying swiftly. The morphological features of oolong tea lie between green and black teas.

In last few decades, tea has attracted the attention of researchers owing to reported health benefits as an antioxidant, anti-carcinogenic, and anti-arteriosclerotic agent. It is contemplated that flavonoids, which comprise one of the largest groups of secondary plant metabolites, are chiefly responsible for these health benefits. The main classes of flavonoids found in tea are flavanols and flavonols. However, based on the structure and conformation of the heterocyclic oxygen ring (C-ring) of the basic molecule, the flavonoids can be further divided into subclasses: flavones, flavanones (hesperetin, naringenin and their glycosides), isoflavones, flavonols, flavanols, anthocyanidins (pelargonidin, cyanidin, and delphinidin), catechins, and biflavans.

Tea Flavonoids

Over the age-old time, tea plant requires sunlight for its normal photosynthesis and growth but excessive light causes stress, thus, the tea plant synthesizes substances known as polyphenols to protect it self from photosynthetic stressors. This group of useful compounds includes flavonoids- the same class of chemicals that give many fruits and vegetables their antioxidant boost. To date more than 4000 flavonoids have been acknowledged.

The natural presence of flavonoids in tea has demonstrated plenty of biochemical and pharmacological effects, including anti-inflammatory and anti-allergic effects. Nearly 95% of tea's polyphenolic compounds are flavonoids. Among plants grown on the earth, tea has been ranked with the highest content of total flavonoids, however, the concentration of flavonoids differs in tea beverages, depending on whether the preparation is blended, decaffeinated, brewed, or iced, *e.g.*, milk protein when added to tea has previously been reported to possibly bind to, and therefore, reduces the flavonoids concentrations.

Catechins

Green tea contains a large amount of polyphenolic substances, which based on their gallate and hydroxylate benzol groups can primarily be divided into four individual catechins, *viz.* (-)-epicatechin (EC), (-)-epicatechin gallate (ECG), (-)-epigallocatechin (EGC) and (-)-epigallocatechin gallate (EGCG), but EGC and EGCG are considered the main constituents of tea catechins. Heat treatment of green tea infusion results in the conversion of these four types of tea catechins into their corresponding epimers, namely (-)-catechin (C), (-)-catechin gallate (CG), (-)-gallocatechin (GC) and (-)-gallocatechingallate (GCG), respectively. The reports show that canned or bottled tea drinks contain high levels of gallocatechin gallate (GCG), an epimer of EGCG. Green tea epicatechins (GTEs) have been shown to have many beneficial health effects, thus, their degradation must be prevented when tea drinks are prepared, stored, and transported.

Tea catechins, which are colorless water-soluble compounds and structurally flavanols, form 20-30% of the dry weight of green tea and impart bitterness and astringency to green tea infusion. More or less all the quality traits of prepared tea, including its color, aroma and taste, are directly or indirectly associated with the alteration in catechins, *e.g.*, a decrease in catechin content during black tea preparation is associated with an increase in the content of monoterpene alcohols, which improve the aroma of tea; degalloation from ester catechins to non-ester catechins decreases bitterness and astringency of green tea, and epimerization-conversion of tea catechins to their corresponding isomers, can occur during preparation and storage of tea. Catechins dominate in green tea. However, black tea contains much lower levels of catechins because they are oxidized during processing.

Flavonols

Kaempferol, quercetin, and myricetin, the main flavonols in tealeaves, make up 2-3% of the water-soluble extractive in tea. In fresh tea shoots, green tea and black tea, these flavonols are largely present in the form of glycosides. To date, about 14 glycosides of myricetin, quercetin and kaempferol have been reported to be present in fresh tea shoots, green tea, and black tea. In tea beverages, the flavonol-aglycones (the non-glycosylated forms) due to their poor solubility in water are not found in significant amount.

Theaflavins and Thearubigins

Theaflavins and thearubigins are the complex flavonoids that predominate in black tea. During preparation of black tea, catechins get oxidized to develop a distinct color pigment and flavor of black tea. Characteristically the pigments of black tea have been divided into orange-colored theaflavins and brownish thearubigins. The black tea contains four major theaflavins, *viz.* theaflavin, theaflavin 3-gallate, theaflavin 3′-gallate, and theaflavin 3,3′-digallate, which are formed through the reaction between quinones derived from a simple catechin and a gallocatechin. These tea flavonoids are thought to have the strongest chemopreventive effects at a concentration usually consumed by human beings.

L-theanine

L-theanine is a unique amino acid found solely in tea plant and the principal component responsible for the exotic taste and flavor of green tea infusion. It is synthesized naturally from glutamic acid and ethylamine in the roots and transferred to the young leaves of tea plant. It constitutes 1-2% of the total dry weight of tealeaves and exists only in free form (non-protein). L-theanine promotes cells tranquility without any weariness. SuntheanineTH, trade name of L-theanine, offers a tremendous opportunity for designing foods and medical foods targeting relaxation and the reduction of stress.

Bioavailability of Flavonoids

The first infusion contains the utmost content of both catechins and flavonols; the catechins content is decreased significantly with the later infusions, however, the decrease in content of flavonols even in third infusion is noticed less marked, indicating relatively slow release of flavonols during infusing. These results suggest the way of brewing tea for higher intake of catechins and flavonols from drinking tea. Consumption of gunpowder tea 5 g per day supplies roughly 600 mg of total catechins and 40 mg of combined flavonols. In one study, it has been shown that addition of milk in tea decreases the bioavailability of catechins. There are many factors that can affect the way how tea compounds are absorbed, metabolized, and excreted.

Therapeutic Significance of Tea Flavonoids

Flavonoids in tea have a protective effect on numerous life supporting systems in human body, and it is easy to conclude that drinking tea has beneficial effects to maintain a good health and to delay the actions of aging process. There is a growing interest in the possible health benefits of tea consumption. The epidemiological studies show that the constituents of tea provide protection against the development of various chronic diseases including cancer and cardiovascular disease.

Quercetin, kaempferol and myricetin, the major flavonoids in black, green and oolong teas, have been shown to inhibit the actions of carcinogens in human body, particularly, the quercetin has been found to inhibit oxidation and cytotoxicity of low-density lipoproteins and to reduce cancer and cardiovascular diseases in humans. All teas, primarily, the semi-fermented teas have been found to inhibit mutagenicity. Some reports have shown that tea inhibits the absorption of carcinogens and mutagens from the diet.

Bioresearchers are finding evidence these days to confirm centuries-old lore regarding the power of tea to prevent illness and prolong human life. Dr. John- a member of the American Health Society says that the tea components may help to reduce the risk of numerous chronic diseases such as stroke, heart attack and some kinds of cancer. The Buddhist monks believe that tea has the preventive and curative powers, and the caffeine in it helps them stay awake. Dr. John Folts, the Director of the University of Wisconsin Medical Schools of Coronary Artery Thrombosis Research and Prevention Center says that the key protective components in tea appear to be flavonoids. Polyphenols in tea are thought to be the disease preventing ingredients, thus, four to five servings of tea a day are essential for the purpose but addition of sugar in tea may promote tooth decay and acidity. The specific functions of tea are being discussed in details under following heads:

Antioxidant Activity

Super oxide generated by a single-electron reduction of oxygen in living cells under physiological conditions plays harmful roles as precursors of more reactive oxygen species. Free radicals (super oxides) occurring in atmosphere can also trigger chain like reactions, which may cause oxidative damage to sensitive biological structures such as DNA or cell membranes, and subsequently result in cancer, heart disease, multiple sclerosis and autoimmune diseases, thus, antioxidants have attracted the attention of bioresearchers for their effects in preventing diseases due to oxidative stress, which leads to disintegration of cell membranes and many pathological diseases. Antioxidants are the substances that significantly diminish the harsh effects of reactive oxygen species, reactive nitrogen species, or both on normal human physiological functions. Flavonoids in tea have very strong antioxidant and free radical scavenging ability, and are much more powerful than vitamin C and E to protect cells from free radical damage.

Green tea is well documented as a source of dietary antioxidants, and catechins having such dietary antioxidant potential are a major group of polyphenolic compounds present in green tea in plenty. Although caffeine is known as to a major ingredient of tea, but its super oxide scavenging potential is much lower than that of catechins. Tea catechins and their epimers serve as powerful antioxidants for eliminating the super oxides directly, and may be useful in prevention of diseases related to *in vivo* oxidative stress. Tea catechins also limit the consumption of α-tocopherol and allow it to act as a scavenger within cell membranes whilst catechins scavenge aqueous peroxyl radicals near the membrane surface. Tea catechins can quench enzymatically generated O_2^- in hypoxanthine (HPX) and xanthine oxidase (XOD) systems, and non-enzymatically in the irradiated riboflavin system. Green tea epicatechins (GTEs) have been shown to be competent scavengers of super oxide-free radicals, hydroxyl radicals and copper mediated LDL oxidation, as the vascular oxidation damage may contribute to endothelial dysfunction in atherosclerosis. GTEs may also improve vascular endothelial functions.

Black tea extracts show strong antioxidant activity, indicating that the theaflavins and thearubigins are biologically active, however, it is tricky to compare the relative inhibitory potency of theaflavins with polyphenols and individual catechins precisely. In one report it has been shown that when an antioxidant activity of a single dose of green tea is compared with that of black tea, both green and black tea significantly and equally increase plasma antioxidant capacity, reaching maximal levels after about one hour, however, the recent reports demonstrate that the increase in plasma antioxidant activity is approximately 1.5 times higher after drinking green tea as compared to black tea. The other reports show that green tea powder is approximately five times more potent than black tea powder for scavenging nitrogen monoxide (NO). Oolong tea extract also exhibits marked antioxidant activity and reducing power.

Anticarcinogenic Activity

The latest foodstuff, which is to be examined as a cancer-fighting agent, is a green tea beverage known to man for thousands of years. The medical doctors at the university of Texas Cancer Center are planning to give powdered green tea extract as daily doses to the patients who have failed to respond to other treatments or whose cancer tumors have been diagnosed as untreatable. Residents of Shizuoka, Japan, where green tea is produced and consumed, were shown to have lower mortality rates from lung, stomach, and liver cancers than comparable populations in non-green tea consuming area.

Drinking green tea in large amounts (ten cups a day) is a possible form of cancer prevention for the general population. Population studies in China and Japan have found a strong link between green tea drinking and the absence of various types of cancer, including esophageal and stomach cancers. Japanese individuals consuming green tea more frequently or in large quantities have been shown to have a lower risk of gastric cancer.

Numerous laboratory studies have demonstrated that green tea inhibits tumor growth and metastases. One of the most important advantages of green tea as a cancer preventive is its low-toxicity, thus, it is believed that an unfractionated green tea extract has synergistic and stronger effects than any single individual tea component. Oral administration of decaffeinated green tea and black tea extracts (0.6%) inhibited lung tumorigenesis. The inhibitory effect appeared to be due to primarily to its major polyphenol, EGCG- the most potent inhibitor among the decaffeinated green and black tea extracts. Scientists at Case Western Reserve University in Cleveland reported that they had isolated two compounds from green tea-epigallocatechin and epigallocatchin-3-gallate that seem to be active in human cancer cells. When human cancer cells are exposed to these compounds, the cells are died. Cancer occurs when cell death is locked in cancerous cells, leading to unchecked proliferation of tissue into malignant tumors.

Tea flavonoids have been linked to benefits in reducing the risk of certain cancers and cardiovascular diseases, and most of these results have been derived from investigations using green tea. Quercetin, kaempferol, and myricetin have been found to be capable to inhibit carcinogen-induced colon and mammary gland tumors, but the former two flavonols have been found more effective in protecting cells from both super oxide and peroxide at concentrations of five and 20 μM while catechin at 500-1000 μM concentrations. Quercetin is undoubtedly the most important flavonol, followed by kaempferol. However, it is noticeable that the content of myricetin is much higher in tea than in many other plants.

Tea catechins inhibit carcinogenesis; these inhibitions are noticed at all stages of cancer progression, *viz.* initiation, promotion, and transformation. This elevates the possibility that tea drinking may reduce the incidence of cancer in humans. The bioresearchers suggested that galloyl esters of catechins are more active than non-galloylated catechins because they have lower redox potentials, thus, it is important to notice that not all the catechins in green tea are equally effective. Some catechins have been found to be less effective against particular cancer, although they show synergistic effects with caffeine and other catechins.

Blood Pressure

It is known that the regulation of blood pressure is highly dependent on catecholaminergic and serotonergic neuron in both the brain and the peripheral nervous system. It has been demonstrated that theanine is absorbed and reduced serotonin levels. Theanine, the major amino acid in green tea, significantly reduces blood pressure and hypertension. The decrease in blood pressure and hypertension appeared to be dose related, and a significant decrease is observed with the administration of a high dose of theanine. Theanine may affect the pathways such as the peripheral nervous system and peripheral blood vessels. Thus, theanine may have a calming effect on the mental state by lowering blood pressure. Tea drinkers have relatively lower levels of cholesterol and blood pressure.

Heart Disease Preventive Activity

Epidemiological studies showed an inverse correlation between tea consumption and concentrations of serum total cholesterol and triacylglycerols. Some studies demonstrated an association between tea consumption and a decreased risk of coronary heart disease. There are numerous reports that demonstrate effects of tea catechins on coronary heart disease and claim that tea drinking may reduce the incidence of heart disease in human beings. The catechins in green tea can lower cholesterol by increasing the quantity of a liver's LDL receptors, which work to clear cholesterol from the blood. Research studies suggest that humans may lower their cholesterol levels up to 10% if they habitually consume green tea 5-10 cups a day.

ARS chemist Joseph T. Judd, who is with the Diet and Human Performance Laboratory, found that drinking tea lowered cholesterol, and therefore, could possibly reduce the risk of coronary heart disease (CHD). While several epidemiological studies show that green and black tea, consumption is associated with reduced risk of CHD. Judd's recent study assessed the effects of black tea consumption on blood lipid and lipoprotein concentrations in adults with mildly high cholesterol, and he found 6-10% lower blood lipids in drinkers of black tea in just 3 weeks. The study's authors concluded that drinking black tea, along with a prudent diet fairly low in fat, cholesterol, and saturated fatty acids, significantly reduces total and LDL cholesterol, and may, therefore, reduce the risk of CHD.

The oxidation of low-density lipoprotein (LDL) causes progressive blockage of arteries or arteriosclerosis and leads to angina pectoris or to coronary heart disease. Tea flavonoids, principally galloctechins, protect LDL against oxidation by aqueous and lipophilic radicals, copper ions, and macrophages. Tea reduces triglycerides, total cholesterol, and LDL-cholesterol. Tea catechins effectively reduce cholesterol absorption by lowering the solubility of cholesterol and total lipids. Tea extracts induce an anti-inflammatory and capillary strengthening effect. It also has the antioxidant activity, thus, can inhibit the LDL oxidation.

Improvement in Learning Ability

It has been shown that administration of theanine has a significant effect on the release or reduction of neurotransmitters, like dopamine and serotonin. It is also known that these transmitters are closely related to memory and learning ability. Since ancient times, it is believed that drinking green tea brings relaxation without inducing drowsiness and improves learning ability. The substance that is responsible for a sense of relaxation is L-theanine. Most Japanese people take several cups of green tea every day while taking rest, welcoming guests, or chatting with friends. Theanine is the chief amino acid in green tea, which acts as a neurotransmitter in brain. It may affect emotions by interacting with neurotransmitters in the brain. It has an equilibrating effect on the control nervous system. Theanine has also been known to act antagonistically against paralysis induced by caffeine.

Anti-microbial Activity

The antimicrobial activity of tannins is well documented worldwide. Tannins have been reported to be bacteriostatic or bacteriocidal, therefore, tannins can theoretically serve as natural regulators of the microbial population in aquatic environments by controlling off flavor production in aquaculture. The presence of tannins in food plants serves as a natural defense mechanism against microbial infections. Thus, tannins in tea serve as natural regulators of the microbial population in different habitats including the human gastrointestinal tract. Green tea catechins have been found to have antibacterial activity against both gram ± ve bacteria, which can be harmful to human but are not effective against *E. coli*, *Pseudomonas aeruginosa* or *Areromonas hydrophila*. Green tea extracts can also kill *Helicobacter pylori* associated with peptic, gastric, and duodenal ulcers.

Fluoride in green tea contributes antibacterial action of tea extracts against Streptococcus. The polyphenols in green tea can prevent teeth from decaying by inhibiting the biological activities of the cariogenic Streptococci (*Streptococcus mutans* and *Streptococcus sobrinus*). Tea extracts not only prevent the growth of *S. mutans* but also deter its adhesion and inhibit glycosyl transferase activity. In addition, green tea polyphenols have a stronger deodorizing activity.

The antimicrobial activity of tannins could be due to their complexation with enzymes or substrates. Many microbial enzymes, cellulase, pectinase, xylanase, peroxidase, laccase and glycosyl transferase are inhibited when mixed with tannins. Tannin toxicity may also be related to their action on the membrane of the microorganisms. Tannins complexation with metal ions could be another possible mechanism responsible for tannin toxicity.

Anti-dental Carries

Dental caries is the commonest of all microbial related diseases. Epidemiological studies also showed that schoolchildren from tea plantation areas had a lower incidence of caries than those from other areas devoid of tea plantation. Levels of maltose, the sugar released from starch, were found to be consistently lower following tea drinking compared to water, thus, it is supposed that under normal situation tea reduces the acid on the tooth enamel for the whole time that starch is trapped in the teeth, and that is why it exerts anti-caries effect. The group of dentistry team has found that compounds in tea destroy cavity-causing bacteria and if included as part of a normal oral hygiene routine, tea drinking may lead to a reduction in dental cavities. Although their work related specifically to black tea, it adds to similar results that had been found previously for green tea. Additionally, tea is one of the few dietary sources of fluoride that helps to strengthen teeth. In addition, green tea polyphenols improve the breath freshness, thus, it acts as a mouth freshener.

Anti-diabetes Activity

Diabetes is associated with high glucose content in blood. Drinking green and black teas infusion can significantly lower down the glucose level in blood by reducing the absorption of glucose. The polyphenols in tea inhibit the activity of α-amylase in saliva and reduce the activity of intestinal α-amylase, which in turn lowers the hydrolysis of starch to glucose and reduces glucose assimilation. Polyphenols can also decrease the activity of other digestive enzymes and reduce glucose absorption. Richard Anderson- ARS chemist with the Beltsville Human Nutrition Research Center (BHNRC), Beltsville, Maryland found that regularly brewed tea, when added to the fat cells of laboratory rats, increased insulin activity by more than 15 times. Anderson noted that this increased insulin activity was found with green, black, and oolong teas, regardless of whether caffeinated or decaffeinated. Further, his research showed that in green and oolong teas, the catechin EGCG was largely responsible for these results.

In black tea, active ingredients included tannins and theaflavins, in addition to EGCG. Ernst J. Schaefer, Director of the Human Nutrition Research Center on Aging (HNRCA's), Lipid Metabolism Laboratory at Tufts University in Boston, Massachusetts found that 8 volunteers with type II diabetes had lower blood sugar levels by 15 to 2% after drinking 6 cups of tea a day constantly for 8 weeks.

Anti-cholera and anti-AIDS Virus Actions

Tea extracts have been used for therapy of cholera in epidemic areas and for the prevention of influenza virus infections. Tannins in tea inhibit the growth of fungi, bacteria, and viruses. Theaflavins extracted from black tea produce inhibitory effect on the AIDS virus but less than that of catechins from green tea. Tannic acid sulfate, (-)-epicatechin sulfate and (-)-epigallocatchin-3-gallate sulfate inhibit the infection of human T-cell lymphotrophic virus type-1-carrying MT-4 cells by human immunodeficiency virus (HIV). Recently it has been demonstrated that epigallocatchin, epicatechin-3-*o*-gallate and EGCG isolated from Chinese green tea are inhibitory to HIV-RT. Tannins have also been reported to inactivate tobacco mosaic virus and herpes simplex viruses.

Controlling Body Weight

Since ancient times the Chinese believe that, oolong tea is effective in controlling body weight and green tea makes people lose weight; it raises metabolic rates and increases fat oxidation rates. Now it is universally accepted that caffeinated tea raises metabolic rate because caffeine is a stimulant. An individual when drinks tea turns on the fat-burning faucet a little bit more than when drinks caffeinated water. Caffeine- a highly active catechin in tea may act together to increase fat oxidation.

Possible Health Hazards of Tea Polyphenols

Tannins in teas are considered nutritionally undesirable as these reduce the nutrition value of foods and affect the utilization of minerals and vitamins. They interfere with the digestion and/or absorption of carbohydrates by interfering with B_{12} utilization, thus, ingestion of a diet containing a much higher content of tannins may result in adverse health effects, particularly, a marked growth depression. There has been a concern about the inhibition of iron absorption through tea polyphenols (tannin) forming insoluble complexes in the gastrointestinal tract, hence, reducing availability of iron in individuals consuming tea, as it is found to be able to inhibit the absorption of non-haem iron to a significant extent, however, the addition of ascorbic acid or milk could counteract the inhibitory effect of tannin in tea on iron absorption. According to research reports, both black and green teas and a green tea extract do not represent a risk of reduced iron availability to animals consuming tea beverages. Other reports demonstrate that when tea and iron both are taken separately no inhibitory influence is observed on iron absorption. An epidemiological investigation also indicates that no significant anaemia is noticed even after drinking an average of 3-7 cups of tea per day. Yet, it is suggested that oral iron drugs should never be taken together with a cup of tea.

Most tea polyphenols (TPs) such as tannic acid, catechin, epicatechin and proanthocyanidin react with starch, proteins, and digestive enzymes, leading to changes in the bio-ability of proteins, *e.g.*, tea polyphenols denature the pectinase, tyrosinase, α-amylase, peroxidase, lipases, xanthine oxidase and proteolytic enzymes, resulting in the formation of precipitates. According to research reports, one molecule of tannin can bind 12 molecules of protein. At very low concentration, the complexes found between polyphenols and proteins are usually soluble, but high concentrations cause precipitation.

This phenomenon occurs frequently in many foods and beverages in which polyphenols are present in large quantities, *e.g.*, tea beverages, beers and fruits and vegetable juices. Chinese green tea contains about 18-25% catechin polyphenols owing to the presence of a hydroxylate benzol group and gallate group structure. TPs exhibit a strong complexing activity with proteins and enzymes. The main mechanism of complexing reaction is considered to be a non-covalent interaction of proline residues of protein molecules with the gallate and hydroxylate benzol groups of TPs. It is believed that the major dietary effect of tannins within digestive tract is due to the formation of less digestive complexes with dietary proteins rather than by inhibition of digestive enzymes.

Occupational Safety and Health Administration has listed tannins as category one carcinogens. Topical application of black tea infusion has been reported to promote Benzo[a]pyrene induced skin carcinogenesis. This carcinogenic activity of tannins may be due to their ability to cause irritation and cellular damage rather than the action on DNA mutation. Tannins are thought to be co-carcinogens or promoters in inducing skin carcinogenesis in the presence of other carcinogens.

Ingestion of high levels of tannins can cause gastroenteritis and congestion of internal wall. From above research reports, it is concluded that tannins seem to be a double-edged sword. On one side, they are beneficial to health due to their chemopreventive activities against carcinogenesis and mutagenesis but on other side they may involve in cancer formation or anti-nutritional activity, thus, there is a possibility that the active components responsible for carcinogenic initiation are not tannins themselves but rather the components associated with tannins.

Types of Herbal Tea

- **Energy tea:** Ginseng, nettles, yerba mate, citrus peels and peppermint
- **Relaxation tea:** Chamomile, lemongrass, yerba mate, Licorice and peppermint
- **Blood pressure tea:** *Apocynum venetum*
- **Sleep tea:** Valerian, lavender, Chamomile and spearmint
- **Skin glow tea:** Red-clover, Sarsapilla, orange peals and anise seeds
- **Head and throat soother tea:** Wild Notopterygium, *Angelica dahurica* and mint
- **Flu tea:** Elder berries, Echinacea, yarrow, ginger and peppermint
- **Immunity tea:** *Astragalus spp.*, Schizandra berries, nettles and peppermint
- **Memory tea:** *Ginkgo biloba*, gotu kola, rosemary, Schizandra berries and spearmint
- **Strong lung tea:** Licorice root, yerba santa, pleurisy root, ginger and sinnamon
- **Kidney fortifier tea:** Alisma and Pyrrosia
- **Tummy upset tea:** Peppermint, catnip, Chamomile and ginger
- **Laxative tea:** Senna leaf
- **Gastroease tea:** *Astragalus* and millet
- **Gas tea:** Fennel seeds, Chamomile and peppermint
- **Diabetes tea:** *Gymnema sylvestre*
- **Sugar control tea:** Mulberry (*Morus* spp.) leaf
- **Sexuality/vigor tea:** Damiana, Ginseng, Sarsapilla, Cinnamon and nutmeg
- **Rejuvenational tea:** Wild *Epimedium* and *Polygoatnum multiflora*
- **Manhood tea:** Broomrape (*Cistanche* spp.), Panax-ginseng and *Astragalus* spp.

Tea Preparations and Applications

Since age-old tea has been consumed worldwide as the most popular beverage after water for thousands of years, the last few decades are beginning to understand the full potential of this widely enjoyed beverage. Its popularity is still increasing because of its potential benefits for human health. Tea extracts and tea beverages, predominantly those based on green tea, have several applications.

The beauty of tea is that it can be consumed in so many ways, depending on individuals' tastes and preferences. Increasing consumer demand for a healthier lifestyle has led to a need to develop tea products, which will accomplish this requirement. In spite of their lack of applications in the field of pharmaceutical, tea flavonoids are now being increasingly added to nutraceuticals/functional foods. Although the consumption of fresh tealeaves extracts is becoming of interest of many people but making a cup of tea by brewing the tealeaves directly is still the most common way to consume tea. It is consumed directly in the forms of strong infusions, soft extracts and powders, and available widely in the form of beverage, toiletries, and cosmetic products. Tea can be consumed either by brewing loose leaves or tea bags or in a ready-to-drink form. In addition, extracts of tea may be prepared in a variety of physical forms in order to cover most application requirements.

Tea extracts range from liquids, such as strong infusions, through soft and dry extracts to purified catechins. Strong infusions are made by soaking tea leaves in alcohol/water mixtures; soft extracts are made by concentrating the strong infusion to a water content of 20-25%, and tea extract powders or dry extracts are made by spray dying the strong infusions after they have been concentrated to 40-50% solids. Tea plant roots have also been extracted and applied as an anti-neoplastic agent.

Consuming a variety of tea preparations adds nutritional benefits to the diet. The inborn characteristic antioxidant properties of catechins in green tea extracts can improve the selling prospective of cereals, biscuits, pastries, cakes, and nutraceuticals as well as conventional health foodstuffs and other dietary supplements. Green tea extracts can also be integrated to give dairy products, confectioneries, ice creams, instantaneous noodles, and snacks a better appeal to the consumer. Besides, many antioxidant-rich health drinks containing green or black tea extracts in combination with other natural extracts are also available in the market. Not only in Japan but also in world over, the consumption of ready-made tea drinks packed in bottles or cans has been increasing remarkably over recent years.

The antibacterial and deodorizing effect of catechins slows down tooth decay process and concomitantly improves breath freshness so offering inherent added value for tooth-pastes, mouthwashes, breath-fresheners and chewing gums with the potential for other body, skin and hair applications. Many cosmetics such as sunscreen lotions, moisturizing creams, perfumes, and shampoos contain tea extracts, as they are believed to have a soothing effect on skin as well as acting as antioxidants to protect the skin and hair from the deleterious effects free radicals. In China, the green tea extract has been added in shampoos, toothpastes and other cosmetics for over 10 years. Tea extracts may also be added in apple juice and other foodstuffs to protect them from any kind of fungal and bacterial infections.

■ ■ ■

40 Coffee: Good or Bad for Health

Kanchan Sandhu

Assistant professor (H.Sc.), Kishi Vigyan Kendra, Nurmahal, Jalandhar, Punjab Agricultural University, Ludhiana, Punjab, India,
E-mail: <kanchansandhu1973@gmail.com>, <kanchansandhu@yahoo.com>

Coffee is probably the most frequently ingested beverage worldwide. In India also after tea, coffee is the popular beverage. Everyone recognizes a roasted coffee bean. A coffee tree is covered with dark-green, waxy leaves growing opposite each other in pairs, pruned short in cultivation but capable of growing more than 30 feet high. Coffee cherries grow along the tree's branches. It takes nearly a year for a cherry to mature after flowering of the fragrant, white blossoms. Coffee is usually sold in a roasted state, and with rare exceptions, coffee is roasted before it is consumed. Roasting process influences the taste of beverage by changing the coffee bean both physically and chemically. During roasting, caramelization occurs as intense heat breaks down starches in the bean, changing them to simple sugar that begin to brown, changing the colour of the bean. Sucrose is rapidly lost during roasting process and may disappear entirely in darker roasts. During roasting, aromatic oils and acids weaken, changing the flavour, and at 205°C, the other oils start to develop. One of these oils is caffeol, created at about 200°C, which is largely responsible for aroma and flavour of the coffee.

The degree of roast has an effect upon coffee flavour and body. Darker roasts are generally bolder because they have less fibre content and a more sugary flavour. Lighter roasts have a more complex and therefore perceived stronger flavour from aromatic oils and acids otherwise destroyed by longer roasting times. Coffee is a brewed beverage with a dark, acidic flavour prepared from the roasted seeds of the coffee plant, colloquially called coffee bean. It is one of the most-consumed beverages in the world. Good quality coffee will commonly have some bitter elements but they should exist in balance with other aspects. Bitterness should not be an overwhelming component. Often people will confuse *bitter* with being *strong*. If your coffee is unpleasantly bitter, one cause may be that your grind is too fine for the steeping duration.

Decaffeinated coffee usually loses some flavour compared to normal coffee. Decaffeinated coffee (also called decaf) is available. This is the coffee from which most of the caffeine has been removed, by the Swiss water process (which involves the soaking of raw beans to remove caffeine) or the use of a chemical solvent such as trichloroethylene (*tri*), or the more popular methylene chloride, in a similar process. Another solvent used is ethyl acetate. The resultant decaffeinated coffee is marketed as *natural decaf* because ethyl acetate is naturally present in fruit.

Therefore, coffee has over a thousand chemicals, many formed during the roasting process. From a physiological point of view, the potential bioactives are caffeine, the diterpenes cafestol and kahweol found in the oil and the polyphenols, are the most notably chlorogenic acid. The commonly consumed preparations of coffee are boiled unfiltered coffee, filtered coffee and decaffeinated coffee.

Effects of Caffeine on Human Body

Coffee contains several compounds, which are known to affect human body chemistry. The coffee bean itself contains chemicals, which are mild psychotropics for humans. These chemicals are toxic in large doses. Coffee can have a stimulating effect on humans due to its caffeine content. Caffeine is similar in structure to adenosine, and therefore, is able to bind to the adenosine receptors in the brain. This results in competitive inhibition of adenosine action. Adenosine is a neurotransmitter and when it binds to its receptors in the brain, it can induce sleep, thus, when caffeine binds to the adenosine receptors, it can prevent sleep. Therefore, physiologically, caffeine has stimulant effect on the body. It increases activity of nerve cells in the brain and triggers the pituitary gland to release more adrenaline (also known as epinephrine, a *fight* or *flight* hormone). Some of the effects include an initial sense of alertness, increased cardiac activity (heart beats faster) and a release of glycogen (stored sugar) giving us an energy boost. The first cup or two of coffee may have what is perceived as being a positive effect on well being, helping us think more clearly and act faster, however, these benefits are reversed with increased intake. If someone is already stressed, the added adrenaline will potentiate this state, increasing the chance of panic attacks, palpitations, insomnia, sweating, *etc*. The effect of caffeine varies from one person to another. Normally, up to four cups of coffee a day is considered as moderate and the caffeine from this quantity of coffee is not considered as harmful.

It is still controversial that coffee is beneficial or harmful as reliable evidence is becoming available supporting its health promoting potential. Some researchers find the association of coffee consumption with cardiovascular complications and cancer insurgence. The health-promoting properties of coffee are often credited to its rich phytochemistry, including caffeine, hydroxyhydroquinone (HHQ), caffeic acid, chlorogenic acid, *etc*. Many research investigations regarding coffee consumption showed its inverse correlation with that of diabetes mellitus, various cancer lines, Parkinsonism and Alzheimer's disease. Furthermore, caffeine and its metabolites help in proper cognitive functionality. Coffee lipid fraction containing cafestol and kahweol act as a safeguard against some malignant cells by modulating the detoxifying enzymes. On the other hand, their higher levels raise serum cholesterol, posing a possible threat to coronary health, *e.g.*, myocardial and cerebral infarction, insomnia and cardiovascular complications. Caffeine also affects adenosine receptors, and its withdrawal is accompanied with muscle fatigue and allied problems in those addicted to coffee. Below are the various health and harmful effects of coffee:

Antioxidant property of coffee

Antioxidants are compounds capable of scavenging the disease causing free radicals produced in the body. One of the most dangerous free radicals is the hydroxyl radical and caffeine has been proved to be effective in removing the hydroxyl radicals. Coffee is the major source of dietary antioxidants. The potential health benefits of these antioxidants ultimately depend on how well they are absorbed and utilized in the body. Coffee contains an anticancer compound methylpyridinium, which is not present in significant amount in other food articles. Methylpyridinium is not present in raw coffee beans but is formed during the roasting process from trigonelline, which is common in raw coffee beans. It is present in both caffeinated and decaffeinated coffee and even in instant coffee. Roast coffee contains more lipophilic antioxidants and chlorogenic acid lactones and is more protective against hydrogen peroxide-induced cell death in primary neuronal cells than green coffee. The espresso method of extraction yields higher antioxidant activity than other brewing methods. Free radicals present in the body can cause damage to body cells. Antioxidants present in the coffee act as free radical scavengers and prevent damage done to the body cells, which in turn reduces the risk of cholesterol-oxidation and heart diseases. Another beneficial feature supported by the antioxidant coffee is that it makes the immune system stronger. Individuals with a stronger immune system are less prone to diseases. Coffee has four times the antioxidants of green tea, which makes it an excellent anti-depressant and enhances performance and memory.

Possible Health benefits of coffee

- Caffeine increases intellectual activity when fatigued or bored.
- Reduced risk of Alzheimer's disease and dementia
- Reduced risk of gallstone disease
- Reduced risk of Parkinson's disease
- Cognitive performance (short term recall)
- Analgesic enhancement (caffeine can help fight pain)
- Reduced risk of some cancers
- Antidiabetic
- Liver protection (coffee can reduce the incidence of liver cirrhosis)
- Cardio protective (protects against cardiovascular disease)
- Laxative/diuretic (coffee is a powerful stimulant, which can help in preventing constipation. Unfortunately, some people experience loose stool as a result).
- Gout (coffee consumption can help in preventing the risk of gout among men over the age of 40 years.)

- Coffee may reduce the risk of developing gallstones, discourage the development of colon cancer, improve cognitive function and reduce the risk of liver damage in people at high risk for liver disease. Coffee has also been shown to improve endurance performance in long-duration physical activities.
- Coffee may reduce the incidence of kidney stones by increasing the flow of urine and decreasing its concentration.

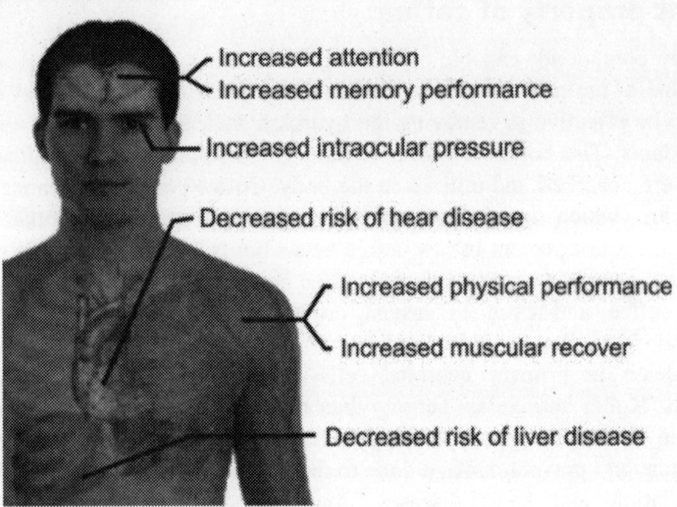

Effects of Caffeine Moderate Consumption

Nutrient absorption: All forms of caffeine can negatively affect absorption or utilisation of a variety of nutrients, especially calcium, protein, iron and water-soluble vitamins, *e.g.*, B and C. Coffee excites more rapid peristaltic movements of the intestines resulting in short transit times and less absorption of nutrients. Keep caffeine consumption away for meal times and supplement taking. Taking vitamins at the same time as a cup of coffee or tea can interfere with the body's absorption of many necessary nutrients.

Calcium: Caffeine inhibits the amount of calcium, which is absorbed through the intestinal tract and depletes the amount retained by the bones. Caffeine causes calcium to be excreted in the urine and faeces. For every 150 mg of caffeine ingested, about the amount in one cup of coffee, 5 mg of calcium is lost. This effect occurs even hours after the consumption of caffeine. A study of postmenopausal women found that those who consumed more than 300 mg of caffeine lost more bone in the spine than women who consumed less. Studies have shown that women with high caffeine intake suffer more hip fractures than those who avoid caffeine or drink in moderation

Vitamin D: Caffeine inhibits vitamin D receptors, which limit the amount of vitamin D absorption. Vitamin D is important in the absorption of calcium. Due to excessive coffee intake vitamin D deficiency can ultimately decrease bone mineral density, resulting in an increased risk for osteoporosis.

Iron: Iron is necessary for red blood cell production. Iron deficiency can lead to many complications. Caffeine interferes with iron absorption. If coffee is consumed along with iron source, the iron absorption can be reduced by up to 80%. Therefore, any beverage containing caffeine should be consumed separately from iron-containing food items or supplements by at least one hour.

B Vitamins: Caffeine has a mild diuretic effect, which increases urination. Water-soluble vitamins, like B-group vitamins, can be depleted as a result of the fluid loss. Caffeine also interferes with the metabolism of vitamin B. On the other hand, caffeine stimulates the HCL production in the stomach, which helps the body to absorb vitamin B_{12}.

Vitamin K: Coffee, including decaf, has high amounts of vitamin K, which affects coagulability of the blood, which is bad for people who are the at risk of heart attack, stroke and blood clots

Minerals: Studies have found that caffeine may reduce the absorption of certain minerals like zinc, manganese and copper. Due to its diuretic effect, it also increases the excretion of the minerals sodium, potassium, magnesium and phosphate.

Acute asthma: Caffeine has overall vasodilation effect on the human body. The airways in lungs are also widened slightly (this effect is known specifically as bronchodilation), enabling someone afflicted with asthma to breathe more easily. The theophylline in coffee may be protective against asthma

Headache relief: Another use of the vasodilation effect of caffeine can be headache relied. Constriction of blood vessels in the brain can cause major headache pain, and caffeine relieves this pain by widening the blood vessels. For this reason, caffeine can be found in many specialized headache medicines, *e.g.* Excedrin.

ADHD: People with ADHD (a hyperactivity disorder) have a lowered ability to focus and a shortened attention span and stimulants like caffeine and ritalin (the most frequently-used treatment for ADHD) allow for longer periods of intense concentration. Unfortunately, the side effects of caffeine make it impractical as a long-term solution to most medical conditions.

Diabetes: Coffee has a significant antioxidant activity and may have an inverse association with the risk of type 2 diabetes mellitus. Coffee also contains minerals such as magnesium and chromium, which help the body in using the hormone insulin, which controls blood sugar (glucose). In type 2 diabetes, the body loses its ability to use insulin and regulate blood sugar effectively.

Blood pressure: Caffeine consumption increases the cortisol levels, which in turn results in constriction of the blood vessels and higher blood pressure.

Coronary heart disease: Coffee may interfere with proper levels of homocysteine and cholesterol by inhibiting vitamins such as folate, vitamin B_{12}, or B_6. A study has shown that cafestol, a substance, which is present in boiled coffee drinks, dramatically increases cholesterol levels, especially in women. Filtered coffee contains only trace amounts of cafestol. Heavy consumption of boiled coffee elevates blood total and LDL cholesterol levels. Unfiltered coffee is a significant source of cafestol and kahweol, which are diterpenes responsible for cholesterol-raising effects of coffee. Diterpenes are extracted by hot water but are retained by a paper filter. This explains why filtered coffee does not affect cholesterol, whereas, Scandinavian boiled, cafetiere, and Turkish coffees do. Coffee may counter several risk factors for heart attack and stroke. In 2009, a study of 83,700 nurses enrolled in the long-term nurses' health study showed a 20% lower risk of stroke in those who reported drinking two or more cups of coffee daily compared to women who drank less coffee or none at all. That pattern held regardless of whether the women had high blood pressure, high cholesterol levels and type 2 diabetes.

Cardiac stimulation: The other effect of caffeine on human body is that of cardiac stimulation. Caffeine gives an 'energy boost' that lasts for a period proportional to the amount of caffeine ingested. Cardiac stimulation is achieved indirectly through the release of epinephrine (adrenaline) by the pituitary gland. However, in order to get the pituitary to release epinephrine, caffeine must first cause an increase in nerve cell activity in the brain. To do this, caffeine takes the place of adenosine, a naturally occurring chemical in the body that slows down nerve cell activity and dilates blood vessels to allow for the increased oxygen uptake necessary during sleep. Caffeine binds to adenosine receptor sites on nerve cells in the brain. Instead of slowing activity, it quickens the pace at which nerve cells fire, warding off drowsiness. It does this by lowering the trigger level for norepinephrine, increasing the likelihood that the firing of one nerve cell in the brain will cause neighbouring cells to fire. This increased activity in the brain alerts the pituitary gland, and it releases epinephrine, setting off the *fight* or *flight* response in the human. One of the functions of the fight or flight response is to release sugar (stored in the liver in the form of glycogen) for immediate use. This is the source of energy used in the 'boost' provided by ingesting caffeine.

Caffeine as laxative/diuretic: Coffee is also a powerful stimulant for peristalsis and is sometimes considered to prevent constipation. However, coffee can also cause excessively loose bowel movements. The simulative effect of coffee consumption on the colon is found in both caffeinated and decaffeinated coffee. Practitioners in alternative medicine often recommend coffee enemas for the *cleansing of colon* due to its stimulus of peristalsis.

Cancer: Coffee contains acknowledged carcinogens such as caffeine and other chemicals produced by the high heat of roasting such as creosote, pymdine, tars and polycyclic aromatic hydrocarbons. Caffeine consumption is also correlated to a reduced risk of oral, esophaheal and pharyngeal cancer. Studies show that coffee protects the liver from cancer, and there is a correlation between coffee consumption and a lowered risk of aggressive prostate cancer.

Liver protection: Coffee can also reduce the incidence of cirrhosis of liver and it has been linked to a reduced risk of hepato cellular carcinoma, a primary liver cancer that usually arises in patients with pre-existing cirrhosis. The exact mechanism and the amount of coffee needed to achieve a beneficial effect have long been unclear. The cytokine transforming growth factor (TGF) beta has long been recognized for promoting fibrosis ability acting through the Smad family of transcription factors. Caffeine may have potent anti-fibrotic capabilities through its ability to antagonize the Smad pathway.

Reduced risk of gallstone disease: Caffeine prevents crystallization of cholesterol and reduces risk of development of gallstones. Drinking caffeinated coffee has been correlated with a lower incidence of gallstone and gallbladder disease in both men and women. Coffee protects primary neuronal cells against hydrogen peroxide-induced cell death.

Appetite suppression and calorie burning: Caffeine may reduce your desire to eat for a brief time but there is not enough evidence to show that long-term consumption aids weight loss. Caffeine may stimulate thermo genesis- one way your body generates heat and energy from digesting food but this is probably not enough to produce significant weight loss.

Water loss: In some people, caffeine can act as a diuretic, which means it increases the amount of urine you excrete. This increase in urine output, mostly water loss, may temporarily decrease your body weight but it does not result in the loss of body fat.

Brain protection: Consuming caffeinated coffee throughout life appears to help preserve cognitive skills, *i.e.*, thinking, memory and comprehension, possibly because of long-term caffeine exposure.

Reduced risk of Parkinson's disease: Coffee may lower the incidence of Parkinson's disease due to high antioxidant activity. Caffeine intake reduces the risk of Parkinson's disease in women who use postmenopausal hormone therapy. In women who did not use hormones, caffeine intake at moderate to high levels decreased the risk of Parkinson's.

Prevention of dental caries: The tannins in coffee may reduce the cariogenic potential of food. In vitro experiments have shown that these polyphenolic compounds may interfere with glucosyl transferase activity of mutans streptococci, which may reduce plaque formation thereby preventing dental decay.

Gout: Coffee consumption decreased the risk of gout in men over the age of 40 years. It is observed that the risk for developing gout in men over 40 was inversely proportional to the amount of coffee consumed.

Analgesic enhancement: Coffee contains caffeine, which increases the effectiveness of painkillers, especially migraie and headache medications. For this reason, many over-the-counter headache drugs include caffeine in their formula. However, coffee can have different effects on individuals that how long it takes to metabolize caffeine, depending on factors such as age, various medications and health of the liver, *e.g.*, the half life (how long it takes the body to clear half of the caffeine from the body) in a healthy adult is about 3-4 hours, in a woman taking the oral contraceptive pill that increases to 5-10 hours, while in a pregnant woman it is up to 11 hours. The half life for a baby in utero or a newborn is even higher so that even on one decent cup of caffeine a day via the mother's milk the neonate may not have got rid of yesterdays intake fully before today's hit.

Taking too much caffeine may cause the following health risks:

- Insomnia, anxiety, tremors and sleeplessness
- Irritability and muscle tremors
- Nervousness and restlessness
- Stomach upset and gastrointestinal symptoms
- Fast heartbeat, increased blood pressure and abnormal heart rhythms
- High levels of caffeine (six or more cups a day) increased the risk of Parkinson's in women who used hormone therapy.
- Coffee contains diterpene cafestol, which increases serum low-density lipoprotein cholesterol concentrations.

Fertility: Numerous studies have also found caffeine consumption by pregnant women to be associated with low birth weight. The American College of Obstetricians and Gynecologists recommends that pregnant women consume no more than 1 to 2 cups of coffee per day. Women who are pregnant or planning to be pregnant should avoid excess caffeine consumption, as it may increase the risk for a variety of fertility problems. High levels of caffeine are bound to accumulate in the baby's body with frequent maternal consumption of caffeine. Just like what it does to adults, caffeine could also send the baby's pulse and breathing rate racing and affect its sleep pattern for an extended duration. A recent study found that the risk of miscarriage doubled with the consumption of 2 or more cups of coffee per day. Another Danish study of 18,478 women linked heavy coffee consumption during pregnancy to significantly increased risk of stillbirths.

Anxiety symptoms: The most important effect of caffeine is on the central nervous system. It acts as a powerful stimulant in the brain, and therefore, increases the alertness levels. Excess caffeine can make one irritable and can increase the anxiety levels. Excess caffeine, because of its ability to stimulate the central nervous system, can prevent sleep. If one wants to sleep normally, one should avoid drinking coffee before going to bed. When you drink coffee on a regular basis, you are open to the appearance of withdrawal symptoms. Coffee consumption also exhausts the adrenal glands over time. Caffeine decreases resistance to stress and increases vulnerability to disease pathogens and other health hazards. Many coffee drinkers are familiar with *coffee jitters*, a nervous condition that occurs when one has had too much caffeine. Coffee can also cause insomnia in some individuals. In others, it can cause narcolepsy.

Osteoporosis: As excess coffee consumption may prevent the full absorption of necessary minerals, it increases the risk of developing osteoporosis. Caffeine may aggravate osteoporosis by leaching calcium from the bones. Coffee intake may induce an extra urinary excretion of calcium. Heavy coffee consumption (4 cups = 600 ml or more) can modestly increase the risk of osteoporosis, especially in women with a low calcium intake. A study on postmenopausal women observed that long-term consumption of two or more cups of caffeinated coffee per day was associated with decreased bone density, regardless of age, alcohol and tobacco consumption, hormone use and other factors.

Weight loss: Caffeine increases the risk of long-term weight gain by increasing stress hormones and creating a greater risk for hypoglycemia, which stimulates appetite. Although caffeine can assist with short-term weight loss, but in long term, heavy consumption is more likely to lead to weight gain. Caffeine speeds up fat metabolism during exercise while conserving glycogen and glucose thereby maintaining brain activity and reducing hunger. This leads to short-term weight loss.

Gastrointestinal problems: Many of the chemicals in coffee and decaf irritate the stomach lining causing an increase of stomach acid leading to digestive disorders. Coffee can damage the lining of gastrointestinal organs, causing gastritis and ulcers. The consumption of coffee is therefore not recommended for people with gastritis, colitis and ulcers.

Caffeine interferes with adenosine, a brain chemical that normally has a calming effect. Therefore, people with any of the following conditions should avoid coffee altogether:

- Bladder or kidney problems
- Bowel problems such as colitis and diverticulitis
- Chronic fatigue syndrome
- Diabetes
- Fibrocystic breast disease
- Generalized anxiety and/or panic attacks
- Heart problems
- High blood pressure
- High cholesterol Hiatal hernia
- Hypoglycemia
- Insomnia
- Irritability
- Indigestion
- Liver disease

- Osteoporosis or family history of osteoporosis
- PMS
- Menopause
- Prostate problems
- Skin irritations
- Ulcers
- Urinary tract irritation

Overall, the decision to consume coffee should be based on personal health profile. If you suffer or are at risk for the conditions that coffee can aggravate, it is better to avoid it.

Excess coffee

Moderate coffee consumption does not cause harm. However, over consumption of coffee and caffeine do have some risks. For most healthy adults, moderate consumption of caffeine (about 2-4 cups of coffee or 200-300 milligrams of caffeine) is not harmful though it can be helpful. However, when we look at people who drink 5 or more cups of coffee (generally over 500 milligrams of caffeine), there are risks that may affect them. Moderation is the key. Drinking coffee in moderation is completely fine.

How Caffeine Becomes Addictive?

Caffeine increases dopamine levels in the brain. Dopamine is a neurotransmitter that activates the pleasure centre in the brain. By manipulating the level of dopamine, caffeine artificially makes the consumer feel good for a short period. This is the same mechanism used by both cocaine and heroin. Caffeine provides an immediate *feel-good* sensation. Caffeine encourages long-term addiction by depriving the consumer of a good night's sleep. Because adenosine reception is essential to deep sleep and it is blocked by caffeine, consumers wake up feeling irritable and use caffeine to mentally *awaken* themselves so that they can function properly since though they had had a good night's sleep. In this way, a positive feedback loop is created, and consumers cannot abstain from caffeine consumption without adverse effects.

Coffee in moderation (a few cups a day) is a safe beverage that may even offer some health benefits. It is also uncovered possible benefits.

Symptoms of Caffeine Withdrawal

Caffeine withdrawal symptoms are generally mild but can be severe in those who have become addicted to it. Caffeine consumption can be reduced and then eliminated gradually. Caffeine withdrawal symptoms include:

- Irritability
- Nausea
- Headaches
- Grogginess
- Vomiting

Coffee consumption rules

- Avoid coffee if you have high blood pressure, heart disease, high cholesterol, gallstones, or a tendency for blood clots or stroke.
- Do not drink coffee in mental illness, insomnia, or anxiety.
- Do not drink coffee if you are pregnant or nursing.
- Use lower caffeine coffees, *i.e.*, Arabica beans have lower caffeine than Robusta beans. Darker roasted beans also have less caffeine.
- Drink plenty of water to avoid dehydration.
- Do not drink coffee with, before or after meals

Few recipes of coffee

Coffee Mocha Soda: Ingredients

- 2 cups milk
- ½ cup hot water
- 8 teaspoons coffee, finely ground to a powder
- 4 scoops chocolate ice cream
- 1 club soda
- Sweetened whipped cream or prepared whipped topping

Method: Place hot water in a medium-sized pitcher. Add coffee and stir until dissolved. Stir in milk and place one scoop of ice cream in each of 4 ice cream soda glasses. Pour coffee/milk mixture equally into each glass, and then, add club soda almost up to brim. Top with sweetened whipping cream or prepared whipped topping.

Mexican coffee Ingredients

- 6 teaspoons chocolate syrup
- ½ cup heavy cream
- ¾ teaspoon cinnamon
- ¼ teaspoon nutmeg
- 1 tablespoon sugar
- 1-½ cups strong hot coffee

Method: Put one and half teaspoon chocolate syrup into each of 4 coffee cups. Mix heavy cream, half of cinnamon, nutmeg and sugar, and whip until it forms soft peaks. Stir remaining ½-teaspoon cinnamon into the hot coffee. Divide the coffee evenly between the 4 cups and stir each to blend the coffee with chocolate syrup. Top each cup with the spiced whipped cream and serve immediately.

Coffee Alexander Ingredients

- 1 cup cold strong coffee
- 1 cup coffee ice cream
- 2 oz. Brandy or Cognac

Method: Pour all ingredients into a blender and blend until smooth and creamy. Serve in wine glasses and garnish each with a coffee bean.

■■■

41 Egg in Medical View

Viroj Wiwanitkit

Wiwanitkit House, Bangkhae, Bangkok, Thailand-10160, Visiting Professor, Hainan Medical College, China, E-mail: <wviroj@yahoo.com>

Egg is a basic thing that everyone has ever seen. In biology, egg is considered as a kind of cell. All vertebrates have egg as a sexual cell for reproductive process. In addition, egg is accepted as the largest cell that can be visible by naked eye. In animal, egg is produced by a sperm gland, which is called egg gland. In human beings, the egg gland is the ovary. A female animal will produce egg for fertilization process with sperm from another male animal. After fertilization, the fertilized egg will further develop into embryo and born to be adult.

Although many animals can produce egg but some animals lay egg and the others do not. The groups of animal that lay egg and the young adult will hatch out from the laid egg include:

- Most fishes
- Reptiles such as turtle and snake
- Amphibians such as frog and toad
- Avian such as chicken, duck, goose and swan
- Some specific mammals

However, in general, egg is known as an important food material, which can be seen in any culture around the world. It is eaten in all countries around the world. Evidences of eating egg can be detectable in historical sites in China, Italian, Egypt, India and Greece. Hence, it can be said that egg is the well-known food source for human beings for many millenniums.

What is egg?

Egg is a specific name of a sexual cell, which is produced by female animal for reproductive purpose. However, in general, the term egg usually means an avian egg. The reason is that a female animal has to produce egg, which includes:

- Egg is considered a safe cell. It contain thick protective layer from the external insult. This is a requirement for the important cell for reproduction or continuation of the species.
- Egg is the time capsule containing genetic data, which is passed from a generation to the next generation. Genetic materials are put into the egg cell.
- Egg is the cell for fertilization and further development into adult animal. Without egg, the new adult animal cannot be formed.

In general, we might divide egg into two groups depending on its structure.

Egg with shell: Egg with shell is usually an egg of animal living on ground, *e.g.*, reptile and avian such as chicken, duck, snake, *etc.*

Egg without shell: Egg without shell is usually egg of animal living in the water, *e.g.*, fish and amphibian such as frog, toad, *etc.* A simple explanation for lack of shell is the fact that there is no need for hard shell since the water can absorb some serious action force on the egg. Though there is no hard shell, there is soft jelly coverage of the egg in this group.

However, it should be noted that the main difference between the two groups is the shell. The internal compositions of eggs in both groups are similar. The two important compositions within the eggs include:

- Yolk (yellow part of the egg)
- Albumen (white part of the egg)

If we divide a chicken egg in a cross sectional direction, we can see the following structures from external to internal sides:

- Cuticle
- Shell
- Outer membrane
- Air cell
- Inner membrane
- Albumen
- Vitelline membrane
- Yolk
- Germination disk

The most-internal structure (germination disk) is the specific structure for reproductive process and is considered as heart of the egg. Therefore, it is well protected.

Development of egg

Egg is internally produced within a female animal's body. The process of egg production is very complex. This process will start only if that female animal reaches puberty. The process is sexually related phenomenon. In chicken, ovulation of egg in a hen is very rapid and it can be seen that a hen can ovulate one egg per day (for human beings 1 egg per month). Based on this fact, chicken is kept as domestic animal for a long time.

Focusing on developmental process of the chicken egg, the process can be described as the followings:

- The process starts when a hen reaches its complete puberty.
- Oocyte will be produced in ovary. At this phase, the yolk is already produced. For yolk formation, the role of low-density lipoprotein receptor family members in the transport of yolk precursors from the plasma compartment into oocytes is mentioned.
- Then the vitelline membrane will be produced for covering yellow part of the egg.

- Yolk with vitelline membrane will be flowed to the infundibulum of oviduct.
- At infundibulum, the yolk with vitelline membrane will stop for a while, about 20 minutes, waiting for fertilization by sperm.
- In case the fertilization occurs, the fertilized egg will be further developed into embryo.
- With or without fertilization, the egg will be further flowed to the next part of the oviduct. At that point, the albumen will be produced and it will cover the yolk. This process is controlled by cellular signaling *via* ovalbumin messenger RNA.
- Then albumen and yolk will be further flowed to final part of the oviduct. At this point, the shell will be formed and the complete structure of an egg can be derived.
- The formation of shell is based mainly on shell gland. At laying cycle, high glycogen particles can be identified at the shell gland and these particles are believed to be the precursors for shell construction.
- Egg will be flowed to the uterus.
- Cuticle will be produced for waxing the shell of the egg.
- The derived complete egg will be flowed to the vagina.
- Finally, egg will reach the cloacae, the termination of the tract.
- The egg will be laid onto the nest.
- If it is fertilized, further embryo developmental process will continue.
- If not, the egg will not be further developed.

The whole process takes only one day for completeness. This seems to be a complex but fast physiological process.

The fully developed chicken egg has its size as the following:

- Length 1.5-2.5 inches
- Diameter 0.5-1.5 inches

Size of the egg is important determinant for the price of egg. Internationally, it is agreed that the egg with smaller number means the larger egg. There are six numbers for classification of eggs based on their size, *i.e.*, 0, 1, 2, 3, 4 and 5.

Based on the nutritional science, egg is accepted as an important protein source. It is grouped in the set of meat, milk and egg [6-7]. It can be easily available around the world. Averagely, the important nutrients within a chicken egg include energy 80 kcal, protein 6.15 g, lipid 5.85 g, carbohydrates 0.7 g, iron 0.8 g, calcium 63 mg, vitamin A 570 IU, vitamin B_1 0.075 mg, vitamin B_2 0.175 mg and water 36.75 g.

Components within egg

Yolk

Yolk seems to be the central part of every egg and seems to be the most important component within egg. The natural characteristics of yolk include:

- Yolk has yellow color.
- It lies within thin membrane called vitelline membrane.
- Yolk is surrounded by a clear fluid namely albumen.
- An egg has only one yolk.
- The genetic content of the egg locates within the yolk.

The size of yolk in each egg varies depending on many factors including health of the hen, nutrients within the yolk and development of the embryo within yolk or not. The yolk of fertilized egg is larger than that of non-fertilized egg. This is because of the content of embryo in fertilized egg.

There are many nutrients within yolk including:

Cholesterol: This is an important source of energy to human beings. The cholesterol can be changed into energy if the body runs off carbohydrates. However, cholesterol can also have bad effect on body if taken excessively. Dyslipedmia can be the common problem in anyone who gets excessive cholesterol. The dyslipidemia is proved to cause many diseases including cardiovascular disease and cerebrovascular disease.

Protein: High content of protein can be seen in yolk. This is useful for the construction building block and repairing of body.

Vitamins: There are many vitamins in yolk including to vitamin A, vitamin B_1, vitamin B_2, folate and vitamin D.

Minerals: There are many minerals in yolk including to sodium, iron and calcium.

Uric acid: Averagely, the important nutrients within a duck egg include energy 59 kcal, protein 2.8 g, cholesterol 213 g, vitamin A 0.6 IU, folate 22.5 μg, calcium 23 mg and iron 0.6 mg.

Albumen

Albumen seems to be less important than yolk. However, if there is no albumen, there can be no egg. Yolk has to be supported by albumen. The natural characteristics of albumen include:

- Albumen has no color.
- It lies between shell and yolk.
- It is the site for interchange between yolk and external environment.

Averagely, the size of albumen is about two-thirds of overall egg. The size depends on the size of yolk. If the yolk is large, albumin will be small. On the other hand, if the yolk is small, albumen will be large.

There are many nutrients within yolk including:

Cholesterol: There is no cholesterol in albumen. Hence, the patients with dyslipidemia can eat albumen without any health precaution.

Protein: Very high content of protein can be seen in albumen. The protein namely albumin can be identified in albumen, and this protein is useful for balancing cellular water within body. Hypoalbuminemia is the important cause of edema and body fluid leakage conditions such as ascites.

Vitamins: The most important vitamin in albumen is folate.

Minerals: There are many minerals in albumen including sodium, iron and calcium but the amounts of those minerals are lower than those of yolk.

Uric acid: There is no uric acid within albumen. Hence, the patients with gout can eat albumen without any problem.

Averagely, the important nutrients within a duck egg include energy 17 kcal, protein 3.5 g, cholesterol 0 g, folate 1 μg, calcium 2 mg and iron 0.01 mg.

Table 1 : Comparison between yolk and albumen

Aspect	Yolk	Albumen
Protein	Lower	Higher
Cholesterol	High	No
Uric acid	High	No
Medical advantage	Tonic, giving energy	Giving protein
Medical precaution	Dyslipidemia, gout	No

Common non-chicken egg

In additional to chicken egg, there are also many eggs, which are eaten daily by human beings. The details of important non-chicken egg are hereby described.

Duck egg

In additional to chicken egg, the second most common eaten egg is the duck egg. Duck is another avian, which is fed as domestic animal. An important difference between chicken and duck is the behaviour of these two animals. Duck prefers swimming in the swamp, whereas, chicken do not. However, the egg production process of duck is not different from that of chicken.

Basically, the shape of duck egg is similar to that of chicken egg. The duck egg has white colour and slightly bigger than chicken egg. It should be noted that duck usually lay their eggs on the mud, not on a grass nest as the chicken. Hence, the duck egg is usually contaminated externally with mud. Cleaning process is required before using the duck egg as food material.

Averagely, the important nutrients within a duck egg include energy 189 kcal, protein 12.1 g, lipid 14.1 g and calcium 40 mg.

It can be seen that duck egg gives more energy and protein but lower calcium than chicken egg. Duck egg seems to be more useful if used as a tonic for refreshment.

Goose egg

Goose egg is less frequently ingested than the chicken and duck eggs. Goose is another avian, which is widely fed in households in developing countries around the world. Goose has the same behaviour to duck. Goose similar to duck prefers swimming in the swamp, whereas, chicken do not. However, the egg production process of goose is the same as that of chicken and duck.

Structurally, the shape of goose egg is similar to that of chicken egg and duck egg. The goose egg has white colour and about 2-3 times bigger than chicken egg and about 1.5 to 2 times bigger than duck egg. It should be noted that geese normally lay their eggs on the mud, similar to duck. Hence, the goose egg is usually contaminated externally with mud and it requires a good cleaning process before being used as food material.

Averagely, the important nutrients within a goose egg include energy 160 kcal, protein 14.6 g, lipid 20.2 g, calcium 42 mg.

It can be seen that goose egg has higher lipid content comparing to chicken egg and duck egg. Although its size is bigger than duck egg but the protein content is lower. Hence, it is not recommended for the patients with problem of dyslipidemia.

Francolin egg

Francolin is a kind of small bird, which can be seen in the tropical countries. This kind of bird is presently fed in farm. Its egg is widely ingested in some cultures such as Chinese culture. The size of francolin egg is very small but it has high nutrients.

Averagely, the important nutrients within a francolin egg include energy 171 kcal, protein 13.1 g, lipid 12 g and calcium 152 mg.

It should be noted that the nutrients content within francolin egg is very high. It gives very high energy despite its small size. This is an important precaution since many francolin eggs can be easily eaten at one time comparing to other larger eggs.

In addition to the already described non-chicken avian eggs, there are also other non-avian eggs, which are ingested by local people in some communities. The important kinds of eggs will be hereby shown:

Turtle egg: Turtle eggs are eaten in some countries. Chinese people love to eat turtle eggs. In natural, turtles creep onto the beach, dig into the sand and lay their eggs. The nutrients content of turtle egg is not high and it is not recommended to eat this specific kind of egg. Since turtle seems to be an animal in danger, high risk for extinction, it is widely proposed for anti-turtle egg eating.

Frog egg: Frog egg is ingested in some Southeast Asian community. This is the local source of protein for the local population. Eating of frog egg is not problematic but the problem is usually due to misunderstanding on toad egg as frog egg. Annually, sporadic case reports on toad egg intoxication can be seen.

Ant egg: Ant egg ingestion is a local practice in Indochina. The local people eat ant eggs expecting to get the acidic taste, not the nutrients. Indeed, this eating behaviour is considered not safe and health education to get rid of this rooted eating behaviour is the topic of concern in local public health.

Usefulness of egg

Egg is accepted as an important food source of protein. It is easily available at any market around the world. The price of the egg is not as high as other protein foodstuffs such as meat and milk. In nutritional science, it is accepted for the usefulness of eggs to the human health. The usefulness is due to the plentiful nutrients within the egg.

The mentioned usefulness of egg includes:

- Giving protein to be the structure of body
- Giving protein for repairing of defects within the body
- Giving energy
- Giving warm
- Stabilizing the biochemical reaction within the body
- Maintaining bone system
- Stabilizing water homeostasis within the body
- Detoxification of some toxins

In addition to the described direct advantages, there are also other indirect advantages of egg. The examples of those indirect advantages of egg include:

- Using as skin tonic
- Using as composition in shampoo
- Using egg shell as decorative arts
- Using as medicine (detoxification)

In medicine, egg is accepted for its usefulness in prevention and treatment of many diseases. It is accepted as an important medicinal food supplementation.

The diseases that the egg can help prevent include:

- Under-nutrition in pediatrics
- Thin, low weight
- Muscle waste, muscle weakness
- Osteopenia and osteoporosis

The diseases the egg can help treat include:

- Protein malnutrition
- Diseases that require egg as nutritional therapy regimen
  - Elderly patients
  - Pediatric patients
  - Pregnant subjects
  - Patients require long hospitalization
- Edema and swelling (use albumin)

An interesting question is whether there is any disadvantage of egg. The answer is *no*. If we eat egg in appropriate amount in correct cooking method, there will be no adverse effect on health status.

However, an important condition is that there is not an actual adverse effect of egg to be concerned. That condition is egg allergy. By definition, egg allergy means immunological reaction to egg in the excessive pattern or hypersensitivity. This is similar to other kinds of food allergy. The important symptoms of egg allergy include:

- Respiratory difficulty
- Palpitation
- Generalized urticaria
- Generalized edema
- Death in the most serious case

It is important to note that anyone with history of egg allergy has to avoid eating egg or egg product since the egg allergy is a deadly condition. Many patients die before reaching the emergency room for proper management.

How to eat egg?

As previously noted, it is needed to eat egg correctly to maximize the advantage of egg to the body and decrease the unwanted effect of egg. Focusing on the principle, the recommendations include:

Eat clean egg: This means egg has to be clean and well cooked before eating.

Eat egg at appropriate amount: Excessive eating or insufficient eating is not recommended since both patterns are not good for our health. In order to judge what the appropriate amount of eating is, there are many factors to be considered as the followings:

- **Age:** The children usually require high protein. Hence, they should eat more egg than the adult should. It is recommended that children should eat up to two eggs per day, whereas, adult should eat only one egg per day
- **Sex:** There is no recommendation depending on sex. However, the pregnant subject should increase the amount of eating from 1 to 2 eggs per day.
- **Personal illness:** Some patients with some specific personal illnesses have to avoid eating egg. Those mentioned illnesses include dyslipidemia and gout. This is due to high content of cholesterol and uric acid in yolk. In addition, there are also some diseases those affect the digestion and absorption of egg. Those diseases include no tooth, gastritis, enteritis, nephritic syndrome and paralysis.
- **Egg contamination:** Generally, egg is classified as a natural clean food material. It is safe and contains no toxin. However, it is not guarantee that there will be no contamination. Egg contamination can be expected and it seems to be very important problem in medicine. The possible causes of egg contamination include sick hens, contamination from hen's gastrointestinal tract (feces), contamination from nest and rotten eggs.

There are many pathogens, which can be contaminated within egg. The important pathogens include:

Virus: Since viral particle is smaller than the pore at shell. It is no doubt that virus can enter into the egg. This fact becomes a present concern because there are many new emerging avian related viruses that might contaminate into egg.

Bacteria: The most important bacteria that can contaminate into the egg are *Salmonella* species. The prevalence of contamination varies from countries to countries. It is reported that fluff and feces on the nest and transport ways are related to the contamination of *Salmonella.*

A serious concern on the egg contamination is presently set. Due to the recent outbreak of new emerging bird flu, egg becomes the food material to be focused. However, until present, there is still no report confirming that egg is the source of infection or can help in transmission of bird flu. However, sanitation in the management of egg material is still required.

Focusing on the sanitation principle, the recommendations are:

- The chicken farm has to be standardized. Closed system is required.
- Buy egg from the standard certified market.
- Reasonable cleaning of egg is needed. This has to be done at farm. The egg should also be repeatedly washed and clean before cooking.
- Washing egg can be done under clean running water (about 3 minutes).
- Avoid eating raw egg since it can be dangerous if there is any contamination of germs.
- Finally, followings are important sign of egg contamination to be remembered:
- Egg with cracking on its shell
- Egg with change of shell colour
- Egg with foul smell (rotten egg)
- Egg from unknown origin

Therapeutic Significance of Egg

Egg can be used as a food supplement for medical therapy. The usage of egg in medical treatment can be in these forms:

Whole egg: This form of using can be seen in any medical wards. It is a routine practice that egg is added into the menu of the patient if the physician orders for increased protein content in the menu. The common egg dish for the patient is the boiled egg. Indeed, boiled egg is considered a safe cooked egg menu since boiling means warming by hot water. This process adds no other components into the egg. It adds no lipid or other components into the egg. In addition, eating boiled egg is easy and can be done by any patients.

Yolk: Yolk is not routinely used in medical treatment. Although some reports confirm the usefulness of yolk on biochemical process in animals but evidences are limited in human beings. Nevertheless, yolk is still indicated as food supplement for increasing cholesterol purpose. It is also reported that blood lutein and zeaxanthine can be raised by yolk supplementation. These two biochemical substances are useful for controlling of good retina function.

Albumen: Albumen is a widely used part of egg in medicine. It is accepted as a drug at present. Purified albumen is available at present and is used for intravenous administration. The indication for using albumen is for increasing oncotic pressure within blood vessels. This is useful for holding water within the blood vessels and helps in decreasing edema. This is applicable to many conditions including renal and liver failure. Another important usage of albumen administration is for prevention of severe ovarian hyper-stimulation syndrome, which is an iatrogenic syndrome due to fertility stimulation, with the symptoms of palpitation, dehydration, abdominal pain and distention. In addition, albumen is also indicated in the patients that cannot eat per oral. The protein supplementation *via* large venous route is primarily based on albumen. Since this administration is considered a dangerous procedure, it has to be performed by expert physician. The access to large vein has to be carefully done with a special medical tool. Nevertheless, a closed observation of the venous cannula has to be done.

■■■

42 Red Meat: Facts and Fantasy

Varsha Rani[1], Sudesh Jood[1] and M.K. Rana[2]

[1]Department of Foods and Nutrition
[2]Department of Vegetable Science,
CCS Haryana Agricultural University, Hisar-125 004, Haryana, India

The primary importance of red meat in human nutrition is due to its high quality proteins and other important essential nutrients including long-chain *n*-3 fatty acids, iron, zinc, selenium, vitamin D, vitamins B_3 and B_{12} that are essential for good health throughout life. In particular, some of these nutrients are more bioavailable in meat than alternative food sources, and their intake is recognized to be below recommendations for some population subgroups. A healthy diet should include lean meat in moderate amounts together with starchy carbohydrates (including wholegrain foods), plenty of fruits and vegetables and moderate amounts of milk and dairy products.

What is red meat?

Meat can broadly be classified as *red* or *white* depending on the concentration of myoglobin in muscle fibre. When myoglobin is exposed to oxygen, reddish oxymyoglobin develops, making the myoglobin-rich meat red in appearance. The redness of meat depends on species, animal age and fibre type. Red meat contains more narrow muscle fibers, which tend to operate over long periods without rest, while white meat contains more broad fibers, which tend to work in short fast bursts. According to traditional culinary terminology, red meat may be defined as the meat, which is red in color when raw and not white when cooked. It includes the meat of most adult mammals and some fowl. It is further classified as beef, veal, pork and lamb (fresh, minced and frozen). According to the United States of America, all meats obtained from livestock are red meats because they contain more myoglobin than chicken or fish.

Energy and red meat

The amount of energy provided by red meat is variable and depends on species and animal. Meat contains no carbohydrate, so the provided energy is composed of protein and fat, which give 4 and 9 kcal of energy per gram of red meat, respectively.

Protein and red meat

Dietary protein is needed for growth, maintenance and repairs of the body and can also provide energy. Protein from different food sources consists of hundreds to thousands of amino acids.

Some of amino acids can be synthesized in human body, therefore, they are termed as nonessential amino acid, whereas, some of them can not be synthesized in human body and are required from diet, therefore, they are termed as essential amino acids. Red meat protein being the excellent source of amino acids provides all essential amino acids, *i.e.*, lysine, threonine, methionine, phenylalanine, tryptophan, leucine, isoleucine, valine, and has no limiting amino acids. In comparison, nearly all plant proteins have one or more limiting amino acid for instance wheat is limited for lysine and pulses are limited for methionine. Eating different plant food articles, in combination, can provide the right balance of essential amino acids, although less efficiently than meat. On an average red meat contains 20-24 g of protein per 100 g raw edible portion, whereas, cooked red meat contains 27-35 g of protein per 100 g of cooked weight. The scientific reason for that is as meat is cooked, the water content decreases and the nutrients become more concentrate. Therefore, the protein content increases while cooking. Lean meat also contains a higher proportion of protein.

Fat and red meat

Fat is the concentrated source of energy and supplies other essential nutrients such as fat-soluble vitamins and essential fatty acids but must be consumed in moderation. Fat makes the foodstuffs more palatable and gives nice flavour to it. Fat is made up of different types of fatty acid, *i.e.*, saturated fatty acids (SFAs), monounsaturated fatty acids (MUFAs) and polyunsaturated fatty acids (PUFAs). The fatty acid profile of a food is determined by the proportions in which each of the fatty acids is present. It is now well accepted that different fatty acids have different effects on blood cholesterol levels, some beneficial and some adverse, and therefore, it is important to think about the fatty acid profile of a food. Fat in meat is present at different allocations, and therefore, termed accordingly. For instance, fat allocated between the muscles called as intermuscular fat, fat within the muscles called as intramuscular fat and fat below the skin called as subcutaneous fat. The subcutaneous fat is considered as visible fat. The fat content of red meat varies widely, depending on the type of red meat, the cut and the degree of trimming. Red meat containing between 5 and 10% fat classified as lean red meat.

Due to its fat composition, regular consumption of red meat is controversially linked with the increased risk of coronary heart disease. However, there is enough evidences stated that a healthy diet containing lean red meat can produce positive changes in lipid biochemistry. It is true that blood cholesterol level increases with the inclusion of high fat red meat but not with the lean red meat. Inclusion of lean red meat with low fat and low saturated fat containing diets reduces the plasma cholesterol and low-density lipoprotein- cholesterol levels in hypercholesterolemic and normocholesterolaemic persons.

Overtime reduction of fat in red meat

Reduction of fat in red meat has been achieved in last 30 years at international market. Thirty years ago, most of the visible (subcutaneous) fat was consumed with the red meat. In the early 1980's, the international red meat industries began to produce meat with less fat. Advances in food processing technologies and breeding practices, as well as feeding designed to increase the carcase lean to fat ratio, and the modern butchery techniques have been the responsible factors in reducing fat content of carcase red meat over the past 20 years at international market. The fat content of carcase meat in the United Kingdom has been reduced by over 30% for pork, 15% for beef and 10% for lamb. New Zealand also has developed quality marked red meat products by introducing trimming of fat from red meat before sale. Therefore, it has been resulted in a 48% reduction of fat in beef, and a 43% reduction in lamb.

Alterations in feeding practices bring the changes not only to total fat content of meat rather it may also effect the fatty acid profile of some carcase meat. The fatty acid profile of non-ruminant meat is an example of that in the diet. Inclusion of vegetable and fish oils in feeds of mono-gastric animals such as pigs increases the *n*-3 PUFAs as compared to traditional cereal based diet, which mainly comprises SFAs and MUFAs. Consumers can also modify the fat content of meat through trimming the fat off meat and using modified preparation and cooking methods, *e.g.*, grilling, roasting, or dry frying can result in fat losses for meat products. Generally, the fat content of traditionally prepared meat products such as sausages, pastry-covered pies and salami is considered to be high, up to 50% but the efforts have been done overtime to reduce the fat content of these products and have been achieved up to remarkable levels. On the other hand, modern products included ready meals and prepared meats are low in fat and have up to 5 percent fat.

Fatty acid composition of red meat

In ruminant animals such as cows and sheep, more than 90% of the dietary unsaturated fatty acids are converted to saturated fatty acids due to their hydrogenation during digestion. Generally, non-ruminant meat, such as pork, contains more unsaturated fatty acids but it is highly influenced by the fatty acid profile of animal feed. Lean red meat contains similar proportions of monounsaturated fatty acids (MUFAs) to saturated fatty acids, though the exact proportion varies depending on the type of meat. The fatty acid profile of meat will also vary depending on the proportion of lean and fat present, as the lean meat is relatively higher in polyunsaturated fatty acids and lower in saturated fatty acids, containing less than 2 g of saturated fatty acids per 100 g of meat. Trimming the fat off meat affects the proportion of fatty acids, as most of the saturated fatty acids present in visible fat are removed. Visible fat in meat contains around 35 g of saturated fatty acids per 100 g of meat. Trimming off visible fat from meat cuts can reduce the fat content by approximately 25-60% after cooking.

Saturated fatty acids (SFAs): Generally, it is assumed that the fat present in meat is saturated fat, which is not true. Meat contains a blend of fatty acids, as the fat content of red meat has reduced due to advancement in food processing technologies and breeding practices, thus, the fatty acid mix has changed with percent saturated fat. Pork, beef, lamb and poultry have 45, 45, 50 and 30 percent saturated fatty acids of total fatty acids blend, respectively.

Palmitic acid (C16:0) and stearic acid (C18:0) are the main saturated fatty acids present in red meat. Myristic acid (C14:0) and lauric acid (C12:0) are also present in red meat but in trace amounts. Not all saturated fatty acids have cholesterol-elevating effect. Stearic acid has no effect on blood cholesterol levels and risk of cardiovascular disease. Although lauric acid (C12:0) increases total serum cholesterol. Myristic acid is the most atherogenic saturated fatty acid and has four times the cholesterol raising potential of palmitic acid.

Monounsaturated fatty acids (MUFAs): Lean red meat has similar proportion of saturated fatty acids and monounsaturated fatty acids. The principal monounsaturated fatty acid in meat is oleic acid (C18:1), and typically, between 30 and 40% of the fat in meat is composed of monounsaturated fatty acid. Dairy products in comparison to red meat contain less monounsaturated fatty acids and more myristic acid, which is worsen health wise. Average whole milk contains 1.1% of monounsaturated fatty acids than 40% in red meat.

Polyunsaturated fatty acids (PUFAs): Omega-6 (ω-6) and omega-3 (ω-3) are the two classes of essential polyunsaturated fatty acid, where ω-6 is represented by linoleic acid (LA) and ω-3 is represented by α-linolenic acid (ALNA). Both, the LA and ALNA must come from diet because synthesis of the ω-6 and ω-3 structures has not been detected in animals. The predominant polyunsaturated fatty acids in red meat are the essential fatty acids, linoleic (ω-6) (LA) and α-linolenic (ω-3). When consumed, the body converts α-linolenic into the long chain *n*-3 polyunsaturated fatty acid, *i.e.*, eicosapentaenoic acid (EPA) and docosahexaenoic acid (DHA), although the rate of synthesis may not be sufficient to meet the requirements. Red meat provides useful amounts of long chain polyunsaturated fatty acids, especially dihomo-γ-linolenic acid (20:3 ω-6), arachidonic acid (20:4 ω-6), eicosapentaenoic acid (20:5 ω-3) and docosapentaenoic acid (22:5 ω-3).

Researches show that EPA and DHA are very helpful in preventing heart attack, stroke, atherosclerosis and cancer and reduce the risk of inflammatory conditions. These long-chain *n*-3 fatty acids have also been linked with improvement in depression, schizophrenia and dementia. Though only a small amount of long-chain *n*-3 PUFAs (EPA and DHA) and *n*-6 PUFAs (arachidonic acid) are found in red meat, there are very few rich sources apart from oily fish, and therefore, meat can usefully contribute to intakes of these important fatty acids for those who consume little or no oily fish

Factors affecting PUFA content

The proportion of polyunsaturated fatty acid in meat is strongly influenced by feeding regime of the animal. The rumen hydrogenates unsaturated dietary fat, which makes it more difficult to modify the fatty acid profile of ruminant meat. Nevertheless, some dietary unsaturated fats bypass the rumen for deposition in fat stores. Feeding linseed and fish oils results in beef with increased ω-3 PUFA levels. The grass fed animals have higher concentrations of α-linolenic acid compared with grain-fed. In addition, beef and lamb have favorable ω-6: ω-3 ratios.

Inclusion of vegetable and fish oil in the feeds of monogastrics such as pig results in significant increase in ω-3 PUFA, and such feeding regimes can produce pork products with nearly half the daily recommendation for eicosapentaenoic acid and docosahexaenoic acid.

Trans-fatty acids

Ruminant meats provide natural trans-fatty acids, contributing around 18% of the total fat. These are formed during biohydrogenation in the rumen. High intakes of trans-fatty acids have been known for their bad health effects because it promotes atherogenesis and adversely influences the ratio of low-density lipoprotein (LDL) to high-density lipoprotein (HDL), a recognized risk factor for cardiovascular disease. The hydrogenated fats from vegetable sources used in bakery goods and other processed foods appear to be more of a concern than the natural trans-fats found in ruminant meat and milk fats. According to food guidelines given by United Kingdom, the trans-fatty acids of 2% of dietary energy intake have been considered safe for human consumption.

Though the ruminant meat and milk contribute to trans-fatty acid intake, these foodstuffs are valuable sources of other nutrients, particularly protein, calcium and iron. Therefore, consumption of trans-fatty acids from industrially produced sources such as fat spreads, biscuits, cakes and pastries, crisps and confectionary should be restricted.

Conjugated linoleic acid (CLA)

Conjugated linoleic acid (CLA), is formed as an intermediate in the biohydrogenation of linoleic acid in ruminants, or a collective term used to describe a mixture of positional and geometric isomers of linoleic acid. Conjugated linoleic acid is only found in useful amounts in ruminant meat and dairy products such as lamb, beef, cheese and milk. Researches on *in vitro* and animal studies indicate that CLA has many potential health benefits on cancer, coronary heart disease (CHD) and diabetes. However, human studies are needed before the effects of CLA on human health are determined. Feeding practices influence the CLA content of meat. Grass-fed beef has been shown to have higher CLA levels than grain-fed beef. Storage of meat and different methods of cooking meat, *e.g.*, frying, grilling, microwaving and baking, have not been found to reduce the conjugated linoleic acid content.

Micronutrient content of red meat

Red meat contains an array of micronutrients, some in substantial amounts, all of which are required for general health and wellbeing. The nutrients discussed in this section are those for which red meat is an average, good, or rich source.

Vitamins and red meat

B group vitamins: Red meat contains a number of B group vitamins, *i.e.*, thiamine (vitamin B_1), riboflavin (vitamin B_2), pantothenic acid, folate, niacin (vitamin B_3) and pyridoxine (vitamins B_6) and cyanocobalamine (vitamin B_{12}). The B group vitamins work as co-factor in different enzyme systems in the body. Low blood levels of several B group vitamins, including vitamin B_{12}, folate and vitamin B_6 (Table-1), have been associated with elevated blood levels of homocysteine, which is a risk factor for cardiovascular disease and stroke.

Thiamine (vitamin B_1): The much known biochemical function of thiamine is to serve as a coenzyme in a number of metabolic processes and mainly the carbohydrates metabolism. Calf muscles and pork including bacon and ham is the richest source of thiamine and fulfill the daily requirement of human body.

Riboflavin (vitamin B_2): Riboflavin is essential for growth and participates in multiple functions, *i.e.*, synthesis of corticosteroids in the adrenal glands, formation of red blood cells, synthesis of glycogen, metabolism of fatty acids and promoting effectiveness of thyroid hormone in regulating enzyme activity. Organ meat, particularly calf liver and kidney, is the excellent source of riboflavin, and a portion size of 100 g provides more than double of daily requirement.

Table 1 : Red meat as average, good and excellent source of vitamins and minerals

Contents	Calf liver	Pork	Beef	Lamb
Vitamin A	Excellent	Average	Average	Average
Thiamine	Excellent	Excellent	Average	Average
Riboflavin	Excellent	Good	Good	–
Niacin	Excellent	Excellent	Excellent	Excellent
Pyridoxine	Excellent	Average	Average	Average
Cyanocobalamine	Excellent	Excellent	Excellent	Excellent
Vitamin C	Good	–	–	–
Vitamin E	Average	Average	Average	Average
Iron	–	–	Good	–
Zinc	Excellent	Good	Excellent	Excellent
Selenium	Excellent	Good	–	–
Potassium	Good	Good	Good	Good
Phosphorus	Excellent	Good	Good	Good

Average: ≤ 10% of the RDA per 100 g; Good: ≥ 15% of the RDA per 100 g; Excellent: ≥ 30% of the RDA per 100 g

Source: Adapted from Wyness *et al.* (2011)

Niacin (vitamin B_3): Niacin is required by all living cells as part of the two coenzymes NAD and NADP, which are essential in the metabolism of carbohydrates, fat and protein. Niacin is also important, as it is the part of glucose tolerance factor of yeast, which enhances the response to insulin. Red meat is the richest source of niacin. Half of the niacin content of meat is derived from tryptophan, which makes it more readily absorbed to the human body than to plant sources where it is bound to glucose.

Pyridoxine (vitamin B_6): Pyridoxine is an essential and important co-enzyme in many of the vital biochemical reactions that involve the amino acids, thus, making it vital for proper protein synthesis. Pyridoxine plays an important role in biochemical conversion of the amino acid tryptophan to the vitamin niacin, which is not possible without the presence of pyridoxine. Calf liver is the excellent source of pyridoxine, whereas, other meats contain average amount of it.

Cyanocobalamine (vitamin B_{12}): Cyanocobalamine is required only in small amounts, *i.e.*, 1.5 µg per day. Food items of animal origin provide the only dietary source of vitamin B_{12} and deficiency does occur in vegetarians. Red meat is the excellent source of vitamin B_{12} and fulfills the daily requirement of human body.

Vitamin A: Vitamin A has its biochemical role in vision, maintenance of epithelial cells, bone development, reproduction and mucopolysaccharide biosynthesis at enzymatic level. Vitamin A may also have a role in the stability of membranes of lysosomes, mitochondria and outer cell membranes. Vitamin A in its active form (retinol) is found only in foods of animal origin. Therefore, red meat provides it directly to the body. As liver and liver products may contain a large amount of vitamin A, these should be avoided in pregnancy, owing to the increased risk of birth defects associated with very high intakes of this vitamin. Large amounts of retinol can also cause liver and bone damage. The amount of vitamin A present in liver can be variable and will depend on age of animal and composition of the feed consumed.

Vitamin D: The main biochemical function of vitamin D is to facilitate and regulate the absorption of calcium, phosphorus and to some extent of magnesium. It is essential for proper growth, development and physical condition of the bone. Earlier, the meat was thought to be a poor source of vitamin D except liver. Later, based on reanalysis of British National Diet and Nutrition surveys of adults and young children, it was confirmed that meat is a major contributor to natural dietary vitamin D intakes, and it is present in both the lean and fat of meat. Both osteomalacia in adults and rickets in children occur due to deficiency of vitamin D and disturbances of its metabolism. Osteomalacia is more common in lacto-vegetarian women than those who consume meat. Infantile rickets is also associated with lacto-vegetarian diet and late introduction of weaning foodstuffs. Recently, vitamin D deficiency has been associated with cardio-vascular diseases, autoimmune disease and cancer.

Dietary sources of vitamin D are relatively insignificant compared with the synthesis in skin from exposure to sunlight. However, housebound people and those who wear concealing clothing are particularly reliant on a dietary supply of vitamin D, and meat usefully contributes to this intake. Vitamin D in meat is derived from the action of sunlight on the skin of animals or from the animals' feed. In particular, the vitamin D metabolite 25-hydroxycholecalciferol, which has high biological activity and faster absorption in body, is found in significant quantities in meat and liver. Furthermore, high protein content of meat may enhance the utilization of vitamin D in humans, particularly where exposure to sunshine is limited. Scientifically, it has been suggested that meat has a *magic meat factor*, as its influence on bone metabolism was found greater that could be expected from the amount of vitamin D present alone.

Vitamin E: Vitamin E acts as an antioxidant and protects the body's cells against damage through reactive oxygen species (free radicals). Meat contains a small amount of vitamin E and contributes around 10% of average daily requirement. Since this is a fat-soluble vitamin, its concentration is higher in fattier cuts of meat. Vegetable oils are particularly high in vitamin E, and therefore, the recent trend to include oil seeds in animals' diets will have contributed to an increase in the vitamin E content of meat. Vitamin E also delays lipid oxidation and browning of meat, which helps in extending the shelf life.

Minerals and Red meat

Meat is an important dietary source of bio-available minerals and trace elements, especially iron and zinc.

Iron: Iron is essential for thc formation of haemoglobin in red blood cells and is an essential component in many enzymatic reactions, *e.g.*, it is a component of the enzyme system, which detoxifies and removes foreign compounds from the body. It also plays an important role in the immune system and is required for normal energy metabolism. Dietary iron exists in two forms, *i.e.*, haem- and non-haem iron, haem iron being more readily absorbed and utilized by the body. Red meat is an excellent source of iron, having 50-60% in the haem form. Approximately, 15-35% of the haem iron is absorbed in the intestine compared with generally less than 5% of non-haem iron. The bioavailability of iron from a food is also extremely important and can be influenced by other dietary factors, especially in case of non-haem iron. The main enhancers of non-haem iron absorption are ascorbic acid and meat. Small amount of meat is recognized to enhance the absorption of non-haem iron from plant food items. The main inhibitors of non-haem iron absorption are calcium, phytates in cereals and legumes and phenolic compounds found in tea, coffee and other beverages. Red meat provides about 15% of total iron intake. Iron deficiency is more prevalent in vegetarians than in non-vegetarians despite the enhanced absorption conferred by high intakes of vitamin C. It has been proven through scientific studies that meat supplements are more effective than iron tablets in maintaining iron status.

Zinc: Zinc is essential for cell division and for growth and tissue repair. It is also necessary for normal reproductive development, a healthy immune system and healing of wounds. Meat contains substantial amounts of zinc, and calf liver, beef and lamb are considered as an excellent source, whereas, pork as a good source of zinc. More importantly, the zinc in meat is present in a highly bio-available form. As with dietary iron, a number of factors affect the bioavailability and absorption of dietary zinc, including the composition of diet, *e.g.*, a small amount of lean beef (75 g/d) has been found to enhance iron and zinc utilization, whereas, phytates in cereals and legumes and oxalates in green leafy vegetables are known to inhibit zinc absorption. This explains the lower plasma zinc in vegetarians despite higher intake.

Other minerals

Meat and meat products contain useful amounts of magnesium, copper, cobalt, phosphorus, chromium and nickel. Red meat in particular contains useful amounts of selenium, although the concentration will depend on diet of the livestock and the soil in which the animal feed was grown.

Antioxidants: Meat as a food has a complex physical structure and chemical composition, which is very susceptible to oxidation. The oxidative stability of meats depends upon the balance and interaction between endogenous antioxidants and pro-oxidants substances (free radicals) and the composition of substrates prone to oxidation including proteins, pigments, cholesterol and polyunsaturated fatty acids (PUFA). Endogenous antioxidants systems are consisted of non-enzymatic compounds, such as vitamin E, vitamin C, carotenoids, ubiquinols, polyphenols, cellular thiols and enzymes like superoxide dismutase, catalase and glutathione peroxidase. Both the enzymatic and non-enzymatic antioxidants counteract the action of pro-oxidants (free radicals) in muscle tissues of living animals, and even, after slaughter. Besides, proteins and peptides of red meat have important antioxidants action due to their ability to scavenge free radicals and chelate pro-oxidative metals.

Chicken has the highest total antioxidant capacity (TAC) followed by pork and beef because it is rich in histidine-containing dipeptides such as carnosine and anserine, which have antioxidant activities.

Selenium: Selenium acts as an antioxidant and is also necessary for the use of iodine in thyroid hormone production and for immune system function. Red meat provides about 10 micrograms selenium per 100 g edible portion, which constitutes almost 25% of the daily requirement. It is considered that bioavailability of selenium from plant foods is greater than that of animal foods, whereas, the raw and cooked red meat provides a highly bio-available selenium.

Glutathione: Meat is considered as the richest source of glutathione among natural sources, which acts as an important reducing agent and maintains ascorbate in a reduced and functional form. Glutathione helps the body to get rid of fat-soluble toxins like pesticides, heavy metals and solvents, and therefore, providing a major cellular defence against toxicological and pathological processes. Consuming glutathione may help in protecting against cancer, formation of mutagens and slow down the process of ageing.

Effect of heat on total antioxidant capacity (TAC) of red meat

The total antioxidant capacity of red meat increases on heating at the beginning of cooking period and reaches its maximum after 5-10 minutes cooking. As the cooking time increases, the TAC levels decrease, although at the longest cooking time, a further increase of TAC may be seen.

Heating can alter the secondary and tertiary structure of proteins, causing modification of their physical properties. On heating, unfolding of the proteins happens, and this is called as denaturation. During denaturation, solvent exposure of antioxidant amino acids, which are located in the core of native protein, increases, and consequently, the capacity of scavenging free radicals also increases. Therefore, it is possible that mild heat treatments could increase the antioxidants activity of red meat proteins due to changes in their quaternary and tertiary structures. On the contrary, it is quite known that prolonged heat treatments can cause thermo-oxidation of different components of muscle foods, resulting in endogenous consumption of antioxidants substances, and consequently, the available TAC from red meat to human body decreases, and the second prolonged heat treatment also promotes the degradation of endogenous anti-oxidative factors such as vitamin E, vitamin C, carotenoids, ubiquinols, polyphenols and cellular thiols. In this way, the pro-oxidant antioxidant balance is dominated by the pro-oxidants properties.

Effect of high temperature red meat cooking on health

When red meat is cooked at high temperature like frying and barbequing, the sugars, creatine, creatinine and amino acids present in it form the meat-related mutagens heterocyclic amines, which may be the causative factor for colorectal carcinogenesis.

Health aspects of red meat

Satiety: Satiety is defined as the sensation of fullness that persists after eating until hunger returns, while satiation is the process that leads us to stop eating, ending an eating occasion. In recent years, there has been great interest in the manipulation of satiation and satiety in order to control energy intake, and thus, bodyweight. Research has been conducted on the effect of different protein sources on satiety, including red meat. Protein tends to be more satiating than the other macronutrients, both at the level of a single eating occasion and over days and weeks. It is considered that higher protein meals and diets may be more satiating in comparison to those with lower protein content and may help to control bodyweight. It should be noted that this effect has not been observed when the energy density of test foods is controlled, and the energy density of a food or diet may be more important in determining its satiating effect than its protein content.

Obesity: Overweight or obesity is an increasing public health problem worldwide, affecting people of all ages and socio-economic groups. Obesity is associated with an increased risk of chronic diseases including cardiovascular disease, type 2 diabetes and some types of cancer. For an individual to become obese, energy intake must be higher than energy expenditure for an extended period. This means that either more energy than needed is consumed or too little energy is used by the body because of a lack of physical activity. In general, weight gain seems to be a result of a combination of both increased energy intake and decreased energy expenditure. Vegetarians have been found to have somewhat lower energy intakes compared with non-vegetarians. The reduced energy intake appears to be mainly caused by higher fibre intake bulking out the diet and to a lesser extent, a higher proportion of energy from carbohydrate and a lower fat intake. Meat makes a relatively high contribution to dietary fat and saturated fat intakes. People following a vegetarian diet tend to weigh less in comparison to non-vegetarians. However, this may be caused by various dietary and lifestyle factors as vegetarians generally adopt healthier behaviour, while protein from various sources, including meat, may promote healthy weight maintenance and weight loss as part of a reduced energy diet because of various factors, including its satiating effects. Diets that are very low in carbohydrates are not recommended. Therefore, people who choose to increase their intake of *lean* meat in order to lose weight are advised to include starchy carbohydrates in their diet.

Cardiovascular diseases: Individuals adopting a vegetarian dietary pattern may be at a lower risk of cardiovascular disease than those adopting other diets, but such data cannot be used to make specific associations about meat consumption. Vegetarians and omnivores differ in a number of respects. Scientific studies have shown a modestly increased risk of cardiovascular disease amongst meat eaters compared with low meat eaters but controlling for potential confounders and other aspects of the diet remains difficult. In recent years, prospective studies have moved towards investigating the evidence for relationships of red (unprocessed) and processed meat separately on risk of cardiovascular disease. Findings are currently inconclusive and more prospective studies of large population groups are needed to establish whether a relationship really exists. Since red meat contains saturated fatty acids, a high intake of it can have adverse effects on cardiovascular disease risk factors such as blood cholesterol levels. It also contains other fatty acids (*n*-3 PUFAs, MUFAs) and nutrients (*e.g.*, B vitamins and selenium) that may offer potential cardio-protective benefits. Intervention studies with lean red meat as part of a healthy diet and in combination with blood pressure-lowering drugs have shown favourable effects on blood cholesterol levels, which suggests that lean red meat can be promoted as part of a healthy diet for primary and secondary cardiovascular disease prevention.

Cancer: Scientific studies have been shown a positive correlation between the red meat consumption and colorectal cancer, although findings of most studies have not reached statistical significance. This could be due to the small cohorts and other issues such as inappropriate and inadequate adjustment for confounders. It is quite difficult to make any conclusions on amounts of meat consumed, which may potentially be associated with colorectal cancer risk, therefore, well-structured scientific studies will be needed to draw clear conclusions. Cancer develops over a long period so it is not possible to assess immediate effects of certain food articles on the risk of cancer. When looking at past dietary patterns, changes throughout life are difficult to account for. Furthermore, it is difficult to assign estimated risks to certain food items, as the intake of all foods is likely to have an impact on cancer risk, in addition to other lifestyle factors including physical activity or drinking habits. Bearing this in mind, findings of associations between foods and cancer always have to be considered with caution, as it is impossible to account for all possible factors influencing the risk of disease. Four components of red meat, *i.e.*, fat, heterocyclic amines, *N*-nitrosation products and iron, provide the basis for potential mechanisms by which it is proposed that meat may play a role in the development of cancer.

Epidemiologically, it is difficult to distinguish the influence of animal fat, protein and meat and no direct link between any dietary factors and human cancers has been found till date. Heterocyclic amines produced on overcooking meat have been shown to be carcinogenic in rats. Although at normal cooking temperature and average-cooking time but the level of these compounds produced is not excessive. Nitrates are added to meat products to prevent microbial proliferation. *N*-nitrosamines, produced from these additives, are present at very low levels in meat products, and their carcinogenic potential is likely to be minor at such levels. It is well known that daily consumption of vegetables and meat reduces the risk of cancer at many sites, whereas, daily meat consumption with less frequent vegetable consumption increases the risk of cancer. Nevertheless, it is sensible to consider that there must be an optimal range for meat intakes in order to ensure a balanced diet, whilst optimal weight is maintained. From this practical point of view, Committee's on Medical Aspects of Food and Nutrition (COMAs) suggested the intake range of 90-140 g cooked meat per day, which is sensible as a public health message.

Bone health: Protein from animal sources is richer in sulphur amino acids, *e.g.*, methionine and cysteine than vegetable protein, and as a result, it produces more sulphuric acid. Consequently, for a number of years, diets high in meat have been suggested to have a negative effect on bone health because of the observations that high protein intakes can increase renal acid and urinary calcium excretion. However, not all studies are in agreement with these findings. Short-term high meat diets have been shown not to affect whole body calcium retention, and high intakes of animal protein have been shown to significantly increase bone mineral density. The inconsistent findings suggest that other components of the diet, such as calcium, potassium, phosphorous, salt, phytates and other components are important and may influence calcium excretion.

Type-2 diabetes: Cohort studies have generally found that the consumption of red meat increases the risk of type-2 diabetes, although the results of most individual studies have been non-significant. However, major inconsistencies regarding consumption levels between the studies, with variable intakes for both *high* and *low* consumers and no indication of absolute amounts of intake in some studies, prevent any conclusions about the amount of meat that may potentially be associated with higher risk. There is no evidence to suggest that lean red meat cannot be recommended as part of a healthy balanced diet in respect to preventing type-2 diabetes and for people with established type-2 diabetes. Intake of protein has been associated with increased satiety. Therefore, lean meat could aid appetite and weight control, including among those with established type-2 diabetes.

■■■

Subject Index

Aflatoxins in Nuts 137
Allspice 200
Almond 114, 118, 120, 121, 122, 123, 140, 191
Alzheimer 29, 121, 134, 156, 233, 300, 303, 353, 355, 369, 377, 458, 460, 461, 466, 474, 483, 495, 572, 573
Amino Acids 18, 317, 318
Anardana 201
Aniseed 201
Anthocyanins 27, 407, 408, 409, 478
Antioxidant Enzymes 490
Antioxidants 31, 35, 57, 107, 376, 382, 457, 458, 459, 460, 461, 462, 463, 465, 473, 484, 485, 487, 488, 489, 490, 491, 492, 494, 563, 573, 600
Aonla 33, 141
Apple 46, 141, 155, 157, 159, 411
Apricot 26, 142
Aquaculture 307
Aquatic Food 341
Arjun tree 230, 231
Arnica 231
Asafoetida 201
Ascorbic Acid 475
Ashwagandha 232, 279
Asiatic Ginseng 233
Asparagus 259
Astaxanthin 300, 301, 302, 303, 304, 305, 306, 307, 483
Babchi Seeds 263
Bael 142
Banana 46, 48, 139, 142, 410
Barley 68, 69, 438
Basil 202, 289, 290, 291, 293, 295, 296, 297, 298, 441
Bathu 160, 161
Beet leaf 33, 48, 161
Betalains 417, 418
Bioactive Compounds 58
Biocolour 445, 446, 452, 454, 455
Bishop's weed 188
Bitter gourd 161, 162, 177, 178, 179, 180, 183, 184, 185, 187
Black bean 73
Black ginger 198
Black musalie 264
Black Oil Tree 261
Black pepper 203
Blue Green Algae 300
Bottle gourd 48, 162
Brazil nut 113, 114, 119, 121
Brinjal 48, 162, 163
Cabbage 26, 33, 46, 163, 318, 443, 518
Caffeine 567, 572, 573, 574, 575, 576, 577, 578, 579, 580
Carbohydrates 3, 4, 8, 13, 14, 15, 23, 32, 50, 60, 63, 65, 68, 69, 71, 72, 73, 76, 77, 79, 82, 96, 119, 319, 321, 346, 372, 375, 377, 460, 545, 546, 553
Cardiovascular Health Benefit 126
Carotenoids 30, 123, 124, 283, 300, 410, 412, 414, 415, 487
Carrot 8, 33, 46, 164, 165, 410, 411
Cashew nut 115, 121
Castor 371, 372
Catechins 561
Celery 165, 166, 204
Charayatah 275
Cherry 143, 151, 409
Chickpea 77
Chilli 166
Chlorella Algae 336

Chlorophyll 309, 312, 338, 415, 417, 434, 435, 436, 437, 438, 440, 442, 443, 501
Citronella grass 266
Cluster bean 167
Coconut 25, 48, 87, 88, 89, 90, 91, 93, 94, 95, 99, 100, 101, 103, 104, 105, 106, 107, 110, 111, 112, 143, 144
Coconut kernel 94
Coconut milk 96
Coconut oil 98, 100, 101, 104, 105, 106, 107, 110
Coconut water 88, 89, 90, 91, 92, 93, 94, 96, 111, 112, 144
Coffee 571, 572, 573, 574, 575, 576, 577, 578, 579, 580, 581, 582
Coleus 230, 255, 262, 263
Common Antioxidant Plants 467
Coral 339
Coriander 33, 190, 191, 441
Cosmetic Industries 448
Cow-itch plant 270
Cucumber 167, 411, 442, 443, 513
Cumin 190
Date palm 144
Detoxification Therapy 503
Development of egg 584
Dietary Fiber Composition of 38
Dietary Fiber Content 45
Dietary fiber 14, 15, 24, 25, 26, 37, 38, 39, 40, 41, 42, 43, 44, 45, 46, 47, 49, 50, 51, 52, 53, 54, 55, 56, 57, 60, 63, 64, 66, 68, 69, 71, 72, 73, 74, 76, 77, 78, 79, 80, 82, 83, 131, 132, 144, 157, 180, 251, 373, 375, 377, 378
Egg 583, 584, 585, 589, 590, 591
Egg Components 585
Ellagic Acid 476
Ephedra 234, 253
Fat and lipids 5
Fats and Oils 16
Fats 5, 6, 16, 17, 23, 24, 31, 32, 98, 120
Fennel 191, 569
Fenugreek 33, 192, 441, 443
Fermentation Process 514
Fermented Food 510, 511, 514
Flavonoids 27, 31, 168, 218, 283, 461, 465, 477, 562, 563
Flavonols 164, 561
Flax seed 359, 360, 362, 363, 365, 366, 367, 368, 369
Flax seed oil 359, 362, 363
Folic Acid 477
Foxglove 234
Free Radicals 34, 457, 458, 459, 461, 463, 485, 491
French bean 75, 168
Fruits and Vegetables 26, 29, 473, 474
Garlic 27, 193, 194
German chamomile 235
Giloe 276
Ginger 195
Glaucoma 303
Glory lily 236
Grape 145, 469, 471, 513
Green gram 48, 71
Green Portals 433
Groundnut 48, 116, 375, 376
Guava 33, 144
Harsingar 244, 245
Hazelnut 115, 118, 120, 122, 123
Herbs for Human Ailments 253
Himalayan fir 238
Holy basil 271, 291
Honey 14, 523, 524, 525, 528, 529, 532, 533, 534, 535, 536, 537, 538, 539, 540, 541, 543
Honey as Food 539
Honey Characteristics 524
Honey Composition 526
Honey Fermentation 542
Horse gram 81
Hydrotherapy 506, 507, 508
Indian aloes 257
Indian Atees 255
Indian Belladonna 238, 239
Indian Cotton 239
Indian hemp 240
Indian Mayapple 273

Indian rhubarb 242
Indian Serpent wood 273
Indian squill 277
Jamun 145, 158, 185, 523
Java glangal 257
Kelp 334, 335
King of bitters 258
Kokam butter tree 267
Kutki 272
Lemon grass 265
Lemon 33, 146, 208, 222, 512, 536
Lentil 82
Lettuce 168, 290
Lima bean 83, 221
Linseed 34, 361, 373, 374
Lipid Peroxidation 459, 465
Litchi 146
Lutein and zeaxanthin 411, 413, 415
Lycopene 26, 175, 410, 411, 412, 415, 474, 483, 496, 497
Macadamia 116, 118, 120, 122, 123, 125
Maize 22, 23, 69, 411, 414
Mango ginger 197
Mango 26, 48, 146, 411
Marine Phytoplankton 331
Marine Plants 329
Medicinal Mushroom 326
Medicinal Plants 462
Medicinal Value of Mushrooms 322
Medicinal 38, 71, 111, 229, 255, 256, 257, 258, 259, 260, 261, 262, 263, 264, 265, 266, 267, 268, 269, 270, 271, 272, 273, 274, 275, 276, 277, 278, 279, 285, 289, 297, 325, 375, 441, 457
Milk Constituents 557
Milk Enzymes 556
Milk Fat 548
Milk Minerals 555
Milk thistle 243
Milk Vitamins 554
Millets 21, 23
Minerals and Red meat 599
Minerals 19, 22, 23, 60, 63, 65, 68, 69, 73, 77, 79, 93, 122, 321, 375, 377, 488, 501, 526, 546, 575, 586, 587
Minor Fruits 151
Miracle fruit 268
Mushroom 316, 317, 318, 323, 324, 326
Musk melon 169
Mustard 34, 379, 443
Neem 272, 382, 383, 467
Nutrition 1, 13, 22, 43, 49, 68, 83, 127, 134, 135, 180, 288, 305, 345, 346, 356, 364, 367, 368, 442, 518, 567, 593, 599, 603
Nutrient Components in Vegetables 3
Nutrients 1, 8, 9, 10, 12, 13, 19, 22, 23, 118, 489
Nutrients and Phytochemicals in Nuts 118
Nutritional Attributes of Mushrooms 317
Nuts and Health Benefits 113, 129
Oats 38, 64, 65, 66, 67
Oilseeds 24, 58, 361, 371, 513
Okra 169, 170
Olive 25, 108, 380, 381, 382, 425
Onion 27, 30, 194, 195, 441
Orange 26, 27, 33, 147, 407, 409, 410, 411, 454
Organic Farming 385, 386, 394, 395, 403, 404, 405
Organic Food Regulations 400
Organic Food 392, 393, 395
Oxidative Stress 458, 486
Papaya 27, 33, 147, 148
Parsley 170, 210, 440
Passionflower 154, 245
Pea 170, 171, 411
Pear 46, 148, 149
Periwinkle 237, 260, 344
Phytochemicals 26, 123, 335, 407, 473
Pine nut 117, 121
Pineapple 149, 343
Pistachio 46, 113, 117, 119, 120, 121, 122, 123, 131
Plankton 331, 335
Plant Enzyme Deficiency 500

Plant Enzyme Therapy 499
Plant Enzymes 498
Plant Food Sources 21
Plant Pigments 406
Plants 13, 28, 229, 290, 343, 355, 406, 436, 457, 463, 500
Polyunsaturated Fats 16, 108
Polyunsaturated Fatty Acids 421, 422
Pomegranate 149, 150
Potato 9, 38, 48, 171, 172, 226
Protective Food 29
Protein 4, 5, 17, 18, 22, 23, 24, 32, 60, 63, 65, 68, 69, 71, 72, 73, 75, 76, 77, 79, 82, 83, 92, 93, 118, 119, 120, 180, 309, 317, 321, 346, 353, 363, 373, 375, 376, 377, 460, 502, 545, 546, 551, 586, 587, 589, 593, 602, 603
Pulses and Beans 71
Pumpkin 172, 411
Radish 33, 172, 409
Red Meat 593
Rice 22, 47, 60, 61, 62, 77, 250, 488
Rice bean 76
Rosha grass 265
Safed Musli 261
Safflower 378, 379
Seabuckthorn 247, 248, 269, 281, 282, 284, 285, 286, 287, 288
Seafood 16, 19, 301, 341, 342, 343, 345, 347, 348, 349, 350, 351, 352, 353, 356, 357, 358, 439, 503
Seaweeds 329, 330, 331, 355
Sesame 46, 377, 378
Shellfish 342, 354, 488
Snakeroot 241
Soybean 9, 34, 78, 80, 106, 517
Spices and Condiments 27
Spinach 26, 29, 30, 46, 173, 174, 318, 319, 410, 411, 415, 440, 443
Spirulina 300, 308, 309, 310, 311, 312, 333, 335, 439
Sponge gourd 173
Stevia 110, 274, 275
Strawberry 150
Sunflower 46, 98, 372, 373
Super Nutrient 308
Sweet corn 174
Sweet flag 213, 256
Sweet potato 48, 174
Tamarind 85, 213
Tea 462, 560, 561, 563, 564, 565, 566, 567, 569, 570
Tea Flavonoids 560, 562
Tea Polyphenols 568
Textile Industries 445, 451
Theaflavins and Thearubigins 562
Therapeutic Significance of Cereals 60
Tomato 26, 46, 175, 176, 410, 411
Tulsi 184, 202, 271, 289, 290, 291, 292, 293, 294, 296, 404
Tulsi in Indian Tradition 294
Turmeric 196, 197, 418, 419, 445, 479
Underutilized Vegetables 215
Uses 40, 61, 177, 181, 255, 256, 257, 258, 259, 260, 261, 262, 263, 264, 265, 266, 267, 268, 269, 270, 271, 272, 273, 274, 275, 276, 277, 278, 279, 285, 286, 295, 297, 299, 353, 361, 443, 523, 543
Vegetable Spices 188
Velvet bean 84, 85, 220
Vetiver 278, 279
Vitamin E 20, 26, 34, 35, 36, 68, 69, 73, 78, 92, 134, 166, 283, 337, 372, 373, 460, 465, 476, 488, 489, 495, 546, 554, 598, 599
Vitamins and red meat 597
Water melon 176
Wheat 21, 22, 34, 35, 38, 47, 55, 58, 62, 63, 64, 438
Wholegrain Food 50
Winter cherry 279
Witch hazel 249
Wormseed 259